Edited by
Richard Dronskowski,
Shinichi Kikkawa, and
Andreas Stein

Handbook of
Solid State Chemistry

Edited by
Richard Dronskowski,
Shinichi Kikkawa, and
Andreas Stein

Handbook of Solid State Chemistry

Volume 6: Functional Materials

WILEY-VCH Verlag GmbH & Co. KGaA

Editors

Richard Dronskowski
RWTH Aachen
Institute of Inorganic Chemistry
Landoltweg 1
52056 Aachen
Germany

Shinichi Kikkawa
Hokkaido University
Faculty of Engineering
N13 W8, Kita-ku
060-8628 Sapporo
Japan

Andreas Stein
University of Minnesota
Department of Chemistry
207 Pleasant St. SE
Minneapolis, MN 55455
USA

Cover Credit: Sven Lidin, Arndt Simon and Franck Tessier

All books published by **Wiley-VCH** are carefully produced. Nevertheless, authors, editors, and publisher do not warrant the information contained in these books, including this book, to be free of errors. Readers are advised to keep in mind that statements, data, illustrations, procedural details or other items may inadvertently be inaccurate.

Library of Congress Card No.: applied for

British Library Cataloguing-in-Publication Data
A catalogue record for this book is available from the British Library.

Bibliographic information published by the Deutsche Nationalbibliothek
The Deutsche Nationalbibliothek lists this publication in the Deutsche Nationalbibliografie; detailed bibliographic data are available on the Internet at http://dnb.d-nb.de.

© 2017 Wiley-VCH Verlag GmbH & Co. KGaA, Boschstr. 12, 69469 Weinheim, Germany

All rights reserved (including those of translation into other languages). No part of this book may be reproduced in any form – by photoprinting, microfilm, or any other means – nor transmitted or translated into a machine language without written permission from the publishers. Registered names, trademarks, etc. used in this book, even when not specifically marked as such, are not to be considered unprotected by law.

Print ISBN: 978-3-527-32587-0
oBook ISBN: 978-3-527-69103-6

Cover Design Formgeber
Typesetting Thomson Digital, Noida, India
Printing and Binding Markono Print Media Pte Ltd, Singapore

Printed on acid-free paper

Preface

When you do great science, you do not have to make a lot of fuss. This oft-forgotten saying from the twentieth century has served these editors pretty well, so the foreword to this definitive six-volume *Handbook of Solid-State Chemistry* in the early twenty-first century will be brief. After all, is there any real need to highlight the paramount successes of solid-state chemistry in the last half century? – Successes that have led to novel magnets, solid-state lighting, dielectrics, phase-change materials, batteries, superconducting compounds, and a lot more? Probably not, but we should stress that many of these exciting matters were derived from curiosity-driven research — work that many practitioners of our beloved branch of chemistry truly appreciate, and this is exactly why they do it. Our objects of study may be immensely important for various applications but, first of all, they are interesting to us; that is, how chemistry defines and challenges itself. Let us also not forget that solid-state chemistry is a neighbor to physics, crystallography, materials science, and other fields, so there is plenty of room at the border, to paraphrase another important quote from a courageous physicist.

Given the incredibly rich heritage of solid-state chemistry, it is probably hard for a newcomer (a young doctoral student, for example) to see the forest for all the trees. In other words, there is a real need to cover solid-state chemistry in its entirety, but only if it is conveniently grouped into digestible categories. Because such an endeavor is not possible in introductory textbooks, this is what we have tried to put together here. The compendium starts with an overview of materials and of the structure of solids. Not too surprisingly, the next volume deals with synthetic techniques, followed by another volume on various ways of (structural) characterization. Being a timely handbook, the fourth volume touches upon nano and hybrid materials, while volume V introduces the reader to the theoretical description of the solid state. Finally, the sixth volume reaches into the real world by focusing on functional materials. Should we have considered more volumes? Yes, probably, but life is short, dear friends.

This handbook would have been impossible to compile for three authors, let alone a single one. Instead, the editors take enormous pride in saying that they managed to motivate more than a hundred first-class scientists living across the globe, each of them specializing in (and sometimes even shaping) a subfield of solid-state chemistry or a related discipline, and all of these wonderful colleagues did their very best to make our dream come true. Thanks to all of you; we sincerely appreciate your contributions. Thank you, once again, on behalf of solid-state chemistry. The editors also would like to thank Wiley-VCH, in particular Dr. Waltraud Wüst and also Dr. Frank Otmar Weinreich, for spiritually (and practically) accompanying us over a few years, and for reminding us here and there that there must be a final deadline. That being said, it is up to the reader to judge whether the tremendous effort was justified. We sincerely hope that this is the case.

A toast to our wonderful science! Long live solid-state chemistry!

Richard Dronskowski
RWTH Aachen, Aachen, Germany

Shinichi Kikkawa
Hokkaido University, Sapporo, Japan

Andreas Stein
University of Minnesota, Minneapolis, USA

Contents

Volume 1: Materials and Structure of Solids

1 **Intermetallic Compounds and Alloy Bonding Theory Derived from Quantum Mechanical One-Electron Models** *1*
Stephen Lee and Daniel C. Fredrickson

2 **Quasicrystal Approximants** *73*
Sven Lidin

3 **Medium-Range Order in Oxide Glasses** *93*
Hellmut Eckert

4 **Suboxides and Other Low-Valent Species** *139*
Arndt Simon

5 **Introduction to the Crystal Chemistry of Transition Metal Oxides** *161*
J.E. Greedan

6 **Perovskite Structure Compounds** *221*
Yuichi Shimakawa

7 **Nitrides of Non-Main Group Elements** *251*
P. Höhn and R. Niewa

8 **Fluorite-Type Transition Metal Oxynitrides** *361*
Franck Tessier

9 **Mechanochemical Synthesis, Vacancy-Ordered Structures and Low-Dimensional Properties of Transition Metal Chalcogenides** *383*
Yutaka Ueda and Tsukio Ohtani

Contents

10 Metal Borides: Versatile Structures and Properties *435*
Barbara Albert and Kathrin Hofmann

11 Metal Pnictides: Structures and Thermoelectric Properties *455*
Abdeljalil Assoud and Holger Kleinke

12 Metal Hydrides *477*
Yaoqing Zhang, Maarten C. Verbraeken, Cédric Tassel, and Hiroshi Kageyama

13 Local Atomic Order in Intermetallics and Alloys *521*
Frank Haarmann

14 Layered Double Hydroxides: Structure–Property Relationships *541*
Shan He, Jingbin Han, Mingfei Shao, Ruizheng Liang, Min Wei, David G. Evans, and Xue Duan

15 Structural Diversity in Complex Layered Oxides *571*
S. Uma

16 Magnetoresistance Materials *595*
Ichiro Terasaki

17 Magnetic Frustration in Spinels, Spin Ice Compounds, $A_3B_5O_{12}$ Garnet, and Multiferroic Materials *617*
Hongyang Zhao, Hideo Kimura, Zhenxiang Cheng, and Tingting Jia

18 Structures and Properties of Dielectrics and Ferroelectrics *643*
Mitsuru Itoh

19 Defect Chemistry and Its Relevance for Ionic Conduction and Reactivity *665*
Joachim Maier

20 Molecular Magnets *703*
J.V. Yakhmi

21 Ge–Sb–Te Phase-Change Materials *735*
Volker L. Deringer and Matthias Wuttig

Index *751*

Volume 2: Synthesis

1 **High-Temperature Methods** *1*
Rainer Pöttgen and Oliver Janka

2 **High-Pressure Methods in Solid-State Chemistry** *23*
Hubert Huppertz, Gunter Heymann, Ulrich Schwarz, and Marcus R. Schwarz

3 **High-Pressure Perovskite: Synthesis, Structure, and Phase Relation** *49*
Yoshiyuki Inaguma

4 **Solvothermal Methods** *107*
Nobuhiro Kumada

5 **High-Throughput Synthesis Under Hydrothermal Conditions** *123*
Nobuaki Aoki, Gimyeong Seong, Tsutomu Aida, Daisuke Hojo, Seiichi Takami, and Tadafumi Adschiri

6 **Particle-Mediated Crystal Growth** *155*
R. Lee Penn

7 **Sol–Gel Synthesis of Solid-State Materials** *179*
Guido Kickelbick and Patrick Wenderoth

8 **Templated Synthesis for Nanostructured Materials** *201*
Yoshiyuki Kuroda and Kazuyuki Kuroda

9 **Bio-Inspired Synthesis and Application of Functional Inorganic Materials by Polymer-Controlled Crystallization** *233*
Lei Liu and Shu-Hong Yu

10 **Reactive Fluxes** *275*

11 **Glass Formation and Crystallization** *287*
T. Komatsu

12 **Glass-Forming Ability, Recent Trends, and Synthesis Methods of Metallic Glasses** *319*
Hidemi Kato, Takeshi Wada, Rui Yamada, and Junji Saida

13 **Crystal Growth Via the Gas Phase by Chemical Vapor Transport Reactions** *351*
Michael Binnewies, Robert Glaum, Marcus Schmidt, and Peer Schmidt

14	**Thermodynamic and Kinetic Aspects of Crystal Growth** *375*
	Detlef Klimm

15	**Chemical Vapor Deposition** *399*
	Takashi Goto and Hirokazu Katsui

16	**Growth of Wide Bandgap Semiconductors by Halide Vapor Phase Epitaxy** *429*
	Yuichi Oshima, Encarnación G. Víllora, and Kiyoshi Shimamura

17	**Growth of Silicon Nanowires** *467*
	Fengji Li and Sam Zhang

18	**Chemical Patterning on Surfaces and in Bulk Gels** *539*
	Olaf Karthaus

19	**Microcontact Printing** *563*
	Kiyoshi Yase

20	**Nanolithography Based on Surface Plasmon** *573*
	Kosei Ueno and Hiroaki Misawa

Index *589*

Volume 3: Characterization

1	**Single-Crystal X-Ray Diffraction** *1*
	Ulli Englert

2	**Laboratory and Synchrotron Powder Diffraction** *29*
	R. E. Dinnebier, M. Etter, and T. Runcevski

3	**Neutron Diffraction** *77*
	Martin Meven and Georg Roth

4	**Modulated Crystal Structures** *109*
	Sander van Smaalen

5	**Characterization of Quasicrystals** *131*
	Walter Steurer

6	**Transmission Electron Microscopy** *155*
	Krumeich Frank

7	**Scanning Probe Microscopy** *183*
	Marek Nowicki and Klaus Wandelt

8	**Solid-State NMR Spectroscopy: Introduction for Solid-State Chemists** *245* Christoph S. Zehe, Renée Siegel, and Jürgen Senker	
9	**Modern Electron Paramagnetic Resonance Techniques and Their Applications to Magnetic Systems** *279* Andrej Zorko, Matej Pregelj, and Denis Arčon	
10	**Photoelectron Spectroscopy** *311* Stephan Breuer and Klaus Wandelt	
11	**Recent Developments in Soft X-Ray Absorption Spectroscopy** *361* Alexander Moewes	
12	**Vibrational Spectroscopy** *393* Götz Eckold and Helmut Schober	
13	**Mößbauer Spectroscopy** *443* Hermann Raphael	
14	**Macroscopic Magnetic Behavior: Spontaneous Magnetic Ordering** *485* Heiko Lueken and Manfred Speldrich	
15	**Dielectric Properties** *523* Rainer Waser and Susanne Hoffmann-Eifert	
16	**Mechanical Properties** *561* Volker Schnabel, Moritz to Baben, Denis Music, William J. Clegg, and Jochen M. Schneider	
17	**Calorimetry** *589* Hitoshi Kawaji	
	Index *615*	

Volume 4: Nano and Hybrid Materials

1	**Self-Assembly of Molecular Metal Oxide Nanoclusters** *1* Laia Vilà-Nadal and Leroy Cronin
2	**Inorganic Nanotubes and Fullerene–Like Nanoparticles from Layered (2D) Compounds** *21* L. Yadgarov, R. Popovitz-Biro, and R. Tenne

3	**Layered Materials: Oxides and Hydroxides** *53* Ida Shintaro
4	**Organoclays and Polymer-Clay Nanocomposites** *79* M.A. Vicente and A. Gil
5	**Zeolite and Zeolite-Like Materials** *97* Watcharop Chaikittisilp and Tatsuya Okubo
6	**Ordered Mesoporous Materials** *121* Michal Kruk
7	**Porous Coordination Polymers/Metal–Organic Frameworks** *141* Ohtani Ryo and Kitagawa Susumu
8	**Metal–Organic Frameworks: An Emerging Class of Solid-State Materials** *165* Joseph E. Mondloch, Rachel C. Klet, Ashlee J. Howarth, Joseph T. Hupp, and Omar K. Farha
9	**Sol–Gel Processing of Porous Materials** *195* Kazuki Nakanishi, Kazuyoshi Kanamori, Yasuaki Tokudome, George Hasegawa, and Yang Zhu
10	**Macroporous Materials Synthesized by Colloidal Crystal Templating** *243* Jinbo Hu and Andreas Stein
11	**Optical Properties of Hybrid Organic–Inorganic Materials and their Applications – Part I: Luminescence and Photochromism** *275* Stephane Parola, Beatriz Julián-López, Luís D. Carlos, and Clément Sanchez
12	**Optical Properties of Hybrid Organic–inorganic Materials and their Applications – Part II: Nonlinear Optics and Plasmonics** *317* Stephane Parola, Beatriz Julián-López, Luís D. Carlos, and Clément Sanchez
13	**Bioactive Glasses** *357* Hirotaka Maeda and Toshihiro Kasuga
14	**Materials for Tissue Engineering** *383* María Vallet-Regí and Antonio J. Salinas

Index *411*

Volume 5: Theoretical Description

1   **Density Functional Theory**   *1*
Michael Springborg and Yi Dong

2   **Eliminating Core Electrons in Electronic Structure Calculations: Pseudopotentials and PAW Potentials**   *29*
Stefan Goedecker and Santanu Saha

3   **Periodic Local Møller–Plesset Perturbation Theory of Second Order for Solids**   *59*
Denis Usvyat, Lorenzo Maschio, and Martin Schütz

4   **Resonating Valence Bonds in Chemistry and Solid State**   *87*
Evgeny A. Plekhanov and Andrei L. Tchougréeff

5   **Many Body Perturbation Theory, Dynamical Mean Field Theory and All That**   *119*
Silke Biermann and Alexander Lichtenstein

6   **Semiempirical Molecular Orbital Methods**   *159*
Thomas Bredow and Karl Jug

7   **Tight-Binding Density Functional Theory: DFTB**   *203*
Gotthard Seifert

8   **DFT Calculations for Real Solids**   *227*
Karlheinz Schwarz and Peter Blaha

9   **Spin Polarization**   *261*
Dong-Kyun Seo

10   **Magnetic Properties from the Perspectives of Electronic Hamiltonian: Spin Exchange Parameters, Spin Orientation, and Spin-Half Misconception**   *285*
Myung-Hwan Whangbo and Hongjun Xiang

11   **Basic Properties of Well-Known Intermetallics and Some New Complex Magnetic Intermetallics**   *345*
Peter Entel

12   **Chemical Bonding in Solids**   *405*
Gordon J. Miller, Yuemei Zhang, and Frank R. Wagner

13	**Lattice Dynamics and Thermochemistry of Solid-State Materials from First-Principles Quantum-Chemical Calculations** *491*	
	Ralf Peter Stoffel and Richard Dronskowski	
14	**Predicting the Structure and Chemistry of Low-Dimensional Materials** *527*	
	Xiaohu Yu, Artem R. Oganov, Zhenhai Wang, Gabriele Saleh, Vinit Sharma, Qiang Zhu, Qinggao Wang, Xiang-Feng Zhou, Ivan A. Popov, Alexander I. Boldyrev, Vladimir S. Baturin, and Sergey V. Lepeshkin	
15	**The Pressing Role of Theory in Studies of Compressed Matter** *571*	
	Eva Zurek	
16	**First-Principles Computation of NMR Parameters in Solid-State Chemistry** *607*	
	Jérôme Cuny, Régis Gautier, and Jean-François Halet	
17	**Quantum Mechanical/Molecular Mechanical (QM/MM) Approaches** *647*	
	C. Richard A. Catlow, John Buckeridge, Matthew R. Farrow, Andrew J. Logsdail, and Alexey A. Sokol	
18	**Modeling Crystal Nucleation and Growth and Polymorphic Transitions** *681*	
	Dirk Zahn	

Index *701*

Volume 6: Functional Materials

1	**Electrical Energy Storage: Batteries** *1*	
	Eric McCalla	
1.1	Motivation *1*	
1.2	Fundamentals of State-of-the-Art Li-Ion Batteries *2*	
1.2.1	$LiCoO_2$ and Graphite Electrodes *2*	
1.2.2	Electrolytes *3*	
1.3	Summary of Structures and Properties of Interest *4*	
1.3.1	Positive Electrode Materials *5*	
1.3.2	Negative Electrode Materials *6*	
1.3.3	Solid-State Electrolytes *7*	
1.4	Synthesis *9*	
1.4.1	All Solid-State Synthesis *9*	
1.4.2	Nanometric Carbon Coating Syntheses *10*	
1.4.3	Soft Chemistry and the Discovery of New Phases *11*	
1.5	Contributions of Solid-State Chemistry to Batteries *12*	
1.5.1	Cationic and Anionic Substitutions *12*	

1.5.2	Unraveling the Mechanisms during Cycling	*13*
1.5.3	Control of Particle Morphology/Coatings	*15*
1.5.4	Making All-Solid Batteries a Reality	*16*
1.6	High-Throughput Chemistry	*18*
1.6.1	Experimental Combinatorial Synthesis	*18*
1.6.2	Computational High-Throughput Calculations	*21*
1.7	Conclusions and Future Direction	*21*
	References	*22*
2	**Electrical Energy Storage: Supercapacitors**	*25*
	Enbo Zhao, Wentian Gu, and Gleb Yushin	
2.1	Introduction	*25*
2.2	Electric Double-layer Capacitors	*27*
2.2.1	Activated Carbon Materials	*30*
2.2.2	Other Carbon Materials	*32*
2.2.3	Carbon–Carbon Composites and Hierarchical Porous Materials	*33*
2.2.4	Quantum or Space Charge Capacitance	*34*
2.3	Pseudocapacitors, Asymmetric Hybrid Capacitors, and Li-ion Capacitors	*36*
2.3.1	Surface Redox Pseudocapacitance	*39*
2.3.1.1	Metals, Metal Oxides, and Metal Nitrides	*39*
2.3.1.2	Conductive Polymers and Functionalized Carbons as Active Materials	*41*
2.3.2	Intercalation Pseudocapacitance	*42*
2.4	Supercapacitor Device Configurations	*43*
2.5	Electrolyte Selection	*44*
2.6	Self-discharge of Supercapacitors	*45*
2.7	Summary and Outlook	*47*
	References	*49*
3	**Dye-Sensitized Solar Cells**	*61*
	Anna Nikolskaia and Oleg Shevaleevskiy	
3.1	Introduction	*61*
3.2	History	*61*
3.3	Operational Principles	*62*
3.4	Materials and Methods for DSSC Fabrication	*64*
3.4.1	Mesoscopic Metal Oxide Electrodes	*64*
3.4.2	Dye Sensitizers	*66*
3.4.3	Counter Electrodes	*68*
3.4.4	Electrolytes	*68*
3.5	New Challenges	*70*
	References	*70*
4	**Electronics and Bioelectronic Interfaces**	*75*
	Seong-Min Kim, Sungjun Park, Won-June Lee, and Myung-Han Yoon	
4.1	Introduction	*75*

4.2	Operation Mechanism of Electrolyte-gated Transistors *76*
4.3	Metal Oxide Transistors and Bioelectronic Interfaces *78*
4.3.1	Sol–Gel Processed Metal Oxide Electronic Materials *79*
4.3.2	Bioelectronic Applications Based on MOx Electronic Materials *81*
4.4	Organic Electrochemical Transistors and Bioelectronic Interfaces *83*
4.4.1	Fundamentals of Organic Electrochemical Transistors *84*
4.4.2	Organic Electrochemical Transistors for Bioelectronic Applications *86*
4.5	Conclusions *88*
	References *90*

5	**Designing Thermoelectric Materials Using 2D Layers** *93*
	Sage R. Bauers and David C. Johnson
5.1	Introduction *93*
5.2	Physical Picture *94*
5.3	Optimizing Thermoelectric Materials: Minimizing Lattice Conductivity *97*
5.3.1	The Phonon Glass/Electron Crystal Approach *97*
5.3.2	Other Methods to Reduce Thermal Conductivity *99*
5.4	Thermoelectric Materials: Maximizing Power Factor *100*
5.5	Bridge to Nanolaminate Structures *103*
5.6	Transition Metal Dichalcogenide Compounds *104*
5.7	Misfit Layer Compounds *104*
5.8	Thin-film Superlattice Materials *106*
5.9	van Der Waals Heterostructures *107*
5.10	Kinetically Trapped Nanolaminates *108*
5.11	$([MSe]_{1+\delta})_1(TiSe_2)_n$ Ferecrystals *111*
5.12	Outlook *116*
	References *117*

6	**Magnetically Responsive Photonic Nanostructures for Display Applications** *123*
	Mingsheng Wang and Yadong Yin
6.1	Introduction *123*
6.2	General Principle of Magnetic Interactions *125*
6.3	Magnetic Tuning of Light Scattering and Reflection *126*
6.4	Magnetic Tuning of Light Interference *129*
6.5	Magnetic Tuning of Bragg Diffraction *133*
6.6	Magnetic Tuning of Light Polarization *143*
6.7	Conclusions *146*
	Acknowledgment *148*
	References *148*

7	**Functional Materials: For Sensing/Diagnostics** *151*
	Rujuta D. Munje, Shalini Prasad, and Edward Graef
7.1	Introduction *151*

7.2	Basic Concept of Label-free Biosensors *152*	
7.2.1	Nanotextured Semiconductor Zinc Oxide as Functional Material *154*	
7.2.2	Synthesis of ZnO Nanostructures for Biosensing Applications *155*	
7.2.3	Functionalization of ZnO Nanostructures for Biosensing Applications *157*	
7.2.4	ZnO Nanostructure-based Wearable Biosensors *159*	
7.3	Environmental Sensors *161*	
7.3.1	Oxygen Sensors *161*	
7.3.2	Hydrocarbon Sensors *164*	
7.4	Conclusions *170*	
	References *171*	

8 Superhard Materials *175*
Ralf Riedel, Leonore Wiehl, Andreas Zerr, Pavel Zinin, and Peter Kroll

8.1	Introduction *175*	
8.2	Hardness and Strength of Materials *176*	
8.3	Superhard Materials *180*	
8.3.1	Carbon Nitride (C–N) Phases *180*	
8.3.1.1	Theoretical Predictions of C–N Structures *180*	
8.3.1.2	Synthesis of Novel C–N-based Compounds under HPHT Conditions *181*	
8.3.2	Novel B–C-based Compounds BC_x *182*	
8.3.2.1	Theory *182*	
8.3.2.2	Synthesis *183*	
8.3.3	B–C–N Compounds *184*	
8.3.3.1	Theoretical Predictions *184*	
8.3.3.2	Synthesis Experiments *185*	
8.3.4	Spinel Nitrides, γ-M_3N_4 where M = Si, Ge, or Sn *186*	
8.3.5	Transition Metal Nitrides of the Group 4 and 5 Elements (c-Zr3N4 and c-Hf3N4, η-Ta2N3) *189*	
8.4	Chances, Expectations, and Prospects of Computational Simulations *192*	
8.5	Conclusions *193*	
	References *194*	

9 Self-healing Materials *201*
Martin D. Hager

9.1	Introduction *201*	
9.2	Definitions and General Principles *203*	
9.3	Self-healing Metals *205*	
9.4	Self-healing Ceramics *207*	
9.5	Self-healing Concrete *209*	
9.6	Self-healing Asphalt *211*	
9.7	Self-healing Polymers *211*	
9.8	Commercial Applications *217*	

9.9	Conclusions and Outlook *217*	
	Acknowledgments *218*	
	References *219*	

10 Functional Surfaces for Biomaterials *227*
Akiko Nagai, Naohiro Horiuchi, Miho Nakamura, Norio Wada, and Kimihiro Yamashita
- 10.1 Introduction *227*
- 10.2 Polarized Hydroxyapatite *229*
- 10.3 Wettability and Surface Free Energy *233*
- 10.4 Initial Cell Behaviors *235*
- 10.5 Crystallization of Calcium Carbonate on Polarized Hydroxyapatite Substrates *237*
- 10.5.1 Theoretical Approaches to Heterogeneous Nucleation in the Presence of Electric Field and the Property of the Electric Field Generated by the Polarized Substrate *237*
- 10.5.2 $CaCO_3$ Crystallization on Polarized HAp Substrates *240*
- References *243*

11 Functional Materials for Gas Storage. Part I: Carbon Dioxide and Toxic Compounds *249*
L. Reguera and E. Reguera
- 11.1 Introduction *249*
- 11.2 Evaluation of Gases' Adsorption *250*
- 11.2.1 Physical Adsorption of Gases and Vapors *250*
- 11.2.2 Gas Adsorption Measurement Techniques *250*
- 11.2.3 Adsorption Data Evaluation *252*
- 11.3 Nature of Adsorbate–Adsorbent Interactions *256*
- 11.4 Adsorption Heats and Gas Adsorption–Desorption *259*
- 11.5 Carbon Dioxide Capture *260*
- 11.5.1 Carbon Dioxide Capture by Chemisorption *261*
- 11.5.2 Carbon Dioxide Capture by Adsorption in Porous Solids *261*
- 11.5.3 Zeolites, Aluminas, and Silicas *262*
- 11.5.4 Carbon-based Materials *263*
- 11.5.5 Prussian Blue and Related Materials *263*
- 11.5.6 Metal–Organic Frameworks *266*
- 11.5.7 Functionalized Porous Solids *268*
- 11.5.8 Porous Organic Polymers *269*
- 11.6 Carbon Dioxide Capture by Clathrates *270*
- 11.7 Air Purification by Adsorption of Toxic Chemicals *273*
- 11.8 Outlook on Materials for CO_2 Capture and Air Purification *275*
- Acknowledgment *275*
- References *276*

12	**Functional Materials for Gas Storage. Part II: Hydrogen and Methane** *281*	
	L. Reguera and E. Reguera	
12.1	Introduction *281*	
12.2	On the Adsorption Forces for H_2 and CH_4 in Porous Solids *282*	
12.3	Adsorbent Materials for H_2 Storage *284*	
12.3.1	Carbon-Based Materials *285*	
12.3.2	Silica, Alumina, and Zeolites *287*	
12.3.3	Metal-Organic Frameworks (MOFs) *288*	
12.3.4	H_2 Storage in MOFs Through Dispersive Interactions *288*	
12.3.5	Ionic MOFs: H_2 Adsorption Through Electrostatic Interactions *289*	
12.3.6	H_2 Adsorption in MOFs with Partially Naked Transition Metal Centers *290*	
12.4	Prussian Blue Analogues and Related Porous Solids *292*	
12.5	Transition Metal Nitroprussides *294*	
12.6	Porous Organic Polymers *295*	
12.7	Current State, Targets, and Perspectives on Materials for H_2 Storage *296*	
12.8	Materials for Methane Storage *298*	
12.9	Methane and Hydrogen Clathrates *301*	
12.9.1	Methane and Natural Gas Hydrates *301*	
12.9.2	Hydrogen Storage in Clathrates *303*	
12.10	Materials for Dissociative Hydrogen Storage *304*	
12.11	Outlook on Materials for H_2 and CH_4 Storage *306*	
	Acknowledgment *307*	
	References *307*	
13	**Supported Catalysts** *313*	
	Isao Ogino, Pedro Serna, and Bruce C. Gates	
13.1	Introduction *313*	
13.2	Classes of Solid Catalysts *314*	
13.2.1	Bulk Solid Catalysts *314*	
13.2.2	Supported Catalysts *314*	
13.3	Preparation of Supported Catalysts *316*	
13.3.1	Industrial Catalyst Preparation Methods *316*	
13.3.2	Synthesis of Supported Metals with Well-Defined Structures *317*	
13.4	Characterization of Supported Catalysts *318*	
13.4.1	Transmission Electron Microscopy *318*	
13.4.2	X-ray Absorption Spectroscopy *319*	
13.4.3	Chemisorption Equilibrium Measurements *321*	
13.4.4	Case Study Illustrating Application of Complementary Characterization Methods *322*	
13.5	Reactivities of Supported Catalysts *322*	

13.5.1	Adsorption	*322*
13.5.2	Restructuring of Supported Catalysts	*324*
13.5.3	Oxidation of Supported Catalysts	*324*
13.5.4	Sulfidation of Supported Catalysts	*325*
13.6	Catalyst Performance: Structure–Property Relationships	*325*
13.6.1	Structure Sensitivity and Structure Insensitivity in Catalysis by Supported Metal Clusters and Particles	*325*
13.6.2	Dependence of Catalytic Activity on Crystal Faces	*326*
13.6.3	Roles of Structural Features such as Steps and Corners	*326*
13.6.4	Effects of Supports as Ligands and Modifiers of Metals	*327*
13.6.5	Effects of Metal Nuclearity	*328*
13.6.6	Effects of Ligands Other Than Supports	*329*
13.6.7	Metal Oxidation State	*330*
13.6.8	Catalyst Deactivation by Sintering and Poisoning by Compounds Containing P, N, S, or As	*330*
13.6.9	Catalyst Regeneration	*331*
	Acknowledgment	*332*
	References	*332*
14	**Hydrogenation by Metals**	***339***
	Xin Jin and Raghunath V. Chaudhari	
14.1	Introduction	*339*
14.2	Classification of Hydrogenation Reactions	*340*
14.2.1	Hydrogenation of Alkenes	*341*
14.2.2	Partial Hydrogenation of Alkynes to Alkenes	*341*
14.2.3	Hydrogenation of Aldehydes and Ketones	*344*
14.2.4	Hydrogenation of Carboxylic Acids, Esters, and Ethers	*348*
14.2.5	Hydrogenation of Nitroaromaticso	*348*
14.2.6	Reductive Amination and Alkylation Reactions	*351*
14.2.7	Hydrogenolysis of C−O Bond	*351*
14.2.8	Hydrodesulphurization of Petroleum Crude	*353*
14.3	Types of Hydrogenation Catalysts	*357*
14.3.1	Nickel, Copper, Iron, and Cobalt	*357*
14.3.2	Platinum Group Catalysts	*358*
14.3.3	Preparation Methods-Conventional	*359*
14.3.3.1	Impregnation Method	*359*
14.3.3.2	Precipitation Method	*359*
14.3.4	Monometallic Nanocatalysts	*360*
14.3.5	Bimetallic Nanocatalysts	*360*
14.3.6	Catalyst Support	*369*
14.3.7	Catalyst Characterization	*370*
14.4	Summary and Conclusions	*373*
	References	*373*

15 Catalysis/Selective Oxidation by Metal Oxides *393*
    Wataru Ueda
15.1   Introduction *393*
15.2   Development of Selective Oxidation Catalysts Based on Complex Metal Oxides *394*
15.3   Distinct Types of Complex Metal Oxide Catalysts *395*
15.3.1 Titanosilicates *397*
15.3.2 Divanadyl Pyrophosphate *397*
15.3.3 Heteropoly Compounds *399*
15.3.4 Crystalline $Mo_3VO_{11.2}$ *400*
15.3.5 Multicomponent Bi–Mo–O *401*
15.4 $Mo_3VO_{11.2}$ as an Unique Selective Oxidation Catalyst with High Dimensionally Structures *403*
15.4.1 Oxide Catalysts for Oxidative Dehydrogenation of Ethane to Ethene *404*
15.4.2 Crystal Structures of $Mo_3VO_{11.2}$ Catalysts *405*
15.4.3 Formation Mechanism of the Crystal Structures *405*
15.4.4 Catalysis Field of Crystalline $Mo_3VO_{11.2}$ for Selective Oxidation of Ethane *407*
15.4.5 Propane (amm)Oxidation Over Mo–V–Te(Sb)–Nb–O Catalysts *409*
15.5   Oxide Catalysts as a New Material Category *411*
15.6   Summary *413*
        References *413*

16 Activity of Zeolitic Catalysts *417*
    Xiangju Meng, Liang Wang, and Feng-Shou Xiao
16.1   Introduction *417*
16.2   Zeolites in Fluid Catalytic Cracking (FCC) *418*
16.2.1   Y Zeolites in FCC *418*
16.2.2   MFI Zeolites in FCC *419*
16.2.3   Rare Earth (RE) in Zeolites in FCC *420*
16.3   Zeolites in MTO *421*
16.3.1   Hydrocarbon Pool Mechanism (HCP) *422*
16.3.2   Direct Mechanism *424*
16.3.2.1 Carbene Mechanism *424*
16.3.2.2 Oxonium Ylide Mechanism *424*
16.3.2.3 Others *424*
16.3.3   Catalyst Deactivation *425*
16.4   Zeolites in FT Synthesis *426*
16.4.1   Hybrid Bifunctional F-T Catalyst *426*
16.4.2   Core–Shell-Structured Catalyst Bifunctional F-T Catalyst *429*
16.5   Zeolites in Biomass Conversion *430*
16.5.1   The Role of the Acidity *430*
16.5.2   Shape-Selectivity of Zeolites in Biomass Conversion *432*
16.5.3   Stability of Zeolites in Biomass Conversion *432*

16.6	Zeolites in Selective Catalytic Reduction (SCR)	433
16.6.1	Cu-Zeolites as SCR Catalyst	433
16.6.2	Fe-Zeolites as SCR Catalyst	435
16.7	Conclusions and Outlook	435
	References	436

17 Nanocatalysis: Catalysis with Nanoscale Materials *443*
Tewodros Asefa and Xiaoxi Huang

- 17.1 Introduction *443*
- 17.1.1 Nanoscience and Nanotechnology *443*
- 17.1.2 Catalysis *444*
- 17.1.3 Nanocatalysts and Nanocatalysts *444*
- 17.2 Types of Nanocatalysts/Nanocatalysis Based on Activation *445*
- 17.2.1 Nanophotocatalysis *445*
- 17.2.2 Thermocatalysis *448*
- 17.2.3 Electrocatalysis *449*
- 17.3 Synthetic Methods Used for Nanocatalysts *449*
- 17.3.1 Zero-Dimensional (0D) Nanocatalysts *450*
- 17.3.2 One-Dimensional (1D) Nanocatalysts *453*
- 17.3.3 Two-Dimensional (2D) Nanocatalysts *453*
- 17.3.4 Three-Dimensional (3D), Supported, Core–Shell, and Multifunctional Nanocatalysts *455*
- 17.3.5 Core–Shell Type Nanostructured Catalysts *460*
- 17.3.6 Metal Organic Framework and Covalent Organic Framework Nanostructured Catalysts *462*
- 17.3.7 Bifunctional- and Multifunctional Nanocatalysts *463*
- 17.4 Effect of Nanostructure (Size, Shape, Grain Boundary, Pores, etc.) on their Catalysis *464*
- 17.4.1 Effect of Size *464*
- 17.4.2 Effect of Shape of Nanocatalysts on Catalysis *465*
- 17.4.3 Effect of Porosity *467*
- 17.5 Characterizations Methods for Nanocatalysts and Nanocatalysis *468*
- 17.6 Applications of Nanocatalysts or Nanocatalysis *468*
- 17.7 Future Prospects and Outlooks and Conclusions *469*
  References *470*

18 Heterogeneous Asymmetric Catalysis *479*
Ágnes Mastalir and Mihály Bartók

- 18.1 Introduction *479*
- 18.2 Aldol Addition *480*
- 18.3 Michael Addition *483*
- 18.4 Diels–Alder Reactions *488*
- 18.5 Diethylzinc Addition *491*

18.6	Epoxidation *494*
18.7	Hydrogenation *498*
18.8	Summary *502*
	Acknowledgment *504*
	References *504*

19 Catalysis by Metal Carbides and Nitrides *511*
Connor Nash, Matt Yung, Yuan Chen, Sarah Carl, Levi Thompson, and Josh Schaidle

19.1	Introduction *511*
19.2	Surface Structure and Active Site Identity *513*
19.3	Catalytic Applications *515*
19.3.1	Biomass Conversion *515*
19.3.1.1	Reforming Reactions for H_2 and CO Production *516*
19.3.1.2	Hydrodeoxygenation Reactions *518*
19.3.1.3	Knowledge Gaps *521*
19.3.2	Syngas and CO_2 Upgrading *521*
19.3.2.1	CO Hydrogenation *522*
19.3.2.2	CO_2 Hydrogenation *525*
19.3.2.3	Knowledge Gaps *528*
19.3.3	Petroleum and Natural Gas Refining *529*
19.3.3.1	Reaction Type and Catalytic Performance *529*
19.3.3.2	Knowledge Gaps *536*
19.3.4	Electrocatalytic Energy Conversion, Energy Storage, and Chemicals Production *537*
19.3.4.1	Electrocatalytic Energy Conversion *537*
19.3.4.2	Energy Storage *541*
19.3.4.3	Chemicals Production *542*
19.3.4.4	Knowledge Gaps *544*
19.4	Conclusions and Focus Areas for Future Research *544*
	References *546*

20 Combinatorial Approaches for Bulk Solid-State Synthesis of Oxides *553*
Paul J. McGinn

20.1	Introduction *553*
20.2	Powder Synthesis from Liquid Precursors *554*
20.2.1	Sol–Gel *554*
20.2.2	Polymer Complex Processing *555*
20.2.3	Combustion Synthesis *555*
20.2.4	Hydrothermal Synthesis *555*
20.2.4.1	Parallel Hydrothermal Synthesis *555*
20.2.4.2	Serial Hydrothermal Synthesis *556*
20.2.5	Flame Spray Pyrolysis *557*
20.3	Combinatorial Inkjetting *558*
20.4	Paste and Slurry Routes *559*

20.5	Combinatorial Synthesis from Dry Powders	*561*
20.5.1	High-Throughput Ball Milling	*561*
20.5.2	Injection Molding	*564*
20.6	Diffusion Couples	*564*
20.7	Summary	*566*
	References	*567*

Index *573*

1
Electrical Energy Storage: Batteries

Eric McCalla

University of Minnesota, Chemical Engineering and Materials Science Department, 421 Washington Ave SE # 250, Minneapolis, MN 55455, USA

1.1
Motivation

Secondary batteries have become the dominant source of energy for personal electronics and are the most promising technology for widespread implementation of electric vehicles. Much of the innovations that have taken place in the last 20 years in the field of Li-ion batteries (and other battery technologies) relate to the design of new materials allowing for greater energy density without compromising long-term performance. This chapter is devoted to components of modern batteries where solid-state chemistry has played and will continue to play a significant role. The batteries of particular interest here will be Li-ion (nearing maturity) and all solid-state batteries (currently used for niche applications only). It should be noted that other battery technologies including Li–S and Li–air batteries are receiving a great deal of attention from battery researchers. However, the major challenges in these systems pertain to reactions taking place in the liquid electrolyte (be it reactions with oxygen gas in Li–air batteries or the formation of dissolved polysulfides in Li–S batteries). By contrast, in metal–ion battery systems the stoichiometry and synthesis conditions of the electrode materials have dramatic effects on the battery properties, and solid-state chemistry has therefore played a vital role in developing these technologies. That being said, the sections pertaining to all solid-state batteries may also impact other battery technologies such as Li–air batteries where solid electrolytes could be revolutionary.

Handbook of Solid State Chemistry, First Edition. Edited by Richard Dronskowski, Shinichi Kikkawa, and Andreas Stein.
© 2017 Wiley-VCH Verlag GmbH & Co. KGaA. Published 2017 by Wiley-VCH Verlag GmbH & Co. KGaA.

1.2 Fundamentals of State-of-the-Art Li-Ion Batteries

1.2.1 LiCoO₂ and Graphite Electrodes

Before delving into how solid-state chemistry has contributed to modern batteries, it is necessary to introduce the functioning of a state-of-the-art Li-ion secondary battery. The most common electrodes used today are $LiCoO_2$ and graphite, and so these will be used as the primary example in this section. Figure 1.1a shows a schematic of such a battery with $LiCoO_2$ as the positive electrode (so named as its electrochemical potential is always above that of the other electrode) and graphite as the negative electrode. When first assembled, the battery is in the discharged state, such that a current must be driven through the external circuit in order to remove Li from $LiCoO_2$ and transfer it to the graphite via the ionic conducting electrolyte that soaks the separator. Then, when the energy is to be provided by the battery, lithium is transferred back to the positive electrode while electrons do work through the external circuit.

The changes in electronic structure during electrochemical cycling govern the voltage that the battery can provide. Figure 1.1b shows a schematic of the energy diagram of a simple battery with $LiCoO_2$ as the positive electrode and lithium metal as the negative. As assembled, the Li 2s bands in the lithium metal are half filled and cobalt is in the 3+ oxidation state such that the 3+/4+ 3d band is fully occupied. During charging, a Li+ ion and an electron are removed from $LiCoO_2$ and recombine at the negative electrode resulting in an energy increase of $\Delta\mu$ shown on the energy diagram. The voltage required to transfer the Li-ion is therefore $V = -\Delta\mu/e$ and is proportional to the difference in energy of the highest filled levels in the two bands (Li 2s and Co 3d in the case of $LiCoO_2$) [3].

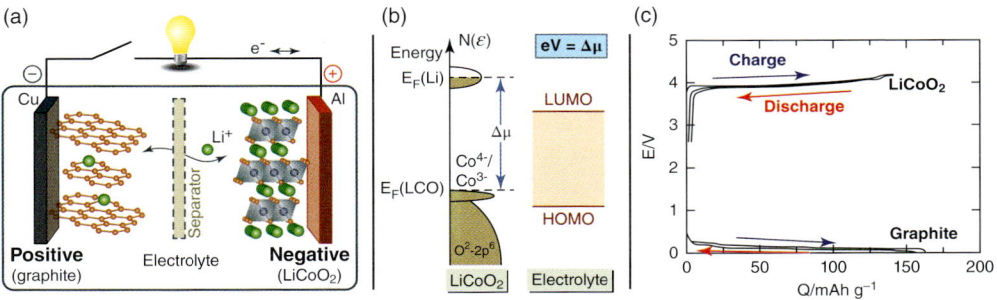

Figure 1.1 (a) Schematic of a state-of-the-art Li-ion battery. (b) Schematic of the band structure of $LiCoO_2$ and Li metal, also shown is the stability window of the electrolyte. The energy scale can also be expressed as Li chemical potential (μ). (c) Voltage versus specific capacity Q for each electrode when cycled against lithium metal (graphite capacity was divided by 2). (Parts (a) and (b) reproduced from Ref. [1] and part (c) from Ref. [2] with permission from the American Chemical Society and Elsevier.)

Thus, as the cell is charged, the shift in Fermi levels increases the voltage until the Co 3+/4+ 3d band is empty. Figure 1.1c shows that for $LiCoO_2$ cycled against lithium, the chemical potentials result in a voltage of 3.9 V. Such a graph is obtained by charging the battery with a constant current until it reaches a maximum cutoff voltage, and then reversing the current until it reaches a lower cutoff and repeating. The figure also shows a graphite electrode cycled versus lithium with a plateau at 0.1 V. Therefore, a $LiCoO_2$/graphite cell typically has an average voltage of 3.8 V. Given the relationship between electronic structure and voltage, one important contribution of solid-state chemistry has been to manipulate the bands in order to increase voltage.

The large push toward greater energy density has motivated much of the research performed to date. The energy density is simply the average voltage multiplied by the charge transferred during discharge. Thus, specific capacity (charge transferred per unit mass, in mAh/g) is an important parameter. There are two limits to the extent to which lithium can be extracted from a positive electrode: the lithium content, and the redox processes available in the material to compensate for the removal of lithium (e.g., oxidation of Co from 3+ to 4+ in the case of $LiCoO_2$). In the case of $LiCoO_2$ the maximum theoretical lithium extraction is 1.0 per formula unit as cobalt is oxidized from 3+ to 4+, giving a theoretical capacity of 272 mAh/g. Thus, in the voltage plots, the key properties to pick out are the total capacity for a given cycle (particularly the discharge capacity) and the voltage at which this capacity is discharged as the energy density is simply the product of these two. Another important property to consider is the irreversible capacity, which is the difference between the charge and discharge capacities in the first cycle. This corresponds to lithium that is lost during the first cycle, most notably in reacting at the surface of the negative electrode where a passivation layer forms (called the solid electrolyte interphase (SEI)). Irreversible capacity must be minimized as it results in inefficiency such that the theoretical energy density can never be reached. Materials with a very high irreversible capacity are, therefore, not easy to integrate in commercial cells.

1.2.2
Electrolytes

The role of the electrolyte is to be electronically insulating while maintaining high ionic conductivity, and be stable under the operating conditions of the battery. Thus, the LUMO of the electrolyte should have a higher energy than the chemical potential of the negative electrode thereby preventing reduction at the negative electrode. However, this is not the case for negative electrodes operating at low potentials. Since the electrolyte is not stable at 0.1 V versus Li/Li+ where graphite operates, the electrolyte decomposes and forms the passivating SEI [4]. This SEI is essential for the safe functioning of the battery, but its formation must be minimized to limit the irreversible capacity. One way to do this is to limit the surface area of electrode materials, such that particle morphology is important in electrode materials. Similarly, at the positive electrode, ideally the

HOMO of the electrolyte should be lower in energy than the electrode chemical potential (as illustrated in Figure 1.1), thereby preventing oxidation of the electrolyte during charging of the cell. For LiCoO$_2$, this is the case up to a voltage of about 4.4 V versus lithium metal [5]. Therefore, in designing positive electrodes, operating potentials should not exceed these levels, unless more stable electrolytes are designed. In commercial cells, the electrolyte is a carbonate-based liquid, but as will be discussed in this chapter there is also a current push toward using solid electrolytes.

1.3
Summary of Structures and Properties of Interest

A few classes of materials will be discussed extensively throughout this chapter and their structures and properties warrant some attention. But first, it is important to have a clear idea of which properties need improving in order to make batteries competitive for new applications such as electric vehicles. Figure 1.2a shows a radar plot of current values for a number of properties of interest for electric vehicles along with targets for 2022 and 2025 [6]. Although safety is always a pressing issue, the current safety is quite excellent primarily due to features at the cell and pack levels. As such, the message to researchers is that new electrode materials need only be as safe as LiCoO$_2$/graphite and if that is the case, overall safety can be managed at the cell and pack levels. The properties that are of highest interest to industry are therefore the energy density, the rate at which the battery can be charged and the cost. A great deal of research continues to be done with these goals in mind and as ever we are guided by the periodic table shown in Figure 1.2b. The task is clearly quite daunting with a great many possible elements that may yet contribute to improving the battery

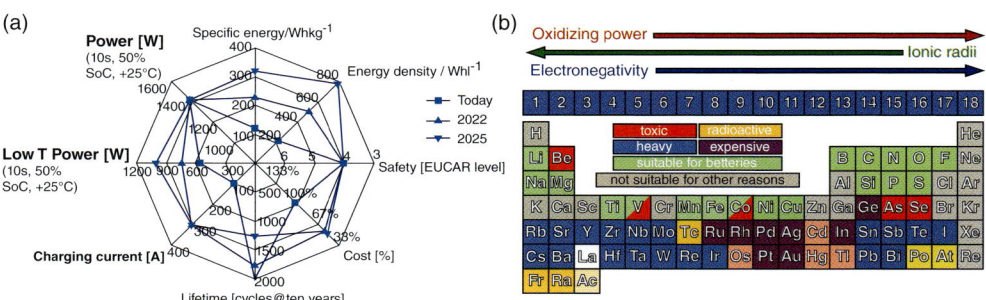

Figure 1.2 (a) Radar plot of the key properties in state-of-the-art Li-ion batteries along with targets for electric vehicles. (b) The playground for solid-state chemistry showing elements that are/may be used in battery technologies. (Parts (a) and (b) reproduced from Refs [6] and [7], respectively, with permission from the Royal Society of Chemistry and the American Chemical Society.)

performance. In the next sections, the properties of currently competitive materials will be outlined in order to establish the benchmarks for ongoing research.

1.3.1
Positive Electrode Materials

Table 1.1 shows key properties for positive electrode materials for Li-ion batteries. The most important properties are energy density and median voltage. Obviously, the objective is to maximize energy density, however it is also necessary to keep the cell operation away from very high potentials as this cannot be sustained with current electrolytes, which is why, for example, $LiNi_{0.5}Mn_{1.5}O_4$ is not currently used commercially. There are two classes of materials that are used significantly at this time: layered oxides, and polyanion ($LiFePO_4$). The former are used because of their high energy density, while the latter is used primarily due its greater rate capability. Of note, the next-generation material $Li[Li_{1/9}Ni_{1/3}Mn_{5/9}]O_2$ is an example of a Li-rich layered oxide and will be discussed throughout this chapter as it yields a significant energy increase compared to the current state of the art.

The structures of these materials are shown in Figure 1.3. The layered oxide structure involves MO_2 slabs (where M is a transition metal) that are hexagonal arrays of MO_6 octahedra. Between each slab is a hexagonal layer of lithium. The oxygen sublattice is cubic close-packed and in layered notation they stack in an ABCABC sequence. The M layer is commonly referred to as the transition metal layer and the properties of the electrode material can be greatly influenced by

Table 1.1 Important properties for commercialized and potential next-generation positive electrode materials [3,6,8].

Material	Commercialized	Specific capacity (mAh/g)	Median voltage (V)	Volumetric energy (Wh/cm³)
Layered oxides				
$LiCoO_2$	Yes	155	3.9	3.05
$Li[Ni_{1-x-y}Mn_xCo_y]O_2$	Yes	140–180	3.8	2.54–3.26
$Li[Ni_{0.8}Co_{0.15}Al_{0.05}]O_2$	Yes	200	3.73	3.54
$Li[Li_{1/9}Ni_{1/3}Mn_{5/9}]O_2$	No	240	3.8	4.06
Spinel				
$LiMn_2O_4$	Yes	120	4.05	2.08
$LiNi_{0.5}Mn_{1.5}O_4$	No	130	4.6	2.63
Polyanion				
$LiFePO_4$	Yes	160	3.45	1.99
$LiFeSO_4F$	No	140	3.6	1.62
Conversion				
FeF_3	No	690	2.0	5.34

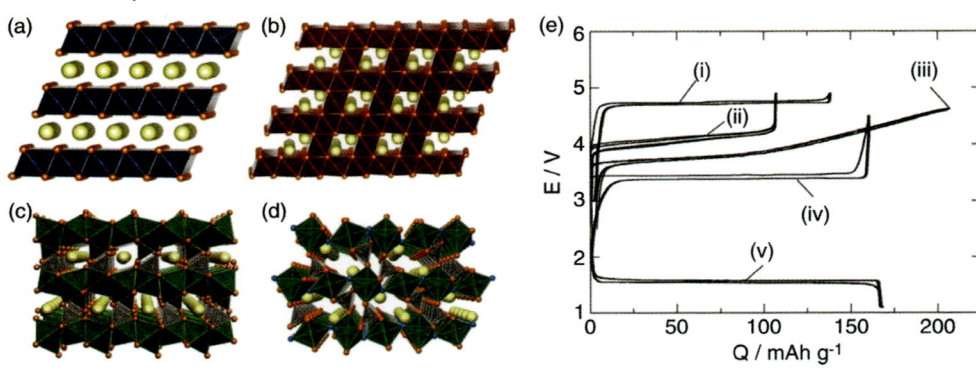

Figure 1.3 Structures of typical intercalation materials used as positive electrodes. (a) Layered LiCoO$_2$. (b) Spinel LiMn$_2$O$_4$. (c) Olivine LiFePO$_4$. (d) Tavorite LiFePO$_4$F. Li atoms are yellow, oxygen are orange. (e) The cycling curves versus lithium metal for (i) LiNi$_{0.5}$Mn$_{1.5}$O$_4$, (ii) LiMn$_2$O$_4$, (iii) Li Ni$_{1/3}$Co$_{1/3}$Mn$_{1/3}$O$_2$, (iv) LiFePO$_4$, and (v) Li$_4$Ti$_5$O$_{12}$. (Parts (a–d) reproduced from Ref. [2] and part (e) from Ref. [7] with permission from the American Chemical Society and Elsevier.)

the composition of this layer (see Table 1.1). This has therefore proved ideal for study by solid-state chemists. The other key feature of this structure for batteries is that lithium can diffuse in 2D along the lithium layer.

The spinel structure is related to the layered one in that oxygen is in a cubic close-packed lattice, but here the overall stoichiometry is 3 metal to 4 oxygen, compared to 1:1 in the layered structure. The resulting structure has lithium in tunnels, which criss-cross such that lithium diffusion during cycling is through this 3D network. The polyanion structures shown in Figure 1.3c and d, by contrast, have 1D tunnels only. One important feature of all intercalation materials is a robust framework that lithium can diffuse into and out of.

The final positive electrode material that warrants some attention is the conversion materials (FeF$_3$ is one example included in the table). Here, the energy is extremely high and therefore of great interest; however, it is fundamentally different in that it is made in the charged state and lithium must be added to the material during charge. It implies that the conversion electrode must be used with a lithium metal negative electrode. Therefore, for the solid-state chemist, there is little interest in conversion electrodes until lithium metal can be made safe in a battery as will be discussed in the next section.

1.3.2
Negative Electrode Materials

Table 1.2 shows the properties of some key negative electrode materials. The use of lithium metal has to date not been possible in commercial Li-ion batteries due to the fact that dendrites form on the surface of the metal during extended cycling and this can result in a short circuit if the dendrite pierces the separator between the two electrodes leading to rapid heating of the cell and the

Table 1.2 Properties of commercialized and potential negative electrode materials.

Material	Lithiated phase	Commercialized	Specific capacity (mAh/g)	Median voltage (V)	Volume change (%)
Lithium	Li	No	3862[a]	0	100
Graphite	LiC_6	Yes	360	0.1	12
$Li_4Ti_5O_{12}$	$Li_7Ti_5O_{12}$	Yes	170	1.5	1
Silicon	$Li_{4.4}Si$	No	4200[a]	0.4	320

Source: Reproduced with permission from Elsevier [9].
a) Theoretical capacity.

electrolyte reacting so rapidly that the cell may combust and explode. This is why graphite is used today, even though the capacity is significantly lower.

The other commercialized material is again used for its stability rather than its contribution to energy density: $Li_4Ti_5O_{12}$ is a spinel material (3 metal per 4 oxygen) that can accept 3 more lithium atoms during a voltage plateau at 1.5 V (shown in Figure 1.3e). This high voltage means that the energy density of the cell is greatly diminished (the voltage is 37% lower for an $LiCoO_2$ cell with $Li_4Ti_5O_{12}$ instead of graphite), but it also means that the electrolyte is stable at the negative electrode and as such cells with $Li_4Ti_5O_{12}$ can cycle longer with negligible energy loss during cycling. The extremely small volume expansion is also advantageous as it results in very little strain in the electrodes and mitigates any cracking of particles that might otherwise occur. This is in very sharp contrast to silicon that has a massive capacity but the volume expansion results in fracturing of particles and rapid capacity loss. In general, the work of solid-state chemists in the realm of negative electrodes has been fairly limited. This can be justified by comparing the properties of positive and negative electrodes: the energy density of current state-of-the-art batteries is limited by the specific capacity at the positive electrode.

1.3.3
Solid-State Electrolytes

The main properties that are typically studied for solid electrolytes are the conductivities (ionic and electronic) and the electrochemical stability window. The ionic conductivity can be measured using electrochemical impedance spectroscopy (EIS) over wide temperature windows from room temperature to at least 300 °C [10]. Liquid electrolytes have ionic conductivities on the order of 10^{-2} S/cm, so it is estimated that a solid electrolyte would require a room-temperature conductivity of 10^{-3} S/cm to be competitive [11]. Figure 1.4 shows the structures of promising solid electrolytes and their corresponding ionic conductivities are included in Table 1.3. The structures are varied in complexity from the relatively simple perovskite structure where Li, La and vacancies share the A sites with 12-fold O coordination to the rather complex structure of the crystalline sulfide

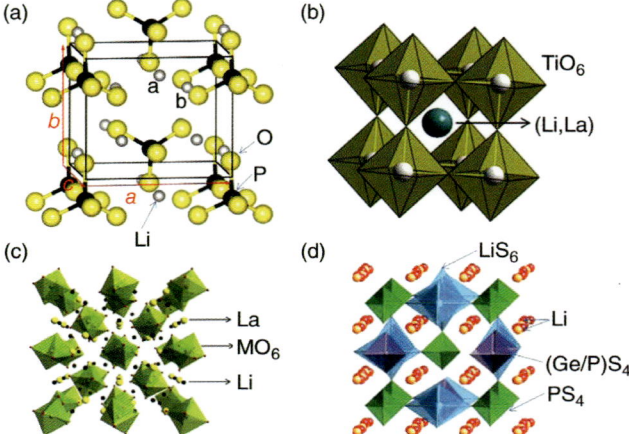

Figure 1.4 Structures of solid electrolytes. (a) Li$_3$PO$_4$ (basis of LiPON). (b) A Li–Ta–Ti perovskite with incomplete occupation of the Li/La site. (c) Garnet Li$_6$La$_3$M$_2$O$_{12}$. (d) The sulfide Li$_{10}$GeP$_2$S$_{12}$. (Part (a) reproduced from Ref. [12], parts (b) and (c) from Ref. [13], and part (d) from Ref. [14] with permission from the American Physical Society, the Royal Society of Chemistry, and the Nature Publishing Group, respectively.)

where Li involved in conduction lies in 1D tunnels. In all cases, the key feature is a rigid framework with sufficient lithium and/or disorder to provide percolation paths throughout the structure.

The complexity of the chemical formulas of these compounds attests to the extensive solid-state chemistry performed in this domain. For example, Li$_{10}$GeP$_2$S$_{12}$ is a composition within the ternary Li$_2$O–GeO$_2$–P$_2$S$_5$ system such that many studies were required to lead to the optimum composition. These studies have been highly successful to the extent that bulk conductivities now rival those of liquid electrolytes. However, this is proving insufficient to give all solid-state batteries that are competitive with their liquid electrolyte containing counterparts. The progress made to date and need for further work will be discussed in Section 5.4.

Table 1.3 Ionic conductivities for a few promising solid electrolytes [11,13,15].

Name	Composition	Ionic conductivity (S/cm)
Garnet	Li$_{6.4}$La$_3$Zr$_{1.4}$Ta$_{0.6}$O$_{12}$	1.0×10^{-3}
Sulfide	Li$_{10}$GeP$_2$S$_{12}$	1.2×10^{-2}
LiPON (amorphous)	Li$_{2.88}$PO$_{3.73}$N$_{0.14}$	3.3×10^{-6}
Perovskite	Li$_{3x}$La$_{2/3-x}$TiO$_3$	1.0×10^{-3}

1.4 Synthesis

There is a wide variety of synthesis techniques that can be used to make the battery materials discussed above. Here, the most common approaches are presented.

1.4.1 All Solid-State Synthesis

The most common way to produce positive electrode and solid electrolyte materials is to use high-temperature solid-state reactions (800–1000 °C is typical). Common starting materials are transition metal oxides reacting with either LiOH or Li_2CO_3. One concern is that Li is lost as Li_2O_2 above 700 °C [16] and is typically compensated for with a small amount of excess Li (2–5%) such that elemental analysis is required to confirm the correct stoichiometry after synthesis.

When multiple transition metals are in the target material, special care must be taken to ensure thorough mixing on atomic length scales. This can be accomplished in a number of different ways. The first is to ball mill the precursors thoroughly in order to ensure maximum mixing in the starting materials [17,18]. Though this is effective, it has the drawback of severely limiting the particle sizes. Sol–gel methods have also been successful in producing a precursor with intimate mixing of transition metals (e.g., $LiNi_{1/3}Mn_{1/3}Co_{1/3}O_2$ was made in Ref. [19]). Although such methods are useful to discover new phases, they are not the method of choice for industrial applications. Instead, a coprecipitation reaction is used to make an intimately mixed precursor containing all of the transition metals. The precursor is then reacted with a lithium source (e.g., LiOH) at high temperatures. For example, sulfates of Ni and Mn can be reacted with NaOH to form a precipitate [20]. The particle size is then easily controlled by varying the reaction time which is useful for battery materials where morphology plays an important role. The coprecipitation method is trivial to scale up for industrial applications as it can be performed in a continuous mode in a vat with gradual sampling away of the particles as they reach the desired size. This method also allows for easy synthesis of particles with complex morphology such as core–shell structures with one phase at the centers of all particles and another at the surfaces. To make such particles, the reactant concentrations added to the reaction must be varied with time. In practice, some care must be taken as ideal pH and stir rates change with the desired precipitate composition [21], however, this can be done to yield the desired particles, as illustrated in Figure 1.5.

Battery materials that can be made by solid-state synthesis include positive electrode materials such as the layered and spinel oxides, polyanion materials, $Li_4Ti_5O_{12}$ negative electrode, and solid-state electrolytes such as the perovskites, garnets, and sulfides.

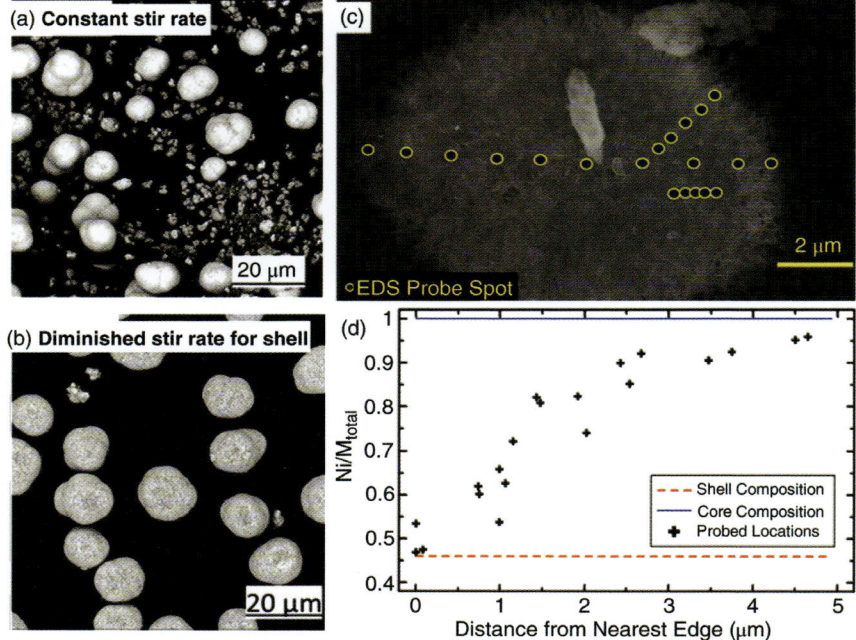

Figure 1.5 SEM images of core–shell particles with (a) poor synthesis conditions making small flaky particles of shell material and (b) proper synthesis where diminished stirring and adjustment of pH prevented the formation of shell particles. SEM image of the particle (c) used for EDX mapping in panel (d). The EDX data confirm the core–shell nature of the particle. (Reproduced from Ref. [21] with permission from the Electrochemical Society.)

1.4.2
Nanometric Carbon Coating Syntheses

Pure LiFePO$_4$ can be made by ceramic methods at high temperatures under flowing nitrogen, although such materials do not show good cycling performance as shown in Figure 1.6a. The poor performance is primarily due to the poor electronic conductivity and the fact that even small amounts of defects can impede the lithium extraction out of the 1D tunnels if the crystallites are large enough. The solution to these problems was to make the material with nanometric particles and coat them with carbon [22]. This solution is only viable because of the fact that the voltage plateau is at 3.45 V versus lithium metal and the electrolyte does not react such that the increased surface area is not detrimental as it would be for high-voltage positive electrodes. This reduction in particle size also vastly improves the rate capability of LiFePO$_4$, which is why it is used commercially today. A variety of methods can be used to make nanosized carbon-coated particles. The simplest is to simply make the battery material first and then ball mill with carbon [23]. However, there also exists a one-step

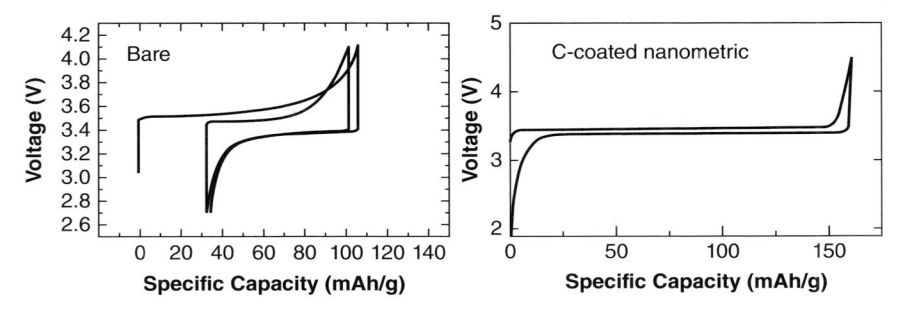

Figure 1.6 Cycling data for pure LiFePO$_4$ made by ceramic synthesis (a) and nanometric particles coated with carbon (b). (Part (a) reproduced from Ref. [2] and part (b) from Ref. [25] with permission from the Electrochemical Society.)

synthesis that gives rise to the desired material: the carbothermal reduction method. In the case of LiFePO$_4$, Barker *et al.* [24] used Fe$_2$O$_3$ with LiH$_2$PO$_4$ and carbon, which, when heated to 650 °C, produce LiFePO$_4$, carbon monoxide, and water. However, with 25% excess carbon, the final product is LiFePO$_4$ with a carbon coating. Their resulting cycling curve showed a capacity of about 97% of the theoretical capacity at room temperature, far better than that obtained in the bare material [24].

1.4.3
Soft Chemistry and the Discovery of New Phases

There are also a wide variety of synthesis routes that have been used in academic studies in order to prepare battery materials and these have led to new phases. Such materials include compounds that form metastable phases at lower temperatures (roughly <500 °C) and these low-temperature syntheses, including solvothermal, hydrothermal, sol–gel, and ionothermal, are commonly referred to as soft chemistry.

A recent example of a promising battery material that was discovered by soft chemistry is LiFeSO$_4$F, which was first synthesized using ionothermal means by reacting iron sulfate hydrate with LiF in the presence of an ionic liquid [8]. Ionic liquids can be readily used as functional solvents in order to promote the growth of crystalline phases [26], but only recently have they been used to make battery materials. Not only did this give a phase never obtained by ceramic methods but this ionothermal synthesis was also obtained at 250 °C, thereby diminishing the cost compared to high-temperature ceramic methods. The product showed electrochemistry comparable to that of LiFePO$_4$ (properties are included in Table 1.1) in micrometer-sized particles without carbon coating as required for LiFePO$_4$, although it has the added challenge of being unstable in water such that it decomposes in air. From a cost perspective, this is still a very attractive process and this work out of the Tarascon group has sparked a great deal of interest in using ionic liquids as the reaction medium. It is worth mentioning that other synthesis routes have since been found to make LiFeSO$_4$F, including solvothermal synthesis [27]. Nonetheless, the ionic liquid medium has shown its worth in the discovery of new phases, and may very well prove to be a green, cost-effective synthesis route for commercial materials as suggested by Eshetu [28].

1.5
Contributions of Solid-State Chemistry to Batteries

There are a number of common strategies used by solid-state chemists in order to improve the performance of a particular battery component with some of them having already been demonstrated in the synthesis section. A complete examination of this topic is beyond the scope of this chapter, but a few examples will be discussed here, with the layered oxides being the primary illustrative example given that this class of materials is the most widely used and studied positive electrode materials. Nonetheless, the methods discussed here are widely applicable for all battery materials studied today.

1.5.1
Cationic and Anionic Substitutions

The most common contribution of solid-state chemistry to the discovery of new battery materials comes about by starting with a promising phase and applying substitutions for the various elements present. The periodic table included in Section 1.3 shows how vast the playing field is for battery materials. In searching through the countless possible compositions, simple rules of thumb are helpful. First, in general, the more ionic the metal–ligand bond, the lower the Fermi level (illustrated in Figure 1.7), and as such the higher the voltage. Thus, a larger difference in electronegativities between the metal and the ligand increases the voltage. This can be seen in the development of LiCoO$_2$ that replaced TiS$_2$ as the positive electrode. TiS$_2$ forms LiTiS$_2$ at a potential of 2.5 V versus lithium metal [29]. The use of LiCoO$_2$ by Goodenough and coworkers [30] can therefore be viewed as a substitution of Co for Ti and O for S, which increases the difference in electronegativity from 1.04 to 1.56 and so the potential increases significantly, up to 3.9 V. For polyanionic materials, the situation becomes more complex. The inductive effect, illustrated in Figure 1.7, describes how, for a polyanion LiM(SX$_4$), the nature of the S—X bonds impact the degree of covalency in the M—X bonds. More specifically, as the S—X bonds become more covalent, there is a charge transfer away from the metal M making the M—X bond more ionic and thus the cell operates at a higher voltage. As shown in Figure 1.7, this explains the trends seen with P, W, and S used as the central S atom. These simple rules help guide us in our choices of substitutions as we manipulate the electronic bands to vary the voltage of the battery.

One can also predict how substitutions affect the capacity of an electrode material by determining the possible redox couples at play. For example, Ni is

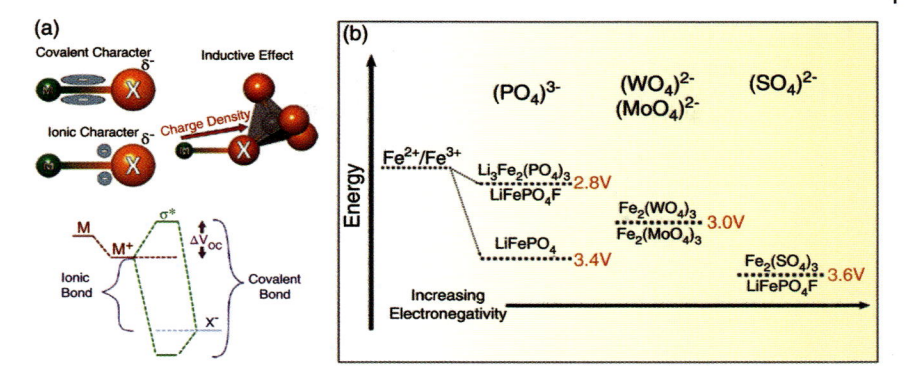

Figure 1.7 Illustration of the inductive effect. (a) For the same redox couple, the more ionic the metal–ligand bond, the higher the voltage. In a polyanionic material, as the central atom S in the polyanion becomes more electronegative, the S—X bonds become more covalent, which induces a charge transfer from the M atom, thereby increasing the ionic character of the M—X bond. (b) The inductive effect is shown for some polyanionic materials: the voltage of the Fe 2+ → 3+ transition increases as the electronegativity of the central atom in the polyanion increases. (Reproduced from Ref. [7] with permission from the American Chemical Society.)

often substituted for Co in layered oxides as Ni can be oxidized from 2+ to 4+, compared to the 3+ to 4+ transition in cobalt. These examples illustrate how substitutions can be used to increase the energy density by either increasing the voltage or the capacity of the cell. In the case of layered oxides, many other substitutions have been attempted, sometimes as an inactive stabilizing element that gives better long-term cycling. An example of this is the use of aluminum in $LiNi_{0.8}Co_{0.15}Al_{0.05}O_2$, where aluminum helps mitigate the poor performance resulting from high nickel content [3]. However, it is difficult to predict how an inactive element will contribute to the cycling, and ultimately such studies rely on trial and error. As such, the most common solid-state chemistry studies of battery materials have involved the synthesis of a particular composition line and its corresponding electrochemistry, with fundamental understanding coming much later as exemplified in the next section.

1.5.2
Unraveling the Mechanisms during Cycling

There are many examples of careful determination of mechanisms involved during cycling of positive electrode materials. Here, focus is on a recent discovery that disproves the long held belief that only the cations undergo oxidation/reduction during cycling. It was shown simultaneously by two groups that for Li-rich oxides, oxygen participates reversibly in the redox process [17,31]. These studies serve here as examples of the typical methods used to unravel the electrochemical mechanisms at work.

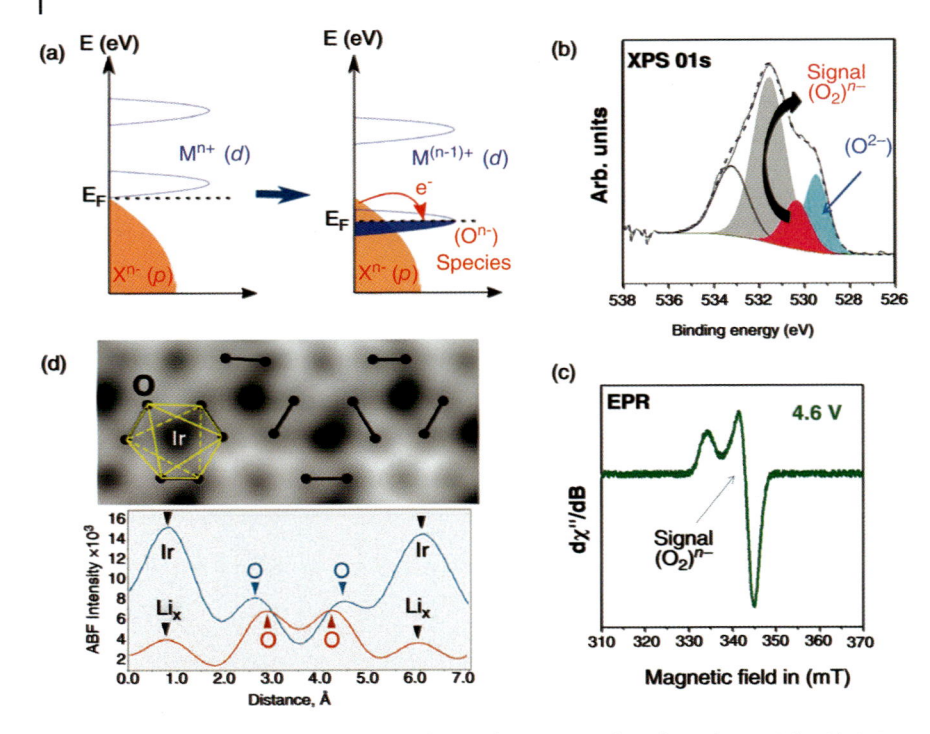

Figure 1.8 Band structure of a Li-rich oxide (a) that undergoes reversible oxygen redox during electrochemical cycling. XPS (b) and EPR (c) data showing the new oxidized oxygen species in Li–Ru–Sn–O materials. (d) Electron microscopy showing the shortened O—O distances resulting from the partial oxidation of oxygen in Li_2IrO_3. (Parts (a–c) reproduced from Ref. [34] and part (d) from Ref. [18] with permission from the Electrochemical Society and American Association for the Advancement of Science, respectively.)

The Tarascon group showed the reversible oxygen redox using model systems: $Li_2Ru_xSn_{1-x}O_3$ and Li_2IrO_3 both of which show the very high capacity obtained for Li-rich oxides. Figure 1.8b and c shows that X-ray photoemission spectroscopy (XPS) and electron paramagnetic resonance (EPR) both showed a new oxygen species in the charged state that disappeared during discharge, while DFT calculations predicted that this species was a peroxo-like O_2^{n-} species with shortened O–O distances [17]. This species appeared after a plateau in the voltage curve where the only element showing changes in oxidation was oxygen, and so it was proposed that the species arose from the fact that the Ir/Ru bands lay below the O 2s band, thereby resulting in a charge transfer from the oxygen back to the metal atoms, as illustrated in the band structure sketch in Figure 1.8a. The shortened O–O distances were also directly observed by aberration corrected transmission electron microscopy as shown in panel (d) [18]. When beam damage can be avoided (which is not always possible in battery materials), high-resolution TEM has proven to be a powerful tool in determining the changes taking place at the atomic scale during electrochemical cycling, with discoveries such as the extensive reversible migration of cations [32] and densification of the surface of particles [33], both taking place in Li-rich oxides. It may be noted that XPS and EPR are only two of the possible techniques used to probe the changes in the electronic states after electrochemical cycling, with X-ray absorption spectroscopy also being commonly used as the leading absorption edge shifts with the average oxidation state.

Croguennec and coworkers also discovered this oxygen redox, but proved it by using a chemical reaction to delithiate the material by reacting it with NO_2BF_4 and then relithiating it by reacting it with LiI [31]. These well-known chemical reactions correspond roughly to potentials of 5.0 and 3.0 V versus lithium. This approach is useful for chemists as the products are not contaminated with carbon and electrolyte decomposition products as they would be after being used in a battery. This greatly simplifies analysis and permitted Koga *et al.* to use redox titrations to determine the average metallic oxidation state and demonstrate that it did not account for the lithium extracted unless some oxygen was oxidized during charge [31]. Although these studies are of a fundamental nature, they have opened up the path for new battery materials that combine both cationic and anionic redox processes in order to maximize the capacity of the material.

1.5.3
Control of Particle Morphology/Coatings

As already discussed in the context of the synthesis of $LiFePO_4$, the morphology of the particles can greatly impact the electrochemical performance, and coatings can be used to great advantage (e.g. to increase conductivity in the case of $LiFePO_4$). Another important example is for $LiCoO_2$ where coatings are used for a different reason. One of the primary limitations of $LiCoO_2$ is that it undergoes phase transformations during charging beyond 4.2 V (i.e., after 0.5 Li extraction), which limits the capacity to about 150 mAh/g, well below the theoretical capacity of 272 mAh/g. Figure 1.9 shows these phase transformations wherein the CoO_2 slabs sheer against each other to change the oxygen stacking from ABCABC to ABAB. The resulting cycling data shows a dramatic decrease in capacity with extended cycling. However, the figure also shows that a thin coating of Al_2O_3 suppresses the phase transformations and the capacity fades far more slowly [35]. This demonstrates the role surfaces of particles can play in stabilizing structures during electrochemical cycling. Coatings can be used for other purposes, including HF scavenging and protection from the electrolyte [36].

Another approach to control the electrode–electrolyte interface is to make particles with a composition gradient as discussed in the synthesis section. This makes it possible to use a material with high energy density as the core and another with low reactivity with the electrolyte as the shell. This has the advantage over a coating in that the shell is electrochemically active. Sun *et al.* demonstrated this nicely using a Ni-rich core stabilized by a Mn-rich shell with a concentration gradient to reduce the internal strain within the particles [38]. Figure 1.10 shows that the resulting material has a capacity close to that of the

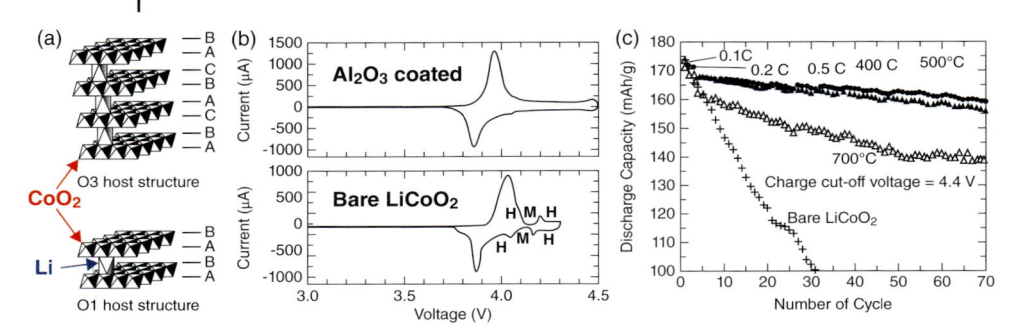

Figure 1.9 (a) During cycling, the LiCoO$_2$ structure changes from an O$_3$ stacking (ABCABC stacking of oxygen layers) to an O1 structure (ABAB stacking). (b) The cycling data of bare LiCoO$_2$ as well as with an Al$_2$O$_3$ coating. The current versus voltage plot can be viewed as the derivative of the voltage versus capacity plots shown throughout this chapter such that peaks here correspond to plateaus in V versus Q. H and M indicate the peaks that are the signature for the phase transitions between the O$_3$ and O1 structures. (c) The discharge capacity for samples with Al$_2$O$_3$ coatings treated at various temperatures. (Part (a) reproduced from Ref. [37] and parts (b) and (c) from Ref. [35] with permission from the American Physical Society and the Electrochemical Society, respectively.)

core material but with the same stability as that of the shell. This proof of concept opens up the path to using higher energy materials such as Li-rich oxides with a protective shell.

1.5.4
Making All-Solid Batteries a Reality

There are high hopes that solid-state electrolytes can be used in many battery technologies, including Li-ion, Li–S, and Li–air. In particular, solid electrolytes

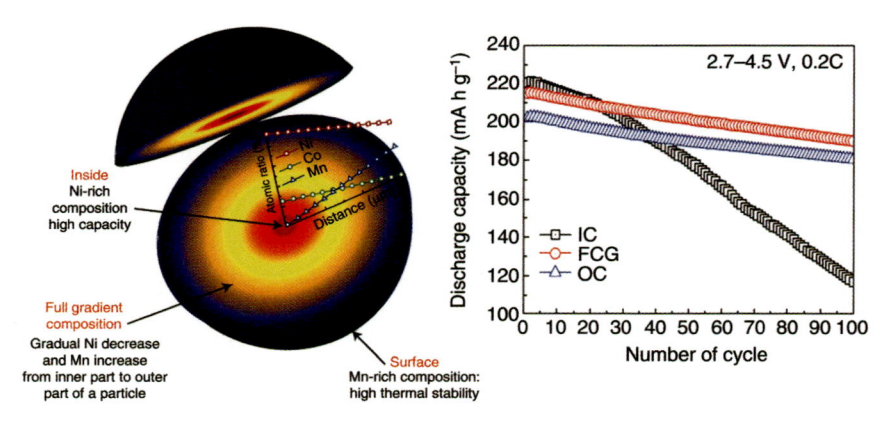

Figure 1.10 The use of gradient particles to mitigate the internal strain in nonuniform particle materials. FCG is the gradient material, IC is the innermost material, and OC is the outermost material. (Reproduced from Ref. [38] with permission from Nature Publishing Group.)

may increase the stability window such that positive electrodes operating above 4.4 V can be used, and they may also suppress dendrite formation in Li metal. These hopes have sparked a great deal of research. As stated earlier, the conductivity of solid electrolytes is no longer the limiting factor with the best being comparable to that of liquid electrolytes. However, this has proved insufficient to make all-solid batteries with good cycling and high energy density a reality. Two strategies are currently used to make a solid battery with good electrochemical cycling. The first is to make a microbattery with thin electrodes and a very thin sputtered electrolyte. This is currently used for niche applications and being developed for applications such as flexible wearable batteries [39]. The electrolyte is typically LiPON that can be sputtered as a thin film such that the short lithium paths through the electrolyte compensate for its moderate conductivity. The greatest issue with microbatteries is now the high resistance that forms at the interface between the positive electrode ($LiCoO_2$) and the electrolyte. The stability window of LiPON is small and is unstable at both at 4 V where $LiCoO_2$ operates and at 0.1 V where Li operates. In fact, all electrolytes currently studied are unstable at one or both electrodes [40,41] and this is illustrated in Figure 1.11. The resulting energy schematic shows that we must rely on the stability of the reaction products that act as an interphase between the electrode and electrolyte. In the case of LiPON, the interphase includes some ionic conducting materials such as Li_3P such that the battery does not shut down completely but the conductivity is diminished by the formation of the interphase [40]. Significant work is now being done in order to minimize the impedance at the interface in such batteries, and one such combinatorial approach will be discussed in the next section. However, it must be pointed out that microbatteries will never be able to compete with state-of-the-art Li-ion batteries in terms of energy density for large-scale applications (the thin electrodes result in a large portion of the cells being current collectors and the casing, thereby reducing the specific energy).

The second approach to make functional solid-state batteries is to make bulk electrodes as those used in Li-ion batteries. However, the poor lithium diffusion in the electrode prevents its operation, and so one typically has to add 40–50% of the electrolyte in the electrodes in order to have percolation through the electrode [42]. This results in an energy density at the battery level that is far diminished from that of the state-of-the-art Li-ion batteries. A number of approaches are being tested to diminish the amount of electrolyte required, but nothing to date has made it to market. Approaches include going to higher energy electrodes (such as conversion electrodes and Li metal) that could partially compensate for the extra electrolyte required [6]. However, it should be recognized that no careful study has yet been performed to see the extent to which a gradient composition within the electrode can be used to minimize the amount of electrolyte required. Such a gradient composition would maximize the amount of active material near the current collector, and ensure a large surface area for the electrode–electrolyte interface. There are therefore strategies to optimize all-solid Li-ion batteries that remain to be tested.

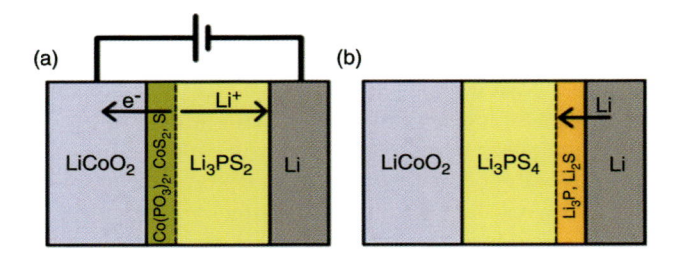

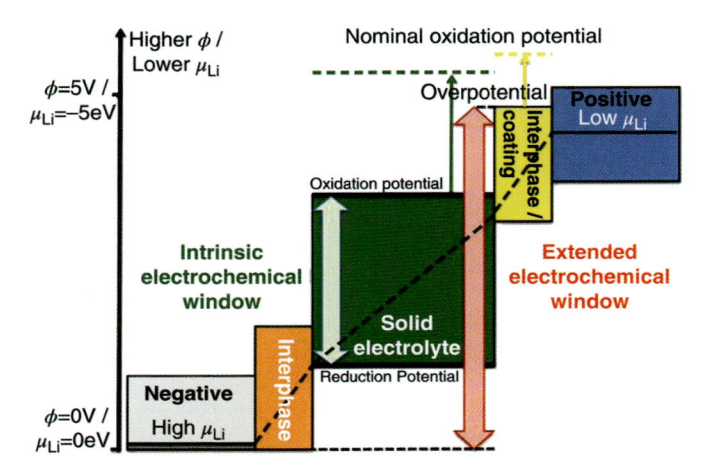

Figure 1.11 (a) Schematic of the interphase formed in a particular cell example (LiCoO$_2$/Li$_3$PS$_4$/Li). (b) A schematic of the chemical potential profile in a cell where interphases grow between the electrolyte and both electrodes. (Top panel reproduced from Ref. [40] and the bottom panel from Ref. [41] with permission from the American Chemical Society.)

1.6 High-Throughput Chemistry

The above discussions make it clear that the materials of interest for battery research lie in complex composition spaces. In the case of layered oxides alone, there exist commercial materials with Co, Ni, Mn, Al, Li, and O, which represents a six-element composition space. Given this complexity, combinatorial approaches are promising tools to efficiently screen battery materials. Such studies have covered all components of the battery, and a few examples will be used here to illustrate the usefulness of high-throughput approaches in battery research.

1.6.1 Experimental Combinatorial Synthesis

Two recent experimental studies demonstrate the value of rapidly synthesizing samples across wide composition spaces. The first is the mapping of the entire

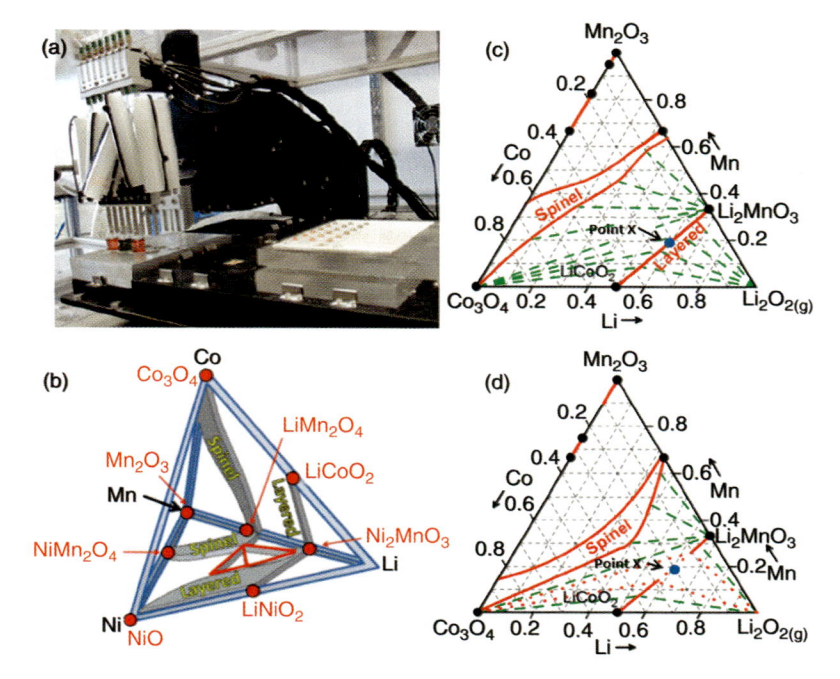

Figure 1.12 The solution-dispensing robot used to prepare milligram-scale mixed metal oxides throughout the Li–Mn–Co–Ni–O pseudo-quaternary system (a). The resulting phase diagram along the Li–Co–Mn and Li–Mn–Ni faces (b). The Li–Mn–Co face is shown for both quenched (c) and slow cooled (d) samples. Point X is discussed in the text. (Parts (a) and (b) reproduced with permission from Ref. [44] and parts (c) and (d) from Ref. [48] with permission from Dalhousie University and the American Chemical Society, respectively.)

Li–Co–Mn–Ni–O pseudoquaternary system of extreme interest as positive electrode materials [43]. Figure 1.12a shows the solution-dispensing robot used to make thousands of samples that were then characterized by XRD [44]. The synthesis was a coprecipitation reaction similar to that used to make commercial materials. The careful analysis of the XRD patterns resulted in phase diagrams that proved quite complex. Of particular interest to solid-state chemists, many compositions were found to transform during the cooling process, as illustrated with point X in Figure 1.12c, which was a single phase if quenched but phase-separated when slow cooled (thermodynamically, the equilibrium is driven by entropy at high temperatures and internal energy at lower temperatures). This has consequences for the synthesis of these materials, such that compositions must be chosen carefully to avoid such phase separation [45]. Another example of how useful the phase diagram is relates to the characterization of core–shell particles after lithiation. XRD yields two sets of lattice parameters that, thanks to the phase diagram, can be used to calculate the lithium content in both the core and the shell [46]. It should be noted that this combinatorial study would be enhanced if combinatorial electrochemistry were coupled to the structural studies, although to date this has only been done on sputtered samples that show very different electrochemistry than bulk samples [47].

The second example of experimental combinatorial science in batteries deals with solid electrolytes [49]. A sputtering system was used to make samples within a ternary system. This requires the use of wedge-shaped shutters to make concentration gradients within a single sputtered film, as shown in Figure 1.13. The problem the researchers were addressing was that in $LiCoO_2$–LiPON–graphite microbatteries, there remains a high impedance at the $LiCoO_2$–LiPON interface due in part to decomposition products when the LiPON reacts. Their objective was therefore to screen for dielectric materials with high ionic conductivity for use as the interphase between $LiCoO_2$ and the LiPON: a proven method to improve the interface impedance [50]. After sputtering Pt contacts onto the plate with the Li–Nb–Ta–O thin films, EIS measurements were made on the samples lying on the 14×14 grid. The resulting 156 conductivity values are shown in the phase diagram for various sintering temperatures, and the plots show a clear optimum composition and sintering temperature. This is a classic example of how combinatorial science can be used for efficient screening of potential battery components and this system can be used to search for new solid electrolytes.

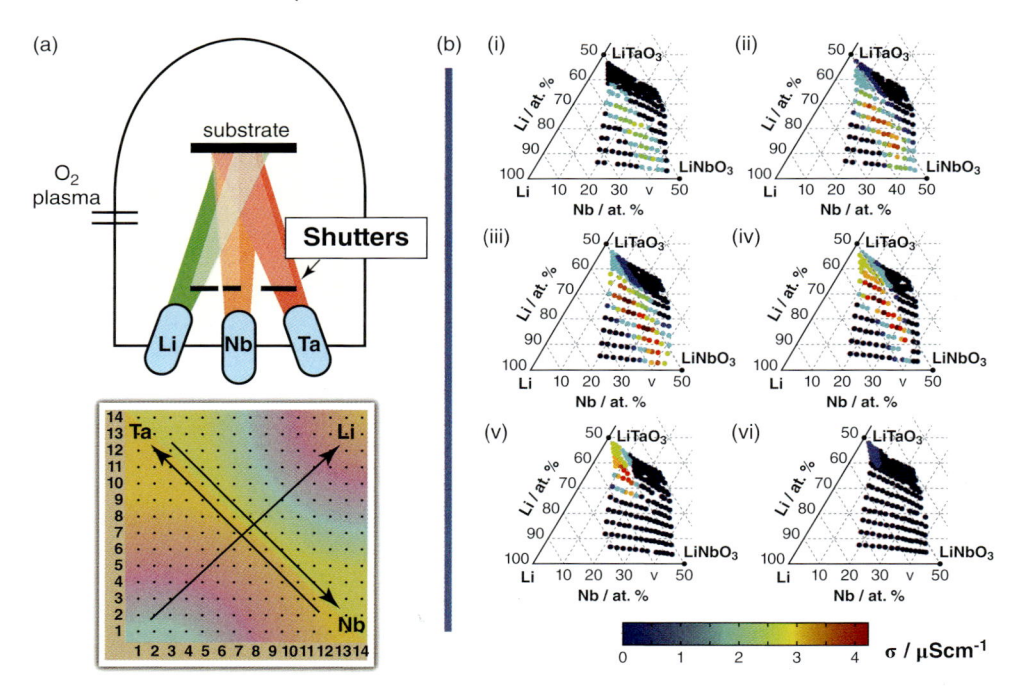

Figure 1.13 (a) The use of combinatorial sputtering in order to search for improved solid electrolytes. (b) Conductivity throughout the ternary system for various sintering temperatures varying from 300 °C (i) to 550 °C (vi). (Reproduced from Ref. [49] with permission from the Electrochemical Society.)

1.6.2
Computational High-Throughput Calculations

Ceder and coworkers at MIT have developed a high-throughput computational platform wherein the objective is to use DFT calculations based on known structures in order to determine which compositions are stable and calculate the properties of the resulting materials [51]. Although this has a wide array of possible applications, an important one is for the design of new battery materials. As of 2016, over 42 000 band structures have been calculated and over 3600 Li intercalation compounds have been studied[1]. Such a rapid screening across so many compositions has proved invaluable in the discovery of new electrode materials, and is very powerful when coupled with the experimental work discussed throughout this chapter [52]. Furthermore, more recently the group has also applied the approach to screen dopants across virtually the whole periodic table in solid electrolyte garnets in order to optimize the electrolyte–electrode interface[1]. This work is truly the computational analog to all the experimental work described throughout this chapter. The results are publicly available at www.materialsproject.org, and this serves as a useful starting point for experimentalists wanting to begin work in a certain composition space in order to quickly determine the lay of the land.

1.7
Conclusions and Future Direction

The techniques outlined here and used for the past 30 years by solid-state chemists have helped establish lithium–ion batteries as the premier technology for energy storage in personal devices and have made it prime for implementation in full electric vehicles. Beyond the final optimizations of Li-ion batteries, solid-state chemistry will also play significant roles in the development of all solid-state batteries by improving the properties at the electrolyte–electrode interfaces and potentially suppressing dendrite formation in lithium metal. There are also high hopes for both Na and Mg–ion batteries. Though much work has already been performed in Na–ion batteries, there are no commercialized products to date, and further studies are required to develop better electrode materials, such that all the approaches outlined here for Li-ion batteries will be crucial in bringing Na–ion batteries to market. Mg–ion batteries, on the other hand, are still in their infancy and are of significant interest because of the fact that magnesium does not form dendrites, such that the primary challenge is to find a positive electrode and electrolyte. Also, given that Mg ions carry a charge of +2, there is the potential for particularly high capacities if new structures can be found with sufficient redox couples to accommodate the same amount of Mg as Li in Li-ion batteries (to date this has not been the case, it will therefore be necessary to go

---

1) www.materialsproject.org/(accessed February 11, 2016).

beyond the common structures used for Li-ion batteries). A wide array of solid-state chemistry is therefore still needed in order to further develop battery technologies, particularly for widespread electric vehicle implementation.

References

1. Goodenough, J.B. and Park, K.-S. (2013) The Li-ion rechargeable battery: a perspective. *J. Am. Chem. Soc.*, **135** (4), 1167–1176.
2. Ohzuku, T. and Brodd, R.J. (2007) An overview of positive-electrode materials for advanced lithium-ion batteries. *J. Power Sources*, **174** (2), 449–456.
3. Ehrlich, G.M. and Dahn, J.R. (2002) Li-ion batteries, in *Handbook of Batteries* (eds D. Linden and T.B. Reddy), 3rd edn, McGraw-Hill, New York.
4. Peled, E., Golodnitsky, D., and Ardel, G. (1997) Advanced model for solid electrolyte interphase electrodes in liquid and polymer electrolytes. *J. Electrochem. Soc.*, **144** (8), L208–L210.
5. Wang, D.Y., Sinha, N.N., Petibon, R., Burns, J.C., and Dahn, J.R. (2014) A systematic study of well-known electrolyte additives in $LiCoO_2$/graphite pouch cells. *J. Power Sources*, **251**, 311–318.
6. Andre, D., Kim, S.-J., Lamp, P., Lux, S.F., Maglia, F., Paschos, O., and Stiaszny, B. (2015) Future generations of cathode materials: an automotive industry perspective. *J. Mater. Chem. A*, **3** (13), 6709–6732.
7. Melot, B.C. and Tarascon, J.-M. (2013) Design and preparation of materials for advanced electrochemical storage. *Acc. Chem. Res.*, **46** (5), 1226–1238.
8. Recham, N., Chotard, J.-N., Dupont, L., Delacourt, C., Walker, W., Armand, M., and Tarascon, J.-M. (2010) A 3.6 V lithium-based fluorosulphate insertion positive electrode for lithium-ion batteries. *Nat. Mater.*, **9** (1), 68–74.
9. Zhang, W.-J. (2011) A review of the electrochemical performance of alloy anodes for lithium-ion batteries. *J. Power Sources*, **196** (1), 13–24.
10. Thangadurai, V., Pinzaru, D., Narayanan, S., and Baral, A.K. (2015) Fast solid-state Li ion conducting garnet-type structure metal oxides for energy storage. *J. Phys. Chem. Lett.*, **6** (2), 292–299.
11. Takada, K. (2013) Progress and prospective of solid-state lithium batteries. *Acta Mater.*, **61** (3), 759–770.
12. Lepley, N.D., Holzwarth, N.A.W., and Du, Y.A. (2013) Structures, Li^+ mobilities, and interfacial properties of solid electrolytes Li_3PS_4 and Li_3PO_4 from first principles. *Phys. Rev. B*, **88** (10), 104103.
13. Thangadurai, V., Narayanan, S., and Pinzaru, D. (2011) Garnet-type solid-state fast Li ion conductors for Li batteries: critical review. *Chem. Soc. Rev.*, **43** (13), 4714–4727.
14. Kamaya, N., Homma, K., Yamakawa, Y., Hirayama, M., Kanno, R., Yonemura, M., Kamiyama, T., Kato, Y., Hama, S., Kawamoto, K., and Mitsui, A. (2011) A lithium superionic conductor. *Nat. Mater.*, **10** (9), 682–686.
15. Knauth, P. (2009) Inorganic solid Li ion conductors: an overview. *Solid State Ion.*, **180** (14), 911–916.
16. McCalla, E., Carey, G.H., and Dahn, J.R. (2012) Lithium loss mechanisms during synthesis of layered $Li_xNi_{2-x}O_2$ for lithium ion batteries. *Solid State Ion.*, **219**, 11–19.
17. Sathiya, M., Rousse, G., Ramesha, K., Laisa, C.P., Vezin, H., Sougrati, M.T., Doublet, M.-L., Foix, D., Gonbeau, D., Walker, W., Prakash, A.S., Ben Hassine, M., Dupont, L., and Tarascon, J.-M. (2013) Reversible anionic redox chemistry in high-capacity layered-oxide electrodes. *Nat. Mater.*, **12** (9), 827–835.
18. McCalla, E., Abakumov, A.M., Saubanère, M., Foix, D., Berg, E.J., Rousse, G., Doublet, M.-L., Gonbeau, D., Novák, P., Tendeloo, G.V., Dominko, R., and Tarascon, J.-M. (2015) Visualization of O–O peroxo-like dimers in high-capacity layered oxides for Li-ion batteries. *Science*, **350** (6267), 1516–1521.

19 Liu, X.-M., Gao, W.-L., and Ji, B.-M. (2011) Synthesis of $LiNi_{1/3}Co_{1/3}Mn_{1/3}O_2$ nanoparticles by modified Pechini method and their enhanced rate capability. *J. Sol-Gel Sci. Technol.*, **61** (1), 56–61.

20 van Bommel, A. and Dahn, J.R. (2009) Analysis of the growth mechanism of co-precipitated spherical and dense nickel, manganese, and cobalt-containing hydroxides in the presence of aqueous ammonia. *Chem. Mater.*, **21** (8), 1500–1503.

21 Camardese, J., Abarbanel, D.W., McCalla, E., and Dahn, J.R. (2014) Synthesis of spherical core–shell $Ni(OH)_2$-$Ni_{1/2}Mn_{1/2}(OH)_2$ particles via a continuously stirred tank reactor. *J. Electrochem. Soc.*, **161** (6), A890–A895.

22 Yuan, L.-X., Wang, Z.-H., Zhang, W.-X., Hu, X.-L., Chen, J.-T., Huang, Y.-H., and Goodenough, J.B. (2011) Development and challenges of $LiFePO_4$ cathode material for lithium-ion batteries. *Energy Environ. Sci.*, **4** (2), 269–284.

23 Jin, B., Gu, H.-B., Zhang, W., Park, K.-H., and Sun, G. (2008) Effect of different carbon conductive additives on electrochemical properties of $LiFePO_4$-C/Li batteries. *J. Solid State Electrochem.*, **12** (12), 1549–1554.

24 Barker, J., Saidi, M.Y., and Swoyer, J.L. (2003) Lithium iron(II) phospho-olivines prepared by a novel carbothermal reduction method. *Electrochem. Solid-State Lett.*, **6** (3), A53–A55.

25 Andersson, A.S., Thomas, J.O., Kalska, B., and Häggström, L. (2000) Thermal stability of $LiFePO_4$-based cathodes. *Electrochem. Solid-State Lett.*, **3** (2), 66–68.

26 Morris, R.E. (2009) Ionothermal synthesis: ionic liquids as functional solvents in the preparation of crystalline materials. *Chem. Commun.*, **21**, 2990–2998.

27 Tripathi, R., Popov, G., Ellis, B.L., Huq, A., and Nazar, L.F. (2012) Lithium metal fluorosulfate polymorphs as positive electrodes for Li-ion batteries: synthetic strategies and effect of cation ordering. *Energy Environ. Sci.*, **5** (3), 6238–6246.

28 Gebresilassie Eshetu, G., Armand, M., Scrosati, B., and Passerini, S. (2014) Energy storage materials synthesized from ionic liquids. *Angew. Chem., Int. Ed.*, **53** (49), 13342–13359.

29 Whittingham, M.S. (2012) History, evolution, and future status of energy storage. *Proc. IEEE*, **100**, 1518–1534.

30 Mizushima, K., Jones, P.C., Wiseman, P.J., and Goodenough, J.B. (1980) Li_xCoO_2 ($0 < x < 1$): a new cathode material for batteries of high energy density. *Mater. Res. Bull.*, **15** (6), 783–789.

31 Koga, H., Croguennec, L., Ménétrier, M., Douhil, K., Belin, S., Bourgeois, L., Suard, E., Weill, F., and Delmas, C. (2013) Reversible oxygen participation to the redox processes revealed for $Li_{1.20}Mn_{0.54}Co_{0.13}Ni_{0.13}O_2$. *J. Electrochem. Soc.*, **160** (6), A786–A792.

32 Sathiya, M., Abakumov, A.M., Foix, D., Rousse, G., Ramesha, K., Saubanère, M., Doublet, M.L., Vezin, H., Laisa, C.P., Prakash, A.S., Gonbeau, D., VanTendeloo, G., and Tarascon, J.-M. (2015) Origin of voltage decay in high-capacity layered oxide electrodes. *Nat. Mater.*, **14** (2), 230–238.

33 Boulineau, A., Simonin, L., Colin, J.-F., Bourbon, C., and Patoux, S. (2013) First evidence of manganese–nickel segregation and densification upon cycling in Li-rich layered oxides for lithium batteries. *Nano Lett.*, **13** (8), 3857–3863.

34 Rozier, P. and Tarascon, J.M. (2015) Review: Li-rich layered oxide cathodes for next-generation Li-ion batteries: chances and challenges. *J. Electrochem. Soc.*, **162** (14), A2490–A2499.

35 Cho, J., Kim, Y.J., and Park, B. (2001) $LiCoO_2$ cathode material that does not show a phase transition from hexagonal to monoclinic phase. *J. Electrochem. Soc.*, **148** (10), A1110–A1115.

36 Chen, Z., Qin, Y., Amine, K., and Sun, Y.-K. (2010) Role of surface coating on cathode materials for lithium-ion batteries. *J. Mater. Chem.*, **20** (36), 7606–7612.

37 van der Ven, A., Aydinol, M.K., Ceder, G., Kresse, G., and Hafner, J. (1998) First-principles investigation of phase stability in Li_xCoO_2. *Phys. Rev. B*, **58** (6), 2975–2987.

38 Sun, Y.-K., Chen, Z., Noh, H.-J., Lee, D.-J., Jung, H.-G., Ren, Y., Wang, S., Yoon, C.S., Myung, S.-T., and Amine, K. (2012)

Nanostructured high-energy cathode materials for advanced lithium batteries. *Nat. Mater.*, **11** (11), 942–947.

39 Goodenough, J.B. and Singh, P. (2015) Review: solid electrolytes in rechargeable electrochemical cells. *J. Electrochem. Soc.*, **162** (14), A2387–A2392.

40 Richards, W.D., Miara, L.J., Wang, Y., Kim, J.C., and Ceder, G. (2016) Interface stability in solid-state batteries. *Chem. Mater.*, **28** (1), 266–273.

41 Zhu, Y., He, X., and Mo, Y. (2015) Origin of outstanding stability in the lithium solid electrolyte materials: insights from thermodynamic analyses based on first-principles calculations. *ACS Appl. Mater. Interfaces*, **7** (42), 23685–23693.

42 Seino, Y., Takada, K., Kim, B.-C., Zhang, L., Ohta, N., Wada, H., Osada, M., and Sasaki, T. (2005) Synthesis of phosphorous sulfide solid electrolyte and all-solid-state lithium batteries with graphite electrode. *Solid State Ionics*, **176** (31), 2389–2393.

43 Brown, C.R., McCalla, E., Watson, C., and Dahn, J.R. (2015) Combinatorial study of the Li–Ni–Mn–Co oxide pseudoquaternary system for use in Li-ion battery materials research. *ACS Comb. Sci.*, **17** (6), 381–391.

44 McCalla, E. (2013) Structural and electrochemical studies of the Li–Mn–Ni–O and Li–Co–Mn–O pseudo-ternary systems. Thesis, Dalhousie University.

45 McCalla, E., Rowe, A.W., Brown, C.R., Hacquebard, L.R.P., and Dahn, J.R. (2013) How phase transformations during cooling affect Li–Mn–Ni–O positive electrodes in lithium ion batteries. *J. Electrochem. Soc.*, **160** (8), A1134–A1138.

46 Camardese, J., Li, J., Abarbanel, D.W., Wright, A.T.B., and Dahn, J.R. (2015) The effect of lithium content and core to shell ratio on structure and electrochemical performance of core–shell $Li_{(1+x)}[Ni_{0.6}Mn_{0.4}]_{(1-x)}O_2$ $Li_{(1+y)}[Ni_{0.2}Mn_{0.8}]_{(1-y)}O_2$ positive electrode materials. *J. Electrochem. Soc.*, **162** (3), A269–A277.

47 Fleischauer, M.D. and Dahn, J.R. (2004) Combinatorial investigations of the Si–Al–Mn system for Li-ion battery applications. *J. Electrochem. Soc.*, **151** (8), A1216–A1221.

48 McCalla, E., Lowartz, C.M., Brown, C.R., and Dahn, J.R. (2013) Formation of layered–layered composites in the Li–Co–Mn oxide pseudoternary system during slow cooling. *Chem. Mater.*, **25** (6), 912–918.

49 Yada, C., Lee, C.E., Laughman, D., Hannah, L., Iba, H., and Hayden, B.E. (2015) A high-throughput approach developing lithium–niobium–tantalum oxides as electrolyte/cathode interlayers for high-voltage all-solid-state lithium batteries. *J. Electrochem. Soc.*, **162** (4), A722–A726.

50 Yada, C., Ohmori, A., Ide, K., Yamasaki, H., Kato, T., Saito, T., Sagane, F., and Iriyama, Y. (2014) Dielectric modification of 5V-class cathodes for high-voltage all-solid-state lithium batteries. *Adv. Energy Mater.*, **4** (9), 1301416–1301421.

51 Jain, A., Ong, S.P., Hautier, G., Chen, W., Richards, W.D., Dacek, S., Cholia, S., Gunter, D., Skinner, D., Ceder, G., and Persson, K.A. (2013) Commentary: The Materials Project: a materials genome approach to accelerating materials innovation. *APL Mater.*, **1** (1), 011002.

52 Meng, Y.S., Wu, Y.W., Hwang, B.J., Li, Y., and Ceder, G. (2004) Combining *ab initio* computation with experiments for designing new electrode materials for advanced lithium batteries: $LiNi_{1/3}Fe_{1/6}Co_{1/6}Mn_{1/3}O_2$. *J. Electrochem. Soc.*, **151** (8), A1134–A1140.

2
Electrical Energy Storage: Supercapacitors

Enbo Zhao,[1] *Wentian Gu,*[2] *and Gleb Yushin*[2]

[1] *Georgia Institute of Technology, Department of Chemistry and Biochemistry, 771 Ferst Drive NW, Atlanta, GA 30332, USA*
[2] *Georgia Institute of Technology, Department of Materials Science and Engineering, 771 Ferst Drive NW, Atlanta, GA 30332, USA*

2.1
Introduction

Capacitor is a high-power charge storage device, which stores charges on the surface of its electrodes. It was first invented in the eighteenth century and by now has become indispensable for industry. One of the key performance characteristics of such a device is the capacitance, with the unit Farad (F), defined as the ratio of the charge accommodated by one of the two electrodes divided by the voltage applied across the device that induces this charge. Capacitance C is typically proportional to the electrode surface area S and inversely proportional to the charge separation distance d, according to Eq. (2.1), where the absolute permittivity ε is the product of $\varepsilon_r \varepsilon_0$, where ε_r is the relative permittivity of the dielectric (charge separating layer) and ε_0 is the permittivity of the vacuum.

$$C = \varepsilon S/d. \tag{2.1}$$

The oldest and the most conventional type of a capacitor is a dielectric capacitor, which consists of two parallel conductors separated by dielectric layer in between (Figure 2.1a). Dielectric capacitors typically offer limited capacitance (on the order of micro-Farads, µF), but they benefit from a very fast response to an externally applied voltage. Smaller versions of these devices are typically used in various electronic equipment components (such as current controls in microcircuits, signal coupling devices in amplifiers, electric filters, and power supply systems, to name a few), while larger dielectric capacitors are used in many alternating current power distribution systems, electric motors, and ultrahigh-power and high-voltage energy storage devices.

An electrolytic capacitor is another common type of commercial capacitors developed in the twentieth century (Figure 2.1b). The anode of an electrolytic

Handbook of Solid State Chemistry, First Edition. Edited by Richard Dronskowski, Shinichi Kikkawa, and Andreas Stein.
© 2017 Wiley-VCH Verlag GmbH & Co. KGaA. Published 2017 by Wiley-VCH Verlag GmbH & Co. KGaA.

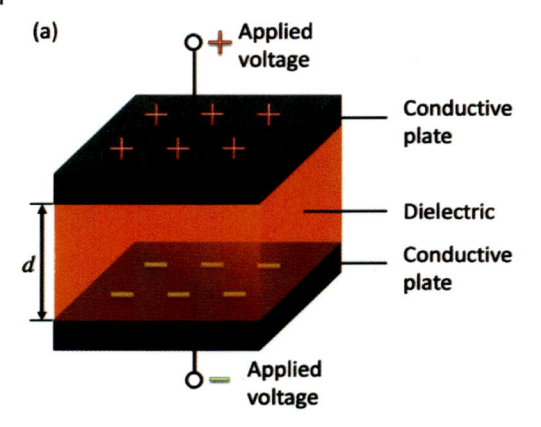

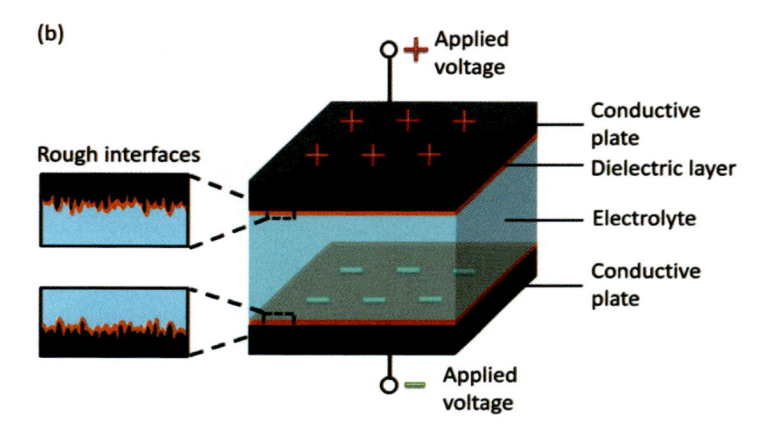

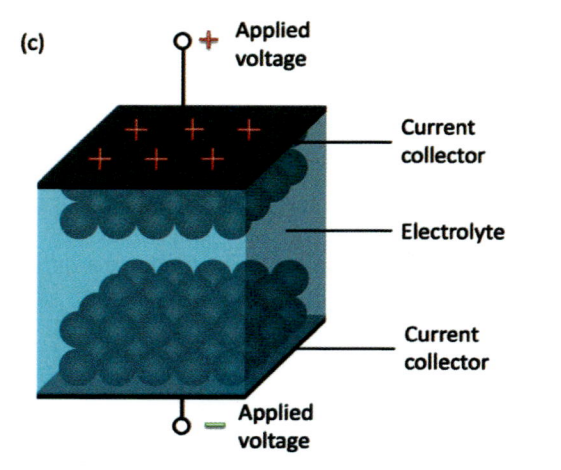

Figure 2.1 Schematic illustration of (a) dielectric capacitor, (b) electrolytic capacitor, and (c) electrochemical capacitor (supercapacitor).

capacitor is a high surface area porous (or roughened by etching) metal foil with a thin dielectric surface oxide layer, which separates the stored charge. The counterelectrode (cathode) is another metal (or carbon) conductor. The space in between an anode and a cathode is filled with ionically conductive liquid or solid electrolyte, which forms a high quality continuous interface with the dielectric oxide layer. Electrolytic capacitors typically offer higher capacitance than dielectric capacitors due to the larger electrode surface area S and a thinner charge separating dielectric layer d.

The latest developments in the field of capacitors include the inventions and commercialization of high-capacitance electrochemical capacitors (EC, as shown in Figure 2.1c), often called supercapacitors, which are typically divided into three categories: (i) electric double-layer capacitors (EDLC), (ii) pseudocapacitors, and (iii) asymmetric hybrid capacitors. While the initial concept (for EDLC) was introduced in the 1950s, their commercialization started only in the 1970s and the substantial market penetration began in the 2000s. A significant market growth in the past decade was stimulated by increasing demand for long life cycle, high-power and high-energy storage devices in several growing applications, such as digital electronic devices; implantable medical devices; uninterruptable power supplies; hybrid motors in transportation vehicles and forklifts, industrial lifts, and cranes; wind power generation systems; and electric grid, among others [1].

In this chapter, we describe the classification, construction, electrode materials, electrolyte compositions, and performance characteristics of various types of supercapacitors.

2.2
Electric Double-layer Capacitors

In a typical EDLC, a nanometer scale double layer comprising two parallel layers of opposite charge is formed at the interface of a solid electrode and a liquid electrolyte (Figure 2.2a with electrodes composed of porous carbon with high internal surface area and Figure 2.2b, with electrodes composed of carbon nanotube (CNT) or carbon onions or carbon aerogels or graphene having high external surface area). EDLCs store energy in this double layer via reversible ion adsorption on the electrode surface [2]. Electrode materials of high surface area (up to $\sim 3000\,m^2/g$) are utilized in EDLCs in order to increase the surface area of the double layer. This large surface area in combination with the ultrathin (nanometer or subnanometer) charge-separating distance contributes to the much higher capacitance (and energy density) of EDLCs, when compared with dielectric and electrolytic capacitors.

EDLCs are the most successfully commercialized supercapacitors by now [4]. Due to the fast establishment of the surface double layer, the charge speed of EDLCs is among the highest of all electrochemical energy storage devices (only electrolytic capacitors can presently exceed the EDLC rate performance).

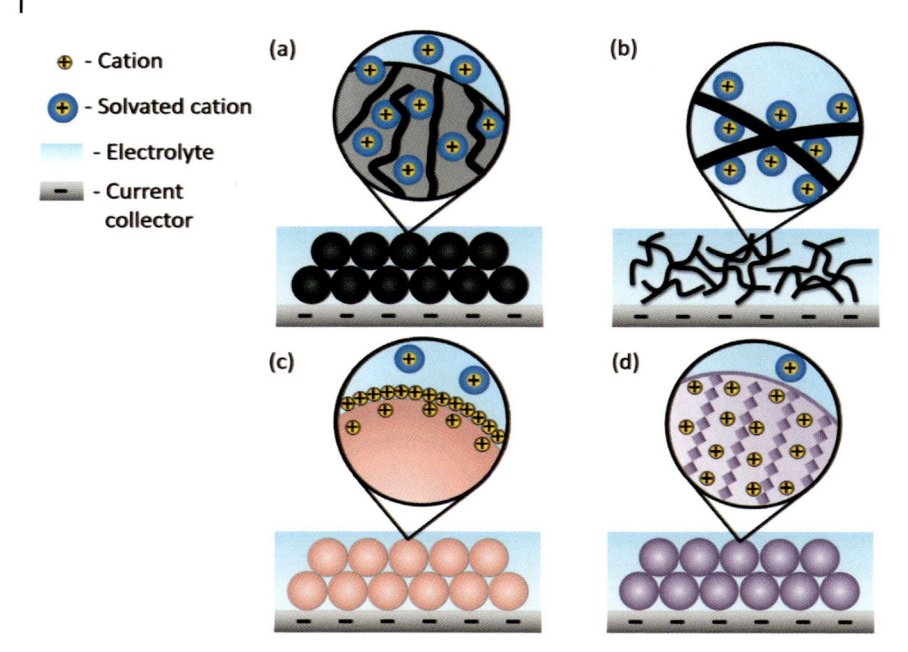

Figure 2.2 Schematic illustration of different charge storage mechanisms in electrochemical capacitors: electrosorption of ions in EDLCs with (a) porous carbon electrode particles or (b) electrode nanoparticles having high external surface area, such as CNT or grapheme, (c) electrochemical adsorption and absorption of ions in pseudocapacitors with electrodes exhibiting classical redox pseudocapacitance, as in hydrous RuO_2, and (d) electrodes exhibiting intercalation-type pseudocapacitance, where ions (e.g., Li^+ or Na^+) are inserted into the active material [3]. For simplicity, only negative electrodes are shown in this schematic.

Besides, the life cycle of EDLCs has been demonstrated to be quite remarkable, in many configurations exceeding 500 000 times [1].

The double-layer capacitance of an EDLC is often also approximated by Eq. (2.1), where the so-called "effective" relative permittivity of the electrolyte is used instead of the actual one. In this case, d is the average thickness of the electrical double layer, and S is the electrode "effective" surface area (accessible by electrolyte ions).

With the capacitance of positive, C_+, and negative, C_-, electrodes and the maximum voltage V_{max}^{EDLC}, at which the device can operate, the energy density of an EDLC could be calculated following Eq. (2.2):

$$E^{EDLC} = \left(\frac{C_- \cdot C_+}{C_- + C_+}\right) \cdot \left(V_{max}^{EDLC}\right)^2. \tag{2.2}$$

The energy density of EDLC is maximized in case the capacitance of both electrodes is identical:

$$E^{EDLC} = \frac{1}{2}C \cdot \left(V_{max}^{EDLC}\right)^2. \tag{2.3}$$

2.2 Electric Double-layer Capacitors | 29

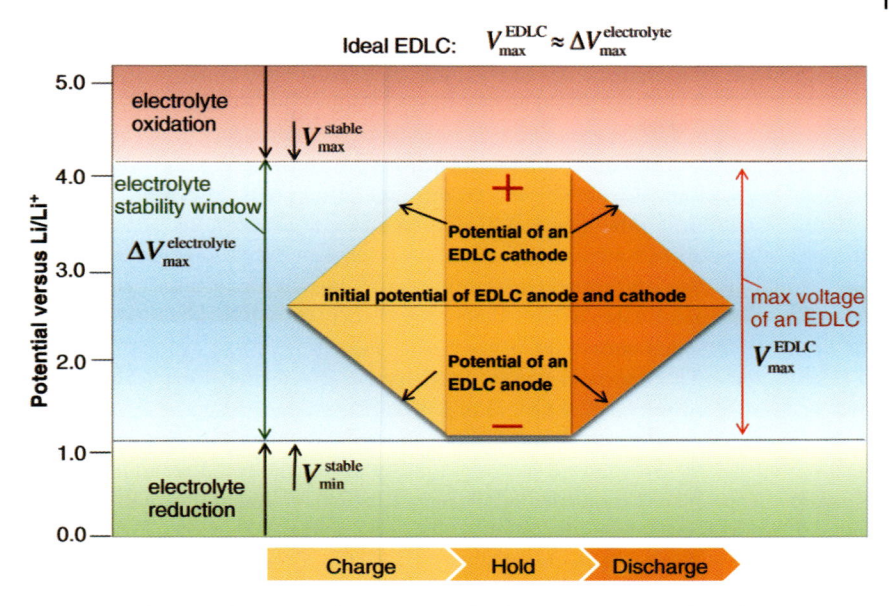

Figure 2.3 Schematic illustration of an EDLC operation. The orange gradient blocks stand for the voltage change of a cathode and an anode in an EDLC. A typical carbonate-based organic electrolyte is used as an example for illustrating electrolyte stability window. Both electrodes may be composed of the same active material, leading to the same initial potential. When charged, the potential of cathode increases and that of anode decreases. The overall voltage of the EDLC increases as a result. During storage, the potential of each electrode should not change significantly (zero self-discharge is assumed). During discharge, the potential of the cathode decreases and that of the anode increases, resulting in a reduced cell voltage.

With the guidance of Eq. (2.3), to achieve high energy density of EDLCs, one shall maximize the capacitance of each electrode and increase the maximum allowed operational voltage of EDLCs. As shown in Figure 2.3, the maximum voltage is generally determined by the stability window of the selected electrolyte. Organic electrolytes could provide larger stability voltage window (typically from about 1.2 V versus Li/Li$^+$ to about 4.3 V versus Li/Li$^+$ for carbonate-based electrolytes) than aqueous electrolytes (typically from about 2.8 versus Li/Li$^+$ to about 3.9 V versus Li/Li$^+$ or smaller for acidic electrolytes, from about 1.7 versus Li/Li$^+$ to about 3.1 V versus Li/Li$^+$ or smaller for basic electrolytes, and from about 2.4 versus Li/Li$^+$ to about 3.7 V versus Li/Li$^+$ or smaller for neutral electrolytes). Ionic liquids may exhibit the broadest stability window (from as low as 0.7 V versus Li/Li$^+$ up to about 4.5 V versus Li/Li$^+$ or smaller) [5]. We will discuss the selection of electrolytes in more detail in Section 2.5.

2.2.1
Activated Carbon Materials

According to Eq. (2.1), an increase in electrode surface area S to achieve large capacitance C per unit mass of the device should motivate engineers to utilize high specific surface area (SSA) electrode materials in EDLCs. Such a requirement, however, should be considered only as a rough guideline because in most applications volumetric (volume-normalized) device characteristics are more important than specific or gravimetric (mass-normalized) ones and since SSA (typically measured by gas sorption techniques) may deviate from the "effective" electrode surface area accessible by electrolyte ions.

Highly porous activated carbons (ACs) have become the most widely used materials in commercial EDLCs. This is because of their large SSA, relatively high chemical stability, reasonable cost, tunable properties, abundance and diversity of the AC precursors, biocompatibility, and established scalable synthesis methodologies. ACs could be produced in various shapes and forms, such as powders and fibers of various size and pore size distributions, mats, monoliths, films, foils, chunks, and irregular shapes (Figure 2.4). Commercial EDLCs utilize irregularly shaped powders due to their lower material costs and ability to process powders into electrodes at a high rate and a low cost.

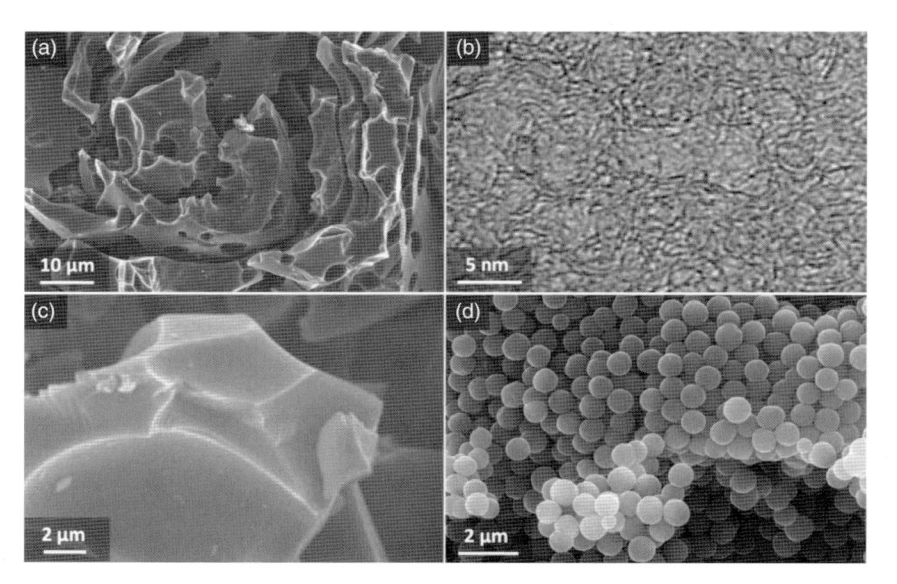

Figure 2.4 Scanning electron microscopy (SEM) and transmission electron microscopy (TEM) micrographs of selected ACs: (a) SEM of polypyrrole-derived AC [38], (b) TEM of porous carbons obtained via pyrolysis of seaweeds [40], (c) SEM of AC derived from hydrothermally treated biomass [39], and (d) SEM of spherical synthetic porous carbon [41].

ACs are typically prepared by using a combination of thermal treatment(s) and partial oxidation of organic compounds, including a very wide selection of natural and synthetic precursors. Some of the most common natural precursors for AC synthesis include seedshells or nutshells (mostly coconut shells [6–13]), waste wood products, coal, petroleum coke, pitch, peat, and lignite, while other precursors, such as starch, sucrose, sugar cane, leaves, seaweed, alginate, straw and various wastes resulting from the production of corn grain, coffee grounds, rice, soybean, maize, and so on are used only occasionally [14–31]. ACs prepared with natural raw materials are of low price, but they are usually lacking precise control on their SSAs, pore size distributions, and chemical purity. ACs with reproducible properties, more uniform microstructure and pores, more controllable and tunable size and shape, and often more tunable porosity (higher "effective" surface area) can be produced from synthetic polymers, such as polyacrylonitrile (PAN), polyvinylidene chloride (PVDC), polyfurfuryl alcohol (PFA), polyvinyl chloride (PVC), polypyrrole (PPy), polydivinylbenzene (PDVB), polyaniline (PANI), and so on [14,32–39].

Current methods for the preparation of ACs are typically classified into two categories: physical (or thermal) activation and chemical activation. The production of ACs by physical activation may involve two steps: carbonization of a precursor (removal of the majority of noncarbon species by thermal decomposition in inert or partially inert atmosphere) and gasification (enhancement of the porosity by partial etching of carbon during the heat treatment in the presence of an oxidizing agent, typically CO_2, H_2O, or a mixture of both) [42,43]. In some cases, low-temperature oxidation in air (typically at temperatures of 250–350 °C) is occasionally performed on some polymer precursors (such as PDVB) to increase carbon yield. Chemical activation is now often conducted in one step (mixing the precursors with an activation agent and heating). Pore formation and removal of the noncarbon species proceed simultaneously. This process typically requires extensive post-activation cleaning (particularly critical for preventing self-discharge and degradation in EDLC applications), which increases production costs. Chemical activation is often modeled to proceed through nanoscale "microexplosions" of an activating agent reacting with the precursor at elevated temperatures, which induce splitting/exfoliation of graphene segments, thus forming high volumes of micropores (pores < 2 nm) in carbons [13,44–54]. Typical examples of activation agents include KOH [44–48], H_3PO_4 [49], $ZnCl_2$ [50], H_2SO_4 [51], NaOH [52], and K_2CO_3 [13,53,54], among others. Compared to the physical activation, chemical activation generally results in larger volume of micropores, higher carbon yield, and more uniform pore size distribution [55–57].

Commercial AC powders commonly offer SSA in the range of 700–2200 m^2/g and moderately high specific capacitance in the range of 70–200 F/g in aqueous and 50–120 F/g in organic electrolytes [58–62]. However, some recent developments in the synthesis of ACs having greatly enhanced specific capacitance (up to 250–300 F/g in aqueous, organic, and ionic liquid (IL)-based electrolytes) demonstrate that for a significant portion of EDLC applications ACs may remain the material of choice [38–40,63–70].

2.2.2
Other Carbon Materials

Many high SSA and highly conductive carbon materials other than ACs have been applied as electrode materials for EDLCs, such as onion-like carbon (OLC), CNTs, graphene, zeolite-templated carbons (ZTC), silica-templated porous carbons, carbide-derived carbons (CDC), and so on (Figure 2.5). Templated and carbide-derived carbons may offer good pore size control than ACs and exhibit more attractive properties, but their production is extremely limited. AC fibers/fabrics commonly exhibit high electrical conductivity [71–86]. AC fabric electrodes, in addition, could offer very high mechanical flexibility. Their often higher power characteristics originate from the smaller electrode thickness, the presence of high volume of mesopores (>2 nm) and macropores (>50 nm) between the individual fibers, and higher electrical conductivity. Unfortunately, the higher price limits their use in price-sensitive applications. CNTs and macroporous carbon aerogels with large external surface area typically offer high rate capability but lower gravimetric and volumetric capacitance. Graphene has received lots of attention as a candidate electrode material for supercapacitors as it possesses superior electrical conductivity, a high theoretical surface area of over $2600\,\text{m}^2/\text{g}$ and chemical tolerance. However, due to the unavoidable aggregation of graphene nanosheets, the measured surface area of graphene may be much lower than the theoretical value. Other than single layer graphene, researchers also studied the electrochemical performance of graphene-based materials (GBMs),

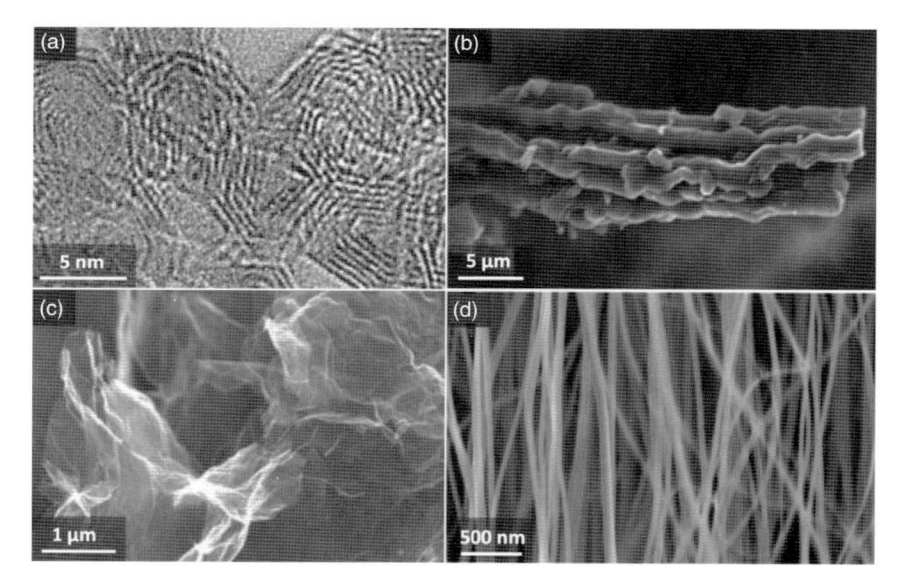

Figure 2.5 Electron microscopy micrographs of selected high SSA conductive carbon materials: (a) TEM of onion-like carbon [92], (b) SEM of micro/mesoporous silicon carbide-derived carbon [93], (c) TEM of graphene [94], and (d) SEM of aligned CNTs [95].

such as graphene nanosheets, graphene oxide, doped graphene oxide, and so on. Many strategies have been applied to increase the porosities or SSAs of GBMs (including routes to prevent the multilayer stacking of graphene sheets), and the capacitance of GBM electrodes were reported to reach 300 F/g in aqueous electrolyte [87–91]. The presence of functional groups and dopants in graphene may increase the measured capacitance value (e.g., due to the contributions of a pseudocapacitance and possibly an increase in the electron density of states, to be discussed in other sections), but typically at the cost of a faster self-discharge. In addition, a relatively low packing density of graphene in EDLC electrodes typically limits its volumetric capacitance to below 40 F/cm^3 (for higher purity graphene without reduction–oxidation (redox) active functional groups).

2.2.3
Carbon–Carbon Composites and Hierarchical Porous Materials

To combine advantages of the low-dimension carbons (such as high performance rate) with practicality of AC powders used in EDLCs, carbon-based nanocomposites (usually with three dimensions) have been proposed and developed. Several categories of such composites may be identified: graphene-based and CNT-based architectures, hierarchical porous carbon-based nanostructures, and those with well-structured pores of different sizes, sometimes ordered and precisely defined, and so on, as shown in Figure 2.6 [96].

One of the limitations of using pure graphene in EDLC electrodes is its tendency of restacking or agglomeration, due to the strong cohesive van der Waals force in between neighboring graphene layers [88]. To overcome this challenge,

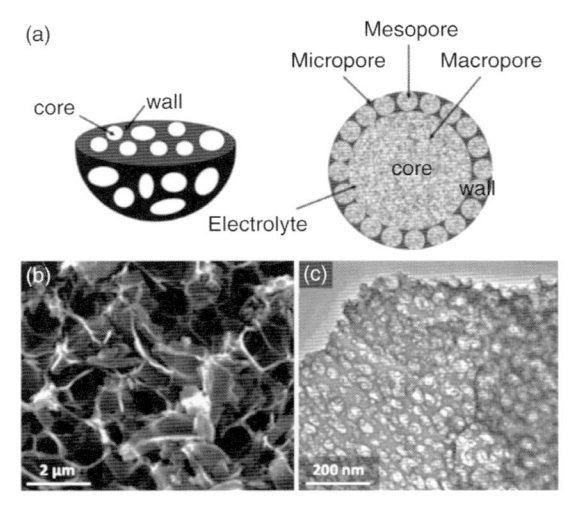

Figure 2.6 Schematic (a), SEM (b), and TEM (c) micrographs of 3D aperiodic hierarchical porous graphitic carbon material [97], which combines macroporous core with mesoporous/microporous pore walls.

CNTs and carbon particles have been applied, respectively, as "stabilizers" or "spacers" to separate the graphene layers and maintain the high surface area of the electrode. Such graphene-based 3D architectures demonstrate much higher capacitance than pure graphene samples. For example, Wei and coworkers have reported a novel strategy to prepare 3D CNT-graphene sandwich structures with CNT pillars grown in between the graphene layers [98]. The produced supercapacitor based on this material achieved a very high specific capacitance of 385 F/g^1 in 6 M KOH solution. In addition to carbon materials, redox-active metal oxides [99–101] and conductive polymers [102,103] were also applied as "spacers" in between graphene layers, which not only maintained the 3D architecture but also added pseudocapacitance and increased the resulting energy density.

The design of highly porous carbon with fine-tuned pore size distributions for EDLCs often suffers intrinsic dilemma, because while micropores are essential for maximizing energy density (energy per unit volume), mesopores allow a faster ion transport within the porous structure (particularly in case of organic and IL electrolytes) and contribute to higher power. To address this issue, researchers worked to introduced pores of various sizes into the electrode materials. For example, Yamada *et al.* reported the synthesis of hierarchical porous carbons consisting of meso/macro/micropores with large surface areas by a colloidal crystal templating technique [104]. A high capacitance of 350 F/g^1 was achieved in 1 M aqueous H_2SO_4 solution. Rojo and coworkers fabricated ultralightweight and highly conductive monolithic carbon aerogels with a 3D continuous hierarchical porous structure through a PPO-PEO-PPO block copolymer assisted route [105]. The resulting carbon monoliths exhibited capacitance of up to 225 F/g^1 in H_2SO_4 solution.

2.2.4
Quantum or Space Charge Capacitance

In our prior discussions, we assumed that the properties of carbon electrodes in EDLCs (such as electrical conductivity and others) do not impact the capacitance value (see Eq. (2.1)). However, even some of the early studies on graphite revealed that double-layer capacitance provided by basal plane of pyrolytic graphite is unusually small compared to metal electrodes with identical dimensions [106–109]. Randin and Yeager proposed that capacitance of a graphite/electrolyte interface (let us call it a double-layer capacitance, C_{DL}, for consistency) may be approximated by two components in a series: the capacitance of a space charge within graphite (C_{SC}) and the capacitance of an electrolyte layer (C_e) [107]. Assuming that the latter should be independent on the electrode itself (and thus could be approximated by a relatively high values of ~20 µF/cm^2 observed on the metal surfaces), Randin and Yeager suggested that the overall capacitance must be governed by that of the C_{SC}, as the smallest of the two, according to their estimations. It may be hypothesized that in a broad range of carbon materials, C_{SC} may affect the total measured capacitance.

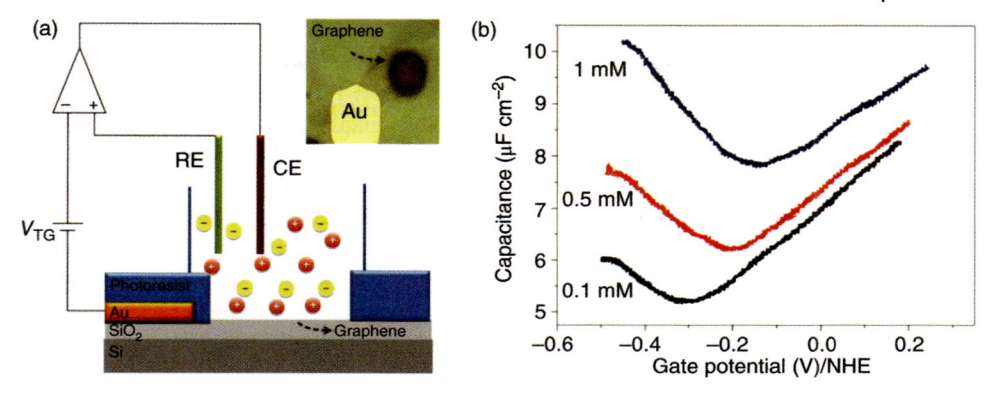

Figure 2.7 (a) Schematic of the quantum capacitance measurement setup and (b) capacitance of graphene measured in a NaF aqueous solution at different ionic concentrations [112].

Another interesting and related observation (made both in graphite and in various other types of carbon electrodes in a broad range of different electrolytes) is increasing capacitance at higher values of the applied voltage, thus forming the so-called "butterfly wing" or "V" or "U" or "dumbbell" shape of the CV curves (Figure 2.7b) [110–112]. Various research groups offered different explanation to the observed deviations from the rectangular shape, but some of the earliest proposals linked this phenomenon to a low density of electronic energy states (DOS) around the Fermi level of graphite [113] and other carbon materials. Indeed, the Fermi level of semiconductive electrodes (such as graphite) cannot accommodate as many charge carriers as conductive electrodes, such as metals. When a graphite-based electrode is charged, electrons are filled in the negative side and depleted from the positive side. The energy of electrons on both sides deviates from the Fermi level, which results in an extra voltage change over the galvani potential difference. Graphene and semiconductive CNTs are often selected as the research objects for this phenomenon because of their well-defined band structures. One of the suitable measurement setups for the characterizations of quantum (or space charge) capacitance is shown in Figure 2.7a. The details of such technique are beyond the scope of this chapter. Interested readers are encouraged to look at some of the representative studies [112,114,115].

To increase the DOS in carbon electrodes around the Fermi level (and thus increase C_{SC}), one may induce doping of electrodes by heteroatoms. In one example, Jang et al. modified the surface of double-walled CNTs via oxidation using nitric acid [116]. Despite their smaller SSAs, the oxidized CNTs exhibited higher capacitances than pristine CNTs in both aqueous and nonaqueous systems. This could be explained, unfortunately, by multiple factors, including pseudocapacitance contribution, changes in the electrolyte/carbon surface energy, and changes in the electrolyte desolvation behavior, in addition to possible changes in the carbon DOS. Zhang et al. enhanced capacitance of graphene via nitrogen doping [117], which may similarly be caused by multiple factors.

With 2.3 at% nitrogen doping, the area-normalized capacitance of graphene increased from 6 to 22 µF/cm^2.

2.3
Pseudocapacitors, Asymmetric Hybrid Capacitors, and Li-ion Capacitors

In spite of the significant recent developments and commercial success, EDLCs are still limited by a relatively low energy density, which is 20–200 times smaller than that of rechargeable Li-ion batteries (LIBs). Therefore, the incorporation of materials that exhibit reversible redox reactions in the electrodes' potential range, such as transition metal oxides and conductive polymers, may offer an attractive solution to boost capacitance and energy density of the supercapacitors (Figure 2.2c and d). Experimentally, the reversible redox process may enhance the specific and volumetric capacitance by 2–100 times compared to plain EDLC electrodes without redox reactions, depending on the nature of the electrode and electrolyte system (Figure 2.8). These capacitors are typically classified as pseudocapacitor [118–120].

The pseudocapacitors store electric charge via reduction and oxidation of electrolyte ions in reversible charge transfer reactions at the electrode. Different from an EDLC, which is free from charge transfer through the double layer in an ideal case, pseudocapacitors perform a net ion exchange between the electrode and the electrolyte during the charge and discharge processes. This is very similar to the analogous process in batteries, except that in pseudocapacitors operating at fast rates such processes are typically limited to the surface of active materials, for example, restricted to the 5–100 nm from the solid surface/electrolyte contact.

The distinction between the pseudocapacitor electrodes and the high-rate battery electrodes based on materials with small particle size may become slightly vague. Historically, electrodes are considered to be pseudocapacitive (as opposed to being called battery or battery-like electrodes) if cyclic voltammetry (CV) curves of such electrodes are either flat (as in EDLCs, Figure 2.9a) or exhibit very broad redox peaks (Figure 2.9b) and if the CV current is linearly dependent on the sweep rate (within a suitable range). The flatness of the CVs is somewhat analogous to the lack of plateaus in the charge–discharge (C–D) curves. In contrast, many battery electrodes exhibit sharp peaks in CVs (particularly at slow rates) and characteristic plateaus in the C–D due to the phase transformations taking place during the redox reactions. Such phase transformations may be considered as one of the distinctions between pseudocapacitor and battery electrodes. Furthermore, in a diffusion-limited regime of battery electrode operations, the CV current is typically proportional to the square root of the sweep rate. However, some of the battery electrodes (particularly if nanosized, nanostructured, or amorphous) also do not exhibit very sharp peaks in CV (such as in Figure 2.9c) and do not show clear plateaus. Instead, they may show sloping C–D profiles (such as in Figure 2.9c), more typical for supercapacitors and

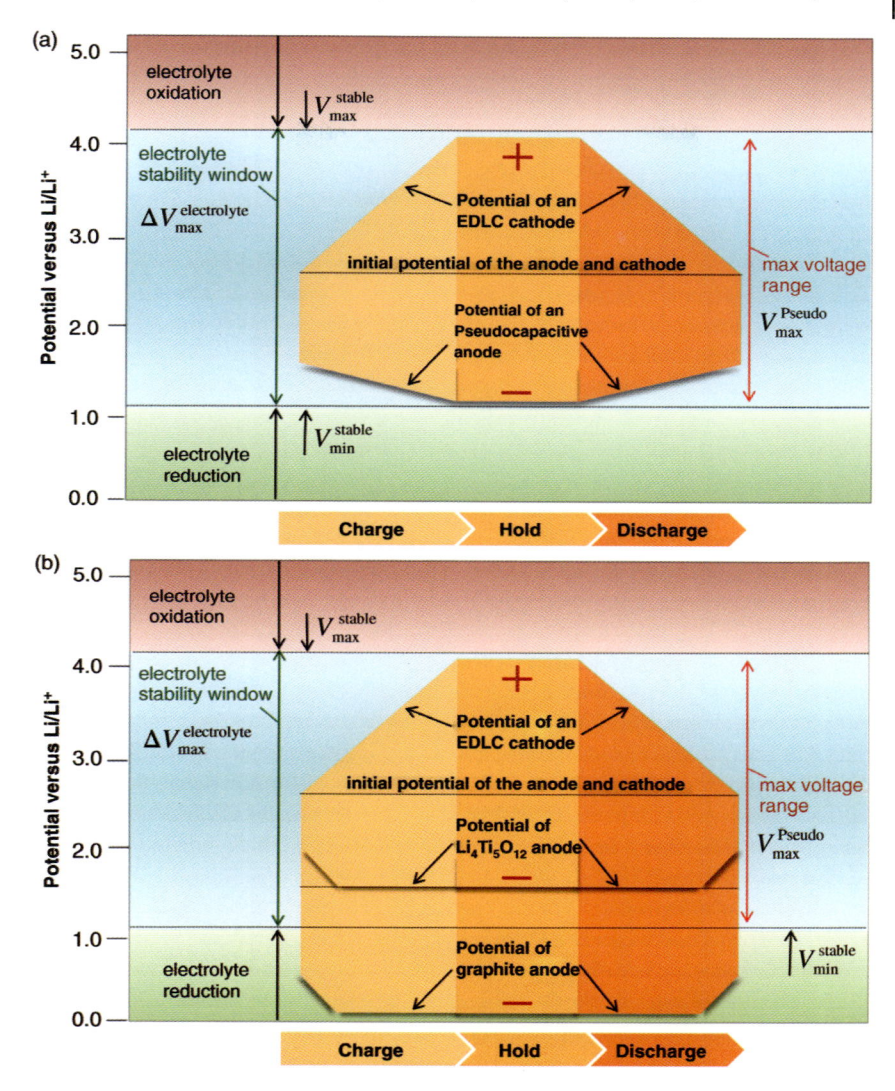

Figure 2.8 Schematics of three types of pseudocapacitors with different combinations of cathode and anode materials. The orange gradient blocks represent an EDLC for comparison. The blue gradient boxes stand for a pseudocapacitor consisting of (a) an EDLC-type cathode and a pseudocapacitive anode, (b) an EDLC-type cathode and a battery-type anode, (c) a pseudocapacitive cathode and an EDLC-type anode, and (d) a pair of pseudocapacitive cathode and anode. Pseudocapacitors typically cover wider voltage window and deliver more gravimetric and volumetric capacitance than ELDCs.

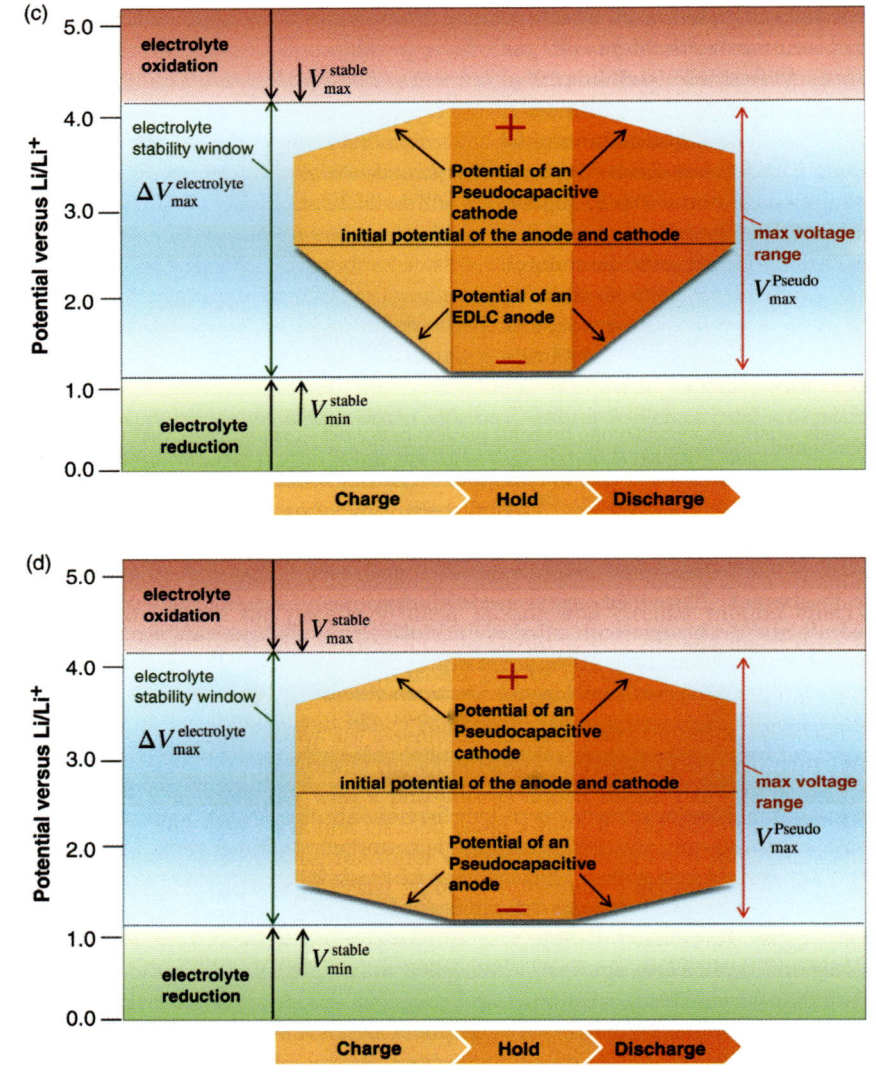

Figure 2.8 *(Continued)*

EDLCs. Some of the materials may also exhibit mixed dependence of the CV current on the sweep rate [3,121].

To make a distinction between a battery and a supercapacitor even more confusing, some devices utilize battery or battery-like material in one electrode and EDLC or pseudocapacitive material in another electrode. Such high-power devices belong to the class of asymmetric hybrid capacitors. Some of such capacitors use the same or almost the same electrodes as in nonaqueous Li-ion

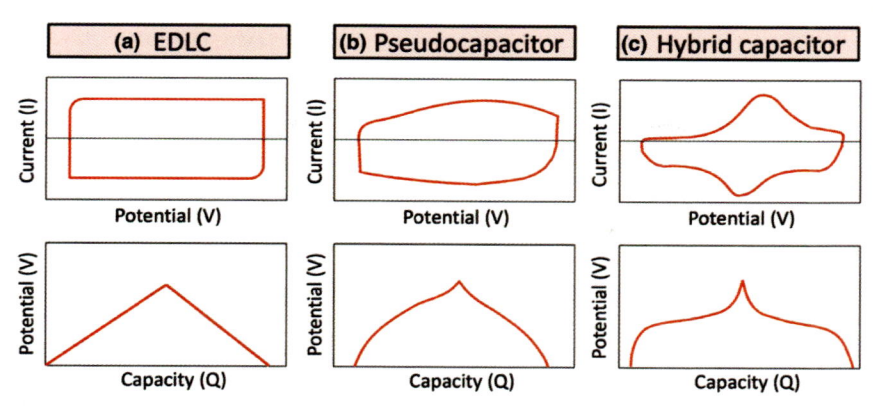

Figure 2.9 Typical CV (upper) and C–D (lower) curves of three types of electrochemical capacitors: (a) EDLC, (b) pseudocapacitor, and (c) hybrid capacitor using a battery-like electrode and an EDLC-type (or pseudocapacitor-type) electrode.

batteries (e.g., graphite or lithium titanate, LTO, anodes). In this case, such devices are often referred to as Li-ion capacitors. In order to achieve high performance rate in such devices, their battery-like electrodes may contain higher content of electrically conductive carbon (to achieve faster electron transport), larger pore volume, and smaller particle or grain size (to achieve faster ion transport) compared to analogous electrodes used in Li-ion batteries.

Because both pseudocapacitive electrodes and battery-like electrodes used in asymmetric capacitors typically exhibit small particle size and large specific surface area, substantial double-layer capacitance in such electrodes coexists along with the redox-induced capacitance. Several types of charge transfer processes may occur in such electrodes, which will be discussed in the next sections of the chapter. Pseudocapacitors and asymmetric capacitors outperform EDLCs in terms of energy density, but they typically suffer from slower performance rate and diminished stability during long-term cycling. Thus, this type of supercapacitors continue to attract significant research interest to overcome these remaining challenges.

2.3.1
Surface Redox Pseudocapacitance

2.3.1.1 Metals, Metal Oxides, and Metal Nitrides
One of the several mechanisms of reversible surface (and/or near-surface) redox reactions in pseudocapacitors involve electrosorption or desorption of ions at the electrode, with simultaneous charge transfer (Figure 2.2c). For example, hydrogen adsorption on platinum and lead adsorption on gold is a typical process of this kind, which has been well studied [122–124].

A variety of metal oxides, such as $RuO_2 \cdot nH_2O$ [129,130], MnO_2 [131–134], NiO [135–137], In_2O_3 [138–140], Co_3O_4 [141,142], V_2O_5 [126,143,144],

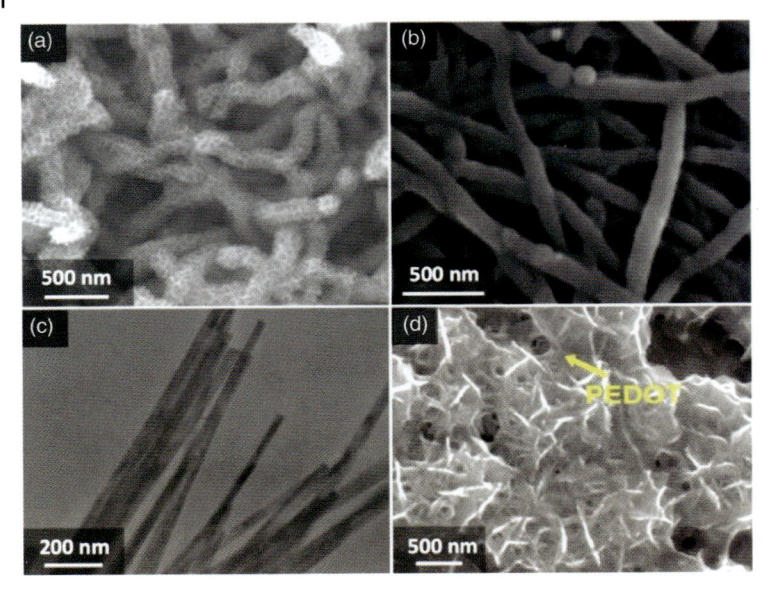

Figure 2.10 SEM micrographs of selected pseudocapacitive metal oxides: (a) Co_3O_4 nanowire@MnO_2 ultrathin nanosheets core/shell arrays [125], (b) vanadium oxide on CNTs [126], (c) TiO_2-B nanowires [127], and (d) graphene/MnO_2/PEDOT:PSS nanostructures [128].

Fe_3O_4 [145,146], Bi_2O_3 [147,148], IrO_2 [149], $NiFe_2O_4$ [150], $BiFeO_3$ [151], and so on, have also been exploited as active materials for redox reactions in pseudocapacitors (see examples in Figure 2.10). For example, hydrated ruthenium oxide ($RuO_2 \cdot nH_2O$) is one of the popular pseudocapacitor electrode materials and has been widely studied for its high conductivity and three accessible distinct oxidation states within 1.2 V [58]. The pseudocapacitance of $RuO_2 \cdot nH_2O$ in aqueous acidic solutions is provided by a fast, reversible electron transfer together with an electroadsorption of protons on the surface of $RuO_2 \cdot nH_2O$ particles. Specific capacitance of more than 600 F/g^1 has been achieved by $RuO_2 \cdot nH_2O$ [130]. However, the high cost of $RuO_2 \cdot nH_2O$ limits its application as electrode material for mass production. Among the less expensive oxides, manganese oxide was found to be a promising candidate [129,152]. The charge storage mechanism based on surface adsorption of both electrolyte cations (K^+, Na^+, . . .) and protons has been proposed in this material [153–157].

Pseudocapacitance in most metal oxides is majorly achieved at the very vicinity (typically within a few or few tens of nanometers) of the electrode surface. Therefore, in order to further increase the utilization rate of the active materials, nanostructured transition metal oxides and nitrides have been studied to enhance the capacitance [125,158–161]. Thin MnO_2 deposits of tens to hundreds of nanometers were produced on various substrates such as metal collectors, carbon nanotubes, and activated carbons. Nanostructured MnO_2 was reported to achieve as high specific capacitances as 1300 F/g^1 in aqueous electrolytes [162]. To improve the cycling stability and the specific capacitance of transition metal oxide nanoparticles, thin conductive polymer deposition on the surface of the electrode was proposed to help enhance proton exchange at the surface [163]. In some of the extensively studied nanostructured metal oxides, metal nitrides have been proposed as alternative active materials due to their higher electrical conductivity [164–167]. For example, nanocrystalline vanadium nitride (VN) exhibits a faradaic reaction (II/IV) at the surface of the nanoparticles, leading to specific capacitance up to 1200 F/g at a scan rate of 2 mV/s in 1 M KOH electrolyte [164].

2.3.1.2 Conductive Polymers and Functionalized Carbons as Active Materials

In addition to using metals, metal oxides, and metal nitrides as electrode materials, the use of conductive polymers is another approach to utilize redox reactions. The pseudocapacitance provided by conductive polymers is based on the reversible redox reactions of their electrochemical doping–dedoping [119,168]. The process of reversible electrochemical doping–dedoping could be represented by the following reactions:

$$P_m - xe^- + xA^- \leftrightarrow P_m^{x+} A_x^-, \tag{2.4}$$

$$P_m + ye^- + xM^+ \leftrightarrow P_m^{y-} M_y^+, \tag{2.5}$$

where P_m is a polymer with a network of conjugated double bonds, m is the polymerization degree, A^- denotes anions, and M^+ represents cations. In both Eqs. (2.4) and (2.5), the doping reactions proceed from the left to the right, and the dedoping reactions the other way around. Reaction (2.4) is a reaction of oxidative p-doping (acceptor doping) and reaction (2.5) is a reaction of reductive n-doping (donor doping). While most conductive polymers can only be p-doped, some of them (polyacetylene, polythiophene, and their derivatives) may be reversibly p-doped and n-doped. In a doped state, the conductive polymers have good intrinsic conductivities, from a few Siemens per centimeters to 500 S/cm [169]. They also have relatively fast charge–discharge kinetics. Conductive polymers offer many advantages as supercapacitor electrodes. They are flexible and highly conductive, easy to process, can be made into thin and thick porous films, and are often inexpensive. Many conductive polymers exhibit high specific capacities and total capacitance, while being able to deliver energy at a relatively rapid rate. The major disadvantage of the conductive polymers used as supercapacitor electrode is a typically poor life cycle compared to EDLCs. One of the known mechanisms of polymer electrode degradation is a substantial volume change in the polymer, induced by the doping or dedoping processes [169–171]. In order to improve life cycle by minimizing particle separation in the electrode, current research efforts on conductive polymers for supercapacitor applications are often focused on polymer–ceramic [172–175] and polymer–carbon [92,176–188] composite materials as well as on hybrid systems where carbon or ceramic components provide structural support (Figure 2.11). High specific capacitance of over 600 F/g

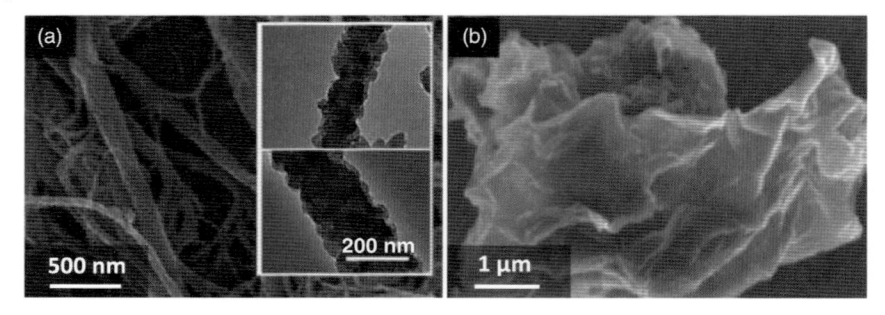

Figure 2.11 SEM images of (a) PANI-coated individual CNT [188] and (b) graphene/polyaniline nanofiber composites [185].

has been achieved by selected composite materials in symmetric two-electrode configuration with no reduction in the capacitance after 10 000 charge–discharge cycles.

Besides, surface functionalization on high SSA carbon materials also often leads to pseudocapacitance [63,64,92,189–192]. A variable amount of heteroatoms (oxygen, nitrogen, sulfur, etc.) can be added to the structures of carbon materials and functionalize their surfaces. Functionalized carbons can be synthesized by applying precursors containing the heteroatoms [193–195] or post-treatment of carbons in heteroatom-enriched atmosphere [196,197]. The resulted functional groups can considerably enhance the gravimetric capacitance of the electrodes by contributing additional pseudocapacitance, which involves rapid charge transfer reactions between the electrolyte and the functional moieties. Even when such carbons exhibit smaller SSA, they may provide higher volumetric capacitance than more porous AC (pure EDLC) electrodes free from functional groups. However, pseudocapacitance based on surface functionalization also suffers from intrinsic shortcomings, including relatively sluggish charge transfer process, substantial self-discharge, and high leakage currents.

2.3.2
Intercalation Pseudocapacitance

Pseudocapacitance is commonly associated with surface and/or near-surface reversible redox reactions, as previously discussed. However, Dunn and coworkers recently demonstrated that a pseudocapacitive mechanism takes place when Li ions are inserted into mesoporous and nanocrystal films of orthorhombic Nb_2O_5 (T-Nb_2O_5) [198,199] (Figures 2.2d and 2.12). The crystalline network of T-Nb_2O_5 offers two-dimensional transport pathways and little structural change on intercalation. The principal benefit realized from intercalation pseudocapacitance is that high levels of charge storage are achieved within short periods of time because there are no limitations from solid-state diffusion.

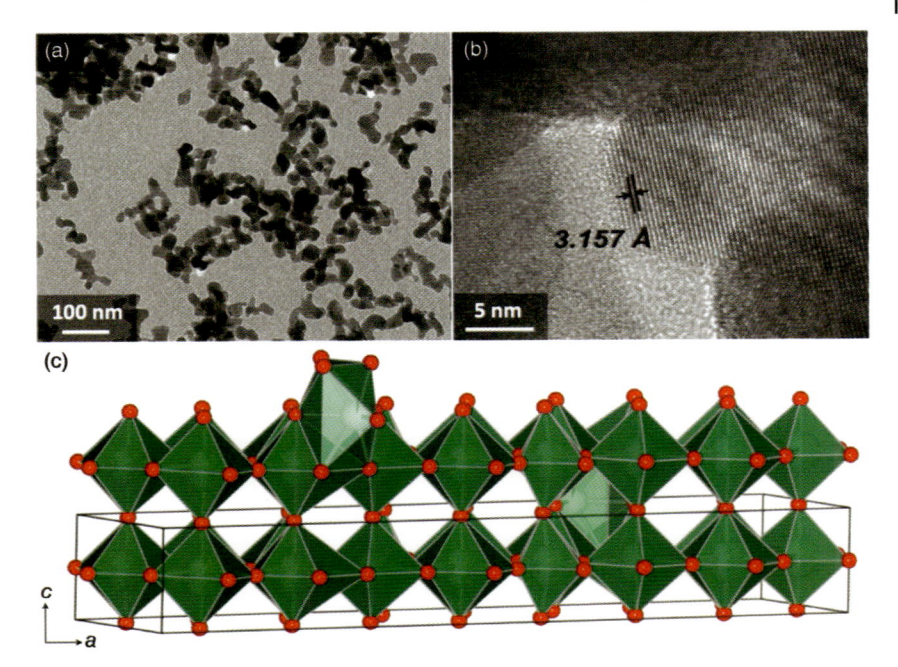

Figure 2.12 Orthorhombic Nb$_2$O$_5$ as interaction pseudocapacitive materials: (a) SEM image, (b) HR-TEM image [199], and (c) crystal structure [198].

2.4
Supercapacitor Device Configurations

Two broad device configurations could be distinguished: (i) symmetric capacitor (with typical C–D profiles of electrodes similar to that schematically shown in Figure 2.3) and (ii) asymmetric capacitor (with typical C–D profiles of electrodes similar to that schematically shown in Figure 2.8). In case of a symmetric capacitor, the same (or nearly the same) materials are used for both anodes and cathodes. These could be pure double-layer carbon electrodes or electrodes comprising the same pseudocapacitive material (e.g., vanadium oxide or functionalized porous carbon). Asymmetric capacitor is a very broad definition referring to that there is a substantial difference between positive and negative electrodes. This difference may be related to chemistry, morphology, weight, density, and charge storage contributions. When an asymmetric capacitor electrodes exhibit substantially different charge storage mechanisms (e.g., double-layer capacitance in one and pseudocapacitance in another or different types of pseudocapacitance in each electrode), the devices are often called hybrid capacitors. As previously discussed, some types of hybrid capacitors (e.g., Li-ion capacitors) use battery-like materials in one of the electrodes. Figure 2.8a and b illustrate C–D of asymmetric capacitors with an EDLC cathode and either a pseudocapacitor-like (Figure 2.8a) or a battery-like (Figure 2.8b) anode.

Figure 2.8c and d illustrate C–D of devices with a pseudocapacitor cathode and either an EDLC anode (Figure 2.8c) or an anode made of another pseudocapacitive material (Figure 2.8d).

2.5
Electrolyte Selection

As discussed in Section 2, electrolytes used in supercapacitors may be divided into three classes: (i) aqueous (solutions of acids, bases, or salts), (ii) organic (salt solutions in organic solvents), (iii) ionic liquids (ILs), and (vi) mixed (e.g., IL with a solvent addition). The first three types of electrolytes have been studied intensively and their pros and cons are known [200]. ILs are nonflammable and nontoxic, and they offer higher operating voltage than their counterparts. Serious shortcomings of ILs are their very high (often prohibitively high) current cost and relatively low ionic mobility at room temperature and below, which limits the charge/discharge rate of IL-based supercapacitors. The advantages of using aqueous electrolytes include their very low cost, safety, and high ionic conductivity. Their disadvantages, however, include their low narrow electrochemical window and corrosion of supercapacitor electrodes observed at higher temperatures and voltages (particularly for acid-based electrolytes, such as H_2SO_4 solutions), which limits the life cycle of the supercapacitors and contributes to self-discharge. Organic electrolytes are somewhat in between aqueous ones and ILs in terms of the price, voltage, and charge–discharge time. If properly selected, organic electrolytes may offer efficient operation of supercapacitors at very low temperatures, down to −70 °C [201]. Organic electrolyte-based supercapacitors offer a life cycle in excess of 500,000 and are used in the vast majority of commercial supercapacitors. We would also like to point out that supercapacitors with organic electrolytes are typically much less flammable than Li-ion batteries comprising similar solvents [202,203].

Besides the selection of electrolyte solvent, it may be important to properly match the size of the electrolyte ions and the pores of the carbon (EDLC) electrode materials to achieve high energy density. Several studies revealed enhanced ion adsorption in subnanometer carbon pores [204–206], where the size of desolvated ions closely approach the size of pores. While the universality of higher gravimetric capacitance in the smallest accessible pores remains a topic of debate in the scientific community, there are no doubts that smaller (yet fully accessible) pores result in higher volumetric capacitance (and thus energy density at the EDLC device level). Too small pores, however, may not be fully accessible, as demonstrated by the studies on the pore size-selective ion sorption [207–209]. For microporous carbons with pore size matching the size of bivalent ions, a classical rectangular shape of CV diagram was obtained, while for microporous carbons modified to exhibit just 0.1–0.2 nm smaller pore size, negligible current was recorded in all the potential range. Then, the bivalent cations and anions in the electrolyte were replaced with slightly smaller monovalent ions, pore size access, and high CV current in some potential range were detected, confirming the necessity to optimize the pore size distribution for a given electrolyte composition. Therefore, in the manufacturing of EDLC electrodes, ion-sieving effect needs to be avoided for an efficient access of electroadsorption sites to electrolyte ions. Some of the recent efforts to prevent the formation of the smallest pores in microporous carbons additionally showed significantly higher EDLC performance rate and power density.

2.6
Self-discharge of Supercapacitors

For many supercapacitor applications, such as alternative and emergency power supply, long shelf life and low self-discharge of supercapacitors is required. The self-discharge is a spontaneous (thermodynamically favored) process and is known to depend on the initial voltage, purity of carbon, and electrolyte as well as electrolyte acidity (Figure 2.13).

Several theories have been developed to distinguish the key mechanisms that causes the self-discharge of a supercapacitor via electrochemical characterizations. According to Conway's model [211,212], self-discharge processes mainly follow three different mechanisms at different potential fading rate. First, when the electrode is polarized to a potential that exceeds the electrochemical window of the electrolyte, the decomposition of electrolyte happens at the electrode/electrolyte interface. This Faradic process reduces the cell potential continuously until the electrode potential falls into the electrochemical window of the electrolyte or the electrolyte is totally consumed. Besides, this process usually produces gases, which may block the pores of the electrodes and the separator and even induce a separation of individual particles within the electrode. This, in turn, may result in the capacitance fading (due to reduction of accessible SSA of the

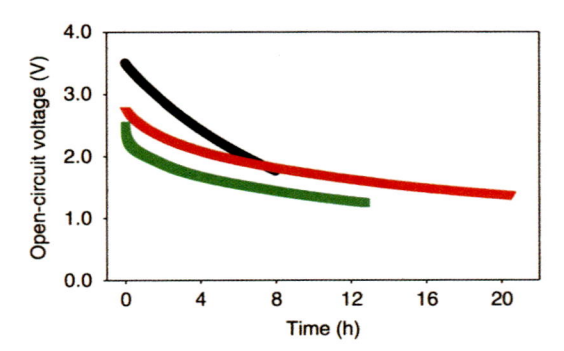

Figure 2.13 Selected examples of self-discharge curves of 3.5 V/25 mF commercial supercapacitor (black), 2.75 V/44 mF commercial supercapacitor (red) and a microsupercapacitor assembled using ionogel electrolyte (green), all demonstrating significant leakage currents [195,210].

electrodes) and in the reduction of power performance (due to increased separator resistance). It also leads to an increase in the ESR.

Second, some redox-active impurities on the surface of electrodes or in the electrolytes (such as O_2, H_2O, H_2O_2, metal ions, etc.) may be involved in undesirable (parasitic) Faradic processes, which may consume the charge stored in a supercapacitor or lead to electrolyte degradation. Other than impurities, the intentionally grafted surface functionalities on the carbon electrodes are believed to be often responsible for the self-discharge and capacitance fading [213,214]. Although these surface functionalities contribute to pseudocapacitance, they are often thermodynamically unstable within the potential range and either readily decompose themselves or induce electrolyte decomposition, thus producing various gases (such as NO_2, SO_2, SO_3, etc.) [61].

Some researchers speculate that selected functional groups on carbon may possibly be stable in the operating potential range. Unfortunately, selective functionalization of carbon surface only with some particular functional groups is a challenging task. Furthermore, commonly available surface chemistry characterization tools (Fourier transform infrared spectroscopy, X-ray photoelectron spectroscopy, chemical titration, etc.) have difficulty unambiguously interpreting the presence of particular functional groups, because the peaks corresponding to different groups are relatively broad and their position may be sufficiently close. At present, no comprehensive understanding exists on the specific contributions of various functional groups to both the desirable and the undesirable performance characteristics of supercapacitors in various organic and aqueous electrolytes. The value of leakage current often correlates with the fraction of functional groups in carbon (and thus with pseudocapacitance) and depends on the electrolyte utilized (pH-neutral aqueous electrolytes commonly induce lower leakage) [195]. So far, the feasibility of the concept to induce pseudocapacitance on carbon without penalties in leakage and degradation is unclear [215]. According to one recently proposed model by Yushin and coworkers [195], functional groups on the AC electrode surface locally change pH in their vicinity. Since electrolyte stability is pH dependent, higher pH at local areas of the cathode (positive electrode) may lead to electrolyte oxidation even when electrolyte should generally be stable at the cathode potential. Similarly, lower pH at local areas of the anode (negative electrode) may lead to electrolyte reduction [195].

The successful formation of stable functional groups on carbon that do not induce leakage has not been achieved in commercial devices. Carbon electrodes utilized in commercial supercapacitors are nonfunctionalized and purified to a high degree. Historically, improvements in the purification procedures allowed commercial devices to increase their life cycle from tens of thousands to several million cycles. In addition, such improvements in the purifications of carbons and electrolytes allowed some of the supercapacitor manufacturers to reduce the characteristic self-discharge time constant from a few hours to days and, in some cases, months.

Third, besides surface impurities and functionalities, self-discharge may also be attributed to Ohmic leakage from unintended interelectrode contacts or leaky bipolar electrodes.

Other than these work, studies on proper modeling and simulation of self-discharge have been performed by many, including some supplement to Conway's model [72,216–218].

2.7
Summary and Outlook

In this chapter, we reviewed the state-of art capacitor technology, including its development, classification, construction materials, and performance characteristics. Significant progress has been made in the past two decades in both science and technology of electrochemical capacitors, which made them attractive for a wide range of growing applications. Supercapacitors are now available in multiple designs that cover a significant portion of the performance range between conventional capacitors and batteries, as shown in Figure 2.14.

There are a number of desirable qualities that make supercapacitors commercially valuable high-power energy storage devices. First, the charge capacity of supercapacitors is contributed by fast double-layer establishment or facile redox reactions on the surface of electrodes, in contrast to batteries, which require

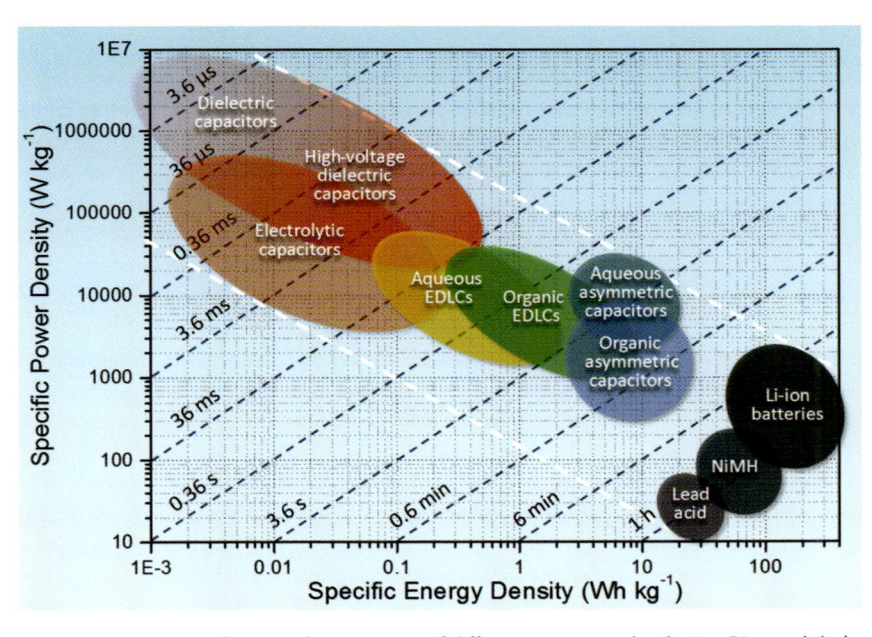

Figure 2.14 Energy and power characteristics of different storage technologies. Diagonal dash lines represent average characteristic charge (or discharge) time of the devices. Characteristics of both commercial and experimental devices are considered.

substantial solid-state diffusion of ions. Such fast charge storage processes lead to quick charge/discharge rates and, therefore, high power densities of these devices even at very low temperatures. Second, because supercapacitors typically operate within a stability range of the utilized electrolyte and because commercial supercapacitor materials (such as ACs) do not undergo phase transformations and do not experience significant volume changes during charge storage, supercapacitors may achieve long life cycle (> 100,000 cycles, some systems up to 10^6 in case of EDLCs) and very reliable (maintenance-free) operation even if environment temperature variations are substantial. Third, the construction of supercapacitors is robust and relatively simple and raw materials (at least carbon electrodes and aqueous electrolytes) are abundant and may be inexpensive. The similarity of their construction to batteries allows fabrication of hybrid devices, where battery materials are used in one of the electrodes (typically anode) to achieve higher energy density for some of the novel applications.

Meanwhile, supercapacitors also suffer from several limitations. First and most important, supercapacitors (particularly EDLCs) suffer from low energy density (energy per unit volume). Because supercapacitors rely only on ion adsorption or redox reactions on the electrode surface, the supercapacitor electrode materials typically exhibit high SSA and pore volume and thus low density and low volumetric charge storage capacity. Large pore volume is also often needed, because active ions are typically stored in electrolyte in a discharged state and the concentration of ions in suitable (organic or aqueous) liquid electrolytes is not nearly as high as possible in the solid battery electrodes. In addition, relatively low average operating voltage of supercapacitors and their sloping discharge profile further lower its energy density. Second, due to the low operating voltage of supercapacitors, stacking of these devices is necessary for high-voltage applications. This requires good matching of cell units and increases complexity of the energy storage system. Similarly, sloping discharge profile is not suitable for some applications, which necessitates the use of an additional rectifying device, which adds complexity, weight, volume, and cost. Finally, the cost of high-performance supercapacitors based on organic electrolytes and synthetic activated carbons is still rather high, which also limits the addressable market for this technology.

In spite of the shortcomings discussed above, we believe the demands for supercapacitors will keep growing. Improvements and novel developments in the material synthesis techniques may allow increase in the energy density of the supercapacitor cells and reduction of their prices. Asymmetric and hybrid capacitors will be competing in price and performance with ultrahigh-power Li-ion batteries and the time will show what technology is most viable in particular applications. The market for EDLCs will keep growing due to their unique combination of properties. The overall prospects and the associated advantages of using supercapacitor devices in new applications will surely power near-term and future efforts in research, development, and, even, in production processes to further improve technical performances in a larger spectrum of applications and significantly reduce costs.

References

1 Gu, W. and Yushin, G. (2014) Review of nanostructured carbon materials for electrochemical capacitor applications: advantages and limitations of activated carbon, carbide-derived carbon, zeolite-templated carbon, carbon aerogels, carbon nanotubes, onion-like carbon, and graphene. *WIREs Energy Environ.*, **3**, 424–473.

2 Sharma, P. and Bhatti, T.S. (2010) A review on electrochemical double-layer capacitors. *Energy Convers. Manage.*, **51**, 2901–2912.

3 Simon, P., Gogotsi, Y., and Dunn, B. (2014) Where do batteries end and supercapacitors begin? *Supramol. Sci.*, **343**, 1210–1211.

4 Simon, P. and Gogotsi, Y. (2013) Capacitive energy storage in nanostructured carbon-electrolyte systems. *Acc. Chem. Res.*, **46** (5), 1094–1103.

5 Galiński, M., Lewandowski, A., and Stępniak, I. (2006) Ionic liquids as electrolytes. *Electrochim. Acta*, **51**, 5567–5580.

6 Laine, J. and Yunes, S. (1992) Effect of the preparation method on the pore size distribution of activated carbon from coconut shell. *Carbon*, **30**, 601–604.

7 Daud, W.M.A.W. and Ali, W.S.W. (2004) Comparison on pore development of activated carbon produced from palm shell and coconut shell. *Bioresour. Technol.*, **93**, 63–69.

8 Yang, K. *et al.* (2010) Preparation of high surface area activated carbon from coconut shells using microwave heating. *Bioresour. Technol.*, **101**, 6163–6169.

9 Hu, Z. and Srinivasan, M. (1999) Preparation of high-surface-area activated carbons from coconut shell. *Micropor. Mesopor. Mat.*, **27**, 11–18.

10 Kumagai, S., Ishizawa, H., and Toida, Y. (2010) Influence of solvent type on dibenzothiophene adsorption onto activated carbon fiber and granular coconut-shell activated carbon. *Fuel*, **89**, 365–371.

11 Azevedo, D. *et al.* (2007) Microporous activated carbon prepared from coconut shells using chemical activation with zinc chloride. *Micropor. Mesopor. Mat.*, **100**, 361–364.

12 Hu, Z. and Srinivasan, M. (2001) Mesoporous high-surface-area activated carbon. *Micropor. Mesopor. Mat.*, **43**, 267–275.

13 Hayashi, J.i., Horikawa, T., Takeda, I., Muroyama, K., and Nasir Ani, F. (2002) Preparing activated carbon from various nutshells by chemical activation with K_2CO_3. *Carbon*, **40**, 2381–2386.

14 Marsh, H. (2001) *Activated Carbon Compendium: A Collection of Papers from the Journal Carbon 1996–2000*, Gulf Professional Publishing.

15 Subramanian, V. *et al.* (2007) Supercapacitors from activated carbon derived from banana fibers. *J. Phys. Chem. C*, **111**, 7527–7531.

16 Wei, L. and Yushin, G. (2011) Electrical double layer capacitors with sucrose derived carbon electrodes in ionic liquid electrolytes. *J. Power Sources*, **196**, 4072–4079.

17 Li, Q.Y., Wang, H.Q., Dai, Q.F., Yang, J.H., and Zhong, Y.L. (2008) Novel activated carbons as electrode materials for electrochemical capacitors from a series of starch. *Solid State Ionics*, **179**, 269–273.

18 Balathanigaimani, M.S. *et al.* (2008) Highly porous electrodes from novel corn grains-based activated carbons for electrical double layer capacitors. *Electrochem. Commun.*, **10**, 868–871.

19 Kierzek, K., Frackowiak, E., Lota, G., Gryglewicz, G., and Machnikowski, J. (2004) Electrochemical capacitors based on highly porous carbons prepared by KOH activation. *Electrochim. Acta*, **49**, 515–523.

20 Zhai, D., Li, B., Du, H., Wang, G., and Kang, F. (2011) The effect of pre-carbonization of mesophase pitch-based activated carbons on their electrochemical performance for electric double-layer capacitors. *J. Solid State Electrochem.*, **15**, 787–794.

21 Rufford, T.E., Hulicova-Jurcakova, D., Khosla, K., Zhu, Z.H., and Lu, G.Q.

22. Lozano-Castello, D. et al. (2003) Influence of pore structure and surface chemistry on electric double layer capacitance in non-aqueous electrolyte. *Carbon*, **41**, 1765–1775.
23. Xu, B. et al. (2010) Activated carbon with high capacitance prepared by NaOH activation for supercapacitors. *Mater. Chem. Phys.*, **124**, 504–509.
24. Rufford, T.E., Hulicova-Jurcakova, D., Zhu, Z., and Lu, G.Q. (2008) Nanoporous carbon electrode from waste coffee beans for high performance supercapacitors. *Electrochem. Commun.*, **10**, 1594–1597.
25. Li, X., Han, C., Chen, X., and Shi, C. (2010) Preparation and performance of straw based activated carbon for supercapacitor in non-aqueous electrolytes. *Micropor. Mesopor. Mat.*, **131**, 303–309.
26. Wang, R. et al. (2012) Promising porous carbon derived from celtuce leaves with outstanding supercapacitance and $CO_{(2)}$ capture performance. *ACS Appl. Mater. Interfaces*, **4**, 5800–5806.
27. Lee, S.G., Park, K.H., Shim, W.G., balathanigaimani, M.S., and Moon, H. (2011) Performance of electrochemical double layer capacitors using highly porous activated carbons prepared from beer lees. *J. Ind. Eng. Chem.*, **17**, 450–454.
28. Zhang, J., Gong, L., Sun, K., Jiang, J., and Zhang, X. (2012) Preparation of activated carbon from waste Camellia oleifera shell for supercapacitor application. *J. Solid State Electrochem.*, **16**, 2179–2186.
29. Elmouwahidi, A., Zapata-Benabithe, Z., Carrasco-Marin, F., and Moreno-Castilla, C. (2012) Activated carbons from KOH-activation of argan (*Argania spinosa*) seed shells as supercapacitor electrodes. *Bioresour. Technol.*, **111**, 185–190.
30. Ding, L. et al. (2012) A novel hydrochar and nickel composite for the electrochemical supercapacitor electrode material. *Mater. Lett.*, **74**, 111–114.
31. Li, Z. et al. (2012) Carbonized chicken eggshell membranes with 3D architectures as high-performance electrode materials for supercapacitors. *Adv. Energy Mater.*, **2**, 431–437.
32. Yan, J. et al. (2010) A high-performance carbon derived from polyaniline for supercapacitors. *Electrochem. Commun.*, **12**, 1279–1282.
33. Xiang, X. et al. (2011) Activated carbon prepared from polyaniline base by $K_{(2)}CO_{(3)}$ activation for application in supercapacitor electrodes. *J. Solid State Electrochem.*, **15**, 579–585.
34. Eliad, L. et al. (2006) Assessing optimal pore-to-ion size relations in the design of porous poly(vinylidene chloride) carbons for EDL capacitors. *Appl. Phys. A*, **82**, 607–613.
35. Hasegawa, G. et al. (2011) Monolithic electrode for electric double-layer capacitors based on macro/meso/microporous S-containing activated carbon with high surface area. *J. Mater. Chem.*, **21**, 2060–2063.
36. Kim, K.-S. and Park, S.-J. (2012) Easy synthesis of polyaniline-based mesoporous carbons and their high electrochemical performance. *Micropor. Mesopor. Mat.*, **163**, 140–146.
37. Guo, S., Wang., F., Chen, H,, Ren, H., Wang, R., and Pan, X. (2012) Preparation and performance of polyvinyl alcohol-based activated carbon as electrode material in both aqueous and organic electrolytes. *J. Solid State Electrochem.*, **16**, 3355–3362.
38. Wei, L., Sevilla, M., Fuertes, A.B., Mokaya, R., and Yushin, G. (2012) Polypyrrole-derived activated carbons for high-performance electrical double-layer capacitors with ionic liquid electrolyte. *Adv. Funct. Mater.*, **22**, 827–834.
39. Wei, L., Sevilla, M., Fuertes, A.B., Mokaya, R., and Yushin, G. (2011) Hydrothermal carbonization of abundant renewable natural organic chemicals for high-performance supercapacitor electrodes. *Adv. Energy Mater.*, **1**, 356–361.
40. Raymundo-Pinero, E., Cadek, M., and Beguin, F. (2009) Tuning carbon materials for supercapacitors by direct pyrolysis of seaweeds. *Adv. Funct. Mater.*, **19**, 1032–1039.

41 Gu, W., Magasinski, A., Zdyrko, B., and Yushin, G. (2015) Metal fluorides nanoconfined in carbon nanopores as reversible high capacity cathodes for Li and Li-ion rechargeable batteries: FeF_2 as an example. *Adv. Energy Mater.*, **5** (4). doi: 10.1002/aenm.201401148

42 Jänes, A., Kurig, H., and Lust, E. (2007) Characterisation of activated nanoporous carbon for supercapacitor electrode materials. *Carbon*, **45**, 1226–1233.

43 Yang, H., Yoshio, M., Isono, K., and Kuramoto, R. (2002) Improvement of commercial activated carbon and its application in electric double layer capacitors. *Electrochem. Solid State Lett.*, **5**, A141. doi: 10.1149/1.1477297

44 Teng, H., Chang, Y.-J., and Hsieh, C.-T. (2001) Performance of electric double-layer capacitors using carbons prepared from phenol-formaldehyde resins by KOH etching. *Carbon*, **39**, 1981–1987.

45 Weng, T.-C. and Teng, H. (2001) Characterization of high porosity carbon electrodes derived from mesophase pitch for electric double-layer capacitors. *J. Electrochem. Soc.*, **148**, A368. doi: 10.1149/1.1357171

46 Kim, Y.J. *et al.* (2004) Correlation between the pore and solvated ion size on capacitance uptake of PVDC-based carbons. *Carbon*, **42**, 1491–1500.

47 Alonso, A. *et al.* (2006) Activated carbon produced from Sasol–Lurgi gasifier pitch and its application as electrodes in supercapacitors. *Carbon*, **44**, 441–446.

48 Huang, C.-W., Hsieh, C.-T., Kuo, P.-L., and Teng, H. (2012) Electric double layer capacitors based on a composite electrode of activated mesophase pitch and carbon nanotubes. *J. Mater. Chem.*, **22**, 7314.

49 Hulicova-Jurcakova, D. Puziy, A.M., Poddubnaya, Olga. I., Suarez-Garcia, F., Tascon, Juan M.D., and Lu, G.Q. (2009) Highly stable performance of supercapacitors from phosphorus-enriched carbons. *J. Am. Chem. Soc.*, **131**, 5026–5027.

50 Xiang, X., Liu, E., Huang, Z., Shen, H., Tian, Y., Xiao, C., Yang, J., and Mao, Z. (2011) Preparation of activated carbon from polyaniline by zinc chloride activation as supercapacitor electrodes. *J. Solid State Electrochem.*, **15**, 2667.

51 Nandhini, R., Mini, P.A., Avinash, B., Nair, S.V., and Subramanian, K.R.V. (2012) Supercapacitor electrodes using nanoscale activated carbon from graphite by ball milling. *Mater. Lett.*, **87**, 165–168.

52 Lillo-Ródenas, M.A., Lozano-Castelló, D., Cazorla-Amorós, D., and Linares-Solano, A. (2001) Preparation of activated carbons from Spanish anthracite: II. Activation by NaOH. *Carbon*, **39**, 751–759.

53 Erdoğan, S. *et al.* (2005) Optimization of nickel adsorption from aqueous solution by using activated carbon prepared from waste apricot by chemical activation. *Appl. Surf. Sci.*, **252**, 1324–1331.

54 Hayashi, J., Yamamoto, N., Horikawa, T., Muroyama, K., and Gomes, V.G. (2005) Preparation and characterization of high-specific-surface-area activated carbons from K_2CO_3-treated waste polyurethane. *J. Colloid Interface Sci.*, **281**, 437–443.

55 Lillo-Ródenas, M., Cazorla-Amorós, D., and Linares-Solano, A. (2003) Understanding chemical reactions between carbons and NaOH and KOH: an insight into the chemical activation mechanism. *Carbon*, **41**, 267–275.

56 Gonzalez-Serrano, E., Cordero, T., Rodriguez-Mirasol, J., and Rodriguez, J. (1997) Development of porosity upon chemical activation of kraft lignin with $ZnCl_2$. *Ind. Eng. Chem. Res.*, **36**, 4832–4838.

57 Evans, M., Halliop, E., and MacDonald, J. (1999) The production of chemically-activated carbon. *Carbon*, **37**, 269–274.

58 Simon, P. and Gogotsi, Y. (2008) Materials for electrochemical capacitors. *Nat. Mater.*, 7, 845–854.

59 Taberna, P.L., Simon, P., and Fauvarque, J.F. (2003) Electrochemical characteristics and impedance spectroscopy studies of carbon–carbon supercapacitors. *J. Electrochem. Soc.*, **150**, A292–A300.

60 Gamby, J., Taberna, P.L., Simon, P., Fauvarque, J.F., and Chesneau, M. (2001) Studies and characterisations of various activated carbons used for carbon/carbon supercapacitors. *J. Power Sources*, **101**, 109–116.

61 Frackowiak, E. and Beguin, F. (2001) Carbon materials for the electrochemical storage of energy in capacitors. *Carbon*, **39**, 937–950.

62 Azaïs, P. et al. (2007) Causes of supercapacitors ageing in organic electrolyte. *J. Power Sources*, **171**, 1046–1053.

63 Hulicova-Jurcakova, D., Seredych, M., Lu, G.Q., and Bandosz, T.J. (2009) Combined effect of nitrogen- and oxygen-containing functional groups of microporous activated carbon on its electrochemical performance in supercapacitors. *Adv. Funct. Mater.*, **19**, 438–447.

64 Hulicova-Jurcakova, D. et al. (2009) Nitrogen-enriched nonporous carbon electrodes with extraordinary supercapacitance. *Adv. Funct. Mater.*, **19**, 1800–1809.

65 Liang, Y.Y. et al. (2010) Direct access to metal or metal oxide nanocrystals integrated with one-dimensional nanoporous carbons for electrochemical energy storage. *J. Am. Chem. Soc.*, **132**, 15030–15037.

66 Balducci, A. et al. (2007) High temperature carbon–carbon supercapacitor using ionic liquid as electrolyte. *J. Power Sources*, **165**, 922–927.

67 Sun, X., Zhang, X., Zhang, H., Zhang, D., and Ma, Y. (2012) A comparative study of activated carbon-based symmetric supercapacitors in Li_2SO_4 and KOH aqueous electrolytes. *J. Solid State Electrochem.*, **16**, 2597–2603.

68 Han, S.-J., Kim, Y.-H., Kim, K.-S., and Park, S.-J. (2012) A study on high electrochemical capacitance of ion exchange resin-based activated carbons for supercapacitor. *Curr. Appl. Phys.*, **12**, 1039–1044.

69 Timperman, L. et al. (2012) Triethylammonium bis(tetrafluoromethylsulfonyl)amide protic ionic liquid as an electrolyte for electrical double-layer capacitors. *Phys. Chem. Chem. Phys.*, **14**, 8199–8207.

70 Raymundo-Pinero, E., Leroux, F., and Beguin, F. (2006) A high-performance carbon for supercapacitors obtained by carbonization of a seaweed biopolymer. *Adv. Mater.*, **18**, 1877–1880.

71 Lewandowski, A., Olejniczak, A., Galinski, M., and Stepniak, I. (2010) Performance of carbon–carbon supercapacitors based on organic, aqueous and ionic liquid electrolytes. *J. Power Sources*, **195**, 5814–5819.

72 Zhang, Q., Rong, J.P., Ma, D.S., and Wei, B.Q. (2011) The governing self-discharge processes in activated carbon fabric-based supercapacitors with different organic electrolytes. *Energy Env. Sci.*, **4**, 2152–2159.

73 Hung, K.S., Masarapu, C., Ko, T.H., and Wei, B.Q. (2009) Wide-temperature range operation supercapacitors from nanostructured activated carbon fabric. *J. Power Sources*, **193**, 944–949.

74 Seo, M.K. and Park, S.J. (2009) Electrochemical characteristics of activated carbon nanofiber electrodes for supercapacitors. *Mater. Sci. Eng. B*, **164**, 106–111.

75 McDonough, J.R. et al. (2009) Carbon nanofiber supercapacitors with large areal capacitances. *Appl. Phys. Lett.*, **95**. doi: 10.1063/1.3273864

76 Xu, B. et al. (2010) Mesoporous activated carbon fiber as electrode material for high-performance electrochemical double layer capacitors with ionic liquid electrolyte. *J. Power Sources*, **195**, 2118–2124.

77 Nakagawa, H., Shudo, A., and Miura, K. (2000) High-capacity electric double-layer capacitor with high-density-activated carbon fiber electrodes. *J. Electrochem. Soc.*, **147**, 38–42.

78 Kim, Y.J. et al. (2005) High energy-density capacitor based on ammonium salt type ionic liquids and their mixing effect by propylene carbonate. *J. Electrochem. Soc.*, **152**, A710–A715.

79 Okajima, K., Ohta, K., and Sudoh, M. (2005) Capacitance behavior of activated carbon fibers with oxygen-plasma treatment. *Electrochim. Acta*, **50**, 2227–2231.

80 Ishikawa, M., Sakamoto, A., Morita, M., Matsuda, Y., and Ishida, K. (1996) Effect of treatment of activated carbon fiber cloth electrodes with cold plasma upon

performance of electric double-layer capacitors. *J. Power Sources*, **60**, 233–238.
81 Kim, C. (2005) Electrochemical characterization of electrospun activated carbon nanofibres as an electrode in supercapacitors. *J. Power Sources*, **142**, 382–388.
82 Kim, B.H., Yang, K.S., Woo, H.G., and Oshida, K. (2011) Supercapacitor performance of porous carbon nanofiber composites prepared by electrospinning polymethylhydrosiloxane (PMHS)/polyacrylonitrile (PAN) blend solutions. *Synth. Met.*, **161**, 1211–1216.
83 Endo, M. et al. (2001) Capacitance and pore-size distribution in aqueous and nonaqueous electrolytes using various activated carbon electrodes. *J. Electrochem. Soc.*, **148**, A910.
84 Babel, K. and Jurewicz, K. (2004) KOH activated carbon fabrics as supercapacitor material. *J. Phys. Chem. Solids*, **65**, 275–280.
85 Endo, M., Kim, Y.J., Maeda, T., Koshiba, K., and Katayam, K. (2001) Morphological effect on the electrochemical behavior of electric double-layer capacitors. *J. Mater. Res.*, **16**, 3402–3410.
86 Endo, M. et al. (2001) Poly(vinylidene chloride)-based carbon as an electrode material for high power capacitors with an aqueous electrolyte. *J. Electrochem. Soc.*, **148**, A1135.
87 Xu, B. et al. (2011) What is the choice for supercapacitors: graphene or graphene oxide? *Energy Environ. Sci.*, **4**, 2826–2830.
88 Sun, Y., Wu, Q., and Shi, G. (2011) Graphene based new energy materials. *Energy Environ. Sci.*, **4**, 1113–1132.
89 Pumera, M. (2011) Graphene-based nanomaterials for energy storage. *Energy Environ. Sci.*, **4**, 668–674.
90 Zhang, L.L., Zhou, R., and Zhao, X.S. (2010) Graphene-based materials as supercapacitor electrodes. *J. Mater. Chem.*, **20**, 5983–5992.
91 Liang, M., Luo, B., and Zhi, L. (2009) Application of graphene and graphene-based materials in clean energy-related devices. *Int. J. Energy Res.*, **33**, 1161–1170.
92 Kovalenko, I., Bucknall, D.G., and Yushin, G. (2010) Detonation nanodiamond and onion-like-carbon-embedded polyaniline for supercapacitors. *Adv. Funct. Mater.*, **20**, 3979–3986.
93 Korenblit, Y. et al. (2010) High-rate electrochemical capacitors based on ordered mesoporous silicon carbide-derived carbon. *ACS Nano*, **4**, 1337–1344.
94 Evanoff, K., Magasinski, A., Yang, J.B., and Yushin, G. (2011) Nanosilicon-coated graphene granules as anodes for Li-ion batteries. *Adv. Energy Mater.*, **1**, 495–498.
95 Evanoff, K. et al. (2012) Towards ultrathick battery electrodes: aligned carbon nanotube-enabled architecture. *Adv. Mater.*, **24**, 533–540.
96 Jiang, H., Lee, P.S., and Li, C. (2013) 3D carbon based nanostructures for advanced supercapacitors. *Energy Environ. Sci.*, **6**, 41–53.
97 Wang, D.W., Li, F., Liu, M., Lu, G.Q., and Cheng, H.M. (2008) 3D aperiodic hierarchical porous graphitic carbon material for high-rate electrochemical capacitive energy storage. *Angew. Chem. Int. Ed.*, **47**, 373–376.
98 Fan, Z. et al. (2010) A three-dimensional carbon nanotube/graphene sandwich and its application as electrode in supercapacitors. *Adv. Mater.*, **22**, 3723–3728.
99 Jiang, H., Sun, T., Li, C., and Ma, J. (2012) Hierarchical porous nanostructures assembled from ultrathin MnO_2 nanoflakes with enhanced supercapacitive performances. *J. Mater. Chem*, **22**, 2751–2756.
100 Jiang, H., Zhao, T., Ma, J., Yan, C., and Li, C. (2011) Ultrafine manganese dioxide nanowire network for high-performance supercapacitors. *Chem. Commun.*, **47**, 1264–1266.
101 Yan, J. et al. (2010) Fast and reversible surface redox reaction of graphene–MnO_2 composites as supercapacitor electrodes. *Carbon*, **48**, 3825–3833.
102 Wu, Q., Xu, Y., Yao, Z., Liu, A., and Shi, G. (2010) Supercapacitors based on flexible graphene/polyaniline nanofiber

composite films. *ACS Nano*, **4**, 1963–1970.
103 Xu, J., Wang, K., Zu, S.-Z., Han, B.-H., and Wei, Z. (2010) Hierarchical nanocomposites of polyaniline nanowire arrays on graphene oxide sheets with synergistic effect for energy storage. *ACS Nano*, **4**, 5019–5026.
104 Yamada, H., Nakamura, H., Nakahara, F., Moriguchi, I., and Kudo, T. (2007) Electrochemical study of high electrochemical double layer capacitance of ordered porous carbons with both meso/macropores and micropores. *J. Phys. Chem. C*, **111**, 227–233.
105 Gutiérrez, M.C. *et al.* (2009) PPO 15-PEO 22-PPO 15 block copolymer assisted synthesis of monolithic macro- and microporous carbon aerogels exhibiting high conductivity and remarkable capacitance. *J. Mater. Chem.*, **19**, 1236–1240.
106 Randin, J.-P. and Yeager, E. (1971) Differential capacitance study of stress-annealed pyrolytic graphite electrodes. *J. Electrochem. Soc.*, **118** (5), 711–714,
107 Randin, J.-P. and Yeager, E. (1972) Differential capacitance study on the basal plane of stress-annealed pyrolytic graphite. *J. Electroanal. Chem.*, **36**, 257–276.
108 Randin, J.-P. and Yeager, E. (1975) Differential capacitance study on the edge orientation of pyrolytic graphite and glassy carbon electrodes. *J. Electroanal. Chem.*, **58**, 313–322.
109 Gerischer, H., McIntyre, R., Scherson, D., and Storck, W. (1987) Density of the electronic states of graphite: derivation from differential capacitance measurements. *J. Phys. Chem.*, **91**, 1930–1935.
110 Hahn, M. *et al.* (2004) Interfacial capacitance and electronic conductance of activated carbon double-layer electrodes. *Electrochem. Solid-State Lett.*, **7**, A33–A36.
111 Gerischer, H., McIntyre, R., Scherson, D., and Storck, W. (1987) Density of the electronic states of graphite-derivation from differential capacitance measurements. *J. Phys. Chem.*, **91**, 1930–1935.
112 Xia, J.L., Chen, F., Li, J.H., and Tao, N.J. (2009) Measurement of the quantum capacitance of graphene. *Nat. Nanotechnol.*, **4**, 505–509.
113 Gerischer, H. (1985) An interpretation of the double layer capacity of graphite electrodes in relation to the density of states at the Fermi level. *J. Phys. Chem.*, **89**, 4249–4251.
114 Dröscher, S. *et al.* (2010) Quantum capacitance and density of states of graphene. *Appl. Phys. Lett.*, **96**, 152104.
115 Ponomarenko, L. *et al.* (2010) Density of states and zero Landau level probed through capacitance of graphene. *Phys. Rev. Lett.*, **105**, 136801.
116 Jang, I.Y., Muramatsu, H., Park, K.C., Kim, Y.J., and Endo, M. (2009) Capacitance response of double-walled carbon nanotubes depending on surface modification. *Electrochem. Commun.*, **11**, 719–723.
117 Zhang, L.L. *et al.* (2012) Nitrogen doping of graphene and its effect on quantum capacitance, and a new insight on the enhanced capacitance of N-doped carbon. *Energy Environ. Sci.*, **5**, 9618–9625.
118 Jayalakshmi, M. and Balasubramanian, K. (2008) Simple capacitors to supercapacitors: an overview. *Int. J. Electrochem. Sci.*, **3**, 1196–1217.
119 Vol'fkovich, Y.M. and Serdyuk, T.M. (2002) Electrochemical capacitors. *Russ. J. Electrochem.*, **38**, 935–958.
120 Janet, H. and Jow, T.R. (2010) Historical introduction to capacitor technology. *IEEE Electr. Insul. M*, **26**. doi: 10.1109/MEI.2010.5383924
121 Augustyn, V., Simon, P., and Dunn, B. (2014) Pseudocapacitive oxide materials for high-rate electrochemical energy storage. *Energy Environ. Sci.*, **7**, 1597–1614.
122 Attard, G.S. *et al.* (1997) Mesoporous platinum films from lyotropic liquid crystalline phases. *Supramol. Sci.*, **278**, 838–840.
123 Will, F.G. and Knorr, C.A. (1960) Investigation of formation and removal of hydrogen and oxygen coverage on platinum by a new nonstationary method. *Zeit. Elektrochem*, **64**, 258.

124 Conway, B.E., Birss, V., and Wojtowicz, J. (1997) The role and utilization of pseudocapacitance for energy storage by supercapacitors. *J. Power Sources*, **66**, 1–14.

125 Liu, J.P. et al. (2011) Co_3O_4 nanowire@MnO_2 ultrathin nanosheet core/shell arrays: a new class of high-performance pseudocapacitive materials. *Adv. Mater.*, **23**, 2076–2080.

126 Boukhalfa, S., Evanoff, K., and Yushin, G. (2012) Atomic layer deposition of vanadium oxide on carbon nanotubes for high-power supercapacitor electrodes. *Energy Environ. Sci.*, **5**, 6872–6879.

127 Wang, Q., Wen, Z.H., and Li, J.H. (2006) A hybrid supercapacitor fabricated with a carbon nanotube cathode and a TiO_2-B nanowire anode. *Adv. Funct. Mater.*, **16**, 2141–2146.

128 Yu, G.H. et al. (2011) Enhancing the supercapacitor performance of graphene/MnO_2 nanostructured electrodes by conductive wrapping. *Nano Lett.*, **11**, 4438–4442.

129 Park, B.-O., Lokhande, C.D., Park, H.-S., Jung, K.-D., and Joo, O.-S. (2004) Performance of supercapacitor with electrodeposited ruthenium oxide film electrodes: effect of film thickness. *J. Power Sources*, **134**, 148–152.

130 Patake, V.D., Pawar, S.M., Shinde, V.R., Gujar, T.P., and Lokhande, C.D. (2010) The growth mechanism and supercapacitor study of anodically deposited amorphous ruthenium oxide films. *Curr. Appl. Phys.*, **10**, 99–103.

131 Chen, S., Zhu, J., Wu, X., Han, Q., and Wang, X. (2010) Graphene oxide: MnO_2 nanocomposites for supercapacitors. *ACS Nano*, **4**, 2822–2830.

132 He, Y. et al. (2013) Freestanding three-dimensional graphene/MnO_2 composite networks as ultralight and flexible supercapacitor electrodes. *ACS Nano*, **7**, 174–182.

133 Fan, Z. et al. (2011) Asymmetric supercapacitors based on graphene/MnO_2 and activated carbon nanofiber electrodes with high power and energy density. *Adv. Funct. Mater.*, **21**, 2366–2375.

134 Yuan, L. et al. (2012) Flexible solid-state supercapacitors based on carbon nanoparticles/MnO_2 nanorods hybrid structure. *ACS Nano*, **6**, 656–661.

135 Zhao, B. et al. (2011) Monolayer graphene/NiO nanosheets with two-dimension structure for supercapacitors. *J. Mater. Chem.*, **21**, 18792–18798.

136 Lu, Q. et al. (2011) Supercapacitor electrodes with high-energy and power densities prepared from monolithic NiO/Ni nanocomposites. *Angew. Chem.*, **123**, 6979–6982.

137 Liang, K., Tang, X., and Hu, W. (2012) High-performance three-dimensional nanoporous NiO film as a supercapacitor electrode. *J. Mater. Chem.*, **22**, 11062–11067.

138 Prasad, K.R., Koga, K., and Miura, N. (2004) Electrochemical deposition of nanostructured indium oxide: high-performance electrode material for redox supercapacitors. *Chem. Mater.*, **16**, 1845–1847.

139 Bastakoti, B.P. et al. (2013) Mesoporous carbon incorporated with In_2O_3 nanoparticles as high-performance supercapacitors. *Eur. J. Inorg. Chem.*, **2013**, 1109–1112.

140 Chen, P.-C., Shen, G., Sukcharoenchoke, S., and Zhou, C. (2009) Flexible and transparent supercapacitor based on In_2O_3 nanowire/carbon nanotube heterogeneous films. *Appl. Phys. Lett.*, **94**, 043113.

141 Xia, X.-h. et al. (2012) Freestanding Co_3O_4 nanowire array for high performance supercapacitors. *RSC Adv.*, **2**, 1835–1841.

142 Xia, X.-h. et al. (2011) Self-supported hydrothermal synthesized hollow Co_3O_4 nanowire arrays with high supercapacitor capacitance. *J. Mater. Chem.*, **21**, 9319–9325.

143 Chen, Z. et al. (2011) High-performance supercapacitors based on intertwined CNT/V_2O_5 nanowire nanocomposites. *Adv. Mater.*, **23**, 791–795.

144 Qu, Q., Zhu, Y., Gao, X., and Wu, Y. (2012) Core–shell structure of polypyrrole grown on V_2O_5 nanoribbon as high performance anode material for

supercapacitors. *Adv. Energy Mater.*, **2**, 950–955.

145 Radhakrishnan, S., Rao, C.R.K., and Vijayan, M. (2011) Performance of conducting polyaniline-DBSA and polyaniline-DBSA/Fe_3O_4 composites as electrode materials for aqueous redox supercapacitors. *J. Appl. Polym. Sci.*, **122**, 1510–1518.

146 Du, X., Wang, C., Chen, M., Jiao, Y., and Wang, J. (2009) Electrochemical performances of nanoparticle Fe_3O_4/activated carbon supercapacitor using KOH electrolyte solution. *J. Phys. Chem. C*, **113**, 2643–2646.

147 Gujar, T.P., Shinde, V.R., Lokhande, C.D., and Han, S.-H. (2006) Electrosynthesis of Bi_2O_3 thin films and their use in electrochemical supercapacitors. *J. Power Sources*, **161**, 1479–1485.

148 Zheng, F.-L. et al. (2010) Synthesis of hierarchical rippled Bi_2O_3 nanobelts for supercapacitor applications. *Chem. Commun.*, **46**, 5021–5023.

149 Chen, Y.M., Cai, J.H., Huang, Y.S., Lee, K.Y., and Tsai, D.S. (2011) Preparation and characterization of iridium dioxide–carbon nanotube nanocomposites for supercapacitors. *Nanotechnology*, **22**, 115706.

150 Yu, Z.-Y., Chen, L.-F., and Yu, S.-H. (2014) Growth of $NiFe_2O_4$ nanoparticles on carbon cloth for high performance flexible supercapacitors. *J. Mater. Chem. A*, **2**, 10889–10894.

151 Lokhande, C.D., Gujar, T.P., Shinde, V.R., Mane, R.S., and Han, S.-H. (2007) Electrochemical supercapacitor application of pervoskite thin films. *Electrochem. Commun.*, **9**, 1805–1809.

152 Long, J.W., Young, A.L., and Rolison, D.R. (2003) Spectroelectrochemical characterization of nanostructured mesoporous manganese oxide in aqueous electrolytes. *J. Electrochem. Soc.*, **150**, A1161.

153 Pang, S.-C. and Anderson, M.A. (2000) Novel electrode materials for electrochemical capacitors. Part II. Material characterization of sol–gel-derived and electrodeposited manganese dioxide thin films. *J. Mater. Res.*, **15**, 2096–2106.

154 Lee, H.Y. and Goodenough, J.B. (1999) Supercapacitor behavior with KCl electrolyte. *J. Solid State Chem.*, **144**, 220–223.

155 Pang, S.C., Anderson, M.A., and Chapman, T.W. (2000) Novel electrode materials for thin-film ultracapacitors: comparison of electrochemical properties of sol–gel-derived and electrodeposited manganese dioxide. *J. Electrochem. Soc.*, **147**, 444–450.

156 Kanoh, H., Tang, W., Makita, Y., and Ooi, K. (1997) Electrochemical intercalation of alkali-metal ions into birnessite-type manganese oxide in aqueous solution. *Langmuir*, **13**, 6845–6849.

157 Lee, H.Y., Kim, S.W., and Lee, H.Y. (2001) Expansion of active site area and improvement of kinetic reversibility in electrochemical pseudocapacitor electrode. *Electrochem. Solid-State Lett.*, **4**, A19–A22.

158 Yan, J. et al. (2009) Carbon nanotube/$MnO_{(2)}$ composites synthesized by microwave-assisted method for supercapacitors with high power and energy densities. *J. Power Sources*, **194**, 1202–1207.

159 Hu, J., Yuan, A.-B., Wang, Y.-Q., and Wang, X.-L. (2009) Improved cyclability of nano-$MnO_{(2)}$/CNT composite supercapacitor electrode derived from room-temperature solid reaction. *Acta Phys. Chim. Sin.*, **25**, 987–993.

160 Sugimoto, W., Iwata, H., Yasunaga, Y., Murakami, Y., and Takasu, Y. (2003) Preparation of ruthenic acid nanosheets and utilization of its interlayer surface for electrochemical energy storage. *Angew. Chem. Int. Ed.*, **42**, 4092–4096.

161 Brezesinski, T., Wang, J., Tolbert, S.H., and Dunn, B. (2010) Next generation pseudocapacitor materials from sol–gel derived transition metal oxides. *J. Sol-Gel Sci. Technol.*, **57**, 330–335.

162 Toupin, M., Brousse, T., and Belanger, D. (2004) Charge storage mechanism of MnO_2 electrode used in aqueous electrochemical capacitor. *Chem. Mater.*, **16**, 3184–3190.

163 Yu, G. et al. (2011) Enhancing the supercapacitor performance of graphene/

MnO$_2$ nanostructured electrodes by conductive wrapping. *Nano Lett.*, **11**, 4438–4442.

164 Choi, D., Blomgren, G.E., and Kumta, P.N. (2006) Fast and reversible surface redox reaction in nanocrystalline vanadium nitride supercapacitors. *Adv. Mater.*, **18**, 1178–1180.

165 Lu, X. et al. (2014) Improving the cycling stability of metal–nitride supercapacitor electrodes with a thin carbon shell. *Adv. Energy Mater.*, **4**. doi: 10.1002/aenm.201300994

166 Fischer, A., Müller, J.O., Antonietti, M., and Thomas, A. (2008) Synthesis of ternary metal nitride nanoparticles using mesoporous carbon nitride as reactive template. *ACS Nano*, **2**, 2489–2496.

167 Lu, X. et al. (2013) High energy density asymmetric quasi-solid-state supercapacitor based on porous vanadium nitride nanowire anode. *Nano Lett.*, **13**, 2628–2633.

168 Conway, B.E. (1991) Transition from supercapacitor to battery behavior in electrochemical energy-storage. *J. Electrochem. Soc.*, **138**, 1539–1548.

169 Snook, G.A., Kao, P., and Best, A.S. (2011) Conducting-polymer-based supercapacitor devices and electrodes. *J. Power Sources*, **196**, 1–12.

170 Tu, L.L. and Jia, C.Y. (2010) Conducting polymers as electrode materials for supercapacitors. *Prog. Chem.*, **22**, 1610–1618.

171 Aradilla, D., Estrany, F., and Aleman, C. (2011) Symmetric supercapacitors based on multilayers of conducting polymers. *J. Phys. Chem. C*, **115**, 8430–8438.

172 Jaidev, Jafri, R.I., Mishra, A.K., and Ramaprabhu, S. (2011) Polyaniline–MnO$_2$ nanotube hybrid nanocomposite as supercapacitor electrode material in acidic electrolyte. *J. Mater. Chem.*, **21**, 17601.

173 Chen, L. et al. (2010) Synthesis and pseudocapacitive studies of composite films of polyaniline and manganese oxide nanoparticles. *J. Power Sources*, **195**, 3742–3747.

174 Liu, F.-J. (2008) Electrodeposition of manganese dioxide in three-dimensional poly(3,4-ethylenedioxythiophene)–poly (styrene sulfonic acid)–polyaniline for supercapacitor. *J. Power Sources*, **182**, 383–388.

175 Zhang, X., Ji, L., Zhang, S., and Yang, W. (2007) Synthesis of a novel polyaniline-intercalated layered manganese oxide nanocomposite as electrode material for electrochemical capacitor. *J. Power Sources*, **173**, 1017–1023.

176 Gupta, V. and Miura, N. (2006) Polyaniline/single-wall carbon nanotube (PANI/SWCNT) composites for high performance supercapacitors. *Electrochim. Acta*, **52**, 1721–1726.

177 Zhou, G. et al. (2011) The effect of carbon particle morphology on the electrochemical properties of nanocarbon/polyaniline composites in supercapacitors. *New Carbon Mater.*, **26**, 180–186.

178 Zheng, L. et al. (2010) The preparation and performance of flocculent polyaniline/carbon nanotubes composite electrode material for supercapacitors. *J. Solid State Electrochem.*, **15**, 675–681.

179 Liu, W., Liu, N., Song, H., and Chen, X. (2011) Properties of polyaniline/ordered mesoporous carbon composites as electrodes for supercapacitors. *New Carbon Mater.*, **26**, 217–223.

180 Konwer, S., Boruah, R., and Dolui, S. K. (2011) *J. Electronic. Mater.*, **40**, 2248–2255.

181 Ge, J., Cheng, G., and Chen, L. (2011) Transparent and flexible electrodes and supercapacitors using polyaniline/single-walled carbon nanotube composite thin films. *Nanoscale*, **3**, 3084–3088.

182 Yuan, C., Shen, L., Zhang, F., Lu, X., and Zhang, X. (2010) Reactive template fabrication of uniform core–shell polyaniline/multiwalled carbon nanotube nanocomposite and its electrochemical capacitance. *Chem. Lett.*, **39**, 850–851.

183 Yang, M., Cheng, B., Song, H., and Chen, X. (2010) Preparation and electrochemical performance of polyaniline-based carbon nanotubes as electrode material for supercapacitor. *Electrochim. Acta*, **55**, 7021–7027.

184 Gómez, H. et al. (2011) Graphene-conducting polymer nanocomposite as novel electrode for supercapacitors. *J. Power Sources*, **196**, 4102–4108.

185 Zhang, K., Zhang, L.L., Zhao, X.S., and Wu, J. (2010) Graphene/polyaniline nanofiber composites as supercapacitor electrodes. *Chem. Mater.*, **22**, 1392–1401.

186 Yan, J. et al. (2010) Preparation of a graphene nanosheet/polyaniline composite with high specific capacitance. *Carbon*, **48**, 487–493.

187 Benson, J. et al. (2013) Multifunctional CNT-polymer composites for ultra-tough structural supercapacitors and desalination devices. *Adv. Mater.*, **25**, 6625–6632.

188 Meng, C., Liu, C., Chen, L., Hu, C., and Fan, S. (2010) Highly flexible and all-solid-state paperlike polymer supercapacitors. *Nano Lett.*, **10**, 4025–4031.

189 Seredych, M., Hulicova-Jurcakova, D., Lu, G.Q., and Bandosz, T.J. (2008) Surface functional groups of carbons and the effects of their chemical character, density and accessibility to ions on electrochemical performance. *Carbon*, **46**, 1475–1488.

190 Hulicova, D., Kodama, M., and Hatori, H. (2006) Electrochemical performance of nitrogen-enriched carbons in aqueous and non-aqueous supercapacitors. *Chem. Mater.*, **18**, 2318–2326.

191 Ania, C.O., Khomenko, V., Raymundo-Piñero, E., Parra, J.B., and Béguin, F. (2007) The large electrochemical capacitance of microporous doped carbon obtained by using a zeolite template. *Adv. Funct. Mater.*, **17**, 1828–1836.

192 Benaddi, H., Bandosz, T.J., Jagiello, J., Schwarz, J.A., Rouzaud, J.N., Legras, D., and Beguin, F. (2000) Surface functionality and porosity of activated carbons obtained from chemical activation of wood. *Carbon*, **38**, 669–674.

193 Tanahashi, I., Yoshida, A., and Nishino, A. (1990) Electrochemical characterization of activated carbon-fiber cloth polarizable electrodes for electric double-layer capacitors. *J. Electrochem. Soc.*, **137**, 3052–3057.

194 Wang, H.L., Gao, Q.M., and Hu, J. (2010) Preparation of porous doped carbons and the high performance in electrochemical capacitors. *Micropor. Mesopor. Mat.*, **131**, 89–96.

195 Gu, W., Sevilla, M., Magasinski, A., Fuertes, A.B., and Yushin, G. (2013) Sulfur-containing activated carbons with greatly reduced content of bottle neck pores for double-layer capacitors: a case study for pseudocapacitance detection. *Energy Environ. Sci.*, **6**, 2465.

196 Hsieh, C.T. and Teng, H. (2002) Influence of oxygen treatment on electric double-layer capacitance of activated carbon fabrics. *Carbon*, **40**, 667–674.

197 Jurewicz, K., Babeł, K., Ziółkowski, A., and Wachowska, H. (2003) Ammoxidation of active carbons for improvement of supercapacitor characteristics. *Electrochim. Acta*, **48**, 1491–1498.

198 Augustyn, V. et al. (2013) High-rate electrochemical energy storage through Li^+ intercalation pseudocapacitance. *Nat. Mater.*, **12**, 518–522.

199 Kim, J.W., Augustyn, V., and Dunn, B. (2012) The effect of crystallinity on the rapid pseudocapacitive response of Nb_2O_5. *Adv. Energy Mater.*, **2**, 141–148.

200 Gu, W. and Yushin, G. (2013) Review of nanostructured carbon materials for electrochemical capacitor applications: advantages and limitations of activated carbon, carbide-derived carbon, zeolite-templated carbon, carbon aerogels, carbon nanotubes, onion-like carbon, and graphene. *WIREs Energy Environ.* doi: 10.1002/wene.102

201 Korenblit, Y. et al. (2012) *In situ* studies of ion transport in microporous supercapacitor electrodes at ultralow temperatures. *Adv. Funct. Mater.*, **22**, 1655–1662.

202 Fic, K., Lota, G., Meller, M., and Frackowiak, E. (2012) Novel insight into neutral medium as electrolyte for high-voltage supercapacitors. *Energy Environ. Sci.*, **5**, 5842–5850.

203 Béguin, F., Presser, V., Balducci, A., and Frackowiak, E. (2014) Carbons and electrolytes for advanced supercapacitors. *Adv. Mater.*, **26**, 2219–2251.

204 Raymundo-Pinero, E., Kierzek, K., Machnikowski, J., and Beguin, F. (2006) Relationship between the nanoporous

texture of activated carbons and their capacitance properties in different electrolytes. *Carbon*, **44**, 2498–2507.

205 Chmiola, J., Yushin, G., Gogotsi, Y., Portet, C., and Simon, P. (2006) Anomalous increase in carbon capacitance at pore size below 1 nm. *Supramol. Sci.*, **313**, 1760–1763.

206 Salitra, G., Soffer, A., Eliad, L., Cohen, Y., and Aurbach, D. (2000) Carbon electrodes for double-layer capacitors: I. Relations between ion and pore dimensions. *J. Electrochem. Soc.*, **147**, 2486–2493.

207 Avraham, E., Yaniv, B., Soffer, A., and Aurbach, D. (2008) Developing ion electroadsorption stereoselectivity, by pore size adjustment with chemical vapor deposition onto active carbon fiber electrodes. Case of Ca_2^+/Na^+ separation in water capacitive desalination. *J. Phys. Chem. C*, **112**, 7385–7389.

208 Eliad, L. *et al.* (2006) Assessing optimal pore-to-ion size relations in the design of porous poly (vinylidene chloride) carbons for EDL capacitors. *Appl. Phys. A*, **82**, 607–613.

209 Noked, M., Avraham, E., Bohadana, Y., Soffer, A., and Aurbach, D. (2009) Development of anion stereoselective, activated carbon molecular sieve electrodes prepared by chemical vapor deposition. *J. Phys. Chem. C*, **113**, 7316–7321.

210 El-Kady, M.F. and Kaner, R.B. (2013) Scalable fabrication of high-power graphene micro-supercapacitors for flexible and on-chip energy storage. *Nat. Commun.*, **4**. doi: 10.1038/ncomms2446

211 Conway, Brian E., Pell, W.G.P., and Liu, T.-C. (1997) Diagnostic analyses for mechanisms of self-discharge of electrochemical capacitors and batteries. *J. Power Sources*, **65** (1–2), 53–59.

212 Liu, T., Pell, W.G.P., and Conway, B.E. (1996) Self-discharge and potential recovery phenomena at thermally and electrochemically prepared RuO_2 supercapacitor electrodes. *Electrochim. Acta*, **42** (23–24), 3541–3552.

213 Morimoto, T., Hiratsuka, K., Sanada, Y., and Kurihara., K. (1996) Electric double-layer capacitor using organic electrolyte. *J. Power Sources*, **60** (2), 239–247.

214 Qiao, W., Koraia, Y, Mochida, I., Hori, Y., and Maeda, T. (2002) Preparation of an activated carbon artifact oxidative modification of coconut shell-based carbon to improved the strength. *Carbon*, **40** (3), 351–358.

215 Nishihara, H. *et al.* (2009) Investigation of the ion storage/transfer behavior in an electrical double-layer capacitor by using ordered microporous carbons as model materials. *Chem. Eur. J.*, **15**, 5355–5363.

216 Kaus, M., Kowal, J., and Sauer, D.U. (2010) Modelling the effects of charge redistribution during self-discharge of supercapacitors. *Electrochim. Acta*, **55**, 7516–7523.

217 Diab, Y, Pascal, V., Gualous, H., and Rojat, G. (2009) Self-discharge characterization and modeling of electrochemical capacitor used for power electronics applications. *IEEE Trans. Power Electr.*, **24**, 510–517.

218 Zhang, X. *et al.* (2012) The effects of surfactant template concentration on the supercapacitive behaviors of hierarchically porous carbons. *J. Power Sources*, **199**, 402–408.

3
Dye-Sensitized Solar Cells

Anna Nikolskaia and Oleg Shevaleevskiy

Russian Academy of Sciences, Institute of Biochemical Physics, Solar Photovoltaics Department, Kosygin St. 4, 119334 Moscow, Russia

3.1
Introduction

In the modern world, solar photovoltaics (PV) have become some of the most promising renewable energy technologies that are capable to play a major role in the future global energy supply [1]. However, the high cost of conventional silicon-based solar cells, dominating the market, is still a critical issue that limits the industrialization of PV technologies [2].

Great efforts that were made over the past two decades toward the search for alternative photovoltaic devices have resulted in the development of the new physicochemical principles of solar energy conversion in PV cells [3,4]. Due to this, dye-sensitized nanocrystalline solar cells (DSSCs) emerged as a new class of PV devices with high-potential possibilities [5]. The DSSCs, considered as third- or next-generation solar cells, attracted a significant interest as a possible cost-attractive alternative to the conventional silicon-based counterparts. DSSCs can achieve up to nearly 13% of certified power conversion efficiency and show great potential for large-scale production due to using cheap materials and an economically friendly and easy fabrication process [6,7].

3.2
History

The early 1990s were marked by the emergence and successful development of the first competitive solar cell that exploited the principles and materials that differed from traditional physics [4]. Grätzel and O'Regan at École Polytechnique Fédérale de Lausanne designed a nanocrystalline dye-sensitized solar cell, also called mesoscopic solar cell [8]. In that system, a TiO_2 mesoscopic layer sensitized with a dye was sandwiched between transparent electrodes and an

Handbook of Solid State Chemistry, First Edition. Edited by Richard Dronskowski, Shinichi Kikkawa, and Andreas Stein.
© 2017 Wiley-VCH Verlag GmbH & Co. KGaA. Published 2017 by Wiley-VCH Verlag GmbH & Co. KGaA.

electrolyte containing a redox couple was used. To improve the DSSC performance, during the 1990s and 2000s, many dyes and electrolytes have been developed and investigated, a variety of mesoporous films with different morphologies and compositions were observed [9]. At the moment the record power conversion efficiencies of DSSC have reached 13% for laboratory cells and about 11% for minimodules [7,10]. DSSCs have demonstrated promising stability data, passing, for example, the critical 1000 h stability test at 80 °C with a durable efficiency of 8–9% [9].

3.3
Operational Principles

Figure 3.1 schematically illustrates the structure and the energy level diagram of a dye-sensitized solar cell. The conventional DSSC is composed of four main elements, including the transparent conducting and counter electrodes (CE), the mesoscopic layer of nanostructured wide-bandgap metal oxide nanoparticles, the dye photosensitizer, and the electrolyte. Nanocrystalline particles are deposited on the conducting electrode coated with fluorine-doped tin oxide (FTO) and create a nanoporous mesoscopic layer with a large surface area to absorb dye molecules and provide good electrically conductive contact between the individual particles. After the photon is absorbed by the dye molecule, the electron is

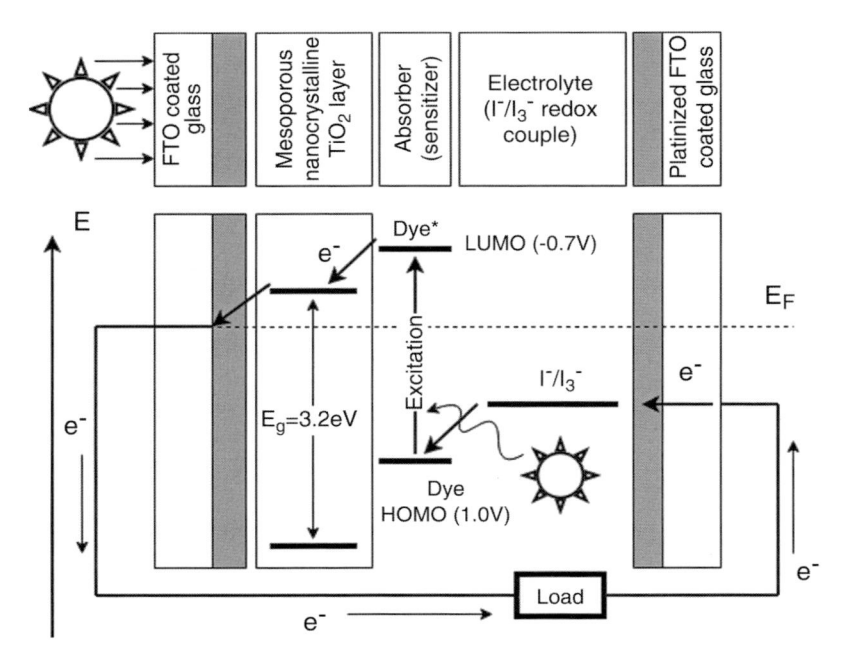

Figure 3.1 The schematic energy level diagram of a DSSC device.

excited from the ground (oxidized) state to the excited state. At the energy diagram, this process is presented in terms of the molecule photoexcitation from the highest occupied molecular orbital (HOMO) to the lowest unoccupied molecular orbital (LUMO) [9,11].

In the next step, the excited electrons are injected into the conduction band of the semiconductor electrode, resulting in the oxidation of the sensitizer, followed by electron diffusion through the nanoporous mesoscopic network toward the conducting electrode. From this electrode, the electrons are transmitted to the external load and reach the platinum counter electrode. In the final step, the oxidized dye is regenerated by accepting the electrons from the redox electrolyte. In parallel, I^- is oxidized to the I_3^- state, followed by ion diffusion toward the Pt counter electrode where it is again reduced to the I^- state [9,11].

Sensitized dyes for DSSCs need to be strongly bound to the semiconductor surface by means of an anchoring group, to provide the electron transfer into the conduction band of the metal oxide semiconductor. For this purpose, the LUMO energy level of the photosensitizer should be sufficiently high. The HOMO energy level should be sufficiently low to provide the regeneration of the oxidized dye by the redox electrolyte.

The optical absorption spectrum of a DSSC photosensitizer must be wide and should closely match the solar spectrum. Photoinduced electron transfer from dye molecules to the conduction band of TiO_2 should be rapid (in the femtosecond time range) and be much faster than the time decay to the ground state of the dye.

The performance of the DSSC device is characterized by the value of power conversion efficiency (η) that can be calculated from a current–voltage (I–V) characteristic:

$$\eta(\%) = \frac{J_{SC} \times V_{OC} \times FF}{P_{in}} \times 100 \ (\%) \tag{3.1}$$

where J_{sc} is the photocurrent density measured at short circuit, V_{oc} is the open-circuit potential, FF is the fill factor of the cell (the ratio between the maximum obtainable power and the theoretical power, that is, the product of J_{sc} and V_{oc}), and P_{in} is the intensity of the incident light [12].

The other important parameter of a DSSC device is the incident monochromatic photon-to-current conversion efficiency (IPCE) spectrum. IPCE is defined as the number of electrons generated by the light in the external circuit divided by the number of incident photons as a function of excitation wavelength [13]:

$$\text{IPCE}(\lambda) = \frac{\text{photocurrent density}}{\text{wavelength} \times \text{photon flux}} \tag{3.2}$$

A typical characteristic I–V curve of and the IPCE spectrum of a DSSC based on nanocrystalline mesoporous TiO_2 layer sensitized with black dye (see Section 3.4.2) are shown in Figures 3.2 and 3.3 [5,12]. The photoelectric parameters of the certified DCCS device with a cell size of 0.186 cm^2 and an efficiency $\eta = 10.4$ are as follows: $J_{sc} = 20.53$ mA/cm^2, $V_{oc} = 0.72$ V, and FF = 0.704.

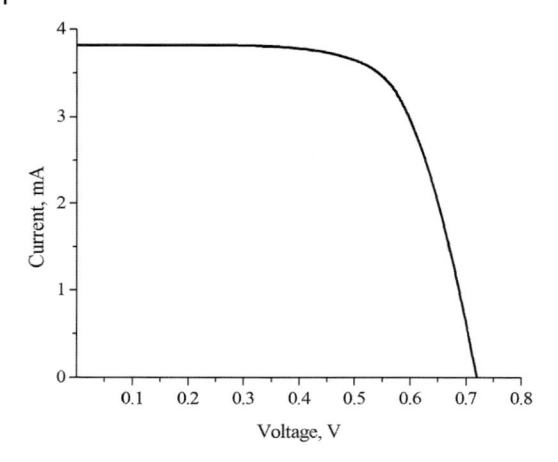

Figure 3.2 Characteristic I–V curve of a DSSC sensitized with black dye (see Section 3.4.2). (Reproduced with permission from Ref. [5]. Copyright 2005, American Chemical Society; also from Ref. [12]. Copyright 2011, John Wiley & Sons, Inc.)

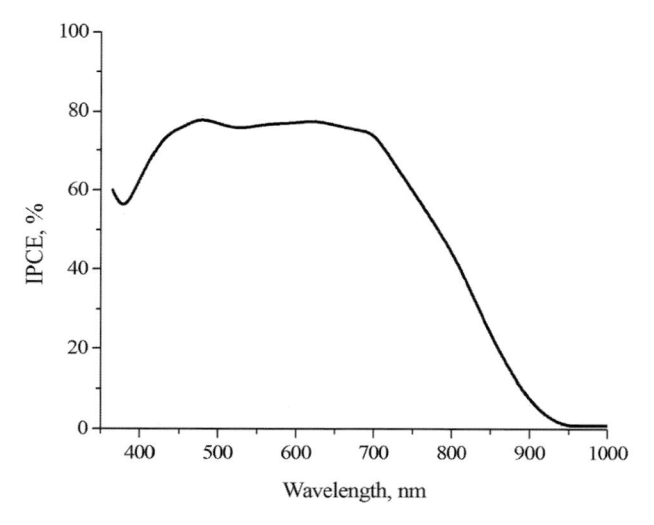

Figure 3.3 IPCE spectrum of a DSSC sensitized with black dye (see Section 3.4.2). (Reproduced with permission from Ref. [5]. Copyright 2005, American Chemical Society; also from Ref. [12]. Copyright 2011, John Wiley & Sons, Inc.)

3.4
Materials and Methods for DSSC Fabrication

3.4.1
Mesoscopic Metal Oxide Electrodes

A variety of wide-bandgap metal oxides, such as TiO_2, ZnO, SnO_2, and Nb_2O_5, can be used as n-type semiconductor materials for fabricating photoanodes for DSSCs. Besides simple oxides, various complex oxides, such as $SrTiO_3$,

Zn_2SnO_4, and Sm_2ScTaO_7, have also been investigated, as well as core-shell structures, such as ZnO-coated SnO_2 [14,15].

Among all known candidates for photoelectrode materials, TiO_2 still remains the most appropiate material for DSSC fabrication, since it has demonstrated the best performance compared to other oxide materials. The main advantages over other metal oxides include excellent thermal stability, nontoxity, and passivity to the chemicals or electrolytes used in DSSCs [1,16]. The anatase form of TiO_2 is mainly used in DSSC because it shows higher electron conductivity compared to the rutile structure of TiO_2 [17].

The morphology (microstructure, particle size, porosity) and the thickness of the TiO_2 layer have an important influence on DSSC efficiency. This suggests that the technology of metal oxide fabrication and the method for coating the metal oxide thin film on conductive glass should be carefully chosen [18]. Hydrothermal growth, screen-printing, and doctor-blade methods are the most common techniques used for the fabrication of nanoporous mesoscopic TiO_2 layers [9]. Aqueous solutions are used for the hydrothermal method, while for screen-printing and doctor-blade techniques a TiO_2 paste should be prepared [19].

Different nanostructures of TiO_2, such as nanowires, nanotubes, nanorods, can be used for DSSCs. A thin film based on regular TiO_2 nanowires can be obtained by spin-coating method [18,19]. A mesoporous TiO_2 structure with a narrow pore size distribution can be formed using a polymer-templated synthesis [20]. A new trend in constructing mesoscopic electrodes is based on the application of the single-walled TiO_2 nanotube arrays, grown by potentiostatic anodization of Ti metal in fluoride-based electrolytes [21]. DSSC structures based on single-walled TiO_2 nanotube arrays show a solar light conversion efficiency of 8% [22].

Doping of metal oxides with suitable cations/anions can be used to improve their electronic properties [18]. For example, an efficiency increase of DSSCs was demonstrated by doping TiO_2 mesoscopic layers with niobium. It was found that Nb doping plays a beneficial role on the electron transport in DSSCs charge injection and transport by tailoring the electronic structure of the TiO_2-dye-electrolyte interface [23,24]. TiO_2 layers doped with Sn^{4+}, Sc^{3+}, Cr^{3+}, Sb^{3+}, and Mg^{2+} are also described in the literature [25].

Besides cation doping, various anions such as N, B, C, S, and F can also be used for enhancing DSSC efficiency [18]. Incorporation of carbon nanomaterials, such as carbon nanotubes, graphene, graphene oxide, and nitrogen-doped graphene, in a TiO_2 layer and their influence on the electronic transfer and charge recombination in photoelectrode was investigated in detail [26].

Recombination processes occur both at the TiO_2/dye/electrolyte interface and at the conducting substrate/TiO_2 interface. To prevent the recombination at the conducting substrate, a compact TiO_2 film with a thickness of 20–50 nm is introduced between a FTO coated glass substrate and a TiO_2 mesoscopic layer [9,27]. A TiO_2 compact layer can be prepared by different methods, such as chemical bath deposition, spray pyrolysis, and sputtering [9]. The most common method involves the deposition of a titanium diisopropoxide bis(acetylacetonate) solution in 2-propanol and ethanol on FTO-coated glass by aerosol spray pyrolysis at 450 °C [28].

During the past few years, p-type semiconductors, such as NiO, Cu_2O, $CuCrO_2$, and $CuGaO_2$, have attracted significant interest as materials for photocathodes [29]. Performance of p-type DSSC is not high. For example, a solar cell based on an Al-doped photoelectrode sensitized with N719 dye provides 5% efficiency [30]. However, integrating of p-type DSSC with n-type DSSC into a tandem structure significantly improves the conversion efficiency [29,30].

3.4.2
Dye Sensitizers

The central component of the DSSC is a dye sensitizer that is capable of harvesting the sun's energy by absorption of photons in the visible and near-infrared spectral regions and generates photo-excited electrons, followed by injection into TiO_2 nanoparticles. The dye sensitizer should have high photostability in a final device and optimal HOMO/LUMO energy level composition in the TiO_2/dye/electrolyte system. The dye should be uniformly distributed and not aggregate on the TiO_2 surface or in solution. Anchoring groups in dye molecules, such as —COOH, —H_2PO_3, and —SO_3H, should be strongly bound to the semiconductor surface [9].

Ruthenium-complex dyes (Figure 3.4) have been intensively investigated for DSSCs. In 1993, Grätzel and co-workers developed a number of mononuclear complexes such as *cis*-(X)$_2$bis(2,2′-bipyridyl-4,4′-dicarboxylate)ruthenium(II) (where X = Cl, Br, I, CN, SCN) [31]. As a result the thiocyanato derivative coded as N3 was synthesized that exhibited the best optical absorption and photoelectrical properties including a broad IPCE spectrum, long excited-state lifetime (~20 ns), and power conversion efficiency exceeding 10% [9,31]. The doubly protonated form of N3 (coded as N719) demonstrated an improved photovoltaic performance in a DSSC (11.2%) [9,32]. The use of a terpyridyl ligand in so-called "black dye" led to an IPCE spectrum over the whole visible range extending up

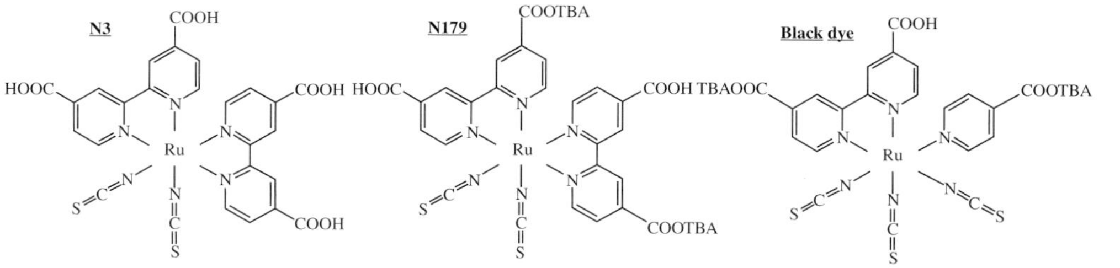

Figure 3.4 Chemical structures of ruthenium-complex dyes for DSSCs (TBA – tetrabutylammonium cation).

to 920 nm and a 10.4% conversion efficiency under AM 1.5 G (100 mW/cm^2) irradiation [12,32]. The N3, N719, and the black dyes are now used as a base for designing new types of Ru-complex dyes [9].

However, the high cost of Ru-based dyes and potential toxicity of these structures initiated the search of cheaper and eco-friendly organic photosensitizers. From this point, the developed metal-complex porphyrin reached significant energy conversion efficiencies [33]. DSSCs with zinc–porphyrin dyes with the cobalt(II/III) redox couple achieved a high conversion efficiency of 13% [10].

Phthalocyanines possess an intense absorption around 700 nm in the Q-band and are also of great interest for use as NIR photosensitizers for DSSCs. The phthalocyanine dyes, however, have the low solubility, which needs to be improved by structural optimization and adding a co-adsorber for suppressing the dye aggregation [9].

To reduce manufacturing costs and to overcome the toxicity problem, metal-free triphenylamine-based organic dyes attain considerable attention. In this case, the efficiency of the DSSC can reach ∼10% comparable with the efficiency of the DSSC based on ruthenium complexes [34]. Application of metal-free tetrathienoacene sensitizers in DSSCs provides ∼10.1% efficiency [35,36].

It is also possible to use natural dyes as photosensitizers for DSSCs despite much lower efficiencies than in synthetic dyes [1]. Natural dyes can be derived from flowers, fruits, vegetables, invertebrates, or minerals; they are nontoxic and readily available at low cost.

The increase of the spectral response in a DSSC can be reached by cosensitization of a mesoscopic layer using a variety of dyes with different absorption spectra. Ozawa and co-workers achieved the conversion efficiency of 11.6% by cosensitizing a DSSC with a mixture of black dye and an organic sensitizer [37]. The use of cosensitization of silyl-anchor and carboxy-anchor dyes in DSSC with the electrolyte containing a cobalt(II/III) redox couple has provided an efficiency of over 14% [38], still remaining the best-known result for DSSCs.

A new trend, started in 2005, is sensitization of DSSCs by inorganic low-bandgap semiconductor particles also called quantum dots (QDs) [39,40]. Quantum dots exhibit a strongly size-dependent optical and electrical property that opens a possibility of adjusting them by tuning QD size. QDs of InP, CdSe, CdS, and PbS can be used instead of dyes [39]. QDs can be planted, using either chemical bath deposition (CBD) or successive ionic layer adsorption and reaction (SILAR) methods [41,42]. However, the reported efficiencies of 4–5% for DSSCs sensitized by QDs are still not satisfactory [43]. Cosensitization of TiO$_2$ by semiconductor quantum dots and dye molecules may enhance the cell performance, for example, cosensitization of TiO$_2$ electrodes with CdS QDs and N719 dye allows DSSCs to reach 5.6% efficiency [44]. Cosensitization with CdSe and CdTe resulted in a conversion efficiency of 6.36% [32,45].

Recent application of a Ti mesh-supported mesoporous carbon counter electrode in QD-sensitized solar cells made it possible to increase the efficiency up to 11% [46].

3.4.3
Counter Electrodes

The counter electrode is an important component in a DSSC, whose main function is to complete the electrical circuit of the solar cell and to provide efficient reduction of the oxidized ions at the FTO/electrolyte interface. For this purpose, the counter electrode should be covered by a catalyst with a low sheet resistance and good chemical stability. The standard CE used in DSSCs is based on a thin platinum (Pt) layer coated on the FTO glass [39,47]. A Pt layer can be obtained on FTO glass substrate by different techniques such as thermal vapor deposition, sputtering, electrodeposition, spray pyrolysis, sputtering, and others. The best performance and long-term stability were obtained with nanoscale Pt clusters prepared by thermal decomposition of platinum chloride compounds deposited on the surface of FTO glass. In this case, the CE remains transparent with resistances of less than $1\,\Omega\,cm^2$ [9].

An alternative technique is based on the synthesis of AuPt bimetallic nanoparticles on an FTO glass substrate by dry plasma reduction at near room temperature under atmospheric pressure. This method has the advantages of avoiding high temperatures, operation under low pressure, and absence of a toxic liquid environment [48,49]. However, because of the high cost of Pt, other conductive materials, such as graphite, graphene, activated carbon, poly(3,4-ethylenedioxythiophene), polypyrrole, and polyaniline, were also used as good alternatives, though with a certain decrease in the device efficiency [50,51]. For example, a DSSC with CE based on graphene oxide nanofibers has shown an efficiency of 7.9% at 0.25 sun illumination [52].

Electrodeposited CoS has been identified as a promising alternative to high-cost Pt catalysts. A DSSC based on TiO_2 sensitized with an organic dye and using a CoS counter electrode reached 6.9% efficiency with the electrolyte containing a cobalt(II/III) redox couple [53].

3.4.4
Electrolytes

Redox electrolytes in DSSCs act as a charge transport medium to transfer electrons from the counter electrode to the oxidized dye. The effective operation of a DSSC requires an electrolyte with high conductivity and stability up to ~80 °C. The redox potential of the electrolyte should be sufficiently negative to match the oxidation potential of the dye sensitizer [39].

Generally, electrolytes can be classified into three categories, including liquid, solid, and quasi-solid electrolytes. Up to now, the most efficient DSSCs are still based on liquid electrolytes [54]. The first DSSC reported in 1991 used an organic liquid electrolyte containing LiI/I_2, with a conversion efficiency of 7.1% [8]. The I^-/I_3^- couple with a standard potential of 0.35 V versus the normal hydrogen electrode (NHE) is the most widely used redox couple in DSSCs due to its very good performance [55]. The most remarkable property of the I^-/I_3^-

couple is slow reduction kinetics of I_2 or I_3^-. Due to this the process of the I_3^- diffusion to the counter electrode is not suppressed [56]. The diffusion coefficient of I_3^- ions in the porous TiO_2 structure was reported to be 7.6×10^{-6} cm^2/s [57].

Other redox couples, such as Br^-/Br_2, $SCN^-/(SCN)_2$, $SeCN^-/(SeCN)_2$, and ferrocene/ferrocenium, have been investigated because of their more positive potential as compared to I^-/I_3^-. However, all of them have a lower driving force for dye regeneration that decreases the photocurrent [39,58]. The most promising alternative redox couple is based on a Co(II/III)-containing redox electrolyte. Cobalt-containing complexes have two main advantages: weak light absorption in the visible light region as compared with iodide/triiodide-based electrolyte, and the possibility of reduction potential tuning through the variation of ligands [59,60]. A record efficiency of 13% was reached using a $[Co(bpy)_3]^{2+/3+}$ redox couple (redox potential 0.56 V versus NHE) in a DSSC combined with a porphyrin-based dye [10,61]. A DSSC with cosensitized organic dyes and with an electrolyte based on a $[Co(phen)_3]^{3+/2+}$ redox couple (redox potential 0.62 V versus NHE) has exhibited a conversion efficiency over 14% [38,61].

The main disadvantage of the cobalt redox couple components is the large molecular size, which limits mass and charge transport and leads to recombination losses. A copper redox system such as $[Cu(dmp)_2]^{+/2+}$ was introduced to overcome this issue. The most efficient DSSC based on an electrolyte containing Cu^+/Cu^{2+} redox couple reached ~10% efficiency [62,63].

Ionic liquids can serve as both iodide source and solvent in DSSCs. The most widely used ionic liquids are alkyl-substituted imidazolium iodides [64]. The major drawbacks of ionic liquids are toxicity and high cost [65].

Liquid electrolytes possess the sealing and long-term stability problems in the practical application of DSSC devices. For this reasons, quasi-solid-state electrolytes and solid-state electrolytes are considered to be the essential alternatives for achievement of a better DSSC stability and to simplify the fabrication process of the device.

To produce a quasi-solid electrolyte, the liquid electrolyte is gelled by using a suitable gelator that can be based either on polymers or on nanopowders [39]. Poly(ethylene oxide) (PEO) polymers are the most popular matrix materials [66]. Poly(methyl methacrylate), polyethylene glycol, poly(vinyl chloride), and other related substances can be used. Different polysaccharides and modified polysaccharide-based materials, such as chitosan, cellulose, and carrageenan, attracted attention due to their higher ionic conductivities at room temperature [67]. Various composite electrolytes have been used involving nanocomposites with silica, titania, and many other inorganic additives [9]. To date, the efficiencies of DSSCs based on quasi-solid-state electrolytes have reached 8–9% [68].

Transparent hole transporting p-type materials (HTM) having a valence band compatible with the HOMO of dye for regeneration of dye and with high charge mobility can be used as solid electrolytes in DSSCs. The important requirement is uniform permeation of HTM molecules into the pores of a mesoporous semiconductor layer [39]. CuI, CuSCN, $CuBr_3SR_2$ (where R = alkyl derivative), NiO and $CuAlO_2$, poly(3,4-ethylenedioxythiophene) (PEDOT), and poly(3-hexylthiophene) (P3HT) were successfully applied in a DSSC [69]. The most studied material is 2,2,7,7-tetrakis(*N*,*N*-di-*p*-methoxyphenylamine)-9,9-spirobifluorene (Spiro-OMeTAD) that has shown the best performance [70].

Despite the fact that solid electrolytes did not provide high efficiencies of PV cells compared to liquid electrolytes, they are still promising candidates for the future research in DSSCs due to their stability.

3.5
New Challenges

DSSCs have a number of attractive features and advantages as compared to the conventional solid-state silicon-based solar cells. One of the main advantages is their high power conversion efficiency under low solar radiation and in cloudy weather conditions. However, until now their overall efficiency remains low compared to silicon-based solar cells. Recent developments in the use of organic–inorganic layer structured perovskites as light absorbers and as electron or hole transporting materials allow reduction in the thickness of photoanodes to the submicron level and have raised the power conversion efficiency of solid-state mesoscopic solar cells above the 10% level. Organo-inorganic hybrid perovskites, such as $CH_3NH_3PbBr_3$ and $CH_3NH_3PbI_3$, have a unique electrical conductivity, optical absorption, and excitonic properties. High efficiencies up to 15–17% for perovskite solar cells (PSC) based on mesoscopic TiO_2 layer with Spiro-OMeTAD as electrolyte and Au as counter electrode were published in recent papers [71,72]. It is expected that such devices may rival thin-film solar cells in terms of the efficiency and stability. The progress on the PSC is fast and impressive, now with efficiencies exceeding 20% [73,74].

The development of the third-generation solar cells has significantly changed the status of PV science and caused a powerful shift in the evolution of alternative energy. The improvement of PV cell properties, optimization of their design, and their commercialization are primary objectives of the future. It should also be noted that not all the scientific problems have been solved while the clear understanding of the energy conversion principles in the systems listed above is a real challenge, but its realization promises a bright future for solar photovoltaics.

References

1 Reddy, K.G., Deepak, T.G., Anjusree, G.S., Thomas, S., Vadukumpully, S., Subramanian, K.R.V., Nair, S.V., and Nair, A.S. (2014) On global energy scenario, dye-sensitized solar cells and the promise of nanotechnology. *Phys. Chem. Chem. Phys.*, **16**, 6838–6858.

2 Soga, T. (2006) Fundamentals of solar cell, in *Nanostructured Materials for Solar Energy Conversion* (eds T. Soga and U. Planning) Elsevier, Amsterdam, pp. 3–43.

3 Kirchartz, T., Bisquert, J., Mora-Sero, I., and Garcia-Belmonte, G. (2015) Classification of solar cells according to

mechanisms of charge separation and charge collection. *Phys. Chem. Chem. Phys.*, **17**, 4007–4014.
4 Shevaleevskiy, O. (2008) The future of solar photovoltaics: a new challenge for chemical physics. *Pure Appl. Chem.*, **80** (10), 2079–2089.
5 Gratzel, M. (2005) Solar energy conversion by dye-sensitized photovoltaics cells. *Inorg. Chem.*, **44** (20), 6841–6851.
6 Hardin, B.E., Snaith, H.J., and McGehee, M.D. (2012) The renaissance of dye-sensitized solar cells. *Nat. Photonics*, **6**, 162–169.
7 Green, M.A., Emery, K., Hishikawa, Y., Warta, W., and Dunlop, E.D. (2016) Solar cell efficiency tables (version 47). *Prog. Photovolt. Res. Appl.*, **24**, 3–11.
8 O'Regan, B. and Gratzel, M. (1991) A low-cost, high-efficiency solar cell based on dye-sensitized colloidal TiO_2 films. *Nature*, **353**, 737–740.
9 Hagfeldt, A., Boschloo, G., Sun, L., Kloo, L., and Pettersson, H. (2010) Dye-sensitized solar cells. *Chem. Rev.*, **110**, 6595–6663.
10 Mathew, S., Yella, A., Gao, P., Humphry-Baker, R., Curchod, B.F.E., Ashari-Astani, N., Tavernelli, I., Rothlisberger, U., Nazeeruddin, M.K., and Gratzel, M. (2014) Dye-sensitized solar cells with 13% efficiency achieved through the molecular engineering of porphyrin sensitizers. *Nat. Chem.*, **6**, 242–247.
11 Gratzel, M. and Durrant, J.R. (2014) Dye- and perovskite-sensitized mesoscopic solar cells, in *Clean Electricity from Photovoltaics*, 2nd edn (eds M.D. Archer and M.A. Green), Imperial College Press, London, pp. 413–452.
12 Luque, A. and Hegedu, S. (eds) (2011) *Handbook of Photovoltaic Science and Engineering*, John Wiley & Sons, Ltd., Chichester.
13 Gratzel, M. and Hagfeldt, A. (1995) Light-Induced redox reactions in nanocrystalline systems. *Chem. Rev.*, **95** (1), 49–68.
14 Suzuki, Y., Okamoto, Y., and Ishii, N. (2015) Dye-sensitized solar cells using double-oxide electrodes: a brief review. *J. Phys. Conf. Ser.*, **596**, 1–4.
15 Todinova, A., Vildanova, M., Aleksandrov, M., Vishnev, A., and Shevaleevskiy, O. (2014) Development of nanostructured TiO_2-based rare earth materials for dye-sensitized solar cells. Proceedings of the 29th European Photovoltaic Solar Energy Conf. and Exhibition, September 22–26, 2014, Amsterdam, Netherlands, pp. 1554–1557.
16 Grätzel, M. (2004) Conversion of sunlight to electric power by nanocrystalline dye-sensitized solar cells. *J. Photochem. Photobiol. Chem.*, **164**, 3–14.
17 Park, N.-G., van de Lagemaat, J., and Frank, A.J. (2000) Comparison of dye-sensitized rutile- and anatase-based TiO_2 solar cells. *J. Phys. Chem. B*, **104**, 8989–8994.
18 Sengupta, D., Dasa, P., Mondal, B., and Mukherjee, K. (2016) Effects of doping, morphology and film-thickness of photo-anode materials for dye sensitized solar cell application. *Renew. Sustain. Energy Rev.*, **60**, 356–376.
19 Clement Raj, C. and Prasanth, R. (2016) A critical review of recent developments in nanomaterials for photoelectrodes in dye sensitized solar cells. *J. Power Sources*, **317**, 120–132.
20 Zukalová, M., Zukal, A., Kavan, L., Nazeeruddin, M.K., Liska, P., and Grätzel, M. (2005) Organized mesoporous TiO_2 films exhibiting greatly enhanced performance in dye-sensitized solar cells. *Nano Lett.*, **5** (9), 1789–1792.
21 Jennings, J.R., Ghicov, A., Peter, L.M., Schmuki, P., and Walker, A.B. (2008) Dye-sensitized solar cells based on oriented tio_2 nanotube arrays: transport, trapping, and transfer of electron. *J. Am. Chem. Soc.*, **130**, 13364–13372.
22 So, S., Hwang, I., and Schmuki, P. (2015) Hierarchical DSSC structures based on "single walled" TiO_2 nanotube arrays reach a back-side illumination solar light conversion efficiency of 8%. *Energy Environ. Sci.*, **8**, 849–854.
23 Tsvetkov, N., Larina, L., Shevaleevskiy, O., and Ahn, B.T. (2011) Electronic structure study of lightly Nb-doped TiO_2 electrode for dye-sensitized solar cells. *Energy Environ. Sci.*, **4**, 1480–1486.
24 Kozlov, S., Nikolskaia, A., Larina, L., Vildanova, M., Vishnev, A., and Shevaleevskiy, O. (2016) Rare-earth and

Nb doping of TiO_2 nanocrystalline mesoscopic layers for high-efficiency dye-sensitized solar cells. *Phys. Status Solidi A*, **213** (7), 1801–1806.

25 Sengupta, D., Das, P., Mondal, B., and Mukherjee, K. (2016) Effects of doping, morphology and film-thickness of photo-anode materials for dye sensitized solar cell application – A review. *Renew. Sustain. Energy Rev.*, **60**, 356–376.

26 Liu, X., Fang, J., Liu, Y., and Lin, T. (2016) Progress in nanostructured photoanodes for dye-sensitized solar cells. *Front. Mater. Sci.*, **10**, 225–237.

27 Burke, A., Ito, S., Snaith, H., Bach, U., Kwiatkowski, J., and Grätzel, M. (2008) The Function of a TiO_2 compact layer in dye-sensitized solar cells incorporating "planar" organic dyes. *Nano Lett.*, **8** (4), 977–981.

28 Fabregat-Santiago, F., Bisquert, J., Cevey, L., Chen, P., Wang, M., Zakeeruddin, S.M., and Gratzel, M. (2009) Electron transport and recombination in solid-state dye solar cell with Spiro-OMeTAD as hole conductor. *J. Am. Chem. Soc.*, **131**, 558–562.

29 Zhang, L., Boschloo, G., Hammarstrom, L., and Tian, H. (2016) Solid state p-type dye-sensitized solar cells: concept, experiment and mechanism. *Phys. Chem. Phys. Chem.*, **18**, 5080–5085.

30 Ursu, D., Vaszilcsin, N., Banica, R., and Miclau, M. (2016) Effect of Al doping on performance of $CuGaO_2$ p-type dye-sensitized solar cells. *J. Mater. Eng. Perform.*, **25** (1), 59–63.

31 Nazeeruddin, M.K., Kay, A., Rodicio, I., Humphry-Baker, R., Mueller, E., Liska, P., Vlachopoulos, N., and Graetzel, M. (1993) Conversion of light to electricity by cis-X_2bis(2,2′-bipyridyl-4,4′-dicarboxylate) ruthenium(II) charge-transfer sensitizers (X = Cl-, Br-, I-, CN-, and SCN-) on nanocrystalline titanium dioxide electrodes. *J. Am. Chem. Soc.*, **115** (14), 6382–6390.

32 Shalini, S., Balasundaraprabhu, R., Kumar, T.S., Prabavathy, N., Senthilarasu, S., and Prasanna, S. (2016) Status and outlook of sensitizers/dyes used in dye sensitized solar cells (DSSC): a review. *Int. J. Energy Res.*, **40** (10), 1303–1320.

33 Yella, A., Lee, H.-W., Tsao, H.N., Yi, C., Chandiran, A.K., Nazeeruddin, M.K., Diau, E.W.-G., Yeh, C.-Y., Zakeeruddin, S.M., and Grätzel, M. (2011) Porphyrin-sensitized solar cells with cobalt (II/III)-based redox electrolyte exceed 12 percent efficiency. *Science*, **334** (6056), 629–634.

34 Mahmood, A. (2016) Triphenylamine based dyes for dye sensitized solar cells: a review. *Sol. Energy*, **123**, 127–144.

35 Zhou, N., Prabakaran, K., Lee, B., Chang, S.H., Harutyunyan, B., Guo, P., Butler, M.R., Timalsina, A., Bedzyk, M.J., Ratner, M.A., Vegiraju, S., Yau, S., Wu, C.-G., Chang, R.P.H., Facchetti, A., Chen, M.-C., and Marks, T.J. (2015) Metal-free tetrathienoacene sensitizers for high-performance dye-sensitized solar cells. *J. Am. Chem. Soc.*, **137**, 4414–4423.

36 Obotowo, I.N., Obot, I.B., and Ekpe, U.J. (2016) Organic sensitizers for dye-sensitized solar cell (DSSC): properties from computation, progress and future perspectives. *J. Mol. Struct.*, **1122**, 80–87.

37 Ozawa, H., Shimizu, R., and Arakawa, H. (2012) Significant improvement in the conversion efficiency of black-dye-based dye-sensitized solar cells by cosensitization with organic dye. *RSC Adv.*, **2**, 3198–3200.

38 Kakiage, K., Aoyama, Y., Yano, T., Oya, K., Fujisawa, J., and Hanaya, M. (2015) Highly-efficient dye sensitized solar cells with collaborative sensitization by silyl-anchor and carboxy-anchor dyes. *Chem. Commun.*, **51**, 15894–15897.

39 Jena, A., Mohanty, S.P., Kumar, P., Naduvath, J., Gondane, V., Lekha, P., Das, J., Narula, H.K., Mallick, S., and Bhargava, P. (2012) Dye sensitized solar cells: a review. *Trans. Indian Ceram. Soc.*, **71** (1), 1–16.

40 Duan, J., Zhang, H., Tang, Q., He, B., and Yu, L. (2015) Recent advances of critical materials in quantum dot–sensitized solar cells. A review. *J. Mater. Chem. A*, **3**, 17497–17510.

41 Choi, Y., Seol, M., Kim, W., and Yong, K. (2014) Chemical bath deposition of stoichiometric CdSe quantum dots for efficient quantum-dot-sensitized solar cell application. *J. Phys. Chem. C*, **118** (11), 5664–5670.

42 Lee, H., Wang, M., Chen, P., Gamelin, D.R., Zakeeruddin, S.M., Grätzel, M., and Nazeeruddin, M.K. (2009) Efficient CdSe quantum-dot solar cells prepared by successive ionic layer adsorption and reaction process. *Nano Lett.*, **9** (12), 4221–4227.

43 Kouhnavard, M., Ikeda, S., Ludin, N.A., Ahmad Khairudin, N.B., Ghaffari, B.V., Mat-Teridi, M.A., Ibrahim, M.A., Sepeai, S., and Sopian, K. (2014) A review of semiconductor materials as sensitizers for quantum dot-sensitized solar cells. *Renew. Sustain. Energy Rev.*, **37**, 397–407.

44 Li, J., Zhao, L., Wang, S., Hu, J., Dong, B., Lu, H., Wan, L., and Wang, P. (2013) Great improvement of photoelectric property from co-sensitization of TiO_2 electrodes with CdS quantum dots and dye N719 in dye-sensitized solar cells. *Mater. Res. Bull.*, **48**, 2566–2570.

45 Pan, Z., Zhao, K., Wang, J., Zhang, H., Feng, Y., and Zhong, X. (2016) Near infrared absorption of $CdSe_xTe_{1-x}$ alloyed quantum dot sensitized solar cells with more than 6% efficiency and high stability. *ACS Nano*, **7** (6), 5215–5222.

46 Du, Z., Pan, Z., Fabregat-Santiago, F., Zhao, K., Long, D., Zhang, H., Zhao, Y., Zhong, X., Yu, J.-S., and Bisquert, J. (2016) Carbon counter electrode-based quantum dot sensitized solar cells with certified efficiency exceeding 11%. *J. Phys. Chem. Lett.*, **7** (16), 3103–3111.

47 Khelashvili, G., Behrens, S., Weidenthaler, C., Vetter, C., Hinsch, A., Kern, R., Skupien, K., Dinjus, E., and Bönnemann, H. (2006) Catalytic platinum layers for dye solar cells: a comparative study. *Thin Solid Films*, **511–512**, 342–348.

48 Dao, V.-D., Tran, C.Q., Ko, S.-H., and Choi, H.-S. (2013) Dry plasma reduction to synthesize supported platinum nanoparticles for flexible dye-sensitized solar cells. *J. Mater. Chem. A*, **1**, 4436–4443.

49 Dao, V.-D., Choi, Y., Yong, K., Larina, L.L., Shevaleevskiy, O., and Choi, H.-S. (2015) A facile synthesis of bimetallic AuPt nanoparticles as a new transparent counter electrode for quantum-dot-sensitized solar cells. *J. Power Source*, **274**, 831–838.

50 Chen, M. and Shao, L.-L. (2016) Review on the recent progress of carbon counter electrodes for dye-sensitized solar cells. *Chem. Eng. J.*, **304**, 629–645.

51 Thomas, S., Deepak, T.G., Anjusree, G.S., Arun, T.A., Nair, S.V., and Nair, A.S. (2014) A review on counter electrode materials in dye-sensitized solar cells. *J. Mater. Chem. A*, **2**, 4474–4490.

52 Kavan, L., Liska, P., Zakeeruddin, S.M., and Grätzel, M. (2016) Low-temperature fabrication of highly-efficient, optically-transparent (FTO-free) graphene cathode for co-mediated dye-sensitized solar cells with acetonitrile-free electrolyte solution. *Electrochim. Acta*, **195**, 34–42.

53 Swami, S.K., Chaturvedi, N., Kumar, A., Kapoor, R., Dutta, V., Frey, J., Moehl, T., Gratzel, M., Mathew, S., and Nazeeruddin, M.K. (2015) Investigation of electrodeposited cobalt sulphide counter electrodes and their application in next-generation dye sensitized solar cells featuring organic dyes and cobalt-based redox electrolytes. *J. Power Source*, **275**, 80–89.

54 Wu, J., Lan, Z., Lin, J., Huang, M., Huang, Y., Fan, L., and Luo, G. (2015) Electrolytes in dye-sensitized solar cells. *Chem. Rev.*, **115** (5), 2136–2173.

55 Boschloo, G. and Hagfeldt, A. (2009) Characteristics of the iodide/triiodide redox mediator in dye-sensitized solar cells. *Acc. Chem. Res.*, **42** (11), 1819–1826.

56 Ondersma, J.W. and Hamann, T.W. (2013) Recombination and redox couples in dye-sensitized solar cells. *Coord. Chem. Rev.*, **257**, 1533–1543.

57 Huang, S.Y., Schlichthorl, G., Nozik, A.J., Gratzel, M., and Frank, A.J. (1997) Charge recombination in dye-sensitized nanocrystalline TiO_2 solar cells. *J. Phys. Chem.*, **101** (14), 2576–2582.

58 Daeneke, T., Kwon, T.-H., Holmes, A.B., Duffy, N.W., Bach, U., and Spiccia, L. (2011) High-efficiency dye-sensitized solar cells with ferrocene-based electrolytes. *Nat. Chem.*, **3**, 211–215.

59 Hamann, T.W. (2012) The end of iodide? Cobalt complex redox shuttles in DSSCs. *Dalton Trans.*, **41**, 3111–3115.

60 Bella, F., Galliano, S., Gerbaldi, C., and Viscardi, G. (2016) Cobalt-based

electrolytes for dye-sensitized solar cells: recent advances towards stable devices. *Energies*, **9**, 384.

61 Feldt, S.M., Wang, G., Boschloo, G., and Hagfeldt, A. (2011) Effects of driving forces for recombination and regeneration on the photovoltaic performance of dye-sensitized solar cells using cobalt polypyridine redox couples. *J. Phys. Chem. C*, **115**, 21500–21507.

62 Freitag, M., Giordano, F., Yang, W., Pazoki, M., Hao, Y., Zietz, B., Gratzel, M., Hagfeldt, A., and Boschloo, G. (2016) Copper phenanthroline as a fast and high-performance redox mediator for dye-sensitized solar cells. *J. Phys. Chem. C*, **120**, 9595–9603.

63 Cong, J., Kinschel, D., Daniel, Q., Safdari, M., Gabrielsson, E., Chen, H., Svensson, P.H., Sun, L., and Kloo, L. (2016) Bis(1,1-bis(2-pyridyl)ethane)copper(I/II) as an efficient redox couple for liquid dye-sensitized solar cells. *J. Mater. Chem. A.* **4**, 14550–14554.

64 Gorlov, M. and Kloo, L. (2008) Ionic liquid electrolytes for dye-sensitized solar cells. *Dalton Trans.*, 2655–2666.

65 Ghavre, M., Byrne, O., Altes, L., Surolia, P.K., Spulak, M., Quilty, B., Thampi, K.R., and Gathergood, N. (2014) Low toxicity functionalised imidazolium salts for task specific ionic liquid electrolytes in dye-sensitised solar cells: a step towards less hazardous energy production. *Green Chem.*, **16**, 2252–2265.

66 Benedetti, J.E., Freitas, F.S., Fernandes, F.C., Gonçalves, A.S., Magalhães, A., and Nogueira, A.F. (2015) Investigation of the structural properties of poly(ethylene oxide) copolymer as gel polymer electrolyte and durability test in dye-sensitized solar cells. *Ionics*, **21** (6), 1771–1780.

67 Su'ait, M.S., Rahman, M.Y.A., and Ahmad, A. (2015) Review on polymer electrolyte in dye-sensitized solar cells (DSSCs). *Sol. Energy*, **115**, 452–470.

68 Achari, M.B., Elumalai, V., Vlachopoulos, N., Safdari, M., Gao, J., Gardner, J.M., and Kloo, L. (2013) A quasi-liquid polymer-based cobalt redox mediator electrolyte for dye-sensitized solar cells. *Phys. Chem. Chem. Phys.*, **15**, 17419–17425.

69 Ye, M., Wen, X., Wang, M., Iocozzia, J., Zhang, N., Lin, C., and Lin, Z. (2015) Recent advances in dye-sensitized solar cells: from photoanodes, sensitizers and electrolytes to counter electrodes. *Mater. Today*, **18** (3), 155–162.

70 Hsu, C.-Y., Chen, Y.-C., Lin, R.Y.-Y., Ho, K.-C., and Lin, J.T. (2012) Solid-state dye-sensitized solar cells based on spirofluorene (spiro-OMeTAD) and arylamines as hole transporting materials. *Phys. Chem. Chem. Phys.*, **14**, 14099–14109.

71 Gratzel, M. (2014) The light and shade of perovskite solar cells. *Nat. Mater.*, **13**, 838–842.

72 Green, M.A., Ho-Baillie, A., and Snaith, H.J. (2014) The emergence of perovskite solar cells. *Nat. Photonics*, **8**, 506–514.

73 Yang, W.S., Noh, J.H., Jeon, N.J., Kim, Y.C., Ryu, S., Seo, J., and Seok, S.I. (2015) High-performance photovoltaic perovskite layers fabricated through intramolecular exchange. *Science*, **348**, 1234–1237.

74 Saliba, M., Orlandi, S., Matsui, T., Aghazada, S., Cavazzini, M., Correa-Baena, J.-P., Gao, P., Scopelliti, R., Mosconi, E., Dahmen, K.-H., De Angelis, F., Abate, A., Hagfeldt, A., Pozzi, G., Graetzel, M., and Nazeeruddin, M.K. (2016) A molecularly engineered hole-transporting material for efficient perovskite solar cells. *Nat. Energy*, **1**, 15017, 1–7.

4
Electronics and Bioelectronic Interfaces

Seong-Min Kim, Sungjun Park, Won-June Lee, and Myung-Han Yoon

Gwangju Institute of Science and Technology, School of Materials Science and Engineering, 123 Cheomdan-gwagiro, Buk-gu, 61005 Gwangju, Republic of Korea

4.1
Introduction

Electronics for biological interfaces, known as bioelectronics, is a converging field of electronics and biology. Bioelectronics has great potential for biomedical applications, such as diagnostic and therapeutic devices. The prerequisites for the development of bioelectronics are as follows: (i) high-performance electronic devices operated stably under aqueous conditions based on high processing throughput, (ii) decent biocompatibility of electronic materials, even during the operation of electronic devices, and (iii) high conversion efficiency from ion-based signals (biological signals) to electron-based signals (electrical signals) or vice versa [1]. An efficient signal conversion at direct biointerfaces is necessary to detect low concentrations of biological analytes, record low amplitude signals in real-time, and electrically stimulate electroactive organs with the minimized voltage per current injection. Along with these prerequisites, the next-generation bioelectronic devices should be designed based on the purpose of the applications. For biomolecule sensors, inexpensive materials and fabrication processes are required to make a device sufficient for single-use disposable sensors [2,3]. Biodegradable and flexible electronics are also promising candidates for developing implantable or wearable devices to interact with bioelectric signals from the brain, nerves, and heart [4,5]. Therefore, exploring novel biocompatible electronic materials and fabrication methods for functional devices is essential.

Recently, high-performance solution-processed electronic materials, such as sol–gel synthesized metal oxide (MOx) semiconductors and chemically synthesized conducting polymers (CPs), have gained tremendous momentum because of flexibility, high transparency, fabrication capability on various substrates, and availability for low-cost fabrication. Because their biocompatibility has also been guaranteed with minimized toxicological effects, the devices based on the solution-processed electronic materials can be utilized to construct direct biological

---

Handbook of Solid State Chemistry, First Edition. Edited by Richard Dronskowski, Shinichi Kikkawa, and Andreas Stein.
© 2017 Wiley-VCH Verlag GmbH & Co. KGaA. Published 2017 by Wiley-VCH Verlag GmbH & Co. KGaA.

interfaces for the recording of biological signals or the stimulation of living entities from cells to organs (e.g., cardiac tissues and brain).

With a view of electronic devices, another option is to choose electrolyte-gated transistors as a platform to meet the above-mentioned prerequisites. The modified structures from conventional solid-state thin-film transistors using electrolytes as dielectric layers offer enormous advantages, such as low operational voltage without water electrolysis under physiological aqueous conditions and self-amplifying availability controlled by gate bias modulation. Moreover, the modified structures convert encountered biological signals (ion-based signals) to electrical signals (electron-based signals) with much higher sensitivity than conventional electrochemical electrode-based biosensors because they directly interface with specimens [6].

In this chapter, we present a comprehensive review of the potential applications of bioelectronic sensors based on electrolyte-gated transistors (EGTs) using emerging solution-processed electronic materials: sol–gel processed metal oxide semiconductors (MOx) and chemically synthesized CPs. In detail, the fundamentals of two different devices based on inorganic semiconductors ("electrolyte-gated thin-film transistors," EGTFTs) and chemically synthesized CPs ("organic electrochemical transistor," OECTs), such as their chemical formation, charge transportation, and electrical performance operated under aqueous solution, will be discussed, and various examples for bioelectronic applications will be presented.

We forecast that electrolyte-gated transistors based on solution-processed electronics material will provide new possibilities for developing next-generation biosensors, from disposable biomolecule sensors to wearable neural recording probes, in an economical fashion.

4.2
Operation Mechanism of Electrolyte-gated Transistors

In general, electrolytes are ionically conductive liquids containing free anions and cations by dissociation of a fully dissolved salt. The different types of electrolytes used as dielectrics in the structure of EGTs are illustrated in Figure 4.1a. The electrolytes can be categorized as electrolyte solutions, ionic liquids (ILs), ion gels, polyelectrolytes, and polymer electrolytes (from left to right) [7]. Among them, electrolyte solutions and ILs have been extensively applied in electronic and electrochemical applications due to their high ionic conductivity, wide potential windows, and various selections of anion and cations because the aqueous solvents have enough space for movement of ionic species. However, early vaporization of solvents is a potential problem for the purpose of functional (switching, memory, and diode) electronic devices operated under ambient conditions. Therefore, other types of nonvolatile electrolytes have been studied for the development of device applications.

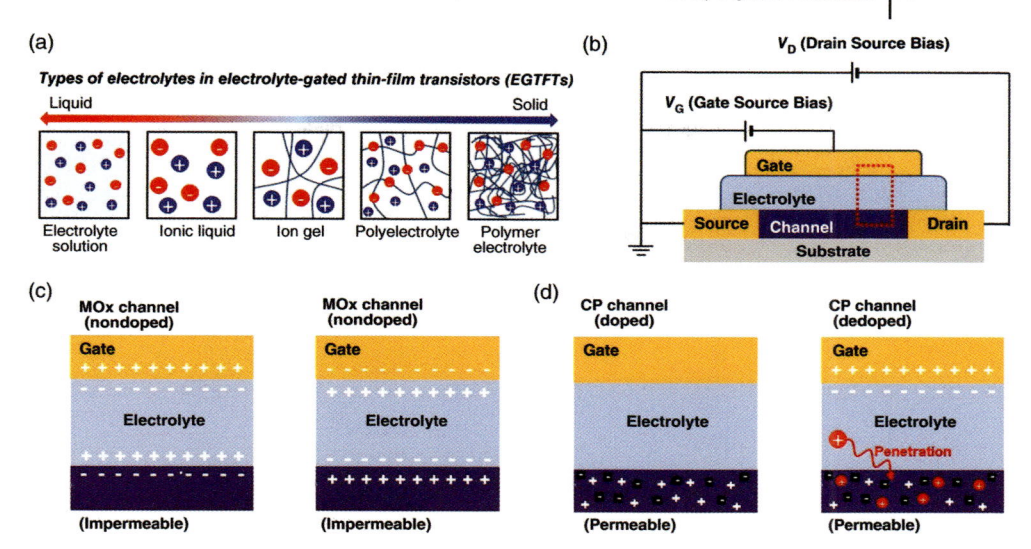

Figure 4.1 (a) Schematic cross-sectional illustration of the EGTFTs. (b) Schematic illustration of different types of electrolytes, ordered from left to right by their physical appearance. (c) Carrier accumulation under positive (left) and negative (right) applied gate biases on impermeable channel surfaces. (d) Doped and dedoped behavior of permeable channel layers without (left) and with (right) a gate bias.

Figure 4.1b shows a top-gate and bottom-contact (TGBC) structured schematic image of EGTs. To understand the operation mechanism of EGTs, the most interesting factor is changes of the charge distributions during applied biases at two interfaces: (1) between the gate electrodes and electrolytes and (2) between the electrolytes and the semiconducting materials.

In the case of a surface between gate electrodes and electrolytes (case 1), when the gate bias is applied at the electrode, the electron charges move into the outermost surface of metal electrodes. For electrical compensation, when electrochemical reference electrodes (e.g., Ag/AgCl electrode) are used as gate electrodes, all of the charges are compensated by faradaic reactions on the electrodes [8]. However, for noble metal gate electrodes, a similar amount of opposite charge to the applied biases will be located in the electrolyte with Angstrom-size neutral regions close to the interface. The structure is called an electrical double layer (EDL) and is composed of two negative and positive layers in parallel with subnanogap neutral regions. The charge distribution and potential drops at the interfaces can be described by the Gouy–Chapman–Stern (GCS) model [9,10]. As a function of the distances from the electrodes, the steep and linear strong potential drops occur across the Helmholtz layer in EDL regions and relatively alleviative the potential drops from the EDL to the bulk electrolytes, called the diffuse layer. The generated total capacitance values at the interface between the metal electrodes and the electrolytes are typically on the order of a few microfarads per square centimeter.

In the case of the surface between the electrolyte and the semiconducting materials (case 2), the charge distribution and operation mechanism can be described based on whether the semiconducting materials are ion impermeable or permeable [6]. As shown in Figure 4.1c and d, for ion-impermeable semiconductors, the charge accumulation of semiconductors has a similar behavior to solid-state thin-film transistors under an applied gate bias. When the EDL between the gate metal and the electrolyte is formed by applied gate biases, the EDL at the opposite electrolyte–semiconductor interface is also formed by charges accumulated in the semiconductors and opposite charges in the electrolytes close to semiconductors. The polarization induced by the potential drops between EDL near the semiconducting surfaces generates a large amount of capacitance, enabling device operation in moderate driving voltage windows. This phenomenon normally occurs in inorganic semiconducting materials, such as silicon and metal oxide semiconductors.

In terms of the ion-permeable semiconductors, electrolyte-gated transistors employing ion-permeable organic semiconductors are called OECTs. In OECTs, the operational mechanisms are based on the reversible electrochemical doping and dedoping reactions at the interface between electrolytes and degenerately doped organic semiconductors. In detail, the cations in the electrolytes penetrate into the CP layer and eliminate mobile carriers, depending on the gate bias, the so-called dedoping process. By these processes, the current can be modulated by carrier density control in the permeable channel layer.

Recently, EGTs have been extensively studied for the application of biosensors and implantable electronics. In the next section, solution-processed organic and inorganic semiconducting materials for EGTs and their application will be introduced.

4.3
Metal Oxide Transistors and Bioelectronic Interfaces

In this section, we review the properties of MOx semiconducting materials and their solution process for the application of bioelectronics. In Section 4.3.1, the unique characteristics, such as high electron mobility, optical transparency, environmental durability, and device operational stability of MOx materials are introduced in view of the atomic bond and structure. In parallel, the advantages and processing mechanism of sol–gel approaches to fabricate high-quality MOx semiconductors and up-to-date developments are also presented. In Section 4.3.2, as a bioelectronic applications based on sol–gel MOx materials, the DNA detection techniques under nonaqueous conditions, and the electrolyte-gated thin-film transistors operated under aqueous conditions are introduced.

4.3.1
Sol–Gel Processed Metal Oxide Electronic Materials

After the first demonstration of room-temperature, vacuum-deposited and amorphous In–Ga–Zn–O metal oxide TFT was reported by Hosono's group [11,12], a significant development in amorphous metal oxide (a-MOx) semiconducting materials has been achieved in industrial and academic areas. MOx have unique electron transport properties, such as large electron mobility exceeding $10\,\text{cm}^2/(\text{V s})$, which is five times higher than that of amorphous Si (a-Si) semiconductors, and high optical transparency with a larger energy gap than 3.1 eV. In addition, stable phase sustainability at high temperature, normally below 600 °C, and compatibility with various inorganic/organic gate insulating materials have been proved [12].

As illustrated in Figure 4.2a, the different characteristics of the electronic structure of both covalent (Si) and ionic (MOx) atomic bonds can explain why MOx semiconducting materials have high electron mobility values despite the amorphous state. In the case of crystalline Si semiconductors, the conduction band minimum (CBM) and valence band maximum (VBM) are placed by tetrahedral covalent bonds with four neighboring silicon atoms from sp^3 hybridized orbitals, and the energy bandgap is formed by the splitting of $\sigma^*-\sigma$ levels, indicating strong directivity for the electronically spatial conduction path (Figure 4.2a, left top). On the other hand, in the case of MOx semiconductors, the CBM of MOx is placed by unoccupied spherical s orbitals of metal cations, and VBM is placed by 2p orbitals of oxygen atoms, indicating that a larger spatial electron conduction path and electronic structure of MOx are formed by the Madelung potential and strong ionic bond tendency originated from the charge transfer from the metal to the oxygen atoms (Figure 4.2a, right top). However, in the case of the amorphous state, the silicon semiconductors with incomplete tetrahedral covalent bonds (Figure 4.2a, left bottom) exhibit much lower electron mobility values by the generation of a localized state between CBM and VBM as a trapping site of charge carriers. In contrast to a-Si, the disordered local structure randomness does not impact the electrical conduction because of overlaps with neighboring spherically bounded s orbitals of metal cations and the fewer defects originated from ionic M—O bonds.

There have been rapid advances in the fabrication of high-quality MOx materials using solution (wet-chemistry) processes based on sol–gel chemistry to form "ceramics from chemical solutions." In comparison with the vacuum deposition process, the solution process has several advantages, such as low-cost production from materials and processes, easy accessibility to large areas, rapid deposition by mass-productive printing techniques, and easy compositional control of binary and ternary systems.

Figure 4.2b shows a series of solution-processed MOx film fabrication steps. After solution synthesis and deposition of xerogel-like oxide films, the following steps are required for the fully densified MOx films: (i) hydrolysis reaction to form a metal hydroxide (M—OH) by exchange of metal ligands, (ii) condensation

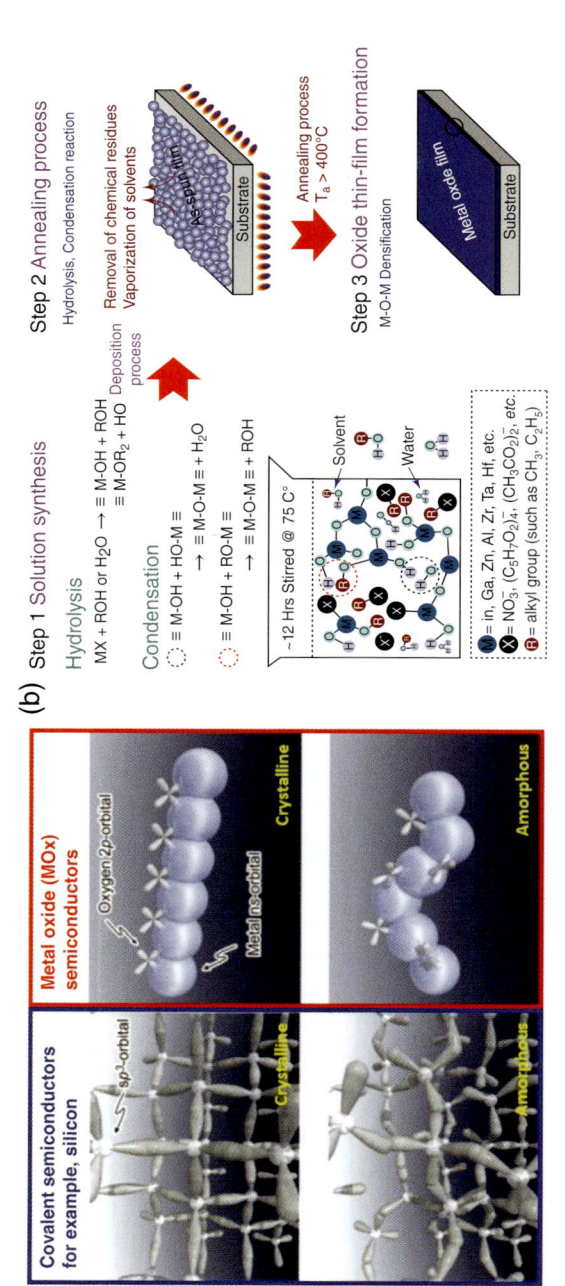

Figure 4.2 (a) Schematic electronic structures of silicon and ionic oxide semiconductors indicating bandgap formation mechanisms in covalent and ionic semiconductors; (left top: crystalline Si; left bottom: amorphous Si; right top: crystalline MOx; and right bottom: amorphous MOx). (Reproduced with permission from Ref. [11]. Copyright 2004 Macmillan Publisher Ltd.). (b) Overall schematic illustration of the steps for the fabrication of metal oxide films from sol–gel solution. (Reproduced with permission from Ref. [13]. Copyright 2015 Wiley-VCH Verlag GmbH.)

reaction to form metal–oxygen–metal (M—O—M) networks, and (iii) film densification by the removal of chemical impurities and by-products. During the above-mentioned steps, a lengthy (>1 h), high-temperature activation energy (>400 °C) is inevitably consumed to fabricate MOx electronic materials. Such long and high-temperature processes deter the development of cost-effective rapid production and future application of plastic-compatible electronics. Various approaches based on predeposition [14–17] and postdeposition [13,18–20] treatment have been proposed to achieve a low-temperature and short annealing process.

4.3.2
Bioelectronic Applications Based on MOx Electronic Materials

The thin-film transistor-based biosensors (bio-TFTs) recognize biological elements (e.g., receptors or biomolecules) using biosensitive semiconducting layers. Once target molecules bind to the surface of the semiconductors, the modulated charge distribution around the surfaces changes the electrostatic potential of the semiconductors. Therefore, the amount of electrical current flowing from the drain to source will be influenced.

Kim et al. [21] developed a DNA detection method using sol–gel-processed MOx semiconductors. They fabricated TFTs with bottom-gate and top-contact structures and performed a dry-wet method by immobilization of DNA on the In–Ga–Zn–O surface, as shown in Figure 4.3a. The prepared artificial double-crossover (DX) DNA was hybridized for high-selective binding ability and the droplets of DNA solution were directly placed on the semiconducting surfaces and maintained for 24 h for physical immobilization and removal of solvents. Figure 4.3b indicates that DX DNA immobilized TFTs showed lower field-effect mobility and on-current values from the transfer characteristics as a result of electrostatic interactions between the negatively charged phosphate groups in the DNA backbone and the MOx channel surface, which decreased the charge transfer by the narrowness of the conduction path (Figure 4.3c). However, the prolonged preparation time for the target molecule immobilization and unsuitability of the electrical characterization against aqueous biological systems need to be further investigated.

EGTFTs are desirable tools for biosensor applications. At the early stage, various cation and anion electrolytes are utilized as gate insulators for MOx EGTFTs (Figure 4.4a), and high capacitance values ($1\sim10\,\mu F/cm^2$) of nanoscale EDL-generated between the electrolyte and the semiconducting surface by the applied gate bias (Figure 4.4b) allow the transistors to operate at extremely low voltages. As shown in Figure 4.4c, there is no serious distance dependency in the dielectric film thickness and EGTFTs can be operated by simple contact of gate metal electrodes on electrolytes, compared to the typical solid-state dielectric materials [6,22–28]. As shown in Figure 4.4d, sub-2 V operation of MOx EGTFTs [27] and CMOS EGTFTs [28] were successfully fabricated by the aerosol jet printing technique by the Frisbie group. Other MOx-based low-voltage

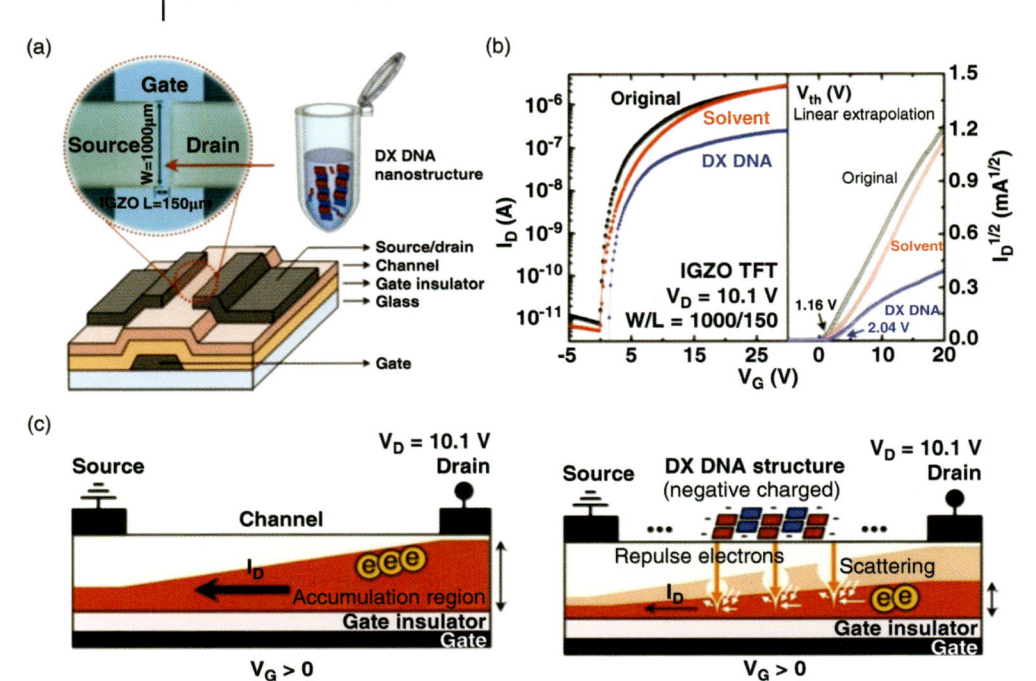

Figure 4.3 (a) Schematic diagram of an artificial DX DNA nanostructure and its immobilization on the IGZO channel surface. (b) Transfer characteristics of the IGZO TFT with and without DX DNA immobilization and with solvent. (c) Schematic diagram of the DNA detection mechanism before and after DX DNA immobilization when a positive gate bias was applied. (Figure 4.3a–c reproduced with permission from Ref. [21]. Copyright 2012 American Institute of Physics.)

operations of EGTFTs using ionic liquids or polymeric electrolytes were also reported by the Dasgupta [25], Iwasa [24], and Zaumseil [26] groups.

For the real application of biosensors using MOx EGTFTs, it is a prerequisite to fabricate high-performance electronic devices operated under physiological aqueous conditions with long operational stability. Park et al. [29] fabricated sol–gel-based In–Ga–Zn–O amorphous MO_x EGTFTs operated with various aqueous ionic solutions, as shown in Figure 4.5a. The fully dissociated NaCl, KCl, and KBr ionic aqueous solutions were used as dielectric media, because they are abundantly composed of ions in human body fluids. The IGZO-EGTFTs exhibited high-performance electrical characteristics, that is, low operational voltage below 0.5 V, high on/off current ratio above 10^7, and low subthreshold below 100 mV. In addition, IGZO-EGTFTs can be stably operated with phosphate buffered saline (PBS) solution (Figure 4.5b) and devices are not degraded by isotonic condition and phosphate ions. Finally, the IGZO-EGTFTs originated from high EDL capacitance induced by charge carrier accumulation without faradaic chemical reaction between semiconducting channel and dielectric media had good characteristics, as indicated in Figure 4.5c.

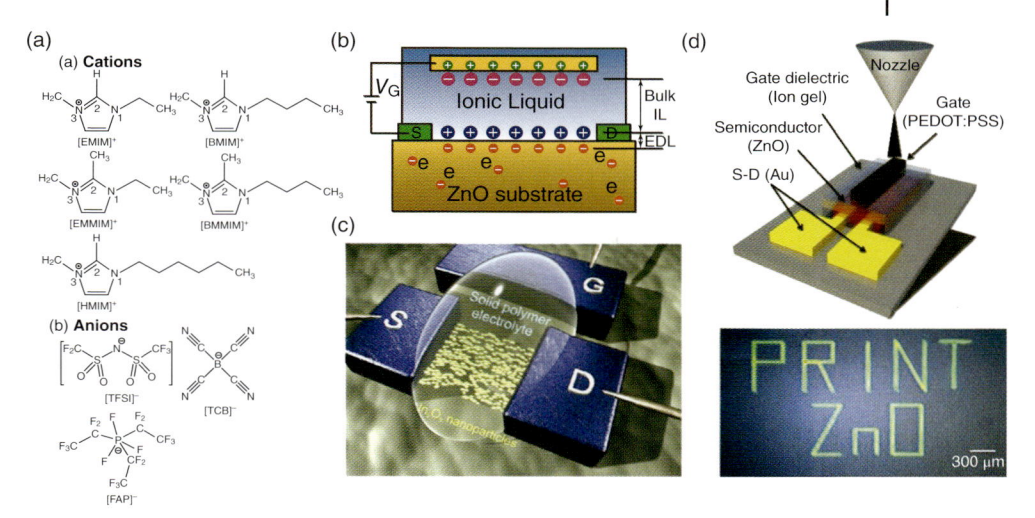

Figure 4.4 (a) Molecular structures of the cations and anions of ILs (reproduced with permission from Ref. [26]. Copyright 2012 American Chemical Society). (b) A schematic cross section of an IL-gated ZnO EGTFTs. Blue, pink, green, and red spheres represent the cation, anion, hole, and electron, respectively (reproduced with permission from Ref. [24]. Copyright 2009 Wiley-VCH Verlag GmbH). (c) A schematic presentation of the EGTFT devices with the solid polymeric electrolyte (reproduced with permission from Ref. [25]. Copyright 2011 American Chemical Society). (d) Scheme of printed ZnO EGTFTs with ion–gel gate insulator (top), optical micrograph of an aerosol jet printed ZnO pattern (reproduced with permission from Ref. [27]. Copyright 2013 Wiley-VCH Verlag GmbH).

Rim and coworkers [30] also demonstrated dopamine-sensitive In-O TFTs fabricated by chemical liftoff lithography (CLL). The nanometer-scale MOx semiconducting channels were formed with high spatial precision of patterned metal electrodes by the CLL method for high-sensitive activity and recognition of the dopamine tested using DNA aptamer immobilized on In-O surfaces to have selective binding characteristics (Figure 4.5d). The real-time sensing measurement of dopamine in PBS solution was successfully recorded with sharp increases in the drain current as shown in Figure 4.5e.

4.4
Organic Electrochemical Transistors and Bioelectronic Interfaces

Since Berggren and coworkers reported organic electrochemical logic circuits as first demonstrations for OECTs based on poly(3,4-ethylenedioxythiophene):poly(styrenesulfonate) (PEDOT:PSS) [31], OECTs have been intensely investigated with respect to their operational mechanisms and applications for bioelectronic interfaces because of their environmental stability compared to other organic-based electrical double-layer transistors, high amplifying efficiency for signal

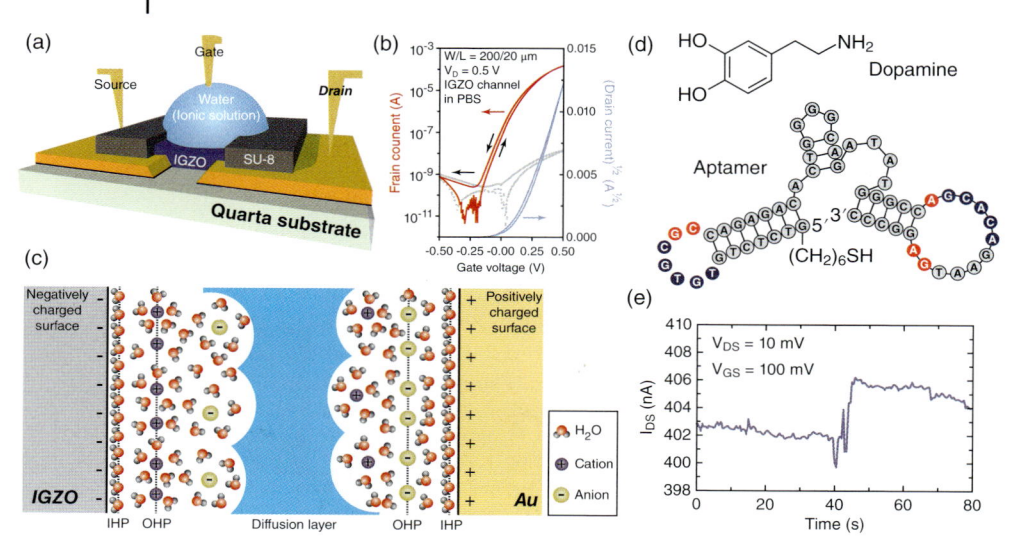

Figure 4.5 (a) A schematic of the aqueous salt EGTFT structure. (b) Representative bidirectional sweep of the transfer curves (red), square root of the channel current (blue), and gate leakage current (gray) at $V_D = +0.5$ V. (c) An illustration of the EDLs formed at the IGZO and Au surface interfacing with an aqueous salt solution. The negative- (positive-) charged surface exhibits an inner Helmholtz plane (IHP) of a compact water layer and an outer Helmholtz plane (OHP) of a diffusive hydrated cation (anions) layer (Figure 4.5a–c reproduced with permission from Ref. [29]. Copyright 2015 Macmillan Publisher Ltd). (d) A dopamine (top) and DNA aptamer (bottom) that had specific binding selectivity on the oxide surface. (e) Real-time sensing recording of 100 pM dopamine in PBS, showing an increase in the current upon exposure (Figure 4.5d and e reproduced with permission from Ref. [30]. Copyright 2015 American Chemical Society).

transducers, and ease of fabrication. OECTs are devices that use degenerately doped p-type semiconductors, such as PEDOT:PSS, as the active channel modulates currents according to a gate bias in an electrochemical manner through faradaic processes. In this section, we focus on the recent progress in investigating the operational mechanisms of organic electrochemical transistors and developing organic electrochemical transistor-based bioelectronics.

4.4.1
Fundamentals of Organic Electrochemical Transistors

As the most frequently used material for a channel of OECTs, PEDOT:PSS is a p-type semiconducting material that is degenerately doped with the sulfonate anionic groups of PSS as the counterions used to balance and stabilize the doping charges (Figure 4.6a) [32]. Generally, the basic operation of OECTs proceeds based on the reversible electrochemical reaction, doping, and dedoping processes of a charge carrier in whole channel volume (faradaic processes) [6]. In other words, the drain current of OECTs varies with applied gated bias by

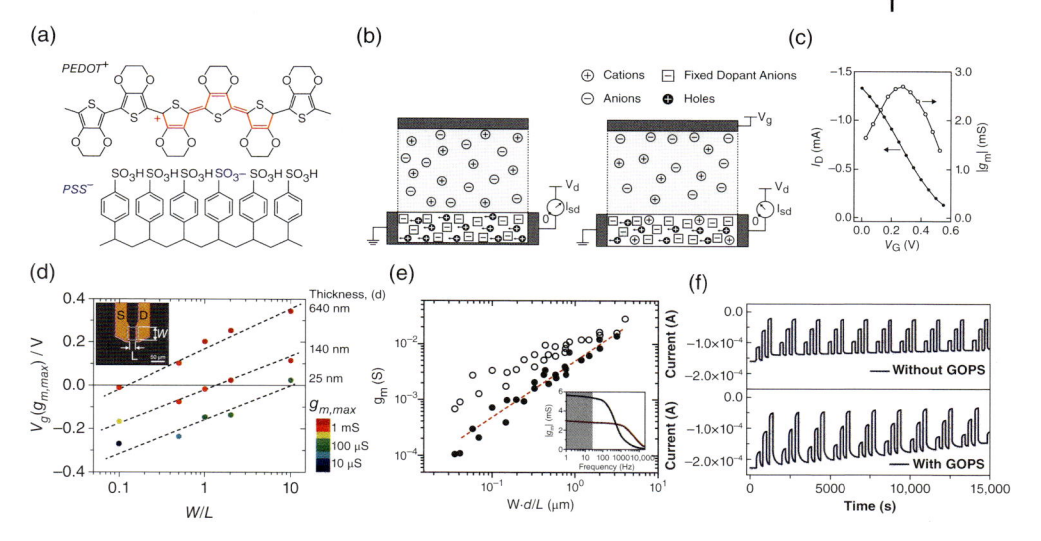

Figure 4.6 (a) A chemical structure of PEDOT:PSS. (b) Schematic diagrams of the operational principle of OECTs (reproduced with permission from Ref. [33]. Copyright 2007 Wiley-VCH Verlag GmbH.). (c) Representative transfer curves for OECTs (Figure 4.5a–c reproduced with permission from Ref. [34]. Copyright 2013 Macmillan Publisher Ltd). The trend curve of (d) the gate bias changes for the maximum transconductance depending on the W/L of the OECTs and (e) the maximum trance conductance changes depending on the $W \cdot d/L$ of the OECTs. (d, inset) The representative photographic images of OECTs, where W, L, and d denote the width, length, and thickness of the channel layer, respectively. (e, inset) The graph shows the frequency dependence of the maximum transconductance of OECTs with thick PEDOT:PSS (blue line) and thin PEDOT:PSS (red line) (Figure 4.5d reproduced with permission from Ref. [35]. Copyright 2013 Wiley-VCH Verlag GmbH; Figure 4.5e reproduced with permission from Ref. [36]. Copyright 2015 American Association for the Advancement of Science). (f) Transient electrical measurement of OECTs using PEDOT:PSS with and without 1 v/v % GOPS (reproduced with permission from Ref. [37]. Copyright 2015 American Institute of Physics).

controlling the charge concentration. In detail, as shown in Figure 4.6b, whereas most charge carriers (holes) flow under applied drain voltage without gate bias, when a positive gate bias is applied, the metal cations, M^+, in the electrolytes (e.g., Na^+, K^+, and Ca^{2+}) penetrate into the polymer channels and dedoped charge carriers in PEDOT [31,33]:

$$PEDOT^+PSS^- + M^+ + e^- \rightarrow PEDOT^0 + M^+PSS^- \quad [31]$$

Based on the operational mechanisms, the practical transfer curves show that the drain currents gradually decreases according to the gate bias increase, indicating that transistors convert the drain current by differences in the applied gate potentials [34]. This conversion efficiency is determined by the figure-of-merit, which is defined as the transconductance, $g_m = \Delta I_D / \Delta V_G$ (Figure 4.6c). In most applications in bioelectronic interfaces, the transconductance is the most important transistor parameter because transistors serve as an amplifying transducer for ionic-based signals from biological specimens. Therefore, higher transconductance values with reliable operational stability of electronic devices are required to develop highly sensitive bioelectronic interfaces.

The transconductance varies with the components of OECTs, such as the electrolyte properties [38], the type of gate electrode [8,38], and the dimensions of the channel [35,36]. By controlling the channel dimensions, transconductance can be delicately modulated. As shown in Figure 4.6d and e, the applied gate bias for maximum transconductance can also be controlled by changing the width–length ratio (W/L) of the channel [35], and the values are proportional to $W \cdot d/L$, where d is the thickness of the CP layers [36]. However, the control of the physical dimensions results in an increase in the time constant, indicating that OECTs with high $W \cdot d/L$ are unsuitable for high-speed responses in the application of recording neuronal signals (Figure 4.6e inset) [36].

In terms of operational stability, electrochemical doping/dedoping processes easily induce the structural changes of the PEDOT:PSS channels, which negatively affect the operational stability of OECTs [6]. Therefore, the silane-based cross-linker, (3-glycidyloxypropyl)trimethoxysilane (GOPS), is added to the PEDOT:PSS channel to enhance the stability in the aqueous environment [35,39,40]. However, GOPS addition induces ion mobility reduction [41] in the channel and results in an increased time constant, as shown in Figure 4.6f [37].

As we described, to design high-performance OECTs, it is important to minimize the trade-off between high transconductance values and low time constant to convert biological signals to electrical signals recorded in real-time with fast response. Therefore, the development of a new class of degenerately doped semiconductors, such as crystalline conducting polymer, is needed to satisfy the high operational stability, high amplification, and low time constant, simultaneously.

4.4.2
Organic Electrochemical Transistors for Bioelectronic Applications

As previously described, OECTs sense chemical and biological specimens by converting an electrical potential change into a current change. The OECTs for bioelectronic applications can be categorized as biomolecular sensors and tissue- (or cell-) based sensors, which record bioelectric signals from target specimens.

In terms of biomolecular sensors, modified metal electrodes with biomolecules, such as an enzyme or single-stranded DNA, have been used as gate electrodes to modulate the electrolyte potentials by reacting with target molecules [42]. Malliaras and coworkers reported the enzymatic sensor using OECTs. They measured glucose concentrations using a platinum gate electrode and phosphate-buffered saline solution with a specific concentration of glucose and glucose oxidase. Through this research, they determined that the drain current changes in OECTs-based enzymatic sensors were induced by potential changes in the electrolytes depending on the concentration of the reactant, glucose [42]. Based on this knowledge, several studies have reported OECT-based enzymatic sensors. Yan and coworkers reported gate electrodes modified with

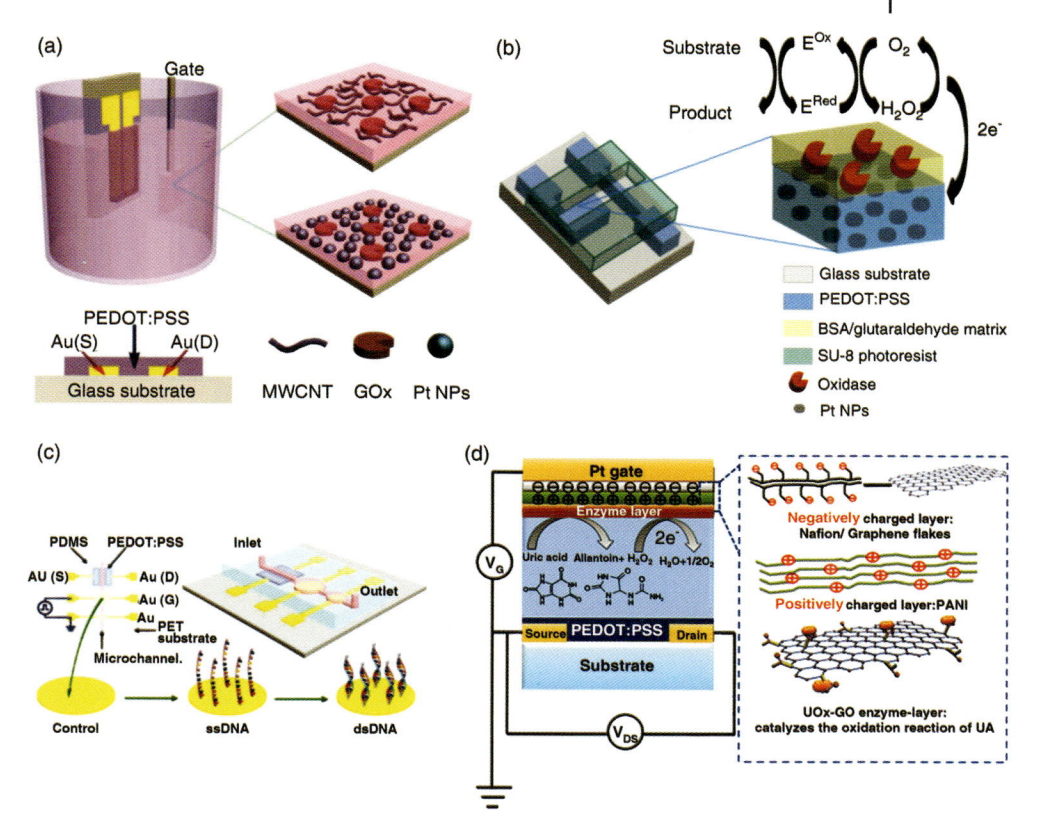

Figure 4.7 The schematic diagrams of OECTs-based enzymatic sensor for sensing (a) glucose, (b) neurotransmitter (e.g., glutamate, acetylcholine) and (c) various substances in saliva (e.g., uric acid, cholesterol and glucose). (d) A schematic diagram of the OECTs-based microfluidic label-free DNA sensor. ((a) reproduced with permission from Ref. [43]. Copyright 2011 Wiley-VCH Verlag GmbH, (b) reproduced with permission from Ref. [44]. Copyright 2014 Wiley-VCH Verlag GmbH, (c) reproduced with permission from Ref. [45]. Copyright 2015 Wiley-VCH Verlag GmbH, (d) reproduced with permission from Ref. [46]. Copyright 2011 Wiley-VCH Verlag GmbH.)

nanomaterials, such as multi-wall carbon nanotubes and platinum nanoparticles (Pt NPs), and they developed glucose biosensors with a detection limit of 5 nM using this electrode (Figure 4.7a) [43]. In addition, PEDOT:PSS/Pt NP- based gate electrodes for sensing glutamate and acetylcholine were reported (Figure 4.7b) [44]. Recently, it was reported that OECT-based enzymatic biosensors for saliva testing can monitor various substances, such as uric acid, cholesterol, and glucose (Figure 4.7c) [45]. In addition to enzymatic sensors, label-free DNA sensors were also developed by designing gate electrodes to capture DNA with specific sequences by hybridization of the DNA. The device can detect complimentary DNA targets with a detection limit of 1 nM (Figure 4.7d) [46].

Because PEDOT:PSS is known as a biocompatible material, OECT-based sensors for recording bioelectric signals from living things, such as bacteria, mammalian cells, tissues, and organs, have also been intensively investigated. In these applications, most living things closely adhere to the channel areas, and their ionic potential changes are reflected in the current changes. Yan and coworkers developed an OECT-based bacteria sensor that captures *Escherichia coli* by immobilized *E. coli* antibodies on the PEDOT:PSS channels and shows a voltage shift of 55 mV after the capture of the bacteria. In terms of cell-based sensors, several studies reported measuring cellular adhesion [47], barrier tissue integrity [48,49], and action potentials of electroactive cells [40]. Hsing and coworkers reported OECT arrays for the action potential recording of cardiomyocytes. The arrays successfully recorded the action potentials of cardiomyocytes adhered to active channels and showed high signal-to-noise ratios larger than 10 dB (Figure 4.8a) [40]. In biomedical applications, OECT-based sensors are also promising platforms for recording bioelectric signals from tissues and organs. Malliaras and coworkers developed OECT-based electrocorticography (ECoG) probes for neural signal recording. OECT-based ECoG probes successfully recorded epileptic neural signals and had a signal-to-noise ratio of 44 dB, which is a higher value than that of PEDOT:PSS surface electrode, 24.2 dB (Figure 4.8b) [50]. Furthermore, it was recently reported that OECTs can record electroencephalography by tuning the channel thickness [36].

The development of a functional platform depending on the purpose of OECT-based bioelectronics is important for the commercialization of the device. Coppedè and coworkers reported a cotton fiber functionalized with PEDOT:PSS by a simple soaking process and demonstrated that the fiber can be used as an active channel for OECTs. Fabricating OECTs with this fiber is one of the simplest and most inexpensive ways to develop wearable sensors (Figure 4.8c) [51]. OECTs on biodegradable polymers were also developed by Biscarini and coworkers. The biodegradable OECTs successfully recorded electrocardiograms by attaching channels to the skin (Figure 4.8d) [52].

4.5
Conclusions

In this chapter, we reviewed various types of electrolyte-gated thin-film transistors and biosensors based on solution-processed metal oxides and organic conducting polymers for the application of future bioelectronics. The exhaustive studies based on the formation, electronic properties, operational mechanisms, and the state-of-the-art research for bioelectronic applications have been introduced. The present solution-processed electronic materials for bioelectronics will serve as an attractive methodology to enable the preparation of all-solution, large-area biosensors and human-implantable electronics in a rapid, scalable, and economic manner for the realization of future human-implantable electronic devices.

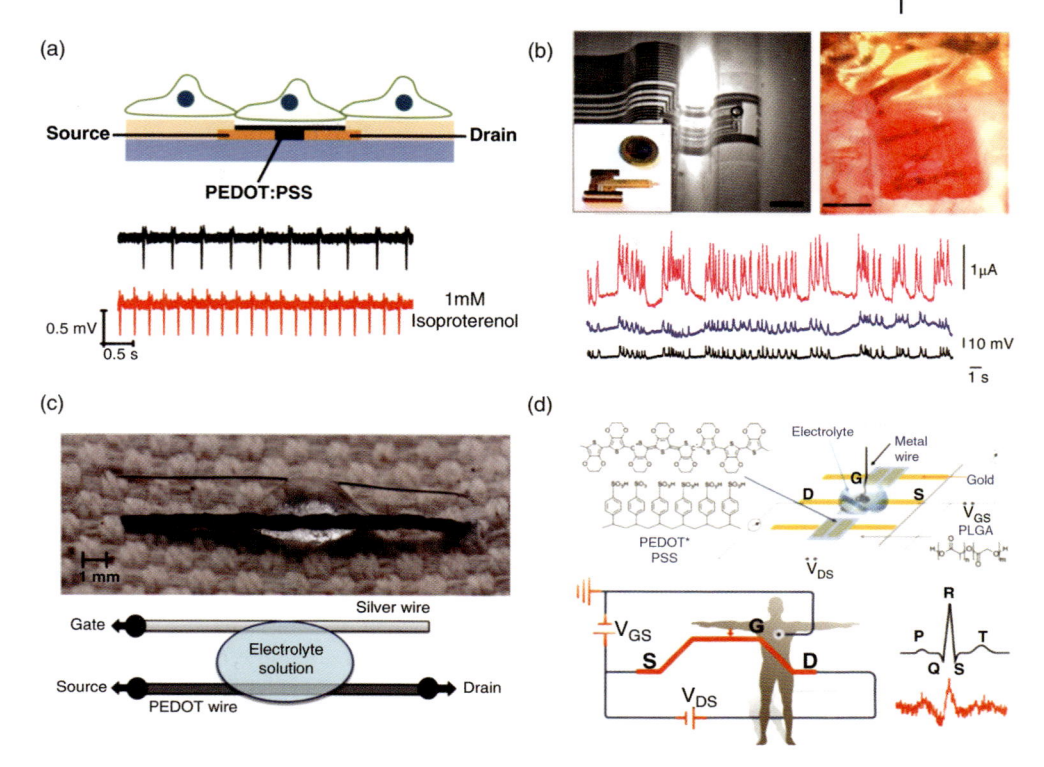

Figure 4.8 (a) A schematic image of OECTs integrated with HL-1 cells (top) and representative traces of spikes from cardiomyocytes before and after isoproterenol treatment (bottom) (reproduced with permission from Ref. [40]. Copyright 2014 Wiley-VCH Verlag GmbH). (b) Photographic images of OECT-based electrocorticography probe (top left, inset) attached to a curvilinear surface (top left) and placed over the cortex (top right). Traces of neural spikes recorded from OECT-based probe (pink), PEDOT:PSS surface electrode (blue) and Ir-penetrating electrode (black) (bottom) (reproduced with permission from Ref. [50]. Copyright 2013 Macmillan Publisher Ltd). (c) A photographic image (top) and a schematic diagram (bottom) of single cotton fiber-based OECTs (reproduced with permission from Ref. [51]. Copyright 2012 Royal Society of Chemistry). (d) A schematic diagram of the OECT-based electrocardiographic sensor fabricated on a biodegradable thin film (top) and a wiring diagram of the sensor in direct contact with skin (bottom left). The acquired drain current trace through the ECG recording (bottom right, red) is analogous to a normal potentiometric recording with standard disposable leads (bottom right, black). (reproduced with permission from Ref. [52]. Copyright 2014 Wiley-VCH Verlag GmbH.).

References

1 Berggren, M. and Richter-Dahlfors, A. (2007) Organic bioelectronics. *Adv. Mater.*, **19** (20), 3201–3213.

2 Chen, C., Xie, Q., Yang, D. et al. (2013) Recent advances in electrochemical glucose biosensors: a review. *RSC Adv.*, **3** (14), 4473–4491.

3 Wang, J. (2008) Electrochemical glucose biosensors. *Chem. Rev.*, **108** (2), 814–825.

4 Hwang, S.-W., Park, G., Cheng, H. et al. (2014) 25th anniversary article: materials for high-performance biodegradable semiconductor devices. *Adv. Mater.*, **26** (13), 1992–2000.

5 Windmiller, J.R. and Wang, J. (2013) Wearable electrochemical sensors and biosensors: a review. *Electroanalysis*, **25** (1), 29–46.

6 Kim, S.H., Hong, K., Xie, W. et al. (2013) Electrolyte-gated transistors for organic and printed electronics. *Adv. Mater.*, **25** (13), 1822–1846.

7 Herlogsson, L. (2011) *Electrolyte-Gated Organic Thin-Film Transistors*, Linköping University Electronic Press, Linköping.

8 Tarabella, G., Santato, C., Yang, S.Y. et al. (2010) Effect of the gate electrode on the response of organic electrochemical transistors. *Appl. Phys. Lett.*, **97** (12), 123304.

9 Grahame, D.C. (1947) The electrical double layer and the theory of electrocapillarity. *Chem. Rev.*, **41** (3), 441–501.

10 Parsons, R. (1990) The electrical double layer: recent experimental and theoretical developments. *Chem. Rev.*, **90** (5), 813–826.

11 Nomura, K., Ohta, H., Takagi, A. et al. (2004) Room-temperature fabrication of transparent flexible thin-film transistors using amorphous oxide semiconductors. *Nature*, **432** (7016), 488–492.

12 Kamiya, T. and Hosono, H. (2010) Material characteristics and applications of transparent amorphous oxide semiconductors. *NPG Asia Mater.*, **2** (1), 15–22.

13 Park, S., Kim, K.-H., Jo, J.-W. et al. (2015) In-depth studies on rapid photochemical activation of various sol–gel metal oxide films for flexible transparent electronics. *Adv. Funct. Mater.*, **25** (19), 2807–2815.

14 Meyers, S.T., Anderson, J.T., Hung, C.M. et al. (2008) Aqueous inorganic inks for low-temperature fabrication of ZnO TFTs. *J. Am. Chem. Soc.*, **130** (51), 17603–17609.

15 Banger, K.K., Yamashita, Y., Mori, K. et al. (2011) Low-temperature, high-performance solution-processed metal oxide thin-film transistors formed by a "sol–gel on chip" process. *Nat. Mater.*, **10** (1), 45–50.

16 Hwan Hwang, Y., Seo, J.-S., J., MoonYun. et al. (2013) An "aqueous route" for the fabrication of low-temperature-processable oxide flexible transparent thin-film transistors on plastic substrates. *NPG Asia Mater.*, **5** (4), e45.

17 Jeong, S., Ha, Y.-G., Moon, J. et al. (2010) Role of gallium doping in dramatically lowering amorphous-oxide processing temperatures for solution-derived indium zinc oxide thin-film transistors. *Adv. Mater.*, **22** (12), 1346–1350.

18 Jun, T., Song, K., Jeong, Y. et al. (2011) High-performance low-temperature solution-processable ZnO thin film transistors by microwave-assisted annealing. *J. Mater. Chem.*, **21** (4), 1102–1108.

19 Han, S.-Y., Herman, G.S., and Chang, C. (2011) Low-temperature, high-performance, solution-processed indium oxide thin-film transistors. *J. Am. Chem. Soc.*, **133** (14), 5166–5169.

20 Kim, Y.-H., Heo, J.-S., Kim, T.-H. et al. (2012) Flexible metal-oxide devices made by room-temperature photochemical activation of sol–gel films. *Nature*, **489** (7414), 128–132.

21 Kim, S.J., Kim, B., Jung, J. et al. (2012) Artificial DNA nanostructure detection using solution-processed In–Ga–Zn–O thin-film transistors. *Appl. Phys. Lett.*, **100** (10), 103702.

22 Lee, J., Panzer, M.J., He, Y. et al. (2007) Ion gel gated polymer thin-film transistors. *J. Am. Chem. Soc.*, **129** (15), 4532–4533.

23 Cho, J.H., Lee, J., Xia, Y. et al. (2008) Printable ion–gel gate dielectrics for low-

voltage polymer thin-film transistors on plastic. *Nat. Mater.*, **7** (11), 900–906.
24 Yuan, H., Shimotani, H., Tsukazaki, A. et al. (2009) High-density carrier accumulation in ZnO field-effect transistors gated by electric double layers of ionic liquids. *Adv. Funct. Mater.*, **19** (7), 1046–1053.
25 Dasgupta, S., Kruk, R., Mechau, N., and Hahn, H. (2011) Inkjet printed, high mobility inorganic-oxide field effect transistors processed at room temperature. *ACS Nano*, **5** (12), 9628–9638.
26 Thiemann, S., Sachnov, S., Porscha, S. et al. (2012) Ionic liquids for electrolyte-gating of ZnO field-effect transistors. *J. Phys. Chem. C*, **116** (25), 13536–13544.
27 Hong, K., Kim, S.H., Lee, K.H., and Frisbie, C.D. (2013) Printed, sub-2V ZnO electrolyte gated transistors and inverters on plastic. *Adv. Mater.*, **25** (25), 3413–3418.
28 Hong, K., Kim, Y.H., Kim, S.H. et al. (2014) Aerosol jet printed, sub-2 V complementary circuits constructed from p- and n-type electrolyte gated transistors. *Adv. Mater.*, **26** (41), 7032–7037.
29 Park, S., Lee, S., Kim, C.-H. et al. (2015) Sub-0.5 V highly stable aqueous salt gated metal oxide electronics. *Sci. Rep.*, **5**, 13088.
30 Kim, J., Rim, Y.S., Chen, H. et al. (2015) Fabrication of high-performance ultrathin In_2O_3 film field-effect transistors and biosensors using chemical lift-off lithography. *ACS Nano*, **9** (4), 4572–4582.
31 Nilsson, D., Robinson, N., Berggren, M., and Forchheimer, R. (2005) Electrochemical logic circuits. *Adv. Mater.*, **17** (3), 353–358.
32 Groenendaal, L., Jonas, F., Freitag, D. et al. (2000) Poly(3,4-ethylenedioxythiophene) and its derivatives: past, present, and future. *Adv. Mater.*, **12** (7), 481–494.
33 Bernards, D.A. and Malliaras, G.G. (2007) Steady-state and transient behavior of organic electrochemical transistors. *Adv. Funct. Mater.*, **17** (17), 3538–3544.
34 Khodagholy, D., Rivnay, J., Sessolo, M. et al. (2013) High transconductance organic electrochemical transistors. *Nat. Commun.*, **4**. doi: 10.1038/ncomms3133

35 Rivnay, J., Leleux, P., Sessolo, M. et al. (2013) Organic electrochemical transistors with maximum transconductance at zero gate bias. *Adv. Mater.*, **25** (48), 7010–7014.
36 Rivnay, J., Leleux, P., Ferro, M. et al. (2015) High-performance transistors for bioelectronics through tuning of channel thickness. *Sci. Adv.*, **1** (4), e1400251–e1400251.
37 Zhang, S., Kumar, P., Nouas, A.S. et al. (2015) Solvent-induced changes in PEDOT:PSS films for organic electrochemical transistors. *APL Mater.*, **3** (1), 014911.
38 Lin, P., Yan, F., and Chan, H.L.W. (2010) Ion-sensitive properties of organic electrochemical transistors. *ACS Appl. Mater. Interfaces*, **2** (6), 1637–1641.
39 Yao, C., Xie, C., Lin, P. et al. (2013) Organic electrochemical transistor array for recording transepithelial ion transport of human airway epithelial cells. *Adv. Mater.*, **25** (45), 6575–6580.
40 Yao, C., Li, Q., Guo, J. et al. (2015) Rigid and flexible organic electrochemical transistor arrays for monitoring action potentials from electrogenic cells. *Adv. Healthc. Mater.*, **4** (4), 528–533.
41 Stavrinidou, E., Leleux, P., Rajaona, H. et al. (2013) Direct measurement of ion mobility in a conducting polymer. *Adv. Mater.*, **25** (32), 4488–4493.
42 Bernards, D.A., Macaya, D.J., Nikolou, M. et al. (2008) Enzymatic sensing with organic electrochemical transistors. *J. Mater. Chem.*, **18** (1), 116.
43 Tang, H., Yan, F., Lin, P. et al. (2011) Highly sensitive glucose biosensors based on organic electrochemical transistors using platinum gate electrodes modified with enzyme and nanomaterials. *Adv. Funct. Mater.*, **21** (12), 2264–2272.
44 Kergoat, L., Piro, B., Simon, D.T. et al. (2014) Detection of glutamate and acetylcholine with organic electrochemical transistors based on conducting polymer/platinum nanoparticle composites. *Adv. Mater.*, **26** (32), 5658–5664.
45 Liao, C., Mak, C., Zhang, M. et al. (2015) Flexible organic electrochemical transistors for highly selective enzyme biosensors and used for saliva testing. *Adv. Mater.*, **27** (4), 676–681.

46 Lin, P., Luo, X., Hsing, I.-M., and Yan, F. (2011) Organic electrochemical transistors integrated in flexible microfluidic systems and used for label-free DNA sensing. *Adv. Mater.*, **23** (35), 4035–4040.

47 Lin, P., Yan, F., Yu, J. *et al.* (2010) The application of organic electrochemical transistors in cell-based biosensors. *Adv. Mater.*, **22** (33), 3655–3660.

48 Jimison, L.H., Tria, S.A., Khodagholy, D. *et al.* (2012) Measurement of barrier tissue integrity with an organic electrochemical transistor. *Adv. Mater.*, **24** (44), 5919–5923.

49 Tria, S.A., Ramuz, M., Huerta, M. *et al.* (2014) Dynamic monitoring of *Salmonella typhimurium* infection of polarized epithelia using organic transistors. *Adv. Healthc. Mater.*, **3** (7), 1053–1060.

50 Khodagholy, D., Doublet, T., Quilichini, P. *et al.* (2013) *In vivo* recordings of brain activity using organic transistors. *Nat. Commun.*, **4**, 1575.

51 Tarabella, G., Villani, M., Calestani, D. *et al.* (2012) A single cotton fiber organic electrochemical transistor for liquid electrolyte saline sensing. *J. Mater. Chem.*, **22** (45), 23830.

52 Campana, A., Cramer, T., Simon, D.T. *et al.* (2014) Electrocardiographic recording with conformable organic electrochemical transistor fabricated on resorbable bioscaffold. *Adv. Mater.*, **26** (23), 3874–3878.

5
Designing Thermoelectric Materials Using 2D Layers

Sage R. Bauers and David C. Johnson

University of Oregon, Department of Chemistry, 1253 Eugene, OR, USA 97403-1299

5.1
Introduction

Thermoelectrics have long been a tantalizing class of devices – solid-state junctions with no moving parts and the ability to reversibly convert between thermal and electric field gradients and only fundamentally limited by Carnot efficiency. Since almost all industrial processes produce waste heat, accounting for an estimated 20–50% of the initial energy input, capturing this heat as a usable form of energy could significantly reduce both cost of operation and offset the total environmental cost of said processes. This has never been a more important consideration than now. The United Nations' 2015 revision of world population expected it to reach 8.5 billion by 2030 and over 11 billion by 2100. Climate change makes reducing carbon dioxide emissions critically important to maintain the livability of the planet, and producing additional power from waste heat obviously reduces the amount of fossil fuels that need to be consumed. To sustain our growing population, and indeed support the current numbers, every effort must be made to maximize our resources, and thermoelectric power generation may be one viable option. However, the discovery of materials with high enough thermoelectric performance to provide either power or cooling on a cost competitive basis has proved difficult except in niche applications. Even after several concentrated waves of concerted research efforts in the United States and around the world, most current commercial devices still rely on decades old material technologies. The crux of this problem lies in the difficulty of balancing the several interdependent and contradictory material properties.

This discussion is for an audience familiar with the basic tenets of solid-state physics and chemistry, but not necessarily with thermoelectric materials. Discussion is centered on optimization at the material level as opposed to modules. We begin with a colloquial discussion of how the basic transport physics that governs thermoelectric performance can be used to guide materials research. Also

Handbook of Solid State Chemistry, First Edition. Edited by Richard Dronskowski, Shinichi Kikkawa, and Andreas Stein.
© 2017 Wiley-VCH Verlag GmbH & Co. KGaA. Published 2017 by Wiley-VCH Verlag GmbH & Co. KGaA.

presented are the common strategies for finding high-performance materials that have been used over the past two decades. The body of the work outlines the current and recent strategies both for finding new materials and for further optimizing the existing library of high-performance materials. Throughout, we include examples and discussions of some of the current "state of the art" materials. We especially emphasize the role of low-dimensional composite structures and give our perspective on future directions of research in this realm.

5.2
Physical Picture

Understanding the flow of charge and heat has been the focus of both fundamental and applied transport studies in solid-state materials for decades. Nontrivial thermoelectric materials display an entanglement of the processes by which these flows proceed – an electronic or heat current may induce a current of the other property. This behavior is captured by the Seebeck coefficient – a spatially independent material property defined as the electric field generated by a thermal gradient across a material, which is obtained experimentally by measuring a voltage as a function of the magnitude of an applied temperature difference.

Conceptually, the Seebeck coefficient is a reasonable material parameter to define when considering any solid material outside of thermal equilibrium. The larger thermal motion from carriers at the hot end of a material will cause diffusion of these carriers toward the cold side of the material. The open-circuit equilibrium between this diffusive separation and the restoring electric field defines the Seebeck coefficient of the material. When separately considering the flow of charge or heat in a particular material, we have a strong intuitive base we can draw upon to roughly determine the properties, but the same is not immediately obvious for the Seebeck coefficient. However, an expression for the Seebeck coefficient may be derived from Boltzmann transport theory, which describes both heat and charge flow in most solids. The full integral expression reduces to a simple relationship in the degenerate carrier regime, which is a reasonable regime for functional thermoelectric materials. This relationship is called the Mott formula [1]:

$$S = \frac{\pi^2}{3} \frac{k^2 T}{e} \frac{d \ln[\sigma(E)]}{dE} \bigg|_{E=E_F},$$

where S is the Seebeck coefficient, k is Boltzmann's constant, σ is the electrical conductivity, T is the absolute temperature, e is the elementary charge, and E_F is the Fermi energy. Over a small energy range, it is reasonable to consider mobility, μ, to be energy independent and conductivity to be monotonic with the density of states (DOS). Thus, a convenient metric to find materials with large Seebeck coefficients is to begin the search in materials with large energy derivatives in the density of states function, $g(E)$, near the chemical potential.

However, the functional implementation of thermoelectric materials depends on more than the Seebeck coefficient. In addition to the maximization of the Seebeck voltage, it is also important to consider materials such that energy may be effectively transferred to (and from) a load (source) (high electronic conductivity) and that a temperature gradient may be maintained (low thermal conductivity). The combination of these parameters leads to a temperature-dependent dimensionless figure of merit defined for thermoelectric materials as follows:

$$zT = \frac{S^2 \sigma}{\kappa} T,$$

where κ is the total thermal conductivity, which is the sum of contributions from both the lattice (κ_L) and the charge carriers (κ_e). The codependence of the terms in the figure of merit makes finding materials with high zT values a challenging task. The figure of merit for useful thermoelectric materials typically has a value of approximately 1. Given a promising starting material, there are two obvious routes for obtaining a higher zT value: either maximizing the numerator, which contains the electronic contributions (called the power factor), while maintaining a low thermal conductivity, or minimizing the thermal conductivity without affecting the power factor. However, the interplay between the various parameters has shown either task to be far from trivial. First, let us look at the power factor.

The higher dimensionality of the Seebeck coefficient in the figure of merit of thermoelectric materials has given a historic precedent to research focused on finding a high Seebeck coefficient. If one considers the approximations of a parabolic band and energy-independent scattering, the Mott formula reduces to the Pisarenko relationship [2]:

$$S = \frac{8\pi^2 k^2}{3eh^2} m^* T \left(\frac{\pi}{3n}\right)^{\frac{3}{2}},$$

where h is Planck's constant, m^* is the carrier effective mass from the shape of $g(E)$, and n is the carrier density. We see that high Seebeck coefficients result from heavy bands with low carrier concentrations, similar to, as expected, a chemical potential at the μ' or μ'' positions as shown in Figure 5.1a. This is consistent with our previous metric of finding high Seebeck coefficients in regions with rapidly changing DOS, as flat bands (high m^*) will integrate to a high DOS over narrow ΔE, and the edge of these bands, where carrier density is lowest, will give the largest energy derivatives. However, recent works have discussed at length that many of the efforts to find high Seebeck coefficients have been stymied by these heavy bands, which negatively affect mobility, and thus the electrical conductivity [3,4]. Instead, a balance of a high Seebeck coefficient and carrier mobility can be achieved in a material with several low-mass pockets in the band structure, as shown in Figure 5.1b, which when integrated result in a high effective mass as perceived in the density of states [5]. This case of several degenerate band extrema is most likely to occur in compounds with high symmetry, further guiding our search toward materials with these structures [6].

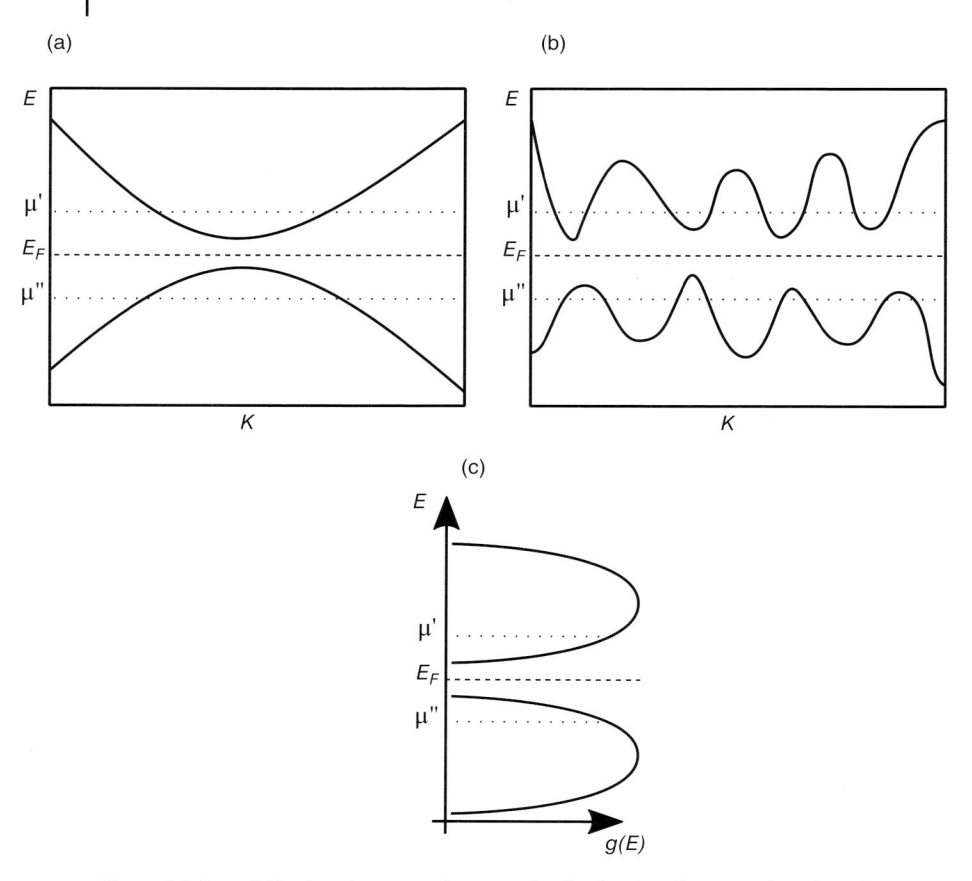

Figure 5.1 Speculative band structure from a compound with (a) a single heavy carrier pockets and a compound with (b) several lighter carrier pockets. However, both compounds might integrate to similar profiles (c) in $g(E)$ and have similar effective masses as perceived by the density of states. When doped, n-doped (p-doped), the chemical potential in each moves toward μ' (μ''), populating the respective pockets with carriers. The light degenerate bands maintain their high mobility, despite the high perceived mass.

The second approach is to focus on the denominator of zT, the thermal conductivity. The total thermal conductivity is the sum of heat moved by the lattice, or phonons, and heat flow associated with the flow of charge ($\kappa = \kappa_L + \kappa_e$). A coupling in the mechanisms by which heat and charge are moved through a material by the electronic component leads to compounds with a high electrical conductivity also having a high thermal conductivity. This is often estimated using the Wiedemann–Franz law, which states that at a certain temperature κ_e is proportionally related to the electronic conductivity of the material by a constant called the Lorenz number, L ($\kappa_e = L\sigma T$). Even when this coupling is taken into account, the maximization of the electronic conductivity is still always favorable to achieve a high value of zT due to the figure of merit's functional form. Thus, the total thermal conductivity must be lowered by either minimizing the Lorenz number, L, or the lattice contribution to the thermal conductivity, κ_L.

While there is no explicit necessity for it to be so, there is often an observed coupling of high κ_L in materials with high σ. To be very general, crystalline materials might conduct both well, whereas glasses would not. Still, the general strategy of finding materials with structures conducive to inherently low thermal conductivities at the expense of the electronic component has been fruitful. For example, compounds with octahedrally coordinated metal centers, such as the various IV–VI metal dichalcogenides, generally have a large degree of phonon–phonon interactions and show much better thermoelectric performance than III–V compounds with higher mobilities [7]. Outside of these few cases, materials with low thermal conductivities have historically been found by searching for complex inorganic structures, typically ternary or quaternary compounds with large and sometimes highly anisotropic unit cells. More recently, several other approaches and extensions to this strategy have been taken to decouple phonon and electronic transport in thermoelectric materials, which will be discussed later.

Armed with the previous discussions, it is valuable to consider limits to a material's thermoelectric performance. Assuming a material that behaves according to the Wiedemann–Franz law has a zero thermal conductivity of the lattice, and a Lorenz number for a typical metal (2.45×10^{-8} V^2/K^2), we can determine a minimum Seebeck coefficient that will define an upper limit to the figure of merit. The expression for the minimum Seebeck coefficient is simply $S_{Min} = (L \times zT_{Target})^{1/2}$. For example, if we want to target a zT value of 1 in a particular compound, the absolute lowest value of Seebeck coefficient for the compound to hit the target is 157 µV/K. Similarly, to obtain a zT value of 2, S must be 221 µV/K and for a value of 4 this becomes 313 µV/K. This reinforces the earlier discussion about the importance of the magnitude of the Seebeck coefficient in determining zT. As a material's carrier density decreases and carriers move away from the degenerate limit, L decreases and the S_{Min} slightly decreases. This suggests that the Seebeck coefficient serves as a useful guide for evaluating a material's prospects as a functional thermoelectric with a single measurement, provided a model exists that extrapolates Seebeck coefficients for different carrier concentrations and temperatures. It also emphasizes the need to minimize lattice thermal conductivity, which we will discuss next.

5.3
Optimizing Thermoelectric Materials: Minimizing Lattice Conductivity

5.3.1
The Phonon Glass/Electron Crystal Approach

Since its introduction by Slack in the 1990s, much of the research in bulk thermoelectric materials has centered around the pursuit of his phonon glass/electron crystal (PGEC) concept [8]. Two approaches to lowering thermal conductivity have dominated: the development of new bulk single-phase structures, which typically have large and very complex unit cells, and the synthesis of nanocomposites, which consist of nanoscale inclusions within a bulk material, most often with both constituents having more conventional and simple structures. Both these approaches were inspired in part by Slack's framing of the challenge in finding an improved thermoelectric material. Slack argued that by finding structures with rigid covalent cages that incorporate loosely bound "rattling" atoms with large displacement parameters into the cage voids, one could create a material with a high figure of merit. He reasoned that electrical mobility would remain high by conduction via the cage behaving as an electron crystal, but the material would effectively scatter acoustic phonons via interactions with the "rattling" atom, lowering the phonon mean free path. The material would behave as a phonon glass due to the static and dynamic displacement of the "rattling" atom. In general, to maximize the effects from the structure, the host cage should be designed as a narrow bandgap semiconductor with small differences in electronegativity between atoms, whereas the rattling ion should be a small but heavy atom to maximize the thermal displacement and disorder from the rattling. Several classes of materials fit Slack's criteria, and a large body of work exists investigating the thermoelectric properties of clathrates [9], zintl phases [10], skutterudites [11], and other structures. A few specific examples are briefly discussed next.

Skutterudites are typically made of MX_6 metal-pnictide octahedra and have a cubic structure similar to the ReO_3 motif but with the anions along four parallel edges of the unit cell displaced inward resulting in a structure consisting of eight corner sharing MX_6 octahedra per unit cell. This anion displacement creates a fourfold ring and an adjacent open cage, within which a rattling atom may be incorporated. The skutterudite structure is shown in Figure 5.2. For thermoelectric applications, antimonide skutterudites are of the most interest. The

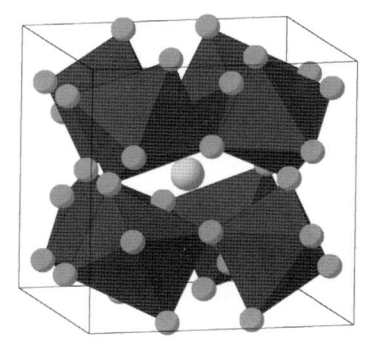

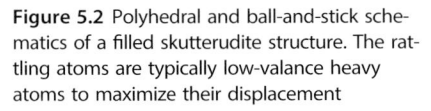

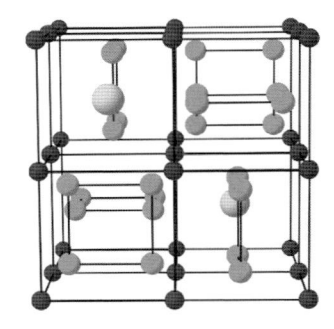

Figure 5.2 Polyhedral and ball-and-stick schematics of a filled skutterudite structure. The rattling atoms are typically low-valance heavy atoms to maximize their displacement parameters and thus their effectiveness at phonon scattering. The host lattice consists of metal centers octahedrally coordinated by a pnictide and should be highly covalent and conductive.

cation composition of the host skutterudite lattice and both the composition and the fill ratio of ions incorporated into the cage are used to tailor and optimize the material's physical properties. Researchers have also focused on tuning the temperature at which an optimal value for zT is achieved via chemical substitutions. To date, the highest performing skutterudite, a $CoSb_3$ host filled with Ba, La, and Yb making an n-type material, exhibits a zT value of 1.7 at 850 K [12]. Skutterudites are among the highest performing materials optimized to date and have been prepared with similar properties by many research groups around the world. Consequently, considerable development work has been done to incorporate them into both working modules and segmented couples.

Zn_4Sb_3 has also been the focus of considerable detective work as researchers focused on initial reports of unusual properties that were greatly dependent on composition and preparation conditions. It was discovered that interstitial Zn atoms in Zn_4Sb_3 greatly reduce the thermal conductivity relative to the ZnSb compound, which has similar stoichiometry, structural features, and transport properties. Multiple interstitial sites in the Zn_4Sb_3 unit cell were shown to have large displacement parameters, which serve to lower the thermal conductivity below those of the best skutterudites, due to the greater degree of disorder available to the lattice [13]. This results in a two- to threefold increase in the figure of merit in between 200–400 °C, demonstrating the power of Slack's original concept.

5.3.2
Other Methods to Reduce Thermal Conductivity

Besides the PGEC concept, other methods have also been used to discover new materials with low thermal conductivities. Among these strategies are using engineered structures and interfaces to scatter phonons, which can include low-dimensional materials, defecting incorporation, or controlling crystalline grain size. Another approach altogether has been finding simple lattices with natural tendencies for phonon–phonon scattering interactions to occur.

By introducing nanoscale precipitates with similar structures into the lattice of a host bulk material, the thermal conductivity may be substantially lowered without severely impacting the power factor. For example, several enhancements to the zT value of PbTe have been seen in the LAST and SALT compounds, which incorporate $AgSbTe_2$ or $NaSbTe_2$ clusters into the rock salt structure [14,15]. These results have led to further work on optimizing PbTe and to the approach of using materials designed with a patterned structure over many length scales to scatter phonons, which has provided us with some of the most promising thermoelectric materials to date. By strategically controlling and introducing point defects, nano-precipitates, and grain boundary size, lead chalcogenide thermoelectric materials have been synthesized with lattice thermal conductivities at or near the amorphous limit [16–18]. By carefully selecting synergistic band structures in the components of the composite material, favorable electrical transport properties between the host and the guest compounds can be tailored, enhancing the power factor as well [19]. This approach has resulted in bulk PbTe with embedded SrTe nanoprecipitates and engineered grain boundaries having an exceptional 2.2 value for zT. The Bi_2Te_3 family has also benefitted from the nanostructuring approach, resulting in BiSbTe alloys having an enhanced room-temperature zT value of 1.2 and a peak value of 1.4 at about 100 °C [20]. These results are extremely promising, though questions about the long-term stability of nanostructured composites at elevated temperatures in large thermal gradients remain.

Perhaps, the most exciting recent development in thermoelectric materials has come from single-crystal SnSe, which does not follow Slack's guidelines. It was found that SnSe displays an exceptionally high (record) value of zT (2.6) along the b-axis at 923 K, due to favorable electronic transport ($S^2\sigma \sim 10\,\mu W/(cm\,K^2)$) and very low thermal conductivity (<0.4 W/(m K) in this direction [21]. SnSe takes on a distorted rock salt structure, which forms layers along the a-axis. One would expect this direction to exhibit both the lowest thermal and the electrical conductivity and while this is true in SnSe, the increased thermal conductivity along the b- and c-lattice directions is negligible compared to the increased electrical conductivity, leading to the exceptional values of zT along these directions. Unlike typical metals and semiconductors, thermal conductivity in SnSe decreases with T in the range of 300–700 K. A thermal conductivity inversely related to temperature is indicative of a material with a high degree of Umklapp scattering. This behavior and the high value of zT in SnSe is highly surprising for both a simple structure and a single crystal and highlights a different strategy – finding materials with a large inherent phonon anharmonicity – as an avenue for achieving a high figure of merit.

Another particularly exciting avenue for bulk materials with low thermal conductivity and high values of zT is in materials that undergo structural instabilities such as Peierls distortions or charge density waves (CDWs). CDWs are structural distortions arising from strong electron–phonon coupling and are found in layered materials whereby an in-plane pairing of atoms serves to stabilize the structure. This stabilization breaks the symmetry of the 2D "sheet" as a modulation of atomic density along one direction in the in-plane structure is formed. This affects both electrical and thermal transport properties in the material. A CDW exists in the layered structure of $In_4Se_{3-\delta}$ and it was shown that the distortion lowers the in-plane thermal conductivity below that of the stacking direction, which is quite surprising as the layers in the material are bonded by weaker van der Waals type forces [22]. The lower thermal conductivity, as well as a more favorable power factor along the in-plane direction, lead to a value of 1.48 for zT, which is ~300% higher than reported for the cross-plane value in the same material [22].

5.4
Thermoelectric Materials: Maximizing Power Factor

Traditionally, maximizing the power factor in a thermoelectric material is done by doping the compound to control carrier concentration. In most cases, the

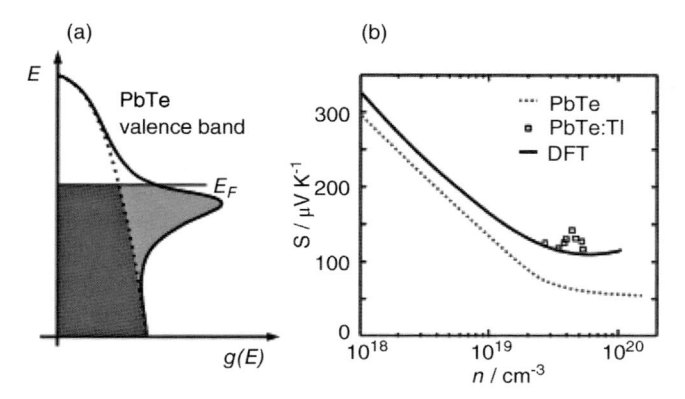

Figure 5.3 (a) Schematic $g(E)$ for PbTe with and without resonant levels (adapted from Ref. [25]). (b) Sharp features in $g(E)$ from resonant states created by Tl doping increase the Seebeck coefficient relative to PbTe that is Na-doped to a similar carrier concentration but without resonant coupling (adapted from Ref. [26]). However, recent density function theory (DFT) calculations of PbTe have predicted a values similar to PbTe:Tl without enhancement from band resonances [26,27].

Seebeck coefficient varies inversely with carrier density, while the conductivity varies proportionally. Neglecting the thermal component, the ideal carrier density optimizes the power factor with respect to these behaviors. In essence, this puts the chemical potential at the ideal level for mixed electronic/thermoelectric transport for a given shape of $g(E)$. Doping any material in this fashion will continue to be an important step in any future work, but several additional approaches have been either proposed or implemented in achieving power factors much greater than possible with carrier concentration optimization alone.

Deliberate band engineering of bulk materials by the introduction of resonance states near the Fermi level has been a promising direction to increase the power factor [23]. By carefully selecting a dopant level with states overlapping those of the host compound, electronic coupling between the host and the dopant can introduce a resonance level in the density of states, which perturbs the electronic structure toward the ideal case of a delta function at the Fermi level, as argued by Mahan and Sofo and seen in Figure 5.3 [24]. Resonant states in Tl-doped PbTe considerably raise the Seebeck coefficient above the value expected by the Pisarenko relationship, resulting in an enhancement of zT at these carrier concentrations [25]. This leads to a zT of about 1.5 at 800 K without nanostructuring the material. It should be noted that while the net gain in zT is appreciable, the introduction of the resonant states did come at the expense of mobility in the form of higher effective carrier mass. Other dopants, such as Na in PbTe, have been shown to enhance the zT value nearly as much, but these results are attributed to moving the chemical potential to sharp regions within the PbTe DOS profile as opposed to the introduction of resonant states [26]. The increase in Seebeck coefficient is less pronounced than the Tl-doped compound, but a higher mobility results in similar overall performance [26]. In each of these cases, the favorable electronic interaction between the host and the added element (Tl or Na) may be optimized by controlling dopant densities.

Another approach to discovering thermoelectric materials by increasing the power factor is based on the ideas and theory that Hicks and Dresselhaus pioneered in the early 1990s. By incorporating promising materials into quantum well or quantum wire nanostructures, Hicks and Dresselhaus noted that an enhancement of zT could be achieved by the introduction of sharp features into the density of states and greatly enhancing the Seebeck coefficient [28,29]. This is consistent with the theory developed by Mahan and Sofo, showing that the power factor is maximized when the density of states at the chemical potential takes the form of a Dirac delta function [24]. The papers published by Hicks and Dresselhaus led to a wave of research on low-dimensional systems. Shortly after the initial predictions by Hicks and Dresselhaus, an enhancement of the Seebeck coefficient and in some cases zT was observed from low-dimensional materials, in both quasi-1D- and 2D-confined structures of bismuth, antimony, or their chalcogenides [30–34]. However, these results have not always been repeatable and increases in the value of zT were attributed to a decrease in the lattice conductivity as much as enhancement of the power factor [35–37].

High power factors have also been discovered in strongly correlated electron systems. In these materials, interactions between carriers become too strong to be neglected. This typically happens when electrons are localized to f- or d-orbitals and strong Coulombic and spin interactions lead to correlated behavior between carriers. This leads to the creation of hybridized heavy bands. The existence of the heavy correlated bands can amplify the thermopower significantly over an uncorrelated material. Depending on filling in these heavy fermion systems, both correlated metals and correlated semiconductors may be formed. The highest energy derivative in $g(E)$ is expected to occur in rare-earth compounds with mixed valence f-level electrons, implying metals containing these electrons near the chemical potential should be promising grounds for finding anomalously high Seebeck coefficients for their carrier densities [38,39]. $YbAl_3$ is an example of such a material, possessing an exceptionally high room-temperature power factor of $180\,\mu W/(K^2\,cm)$ at 300 K [40]. The large peaks in $g(E)$ of $YbAl_3$ near the Fermi energy, shown in Figure 5.4, come from the 4f-levels in Yb [41] and without the background states come very close to approximating the δ-function-like profile. Enhancement of the Seebeck coefficient in correlated semiconductors is especially highlighted by $FeSb_2$, which has a colossal Seebeck coefficient of $45\,000\,\mu V/K$ and an unparalleled power factor of over $10^3\,\mu W/(K^2\,cm)$ at low temperatures [42]. This behavior has been explained by the enhancement of the thermopower by strongly correlated 3d-electrons in the system. While the high thermal conductivity at these temperatures has limited $FeSb_2$ use as a functional thermoelectric material, the results illustrate the exceptional power factors that have been found in correlated electron systems.

Another class of compounds that have also shown great promise as spin-correlated thermoelectric materials are the layered cobalt oxides. The compound $Na_xCo_2O_4$ shows a high Seebeck coefficient for its carrier density, on the order of $100\,\mu V/K$ [43]. A large spin entropy present in the family of materials is

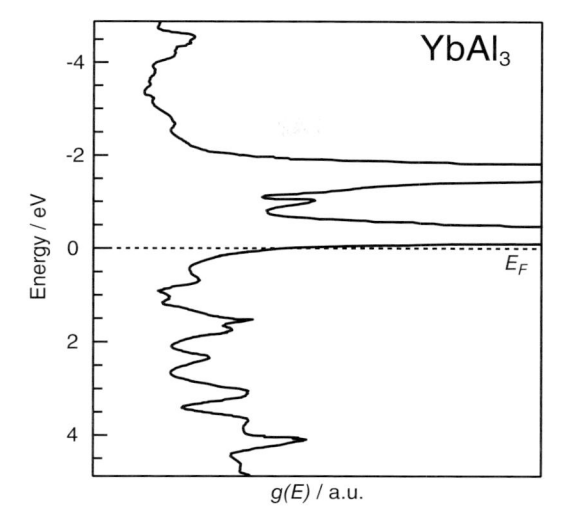

Figure 5.4 Density of states for YbAl$_3$ modeled after the calculated data in Ref. [41]. The delta-function-like features in $g(E)$ near the chemical potential result in high Seebeck coefficients for the compound, despite a metal-like carrier density. This results in unparalleled room-temperature power factors for the material.

responsible for the significantly increasing the Seebeck coefficient. This degeneracy can be removed by making measurements at low temperatures and high magnetic fields, which suppresses the thermopower and indicates that the correlated spin behavior is the source. Careful extrapolation of these results to 300 K for the Na$_x$Co$_2$O$_4$ compound shows that most of the room-temperature thermopower results from the correlated spin enhancement [43].

5.5
Bridge to Nanolaminate Structures

Looking at the past few decades of research, a rational methodology and understanding for systematically reducing the thermal conductivity of materials has been developed well. Gains have also been made in the power factors of materials, but most of the highest gains come outside the classic picture of balancing the contribution of the Seebeck coefficient and conductivity by optimizing the carrier concentration. The current forum is too brief and the collective understanding too narrow to give a full account of how strongly correlated electron behaviors affect thermoelectric properties. However, there is a large research effort in understanding this behavior. Further investigating and developing a deeper understanding of structure–property relationships in systems with quantum well structures or materials where electron–electron, electron–spin, or electron–phonon behavior enhance the power factor are a means by which we might expect to see considerable gains in high-performance thermoelectric materials.

On the other hand, nanostructuring and rational doping of promising bulk materials remains a potential to enhance their thermoelectric properties. For the first time, scalable materials technologies that easily surpass bismuth telluride are coming to the fore, which may lead to more widespread implementation of thermoelectric devices both for power generation and for temperature control. However, the avenues by which these gains have been made result in materials that are inherently difficult to characterize, especially at the scale of the structural details that lead to the enhanced behavior.

While eventual implementation of thermoelectric materials in everyday devices will likely result from progress made in nanostructured bulk materials, further research on more readily characterizable systems remains a valuable tool to understand the underpinning physical phenomena and as a means to discover promising new material systems for further investigation. This is especially true in the case of finding, understanding, and describing cases that are dominated by correlated electron behavior, where simple physical models are often inadequate to describe the enhancement material parameters. For the remainder of the chapter, we will discuss a brief history, the current state, and prospects of thermoelectric materials research in designed layered materials.

5.6
Transition Metal Dichalcogenide Compounds

The simplest form of composite nanolaminate structures investigated as thermoelectric materials are intercalates of transition metal dichalcogenides. While several systems have been investigated, the titanium-based transition metal dichalcogenides TiS_2 and $TiSe_2$ have been the most promising as host materials and for the properties found in their intercalation compounds [44–48]. Several guest atoms (e.g., Mn, Fe, Co, Ni, Cu, Nd, and Bi) have been used in TiX_2 intercalates, with the best results coming from Cu intercalated compounds at about a 0.1 Cu to Ti ratio. Unfortunately, there has not been much improvement from intercalates beyond finding optimal guest atoms and their doping concentrations, which have not shown enough enhancement for implementation. The highest value of zT reported for these materials is 0.45 in $Cu_{0.1}TiS_2$ at 800 K [46].

5.7
Misfit Layer Compounds

A class of naturally occurring nanolaminate materials, called misfit layer compounds (MLCs), has been investigated with respect to potential thermoelectric applications. These compounds consist of two structures that are interleaved as "sheets" such that the constituents layer along the c-axis of the composite crystal. In order to accommodate the different structures in the nanolaminate material, the structures distort relative to bulk constituents. The resulting structures

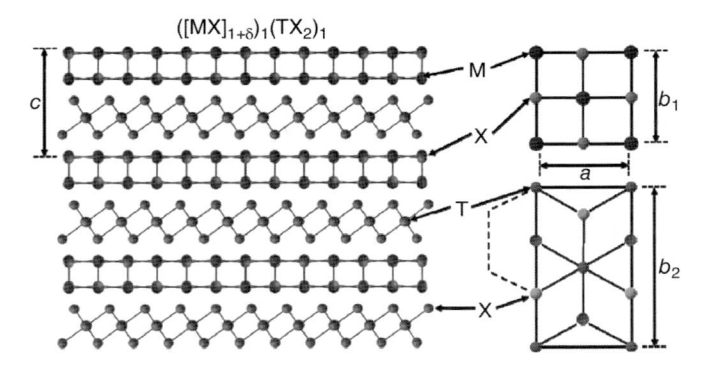

Figure 5.5 Schematic of the structures found in a chalcogenide misfit compound. The basal planes of the rock-salt (MX) and octahedrally coordinated transition metal dichalcogenide (TX$_2$) have a commensurate a-axis, but a distinct b-axes (given by b_1 and b_2, respectively), and the structures are interleaved along the c-axis.

typically have a four-dimensional unit cell, with a commensurate (or common) a-axis between the constituents, incommensurate b-axes (one for each constituent), and a c-axis defined by the layering of the superstructure [49].

There are two main families of these compounds that have been examined as potential thermoelectric materials. The first and the most widely studied are the layered cobalt oxides, which can be considered a subset of the spin-correlated compounds previously mentioned in this chapter [50,51]. The second are chalcogenide composites made of interleaved rock salt and transition metal dichalcogenide structures. The general formula for these materials is $([MX]_{1+\delta})_m(TX_2)_n$, where M = Sn, Pb, Bi or rare earth, T = Ti, V, Cr, Nb, or Ta, and X = S or Se. The subscripts $1 + \delta$, m, and n denote the difference in basal-plane area per cation of the two structures, the number of rock salt bilayers per repeating unit, and the number of dichalcogenide trilayers per repeating unit, respectively. Typically, only the $m = n = 1$ compound can be prepared using high-temperature synthesis techniques. The structure of a chalcogenide MLC is shown in Figure 5.5.

Some of the most notable and thoroughly studied materials in the chalcogenide family are the intergrowths of TiS$_2$ with SnS, PbS, or BiS. The Sn-based materials were found to have the best figure of merit, due to both an increased power factor and a reduced total thermal conductivity relative to the Pb- and Bi-containing compounds [52]. Peak values of zT for pellets pressed from powders of these compounds range from ~0.3 to 0.4, with peak values measured at 700 K [52]. These compounds showed a very low lattice thermal conductivity, with the Bi-containing compounds falling below the theoretical minimum of ~0.5 W/(m K) at and above 350 °C. In addition, hot-pressed pellets of NbS$_2$ and CrS$_2$ compounds interleaved with LaS rock salt layers have shown promising thermoelectric properties with zT values of about 0.15 [53]. Further optimization by further controlling the microstructure, similar to what has been done in the IV–VI dichalcogenide systems, could improve these values.

The interleaved structure of the MLCs has resulted in the literature discussing them as two constituents with weak interactions between the incommensurate layers. However, the traditional high-temperature synthesis approach used to make them from a direct reaction of the constituent elements requires that the MLC is more thermodynamically stable than a bulk mixture of the two constituents. This implies that the interaction between the constituent layers in the MLCs lowers the free energy enough to overcome the truncated structures and incommensurate interface between them. The stabilization interaction has been suggested to be charge transfer between the constituents, which would create a large capacitive energy [54,55]. The rigid band structure picture where the bands of the constituents can be summed to produce the bands of the MLC is certainly an oversimplification of these complicated materials. Understanding the interaction between the constituents and how the bands change as the structures distort is required to control and optimize the band structures of these compounds and to control doping for optimal thermoelectric performance. Due to the method of their synthesis, however, there is little inherent tunability in either composition or layering sequences. To date, all of the MLCs discovered form with $n=1$ or 2 (with one 3) and $m=1$, with the notable exception of $([EuS]_{1.15})_{1.5}(NbS_2)_1$, which forms a trilayer of the rocksalt structure [56]. This is an especially interesting example as it shows a mixed valence rare earth, indicating f-electrons may be available to introduce resonant levels and correlated behavior. Reports of rationally doping misfit layered materials are few and mostly limited to the oxide systems [57,58]. Thus, within the common realms of their synthesis, MLCs do not have the tunability necessary to be functional thermoelectric materials.

5.8
Thin-film Superlattice Materials

Spurred by the work of Hicks and Dresselhaus, the search for low-dimensional confined structures with favorable thermoelectric properties became an active field of research [28,29]. Soon, researchers realized that the inherent anisotropy of superlattices lends itself to the study of properties in two directions – both in-plane and cross-plane relative to the stacking direction of the material. Possible mechanisms for enhancement of the figure of merit are unique along each direction. The cross-plane enhancement of zT in a superlattice would primarily come about by interfaces scattering phonons by reflection while still transmitting electrons. On the other hand, in-plane thermoelectric enhancement can be imagined to be enhanced both in the power factor by band structure changes associated with quantum confinement (see previous section) and in the increased interface scattering of phonons lowering the thermal conductivity.

Almost immediately after the theoretical predictions, promising superlattice compounds began to be reported. Two reports in particular stood out for their large enhancement of zT. After several reports of systematically optimizing Bi_2Te_3–Sb_2Te_3 superlattices from the Venkatasubramanian group at RTI

International, North Carolina, they eventually reported an unprecedented room-temperature zT value of ~2.4 in the material [59]. Harman *et al.* also released a series of papers on PbTe-based superlattices where zT was systematically increased fourfold from the bulk value of ~0.4 to ~1.6 [60]. Unfortunately, neither result has been reproduced despite several efforts [61,62]. Bottner has summarized these and other works on superlattice thermoelectric materials in a short review [63].

The rapid progress in material properties spurred interest in the field, including a quick succession of readily characterizable, thought-provoking structure–property relationships that were developed because materials structures could be varied to test optimization strategies [64–68]. However, it is important to note that these results were confined to materials where epitaxial synthesis was possible. The decades of refinement in various epitaxial growth systems and well-developed growth mechanisms has naturally led to synthesis and optimization of superlattice thermoelectric materials to be confined to thin films within these systems. Little work has been done outside of what is accessible by these means. However, the strong theoretical arguments for enhanced performance and their potential use as a thermoelectric-based on-chip cooler provides motivation for further research in this area.

5.9
van Der Waals Heterostructures

In the past two decades, there has been a surging interest in 2D materials resulting in a Nobel prize in physics for work in graphene [69] and over 15 000 annual publications as properties emerged in 2D layers that were not found in the bulk. While this interest grew out of exceptional properties discovered in graphene, several other 2D and few-layer materials have been discovered or predicted to have properties not found in the bulk. These include planar hexagonal boron nitride, black phosphorous that forms puckered sheets, Bi and Sb chalcogenides, several oxides, including more complex compounds, and several transition metal dichalcogenides [35,70–79]. These low-dimensional materials possess a broad array of unique physical properties not found in the bulk compounds, which result from unique electronic environments, quantum confinement, and surface-to-volume enhancement (by several orders of magnitude). An example of this is MoS_2, where the band structure near the Fermi energy systematically changes as the thickness is reduced down to a single S-Mo-S trilayer [78]. Both calculation and initial results have also found that the thermoelectric properties of isolated layers can be enhanced [35,80–83].

As the major breakthroughs in isolated 2D materials have become less frequent, attention is turning to the development and investigation of vdW heterostructures, where several 2D layers are stacked to form a superstructure. The individual layers exhibit sufficient stability to maintain the distinct layers, but the weak van der Waals bonding between them acts to keep the superstructures intact, hence the name. The promise of these heterostructures is that stacks comprised of two or more complimentary constituents can be potentially created to yield emergent properties that do not exist in the individual constituents. In this way, researchers are actively pursuing new "designed" materials, which truly are, as the old adage says, greater than the sum of their parts. And while the field is new, if given the appropriate 2D building blocks and a method to assemble them into a superstructure, no shortage of big ideas have be conjured up [84].

Heterostructures could be an especially interesting approach to the study of thermoelectric materials, since several designed properties are simultaneously needed for high performance. Building on the ideas from Hicks and Dresselhaus and their suggestions of quantum confinement as an effective means of achieving high power factors, it could be expected that vdW heterostructures would naturally exhibit high zT values by their anisotropic structure alone. However, in addition to possible enhancement by quantum confinement, rationally designed vdW structures may potentially give additional tunability due to the composite properties of the system. For example, if considering designing a heterostructure optimized for in-plane maximization of zT, an unprecedented number of controllable parameters are available. A material or even a composite with a "working" band structure could be chosen that possesses a promising electronic structure (bandgap and shape of $g(E)$) for having a high power factor. A second constituent containing a "distribution" band could then be included. This would be designed to donate or accept carriers as needed for optimized transport in the "working" material, but the structure should maintain high mobility in the working band as no impurities are introduced. An ideal system would not be perturbed by additional constituents, but would instead modulation dope the structure allowing for optimizing materials without perturbing the working band structure. The lack of a strong bonding network along stacking direction should result in inherently low thermal conductivity, which could be further optimized by controlling stacking orders or thicknesses of insulating layers. Alternatively, a third constituent designed to minimize thermal transport could also be added to the heterostructure.

While the potential for van der Waals systems as a platform for composite thermoelectric materials discovery is high, they need to isolate and reassemble several constituent monolayers. This is a considerable synthetic burden, even in a case of pure academic interest. At present, there is a lack of a high-throughput fabrication method and it will be necessary to test large suites of materials to develop the knowledge of how layers in a composite will interact. These together prevent the use of heterostructures as an efficient research vehicle for thermoelectric materials. Previously investigated laminate structures such as misfit layer compounds or dichalcogenide superlattices, however, provide a useful structure and property relationship platform from which the interplay between composition, structure, and properties can be systematically developed.

5.10
Kinetically Trapped Nanolaminates

Recently, a bottom-up synthetic approach has been introduced that can be used to create a large class of materials similar to the chalcogenide misfit layered compounds [85–87]. These materials, called ferecrystals, are based on the chalcogenide MLCs, with the bulk of this family of compounds consisting of interleaved metal chalcogenide rock salt bilayers, MX, and transition metal dichalcogenide trilayers, TX_2. The compounds form over a wide range of material systems, with many of the ternary selenide members such that T = Ti, V, Nb, Mo, Ta, W and M = Sn, Pb, Bi having been reported [88]. The first telluride misfits in the PbTe–$TiTe_2$ system were also prepared using this synthetic approach [89]. Because of the wide range of chemical systems available, a wide range of physical properties have been observed, with transport in materials ranging from superconductivity, n- or p-type metallic to semimetallic to semiconducting.

Ferecrystals are made as thin films by sequentially depositing thin elemental layers from the vapor phase onto a substrate to form an amorphous precursor that, when properly calibrated, has similar composition profiles and nanoarchitecture as the desired final crystalline product [86,90]. With modest heating, this approximate elemental distribution can be kinetically trapped to form a material consisting of crystallographically aligned 2D layers precisely stacked with structurally abrupt interfaces in a layering sequence determined by the local precursor structure. While crystallites are precisely aligned along the c-axis of the composite, they crystallize with an average random in-plane orientation, resembling a systematic layering of 2D powders [88,90,91]. Little registration in the basal planes has been observed between constituents in the structure, leading to a global rotational, or turbostratic, disorder through the compound. This severely impacts phonon propagation in the materials and is structurally distinct relative to MLCs, where a coherent arrangement persists between layers in the composite structure. While the MLCs already have very low thermal conductivities due to the high interface volume ratio of the laminated layers, ferecrystals have even lower thermal conductivities. Disordered WSe_2-layered compounds have the lowest thermal conductivity ever measured perpendicular to the layering direction (~0.05 W/(m K)) [92] with insulating intergrowths being slightly higher along this direction (~0.1 W/(m K)) [93] but lower along the layers (~0.4 W/(m K)) [94]. These measurements were made at room temperature, and in all cases, the values are lower than those of the state-of-the art thermoelectric materials.

Since the sequence of layers in ferecrystals is kinetically controlled by the structure of the precursor, designed composite structures can be prepared. The large breadth of synthetic space afforded by this approach allows for unprecedented accessibility to explore properties of solid-state materials over a wide range of local structures and compositions. As long as materials are synthesized as a superlattice – with several repeating units along the stacking direction – a unique diffraction pattern allows the structure to be determined for the constituent layers that make up the compound. This is highlighted in Figure 5.6 showing the six structural isomers, which have very similar compositions and c-lattice parameters, which can be formed from four repeating units of MX and TX_2. The compounds all have similar c-lattice parameters, given by the positions of the peaks, due to the identical bulk constituents. However, the unique stacking sequences result in unique diffraction patterns as the different locations of atomic planes within the unit cell will scatter with differing intensities. This

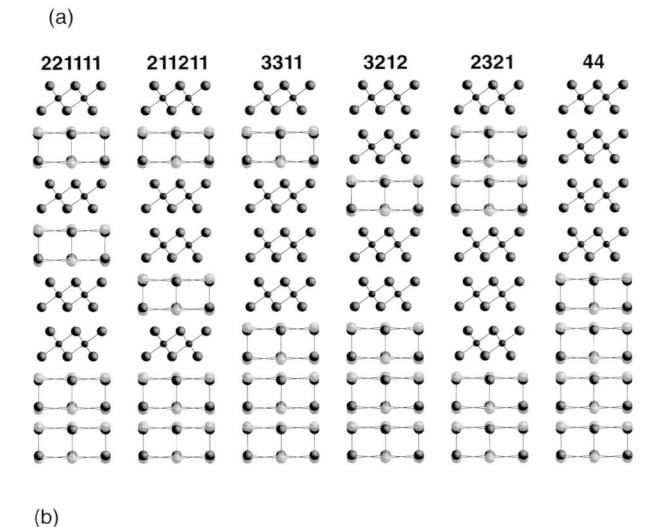

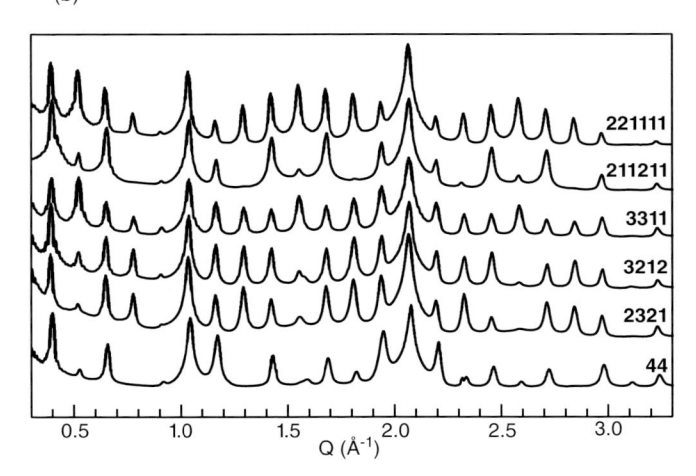

Figure 5.6 Structures of ([PbSe]$_{1+\delta}$)$_4$(TiSe$_2$)$_4$ isomers. (a) The six structural isomers that can be made in the $m=n=4$ layering scheme. (b) Confirmation of consistent c-lattice parameters from samples with these structures, but different peak intensities due to the unique electron density profiles within the superlattice.

approach can be extended further and the number of unique structures, or distinct sequences of layers, is describable by the combinatorics mathematics of a necklace of n beads of m colors. For example, a two-constituent ($m=2$) composite structure comprised of 20 layers ($n=20$), which would be 12 nm thick, could have about 27 000 distinct layering motifs. If this is extended to three constituents ($m=3$), the number of unique structures increases to nearly 90 million, and this goes well into the billions with four constituents ($m=4$). To date, there is nothing to suggest that the synthetic method of kinetically trapping the local structure of a thin-layered precursor into a designed compound could not be applied to each of the cases above in several material systems. Thus, unlike typical hurdles in solid-state chemistry, a point has been reached where the number of metastable compounds that are synthetically accessible vastly exceeds the practical experimental throughput. Developing the theory and intuition of the best layering schemes to test the fundamental interactions in these materials will be a necessary step to move forward with maximum efficiency.

The published body of literature on the chalcogenide misfit compounds universally suggests conduction is localized to the dichalcogenide constituent [49,54]. This behavior is also seen in several experiments in several ferecrystal material systems. Much of the research in these systems also suggests that stabilization of compounds is aided, at least in part, by charge transfer across constituents in the laminate structure. Because of this transfer, layering sequences within a material system may be changed to modulation-dope the layers. Complicated layering structures may be created as necessary for the optimization of material properties, for example, structurally isomeric suites of samples comprised of the same composition and c-lattice parameter, but a distinct layering pattern within the unit cell, such as the $([PbSe]_{1+\delta})_4(TiSe_2)_4$ structural isomers shown above. As long as we do not exceed the depletion width or change the modulation doping efficiency, this may preserve the electronic interactions governed by the integer number of layers of each constituent. Recent work on $([MSe]_{1+\delta})_1(TiSe_2)_n$ ferecrystals [95–99] suggests that they provide a promising platform for systematically making changes to a compound to optimize the figure of merit as discussed next.

5.11 $([MSe]_{1+\delta})_1(TiSe_2)_n$ Ferecrystals

Bulk stoichiometric $TiSe_2$ is a semimetal, but generally the compound forms with a small (2–4%) excess of Ti atoms residing in the van der Waals gaps. The resulting free carrier density is on the order of 10 [21]. The misfit layered compound literature often assumes a rigid band approximation for the interactions of the composite crystal such that the density of states of the laminate structure is simply a superposition of those from each constituent. While the structural distortions and interactions between layers perturb the band structures, this approximation's agreement with data suggests that it is a reasonable initial assumption and it has been made in the TiX_2 MLC literature for both the Pb selenides [100] and the Pb and Sn sulfides [101,102]. In the family of compounds containing $TiSe_2$ dichalcogenide layers, the electrons occupying the Se-4p levels in the rock salt are higher energy than the empty Ti-3d states and charge is donated to the $TiSe_2$ layers, as shown in Figure 5.7. These populated states in the $TiSe_2$ are considered to dominate the charge transport in the composite compound. Evidence for conduction within the $TiSe_2$ layers is given in intergrowths with alloyed rock salts, where compounds with $Pb_xSn_{1-x}Se$ layers show unchanged or even higher mobilities relative to the end members [103]. The donated charge increases the carrier concentration

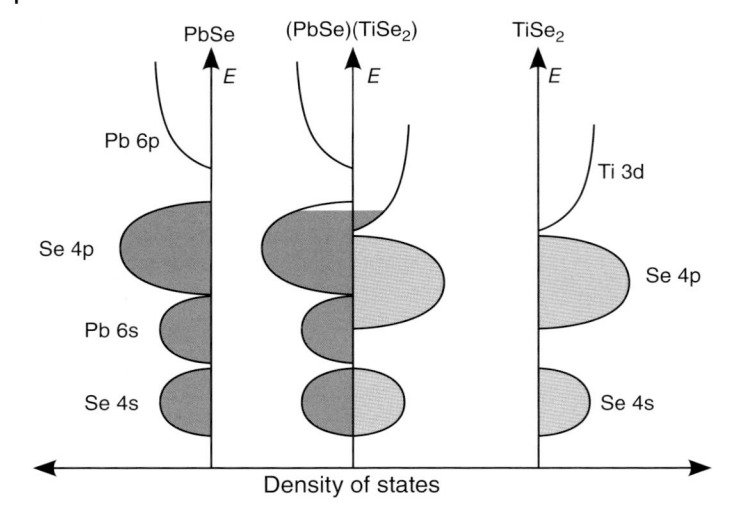

Figure 5.7 Schematic density of states based on electronegativity and coordination environments for $([PbSe]_{1+\delta})_m(TiSe_2)_n$ compounds. The composite band structure is considered to be a superposition of the individual constituents' band structures and conduction is assumed to occur through a single band – in this case by electrons populating the Ti-3d band. In this model, the extent of charge transfer between PbSe and $TiSe_2$ constituents will determine the carrier density.

above the intrinsic values for bulk $TiSe_2$. This simple band-filling picture also suggests that the Bi-containing compounds relative to the isovalent Pb and Sn will have higher carrier concentrations due to more filling of the Ti-3d band. This was observed and discussed both in the n-type $([MS]_{1+\delta})_1(TiS_2)_2$ suite of misfit layered compounds [52] and in the $([MSe]_{1+\delta})_1(TiSe_2)_1$ family of ferecrystal compounds (where M = Sn [97], Pb [95], and Bi [104]) as shown in Figure 5.8. Normalizing these data to the basal plane cation density shows that bismuth substitution results in an increase in carrier concentration of approximately 1 electron per bismuth atom, further agreeing with the rigid band picture and showing high modulation doping efficiency for the $m = n = 1$ structures. This effect appears stable with temperature, though a slight convergence between the Bi and other data sets is observed at low temperatures suggesting more of the donated charge localizes to the Bi layers at low temperatures. The room-temperature Seebeck coefficients of the three $([MSe]_{1+\delta})_1(TiSe_2)_1$ compounds are −75, −66, and −42 μV/K, for Sn, Pb, and Bi, respectively. Calculating the effective carrier mass with the Pisarenko relationship shows the compounds to be similar, with m^* values of 5.8(3) m_e, suggesting the effect in all cases is being dominated at different levels within the same parabolic band in the common $TiSe_2$ constituent.

Especially powerful in these samples for tracking structure–property relationships is the ability to structurally characterize the nanolaminate systems. Rietveld refinement of the 00l composite structure can be performed to obtain planes of atoms in the out-of-plane direction. Structural refinements of the stacking

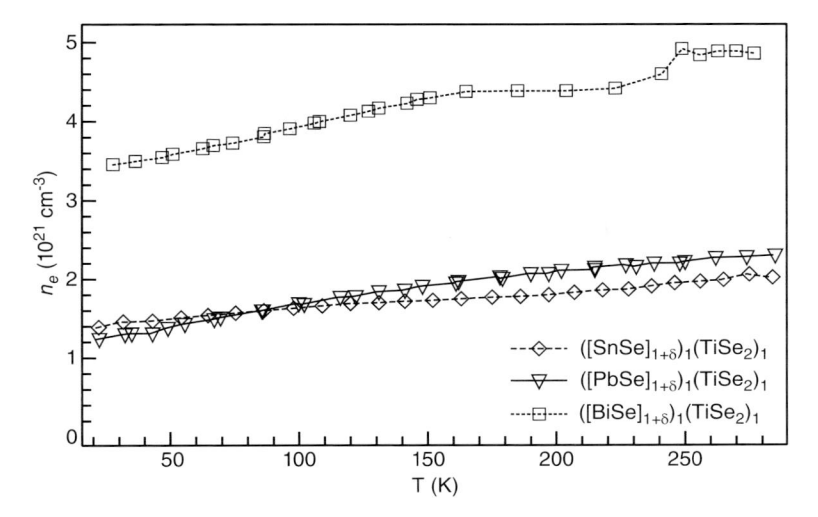

Figure 5.8 Carrier densities as a function of temperature in ($[MSe]_{1+\delta})_1(TiSe_2)_1$ for M= Sn [97], Pb [95], and Bi [104]. The compounds containing isovalent Sn and Pb atoms show very similar carrier concentrations. As expected from the conduction mechanism shown and discussed in Figure 5.7, the compound containing trivalent Bi atoms have an increased carrier concentration of approximately 1 electron per Bi atom.

planes within the $m=n=1$ structures of TiSe$_2$ interleaved with SnSe and PbSe are shown in Figure 5.9 [95,97]. Both compounds exhibit a similar total thickness of the TiSe$_2$ constituent when including the "van der Waals" gap, but the SnSe-containing compound shows a slightly larger Ti–Se distance and a slightly smaller gap between the two structures. A puckering distortion of Se atoms moving inward within the rock salt structure is also observed in both cases; however, the distortion in the SnSe is of much greater magnitude despite the smaller distance between the terminating metal planes. Structural changes can also be

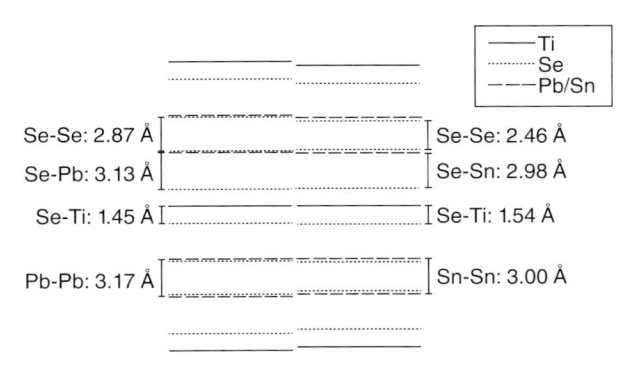

Figure 5.9 Atomic plane positions along the stacking direction of the superlattice in $([MSe]_{1+\delta})_1(TiSe_2)_1$ where M = Sn, Pb as determined from Rietveld refinement. Two unit cells are shown for each superlattice, with the origin at the central Ti plane. Line positions are to scale.

tracked within a material system as the layers are changed, for example, in the ($[BiSe]_{1+\delta})_1(TiSe_2)_n$ ($n = 2–4$) compounds. Here, a systematic increase in the degree of BiSe puckering is observed alongside an offset in Ti planes relative to the Se planes in the TiSe$_2$ trilayers, which trends toward the bulk structures as n is increased [105]. Current work is underway to relate these structural behaviors to transport measurements both within and across homologous series of compounds to build libraries of structure–property relationships in these materials. Future corroboration of band structure calculations will also aid in a deeper understanding of how to optimize nanolaminate materials.

The greater puckering distortion within the SnSe structure alludes to a greater interaction between layers. Some additional insight as to what is gained from the in-plane lattice parameters of the ($[SnSe]_{1+\delta})_1(TiSe_2)_1$ compound relative to other ferecrystal intergrowths. The SnSe a-lattice parameter is much greater when interleaved with TiSe$_2$ relative to the same MX layer in other dichalcogenides [91]. On the other hand, the TiSe$_2$ a-lattice shows less sensitivity as the MX layer is changed and is typically much larger than the TMD lattice in other intergrowths. Projecting the SnSe lattice onto the TiSe$_2$ lattice, as seen in the inset of Figure 5.10, shows that because of a small lattice mismatch, a regular repeating structure along one dimension can be accommodated between constituents. Perhaps, the structures distort due to an energetic gain by this repeating structure,

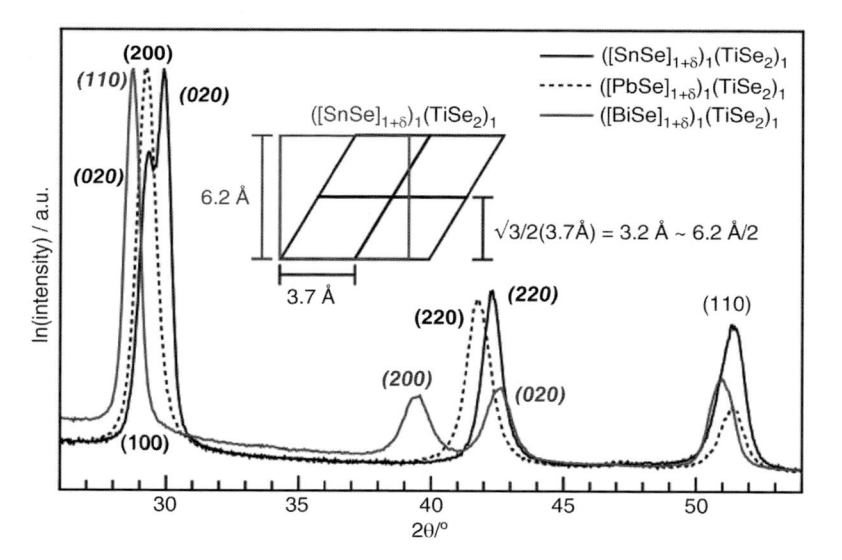

Figure 5.10 In-plane diffraction patterns ($\lambda =$ Cu k_α) of ($[MSe]_{1+\delta})_1(TiSe_2)_1$ for M = Sn, Pb, and Bi adapted from Ref. [91]. The MSe layer indices are shown in bold and match the curve colors, with the SnSe and BiSe indices italicized. TiSe$_2$ peaks are indexed in black and in normal type. The in-plane MSe layers are square for PbSe, slightly tetragonally distorted for SnSe, and highly distorted for BiSe. Also shown in the inset is a schematic of the approximate lattice match of SnSe in TiSe$_2$ in the distorted SnSe layers with lattice parameters taken from Ref. [91].

similar to the structural distortions found in misfit compounds. The Pb- and Bi-containing ($[MSe]_{1+\delta})_1(TiSe_2)_1$ ferecrystals show an entirely different structures, with the PbSe-containing compound having square in-plane rock salt layers and the BiSe-containing compound showing a far more distorted structure than SnSe.

The first step of optimizing typical thermoelectric materials is optimizing the carrier concentration of the existing material by a means that does not severely perturb the band structure. Considering the Pisarenko relationship, which appears valid for these compounds, the synthetic control, and the stability of the $TiSe_2$ bands as the interleaved layers are changed, modulation doping the structure by changing the relative ratio of the layers is a logical step forward. A general trend across the three rock salt systems is that by adding additional $TiSe_2$ layers, the charge donated into the Ti-3d band can be diluted, lowering the carrier concentration and raising both the Seebeck coefficient and the power factor as seen in Figure 5.11 [99,105]. While

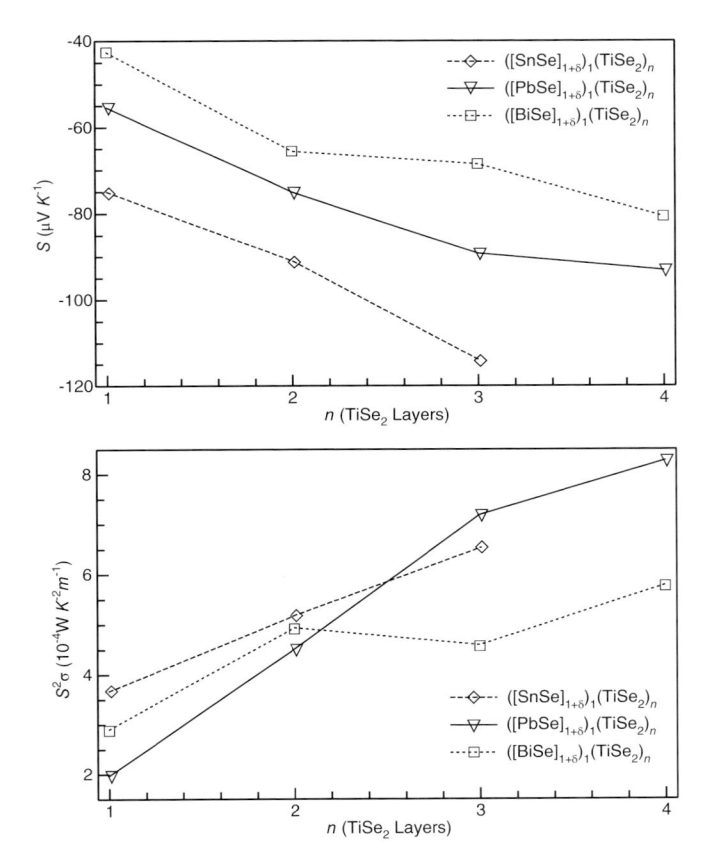

Figure 5.11 Seebeck coefficients and power factors of $([MSe]_{1+\delta})_1(TiSe_2)_n$ for M = Sn, Pb, and Bi. A systematic increase in the magnitude of the Seebeck coefficient is observed for increasing n in all material systems. This also translates to increased power factor, though variation in sample quality, and hence mobility [106], results in slightly more scatter in the data.

thermal conductivity measurements of the structures are necessary for a full validation of the effectiveness, this example illustrates the power of using interlayer interactions within different nanolaminate material systems for enhancing the thermoelectric power factor. However, if we consider the single-band model to be valid, this is still within the classic regime of carrier concentration optimization, albeit by modulation doping. Exploring the more recent strategies of enhancing the power factor in nanolaminate structures by incorporating composite bands, resonant states, or correlated electron behaviors is an available and hopefully fruitful approach for future materials improvement.

Moving forward in the field of nanolaminate thermoelectric materials, combining the synthesis and characterization of strategic sets of compounds with calculations in a feedback loop that quickly develops functional materials has great appeal. Once a more complete understanding of how layers interact is developed, which can be achieved only by synthesis and characterization, these structure–property relationships may be extended by computational work, which could quickly suggest new promising chemical systems and layering sequences. These may then be tested, the theory refined, and new predictions made. In this way, the vast synthetic space afforded by assembling 2D structures may be narrowed to only the most promising suites of compounds, maximizing both throughput and results.

5.12
Outlook

The discussions within this chapter have focused only on the first aspects of developing efficient thermoelectric devices: finding materials with high zT values. However, several additional considerations are necessary: for every promising material found, a countermaterial of opposite carrier type must also be developed for integration into a working device. Each must also be optimized for contact resistances and other parasitic effects within the module and to maximize the efficiency of the couple rather than just the materials themselves. Similarly, each must also share similar zT values and temperature ranges wherein they operate. Furthermore, as devices operate across some temperature difference and zT values are highly dependent on temperature, the average zT value of the device is more critical than the peak values. Thus, within each n-type or p-type leg, segmenting a device into several materials at different points along the operating temperature gradient is often done to achieve the highest device efficiencies. This creates the need to ensure additional compatibility requirements between materials. Depending on the end use, the total device performane might then be expressed as the maximum temperature difference attainable across a thermoelectric cooler or a generator's efficiency, as given by the thermodynamic cycle (Carnot) that governs it.

While the ambitions for an earth abundant in high-performance thermoelectrics for large-scale power conversion may still be a long time coming, the recent advances in materials with high zT values should open up further niche applications for thermoelectric devices, which in turn should hopefully drive further interest in materials and device development. It is rare that one physical phenomenon has the potential to be utilized across such a large scope of human needs. For example, thermoelectric devices have the potential not only to be used as power generators, as the New Horizons spacecraft captures pictures of Pluto in unforeseen detail, but also to recapture considerable energy from waste heat from every industrial cooling tower or automobile exhaust across the world. Similarly, their niche use as automobile seat warmers may be extended to large-scale refrigeration or even cooling to cryogenic temperatures.

Since nearly all of the current state-of-the art materials have thermal conductivities approaching the amorphous limit, reexamining old methods or finding new strategies to enhance the power factor, and specifically the Seebeck coefficient, will be necessary push zT values further. This task has been a far from trivial as shown by the slow progress over half a century of active research. The considerable materials challenge in discovering new compounds and new strategies beyond optimally doping bulk semiconductors will be aided by the synthetic, analytical, and predictive tools that recently have become available. The next decade promises considerable progress!

References

1 Mott, N. and Jones, H. (1936) *The Theory of the Properties of Metals and Alloys*, Oxford University Press, Oxford.
2 Snyder, G.J. and Toberer, E.S. (2008) Complex thermoelectric materials. *Nat. Mater.*, **7** (2), 105–114.
3 Pei, Y., Wang, H., and Snyder, G.J. (2012) Band engineering of thermoelectric materials. *Adv. Mater.*, **24** (46), 6125–6135.
4 Pei, Y., LaLonde, A.D., Wang, H., and Snyder, G.J. (2012) Low effective mass leading to high thermoelectric performance. *Energy Environ. Sci.*, **5** (7), 7963.
5 Pei, Y., Shi, X., LaLonde, A., Wang, H., Chen, L., and Snyder, G.J. (2011) Convergence of electronic bands for high performance bulk thermoelectrics. *Nature*, **473** (7345), 66–69.
6 DiSalvo, F.J. (1999) Thermoelectric cooling and power generation. *Science*, **285** (5428), 703–706.
7 Morelli, D.T., Jovovic, V., and Heremans, J.P. (2008) Intrinsically minimal thermal conductivity in cubic I-V-VI2 semiconductors. *Phys. Rev. Lett.*, **101** (3), 035901.
8 Slack, G.A. (1995) New Materials and performance limits for thermoelectric cooling, in *CRC Handbook of Thermoelectrics* (ed. D.M. Rowe), CRC Press, Boca Raton, FL, USA, pp. 407–440.
9 Kleinke, H. (2010) New bulk materials for thermoelectric power generation: clathrates and complex antimonides. *Chem. Mater.*, **22** (3), 604–611.
10 Toberer, E.S., May, A.F., and Snyder, G.J. (2010) Zintl chemistry for designing high efficiency thermoelectric materials. *Chem. Mater.*, **22** (3), 624–634.
11 Sales, B.C., Mandrus, D., and Williams, R.K. (1996) Filled skutterudite antimonides: a new class of thermoelectric materials. *Science*, **272** (5266), 1325–1328.
12 Shi, X., Yang, J., Salvador, J.R., Chi, M., Cho, J.Y., Wang, H., Bai, S., Yang, J., Zhang, W., and Chen, L. (2011) Multiple-filled skutterudites: high thermoelectric figure of merit through separately optimizing electrical and thermal

transports. *J. Am. Chem. Soc.*, **133** (20), 7837–7846.

13 Snyder, G.J., Christensen, M., Nishibori, E., Caillat, T., and Iversen, B.B. (2004) Disordered zinc in Zn4Sb3 with phonon-glass and electron-crystal thermoelectric properties. *Nat. Mater.*, **3** (7), 458–463.

14 Hsu, K.F., Loo, S., Guo, F., Chen, W., Dyck, J.S., Uher, C., Hogan, T., Polychroniadis, E.K., and Kanatzidis, M.G. (2004) Cubic AgPb$_m$SbTe$_{2+m}$: Bulk thermoelectric materials with high figure of merit. *Science*, **303** (5659), 818–821.

15 Poudeu, P.F.P., D'Angelo, J., Downey, A.D., Short, J.L., Hogan, T.P., and Kanatzidis, M.G. (2006) High thermoelectric figure of merit and nanostructuring in bulk p-type Na$_{1-x}$Pb$_m$SbyTe$_{m+2}$. *Angew. Chem.*, **118** (23), 3919–3923.

16 Biswas, K., He, J., Zhang, Q., Wang, G., Uher, C., Dravid, V.P., and Kanatzidis, M.G. (2011) Strained endotaxial nanostructures with high thermoelectric figure of merit. *Nat. Chem.*, **3** (2), 160–166.

17 Biswas, K., He, J., Wang, G., Lo, S.-H., Uher, C., Dravid, V.P., and Kanatzidis, M.G. (2011) High thermoelectric figure of merit in nanostructured p-type PbTe–MTe (M = Ca, Ba). *Energy Environ. Sci.*, **4** (11), 4675.

18 Biswas, K., He, J., Blum, I.D., Wu, C.-I., Hogan, T.P., Seidman, D.N., Dravid, V.P., and Kanatzidis, M.G. (2012) High-performance bulk thermoelectrics with all-scale hierarchical architectures. *Nature*, **489** (7416), 414–418.

19 Zhao, L.-D., Hao, S., Lo, S.-H., Wu, C.-I., Zhou, X., Lee, Y., Li, H., Biswas, K., Hogan, T.P., Uher, C. *et al.* (2013) High thermoelectric performance via hierarchical compositionally alloyed nanostructures. *J. Am. Chem. Soc.*, **135** (19), 7364–7370.

20 Poudel, B., Hao, Q., Ma, Y., Lan, Y., Minnich, A., Yu, B., Yan, X., Wang, D., Muto, A., Vashaee, D. *et al.* (2008) High-thermoelectric performance of nanostructured bismuth antimony telluride bulk alloys. *Science*, **320** (5876), 634–638.

21 Zhao, L.-D., Lo, S.-H., Zhang, Y., Sun, H., Tan, G., Uher, C., Wolverton, C., Dravid, V.P., and Kanatzidis, M.G. (2014) Ultralow thermal conductivity and high thermoelectric figure of merit in SnSe crystals. *Nature*, **508** (7496), 373–377.

22 Rhyee, J.-S., Lee, K.H., Lee, S.M., Cho, E., Kim, S., Il, Lee, E., Kwon, Y.S., Shim, J.H., and Kotliar, G. (2009) Peierls distortion as a route to high thermoelectric performance in In$_4$Se$_{3-\delta}$ crystals. *Nature*, **459** (7249), 965–968.

23 Heremans, J.P., Wiendlocha, B., and Chamoire, A.M. (2012) Resonant levels in bulk thermoelectric semiconductors. *Energy Environ. Sci.*, **5** (2), 5510–5530.

24 Mahan, G.D. and Sofo, J.O. (1996) The best thermoelectric. *Proc. Natl. Acad. Sci. USA*, **93** (15), 7436–7439.

25 Heremans, J.P., Jovovic, V., Toberer, E.S., Saramat, A., Kurosaki, K., Charoenphakdee, A., Yamanaka, S., and Snyder, G.J. (2008) Enhancement of thermoelectric efficiency in PbTe by distortion of the electronic density of states. *Science*, **321** (5888), 554–557.

26 Pei, Y., LaLonde, A., Iwanaga, S., and Snyder, G.J. (2011) High thermoelectric figure of merit in heavy hole dominated PbTe. *Energy Environ. Sci.*, **4** (6), 2085.

27 Singh, D.J. (2010) Doping-dependent thermopower of PbTe from Boltzmann transport calculations. *Phys. Rev. B*, **81** (19), 195217.

28 Hicks, L.D. and Dresselhaus, M.S. (1993) Effect of quantum-well structures on the thermoelectric figure of merit. *Phys. Rev. B*, **47** (19), 12727–12731.

29 Hicks, L.D. and Dresselhaus, M.S. (1993) Thermoelectric figure of merit of a one-dimensional conductor. *Phys. Rev. B*, **47** (24), 16631–16634.

30 Hicks, L.D., Harman, T.C., Sun, X., and Dresselhaus, M.S. (1996) Experimental study of the effect of quantum-well structures on the thermoelectric figure of merit. *Phys. Rev. B*, **53** (16), R10493–R10496.

31 Harman, T.C., Spears, D.L., and Manfra, M.J. (1996) High thermoelectric figures of merit in PbTe quantum wells. *J. Electron. Mater.*, **25** (7), 1121–1127.

32 Heremans, J. and Thrush, C.M. (1999) Thermoelectric power of bismuth

nanowires. *Phys. Rev. B*, **59** (19), 12579–12583.

33 Rabina, O., Lin, Y.-M., and Dresselhaus, M.S. (2001) Anomalously high thermoelectric figure of merit in $Bi_{1-x}Sb_x$ nanowires by carrier pocket alignment. *Appl. Phys. Lett.*, **79** (1), 81.

34 Heremans, J.P., Thrush, C.M., Morelli, D.T., and Wu, M.-C. (2002) Thermoelectric power of bismuth nanocomposites. *Phys. Rev. Lett.*, **88** (21), 216801.

35 Teweldebrhan, D., Goyal, V., and Balandin, A.A. (2010) Exfoliation and characterization of bismuth telluride atomic quintuples and quasi-two-dimensional crystals. *Nano Lett.*, **10** (4), 1209–1218.

36 Touzelbaev, M.N., Zhou, P., Venkatasubramanian, R., and Goodson, K.E. (2001) Thermal characterization of Bi_2Te_3/Sb_2Te_3 superlattices. *J. Appl. Phys.*, **90** (2), 763.

37 Winkler, M., Liu, X., König, J.D., Buller, S., Schürmann, U., Kienle, L., Bensch, W., and Böttner, H. (2012) Electrical and structural properties of Bi_2Te_3 and Sb_2Te_3 thin films grown by the nanoalloying method with different deposition patterns and compositions. *J. Mater. Chem.*, **22** (22), 11323.

38 Mahan, G.D. (1997) Rare earth thermoelectrics. In XVI ICT '97. Proceedings ICT'97. 16th International Conference on Thermoelectrics (Cat. No. 97TH8291); IEEE, pp 21–27.

39 Mahan, G.D. (1997) Good thermoelectrics. *Solid State Phys.*, **51**, 81–157.

40 Rowe, D.M., Kuznetsov, V.L., Kuznetsova, L.A., and Min, G. (2002) Electrical and thermal transport properties of intermediate-valence $YbAl_3$. *J. Phys. D*, **35** (17), 2183–2186.

41 Lee, S.J., Park, J.M., Canfield, P.C., and Lynch, D.W. (2003) Optical properties and electronic structures of single crystalline R Al_3 (R = Sc, Yb, and Lu). *Phys. Rev. B*, **67** (7), 075104.

42 Bentien, A., Johnsen, S., Madsen, G.K.H., Iversen, B.B., and Steglich, F. (2007) Colossal seebeck coefficient in strongly correlated semiconductor $FeSb_2$. *Europhys. Lett.*, **80** (1), 17008.

43 Wang, Y., Rogado, N.S., Cava, R.J., and Ong, N.P. (2003) Spin entropy as the likely source of enhanced thermopower in $Na_xCo_2O_4$. *Nature*, **423** (6938), 425–428.

44 Imai, H., Shimakawa, Y., and Kubo, Y. (2001) Large thermoelectric power factor in TiS_2 crystal with nearly stoichiometric composition. *Phys. Rev. B*, **64** (24), 241104.

45 Gaby, J.H., DeLong, B., Brown, F.C., Kirby, R., and Lévy, F. (1981) Origin of the structural transition in $TiSe_2$. *Solid State Commun.*, **39** (11), 1167–1170.

46 Guilmeau, E., Bréard, Y., and Maignan, A. (2011) Transport and thermoelectric properties in copper intercalated TiS_2 chalcogenide. *Appl. Phys. Lett.*, **99** (5), 052107.

47 Bhatt, R., Basu, R., Bhattacharya, S., Singh, A., Aswal, D.K., Gupta, S.K., Okram, G.S., Ganesan, V., Venkateshwarlu, D., Surgers, C. et al. (2013) Low temperature thermoelectric properties of Cu intercalated $TiSe_2$: a charge density wave material. *Appl. Phys. A*, **111** (2), 465–470.

48 Bhatt, R., Bhattacharya, S., Patel, M., Basu, R., Singh, A., Sürger, C., Navaneethan, M., Hayakawa, Y., Aswal, D.K., and Gupta, S.K. (2013) Thermoelectric performance of Cu intercalated layered $TiSe_2$ above 300 K. *J. Appl. Phys.*, **114** (11), 114509.

49 Wiegers, G. (1996) Misfit layer compounds: structures and physical properties. *Prog. Solid State Chem.*, **24** (1–2), 1–139.

50 Koumoto, K., Terasaki, I., and Funahashi, R. (2011) Complex oxide materials for potential thermoelectric applications. *MRS Bull.*, **31** (03), 206–210.

51 Limelette, P., Hébert, S., Hardy, V., Frésard, R., Simon, C., and Maignan, A. (2006) Scaling behavior in thermoelectric misfit cobalt oxides. *Phys. Rev. Lett.*, **97** (4), 046601.

52 Wan, C., Wang, Y., Wang, N., and Koumoto, K. (2010) Low-thermal-conductivity $(MS)_{1+x}(TiS_2)_2$ (M = Pb, Bi, Sn) misfit layer compounds for bulk

thermoelectric materials. *Materials (Basel)*, **3** (4), 2606–2617.

53 Jood, P., Ohta, M., Nishiate, H., Yamamoto, A., Lebedev, O.I., Berthebaud, D., Suekuni, K., and Kunii, M. (2014) Microstructural control and thermoelectric properties of misfit layered sulfides $(LaS)_{1+m}TS_2$ (T = Cr, Nb): the natural superlattice systems. *Chem. Mater.*, **26** (8), 2684–2692.

54 Wiegers, G.A. (1995) Charge transfer between layers in misfit layer compounds. *J. Alloys Compd.*, **219** (1–2), 152–156.

55 Fang, C.M., Ettema, A.R.H.F., Haas, C., Wiegers, G.A., van Leuken, H., and de Groot, R.A. (1995) Electronic structure of the misfit-layer compound $(SnS)_{1.17}NbS_2$ deduced from band-structure calculations and photoelectron spectra. *Phys. Rev. B*, **52** (4), 2336–2347.

56 Cario, L., Lafond, A., Palvadeau, P., Deudon, C., and Meerschaut, A. (1999) Evidence of a mixed-valence state for europium in the misfit layer compound $[(EuS)_{1.5}]_{1.15}NbS_2$ by means of a superspace structural determination, Mössbauer spectroscopy, and magnetic measurements. *J. Solid State Chem.*, **147** (1), 58–67.

57 Nong, N.V., Liu, C.-J., and Ohtaki, M. (2010) Improvement on the high temperature thermoelectric performance of Ga-doped misfit-layered $Ca_3Co_{4-x}Ga_xO_{9+\delta}$ ($x = 0$, 0.05, 0.1, and 0.2). *J. Alloys Compd.*, **491** (1–2), 53–56.

58 Shen, J.J., Liu, X.X., Zhu, T.J., and Zhao, X.B. (2009) Improved thermoelectric properties of La-doped $Bi_2Sr_2Co_2O_9$-layered misfit oxides. *J. Mater. Sci.*, **44** (7), 1889–1893.

59 Venkatasubramanian, R., Siivola, E., Colpitts, T., and O'Quinn, B. (2001) Thin-film thermoelectric devices with high room-temperature figures of merit. *Nature*, **413** (6856), 597–602.

60 Harman, T.C., Taylor, P.J., Walsh, M.P., and LaForge, B.E. (2002) Quantum dot superlattice thermoelectric materials and devices. *Science*, **297** (5590), 2229–2232.

61 Winkler, M., Liu, X., Schürmann, U., König, J.D., Kienle, L., Bensch, W., and Böttner, H. (2012) Current status in fabrication, structural and transport property characterization, and theoretical understanding of Bi_2Te_3/Sb_2Te_3 superlattice systems. *Zeit. Anorg. Allg. Chem.*, **638** (15), 2441–2454.

62 Hansen, A.-L., Dankwort, T., Winkler, M., Ditto, J., Johnson, D.C., Koenig, J.D., Bartholomé, K., Kienle, L., and Bensch, W. (2014) Synthesis and thermal instability of high-quality Bi_2Te_3/Sb_2Te_3 superlattice thin film thermoelectrics. *Chem. Mater.*, **26** (22), 6518–6522.

63 Böttner, H., Chen, G., and Venkatasubramanian, R. (2011) Aspects of thin-film superlattice thermoelectric materials, devices, and applications. *MRS Bull.*, **31** (03), 211–217.

64 Cui, H., Bhat, I., O'Quinn, B., and Venkatasubramanian, R. (2001) In-situ monitoring of the growth of Bi_2Te_2 and Sb_2Te_3 films and Bi_2Te_3–Sb_2Te_3 superlattice using spectroscopic ellipsometry. *J. Electron. Mater.*, **30** (11), 1376–1381.

65 Peranio, N., Eibl, O., and Nurnus, J. (2006) Structural and thermoelectric properties of epitaxially grown Bi_2Te_3 thin films and superlattices. *J. Appl. Phys.*, **100** (11), 114306.

66 Wang, G., Endicott, L., and Uher, C. (2011) Recent advances in the growth of Bi–Sb–Te–Se thin films. *Sci. Adv. Mater.*, **3** (4), 539–560.

67 Banga, D., Lensch-Falk, J.L., Medlin, D.L., Stavila, V., Yang, N.Y.C., Robinson, D.B., and Sharma, P.A. (2012) Periodic modulation of Sb stoichiometry in $Bi_2Te_3/Bi_{2-x}Sb_xTe_3$ multilayers using pulsed electrodeposition. *Cryst. Growth Des.*, **12** (3), 1347–1353.

68 Zeipl, R., Walachová, J., Pavelka, M., Jelínek, M., Studnička, V., and Kocourek, T. (2008) Power factor of very thin thermoelectric layers of different thickness prepared by laser ablation. *Appl. Phys. A*, **93** (3), 663–667.

69 Nature Materials (2011) Editorial: It's still all about graphene. *Nat. Mater.*, **10** (1), 1.

70 Novoselov, K.S., Geim, A.K., Morozov, S.V., Jiang, D., Zhang, Y., Dubonos, S.V, Grigorieva, I.V., and Firsov, A.A. (2004) Electric field effect in atomically thin

carbon films. *Science*, **306** (5696), 666–669.

71 Pacilé, D., Meyer, J.C., Girit, C.O., and Zettl, A. (2008) The two-dimensional phase of boron nitride: few-atomic-layer sheets and suspended membranes. *Appl. Phys. Lett.*, **92** (13), 133107.

72 Xia, F., Wang, H., and Jia, Y. (2014) Rediscovering black phosphorus as an anisotropic layered material for optoelectronics and electronics. *Nat. Commun.*, **5**, 4458.

73 Min, Y., Moon, G.D., Kim, B.S., Lim, B., Kim, J.-S., Kang, C.Y., and Jeong, U. (2012) Quick, controlled synthesis of ultrathin Bi_2Se_3 nanodiscs and nanosheets. *J. Am. Chem. Soc.*, **134** (6), 2872–2875.

74 Mas-Ballesté, R., Gómez-Navarro, C., Gómez-Herrero, J., and Zamora, F. (2011) 2D Materials: to graphene and beyond. *Nanoscale*, **3** (1), 20–30.

75 Huang, X., Zeng, Z., and Zhang, H. (2013) Metal dichalcogenide nanosheets: preparation, properties and applications. *Chem. Soc. Rev.*, **42** (5), 1934–1946.

76 Coleman, J.N., Lotya, M., O'Neill, A., Bergin, S.D., King, P.J., Khan, U., Young, K., Gaucher, A., De, S., Smith, R.J. et al. (2011) Two-dimensional nanosheets produced by liquid exfoliation of layered materials. *Science*, **331** (6017), 568–571.

77 Zeng, Z., Sun, T., Zhu, J., Huang, X., Yin, Z., Lu, G., Fan, Z., Yan, Q., Hng, H.H., and Zhang, H. (2012) An effective method for the fabrication of few-layer-thick inorganic nanosheets. *Angew. Chem. Int. Ed. Engl.*, **51** (36), 9052–9056.

78 Splendiani, A., Sun, L., Zhang, Y., Li, T., Kim, J., Chim, C.-Y., Galli, G., and Wang, F. (2010) Emerging photoluminescence in monolayer MoS_2. *Nano Lett.*, **10** (4), 1271–1275.

79 Radisavljevic, B., Radenovic, A., Brivio, J., Giacometti, V., and Kis, A. (2011) Single-layer MoS_2 transistors. *Nat. Nanotechnol.*, **6** (3), 147–150.

80 Goyal, V., Teweldebrhan, D., and Balandin, A.A. (2010) Mechanically-exfoliated stacks of thin films of Bi_2Te_3 topological insulators with enhanced thermoelectric performance. *Appl. Phys. Lett.*, **97** (13), 133117.

81 Wu, J., Schmidt, H., Amara, K.K., Xu, X., Eda, G., and Özyilmaz, B. (2014) Large thermoelectricity via variable range hopping in chemical vapor deposition grown single-layer MoS_2. *Nano Lett.*, **14** (5), 2730–2734.

82 Fei, R., Faghaninia, A., Soklaski, R., Yan, J.-A., Lo, C., and Yang, L. (2014) Enhanced thermoelectric efficiency via orthogonal electrical and thermal conductances in phosphorene. *Nano Lett.*, **14** (11), 6393–6399.

83 Huang, W., Da, H., and Liang, G. (2013) Thermoelectric performance of MX_2 (M = Mo,W; X = S, Se) monolayers. *J. Appl. Phys.*, **113** (10), 104304.

84 Geim, A.K. and Grigorieva, I.V. (2013) Van der Waals heterostructures. *Nature*, **499** (7459), 419–425.

85 Lin, Q., Smeller, M., Heideman, C.L., Zschack, P., Koyano, M., Anderson, M.D., Kykyneshi, R., Keszler, D.A., Anderson, I.M., and Johnson, D.C. (2009) Rational synthesis and characterization of a new family of low thermal conductivity misfit layer compounds $[(PbSe)_{0.99}]$ M (WSe_2) N^+. *Chem. Mater.*, **22** (3), 1002–1009.

86 Atkins, R., Wilson, J., Zschack, P., Grosse, C., Neumann, W., and Johnson, D.C. (2012) Synthesis of $[(SnSe)_{1.15}]$ M $(TaSe_2)$ N ferecrystals: structurally tunable metallic compounds. *Chem. Mater.*, **24** (23), 4594–4599.

87 Beekman, M., Cogburn, G., Heideman, C., Rouvimov, S., Zschack, P., Neumann, W., and Johnson, D.C. (2012) New layered intergrowths in the Sn–Mo–Se system. *J. Electron. Mater.*, **41** (6), 1476–1480.

88 Beekman, M., Heideman, C.L., and Johnson, D.C. (2014) Ferecrystals: non-epitaxial layered intergrowths. *Semicond. Sci. Technol.*, **29** (6), 064012.

89 Moore, D.B., Beekman, M., Disch, S., and Johnson, D.C. (2014) Telluride misfit layer compounds: $[(PbTe)_{1.17}]m(TiTe_2)n$. *Angew. Chem. Int. Ed. Engl.*, **53** (22), 5672–5675.

90 Heideman, C.L., Tepfer, S., Lin, Q., Rostek, R., Zschack, P., Anderson, M.D., Anderson, I.M., and Johnson, D.C. (2013) Designed synthesis, structure, and properties of a family of ferecrystalline

compounds [(PbSe)$_{1.00}$]$_m$(MoSe$_2$)$_n$. *J. Am. Chem. Soc.*, **135** (30), 11055–11062.

91 Falmbigl, M., Alemayehu, M.B., Merrill, D.R., Beekman, M., and Johnson, D.C. (2015) In-plane structure of ferecrystalline compounds. *Cryst. Res. Technol.*, **50** (6), 464–472.

92 Chiritescu, C., Cahill, D.G., Nguyen, N., Johnson, D., Bodapati, A., Keblinski, P., and Zschack, P. (2007) Ultralow thermal conductivity in disordered, layered WSe$_2$ crystals. *Science*, **315** (5810), 351–353.

93 Chiritescu, C., Cahill, D.G., Heideman, C., Lin, Q., Mortensen, C., Nguyen, N.T., Johnson, D., Rostek, R., and Böttner, H. (2008) Low thermal conductivity in nanoscale layered materials synthesized by the method of modulated elemental reactants. *J. Appl. Phys.*, **104** (3), 033533.

94 Mavrokefalos, A., Lin, Q., Beekman, M., Seol, J.H., Lee, Y.J., Kong, H., Pettes, M.T., Johnson, D.C., and Shi, L. (2010) In-plane thermal and thermoelectric properties of misfit-layered [(PbSe)$_{0.99}$]$_x$(WSe$_2$)$_x$ superlattice thin films. *Appl. Phys. Lett.*, **96** (18), 181908.

95 Moore, D.B., Jones, Z., Stolt, M.J., Atkins, R., and Johnson, D.C. (2012) Structural and electrical properties of (PbSe)$_{1.16}$TiSe$_2$. *Emerg. Mater. Res.*, **1** (6), 292–298.

96 Moore, D.B., Beekman, M., Disch, S., Zschack, P., Hausler, I., Neumann, W., and Johnson, D.C. (2013) Synthesis, structure, and properties of turbostratically disordered (PbSe)$_{1.18}$(TiSe$_2$)$_2$. *Chem. Mater.*, **25** (12), 2404–2409.

97 Merrill, D.R., Moore, D.B., Ditto, J., Sutherland, D.R., Falmbigl, M., Winkler, M., Pernau, H.-F., and Johnson, D.C. (2014) The synthesis, structure, and electrical characterization of (SnSe)$_{1.2}$TiSe$_2$. *Eur. J. Inorg. Chem.* doi: 10.1002/ejic.201402814

98 Merrill, D., Moore, D., Bauers, S., Falmbigl, M., and Johnson, D. (2015) Misfit layer compounds and ferecrystals: model systems for thermoelectric nanocomposites. *Materials (Basel)*, **8** (4), 2000–2029.

99 Bauers, S.R., Merrill, D.R., Moore, D.B., and Johnson, D.C. (2015) Carrier dilution in TiSe$_2$ based intergrowth compounds for enhanced thermoelectric performance. *J. Mater. Chem. C*, **3** (40), 10451–10458.

100 Giang, N., Xu, Q., Hor, Y.S., Williams, A.J., Dutton, S.E., Zandbergen, H.W., and Cava, R.J. (2010) Superconductivity at 2.3 K in the misfit compound (PbSe)$_{1.16}$(TiSe$_2$)$_2$. *Phys. Rev. B*, **82** (2), 24503.

101 Ohno, Y. (1991) Electronic structure of the misfit-layer compounds PbTiS$_3$ and SnNbS$_3$. *Phys. Rev. B*, **44** (3), 1281–1291.

102 Wiegers, G.A. and Haange, R.J. (1991) Electrical transport properties of the misfit layer compounds (SnS)$_{1.20}$TiS$_2$ and (PbS)$_{1.18}$TiS$_2$. *Eur. J. Solid State Inorg. Chem.*, **28** (5), 1071–1078.

103 Merrill, D.R., Sutherland, D.R., Ditto, J., Bauers, S.R., Falmbigl, M., Medlin, D.L., and Johnson, D.C. (2015) Kinetically controlled site-specific substitutions in higher-order heterostructures. *Chem. Mater.*, **27** (11), 4066–4072.

104 Merrill, D.R., Moore, D.B., Coffey, M.N., Jansons, A.W., Falmbigl, M., and Johnson, D.C. (2014) Synthesis and characterization of turbostratically disordered (BiSe)$_{1.15}$TiSe$_2$. *Semicond. Sci. Technol.*, **29** (6), 064004.

105 Wood, S.R., Merrill, D.R., Falmbigl, M., Moore, D.B., Ditto, J., Esters, M., and Johnson, D.C. (2015) Tuning electrical properties through control of TiSe$_2$ thickness in (BiSe)$_{1+\delta}$(TiSe$_2$)N compounds. *Chem. Mater.*, **27** (17), 6067–6076.

106 Bauers, S.R., Moore, D.B., Ditto, J., and Johnson, D.C. (2015) Phase width of kinetically stable ([PbSe]$_{1+\delta}$)$_1$(TiSe$_2$)$_1$ ferecrystals and the effect of precursor composition on electrical properties. *J. Alloys Compd.*, **645**, 118–124.

6
Magnetically Responsive Photonic Nanostructures for Display Applications

Mingsheng Wang and Yadong Yin

University of California, Department of Chemistry, 501 Big Springs Road, Riverside, CA 92521, USA

6.1
Introduction

Display technology is a truly interdisciplinary field [1]. The evolution of display technology has been pioneered primarily by the desire of general public for entertainment, as it provides people with a virtual window to a remote scene and also satisfies their need for observing distant places and different times without actually being there. The simplest prototype of display that comes to people's mind is probably a matrix of pixels with varying light intensity. If these pixels are small enough and dense enough, they will blend into an image. And if these images are exchanged fast enough, they will merge into a moving scene. With rapid development in the past few decades, display technology has seen significant improvement and people are now able to choose from a large variety of display products with great features: vivid colors, high contrasts, and high refresh rates.

Generally, displays can be divided into emissive and nonemissive types [2]. For emissive displays, each pixel emits light with different intensity and color that stimulates the human eye directly, with examples including cathode ray tubes (CRTs) [3], plasma display panels (PDPs) [4], light-emitting diodes (LEDs) [5], and field emission displays (FEDs) [6]. Owing to the self-emissive properties, they can be used under very low ambient light. They become completely dark after being turned off and hence produce high contrast ratios. On the other hand, nonemissive displays do not emit light themselves. A representative example is the widely commercialized liquid crystal displays (LCDs), in which the liquid crystal molecules in each pixel work as an independent light modulator [7]. The external electric signals reorient the liquid crystal molecules, induce phase retardations, and hence modulate the incident light. The light can be from either a backlight unit or ambience. When the ambient light is used as the light source, the power consumption can be significantly reduced as no backlight is needed [8]. Moreover, the nonemissive displays could have outstanding performance for outdoor applications, as they exhibit an even higher luminance in a very bright environment, while the information in most emissive displays could be washed out under the same condition.

Successful functioning of nonemissive displays requires effective modulation of lights, via the fabrication and design of materials with novel optical properties [9]. When light encounters a material surface or passes through a material medium, interactions between the light and the electrons of the atoms and molecules of the material occur [10]. Generally, the basic process of the interaction of light with matter can be described by means of quantum theory: the electron of an atom, a molecule, or an atomic lattice can absorb a photon and use its energy to jump into an energetically higher state [11]. Conversely, an electron can fall into a state of lower energy, with the energy difference being sent out as a photon. Actually, it is usually not only a single electron but also the whole electron shell of an atom or molecule or their assemblies that experience a change of state in these processes. For example, scattering means that a photon is absorbed and immediately emitted again. The absorbed energy equals the emitted energy and, as a result, it does not change the wavelength of the light. The emitted light is uniformly distributed in all directions in most cases. However, for a smooth piece of a transparent solid where the atoms are arranged in an orderly pattern and are very close to each other, the scattering of light happens mainly in the backward and forward directions. Light scattered in the opposite direction of incident light is named as reflected light, while light scattered in the forward direction combines with the incident beam to give rise to the phenomenon of refraction. The physical effect of this combination is to make the transmitted light appear as though it has traveled more slowly through the materials. In the case that the photon is not emitted, its energy is converted into vibrations of the materials, and as a result the absorption of light occurs.

New methods of light modulation have been intensively investigated [9d,12] and have been greatly propelled by the development of new materials that interact with lights in a large variety of manners, among which magnetically responsive materials have attracted lots of interests and have been considered important candidates for optical modulations due to their several unique advantages [13]. First, the behaviors of magnetic materials are controlled by magnetic dipole–dipole interactions and are, therefore, sensitive to the strength and direction of the external fields, thus allowing significant freedom for experimental design. Second, magnetic dipole–dipole interactions can be effectively initiated by external magnetic fields, providing sufficient driving forces for enabling instant modulation. Third, magnetic interactions could act at a distance, enabling contactless manipulation of magnetically responsive materials, and are largely independent of the changes in environmental conditions, such as temperature and humidity. More importantly, when the strengths and spatial distributions of magnetic fields can be programmed precisely, one can achieve fine control over the light modulation.

In this chapter, we present the recent developments of the most fascinating theme about light modulation by magnetic means. We start with a brief introduction of the general behavior of magnetic materials under external magnetic fields, and then demonstrate the design and fabrication of magnetically responsive photonic structures, specifically with dimensions in the range from several nanometers to a few of micrometers. The ability of these responsive structures for the modulation of the intensity, wavelength, and phase of light via the interaction mechanism of different light materials is discussed. By taking advantage of the unique features of the magnetically responsive system, we highlight their unique practical applications in various types of displays.

6.2
General Principle of Magnetic Interactions

The ability to use magnetic fields to control the behavior of magnetically responsive materials is crucial to the design of magnetically responsive displays. Generally, magnetic particles experience various types of magnetic interactions in external magnetic fields [13]. Upon the application of external magnetic fields, they are magnetized with an induced magnetic dipole moment $m = \chi H V$, where χ is the volume susceptibility of the particle, H is the local magnetic field, and V is the volume of the particle. For a particle with a magnetic moment of m, its induced magnetic field H_1 felt by another particle can be described as

$$H_1 = [3(m \cdot r)r - m]/d^3,$$

where r is the unit vector parallel to the line pointed from the center of the first particle to that of the second one and d is the center-to-center distance. The dipole–dipole interaction energy of the second particle with the same magnetic moment m can thus be written as

$$U_2 = m \cdot H_1 = (3\cos^2\theta - 1)m^2/d^3,$$

where θ ranging from 0 to 90°, is the angle between the direction of external magnetic field and the line connecting the center of the two particles. The dipole force exerted on the second particle induced by the first particle can be expressed as

$$F_{21} = \nabla(m \cdot H_1) = 3r(1 - 3\cos^2\theta)m^2/d^4,$$

which clearly shows the dependence of the dipole–dipole force on the configuration of the two dipoles. At the critical angle of 54.09°, the interaction approaches zero. The dipole–dipole interaction is attractive when $0° \leq \alpha < 54.09°$ and repulsive in cases where $54.09° < \alpha \leq 90°$. When the interaction energy is large enough to overcome thermal fluctuations, the magnetic dipole–dipole force drives the self-assembly of particles into one-dimensional chain-like structures along the dipole moment.

Practically, the magnetic fields for experimental use are generated by a permanent magnet or an electromagnet and are, therefore, not always homogeneous. Magnetic particles in a nonhomogeneous magnetic field will experience a force along the direction of increasing field strength. The magnitude of this magnetophoretic force is monotonically dependent on the difference between the magnetic susceptibility of the particles and that of their surroundings. If the difference is large enough, the magnetophoretic force can be approximated as $F_{phr} = \nabla B^2 \cdot V\chi/2\mu_0$. It drives the movement of particles toward regions with the maximum magnetic field strength and subsequently induces a particle concentration gradient or crystallization.

Magnetic shape anisotropy can also play a very important role in the magnetization of particles of anisotropic shapes [14]. A magnetized particle has induced magnetic poles on its surface, which produces a demagnetizing field in the opposing direction to the magnetization, $B_{demag} = -\mu_0 N \cdot m$, where N is the demagnetizing tensor and m is the induced magnetic moments. Then, the energy of the particle in its own demagnetizing field is given by the integral:

$$E_{demag} = -\frac{1}{2}\int B_{demag} \cdot m dV = \frac{1}{2}\int \mu_0(N \cdot m) \cdot m dV.$$

For an infinitely long cylinder, the expression can be simplified as follows:

$$E_{demag} = \frac{1}{4}\mu_0 m^2 \sin^2\theta,$$

where θ is the angle between the major axis of the cylinder and the field direction. As a result, anisotropic particles spontaneously align their major axis parallel to the direction of external magnetic fields, in order to reduce both the demagnetizing field and the associated energy [15].

For an infinitely expanded and/or very thin ellipsoid, the expression can be simplified as follows:

$$E_{demag} = \frac{1}{2}\mu_0 m^2 \cos^2\theta,$$

where θ is the angle between the major axis of the thin ellipsoid and the field direction. As a result, an in-plane orientation of magnetic moment of thin ellipsoid is energetically favorable, and it spontaneously aligns their plane parallel to the direction of external magnetic fields, in order to reduce both the demagnetizing field and the energy associated with it [16].

6.3
Magnetic Tuning of Light Scattering and Reflection

The scattering by particles with diameters on the order of light wavelengths or larger is known as Mie scattering [17]. As the fast development of nanoscience and nanotechnology, the Mie scattering of nanoparticles has attracted a lot of interest. The degree to which a nanomaterial can scatter incident light is described through its scattering cross section, which is very sensitive to the structure, molecular composition, size, shape, and orientation of the nanomaterial.

Magnetic nanoparticles suspended in a carrier fluid have been intensively studied for their potential use in light transmission modulation [18]. The common idea in these applications is to use external stimulus, such as a magnetic field, to coax and assemble magnetic nanostructures into ordered or aligned systems in a reversible way, which results in the change of their scattering cross section. A simple demonstration for this idea is the use of $Fe_3O_4/SiO_2/TiO_2$ peapod-like nanostructures for magnetically controlled light intensity modulation and chopping [19]. The peapod-like nanostructures were synthesized by a multi-step process. Fe_3O_4 nanoclusters were first synthesized and coated with a layer of SiO_2. Then the $Fe_3O_4@SiO_2$ core/shell nanoparticles were aligned into chain-like intermediate structures by an external magnetic field. These intermediate structures were finally stabilized by being encapsulated in a TiO_2 tubular shell, and their morphology was confirmed by TEM characterization.

In a typical light modulation experiment, an aqueous solution containing the peapod-like nanostructures is placed in a cylindrical glass cell. The glass cell is positioned inside a solenoid that is driven by an audio amplifier that in turn is driven by a function generator. Alternating linearly polarized magnetic fields of a few milliteslas magnitude were facilely obtained at the center of the solenoid with tunable frequency capability. A constant intensity-polarized laser beam is directed at the sample, goes through the sample, and is then detected by a photodetector that in turn is monitored by an oscilloscope. A typical result of the optical intensity modulation under an alternating magnetic field with a strength of 4.5 mT and a frequency of 905 Hz is shown in Figure 6.1. The top signal drives the solenoid magnetic field, while the bottom signal is what is observed at a photodetector due to the modulated light. The light intensity modulation is achieved by a change in the scattering cross sections of the peapod-like nanostructures, due to their alignments with the alternating magnetic field, as schematized in Figure 6.1d. For a full period of the periodic signal driving the magnetic field, the peapod-like nanostructures align themselves twice, once for the +B direction and the other for the −B direction, allowing maximum light passage each time. As a result, the laser beam comes out of the sample with its intensity modulated at twice the frequency of the driving magnetic field signal.

The high fidelity in the waveform of the modulated signal and the constant phase relation between the driving signal and the modulated signal benefit from the small inertia of the particles. It is worth noting that the modulation frequency is comparable with the frequency of commercial mechanical spoked-wheel-based optical choppers. Since no mechanical moving parts are involved, this magnetically controlled light intensity modulation system may find applications in optical circuits with advantages including high reliability, long lifetime, and great flexibility and convenience for a high degree of device integration.

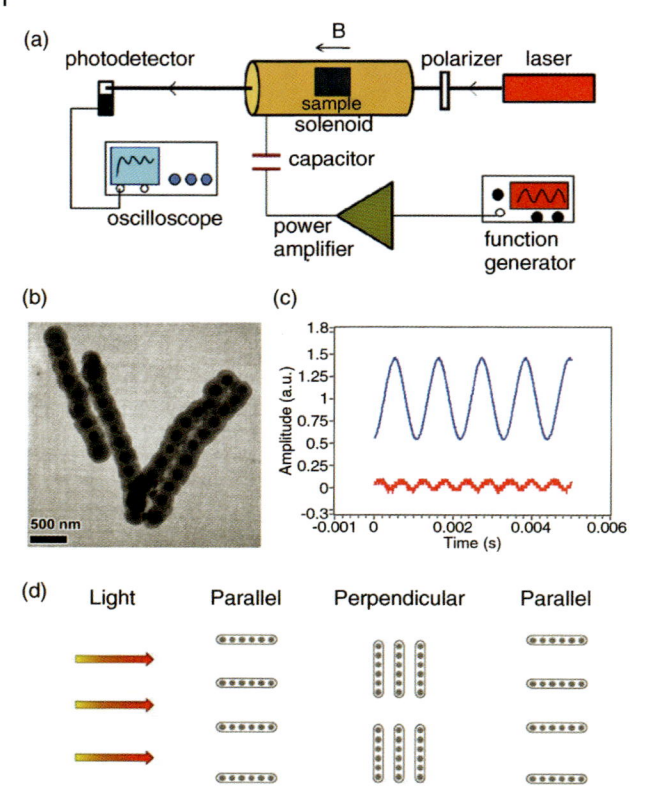

Figure 6.1 (a) Schematic diagram of the experimental setup used for modulation of laser intensity, (b) TEM image of $Fe_3O_4/SiO_2/TiO_2$ peapod-like superstructure, (c) Oscilloscope screenshot of light modulation with peapod superstructures with lengths of 1 ~ 3 µm at 905 Hz, (d) Depiction of how light modulation is caused during a full cycle of the magnetic field oscillation. Peapods are aligned parallel, perpendicular, and antiparallel to the initial orientation of the magnetic field. Thus, peapods orient themselves parallel to the field twice within one full cycle, and hence twice allow maximum light passage causing frequency doubling. (Reproduced with permission from Ref. [19a]. Copyright 2010 Royal Society of Chemistry and Ref. [19b]. Copyright 2010 American Chemical Society.)

When light encounters a metallic surface, the penetrating light causes the cloud of negative charge – consisting of weakly bound electrons of the metal atoms – to vibrate in phase with the light frequency, and this vibrating and charged cloud produces light of the same frequency, specifically reflected light. Therefore, the use of planar metallic materials as scatters not only enables the light transmission modulation but also brings the opportunities for effectively modulating the intensity of reflected light and thus benefits the development of reflective displays [20].

Gold microplates were chosen as an example for the demonstration of this design principle [21]. After being synthesized, they were rendered magnetically functional via conjugation to amine terminated superparamagnetic γ-Fe_2O_3@-SiO_2 nanoparticles. Since only a thin layer of magnetic material is necessary to render the plates magnetic, they retain their high reflectivity and their surrounding solution remains transparent, while their orientation can be readily, rapidly, and reversibly tuned by applying a magnetic field oriented in varying directions. If the direction of the incident light is fixed, the reorientation of microplates with external magnetic fields results in the changes of both its projected area and its scattering cross section, and thus the changes in the light transmittance and reflectance.

The reflectance of a dispersion of gold microplates can be conveniently modulated by external magnetic fields. As shown in Figure 6.2c, the reflectance was strongly dependent on the angle of the applied magnetic field. Almost no change in reflectance occurred when θ was increased from $0° \sim 60°$, while a dramatic increase from 1 to 6% was observed as θ was increased from 60 to 90°, suggesting a sixfold difference between the maximum and the minimum reflectance states. The transmittance changes of gold microplate dispersion exposed to a direction-varying magnetic field was also monitored by the transmittance spectra. As shown in Figure 6.2d, it could be varied from 42 to 72% as θ was decreased from 90 to 0°. This observed increase was largely due to a gradual change in the projected cross section of the plates, which was minimized when the major axis was aligned with the beam path and maximized when the plates were oriented perpendicular to the beam path.

In the absence of a magnetic stimulus, the gold plates dispersion exhibited a shiny, golden color reminiscent of bulk gold. Upon the application of a magnetic field oriented parallel to the viewing direction, the plates rapidly aligned by the applied field and the solution became considerably transparent due to the minimized cross section of the microplates. As the applied magnetic field was rotated such that the angle between the magnetic field and the viewing direction (θ) increased, the plate solution transitioned from clear and transparent to opaque and reflective. This optical change was due to the plates in the solution becoming aligned perpendicular to the viewing angle ($\theta = 90°$), which maximized the projected cross section of the plates to reduce transmitted light, as well as direct more reflected light back to the observer, as suggested by Figure 6.2e.

6.4
Magnetic Tuning of Light Interference

Thin-film interference occurs when incident light reflected by the upper and lower interfaces of planar structures interfere with one another. If the thickness of the planar structures is comparable to the wavelength of visible light, structural colors will be observed as a result of the light interference. Thin-film interference is common in the natural world as exemplified by soap bubbles and oil films. Unlike the colors from organic dyes or pigments, interference colors are free from bleaching and allow for considerably longer lifetime in real

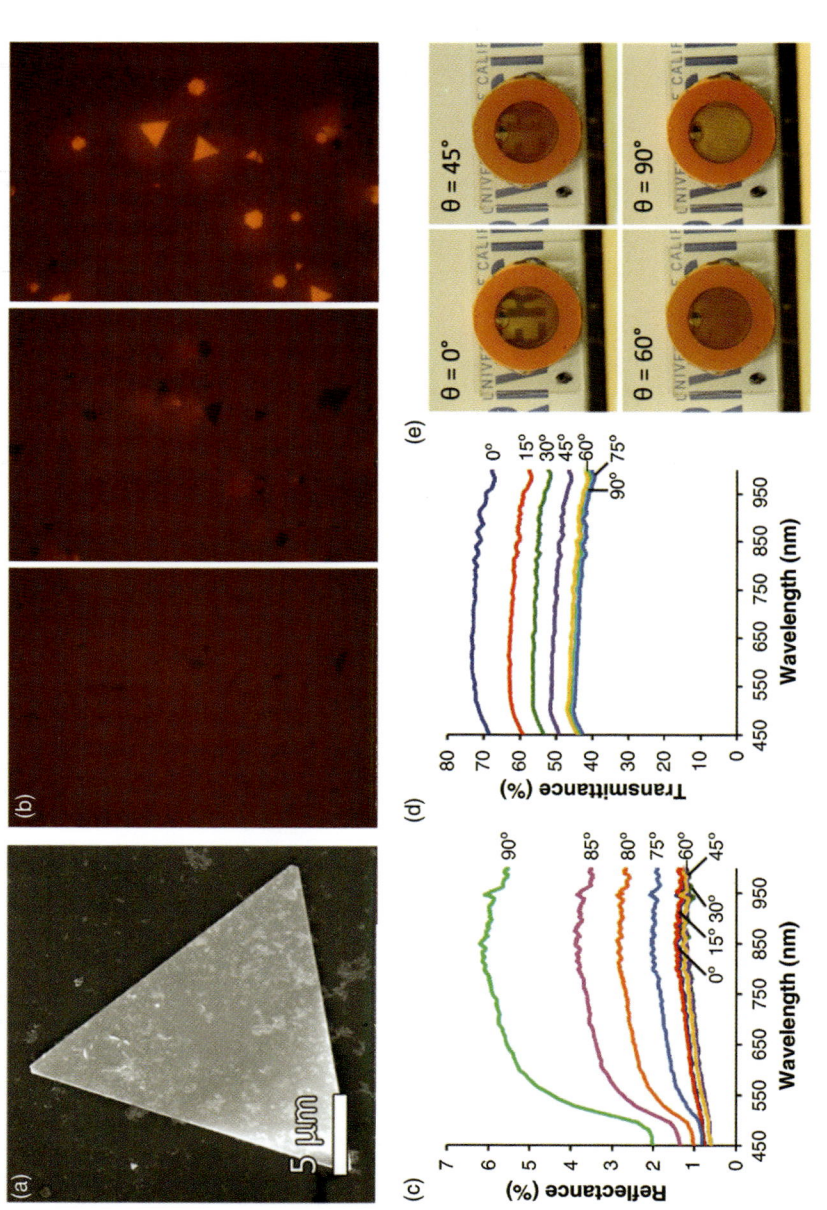

Figure 6.2 (a) SEM images of gold microplates with magnetic nanoparticle deposition. (b) Optical microscopy images of gold microplates under an applied magnetic field with varying orientations relative to the viewing directions: 0°, 45°, and 90°, from left to right. (c) Reflectance spectrum of a dispersion of gold microplates exposed to a magnetic field oriented at various angles (from 0 to 90°) relative to the path of the incident light. (d) Transmittance spectrum of a dispersion of gold microplates exposed to a magnetic field oriented at various angles (from 0 to 90°) relative to the path of the incident light. (e) Digital photographs of a bulk gold microplate solution under an applied magnetic field with various orientations (θ) relative to the viewing direction. (Reproduced with permission from Ref. [21]. Copyright 2016 Royal Society of Chemistry.)

applications associated with colorful displays [22]. The active tuning of interference colors represents one important mechanism for the modulation of light in its wavelengths [23].

Interference of light can happen in a constructive or a destructive manner. The degree of constructive or destructive interference between the two light waves depends on the difference in their phase. Constructive interference occurs with maximum reflection when the interfered waves are in phase, while destructive interference happens with minimum reflection when the interfered waves are out of phase. The conditions for constructive or destructive interference vary depending on the relative refractive indices of the systems. In cases where $n_1 > n_2 > n_3$ or $n_1 < n_2 < n_3$, the interference conditions are $2n_2 d \cos\theta_2 = m\lambda$ for constructive interference of reflected light and $2n_2 d \cos\theta_2 = (m - 0.5)\lambda$ for destructive interference of reflected light, where n_1, n_2, and n_3 are the refractive indices of the upper medium, the thin film, and the bottom medium, respectively; d is the film thickness; θ_2 is the angle of refraction; and m is an integer. In cases where $n_1 \langle n_2 \rangle n_3$ or $n_1 > n_2 < n_3$, the above two conditions switch. The above equations indicate that the interference color depends upon the incident angle, with its relation to the diffraction angle governed by Snell's law, $n_2 \sin\theta_2 = n_1 \sin\theta_1$, underscoring the possibility of modulating the optical properties of the films through orientational manipulation.

The magnetic modulation of light interference has been successfully demonstrated using magnetically responsive Fe_3O_4 microplates [24]. These microplates were fabricated from tiny Fe_3O_4 nanoparticles with diameters less than 10 nm through a template-directed assembly method using topologically patterned polyurethane templates. A ferrofluid droplet containing polyacrylate-capped Fe_3O_4 nanocrystals was applied to the polymer patterns and covered with a glass cover slide. During drying, the capillary force causes the movement of ferrofluid droplets between the polymer bumps and the surrounding areas, resulting in discontinuity of the film at the edge of the patterns. Soaking of the film can release Fe_3O_4 microplates in the freestanding form. After fabrication, interference colors can be observed on these microplates. The interference colors of microplates are highly dependent on their thicknesses and can be tuned from red to blue by modifying the experimental parameters, for instance, the original concentration of Fe_3O_4 nanoparticles, as shown in Figure 6.3c–e.

These microplates exhibit satisfactory response to external magnetic fields after being dispersed into a liquid medium. They spontaneously align their plane parallel to the field directions. The reflectance spectrum of one typical microplate at an incident angle of 0° showed a series of diffraction maxima and minima in the visible and near-infrared range, a phenomenon that is typical for thin-film interference (Figure 6.3f). All the peaks and valleys can be assigned to different orders of constructive and destructive interferences, respectively. With an increase in the incident angle to 3°, the peak and valley positions of the same interference orders shifted to shorter wavelengths, as indicated in Figure 6.3g. The angular-dependent optical properties were further demonstrated by the continued blueshift of the peaks and valleys at even higher incident angles of 9 and 15°.

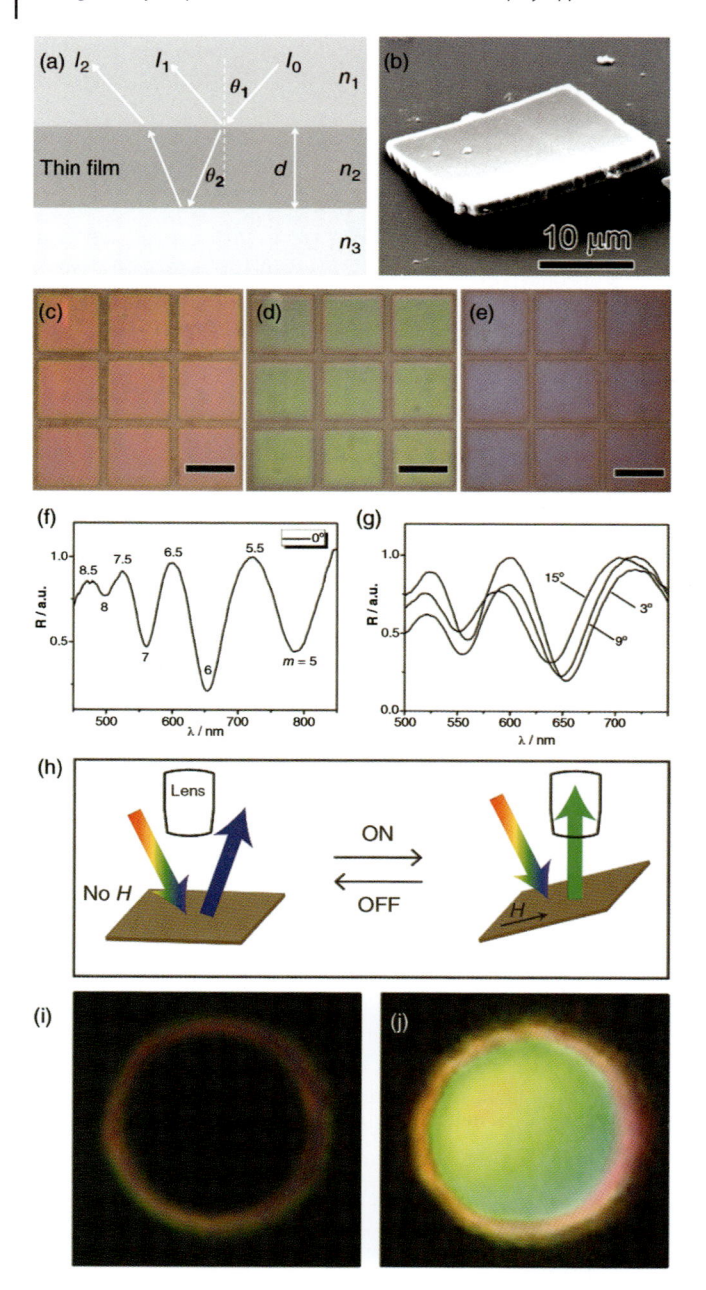

Figure 6.3 (a) Schematic illustration of thin-film interference. The light reflected from the upper and lower interfaces (I_1 and I_2) interfere with one another to form a new wave. θ_1 is the incident angle, while θ_2 responds to the diffractive angle. (b) A representative TEM image of a single microplate. (c–e) Bright-field optical images of dried nanocrystal thin plates with different thicknesses formed against the same square-shape template. Scale bar: 20 μm.

The rotation of one single plate upon the application of an external magnetic field was monitored under a dark-field optical microscope. In the absence of a magnetic field, the plates lie flat with their main axis parallel to the substrate due to gravity. Therefore, the reflected light does not travel along the observation direction and no interference color can be observed. Instead, only randomly scattered light is visible from the plate (Figure 6.3h). When a magnetic field is applied at an angle 15° offset from the horizontal, the plate aligns its main axis parallel to the field to minimize its magnetostatic potential, so that the reflected light goes into the objective lens and its vivid structural color can be observed (Figure 6.3h). Such "ON/OFF" color switching of microplates is very convenient to be achieved by magnetically controlling their orientations. Further optimization of this light interference modulation system and its integration with existing optical devices will enable a wide range of chromatic display applications in various fields.

6.5
Magnetic Tuning of Bragg Diffraction

When light wave incidents on order structures constructed by building units with dimensions comparable to, or even smaller than, its wavelengths, Bragg diffraction occurs. Light is scattered in a specular fashion by the building units of the structures and undergoes constructive interference. Take a crystalline solid as an example, the light waves are scattered from lattice planes separated by the interplanar distance d. When the scattered waves interfere constructively, they remain in phase since the path length of each wave is equal to an integer multiple of the wavelength. The path difference between two waves undergoing interference is given by $2d \sin \theta$, where θ is the scattering angle. The effect of the constructive or destructive interference intensifies as a cumulative effect of reflection in successive crystallographic planes of the crystalline lattice. The condition for the constructive interference to be at its strongest is described by Bragg's law:

$$2d \sin \theta = n\lambda,$$

where n is a positive integer and λ is the wavelength of diffraction light wave.

(f) Reflectance spectrum of a typical nanocrystal thin-film microplate at an incident angle of 0°. (g) Reflectance spectrum of a typical nanocrystal thin-film microplate at an incident angle of 3°, 9°, and 15°. (h) Schematic illustration of the ON/OFF switch off the diffractions of a thin microplate observed under a dark-field optical microscope. (i) The reflected light does not travel along the observation direction, corresponding to an "OFF" state. (j) The observed structural color of the microplate when a magnetic field of 15° offset from the horizontal direction is applied, corresponding to an "ON" state. (Reproduced with permission from Ref. [24]. Copyright 2015 John Wiley & Sons, Inc.)

When nanoparticles with dimensions comparable to the wavelength of visible light were organized into ordered structures, the resultant structures can strongly diffract visible light at certain wavelengths and thus exhibit shiny structural colors, as determined by Bragg's law [25]. Such ordered structures are named as photonic structures and are of special interest as important candidates for potential applications in displays, as the tuning of these ordered structures directly results in the modulation of the wavelength of diffraction light [26]. By using magnetic responsive nanoparticles as building units, the effectiveness of magnetic fields in organizing nanoparticles into photonic structures and in tuning the optical properties of the photonic structures is demonstrated [27].

A representative TEM image of the building block, superparamagnetic magnetite colloidal nanoclusters (CNCs) is shown in Figure 6.4 [13b,27c]. The surface of the original CNCs is grafted with a layer of polyelectrolyte, polyacrylate, during the synthesis, which provides a strong interparticle electrostatic force in aqueous solution. Upon the application of an external magnetic field, magnetic dipole moments are induced in each CNC, and the interactions between magnetic dipoles drive the assembly of them into 1D chain-like ordered structures that strongly diffract incident lights. The key to the successful assembly of CNCs and tuning of the optical diffraction is the coexistence of a highly tunable magnetic dipole–dipole force and comparable long-range electrostatic force, both of which are separation-dependent. Along the chain, the magnetic dipole–dipole attraction is balanced by the electrostatic force, while the interchain magnetic repulsive force, as well as the electrostatic force, keeps the chains away from each other.

The strength of magnetic field determines the magnitude of the magnetic dipole–dipole interactions and, therefore, the interparticle spacing of the structures [27c]. As a result, the optical property of the as-assembled structures can be effectively tuned by the external magnetic field. For example, enhancing the magnetic field strength induces stronger magnetic attraction along the chain that brings the particles closer and the diffraction wavelength blueshifts consequently, while decreasing the magnetic field strength results in the redshifts of the diffraction wavelength, with considerably strong diffraction intensities achieved at a significantly low particle concentration (∼0.1% volume fraction). More importantly, the diffraction intensity of these photonic structures can be periodically modulated by a high-frequency alternating magnetic field. Previous experiments suggested that the optical signals generated by the diffraction can be modulated in a frequency of 30 Hz, which makes them potential candidates for colorful display applications.

Encapsulating the dispersion of CNCs in a solid flexible film makes it convenient to integrate the magnetically responsive photonic structures into complex devices for practical applications [28]. This concept was demonstrated by dispersing an ethylene glycol solution of Fe_3O_4@SiO_2 particles in a liquid prepolymer of polydimethylsiloxane (PDMS) in the form of emulsion droplets with typical sizes of several micrometers, followed by solidification of the polymer matrix. Applying a uniform magnetic field to the film then induces chaining of

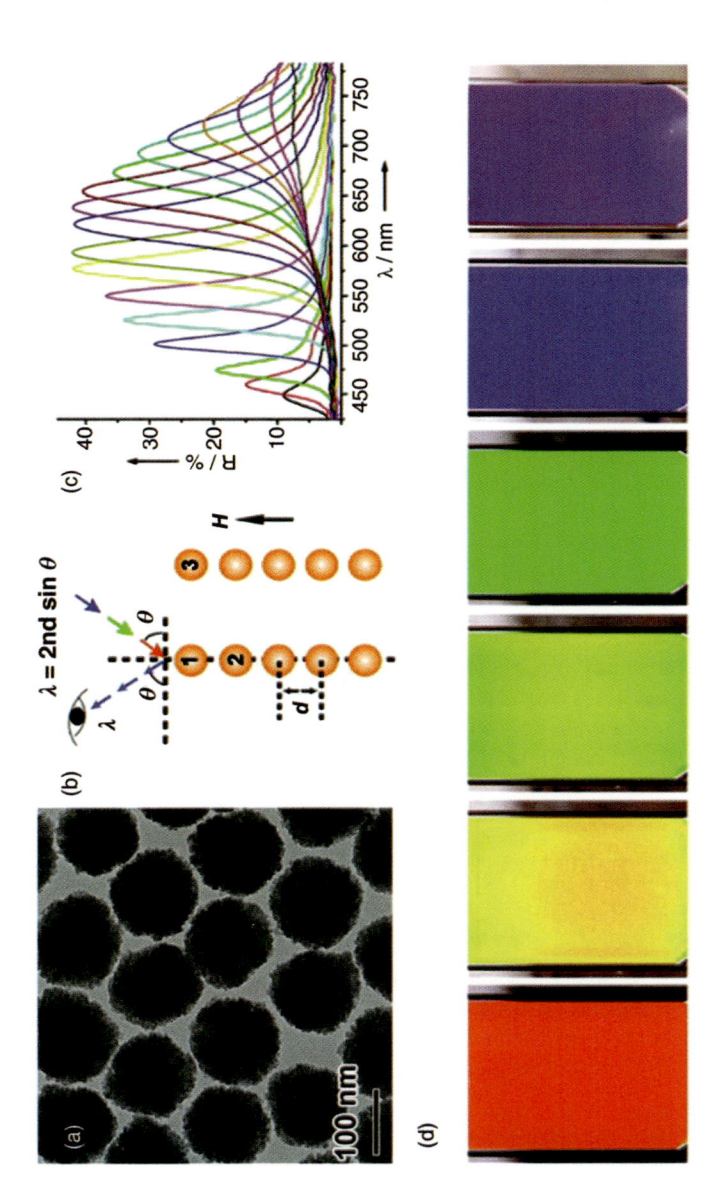

Figure 6.4 (a) A representative TEM image of the CNCs. (b) Scheme of Bragg diffraction from the chains of CNCs. (c) Reflectance spectra of a CNC aqueous dispersion under magnetic fields with different strengths. (d) Digital photos showing the diffraction color change in a typical CNC dispersion encapsulated in a capillary tube with a width of 1 cm in response to a magnetic field with increasing strengths from left to right. (Reproduced with permission from Ref. [13b]. Copyright 2012 American Chemical Society.)

the magnetic particles inside each droplet with the same diffraction wavelength so that the film displays a homogeneous color that can be recognized by the naked eye. The composite film maintains the excellent flexibility of the PDMS matrix and is still able to show a rapid response to external magnetic fields even if folded into various shapes. Color gradient effects can be easily created by controlling the assembly of each droplet in parallel. By incorporation of different types of CNC particle solutions that have similar background colors but show different visible responses to the magnetic field, for example, by choice of CNCs with different sizes, a flexible film with a patterned display is created. Without an external field, the film shows the native brown color of iron oxide with essentially no contrast. Upon the application of the magnetic field, however, a color pattern can be clearly observed due to the diffraction contrast resulting from different sizes of CNC particles. This type of graphic display may find use in applications such as anticounterfeiting devices or switchable signage where prestored information can be hidden unless activated by external magnetic stimuli (Figure 6.5).

In practical applications, a magnetic field is often not uniform and a large gradient usually exists, which induces a magnetic packing force on every particle, attracts them toward the maximum of the local magnetic gradient. As a result,

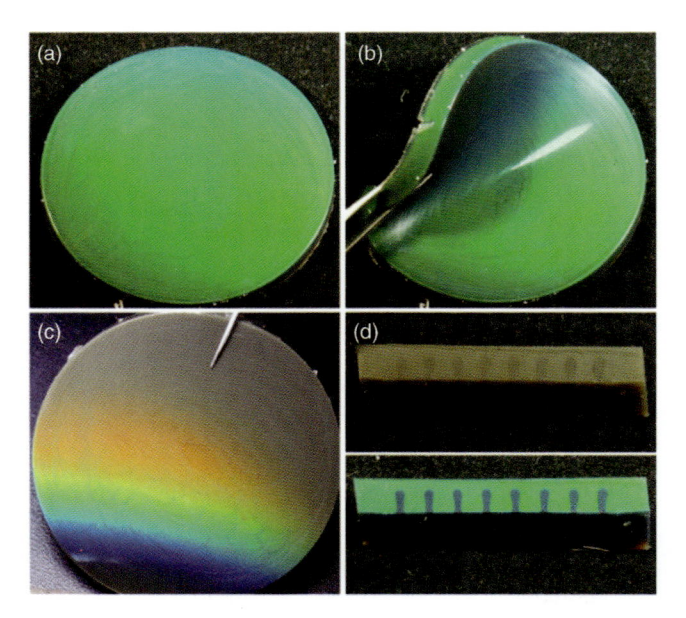

Figure 6.5 (a–b) Photograph showing a flexible PDMS film displaying structural color under a uniform magnetic field; (c) the film displaying "rainbow" structural colors under a nonuniform magnetic field. (d) Digital photos of a PDMS film that can display color patterns upon the application of an external magnetic field. The diameter of the disk in (a–c) is around 47 mm, and the length of the bar in (d) is around 25 mm. (Reprinted with permission from Ref. [13b]. Copyright 2012 American Chemical Society.)

the dynamically ordered assemblies of $Fe_3O_4@SiO_2$ suffer from chain coagulation induced by the magnetic packing force, gradually degrading the colloidal stability of the system and consequently their photonic performance after long time exposure to magnetic fields. The use of a matrix, for example, agarose hydrogel, is able to enhance the stability against the magnetic packing force, while retain the tunability of the photonic structures [29]. The steric effect of an agarose hydrogel network enables the dynamic chains of $Fe_3O_4@SiO_2$ colloids to withstand the packing force. In addition, the hydroxyl groups of the agarose network function as flexible anchors for the $Fe_3O_4@SiO_2$ colloids via hydrogen bonding to provide additional resistance against aggregation.

One remarkable improvement obtained by incorporating agarose hydrogel film is the excellent long-term stability achievable in a magnetic field with a large gradient. The diffraction intensity and the peak position of an agarose/$Fe_3O_4@SiO_2$ composite hydrogel film were found to be nearly constant, after 12 h of exposure to a magnetic field with strength of 360 G and a gradient of 115 G/cm, as plotted in Figure 6.6a. A simple display unit made of two composite agarose films in contact with each other further illustrates the excellent stabilization of the photonic chain structures by the agarose matrix. The digital photos of the display unit exhibit no obvious change after 18 h of exposure to a magnetic field. The boundary between the green and the orange agarose films remained clear and its position did not shift. The dramatically improved structural and photonic stability of the assemblies in magnetic fields greatly broadens the potential of the magnetically responsive photonic structures for practical applications.

Benefiting from the instantaneous nature of the magnetic creation and tuning of structural colors as well as the photopolymerization technique, structural colors can also be fixed in a polymer film with a high spatial resolution [30]. By using a spatially modulated focused UV beam as the printing tool, and a mixture

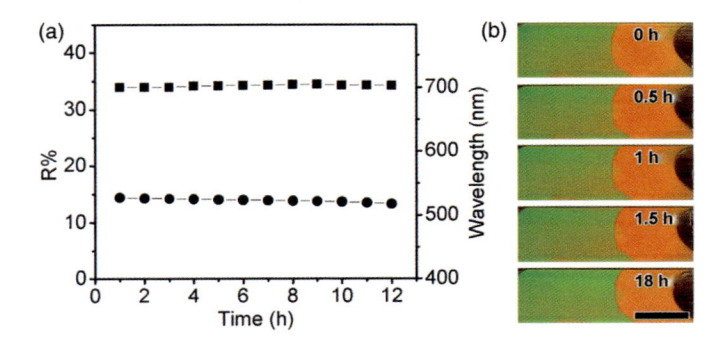

Figure 6.6 (a) Plots of reflectance intensity (■) and peak position (●) versus time for a composite hydrogel film in an external magnetic field of 360 G with a gradient of 115 G/cm. (b) Digital images of the patterned photonic agarose film under a magnetic field of 570 G with a gradient of 198 G/cm over different periods: 0 h, 0.5 h, 1 h, 1.5 h, and 18 h, from top to bottom. The scale bar corresponds to 1 mm. (Reproduced with permission from Ref. [29]. Copyright 2015 Springer.)

of CNC particles and UV curable resin as the ink, we have developed a fast and high-resolution color printing technique for producing various multicolored patterns with a single ink by repeating the "magnetic tuning" and "maskless lithographical fixing" processes, which can be programmed to print different color regions sequentially and eventually yield a large complex multicolored pattern. Since the ultimate resolution can reach 2 μm, it is easy to realize grayscale modulation and spatial color mixing by varying the density and color of primary color units in a pixel that is smaller than the resolution of a human eye. Graphic coding can also be realized by the similar technique, as schematized in Figure 6.7. It is demonstrated that a single particle with up to eight different color encodings can be fabricated within approximately one second, superior to conventional techniques that usually involves binary encoding and requires many cycles of consecutive patterning of colorful materials at high resolution. With the advantages including multicolor production from a single ink, no need to move the substrate or change any physical photomask during printing, and the intrinsic characteristics of structural colors such as iridescence and metallic appearance and high color durability without bleaching, this technique represents a new platform for high-resolution color printing or encoding with immediate applications ranging from forgery protection, structurally colored graphic design to biomedical applications.

In addition to tuning the interparticle separation, the orientation of the photonic chains can also be programmed in each "magnetic tuning and lithographical photopolymerization" cycle to print patterns with angular-dependent color contrast, which shows switchable color distribution when the angle of incident light changes or the samples are tilted. Again, only a single magnetic ink is needed for printing a photonic crystal film with different crystal orientation distribution [31]. As shown in Figure 6.8, upon incidence of light, different structural colors appear in areas with different chain orientations. When the incident angle is close to the longitudinal direction of the particle chains, the corresponding areas diffract longer wavelength light (green, characters in case 1), while the other areas diffract shorter wavelength light (blue, background in case 1). On the contrary, as the incident light is projected close to the chain direction in the background area (case 2), the characters and background instantly switch their colors. With multiple masks, it is expected that more complicated and unique patterns can be fabricated, which may be particularly interesting for anticounterfeiting display applications.

The above angular-dependent photonic structures can also be made into individual unit with much smaller dimensions. Angular-dependent photonic structures were fabricated by either fixing photonic chains in polymer microspheres through the magnetically induced assembly of particles and UV-induced polymerization [32] or wrapping photonic chains with another layer of silica via the sol–gel method [33]. As discussed above, the photonic chains assembled from superparamagnetic particles are magnetically anisotropic in nature. Due to the dipole–dipole interaction between neighboring particles, the chains tend to align along the magnetic field to minimize their magnetostatic energy, which enables

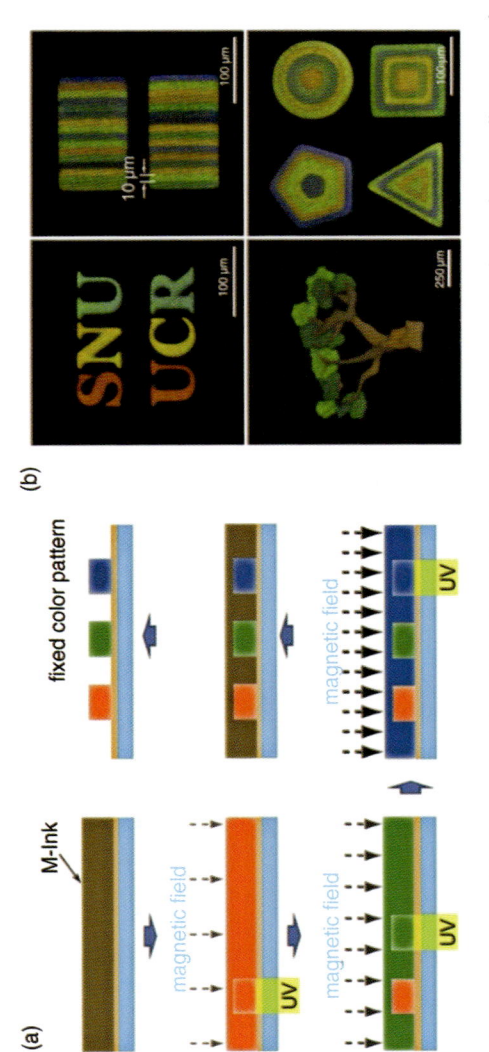

Figure 6.7 High-resolution patterning of multiple structural colors with a single magnetic ink. (a) Schematic illustration of multicolor patterning with a single ink by the sequential steps of "tuning and fixing." (b) High-resolution multicolor patterns produced using magnetic ink. (Reproduced with permission from Ref. [30]. Copyright 2009 Nature Publishing Group.)

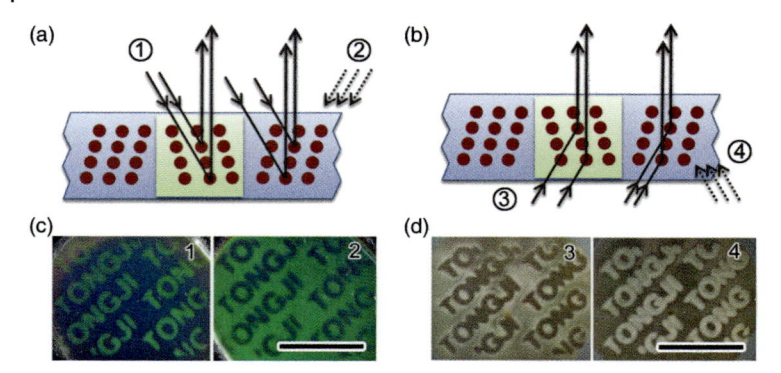

Figure 6.8 (a) Scheme of light paths in photonic films in reflection modes. (b) Scheme of light paths in photonic films in transmission modes. (c) A digital photograph of a piece of photonic films in reflection modes. (d) A digital photograph of a piece of photonic films in transmission modes. The green and blue sections in the scheme correspond to characters and the background, respectively. Photographs 1 and 2 (or 3 and 4) show the same sample as the incident light is projected along opposite directions. The scale bars in (c and d) are 1 cm. (Reproduced with permission from Ref. [31]. Copyright 2011 American Chemical Society.)

the tuning of their orientations and thus their optical properties by external magnetic fields.

Figure 6.9 demonstrates the switching of the diffraction of photonic-chain-blended microspheres between "on" and "off" states by rotating the external magnetic field parallel or perpendicular to the viewing direction. In the "on" state, all the photonic chains embedded in the microspheres are aligned parallel to the viewing angle so that their shiny structural colors can be observed. Since the fabrication of microspheres and magnetic tuning process are separated into individual steps, color mixing can be easily realized by mixing microspheres with different diffraction colors together, as suggested by the presence of both "blue" and "green" microspheres in Figure 6.9b. Such a multistep method has several advantages including long-term stability of optical response, improved tolerance to environmental variances such as ionic strength and solvent hydrophobicity, and greater convenience for incorporation into many liquid or solid matrices.

This fabrication method can also be extended for the design of display units with bistability, which is a commonly desired feature as it allows the recording of visual information for a considerably long period and avoids the energy cost for constant refreshing. In a bistable system, energy is consumed only during occasional refreshing; afterward, the information displayed by the electronic paper will be kept for minutes or even longer without the need of any energy input.

In a simple demonstration, 1D photonic nanochains containing periodically arranged magnetic spheres along their long axis are used as the basic color units,

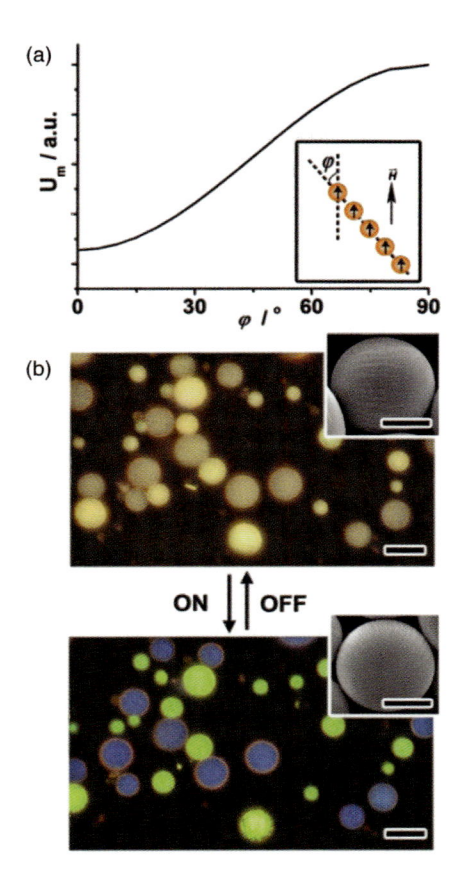

Figure 6.9 (a) Plot of magnetic dipole–dipole energy of a nanochain versus the orientation with respect to the external magnetic field. (b) "ON/OFF" switching of the color of a mixture of two types of microspheres, imaged by dark-field optical microscopy, from the native light brown of iron oxide to blue and green by tuning the magnetic field orientation from horizontal to vertical. Insets are SEM images of the microspheres in horizontal (OFF) and vertical (ON) orientations. All scale bars are 20 μm.

which can be magnetically switched on and off very rapidly [33b]. Only if they are aligned parallel to the viewing angle by the magnetic field, incident light can be diffracted and strong photonic response can be observed, corresponding to an "ON" state, otherwise no diffraction happens and nanochains only show the native brown color of magnetite, corresponding to an "OFF" state. The "ON" state can be maintained for different lengths of period by controlling the length of the nanochains or the viscosity of the solvents. When highly viscous solvents were used, their photonic response could last for a considerably longer time after the removal of external magnetic fields.

A writing tablet fabricated by dispersing nanochains in 95% glycerol solution has been used for demonstration. Upon the application of a refrigerator magnet with patterned magnetic field distribution, alternating green or brown stripes appear within a second. In the areas where the magnetic field is parallel to the viewing angle, green strips can be observed, while brown stripes appear in areas where the direction of magnetic field is perpendicular. The alternating strips remain visually almost unchanged 3 mins after the removal of the magnet, indicating a good bistability. In fact, these strips can stay distinguishable for longer than 10 mins. Patterns or letters can also be directly written in this writing tablet by using a small magnet as a pen. More complicated bistable displays, for example, a display that can show two distinct colors when the external field is switched on and off can also be achieved by the similar "color mixing" technique: mixing two photonic nanochains with different diffraction colors and chain lengths together. The shorter nanochains are much easier to undergo rotations and lose their photonic properties relatively quicker after the removal of external magnetic stimuli, while the longer nanochains can remain aligned and diffract incident light for much longer time (Figure 6.10).

The use of ultrasmall one-dimensional photonic nanostructures in the inks allows colors to be more uniform and highly resolved. Because photonic-ink displays reflect light (rather than emit it, like traditional light-emitting displays), the system is vastly more energy-efficient and also shows no glare under sunlight, which are useful for outdoor purposes such as billboards and e-readers. Given the electrode-less remote control, temporary bistability, energy efficiency, and low toxicity of such photonic inks, they are extremely utilizable in display technologies.

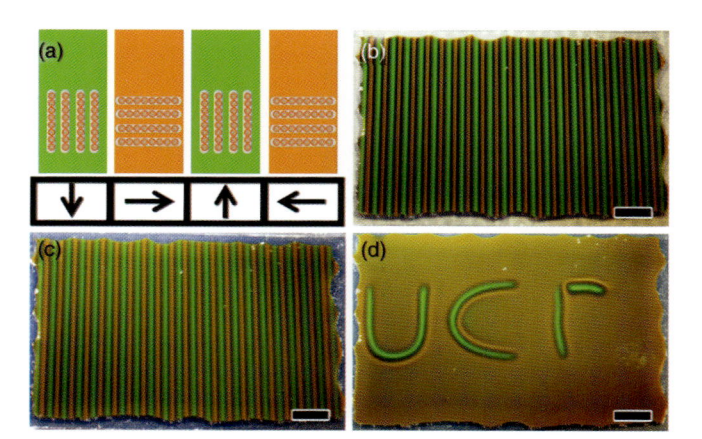

Figure 6.10 (a) Scheme showing how the pattern magnet influences the color of inks. (b) Digital image of the display right after the removal of a pattern magnet. (c) Digital image of the display 3 min after the removal of a pattern magnet. (d) Letters written on the display by using a small piece of magnet as pen. All scale bars are 1 cm. (Reprinted with permission from Ref. [33b]. Copyright 2013 Royal Society of Chemistry.)

6.6
Magnetic Tuning of Light Polarization

Anisotropic materials interact with light in a much more complex manner than their isotropic counterparts, allowing more degrees of freedom for designing new mechanisms of light modulation. When an arbitrary beam of light strikes the surface of an optically anisotropic material, for example, a birefringent material, the polarizations corresponding to the ordinary and extraordinary rays generally take different paths. According to Snell's law of refraction, the angle of refraction will be governed by the effective refractive index that is different between these two polarizations. These two orthogonal polarizations propagate with two different velocities, and thus they develop a phase difference as they propagate. When they exit the birefringent material, the total polarization of the wave changes when these two states recombine.

One important class of birefringent materials is liquid crystals that have been broadly used in commercial displays. The basic units that constitute liquid crystals are often organic molecules with anisotropic rod-like or disk-like shapes. These molecules have two principal refractive indices, ordinary refractive index n_o and extraordinary refractive index n_e. The first one is measured for the light wave where the electric vector vibrates perpendicular to the optical axis (ordinary wave). The index n_e is measured for the light wave where the electric vector vibrates along the optical axis (extraordinary wave). The intensity of light transmitted through a liquid crystal sandwiched between cross polarizers can be typically described as $I = I_0 \sin^2(2\alpha)\sin^2(\pi \Delta n L/\lambda)$, where I_0 is the intensity of light passing through the first polarizer, α is the angle between the transmission axes of the polarizer and the long axis of the liquid crystal, Δn is the difference in n_o and n_e, L is the sample thickness, and λ is the wavelength of incident light [34].

Effective alignments of liquid crystal molecules along desired directions are, therefore, critical for the modulation of light transmittance and the design of liquid crystal displays, and are usually achieved by electric fields in commercial applications. However, recent studies also suggested that magnetic fields can serve the same task when the liquid crystals are composed of magnetically responsive materials instead of commercially nonmagnetic organic molecules. For example, aqueous dispersions of the $Fe_3O_4@SiO_2$ nanorods with an aspect ratio of 7 were found to self-organize into liquid crystal phases under elevated volumetric fractions (>3%) [35]. Upon the application of an external magnetic field, the nanorods spontaneously align their long axis parallel to the field to minimize their magnetostatic potentials, enabling the tuning of the resultant liquid crystals.

The optical tuning of such liquid crystal by a magnetic field has been examined under polarized optical microscope (POM). When the field direction is parallel or perpendicular to the polarizer, α is equal to 0 or 90°, leading to dark optical views (Figure 6.11a and c). As the field direction turns to 45° relative to the polarizer, α changes to 45°, so that the intensity reaches the maximum

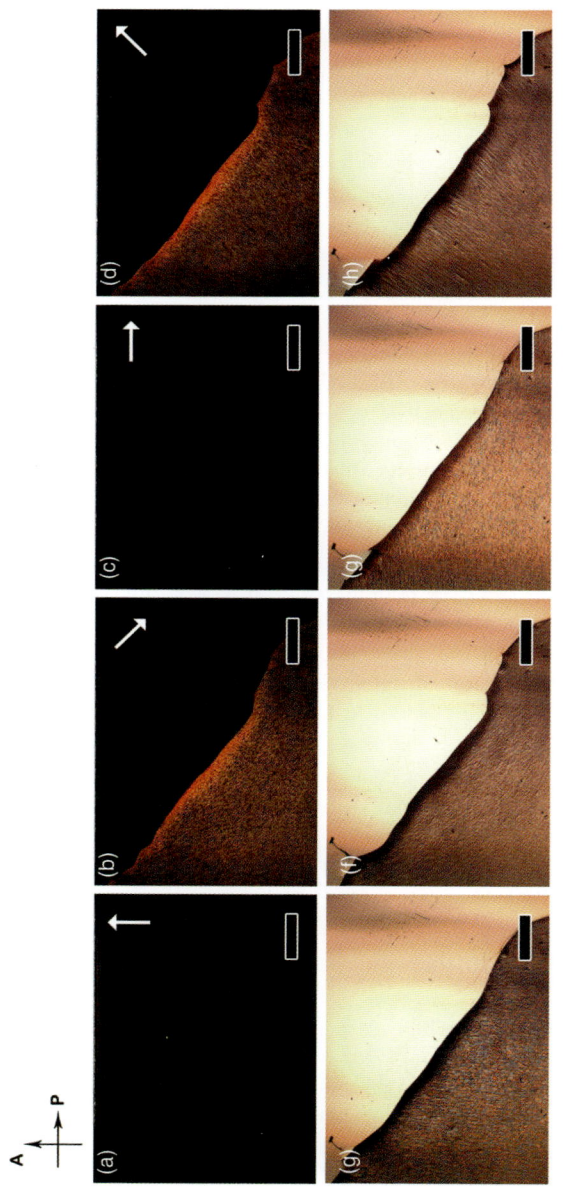

144 | *6 Magnetically Responsive Photonic Nanostructures for Display Applications*

Figure 6.11 (a–d) POM images and (e–h) bright-field OM images of a magnetic liquid crystal film under magnetic fields oriented in different directions. Black arrows at the top-left indicate the transmission axis of the polarizer (P) and analyzer (A). White arrows indicate the field direction. The top-right corner in each image contains no sample. Scale bars: 500 μm. (Reproduced with permission from Ref. [35]. Copyright 2014 American Chemical Society.)

according to Equation 1, resulting in bright views, as shown in Figure 6.11b and d. In contrast, the corresponding bright-field optical microscopy images of the same sample did not show apparent differences in the darkness of the view in response to the changes in the direction of the magnetic field, as indicated in Figure 6.11e–h. Owing to the overwhelming effect of the magnetic torque exerted on the nanorods, such optical tuning happens instantly. Optical modulation characterizations indicated that drastic changes in the transmittance of liquid crystals occur within 0.01 s, under a high-frequency alternating magnetic field. This corresponds to a switching frequency of 100 Hz and is comparable to the refreshing rates of commercial liquid crystal displays.

The facial magnetic control of such liquid crystals brings the opportunity for convenient fixation of the orientational order. It was further demonstrated that thin films patterned with various optical polarizations can be conveniently produced by combining the magnetic liquid crystals with lithography processes. As schematically shown in Figure 6.12a, a liquid crystal solution containing magnetic nanorods and PEGDA resin was first sandwiched between a glass cover slip and a glass slide to form a liquid film. A photomask was then placed on top of the sample, followed by the application of a magnetic field. Upon exposure to UV light, the orientation of the nanorods in the uncovered regions was fixed along a specific direction within the plane of the film. The photomask was then removed and the sample was again exposed to UV light in the presence of a magnetic field rotated 45° (in plane) from the initial field direction. A thin film with polarization patterns showing different transmittances to a polarized light

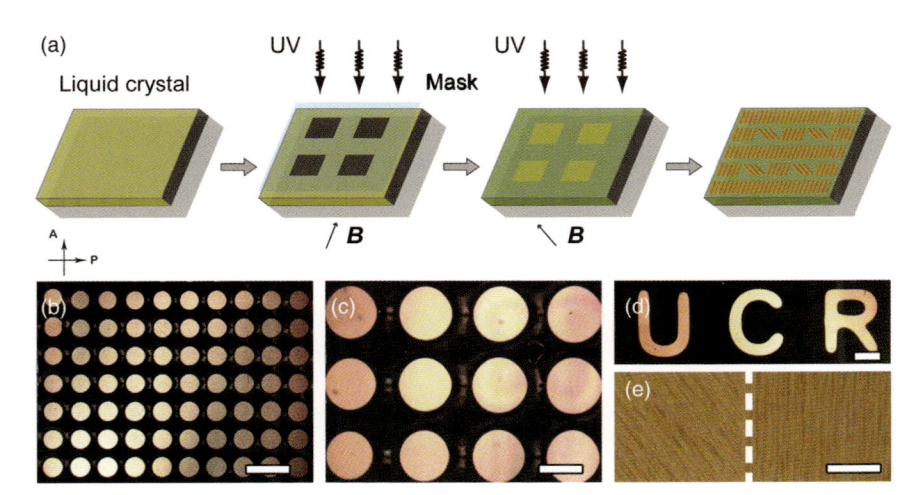

Figure 6.12 (a) Scheme showing the lithography process for the fabrication of thin films with patterns of different polarizations, (b–d) POM images of various polarization-modulated patterns, and (e) enlarged OM image shows the arrangement of nanorods in the pattern (left) and surrounding area (right). Scale bars: b–d) 500 μm; e) 10 μm. (Reproduced with permission from Ref. [35]. Copyright 2014 American Chemical Society.)

was obtained at the end. Figure 6.12b–d display the POM images of as-prepared samples after the application of different patterns. In these cases, the transmission axis of the polarizer was set to be parallel to the initial field direction. The areas cured during the first exposure appear dark under the POM, owing to the parallel arrangement of the nanorods relative to the transmission axis of the polarizer, while the areas cured during the second exposure are bright since all nanorods are oriented 45° relative to the transmission axis of the polarizer. An enlarged bright-field optical microscopy image is shown in Figure 6.12e, which accentuates the alignment of the nanorods at the boundary of the bright (left) and dark (right) areas (separated by the dotted line) and clearly confirms the 45° angle between the two orientations.

Changing the orientation of the nanorods relative to the transmission axis of the polarizer also allows convenient modulation of the transmittance intensity. As depicted in the extreme cases in Figure 6.13a–d, shifting the transmission axis of the polarizer to be parallel to the direction of the second field completely reverses the dark and bright areas, while almost no contrast can be observed under their bright-field optical images (Figure 6.13e and f).

Benefiting from the instantaneous and contactless nature of magnetic manipulation, the magnetically actuated liquid crystals save the use of transparent electrodes and are able to reach a comparable switching frequency as commercial liquid crystal displays. Such ability of using magnetic fields to modulate the light transmittance of liquid crystals is also expected to catalyze new applications associated with displays, for example, optical waveguides, actuators, and anti-counterfeiting devices.

6.7
Conclusions

In this chapter, we introduce the design of magnetically responsive photonic nanostructures and demonstrate their abilities for reversible optical modulation through controlling the strength and the direction of external magnetic fields. The direction of magnetic field determines the orientation of magnetically responsive particles, which usually enables effective tuning of their scattering cross-sections and the intensity of transmitted or reflected light. The variations in the strength of magnetic fields can induce dramatic changes in local concentration and interparticle interaction potentials of particles, thus serving as a convenient tool for tuning the collective optical property of particle assemblies and allow the modulation of light intensity in certain wavelengths. Our future work will be focused both on the optimization of the design of nanostructures, with respect to assembly efficiency, magnetic response rate, optical modulation range, and integration density, and on the miniaturization of the responsive system into smaller scale, which allows pixel-level control of optical signals in parallel toward real high-resolution display applications.

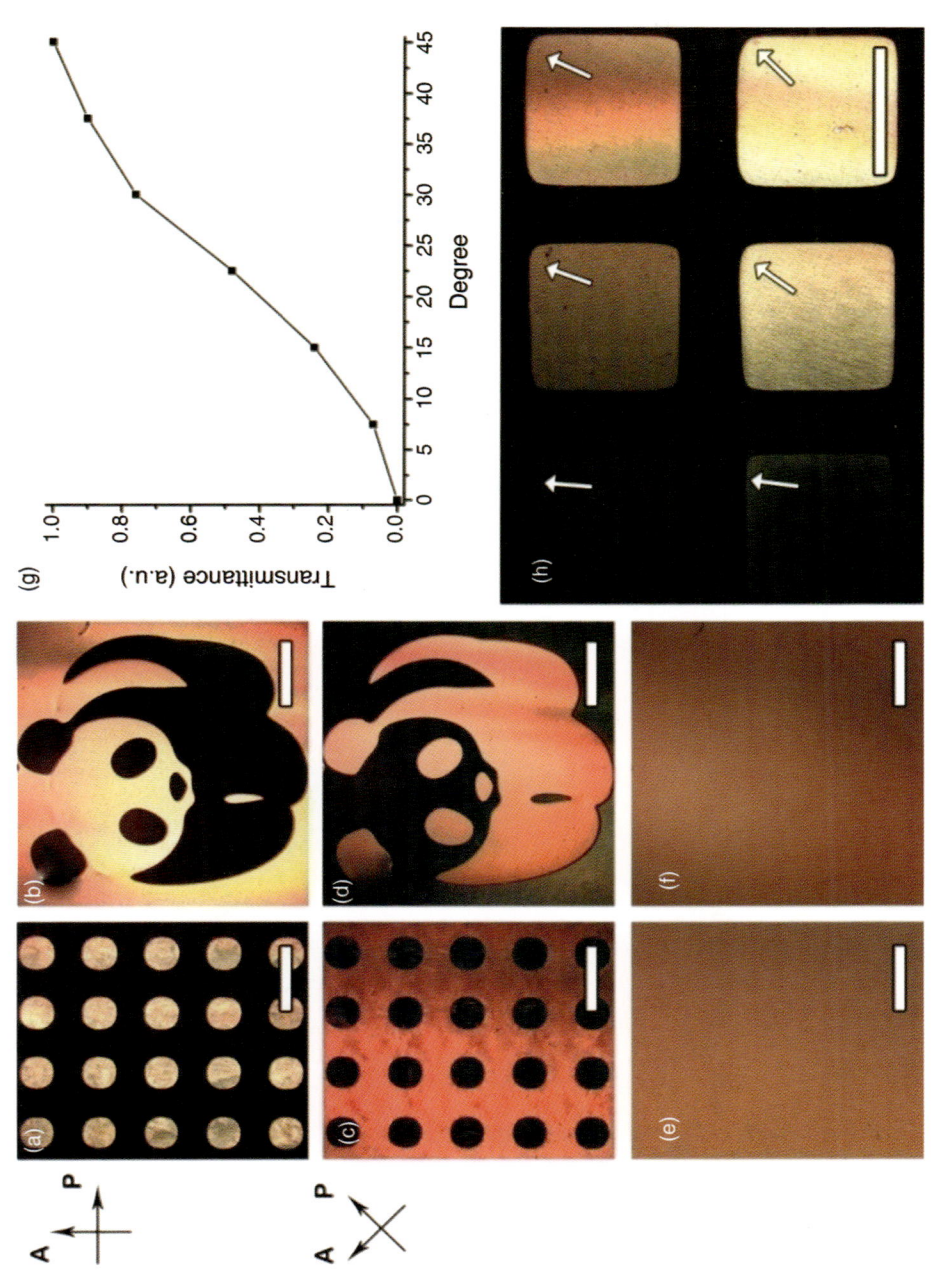

Figure 6.13 (a–d) POM images of two polarization-modulated patterns under cross polarizers before (a–b) and after (c–d) shifting the direction of the transmission axis of the polarizers for 45°. (e–f) Bright-field images of the same patterns. Scale bars: 500 μm. (Reprinted with permission from Ref. [35]. Copyright 2014 American Chemical Society.)

Acknowledgment

We are grateful for the partial financial support from the US National Science Foundation (Award Nos: DMR-0956081 and CHE-1308587), the US Army Research Laboratory (Award No: W911NF-10-1-0484), the Cottrell Scholar Award from the Research Corporation for Science Advancement, the Nontenured Faculty Grant from 3M, and the Young Professor Grant from DuPont.

References

1 Hainich, R.R. and Bimber, O. (2011) *Displays: Fundamentals and Applications*, A K Peters/CRC Press, Boca Raton, FL.
2 Lee, J.H., Liu, D.N., and Wu, S.T. (2008) *Wiley SID Series in Display Technology*, John Wiley & Sons, Ltd, Chichester, U.K, pp. xvi, 262.
3 (a) Soltic, S. and Chalmers, A.N. (2004) *J. Electron. Imaging*, **13**, 688–700; (b) Seats, P. (1981) *J. Electrochem. Soc.*, **128**, C87–C87; (c) Roehrig, H., Hartmann, W.H., Rovinelli, R.J., Capp, M.P., Evanoff, M.G., and Krupinski, E.A. (2001) *Radiology*, **221**, 168–168; (d) Szczyrbowski, J., Rogels, S. and Hartig, K. (1990) *Opt. Thin Films III*, **1323**, 110–121.
4 (a) Brown, F.H. and Zayac, M.T. (1971) Sid International Symposium Digest of Technical Papers, 11, pp. 98–99; (b) Chodil, G. and Dejule, M. (1971) Sid International Symposium Digest of Technical Papers, 11, pp. 102–103; (c) Petty, W.D. and Slottow, H.G. (1971) *IEEE Trans. Electron Devices*, **Ed18**, 654–658; (d) Boeuf, J.P. (2003) *J. Phys. D*, **36**, R53–R79.
5 (a) Jang, E., Jun, S., Jang, H., Llim, J., Kim, B., and Kim, Y. (2010) *Adv. Mater.*, **22**, 3076–3080; (b) Nuese, C.J., Kressel, H. and Ladany, I. (1973) *J. Vac. Sci. Technol.*, **10**, 772–788; (c) Manevitz, A. (1969) *J. Opt. Soc. Am.*, **59**, 479; (d) Gorrn, P., Sander, M., Meyer, J., Kroger, M., Becker, E., Johannes, H.H., Kowalsky, W., and Riedl, T. (2006) *Adv. Mater.*, **18**, 738–741; (e) Geffroy, B., Le Roy, P. and Prat, C. (2006) *Polym. Int.*, **55** 572–582.
6 (a) Choi, W.B., Chung, D.S., Kang, J.H., Kim, H.Y., Jin, Y.W., Han, I.T., Lee, Y.H., Jung, J.E., Lee, N.S., Park, G.S., and Kim, J.M. (1999) *Appl. Phys. Lett.*, **75**, 3129–3131; (b) Talin, A.A., Dean, K.A., and Jaskie, J.E. (2001) *Solid-State Electron.*, **45**, 963–976; (c) Jaskie, J.E. (1996) *MRS Bull.*, **21**, 59–64; (d) Burden, A.P. (2001) *Int. Mater. Rev.*, **46**, 213–231; (e) Yoshida, Y., Ishizuka, A., and Makishima, H. (1995) *Mater. Chem. Phys.*, **40** 267–272.
7 (a) Gray, G.W. and Kelly, S.M. (1999) *J. Mater. Chem.*, **9**, 2037–2050; (b) Castellano, J.A. (2006) *Am. Sci.*, **94**, 438–445; (c) Kawamoto, H. (2002) *Proc. IEEE*, **90**, 460–500; (d) Scheffer, T.J. and Nehring, J. (1984) *Appl. Phys. Lett.*, **45**, 1021–1023; (e) Schadt, M. and Leenhouts, F. (1987) *Appl. Phys. Lett.*, **50**, 236–238.
8 (a) Comiskey, B., Albert, J.D., Yoshizawa, H., and Jacobson, J. (1998) *Nature*, **394**, 253–255; (b) Yang, D.K., Doane, J.W., Yaniv, Z., and Glasser, J. (1994) *Appl. Phys. Lett.*, **64**, 1905–1907; (c) Wu, S.T. and Wu, C.S. (1996) *Appl. Phys. Lett.*, **68**, 1455–1457; (d) Ge, Z.B., Wu, T.X., Zhu, X.Y., and Wu, S.T. (2005) *J. Opt. Soc. Am. A*, **22**, 966–977.
9 (a) Vilardy, J.M., Millan, M.S. and Perez-Cabre, E. (2013) 8th Iberoamerican Optics Meeting and 11th Latin American Meeting on Optics, Lasers, and Applications, 8785; (b) Takizawa, K., Fujii, T., Sunaga, T., and Kishi, K. (1998) *Appl. Opt.*, **37**, 6182–6195; (c) Hattori, T., Mcallister, D.F. and Sakuma, S. (1992) *Opt. Eng.*, **31**, 350–352; (d) Korpel, A., Adler, R., Desmares, P., and Watson, W. (1966) *Appl. Opt.*, **5**, 1667–1675.
10 Born, M. and Wolf, E. (1980) *Principles of Optics: Electromagnetic Theory of Propagation, Interference and Diffraction of Light*, 6th edn, Pergamon Press, Oxford.

11 (a) Klauder, J.R. and Sudarshan, E.C.G. (1968) *Fundamentals of Quantum Optics*, W. A. Benjamin, New York; (b) Agarwal, G.S. (2012) *Quantum Optics*, Cambridge University Press, Cambridge.

12 (a) Yamauchi, M. and Eiju, T. (1995) *Opt. Commun.*, **115**, 19–25; (b) Buhrer, C.F., Bloom, L.R., and Baird, D.H. (1963) *Appl. Opt.*, **2**, 839–846; (c) Lee, M., Katz, H.E., Erben, C., Gill, D.M., Gopalan, P., Heber, J.D., and McGee, D.J. (2002) *Science*, **298**, 1401–1403; (d) Coffa, S., Franzo, G., and Priolo, F. (1996) *Appl. Phys. Lett.*, **69**, 2077–2079; (e) Wang, L.H., Jacques, S.L., and Zhao, X.M. (1995) *Opt. Lett.*, **20**, 629–631; (f) Leutz, W. and Maret, G. (1995) *Physica B*, **204**, 14–19; (g) Ercole, F., Davis, T.P., and Evans, R.A. (2010) *Polym. Chem.*, **1**, 37–54.

13 (a) Wang, M.S., He, L., and Yin, Y.D. (2013) *Mater. Today*, **16**, 110–116; (b) He, L., Wang, M.S., Ge, J.P., and Yin, Y.D. (2012) *Acc. Chem. Res.*, **45**, 1431–1440.

14 (a) Damon, R.W. and Eshbach, J.R. (1961) *J. Phys. Chem. Solids*, **19**, 308–320; (b) Beleggia, M., Tandon, S., Zhu, Y., and De Graef, M. (2004) *J. Magn. Magn. Mater.*, **278**, 270–284; (c) Ignatchenko, V.A., Edelman, I.S., and Petrov, D.A. (2010) *Phys. Rev. B*, **81**, 054419.

15 (a) Kasperski, M. (2014) *J. Appl. Phys.*, **116**, 143904; (b) Joseph, R.I. and Schlomann, E. (1961) *J. Appl. Phys.*, **32**, 1001–1005.

16 (a) Vi, K. and Sigal, M.A. (1992) *Phys. Status Solidi B*, **170**, 569–584; (b) Sigal, M.A. and Kostenko, V.I. (1991) *Phys. Status Solidi A*, **128**, 219–234; (c) Craik, D.J. and Cooper, P.V. (1972) *Phys. Lett. A*, **A 41**, 255–256.

17 (a) Bohren, C.F. and Huffman, D.R. (1983) *Absorption and Scattering of Light by Small Particles*, John Wiley & Sons, Inc., New York; (b) Hulst, H.C.v.d. (1957) *Light Scattering by Small Particles*, John Wiley & Sons, Inc., New York.

18 (a) Li, J., Liu, X.D., Lin, Y.Q., Bai, L., Li, Q., Chen, X.M., and Wang, A.R. (2007) *Appl. Phys. Lett.*, **91**, 253108; (b) Fan, F., Chen, S., Lin, W., Miao, Y.P., Chang, S.J., Liu, B., Wang, X.H., and Lin, L. (2013) *Appl. Phys. Lett.*, **103**, 161115; (c) Chen, S., Fan, F., Chang, S.J., Miao, Y.P., Chen, M., Li, J.N., Wang, X.H., and Lin, L. (2014) *Opt. Express*, **22**, 6313–6321.

19 (a) Ye, M.M., Zorba, S., He, L., Hu, Y.X., Maxwell, R.T., Farah, C., Zhang, Q., and Yin, Y.D. (2010) *J. Mater. Chem.*, **20**, 7965–7969; (b) Zorba, S., Maxwell, R.T., Farah, C., He, L., Ye, M.M., and Yin, Y.D. (2010) *J. Phys. Chem. C*, **114**, 17868–17873.

20 (a) Erb, R.M., Cherenack, K.H., Stahel, R.E., Libanori, R., Kinkeldei, T., Munzenrieder, N., Troster, G., and Studart, A.R. (2012) *ACS Appl. Mater. Interfaces*, **4**, 2860–2864; (b) Mao, Y.W., Liu, J., and Ge, J.P. (2012) *Langmuir*, **28**, 13112–13117; (c) Bubenhofer, S.B., Athanassiou, E.K., Grass, R.N., Koehler, F.M., Rossier, M., and Stark, W.J. (2009) *Nanotechnology*, **20**, 485302.

21 Goebl, J., Liu, Y., Wong, S., Zorba, S., and Yin, Y.D. (2016) *Nanoscale Horiz.*, **1**, 64–68.

22 (a) Gu, Z.Z., Uetsuka, H., Takahashi, K., Nakajima, R., Onishi, H., Fujishima, A., and Sato, O. (2003) *Angew. Chem. Int. Ed.*, **42**, 894–897; (b) Kinoshita, S. and Yoshioka, S. (2005) *ChemPhysChem*, **6**, 1442–1459; (c) Parker, A.R. (2000) *J. Opt. A*, **2**, R15–R28.

23 Lezec, H.J., McMahon, J.J., Nalamasu, O., and Ajayan, P.M. (2007) *Nano Lett.*, 7, 329–333.

24 He, L., Janner, M., Lu, Q.P., Wang, M.S., Ma, H., and Yin, Y.D. (2015) *Adv. Mater.*, **27**, 86–92.

25 Kinoshita, S., Yoshioka, S., and Miyazaki, J. (2008) *Rep. Prog. Phys.*, **71**, 076401.

26 (a) Joannopoulos, J.D., Johnson, S.G., Winn, J.N., and Meade, R.D. (2008) *Photonic Crystals: Molding the Flow of Light*, 2nd edn, pp. 1–286; (b) Joannopoulos, J.D., Villeneuve, P.R. and Fan, S.H. (1997) *Nature*, **386**, 143–149.

27 (a) Ge, J.P., He, L., Goebl, J., and Yin, Y.D. (2009) *J. Am. Chem. Soc.*, **131**, 3484–3486; (b) Ge, J.P., He, L., Hu, Y.X., and Yin, Y.D. (2011) *Nanoscale*, **3**, 177–183; (c) Ge, J.P., Hu, Y.X., and Yin, Y.D. (2007) *Angew. Chem. Int. Ed.*, **46**, 7428–7431; (d) Ge, J.P., Hu, Y.X., Zhang, T.R., Huynh, T., and Yin, Y.D. (2008) *Langmuir*, **24**, 3671–3680; (e) Xu, X.L., Friedman, G., Humfeld, K.D., Majetich, S.A., and Asher, S.A. (2002)

Chem. Mater., **14**, 1249–1256; (f) Xu, X.L., Majetich, S.A., and Asher, S.A. (2002) *J. Am. Chem. Soc.*, **124** 13864–13868.

28 Ge, J.P. and Yin, Y.D. (2008) *Adv. Mater.*, **20**, 3485–3491.

29 Hu, Y.X., He, L., Han, X.G., Wang, M.S., and Yin, Y.D. (2015) *Nano Res.*, **8**, 611–620.

30 Kim, H., Ge, J.P., Kim, J., Choi, S., Lee, H., Lee, H., Park, W., Yin, Y., and Kwon, S. (2009) *Nat. Photonics*, **3**, 534–540.

31 Xuan, R.Y. and Ge, J.P. (2011) *Langmuir*, **27**, 5694–5699.

32 Ge, J.P., Lee, H., He, L., Kim, J., Lu, Z.D., Kim, H., Goebl, J., Kwon, S., and Yin, Y.D. (2009) *J. Am. Chem. Soc.*, **131**, 15687–15694.

33 (a) Hu, Y.X., He, L., and Yin, Y.D. (2011) *Angew. Chem. Int. Ed.*, **50**, 3747–3750; (b) Wang, M.S., He, L., Hu, Y.X., and Yin, Y.D. (2013) *J. Mater. Chem. C*, **1**, 6151–6156.

34 Stephen, M.J. and Straley, J.P. (1974) *Rev. Mod. Phys.*, **46**, 617–704.

35 Wang, M.S., He, L., Zorba, S., and Yin, Y.D. (2014) *Nano Lett.*, **14**, 3966–3971.

7
Functional Materials: For Sensing/Diagnostics

Rujuta D. Munje,[1] Shalini Prasad,[1] and Edward Graef[2]

[1]*University of Texas at Dallas, Department of Bioengineering, 800 W. Campbell Road, EC 39, Richardson, TX 75080, USA*
[2]*University of Texas at Dallas, Department of Physics, 800 W. Campbell Road, EC 39, Richardson, TX 75080, USA*

7.1
Introduction

There has been a tremendous growth in the field of diagnostic sensors and environmental gas sensors in the past 10 years. The global market for commercial sensors was estimated to be around US$12 billion in 2015 [1]. A sensor is an analytical device that converts a biological response or an environmental change into a signal readable through a transducer. Most common transduction methodologies are either electrical or electrochemical. A diagnostic sensor or a biosensor is used for recognition of molecules such as proteins, enzymes, phages, hormones, aptamers, or metabolites from body fluids. Gas sensing has become an active field of research that has far reaching implication into the future. From the viewpoint of global climate change, air quality, and toxic gas detection, gas sensors are being researched both on the industrial and the university levels to provide tools to better understand these applications and more [2–5]. Environmental gas sensors can be used for detection of gas, VOCs, or trace elements from the environment. Both biosensors and environmental sensors have evolved through multiple generations based on the advancements of the functional materials that are being used for detection of the biomolecule or environmental gas of interest. Functional materials can be characterized as materials possessing special properties such as magnetic, piezoelectric, high tensile strength, capacitive energy storage, and so on. Functional materials can be liquid crystals, semiconductors, ceramics, colloids, thin films, lasers, optics, or magnetic materials. The enrichment in the field of nanotechnology in terms of growth, fabrication, and integration of nanomaterials into a sensor device has set a new paradigm for the form and function of sensor devices [6]. This chapter reviews the current state-of-the art nanomaterial technology leveraged both for the

Handbook of Solid State Chemistry, First Edition. Edited by Richard Dronskowski, Shinichi Kikkawa, and Andreas Stein.
© 2017 Wiley-VCH Verlag GmbH & Co. KGaA. Published 2017 by Wiley-VCH Verlag GmbH & Co. KGaA.

development of protein, hormone, and enzyme biosensors and for the development of environmental gas sensors.

This chapter explains the role of nanomaterials as active functional materials in sensor devices. The latest generation of diagnostic biosensors is focused on wearable biosensing. Wearable health-monitoring devices represent an exciting opportunity in health care, promising to be the disruptive technology changing the status quo in patient care. A highly functional wearable biosensor that can provide meaningful diagnostics to guide therapeutics would be extremely valuable to end-user consumers or health-professionals. In order to make wearable biosensors as successful consumer products, it is important to demonstrate enhanced multiplexed functionality, reliability, and ease of use [7,8]. Nanotechnology provides new avenues to incorporate these desired features along with the ability to build small form factor and flexibility of sensor devices into wearable sensing [9]. Similarly, for environmental gas sensing, specificity of sensing, and reversibility of the functional material to its original state and control on diffusion time are crucial to achieve a highly responsive, reliable, and long-lasting sensing device. Nanomaterials with their varying structural and functional characteristics avail a plethora of options to build the sensors of high accuracy and efficacy. Solid-state semiconductor sensors have the largest investment of study. These materials usually consist of oxides of varying configurations. Through the tuning of these semiconductor materials many advances have been made in the detection of various gases. This chapter has been organized in two sections. The first section describes the popular functional nanomaterials used for biosensing and diagnostics. It discusses the properties of nanomaterials that affect biosensing performance, the techniques for functionalization of nanomaterials to achieve specific biomolecule detection and dynamics of nanomaterial and biomolecular fluid interfaces transduced into an electrical/electrochemical signal to achieve ultrasensitivity of detection. The second section focuses on the development of environmental gas sensors. It describes the current state-of-the-art nanomaterials utilized for hydrogen, oxygen, carbon dioxide, and hydrocarbons detection, which are just some of the few important gas species that can be detected utilizing semiconductor materials [10–13].

7.2
Basic Concept of Label-free Biosensors

Electrical biosensors can be divided into subcategories based on the electrical technique used for detecting the change in measurement that can be used as a calibration for sensing of a biomarker or analyte. The popular techniques that exist are voltammetry, impedance spectroscopy, amperometry, transverse current–voltage or orthogonal current–voltage characteristics, and so on. In all of these techniques, the surface functionalization of the sensor substrate carried out in order to detect the biomarker of interest may or may not include labels bound to target molecule. These labels can be in the form of redox polymers, fluorophores, enzymes, or any chemical or nanomaterial that can be bound to a target molecule and used to measure the concentration of the target biomolecule. The sensors with functionalized assay where no label is introduced to measure the concentration of the target molecule at the sensor surface are called label-free biosensors [14]. Label-free biosensors are advantageous especially in case of protein detection, as they do not change the protein conformation on interaction [14]. Also, they reduce the time and expense of assay preparation associated with labels. Label-free sensors enable real-time detection of target-capture probe binding. Impedance spectroscopy is a great tool for label-free biosensing [15]. The changes in the dielectric properties or resistance at the interface of sensor surface and biomarker diluted fluid can be detected using this technique. Impedance spectroscopy applies an AC voltage with a constant DC bias in a range of frequencies and reads the output current that has a different phase angle than the applied voltage [14,15]. Using the output current and input voltage, impedance is calculated. This impedance is the total sensor system impedance that can be related to interactions of target biomolecules on the sensor substrate. Thus, the changes in impedance can be tracked to monitor changes due to binding of different concentrations of target biomolecule to the capture probe without using any label [16,17].

The standard generalized schematic of label-free biosensor used is shown in Figure 7.1. Crucial components and process flow of the sensor system are displayed in Figure 7.1. The sensor substrate is selected based on the end-user application of the biosensor. It can be a silicon wafer, PCB, glass, flexible materials, and so on. The functional materials are generally selected to enable surface biofunctionalization of capture probes or bioreceptors. The functional materials

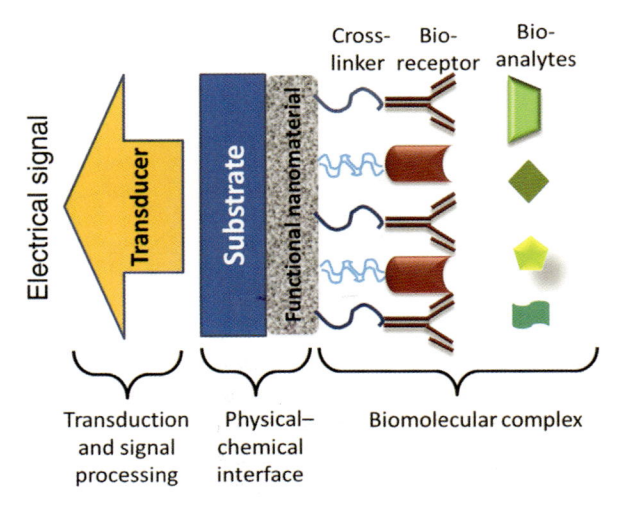

Figure 7.1 Schematic of label-free biosensor. Different shapes of bioanalytes represent varied types of analytes such as enzymes, proteins, and hormones.

also play a significant role in increasing the sensitivity, signal-to-noise ratio, and sometimes selectivity of the biosensor.

The increasing advances in the field of nanotechnology have made it possible to integrate 1D, 2D, and 3D nanomaterials in the sensor design to enhance the performance of electrical biosensor. The ability to tailor the size and shape of nanomaterials to achieve desired functional properties offers multiple avenues to design novel sensing systems. The cross-linker is a chemical molecule that facilitates the chemical or physical linking between functional material and the capture probe [14]. The bioreceptor or capture probe is a molecule that is specific to the target molecule and chemically binds to the target probe as shown in Figure 7.1. The commonly used bioreceptors are antibodies, enzymes, nucleic acids, and so on. The electrical transduction methodology such as impedance spectroscopy translates the biomolecular complex interactions on the functional material and the substrate into a readable electrical signal.

The following sections will focus on the application of the semiconductor functional material zinc oxide (ZnO) with respect to the synthesis and deposition techniques, material properties suitable for biosensing, and its applications.

7.2.1
Nanotextured Semiconductor Zinc Oxide as Functional Material

Zinc oxide is a popular n-type, II–VI semiconductor. It is a compound semiconductor whose ionicity is at the borderline between covalent and ionic semiconductors. ZnO exists mainly in wurtzite and zinc blende crystal structures as shown in Figure 7.2. The thermodynamically stable phase at ambient conditions is wurtzite phase, shown in Figure 7.2a. It has a hexagonal unit cell with two lattice parameters a and c, where $c/a = 1.633$. The structure consists of two closely packed interpenetrating hexagonal sublattices. It is known that the c-axial growth of each columnar ZnO structure within the film results in either zinc or

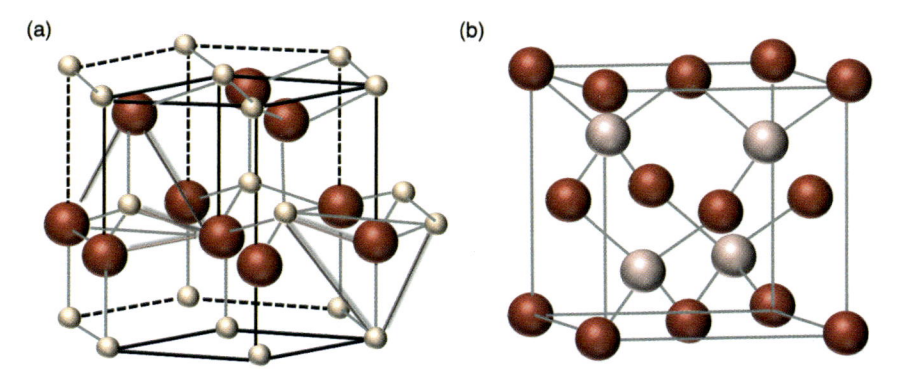

Figure 7.2 Crystal structures of ZnO. (a) Wurtzite, hexagonal packing. (b) Zinc blende, cubic packing.

oxygen surface terminations, which creates either a polar (0001) or (000$\bar{1}$) plane, respectively. The zinc blende crystal lattice is rather unstable and quickly transforms into wurtzite form. It can be stabilized by heteroepitaxial growth on cubic substrates. A schematic of the zinc blende crystal lattice is shown in Figure 7.2b [18].

ZnO is a biocompatible semiconductor material that has been used in numerous studies as functional material for biosensing. ZnO nanostructures are believed to be nontoxic and biosafe for *in vivo* applications and are widely used in cosmetics and drug carriers. ZnO has a wide direct bandgap of 3.37 eV and high isoelectric point (IEP) of ~9.5 along with tunable electrical properties. The ZnO matrix is positively charged at pH 7. A higher bandgap allows for higher breakdown voltages and the ability to sustain larger electric fields. The superior IEP allows for enhanced adsorption of proteins and linker biochemical molecules due to higher electrostatic attraction [19]. The electronic properties of ZnO can be modulated by controlling the extent of oxygen deficiency. Also, the semiconducting properties of ZnO help in tuning the resistivity and surface charge distribution of the thin film to a suitable level in order to improve the signal-to-noise ratio, even in the presence of a complex medium like body fluids. Moreover, ZnO being a polar semiconductor can also address the issues of specificity in the biosensor. The surface terminations of ZnO can be used to achieve favorable linker chemistry. Particular deposition conditions can be used to achieve n-type ZnO material that grows in wurtzite crystal form. The advantage of this type of crystal form in ZnO is the availability of alternate stacked layers with zinc and oxygen terminations. The ability to modulate the surface states can be useful in achieving greater specificity of the sensor.

7.2.2
Synthesis of ZnO Nanostructures for Biosensing Applications

A variety of one-dimensional, two-dimensional, and three-dimensional ZnO nanostructures synthesized by different physical and chemical growth mechanisms have been researched for biosensing and diagnostic applications. The structural, electrical, and chemical properties of ZnO vary according to the shape, size, and surface chemistry of the nanostructures, as well as the growth mechanisms. ZnO nanoparticles, nanowires, nanobelts, nanoflowers, nanotubes, thin films, nanocombs, and nanoflakes are some of the types of nanostructures found to be implemented for biosensing applications. The physical growth mechanisms primarily include evaporation techniques such as RF magnetron sputtering and pulse layer deposition (PLD). The chemical or wet growth mechanisms include hydrothermal growth, solid–liquid–solid (SLS), and atomic layer deposition (ALD). Due to the ease of fabrication, options for cheap growth mechanisms and ability to control the characteristic properties of nanaostructures, ZnO is an attractive choice for development of low-cost biosensing applications. The different synthesis approaches and their applications are summarized in Table 7.1 above.

Table 7.1 Synthesis and applications of ZnO for biosensing.

Nanostructure	Growth mechanism	Application	Reference
ZnO nanotextured thin film	RF magnetron sputtering	Glucose detection	[20]
ZnO nanoparticles	In solution	Triglyceride detection	[21]
ZnO nanowires	In solution	Glucose detection	[22]
ZnO nanoflower	In solution	Cholesterol	[23]
ZnO nanobelt	In solution	Urea	[24]
ZnO nanofork	In solution	Hydrogen peroxide	[25]
ZnO nanotextured thin film	Plasma enhanced chemical vapor deposition	EGFR proteins	[26]

Saha and Gupta studied RF magnetron sputtered ZnO thin films grown at different pressures (20–50 mT) in a gas mixture of argon and oxygen. They demonstrated the effect of variable surface defects in ZnO thin films, on glucose detection. The electrochemical studies were carried out for biosensing of glucose. The surface of a ZnO thin film grown at higher pressure of 50 mT was found to be most favorable for biomolecular orientation on the surface. This resulted in higher affinity of glucose oxidase enzyme toward glucose, improving the stability of the electrodes. Also, the lowered activation energy in the prepared matrix provided enhanced electron transfer at the electrode surface [20]. Narang and Pundir developed a composite of ZnO nanoparticles (NPs) using wet chemical method with reactants zinc nitrate and sodium hydroxide. They used chitosan (CHIT), which is a biopolymer used as a linker for biofunctionalization. ZnO NPs-CHIT composite was prepared for biofunctionalization on platinum (Pt) electrodes. Cyclic voltammetry was performed on the CHIT/Pt and ZnO NPs–CHIT/Pt composite. It was demonstrated that redox potentials of the ZnO NPs-CHIT/Pt electrode shifted toward the higher side as compared to CHIT/Pt electrodes. This study proved that there is an accelerated electron transfer due to the presence of ZnO NPs and ZnO NPs also provide a biocompatible environment [21]. Umar *et al.* fabricated flower-shaped ZnO on gold electrodes using *in situ* preparation with reactants zinc nitrate and hexamethylenetetramine. Cyclic voltammetry measurements were carried out after physical adsorption of cholesterol oxidase (ChOx) on ZnO nanoflower structures. Nafion was used to tightly attach the nanostructures and ChOx on the gold electrodes. The Michaelis–Menten constant was derived using cyclic voltammetry measurements and it was observed to be 2.57 mM, which represents the highest affinity of ChOx toward cholesterol on the ZnO nanoflowers. A very high sensitivity of 61.7 µA/(µM cm^2) was also demonstrated, which was attributed to the active sites of immobilized ChOx and the electrode surface that leads to a sharper and well-defined peak and also the large surface area due to the dimension and shape of the flower-shaped nanostructures [23]. Reyes *et al.* developed a ZnO-based TFT biosensor where the ZnO thin film was deposited using plasma-enhanced chemical vapor deposition and acted as an n-type conduction channel. The ZnO thin film was biofunctionalized with cross-linkers and monoclonal EGFR antibodies to act as capture probe. The drain-source current was observed with unfunctionalized ZnO-FET, where the positive bias at the gate causes a drain-source current purely due to the electron carriers in the ZnO thin film. As the surface was functionalized with EGFR antibodies, it acted as a resistive layer and decreased the current. For the third step, when EGFR antibodies were exposed to EGFR protein, it formed a positively charged tip complex protein that caused accumulation of electrons at the conduction path near gate. This amplified the drain current that was used as a calibration response due to different concentrations of EGFR [26]. This type of ZnO-bio-TFT has potential application in handheld and point-of-care diagnostics.

7.2.3
Functionalization of ZnO Nanostructures for Biosensing Applications

The electrical characteristics and surface states of ZnO can be tuned based on deposition conditions [9]. Due to the nature of its growth mechanism, each columnar ZnO structure terminates with either a polar zinc or an oxygen surface [27]. These surface characteristics of ZnO films can, therefore, be leveraged for biosensing [28]. The ability to tailor and functionalize the terminations of the ZnO surface would leverage the electrochemical properties of the thin film for performing the desired biochemical function, improving the sensor selectivity.

ZnO RF magnetron sputter deposited films with and without oxygen flow were studied by depositing them on PCB-based gold electrodes as shown in Figure 7.3a and b [29]. The linker dithiobis(succinimidyl-propionate) (DSP) contains a thiol group that has specificity in binding to the zinc terminated sites. This linker was functionalized as shown in Figure 7.3d on ZnO thin-film sputter deposited without oxygen flow. The linker 3-aminopropyl triethoxysilane (APTES) contains a silane group, which has specificity in binding to the oxygen-terminated sites. This linker was functionalized as shown in Figure 7.3e on ZnO thin-film sputter deposited with oxygen flow. The immunoassay used for this analysis is shown schematically in Figure 7.3c. We specifically focused on analyzing the methods of selective functionalization of surface terminations of ZnO thin films and investigating the biosensor performance for troponin-T detection from phosphate buffered saline (PBS). Human serum (HS) samples containing variable doses of antigen were also tested to demonstrate the potential of the sensor for point-of-care diagnostics. Evaluation of each linker molecule bound to the ZnO surface and its utility in biosensor performance was studied by using non-Faradaic electrochemical impedance spectroscopy (EIS). This was supported through fluorescence analysis using fluorophores.

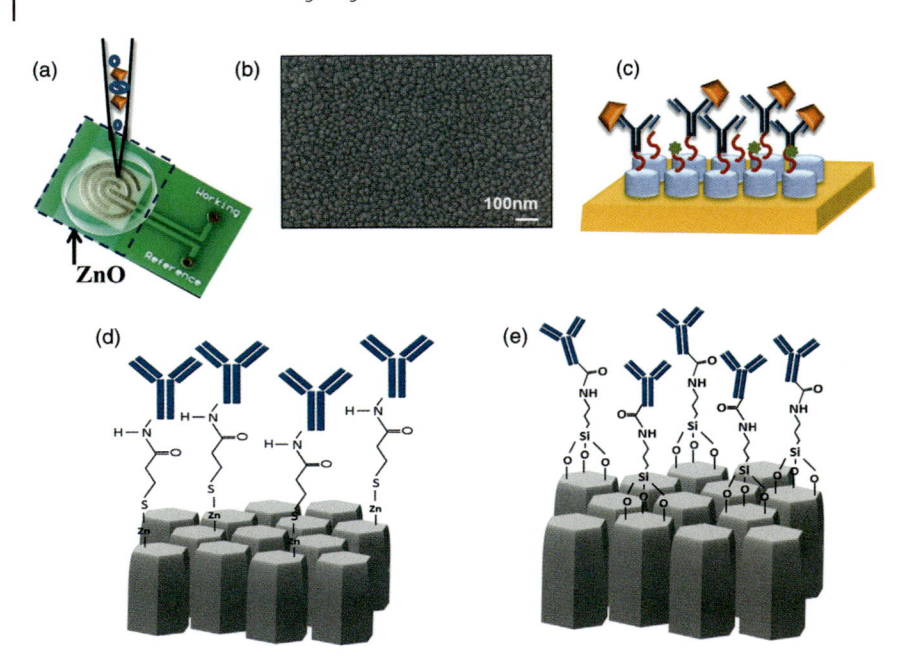

Figure 7.3 (a) PCB substrate with sputtered ZnO over gold electrode pattern. (b) SEM image of nanotextured ZnO thin film. (c) Schematic of immunoassay functionalization on ZnO thin-film-based electrodes for troponin-T. (d) Immunoassay setup on the ZnO surface using DSP bound with α-troponin-T. (e) Immunoassay setup on the ZnO surface using APTES bound with α-troponin-T [29].

The results of the above-mentioned study are shown in Figure 7.4. The equivalent circuit diagrams for DSP and APTES linker functionalized sensor systems are shown in Figure 7.4a and b, respectively. An equivalent circuit diagram is an electrical parameter representation of a system under evaluation, where resistance and capacitance of the system is arranged in particular manner. The Zn–thiol interaction-based sensor was imagined to consist of two resistance-capacitance arranged in parallel (RC) pairs, where one pair was formed due to the binding of thiol in DSP to individual zinc terminations and the other pair is formed due to the binding of biomolecules. The APTES–oxygen interaction-based sensor is imagined to be consisting of only one RC component, assuming that APTES will form a continuous monolayer on the surface establishing a connected platform for antibody–antigen interactions. The difference in the surface chemistries of a DSP-based immunoassay and an APTES-based immunoassay can be interpreted from the shape of Nyquist plots obtained after performing EIS. Nyquist plots represent the dynamics between real impedance and imaginary impedance of the total impedance of the sensor system measured using EIS. The results of fluorescence study are shown in Figure 7.4c. It displays the relative effectiveness of the binding of DSP and APTES linkers in terms of percentage of fluorescent pixel intensity. Higher fluorescent pixel intensity due to

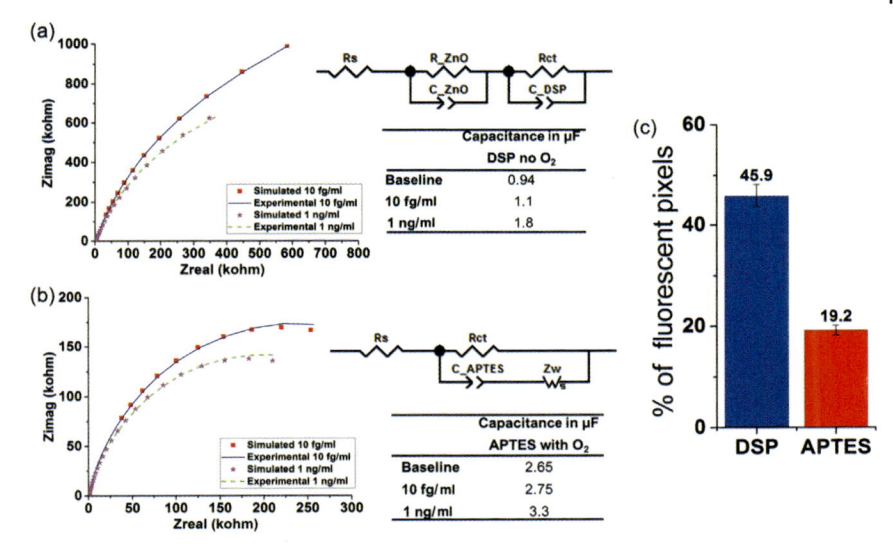

Figure 7.4 (a) C_{dl} for DSP linker represented as the combination of C_{ZnO} and C_{DSP}. Simulated and experimental Nyquist plots in PBS using DSP. (b) C_{dl} for APTES linker symbolized as the combination of C_{ZnO} and C_{APTES}. Simulated and experimental Nyquist plots in PBS using APTES. (c) Fluorescent pixel comparison of DSP and APTES using Rhodamine 123 fluorophore [29].

higher binding of fluorescent molecule Rhodamine 123 can be interpreted as the better coverage of the linker on the ZnO surface. DSP showed approximately three times more binding on the ZnO surface than the APTES linker. Thus, the enhanced performance of the sensor was achieved through leveraging the zinc–thiol interactions using DSP toward designing a robust immunoassay [29].

7.2.4
ZnO Nanostructure-based Wearable Biosensors

Wearable health-monitoring devices represent an exciting opportunity in healthcare. Providing patients with tools to track their own conditions, they could be empowered to take responsibility for their own health. Sweat contains valuable medical information and is perhaps the most suitable body fluid for designing noninvasive diagnostics platforms for consumer healthcare applications as compared to blood or urine [30].

We used n-type semiconductor material zinc oxide (ZnO) as the active region in a three electrode setup for biosensing on polyamide substrates. We used the pulsed laser deposition (PLD) technique to deposit ZnO thin films in the active regions [31]. PLD deposited thin films are highly crystalline and uniform with stable electrical properties even after deposition on flexible substrates [32] (see Table 7.2).

7 Functional Materials: For Sensing/Diagnostics

Table 7.2 Synthesis and applications of ZnO nanostrures on flexible substrates.

Nanostructure	Growth mechanism	Flexible substrate	Application	Reference
ZnO nanowires	In solution	Plastic	pH sensor	[33]
ZnO nanotextured thin film	RF magnetron sputtering	Polyimide	Ethyl glucuronide	[34]
ZnO nanowires	In solution	Polyimide	Breath monitoring	[35]

Electric field control of charge carrier density in a semiconductor has been identified as a method for tuning the electronic states of condensed matter. When a semiconducting material interacts with liquid electrolytes, an electrical double layer (EDL) is formed, which can have significantly high charge carrying capacity, at times as high as $8.0 \times 10^{14}\,cm^{-2}$ at ionic liquid/solid interfaces [36]. This property was leveraged in this case, where biocomplex formed at the ZnO–electrolyte interface was highly capacitive due to the formation of capacitive electrical double layer (C_{EDL}). The variations in C_{EDL} due to variable concentrations of cortisol were used for calibration mapping in this application [31].

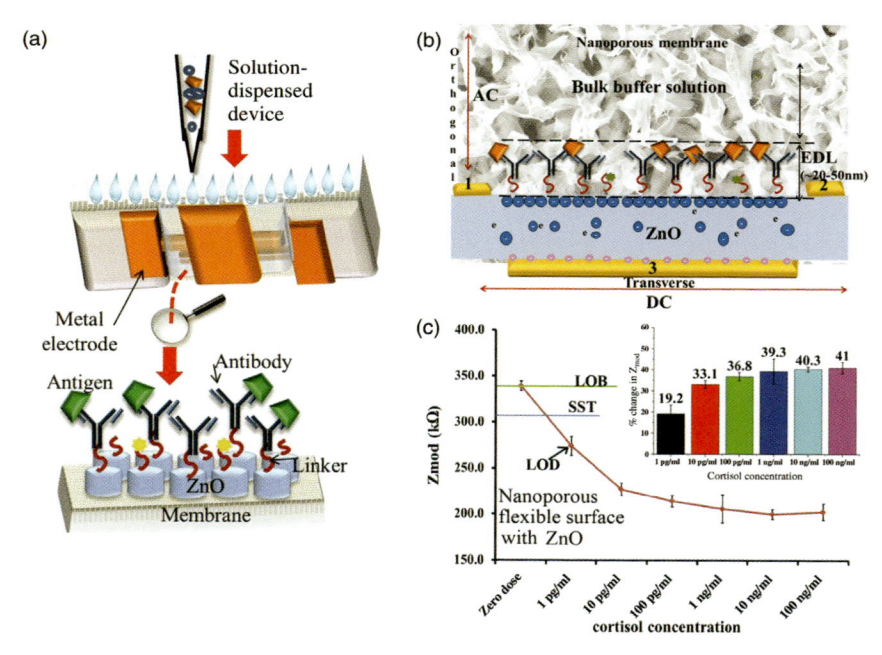

Figure 7.5 (a) Schematic representation of flexible substrate-based biosensor for cortisol detection [37]. (b) Cross-section of ZnO surface on the flexible nanoporous polyamide substrate with modification to EDL during immobilization of biomolecules [31]. (c) Change in total impedance Zmod for varying cortisol concentration of 1 pg/ml to 100 ng/ml. Inset represents percentage change in impedance with respect to baseline, that is, zero dose [31].

ZnO is a versatile semiconductor material that can be used as a promising platform for biosensing. ZnO is one of the prime candidates for ultrasensitive, selective, and low-cost wearable diagnostic applications. Moreover, doping ZnO nanostructures can improve its functional properties and can be an effective approach for tuning the characteristics as per required sensing application (Figure 7.5).

7.3 Environmental Sensors

7.3.1 Oxygen Sensors

Oxygen sensors have use in multiple applications. Enormous interest in these sensors has been shown in the auto industry as well as the life and medical sciences. In the auto industry, oxygen sensors are utilized in the engine management systems to control the air-to-fuel ratio supplied to the engine [5]. In the biological and medical realm, they are used to monitor chemical processes as well as in food processing applications [4]. For all of these cases, they have to be operated at high temperatures and must be capable of detecting large swings in oxygen concentration [4,5,38–40]. Current sensors on the market utilize potentiometric equilibrium, current limiting amperometry, and resistance variation measurements in most applications.

Metal oxide-based semiconducting materials offer a type of resistive gas sensing approach. With their ability to be easily miniaturized, high signal-to-noise ratio, and ease of portability, MOS-based sensors have grown in popularity over the past few years. Ceria is one such material that is capable of all of these requirements. With its low activation energy, this behavior allows for ionic conduction when the material is at high temperatures in the range of 600–800 °C [4,41]. When in pure or doped form, the fluorite crystalline structures of ceria behave like a resistive system. The materials characteristics respond to the ambient oxygen concentration by the capture or release oxygen from its surrounding environment [42].

Due to the elevated temperature operation of these sensors, their detection criteria change. Between the temperatures of 300–500 °C, the material relies on chemisorption of oxygen at the surface of the detector that allows for the production of oxygen ions [43]. The extent of chemisorption is dependent upon the partial pressure of the oxygen within the sampling environment. Once the material reaches around 600–1000 °C diffusion begins to dominate. These vacancies are described in Eq. (7.1). It is important to note that none of this is possible unless the ceria structures are created with a nonstoichiometric low oxygen component to allow for the ease of adsorption of environmental oxygen [44].

$$O_o^X \leftrightarrow V\ddot{o} + 2e' + \frac{1}{2}O_2^{(g)}. \tag{7.1}$$

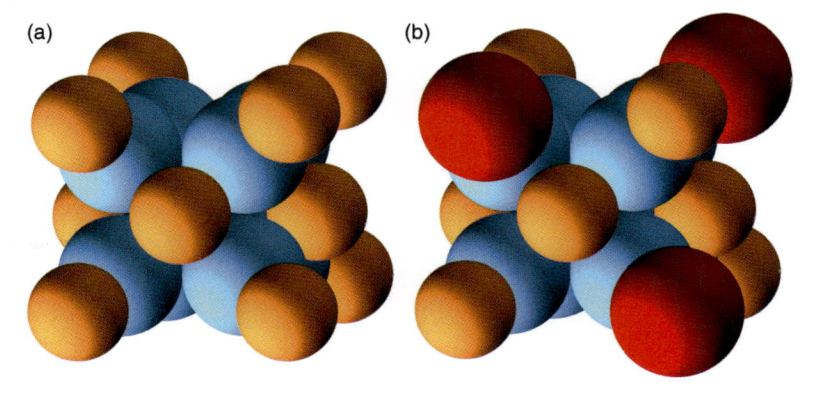

Figure 7.6 (a) Flourite structure of pure cerium oxide (ceria). (b) Substitutional doping of ceria by addition of Samarium.

Doping of the ceria influences the materials behavior when exposed to oxygen. Dopants capable of trivalent states are ideal in aiding in the adsorption of oxygen. Aluminum, chromium, and samarium are just a few of the metals capable of existing in a trivalent state. Samarium in particular allows for a trivalent state that aids in the mobility of the oxygen ions while also relaxing and minimizing the strain that is produced in the fluorite structure in Figure 7.6a of ceria when doped. Upon elevation of the sensors temperature, the oxygen ions at the surface are adsorbed by overcoming the activation energy of the barrier of the material. This adsorption modifies the thin-film current that is measured [44].

The measurement of current through the film is dependent on both the conductivity and the temperature of operation. As either of these increases, the current changes with it. These current values are measured utilizing the constant voltage technique. This involves holding the voltage source constant as the film is characterized, Figure 7.7. As shown in Figure 7.7, conductance and, therefore, resistance varies as one increases the temperature of the thin film as it is tested (Table 7.3).

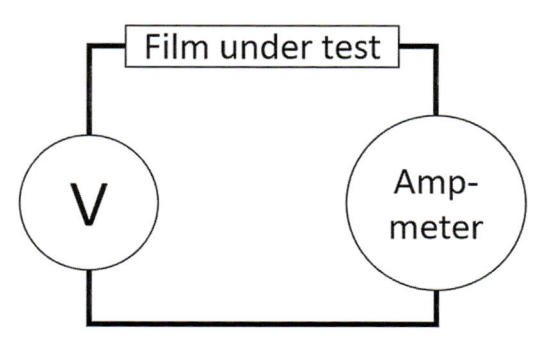

Figure 7.7 Layout of constant voltage setup for the characterization of doped ceria thin films.

Table 7.3 Resistance change between the different adsorption regimes with respect to Samarium doping concentration [44].

Thin film	Resistance (Ω) @673 K	Resistance (Ω) @973 K
Pure ceria	Infinite	3.5×10^9
3% SDC	2.5×10^8	1.11×10^7
6% SDC	2.01×10^8	8.38×10^6
10% SDC	2×10^9	1.95×10^7
14% SDC	1×10^{10}	4.2×10^7

The doped ceria was prepared by growing it on (0001) stoichiometric aluminum oxide (Al_2O_3) utilizing molecular beam epitaxy (MBE). This ultrahigh vacuum (UHV) process utilized a dual chamber system that allowed for the controlled evaporation of selected metals onto the surface of the substrate. An oxygen plasma was generated utilizing electron cyclotron resonance during the deposition process. Growth of the material was monitored utilizing calibrated quartz crystal oscillators while the quality of the growth was monitored utilizing reflection high-energy electron diffraction (RHEED). Additional X-ray photoelectron spectroscopy (XPS), X-ray diffraction (XRD), and Rutherford backscattering (RBS) measurements were performed *ex situ* to determine the orientation of the films as well as the their quality, thickness, and composition. Due to the similar atomic numbers of Sm and Ce, it was difficult to resolve their individual peaks in the RBS spectra. From the XRD results, it was shown that the thin film grew in a (111) alignment. This implies that the growth of the doped ceria thin film was a 2D process throughout the deposition [45].

The testing of the thin film requires the creation of the stack shown in Figure 7.8. This stack was created using predeposited silicon/silicon oxide wafers that are then processed via a standard lithographic process to create a Cr/Au pattern. The doped ceria is then mounted to this substrate for characterization. The doped ceria films were grown to different thickness between 50 and 300 nm with different Sm concentrations from 3 to 14 at% [44,46].

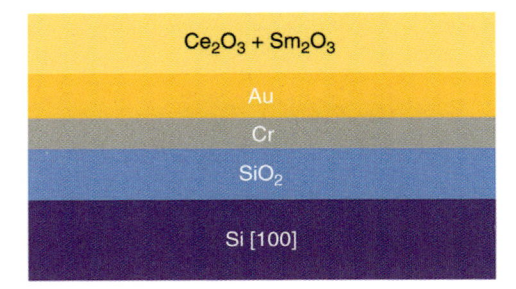

Figure 7.8 Overview of sensor stack to determine oxygen concentration utilizing doped ceria.

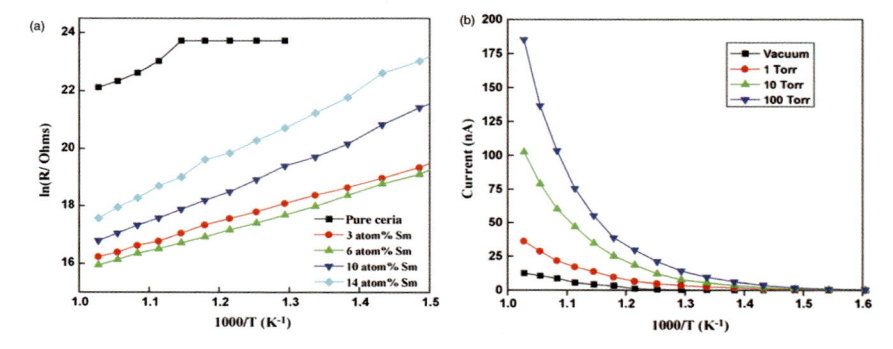

Figure 7.9 (a) Ln(R) versus scaled inverse temperature plot of different Sm-doped ceria thin films. As the temperature of the thin films is increased, the conductivity of the film increases [43]. (b) Current versus inverse temperature with respect to different oxygen pressures. The current was measured utilizing a 10 at% Sm-doped ceria film [44].

The resistive behavior of these thin films is shown in Figure 7.9a. All but the pure ceria show an exponential relationship with respect to the inverse to the temperature. This behavior shows that as the temperature is increased in these doped thin films, the conductivity and therefore the ability of thin film to adsorb oxygen ions increases. This behavior is not dependent upon the environmental oxygen concentrations. From Figure 7.9a, it was determined that a 6% doping of Sm created optimal conditions in the lattice for oxygen diffusion [44]. The increase in resistance at higher and lower doping amounts is due to the amount of defects created in the lattice. These defects increase the activation energy required for oxygen ions to "hope" along the lattice and start to create more polycrystalline features due to the creation of partial grain boundaries in the thin film.

Epitaxially grown ceria films that have been doped with Sm are able to differentiate between the presence and the absence of oxygen at different pressures and elevated temperatures. The pressure variation increases and decreases the amount of oxygen per unit volume such that the material is able to differentiate between different concentrations of oxygen that it is exposed to. A 6% doped thin film shows the best behavior out of the differing doping amounts studied due to the minimization of lattice stress while also providing an elevated number of oxygen binding sites when compared to pure ceria [44]. This causes a change in conductance through the film that allows the film to successfully differentiate between changing oxygen pressures.

7.3.2
Hydrocarbon Sensors

Carbon nanotubes have been researched heavily for the past decade [12]. Due to their strong ballistic conduction properties, ease of functionalization, and well-studied electron transport mechanism, multiwalled carbon nanotubes (MWCNTs) show the most promise in the areas of nanoelectronics, nanofunctionalized composites, and nanointerconnects [12,13,43]. Recently, it has been reported that through chemical functionalization of MWCNTs, they are capable of detecting trace amounts of chemicals [10,47,48]. The ability to detect any chemical with functionalized MWCNTs is through ballistic transportation of electrons [49]. Detection of hydrocarbons, specifically, of different weights is of great importance as a monitor for environmental pollution due to the production of different fuel sources as well as other anthropogenic sources [50].

Currently available sensors on the market consist of micrometer-scale detectors that rely on complicated microfabrication techniques [12,51]. These sensors rely on electrical detection, optical fluorescence, or impedance-based changes to detect the presence of hydrocarbons [3,50,52]. Several different sensors utilizing different sensing techniques have been studied. Utilizing a carbon nanotube-functionalized quartz crystal microbalance, it is possible to detect hydrocarbons that attach themselves to CNTs. The disadvantages of this technique are that it is difficult to reduce nongravimetric effects, especially in environmental applications. They also suffer sensitivity issues due to a lack of selectivity issues. This greatly increases the difficulty of differentiating between any two hydrocarbons. Even with the addition of polymer stabilization, different polymers show different diffusion abilities for different hydrocarbons. There are also computational methods that aid in differentiation, but they have their own set of issues due to their reliance on computationally costly methods (i.e., neural networks) [53]. Another form of sensor relies upon a yttrium-stabilized zirconia sensor. They have been approved for automotive use due to their ability to withstand harsh conditions and high temperatures. Their drawbacks come from their difficulty to differentiate between hydrocarbons due to the different fractions of each in the mixtures that occur in automotive exhausts [54]. A final form of hydrocarbon detector is one that utilizes embedded quantum dots in a polymer-based film. These devices have shown sensitivity to hydrocarbons, but they have inconsistent behavior experiment to experiment. The inconsistency comes from a combination of enhancement and quenching effects due to the loading of the quantum dots are well as increasing the concentration of the hydrocarbons. This inconsistent response limits the practical applications of this sensing material [55]. The limitations of these microscale devices are the ability to trigger false alarm readings, a lack of robustness, and low end detection sensitivity [56]. Only through the utilization of nanoscale structures with microscale systems are we able to eliminate these issues from hydrocarbon sensors [12].

Hydrocarbon sensors using functionalized materials already exist on the market, but they suffer from manipulation difficulties of these ordered functional nanomaterials and their low signal-to-noise ratios [57]. However, utilizing a newly designed microelectrode array, Figure 7.10, along with accurate spatial patterning of functionalized MWCNTs, the signal-to-noise ratio and chemical selectivity are higher and more repeatable than current technology. Due to the physical offset of the linear electrodes combined with the circular electrodes,

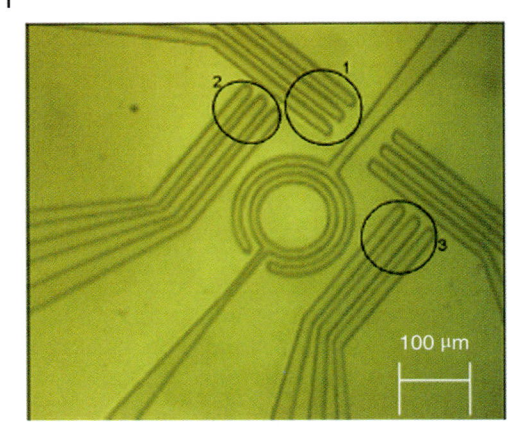

Figure 7.10 Microelectrode array fabricated utilizing standard photolithography. Planer electrodes are 15 µm in width (1,2,3). Circular electrodes were 100 µm in diameter [49].

Figure 7.10, the sensor is able to generate a gradient around the microelectrodes as necessary. This allows the electrode system to be used for electrorotation and electrophoresis experiments by applying signals of varying voltage levels simultaneously to achieve the necessary linear motion and circular motion effects. Another advantage of this microelectrode layout is that for any given experiment, either circular or planar electrodes can be utilized as the measurement electrodes [49].

The microelectrode arrays are assembled utilizing standard lithography methods. The process is schematically shown in Figure 7.11a [50,58,59]. The substrate consists of a silicon wafer with a phosphorus-doped silicon oxide layer that is 1 µm thick. This wafer was first cleaned using a sulfuric acid and hydrogen peroxide piranha solution. After the wafer was washed and dried, it was coated and baked with a positive photoresist 3 µm in thickness. This was then exposed

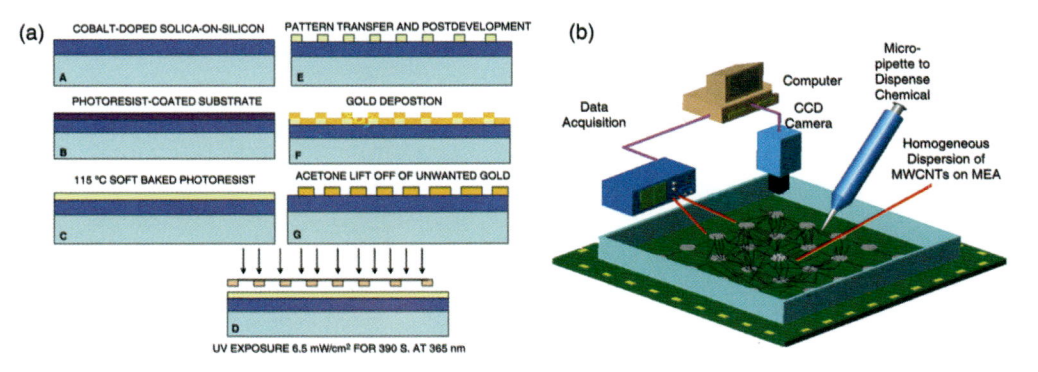

Figure 7.11 (a) Process layout for the creation of microelectrode array [49]. (b) Schematic of the measurement setup utilizing MWCNTs. These nanotubes are distributed in a homogeneous manner [49].

through a contact mask utilizing UV light and then developed. After a wafer inspection step, the wafer is then sputtered with 36 nm of gold. The final step utilized acetone to lift off the developed resist, leaving the microelectrodes shown in Figure 7.11 [49].

Before application of the MWCNTs, they must first undergo functionalization [49,50,60–64]. They are first clumped together utilizing in DI water utilizing a centrifuge. This pellet of MWCNTs is then transferred into sodium dodecyl sulfate and allowed to diffuse into the liquid, creating a suspension of SDS and MWCNTs. This solution allows for a thiol end group to attach to the endcaps that allows for dangling bonds to occur. This solution is then dispersed homogeneously across the microelectrodes to provide a uniform layer of functionalized MWCNTs that will be receptive to different types of hydrocarbon chains.

These functionalized MWCNTs act as an electrochemical transducer that modifies its resistance when exposed to specific chemicals, which are shown in Table 7.4 [65]. When any of these chemicals are present and allowed to react with the MWCNTs, the adsorption causes a shift in the resonance peak of the system [2]. When the shift is compared to the unreacted harmonic peak, it is possible to identify which chemical is present in the system. To record these peaks and their behavior, a set of micropositioners with attached microneedles are connected to a data acquisition board (DAQ) combined with an optical microscope to allow for the collection of *in situ* data. In order to establish and optimize both noise levels and baseline behavior, the microelectrodes and microneedles were first analyzed utilizing this setup without the addition of MWCNTs.

This functionalized sensor was tested utilizing four different hydrocarbons: methanol, ethanol, propanol, and butanol. In each experiment, the hydrocarbon was diluted to the necessary concentration of 100 000 ppm, 10 000 ppm, 1000 ppm, and 1 ppm via DI water. A sampling time of 2 min was used with a rate of measurement at 10 kHz. The voltage changes versus time were measured continuously and interpreted using complex analysis in order to differentiate between the various chemicals studied. This data allowed for the empirical discovery of the ambiguous frequencies listed in Figures 7.12 and 7.13 [49]. The

Table 7.4 Response frequencies, harmonic repetition, and frequency ambiguity for selected hydrocarbons used in experiments.

S. no.	Chemical	Initial frequency response (Hz)	Harmonic repetition rate (Hz)	Frequencies of ambiguity (Hz)	Standard deviation
1	Methanol	100	100	100, 200, 300	0.811
2	Ethanol	60	120	60, 180, 300	0
3	Propanol	60	60	60, 120, 180	0.507
4	Butanol	60	10	60, 90, 100, 120, 190	0.707

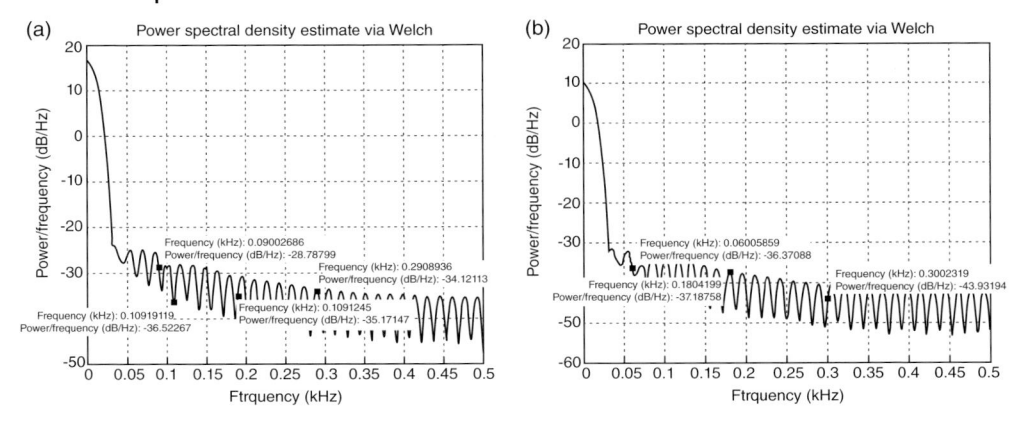

Figure 7.12 (a) Power spectral density plot for methanol at a concentration of 1 ppm. The PSD was calculated at a set of frequencies: 100, 200, and 300 Hz. These frequencies were chosen due to the known response of methanol. There is also an ambiguity at these frequencies between methanol and butanol [49]. (b) PSD of ethanol at 1 ppm concentration. PSD was computed at 60, 180, and 300 Hz. These frequencies were chosen to eliminate the ambiguity of detection between propanol and butanol [49].

time series were used to generate the power spectral density for each of the hydrocarbons, Figures 7.12 and 7.13.

To properly understand and differentiate between these hydrocarbons, a baseline collection of the dry sensor must be performed. Temporal voltage data was collected over 2 min at a sampling rate of 10 kHz. This established the resonant behavior of the functionalized MWCNT material. After baseline harmonics were

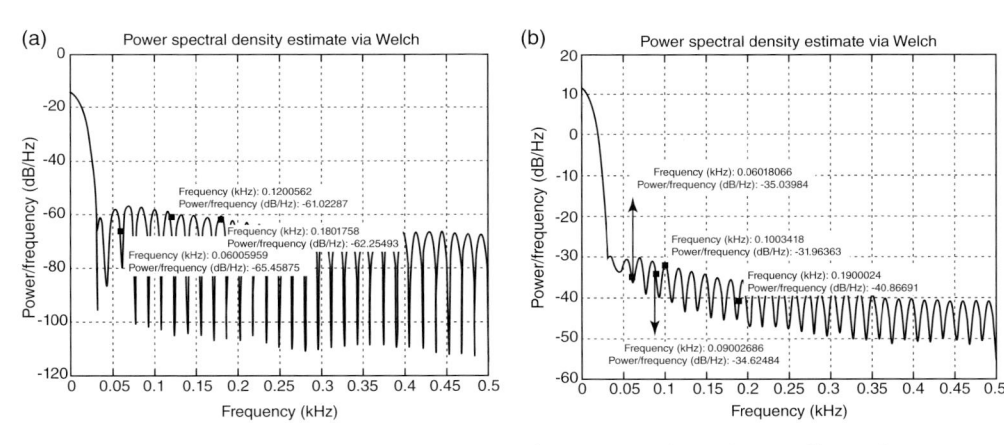

Figure 7.13 (a) Power spectral density of propanol at 1 ppm concentration. The PSD was calculated for ambiguous frequencies of 60, 120, and 180 Hz. The PSD at these frequencies allows for the differentiation of propanol with respect to methanol, ethanol, and butanol [49]. (b) PSD of butanol at 1 ppm concentration. Calculations of the PSD were performed using the frequencies of 60, 90, 100, and 190 Hz to differentiate between ethanol, methanol, and propanol [49].

established, methanol was tested and showed a harmonic frequency of 100 Hz regardless of concentration tested with an initial response frequency of 100 Hz. Figure 7.12a shows the power spectral density of methanol, the calculations for each of the hydrocarbons was produced using the Welch function in the MATLAB analysis toolkit. PSD for ethanol is shown in Figure 7.12b, this hydrocarbon is shown to have a harmonic frequency of 60 Hz that was differentiable from background noise with an initial response frequency of 60 Hz as well. This behavior held true from 100 000 ppm to 1 ppm of ethanol. Propanol is shown in Figure 7.13a. The harmonic frequency found for propanol is 120 Hz with an initial main peak occurring at 60 Hz. Butanol, Figure 7.13b, showed consistent behavior across all concentrations tested with a harmonic frequency of 10 Hz with an initial response frequency of 60 Hz [49].

The ability to differentiate these molecules was dependent upon several factors. Namely, the design geometry of the microelectrodes, the form of detection, the functionalization of the surface, and the analysis techniques employed. All of these factors affect the overall behavior of the detection system. The microelectrode array geometry has been shown to have an effect on the noise behavior of the system in previous literature [51,66]. Noise follows an increasing trend as one shrinks the size of the electrodes. The material used also plays a part in the detection as well due to its influence on the contact resistance between itself and the MWCNTs. The application drives the geometry and material used for these types of sensors, so there is no "one size fits all" sensor design. The form of detection relies on the fact that a functionalized MWCNT acts as a nanoscale transducer when placed across electrode arrays. A method of improvement would work to tune the array design such that the detection method would utilize a cohesive interface utilizing different MWCNT deposition techniques [49].

Functionalization of the MWCNTs is dependent on the use of SDS to create S—OH bonds with the ionization of the specific species of hydrocarbons [60,63]. This form of bonding is the main detection method and relies on the endcaps forming the appropriate form of dangling bonds. To improve upon this detection method, it has been suggested in the literature that a carboxylic group and hydroxylic functionalization approach would create a larger density of reactive binding sites that would, in turn, increase the signal-to-noise ratio of the sensor via increased signal densities during detection [62,64,67].

While analyzing the collected temporal data, one cannot solely rely on the time domain analysis of the different chemicals due to their similarities. Multiple levels of analysis are needed in order to differentiate between the species. Utilizing fast Fourier transformations, the temporal data is analyzed to determine the presence of specific chemicals in the set studied. Power spectral density analysis is then performed to accurately identify the chemical exactly. This was determined by taking the peak frequencies measured via the FFT and measuring the power per unit frequency for them, Figure 7.14. This established the key frequency peak identifiers that could be measured electrically for each of this family of chemicals.

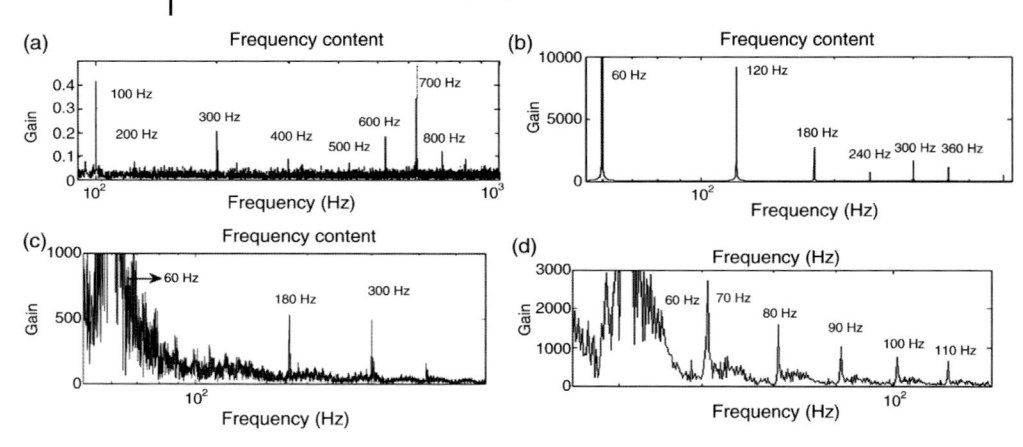

Figure 7.14 Frequency spectra from the studied molecules utilizing FFT analysis. (a) Methanol frequency spectrum with peaks at 100, 200, and 300 Hz [49]. (b) Propanol frequency spectrum with peaks occurring at 60, 120, and 180 [49]. (c) Ethanol with frequency peaks at 60, 180, and 300 Hz [49]. (d) Butanol spectrum with peaks occurring at 60, 90, 120, and 190 [49].

The rate of ionization and the number of binding sites play significant roles in the ability to differentiate these chemicals. Due to the functionalization mechanism being identical for all of the chemicals, there occurs a frequency ambiguity due to them having identical binding mechanisms with the MWCNTs [49]. Only via FFT and PSD analysis can one successfully differentiate the materials from each other.

The future of these hydrocarbon sensors is only limited by the chemistry used to functionalize the MWCNTs. Exploring functionalized single-walled carbon nanotubes, functionalization methods, and different electrode geometries and depositions will help only to increase the applications for these types of hydrocarbon sensors. Through this exploration, these sensors should be able to study and differentiate additional families of chemicals as well.

7.4
Conclusions

In summary, versatile properties of functional materials can be leveraged to improve the overall performance of sensing applications. As a result, sensor functionality is greatly enhanced by enabling the design of sensors that are small, portable, easy to use, low cost, disposable, and highly versatile diagnostic instruments. We have attempted to cover the most exciting research studies on functional materials that are applicable for the current generation of sensing applications and industrial needs. However, there are myriad types of functional materials that are evolving everyday to advance the world of sensors.

References

1 Bahadır, E.B. and Sezgintürk, M.K. (2015) Applications of commercial biosensors in clinical, food, environmental, and biothreat/biowarfare analyses. *Anal. Biochem.*, **478**, 107–120.

2 Bockrath, M., Liang, W., Bozovic, D., Hafner, J.H., Lieber, C.M., Tinkham, M. et al. (2001) Resonant electron scattering by defects in single-walled carbon nanotubes. *Science*, **291**, 283–285.

3 Canziani, G., Zhang, W., Cines, D., Rux, A., Willis, S., Cohen, G. et al. (1999) Exploring biomolecular recognition using optical biosensors. *Methods*, **19**, 253–269.

4 Izu, N., Shin, W., Matsubara, I., and Murayama, N. (2004) Development of resistive oxygen sensors based on cerium oxide thick film. *J. Electroceram.*, **13**, 703–706.

5 Ramamoorthy, R., Dutta, P., and Akbar, S. (2003) Oxygen sensors: materials, methods, designs and applications. *J. Mater. Sci.*, **38**, 4271–4282.

6 Jianrong, C., Yuqing, M., Nongyue, H., Xiaohua, W., and Sijiao, L. (2004) Nanotechnology and biosensors. *Biotechnol. Adv.*, **22**, 505–518.

7 Bandodkar, A.J., Jeerapan, I., and Wang, J. (2016) Wearable chemical sensors: present challenges and future prospects. *ACS Sens.*, **1** (5), 464–482.

8 Shafiee, H., Asghar, W., Inci, F., Yuksekkaya, M., Jahangir, M., Zhang, M.H. et al. (2015) Paper and flexible substrates as materials for biosensing platforms to detect multiple biotargets. *Sci. Rep.*, **5**. doi: 10.1038/srep08719

9 Fan, Z. and Lu, J.G. (2005) Zinc oxide nanostructures: synthesis and properties. *J. Nanosci. Nanotechnol.*, **5**, 1561–1573.

10 Dai, H. (2002) Carbon nanotubes: opportunities and challenges. *Surf. Sci.*, **500**, 218–241.

11 Elibol, O., Morisette, D., Akin, D., Denton, J., and Bashir, R. (2003) Integrated nanoscale silicon sensors using top-down fabrication. *Appl. Phys. Lett.*, **83**, 4613–4615.

12 Hierold, C. (2004) From micro-to nanosystems: mechanical sensors go nano. *J. Micromech. Microeng.*, **14**, S1.

13 Koehne, J., Chen, H., Li, J., Cassell, A.M., Ye, Q., Ng, H.T. et al. (2003) Ultrasensitive label-free DNA analysis using an electronic chip based on carbon nanotube nanoelectrode arrays. *Nanotechnology*, **14**, 1239.

14 Daniels, J.S. and Pourmand, N. (2007) Label-free impedance biosensors: opportunities and challenges. *Electroanalysis*, **19**, 1239–1257.

15 Randviir, E.P. and Banks, C.E. (2013) Electrochemical impedance spectroscopy: an overview of bioanalytical applications. *Anal. Methods*, **5**, 1098–1115.

16 Panneer Selvam, A. and Prasad, S. (2013) Nanosensor electrical immunoassay for quantitative detection of NT-pro brain natriuretic peptide. *Future Cardiol.*, **9**, 137–147.

17 Jacobs, M., Muthukumar, S., Selvam, A.P., Craven, J.E., and Prasad, S. (2014) Ultra-sensitive electrical immunoassay biosensors using nanotextured zinc oxide thin films on printed circuit board platforms. *Biosens. Bioelectron.*, **55**, 7–13.

18 Özgür, Ü. and Morkoç, H. (2006) Optical properties of ZnO and related alloys, in *Zinc Oxide Bulk, Thin Films and Nanostructures* (eds C. Jagadish and S. Pearton), Chapter 5, Elsevier Science Ltd., Oxford, pp. 175–239.

19 Gomez, J.L. and Tigli, O. (2013) Zinc oxide nanostructures: from growth to application. *J. Mater. Sci.*, **48**, 612–624.

20 Saha, S. and Gupta, V. (2011) Influence of surface defects in ZnO thin films on its biosensing response characteristic. *J. Appl. Phys.*, **110**, 064904.

21 Narang, J. and Pundir, C. (2011) Construction of a triglyceride amperometric biosensor based on chitosan–ZnO nanocomposite film. *Int. J. Biol. Macromol.*, **49**, 707–715.

22 Ali, S.M.U., Nur, O., Willander, M., and Danielsson, B. (2010) A fast and sensitive potentiometric glucose microsensor based on glucose oxidase coated ZnO nanowires grown on a thin silver wire. *Sens. Actuator B*, **145**, 869–874.

23 Umar, A., Rahman, M., Al-Hajry, A., and Hahn, Y.-B. (2009) Highly-sensitive

cholesterol biosensor based on well-crystallized flower-shaped ZnO nanostructures. *Talanta*, **78**, 284–289.
24 Ansari, S., Wahab, R., Ansari, Z., Kim, Y.-S., Khang, G., Al-Hajry, A. et al. (2009) Effect of nanostructure on the urea sensing properties of sol–gel synthesized ZnO. *Sens. Actuator B*, **137**, 566–573.
25 Yang, Z., Zong, X., Ye, Z., Zhao, B., Wang, Q., and Wang, P. (2010) The application of complex multiple forklike ZnO nanostructures to rapid and ultrahigh sensitive hydrogen peroxide biosensors. *Biomaterials*, **31**, 7534–7541.
26 Reyes, P.I., Ku, C.-J., Duan, Z., Lu, Y., Solanki, A., and Lee, K.-B. (2011) ZnO thin film transistor immunosensor with high sensitivity and selectivity. *Appl. Phys. Lett.*, **98**, 173702.
27 Wöll, C. (2007) The chemistry and physics of zinc oxide surfaces. *Prog. Surf. Sci.*, **82**, 55–120.
28 Sadik, P.W., Pearton, S.J., Norton, D.P., Lambers, E., and Ren, F. (2007) Functionalizing Zn-and O-terminated ZnO with thiols. *J. Appl. Phys.*, **101**, 104514.
29 Munje, R.D., Jacobs, M., Muthukumar, S., Quadri, B., Shanmugam, N.R., and Prasad, S. (2015) A novel approach for electrical tuning of nano-textured zinc oxide surfaces for ultra-sensitive troponin-T detection. *Anal. Methods*, **7**, 10136–10144.
30 McSwiney, B. (1934) The composition of human perspiration. (Samuel Hyde Memorial Lecture) *J. R. Soc. Med.*, **27**, 839–848.
31 Munje, R.D., Muthukumar, S., Selvam, A.P., and Prasad, S. (2015) Flexible nanoporous tunable electrical double layer biosensors for sweat diagnostics. *Sci. Rep.*, **5**. doi: 10.1038/srep14586
32 Franklin, J., Zou, B., Petrov, P., McComb, D., Ryan, M., and McLachlan, M. (2011) Optimised pulsed laser deposition of ZnO thin films on transparent conducting substrates. *J. Mater. Chem.*, **21**, 8178–8182.
33 Fortunato, G., Maiolo, L., Maita, F., Minotti, A., Mirabella, S., Strano, V. et al. (2014) LTPS TFT technology on flexible substrates for sensor applications. Active-Matrix Flatpanel Displays and Devices (AM-FPD), 2014 21st International Workshop on, pp. 311–314.
34 Selvam, A.P., Muthukumar, S., Kamakoti, V., and Prasad, S. (2016) A wearable biochemical sensor for monitoring alcohol consumption lifestyle through ethyl glucuronide (EtG) detection in human sweat. *Sci. Rep.*, **6**. doi: 10.1038/srep23111
35 Park, H., Ahn, H., Kim, D.-J., and Koo, H. (2013) Nanostructured gas sensors integrated into fabric for wearable breath monitoring system, Proceedings of the 2013 International Symposium on Wearable Computers, pp. 129–130.
36 Yuan, H., Shimotani, H., Tsukazaki, A., Ohtomo, A., Kawasaki, M., and Iwasa, Y. (2009) High-density carrier accumulation in ZnO field-effect transistors gated by electric double layers of ionic liquids. *Adv. Funct. Mater.*, **19**, 1046–1053.
37 R. Munje (2016) Optimized Assembly and Utilization of Functional Materials Towards Designing Stable and Reliable Affinity Based Biosenisng, PhD dissertation.
38 Beie, H.-J. and Gnörich, A. (1991) Oxygen gas sensors based on CeO_2 thick and thin films. *Sens. Actuator B*, **4**, 393–399.
39 Gerblinger, J., Lohwasser, W., Lampe, U., and Meixner, H. (1995) High temperature oxygen sensor based on sputtered cerium oxide. *Sens. Actuator B*, **26**, 93–96.
40 Izu, N., Itoh, T., Shin, W., Matsubara, I., and Murayama, N. (2007) The effect of hafnia doping on the resistance of ceria for use in resistive oxygen sensors. *Sens. Actuator B*, **123**, 407–412.
41 Andersson, D.A., Simak, S.I., Skorodumova, N.V., Abrikosov, I.A., and Johansson, B. (2006) Optimization of ionic conductivity in doped ceria. *Proc. Natl. Acad. Sci. USA*, **103**, 3518–3521.
42 Koch, K.T. and Saraf, L.V. (2004) Synthesis and characterization of pure and doped ceria films by sol–gel and sputtering. *J. Undergraduate Res.*, **4**, 84–90.
43 Xu, Y., Zhou, X., and Sorensen, O.T. (2000) Oxygen sensors based on

semiconducting metal oxides: an overview. *Sens. Actuator B*, **65**, 2–4.

44 Gupta, S., Kuchibhatla, S.V., Engelhard, M.H., Shutthanandan, V., Nachimuthu, P., Jiang, W. *et al.* (2009) Influence of samaria doping on the resistance of ceria thin films and its implications to the planar oxygen sensing devices. *Sens. Actuator B*, **139**, 380–386.

45 Yu, Z., Kuchibhatla, S.V., Engelhard, M.H., Shutthanandan, V., Wang, C.M., Nachimuthu, P. *et al.* (2008) Growth and structure of epitaxial Ce 0.8 Sm 0.2 O 1.9 by oxygen-plasma-assisted molecular beam epitaxy. *J. Cryst. Growth*, **310**, 2450–2456.

46 Thevuthasan, S., Peden, C., Engelhard, M., Baer, D., Herman, G., Jiang, W. *et al.* (1999) The ion beam materials analysis laboratory at the environmental molecular sciences laboratory. *Nucl. Instrum. Methods Phys. Res. A*, **420**, 81–89.

47 Chung, J., Lee, K.-H., Lee, J., Troya, D., and Schatz, G.C. (2004) Multi-walled carbon nanotubes experiencing electrical breakdown as gas sensors. *Nanotechnology*, **15**, 1596.

48 Suehiro, J., Zhou, G., and Hara, M. (2003) Fabrication of a carbon nanotube-based gas sensor using dielectrophoresis and its application for ammonia detection by impedance spectroscopy. *J. Phys. D*, **36**, L109.

49 Padigi, S.K., Reddy, R.K.K., and Prasad, S. (2007) Carbon nanotube based aliphatic hydrocarbon sensor. *Biosens. Bioelectron.*, **22**, 829–837.

50 Prasad, S., Zhang, X., Ozkan, C.S., and Ozkan, M. (2004) Neuron-based microarray sensors for environmental sensing. *Electrophoresis*, **25**, 3746–3760.

51 Yang, M., Prasad, S., Zhang, X., Ozkan, M., and Ozkan, C.S. (2004) Cascaded chemical sensing using a single cell as a sensor. *Sensor Lett.*, **2**, 1–8.

52 Li, C., Chen, W., Yang, X., Sun, C., Gao, C., Zheng, Z. *et al.* (2005) Impedance labelless detection-based polypyrrole protein biosensor. *Front. Biosci*, **10**, 2518–2526.

53 Lucklum, R., Rösler, S., Hartmann, J., and Hauptmann, P. (1996) On-line detection of organic pollutants in water by thickness shear mode resonators. *Sens. Actuator B*, **35**, 103–111.

54 Han, M., Assanis, D.N., Jacobs, T.J., and Bohac, S.V. (2008) Method and detailed analysis of individual hydrocarbon species from diesel combustion modes and diesel oxidation catalyst. *J. Eng. Gas Turb. Power*, **130**, 042803.

55 Zhao, Z., Arrandale, M., Vassiltsova, O.V., Petrukhina, M.A., and Carpenter, M.A. (2009) Sensing mechanism investigation on semiconductor quantum dot/polymer thin film based hydrocarbon sensor. *Sens. Actuator B*, **141**, 26–33.

56 Paddle, B.M. (1996) Biosensors for chemical and biological agents of defence interest. *Biosens. Bioelectron.*, **11**, 1079–1113.

57 Davis, J.J., Coleman, K.S., Azamian, B.R., Bagshaw, C.B., and Green, M.L. (2003) Chemical and biochemical sensing with modified single walled carbon nanotubes. *Chem. Eur. J.*, **9**, 3732–3739.

58 Chan, R.H., Fung, C.K., and Li, W.J. (2004) Rapid assembly of carbon nanotubes for nanosensing by dielectrophoretic force. *Nanotechnology*, **15**, S672.

59 Madou, M.J. (2002) *Fundamentals of Microfabrication: The Science of Miniaturization*, CRC Press.

60 Dyke, C.A. and Tour, J.M. (2003) Unbundled and highly functionalized carbon nanotubes from aqueous reactions. *Nano. Lett.*, **3**, 1215–1218.

61 Lin, Y., Taylor, S., Li, H., Fernando, K.S., Qu, L., Wang, W. *et al.* (2004) Advances toward bioapplications of carbon nanotubes. *J. Mater. Chem.*, **14**, 527–541.

62 Liu, J., Rinzler, A.G., Dai, H., Hafner, J.H., Bradley, R.K., Boul, P.J. *et al.* (1998) Fullerene pipes. *Science*, **280**, 1253–1256.

63 Richard, C., Balavoine, F., Schultz, P., Ebbesen, T.W., and Mioskowski, C. (2003) Supramolecular self-assembly of lipid derivatives on carbon nanotubes. *Science*, **300**, 775–778.

64 Rinzler, A., Liu, J., Dai, H., Nikolaev, P., Huffman, C., Rodriguez-Macias, F. *et al.* (1998) Large-scale purification of single-wall carbon nanotubes: process, product,

and characterization. *Appl. Phys. A*, **67**, 29–37.

65 Tománek, D. (2002) Ballistic conductance in quantum devices: from organic polymers to nanotubes. *Curr. Appl. Phys.*, **2**, 47–49.

66 Martinoia, S., Massobrio, P., Bove, M., and Massobrio, G. (2004) Cultured neurons coupled to microelectrode arrays: circuit models, simulations and experimental data. *IEEE Trans. Biomed. Eng.*, **51**, 859–863.

67 Velasco-Santos, C., Martinez-Hernandez, A., Lozada-Cassou, M., Alvarez-Castillo, A., and Castano, V. (2002) Chemical functionalization of carbon nanotubes through an organosilane. *Nanotechnology*, **13**, 495.

8
Superhard Materials

Ralf Riedel,[1] *Leonore Wiehl,*[2] *Andreas Zerr,*[3] *Pavel Zinin,*[4] *and Peter Kroll*[5]

[1]*Dean of the Department of Materials and Earth Sciences, Technische Universität Darmstadt, FB Material-und Geowissenschaften, Jovanka-Bontschits-Str. 2, 64287 Darmstadt, Germany*
[2]*Technische Universität Darmstadt, FB Material-und Geowissenschaften, Jovanka-Bontschits-Str. 2, 64287 Darmstadt, Germany*
[3]*Université Paris 13, LSPM-CNRS, 99 av. J.B. Clément, 93430 Villetaneuse, France*
[4]*University of Hawaii, 2525 Correa Rd., Honolulu, Hi, 96822, USA*
[5]*University of Texas at Arlington, Department of Chemistry and Biochemistry, 700 Planetarium Place, Arlington, Texas 76019, USA*

8.1
Introduction

The search for superhard materials is motivated by technical applications under extreme conditions. The hardest materials known so far are diamond and cubic boron nitride (c-BN) with Vickers hardness of $H_V = 90\,\text{GPa}/50\,\text{GPa}$ [1] and bulk modulus $B_0 = 445\,\text{GPa}$ [2]/396 GPa [3]) for diamond/c-BN, respectively, but their usability is limited by temperature and/or chemical constraints. Thus, for many applications they need to be replaced by materials with advanced performance. The term "superhard" usually denotes materials with Vickers hardness (H_V) larger than 40 GPa, whereas materials with Vickers hardness larger than 80 GPa are denoted as "ultrahard" [1,4–6].

Synthesis strategies of new materials are guided by theoretical predictions of properties. These predictions do not only have to consider the single-crystal properties of a pure material but also the variation of these properties in multiphase polycrystalline aggregates. The mechanical properties of materials are determined not only by the type and distribution of the constituent phases but to a large extent also by microstructural features, namely, porosity, grain size, and grain size distribution as well as defects. To make the situation even more complex, a lot of different properties are used to characterize the mechanical performance of a material, such as hardness, fracture toughness, strength, yield stress, and last but not least the elastic moduli. From these, only the elastic

Handbook of Solid State Chemistry, First Edition. Edited by Richard Dronskowski, Shinichi Kikkawa, and Andreas Stein.
© 2017 Wiley-VCH Verlag GmbH & Co. KGaA. Published 2017 by Wiley-VCH Verlag GmbH & Co. KGaA.

moduli are described by well-defined parameters, which can be calculated from first principles. Other quantities, for example, hardness, are not well defined on a theoretical basis [1]. Instead, they are defined by a certain measuring process. This means that even for an identical sample, different methods of hardness measurements (e.g., Vickers or Knoop) will result in different numerical values. In consequence, hardness values of different samples are comparable only when measured with the same method. This already indicates the difficulty to find relations between mechanical properties such as hardness and elastic moduli. Section 8.2 will give a short overview of these quantities and the correlation between them.

The main part of this chapter will present some classes of hardest compounds known so far. Here, we will concentrate on novel nitrides and some carbides and carbonitrides. Especially, we will not consider the two best-known hard materials, diamond and c-BN, as there is a huge amount of recent literature (e.g., [7,8].) and specialized books available.

8.2
Hardness and Strength of Materials

Mechanical properties of materials are governed by two factors, namely, elasticity and plasticity. The elastic moduli describe the resistance of a material to small and fully reversible deformations, whereas the hardness is a measure of the material's resistance to irreversible (plastic) deformations. Both properties contribute to the utility of a material under harsh conditions. Elastic moduli and elastic stiffness coefficients are well-defined quantities. They can be calculated in good quality from first principles for homogeneous single-phase materials and they can be measured with very high precision. Hardness, on the other hand, is not as well defined, though it can be measured in a reproducible way. Moreover, it is still not clear if a relation between hardness and elastic moduli can be established. In the following, we want to give a short overview of hardness and elastic properties and some recent attempts to interrelate these quantities.

The *elastic properties* of an isotropic homogeneous solid can be described by any two of the four quantities: Young's modulus (E), shear (or rigidity) modulus (G), bulk modulus (B), Poisson's ratio (ν). They are related [9,10] by

$$\frac{9}{E} = \frac{1}{B} + \frac{3}{G} \quad \text{and} \quad \nu = \frac{(3/2 - G/B)}{(3 + G/B)}.$$

Figure 8.1 shows the dependence of G/B on Poisson's ratio ν. From principle arguments (positive elastic energy and positive bulk modulus), Poisson's ratio can adopt only values in the range $-1 < \nu < 0.5$, but most materials show positive ν values. Negative values, however, are known for certain anisotropic (the so-called "auxetic") materials.

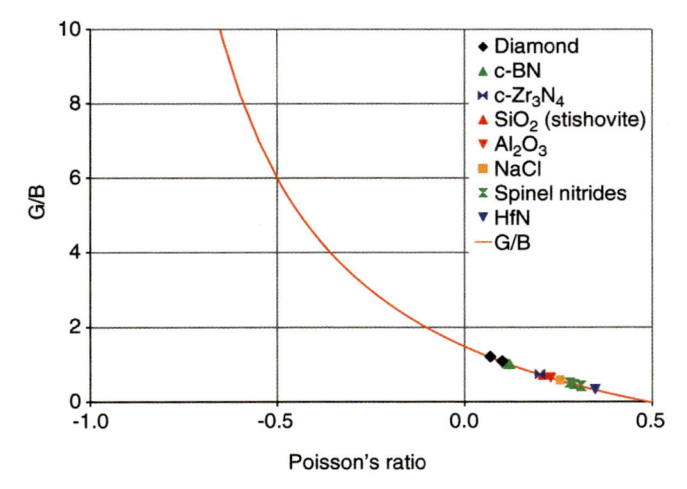

Figure 8.1 Dependence of the G/B ratio on Poisson's ratio for a homogeneous isotropic solid. Included are some experimental values (isotropic averages).

Single crystals, including cubic crystals of the highest symmetry, are not elastically isotropic. They need at least 3 coefficients (cubic case) for the description of their elastic properties and up to 21 in the triclinic case (only translational symmetry). The elastic coefficients form a fourth-rank tensor, which is usually given by a 6×6-matrix (c_{ij}) or (s_{ij}) for the mutual inverse stiffness or compliance tensors, respectively [9]. The above-mentioned parameters, Young's modulus (E), shear modulus (G), bulk modulus (B), and Poisson's ratio (ν), can be calculated, due to their primary definitions, from the elasticity tensor of any anisotropic solid. In consequence, not only E, G, and ν will be dependent on orientation (only B remains a single number) but also the simple relations between E, G, B, and ν are no longer valid. This holds both for single crystals and for textured polycrystals or composites with preferred orientations.

Ceramics are composed of many small crystallites of one or several phases with a (ideally) random distribution of orientations. Such a polycrystal is macroscopically isotropic. The quasi-isotropic elastic moduli of a polycrystal may be computed by averaging the single-crystal coefficients over all phases and all orientations [10–12]. Here, the knowledge of the microstructure is essential. A compromise has to be made on averaging, namely, to assume all crystal grains to share either a common strain state (Voigt) or a common stress state (Reuss). As these two conditions are mutually exclusive, only lower and upper bounds of the averaged elastic coefficients can be calculated and often the mean value of both is used according to Hill [13].

Hardness is defined as the resistance of a material to irreversible deformations and is mainly measured by indentation methods [14,15]. A diamond tip, shaped as four- or three-sided pyramid (Vickers, Knoop, and Berkovich) or as sphere (Brinell) is pressed into the sample surface with a certain force. After releasing the force, there remains a permanent indentation, the size of which is used as a measure of hardness. Forces (loads) in the range from kilonewton to 10 mN are used and the size of the resulting permanent indents vary from hundreds of micrometers to nanometers (nanoindentation).

In practice, hardness is defined by the measuring process. Experimental values are comparable only when measured with the same method [14,16]. Even then, for example, when using the identical Vickers indenter, the resulting hardness values vary with the applied load. This comes from the fact that the permanent indent size is related to the plastic, that is, irreversible deformation, produced by the creation and movement of dislocations and other defects, whereas the elastic deformations relax after load release. And the elastic contribution to the deformation increases at low loads. It is commonly agreed to take the asymptotic value from a series of measurements with increasing loads as the "true hardness" of the material [4,5], whereas the apparent increase in hardness values (decrease in indentation size) at low loads is regarded to be correlated with the elastic properties. By monitoring the indent size during the loading/unloading process of one specific test load, also the elastic (= Young's) modulus may be estimated from the initial slope of the unloading curve [17,18] or by the continuous stiffness measurement technique [15].

Based on empirical data many attempts have been made to correlate hardness with the elastic moduli, especially the bulk modulus or the shear modulus.

Bulk modulus. Earlier predictions of superhard materials were based on the assumption that "hardness is determined by the bulk modulus," because "it is the strength and compressibility of the bond that plays the primary role in a solid's ability to resist deformation" [19]. In contrast to this assumption, however, several compounds are known, which show a high bulk modulus but low hardness, for example, osmium [20], or vice versa, are superhard with low elastic moduli, for example, B_6O [21]. Moreover, a direct correlation of hardness with bulk modulus (B) is not plausible, because $B = -V(dP/dV)$ is the volume change under hydrostatic compression, whereas hardness is related to plastic deformations, which always involve shear.

Shear modulus. Thus, hardness is more likely related to the shear modulus than to the bulk modulus and such empirical findings were emphasized by several authors. For example, Pugh *et al.* [22] reported that the resistance of metals to plastic deformation is proportional to the elastic shear modulus, G. Teter [23] and Haines *et al.* [1] presented collections of empirical hardness and bulk/shear modulus data of materials with covalent bonding, which show a fairly good linear correlation of the shear modulus with hardness, but a large scatter in the bulk modulus versus hardness plot. More detailed discussions supporting a correlation of hardness (and/or yield stress) with shear modulus are reviewed by Zerr and Riedel in the *Handbook of Ceramic Hard Materials* (2000) [14].

G/B. Also other variants are discussed in the literature. Chen *et al.* [24] correlate hardness with the ratio of shear and bulk modulus $k = G/B$, based on the observation that brittle/ductile materials have high/low k-values. And brittle materials are expected to be harder than ductile materials.

On the whole, the empirical correlation of hardness with elastic properties is not satisfying. This is not surprising, however, because here two classes of mechanical properties are compared that involve deformations differing by orders of magnitude and that accordingly are related to different physical phenomena on the atomic scale. The elastic moduli describe the resistance to reversible (and small) deformations, whereas hardness is a measure for an irreversible plastic (and large) deformation. In addition, the experimentally obtained mechanical properties such as shear modulus or hardness are sensitive, and this in different manner, to microstructural properties (e.g., porosity, texture, presence of secondary phases, etc.). Apparently, in the case of hard and superhard materials fabrication of high-quality dense polycrystalline samples, needed for reliable hardness measurements, is extremely difficult without the use of sintering aids that bias the property of interest. Even at high pressures and temperatures (10–20 GPa and 2000 K), high hardness often precludes the formation of dense polycrystalline single-phase samples, and an alternative way to access reliable hardness values could be the production of single crystals of macroscopic size.

Theoretical prediction of hardness needs microscopic models at atomic level. The results from three different theoretical models, which derive hardness from the resistance of chemical bonds, are compared in a review by Tian *et al.* [4]. These models emphasize the "bond resistance," that is, the energy needed to break bonds [25], the "bond strength" [26] or the electronegativity [27,28] of single crystals. The experimental hardness values of numerous covalent as well as ionic compounds (cf. references in [4]) are compared with calculated values from the three microscopic models and in addition with the macroscopic model by Chen *et al.* [24], who relate hardness to the G/B ratio. All the microscopic models are limited to intrinsic hardness of ideal single crystals, that is, the hardness achieved by strong chemical bonds. No microstructural features are considered.

The role of the microstructure. In real materials, however, the microstructure plays an important role with respect to mechanical and other properties. The extrinsic hardness of a polycrystal, achieved by structuring, may be much smaller (e.g., for porous materials) or even higher than that of a single crystal. One well-known effect is the Hall–Petch effect [29,30]. According to the semiempiric Hall–Petch relationship the hardness of polycrystals increases with decreasing crystallite size (in the range of micrometers down to ∼100 nm) and was explained by a suppression of the dislocation activities inside the grain with decreasing grain size. At even smaller grain size (below 10–20 nm) again a softening has been observed in some compounds; however, the existence of the so-called reverse Hall–Petch effect has been doubted by others. For a recent discussion see, for example, Wollmershauser *et al.* [31] or Huang *et al.* [32]. The second and even much stronger influence of the microstructure is observed in nanocrystalline composite materials (NCs), which may be considerably harder than the corresponding single crystals, cf. Veprek *et al.* [33–35].

The search for superhard materials, researchers look for all aspects of "strength," as the needs of applications may be manifold. So the term "superhard" sometimes is used for materials that not only have a high Vickers hardness but also have high bulk and shear moduli [1]. Even more important for industrial applications is the availability of hard materials as coatings or protecting layers exhibiting a good adhesion to substrates or chemical stability in tribological contact. The first of these properties is lacking, for example, for cubic BN and the second for diamond by machining of ferrous alloys such as steels. Other parameters that could influence the choice of hard materials are fracture toughness and chemical-, thermal shock-, or wear resistance. Thus, hardness alone is not the criterion for the choice of a material but rather a combination of properties most suitable for particular applications.

From empirical evidence of materials fulfilling all these requirements, some most promising classes of possible superhard materials were identified: (i) compounds of light elements (B, C, N, and O), having three-dimensional network structures with strong covalent bonds, short bond lengths, and high bond densities, and (ii) compounds of light elements (B, C, N) with (heavy) transition metals, having a high valence electron density.

8.3
Superhard Materials

8.3.1
Carbon Nitride (C—N) Phases

8.3.1.1 Theoretical Predictions of C—N Structures

In 1989, a low-compressibility β-C_3N_4 structure (β-Si_3N_4 type, space group $P6_3/m$) has been suggested by Liu and Cohen [19] using an empirical model and an *ab initio* calculation of the bulk moduli for covalent solids. The bulk modulus of β-C_3N_4 was predicted to be higher than that of diamond. This prediction initiated a hype in the search for superhard C_3N_4 phases. Numerous hypothetical carbon nitrides have been proposed and analyzed in the literature (see review [36]): α-C_3N_4 (P31c) [37]; β-C_3N_4 ($P6_3/m$) [19]; pseudocubic or "defect zincblende"-C_3N_4 (P-C_3N_4) ($P\bar{4}2m$) [38,39]; "cubic" c-C_3N_4 ($I\bar{4}3d$) [39]; spinel-C_3N_4 ($Fd3m$) [40]; and λ-C_3N_4 ($P4_322$) [41]. Up-to-date descriptions of the proposed C_3N_4 phases and reviews of the experimental results can be found elsewhere [1,36,42–45].

The lack of progress in the synthesis of sp^3 bonded C_3N_4 materials pushed researchers to redirect the search toward novel carbon nitrides with stoichiometries other than C_3N_4. Examination of various structures of carbon nitride with 1:1 stoichiometry, led Cote and Cohen [46] to the conclusion that for 1:1 stoichiometry sp^2 bonding of nitrogen is energetically preferable to sp^3. Several stable phases of carbon nitride (CN) with 1:1 stoichiometry were proposed by Kim *et al.* [47–49] and Wang [50,51]. Recently, a novel carbon mononitride with tetragonal $P4_2/m$ symmetry was proposed by Zhang *et al.* [52]. Mechanical calculations suggested that the $P4_2/m$-CN is ultraincompressible and superhard. Behavior and stability under high pressure of the C–N structures proposed before 2015 are summarized by Dong et al. [53]. Detailed enthalpy calculations allowed them to reconstruct the phase diagram of the C—N system in the pressure range from 0 to 300 GPa. According to Dong et al., the first thermodynamically stable carbon nitride, $P4_2/m$-CN, proposed in Ref. [52] appears at a pressure of just 14 GPa and it is stable up to 22 GPa. The $Pnnm$-CN structure proposed by Wang [50] is stable at 22–97 GPa, and α-C_3N_4 proposed by Guo and Goddard [37] (P31c) is the most stable structure in the pressure range from 22 to 68 GPa. The structure bct-CN_2 proposed by Li et al. [54] ($I\bar{4}2d$) is stable at 59–298 GPa. The cubic-C_3N_4 ($I\bar{4}3d$) structure proposed by Teter and Hemley [39] is stable at 187–231 GPa. Dong et al. also predicted three new C_3N_4 phases with space groups (a) Cm (stable at 68–98 GPa), $I\bar{4}2m$ (stable at 98–187 GPa), (c) $Cmc2_1$ (stable at 224–300 GPa).

8.3.1.2 Synthesis of Novel C—N-based Compounds under HPHT Conditions

There are a number of publications where authors claimed the synthesis of new C_3N_4 phases; however, their publications did not include either X-ray diffraction (XRD) data or data on elemental compositions of the new phases [42,43,45]. The first successful synthesis of a dense C—N material was achieved in the laboratory of Prof. Riedel in 2007. The results were published by Horvath-Bordon et al. describing the synthesis of a well-crystallized carbon nitride imide, $C_2N_2(NH)$, in a laser-heated diamond anvil cell (LHDAC) from a single source precursor, $C_2N_4H_4$ at 41 GPa and 1700 °C [55]. The crystal structure of the new compound (defect-wurtzite $dwur$-$C_2N_2(NH)$) was determined in space group $Cmc2_1$ by combining electron diffraction results with first-principles theoretical studies and was later confirmed by Salamat et al. [56]. The ambient-pressure bulk modulus was found to be $B_0 = 258$ GPa, which is significantly lower than those of diamond and the hypothetical dense polymorphs of C_3N_4 ($B_0 = 430$–460 GPa).

Nanoplatelets of a new carbon–nitride-related compound, namely, $C_2N_2(CH_2)$, were synthesized by subjecting nanoparticles of a $C_3N_4H_xO_y$ precursor in a laser-heated diamond anvil cell to the pressure of 40 GPa and temperature of 1200–2000 K [57]. The experimental bulk modulus, B_0, of $C_2N_2(CH_2)$ was determined to be 258 GPa from the analysis of high-pressure (up to 30 GPa) X-ray diffraction patterns obtained using synchrotron radiation [58]. This bulk modulus is 40% lower than that of diamond. Another carbon nitride was investigated under high pressure and temperature, starting from hydrogen-bearing graphitic C_3N_4. In a wide pressure range from 30 to 125 GPa an orthorhombic phase was found to be stable, which has unit cell parameters similar to those of the reported hydrogen-bearing carbon nitride phases $C_2N_2(NH)$ and $C_2N_2(CH_2)$ [59]. The authors concluded that the synthesis of superhard C_3N_4 phases likely will need hydrogen-free starting materials.

8.3.2
Novel B—C-based Compounds BC_x

8.3.2.1 Theory

Theoretical simulations of pressure- and temperature-induced phase transitions in the B—C system conducted by Lowther [60] demonstrated that the incorporation of B atoms into a diamond structure should not lead to a drastic distortion of the cubic cell of diamond. The unit cells obtained in the phases predicted theoretically are slightly larger than that of diamond [60,61]. Both the bulk and shear moduli show a steady decrease with boron concentration. While it is not as hard as diamond, the overall electronic character indicates features different from that of diamond. The higher the concentration of boron, the less hard the material becomes. It was also found that incorporation of large amounts of boron strongly affects the transport properties of the diamond-like phases [60].

Several new structures of dense BC_x phases were proposed recently: (a) a diamond-like t-BC_2 (dlt-BC_2) with a tetragonal lattice originating from the cubic structure [62]; (b) a c-BC_3 [63] with a slightly deformed cubic cell; (c) a tetragonal BC_3 (t-BC_3) phase originating from the cubic structure [64]; (d) three diamond-like BC_3 (dl-BC_3) structures (orthorhombic $Pmma$-a consisting of a sandwich-like "B layer" between C layers, orthorhombic $Pmma$-b with B—B bondings, and tetragonal $P\bar{4}m2$ [65]); (e) a stable dl-BC_5 compound with a slightly distorted cubic cell [66,67] and dl-BC_5 with two possible $Pmma$ structures [68]; and (f) dl-BC_7 [69–71]. These dense BC_x phases were predicted to be superhard [61,62,72] and, more importantly, they should exhibit interesting electrical properties. Unlike diamond, which is an insulator, the dense BC_x phases may be conductors or superconductors: the dlt-BC_2 phase was predicted to be a conductor with three-dimensional conductivity [62]; the t-BC_3 phase should exhibit a sandwich-like metal and/or insulator layered structure with the anisotropic conductivity on the basal planes formed by the metallic CBC blocks [64], c-BC_3, or dl-BC_3 structures that would behave as metals under ambient conditions [63] and were predicted to be superconductors at low temperatures [65,73,74]; the dl-BC_5 phases should exhibit metallic [66] or hole conducting [68] behavior under ambient conditions and superconductor behavior at low temperatures [67]; and the dl-BC_7 phase was predicted to be a superconductor [70,71]. Zhang et al. [75] solved the crystal structure of recently synthesized cubic BC_3 using an unbiased swarm structure search, which identifies a highly symmetric BC_3 phase in the cubic diamond structure (d-BC_3) that contains a distinct B—B bonding network along the body diagonals of a large 64-atom unit cell. Mikhaylushkin et al. [76] concluded that disordered dl-BC_3 is formed at high temperature in the exothermic reaction of graphitic $BC_3 \rightarrow$ dl-BC_3, and that disorder on the C and B sites of dl-BC_3 was responsible for the observed phases. Several novel, dense BC_x phases with a high concentration of boron (e.g., B_4C_3 and B_4C_4) have been predicted to be metastable under ambient conditions [77,78]. A combination of high elastic moduli and various electrical properties that can be tuned by varying the B/C ratio of dense BC_x phases makes them attractive for a broad range of applications and industries, especially the (micro)electronic industry.

8.3.2.2 Synthesis

A direct transformation from the g-$BC_{1.6}$ phase to a new *diamond-like $BC_{1.6}$* phase was obtained at high temperature, 2230 ± 140 K and high pressure, 45 GPa [79]. Later, also in a BC_5 system, the observation of a phase transition from a graphitic to a diamond-like phase was claimed, namely, the synthesis of dl-BC_5 at 24 GPa and temperature of 2200 K [80,81]. Unfortunately, no raw experimental data were presented supporting the synthesis of dl-BC_5.

The synthesis of a cubic BC_3 (c-BC_3) phase by direct transformation from graphitic phases at a pressure of 39 GPa and temperature of 2200 K in a laser-heated diamond anvil cell (DAC) was reported by Zinin *et al.* [82]. Using a combination of X-ray diffraction (XRD), electron diffraction (ED), transmission electron microscopy (TEM) imaging, and electron energy loss spectroscopy (EELS) measurements, they concluded that the obtained c-BC_3 phase is a heteronanodiamond. The zero-pressure lattice parameter of the c-BC_3 calculated from diffraction peaks was found to be $a = (3.589 \pm 0.007)$ Å. A high-resolution TEM image of c-BC_3 demonstrates that c-BC_3 is a single, uniform nanocrystalline phase with a grain size of about 3–5 nm. The B-K and C-K edge EELS spectra of the c-BC_3 phase are similar to those of BC_2N phases [83,84] and diamond and boron containing diamond-like carbon (DLC) films [85]. The B-K and C-K edge spectra are dominated by sp^3 bonding (1s → σ^* transition peaks) and show only weak, vestigial 1s → π^* transition peaks below the primary σ^* features. The EELS quantification of the elemental composition ratio was carried out on the c-BC_3 phase, which corresponds approximately to BC_3 (B/C = 2.8 ± 0.7).

Quite unexpected results were obtained recently during characterization of the BC_x films deposited by chemical vapor deposition and described elsewhere [86–90]. The use of EELS combined with TEM, Raman spectroscopy, surface Brillouin scattering (SBS), laser ultrasonic (LU) technique, and analysis of elastic properties demonstrated that low-pressure synthesis (chemical vapor deposition) of BC_x phases may also lead to the creation of diamond-like boron-rich carbides [91]. The elastic properties of the dl-BC_x phases depend on the carbon sp^2 versus sp^3 content, which decreases with increasing boron concentration, while the boron bonds determine the shape of the Raman spectra of the dl-BC_x after high-pressure high-temperature treatment. Using the estimation of the density value based on the sp^3 fraction and measured velocities of the acoustical waves, the shear modulus G of diamond-like BC_4 (dl-BC_4), containing 10% carbon atoms with sp^3 bonds, and dl-B_3C_2, containing 38% carbon atoms with sp^3 bonds, were found to be $G = 19.3$ GPa and $G = 170$ GPa, respectively. The presented experimental data also imply that boron atoms lead to a creation of sp^3 bonds during the deposition processes. Low-pressure chemical vapor deposition at 950 K leads to the synthesis of "diamond-like boron carbides" with high concentrations of boron ($0.66 < x < 4$) in which sp^2 fraction and elastic properties depend on the boron concentration. This is in line with the theoretical simulations [74,92] demonstrating that the total energies of the g-BC_3 structure should be higher than that of the equivalent dl-BC_3 phase under ambient conditions.

8.3.3
B—C—N Compounds

B—C—N compounds, especially $C_x(BN)_y$, were regarded as the most promising candidates for new superhard compounds, because they are mixed crystals of the hardest single-phase materials known, namely, diamond and cubic boron nitride with sphalerite-type (i.e., diamond-like) crystal structure. The idea was to combine the most favorable properties of both end members, namely, the higher hardness of diamond with the higher temperature resistance and enhanced chemical stability of c-BN (e.g., for machining iron alloys). Starting from the phase diagrams of carbon and BN, many efforts have been made to explore the phase diagram of ternary B—C—N compounds, experimentally and theoretically.

8.3.3.1 Theoretical Predictions
First-principles methods have been used to predict possible crystal structures in the ternary B—C—N system, and their pressure/temperature stability fields. The reported crystal structures largely are related to the known structures of carbon and boron nitride, which may be classified as follows. Networks of hexagonal rings, which may be flat (sp^2 bonded) or "puckered" (sp^3 bonded), are stacked along their threefold axis with a repeat period of two or three per unit cell and different relative shifts of the layers.

	Three layers	Two layers
sp^3	C: diamond, $Fd3m$	C: lonsdaleite, $P6_3/mmc$ [93,94]
	c-BN, sphalerite, $F\bar{4}3m$	w-BN, wurtzite, $P6_3mc$
sp^2	r-BN, rhombohedral, $R\bar{3}m$ [95]	C: graphite, $P6_3mc$
		h-BN, $P6_3/mmc$

In the hypothetical B—C—N structures, additional degrees of freedom arise from the three different atom types to be distributed on the lattice sites and from possible distortions of the unit cell, leading to many different structure models. The term "graphitic" is usually used for all crystal phases built up from flat (sp^2 bonded) layers, irrespective of their topology. This must not be taken literally. Even the binary hexagonal h-BN is not isotypic to graphite, although both structures consist of a stacking of two flat layers per unit cell, but in h-BN the relative shift of these layers is different from that of graphite. In a similar way, the term "diamond-like" sometimes is used synonymously with "sp^3 bonded."

A lot of structure models to describe BCN phases have been proposed. Only a few examples can be given here. One of the first was a graphitic (monolayer) BC_2N compound by Liu et al. [96]. They found that C—C and B—N bonds are preferred in the most stable structure models. Tateyama *et al.* [97] reported a "heterodiamond" β-BC_2N structure, consisting of sphalerite (BN) and diamond (C_2) layers alternating along the *c*-axis. The predicted bulk modulus of $B = 438$ GPa is nearly the same as that of diamond. The preference of C—C and B—N bonds, and the disadvantage of C—N, N—N, and B—B bonds, was confirmed by Yuge *et al.* [98], who explored the phase stability of a solid solution series of diamond-like $(BN)_{(1-x)}(C_2)_x$ ($0 \leq x \leq 1$) compounds at ambient pressure. The results suggested phase separation into c-BN and diamond and they found a complete miscibility only at temperatures larger than ~4800 K, that is, above the melting points of both c-BN and diamond. Other authors calculated specific models in order to explain the experimental XRD patterns of some synthesized B–C–N compounds. Two tetragonal ($z*$-BC_2N and t-BC_2N) [99] and one rhombohedral (rh-BC_2N) [100] crystal structures were suggested as most likely describing the experimentally obtained c-BC_2N reported by Zhao *et al.* [83]. A 3C-BC_4N structure with trigonal symmetry [101] was proposed to describe the experimental XRD pattern of c-BC_4N [83]. Further examples are collected in a recent PhD thesis by Bhat [102].

None of these predicted structures is really stable with respect to decomposition into C and BN phases. The lowest formation enthalpies in the ternary $B_xC_yN_z$ phase diagram have been found along a line C—BN, that is, with nearly equal amounts of boron and nitrogen ($x \approx z$) [103]. Guided by this, most of the synthesis experiments were started with compositions BC_xN, especially BC_2N or BC_4N.

8.3.3.2 Synthesis Experiments

Attempts to synthesize superhard B—C—N compounds were performed in two steps. First, ternary B—C—N compositions were produced at ambient pressure along high-temperature routes from the elements or from simple inorganic compounds (e.g., halides and ammonia-based compounds) or from organic precursors such as piperazine borane and pyridine borane or other organic compounds. These experiments resulted in sp^2-bonded phases containing hexagonal networks of threefold coordinated B, C, and N atoms, which were amorphous or showed layered ("graphitic") structures (Riedel [14,104,105]; Hubacek and Sato [106]). In the second step, either such precursors or ball-milled mixtures of graphite and hexagonal boron nitride (h-BN), were subjected to high pressures and temperatures in order to obtain ternary sp^3-bonded diamond-like phases.

For more than 35 years [107], high-pressure high-temperature (HP-HT) experiments have been performed to synthesize superhard cubic B—C—N phases. Different static and dynamic methods/techniques have been used such as multianvil presses, belt presses, LHDAC, shock waves, or others. Many of these experiments did not yield ternary phases but instead resulted in mixtures of diamond and c-BN (at higher pressures) or graphite and h-BN (at lower pressures) [108,109]. An overview is given by Bhat [102]. Others claimed the successful production of ternary B—C—N phases. For example, Knittle *et al.* [110] reported cubic $C_x(BN)_{(1-x)}$ ($x = 0.3$–0.6) solid solutions, Komatsu *et al.* [111–113] described the production of a heterodiamond c-BC$_{2.5}$N, Solozhenko *et al.* [114] claimed the synthesis of cubic BC$_2$N from graphitic BC$_2$N, and Zhao *et al.* [83] described the synthesis of superhard BC$_2$N and BC$_4$N.

It is, however, a great challenge to verify the successful synthesis of a C$_x$(BN)$_{(1-x)}$ solid solution and to distinguish the product from a simple mixture of diamond and c-BN [84]. Mainly X-ray diffraction and electron microscopy have been used for this purpose. The powder X-ray patterns of diamond and c-BN are very similar due to their similar lattice parameters of 3.5667 and 3.6158 Å (JCPDS 6-0675 and 35-1365). The only obvious differences are a few reflections extinguished in the diamond pattern, for example, (200), (222), and (420), because of the *d*-glide planes that are missing in c-BN, but these reflections should be present in an ordered isotypic C$_x$(BN)$_{(1-x)}$ solid solution as well. The diamond and c-BN lines are clearly separated for well-crystallized samples, having narrow diffraction lines. In contrast, for nanocrystalline samples, also when using synchrotron radiation, the line width (FWHM) is larger than the difference of the diamond/c-BN line positions and a possible line splitting cannot be detected unambiguously. Accordingly, the mean positions of these superimposed lines, and the lattice parameter deduced from them, will shift with the diamond/c-BN ratio, resulting in Vegard's law similar to the case of an ideal solid solution.

Using electron microscopy, the limiting factor is the spatial resolution of diffraction and chemical analysis, that is, the size of the aperture limiting the electron beam in relation to the crystallite size. Nanoparticles have been observed in TEM images of HP-HT-synthesized B–C–N samples (e.g., [83,84].), but the elemental analysis using electron energy loss spectroscopy or the selected area electron diffraction (SAED) covered larger regions containing many particles and could not reveal the composition and crystalline state of an individual particle.

The reports claiming ternary solid solutions are contrasted with a lot of publications (experiments and theory) where the authors explicitly state the phase separation into carbon and BN phases. Also, some earlier claims of ternary phases [84] were later modified by the same authors, stating that "all ternary diamond-like B-C-N phases tend to decompose at high temperature" [115]. In the past few years, only a few reports about B—C—N (e.g., [116].) were published. In view of the difficulties in the B—C—N system to distinguish ternary phases from mechanical mixtures, the interested reader is advised to consult the original literature and judge by himself if he believes in the existence of ternary hard B—C—N phases.

8.3.4
Spinel Nitrides, γ-M$_3$N$_4$ where M = Si, Ge, or Sn

Attempts to synthesize a dense carbon nitride at high pressures and high temperatures initiated the research on other nitrides of group 14 elements, such as Si, Ge, and Sn. It was found that nitrides of Si and Ge crystallize in the spinel structure (Figure 8.2), denoted hereafter as γ-M$_3$N$_4$ (M = Si, Ge) [117–121], at static pressures exceeding 10 GPa and temperatures of 1000–2000 K. Small

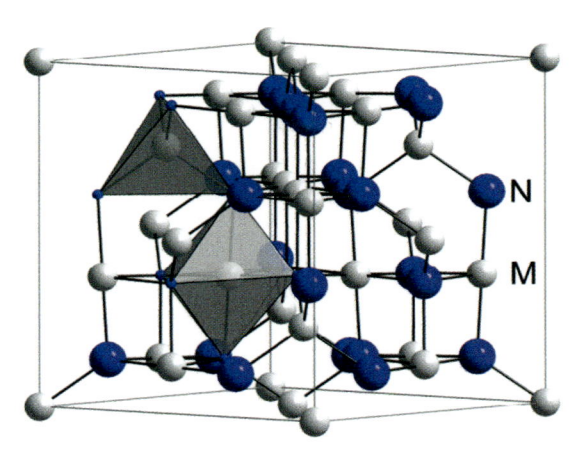

Figure 8.2 Ball-and-stick model of the spinel structure (space group $Fd\bar{3}m$, no. 227) of γ-M_3N_4 (M = Si, Ge, and Sn).

amounts (<10^{-3} mm^3) were obtained via the reactions of the elements, Si or Ge with nitrogen, in a LHDAC and recovered to ambient conditions [119,120]. Larger amounts (~1 mm^3) were obtained using a multianvil HP-HT apparatus starting either from the low-pressure trigonal or hexagonal phases where all cations are fourfold coordinated, α- or β-M_3N_4 [117,121], or from amorphous precursors, a-Si_3N_4 or Si_2N_3H [118]. In the spinel structure two-thirds of the cations are octahedrally coordinated (Figure 8.2) and the remaining one-third are tetrahedrally coordinated. The higher coordination number led to a 20–25% higher density of γ-M_3N_4 (Table 8.1) compared to that of α- and β-M_3N_4. The existence of γ-Sn_3N_4 was first recognized by the analysis of products of a reaction of SnI_4 with KNH_2 in liquid ammonia at 243 K [122], but the material was obtained much earlier in the form of thin films deposited using magnetron sputtering [123,124]. In the latter works, reproduced recently by Caskey et al. [125], the spinel structure was not recognized due to limited efforts spent

Table 8.1 Selection of the most reliable experimental values of the cubic lattice parameter, density, elastic moduli, hardness, bandgap, and thermal stability limit of γ-M_3N_4 (M = Si, Ge, and Sn) and c-A_3N_4 (A = Zr, Hf) reported in the above-cited publications.

	a, Å	ρ, g/cm^3	B_0, GPa	G_0, GPa	H_V, GPa	E_g, eV	$T_{stab.}$, K
γ-Si_3N_4	7.7381	4.03	290	148	35	5.05	>1673
γ-Ge_3N_4	8.2125	6.57	295	141	28	3.5	1000
γ-Sn_3N_4	9.037	7.42	149	64	11	1.55	573
c-Zr_3N_4	6.740	7.15	219	163	>12[a]	1.6	773
c-Hf_3N_4	6.701	13.06	227	n.a.	n.a.	1.8	n.a.

Note: The experimental uncertainties can be also found there.
a) Sample porosity p = 25–30%.

to the structure solution search. γ-Sn_3N_4 was also synthesized at elevated pressures of ~2.5 GPa and moderate temperatures of 623 K via a solid-state metathesis reaction of SnI_4 with Li_3N [126]. Later on, a successful HP-HT synthesis of solid solutions γ-$(Si_{1-x}Ge_x)_3N_4$ was reported and a nearly linear increase in the cubic lattice parameter a [127] with x was recognized. Another solid solution crystallizing in the spinel structure under HP-HT was reported to be γ-Si_2AlON_3 [128]. Mixtures of carbon with Si_3N_4, however, do not react to an admixture of SiC with γ-$(Si_{1-x}C_x)_3N_4$ solid solutions under HP-HT conditions but to well-sintered composites of γ-Si_3N_4 and diamond with H_V values as high as 42 GPa [129].

Binary spinel nitrides of silicon and germanium were obtained via shock wave synthesis, a method that allows economic production of powders of hard materials in large amounts [130]. In experiments where a projectile is accelerated by a propellant gun, β-Si_3N_4 was found to start to transform to γ-Si_3N_4 with a grain size of 10–50 nm at pressures above P ~ 20 GPa, but a maximum yield of 80% was reached at 50 GPa [131]. A nearly 100% yield of nanocrystalline γ-Si_3N_4 was claimed to be obtained in experiments on shock wave loading of powder mixtures of Si_3N_4 and copper or KBr using flyer plates accelerated by explosives [132–134].

As for other high-pressure ceramic compounds with covalent bonding, high elastic moduli and hardness were predicted and later on measured for the spinel nitrides (Table 8.1). Here, we do not consider theoretical studies due to their large number and a limited space this chapter provides, but the interested readers can consult earlier reviews [118,130,135]. Experimental values of the bulk modulus, B_0, of γ-M_3N_4 were derived from HP equations of state (EoS) [136–139], while their shear moduli, G_0, were derived from nanoindentation tests [126,137]. Hardness measurements were performed for all binary spinel nitrides using Vickers-indentation and nanoindentation techniques [126,137,140]. The highest Vickers hardness $H_V = 33–37$ GPa was reported for a transparent sample of γ-Si_3N_4 densified at $P = 17$ GPa by heating to 2100 K for 1 h [140]. The same work reports on an exceptional thermal stability of γ-Si_3N_4 in air up to 1673 K that does not oxidize but starts to transform into a mixture of the low-pressure hexagonal α- and β-phases at 1873 K. γ-Ge_3N_4 appears to be stable at up to 1000 K [141] and γ-Sn_3N_4, in vacuum, up to at least 573 K [122]. The most reliable experimental data on elastic moduli, Vickers hardness, and thermal stability of γ-M_3N_4 are summarized in Table 8.1.

In addition to their advanced mechanical properties and thermal stability, the spinel nitrides, γ-M_3N_4, exhibit interesting optoelectronic properties. Since the first theoretical prediction that γ-Si_3N_4 has a direct bandgap [40], this statement was extended to the other family members but confirmed experimentally only 15 years later [142]. The fundamental bandgap energy E_g of γ-M_3N_4 changes almost linearly from 5.05 eV (for γ-Si_3N_4) to ~1.6 eV (for γ-Sn_3N_4) as experimentally confirmed for the end members (Table 8.1) and the solid solutions γ-$(Si_{1-x}Ge_x)_3N_4$ (where $0 \leq x \leq 1$) [125,142–144] and calculated for the solid solutions γ-$(Ge_{1-y}Sn_y)_3N_4$ (where $0 \leq y \leq 1$) [144]. A more important finding is that γ-M_3N_4 exhibits strong free exciton-binding energies (measured to be

~0.65 eV for γ-Si$_3$N$_4$ [142] and calculated for others in [144]), significantly exceeding those of the competing materials having comparable E_g values (e.g., (Ga,In)(As,N), ZnO, and h-BN). This suggests a high efficiency of the transformation of electric power into light in LEDs based on defect-free crystals of the spinel nitrides [142,144]. Combined with a very high hardness and exceptional thermal stability in air, γ-M$_3$N$_4$ and their solid solutions have a potential for application not only as robust and efficient photonic emitters but also as photovoltaic and transistor materials. Finally, γ-Sn$_3$N$_4$ appears to be one of the very few materials combining E_g values <2 eV with a band alignment similar to water redox potentials, which is needed for an effective water splitting upon visible light irradiation [125].

8.3.5
Transition Metal Nitrides of the Group 4 and 5 Elements (c-Zr3N4 and c-Hf3N4, η-Ta2N3)

At atmospheric pressure, transition metals of group 4 and 5 elements react with nitrogen (if available in excess) to mononitrides where the cations have the oxidation state +3. In contrast, an increase in pressure above 7 GPa leads to nitrides with cations in higher oxidation states. In the case of zirconium and hafnium, the oxidation state reaches the maximum value of +4 resulting in the formation of binary nitrides c-A$_3$N$_4$ (A = Zr or Hf) having the cubic Th$_3$P$_4$-type structure (Figure 8.3) [145,146]. These materials can be recovered to ambient conditions and remain metastable at up to 773 K even when heated in air [147]. Starting

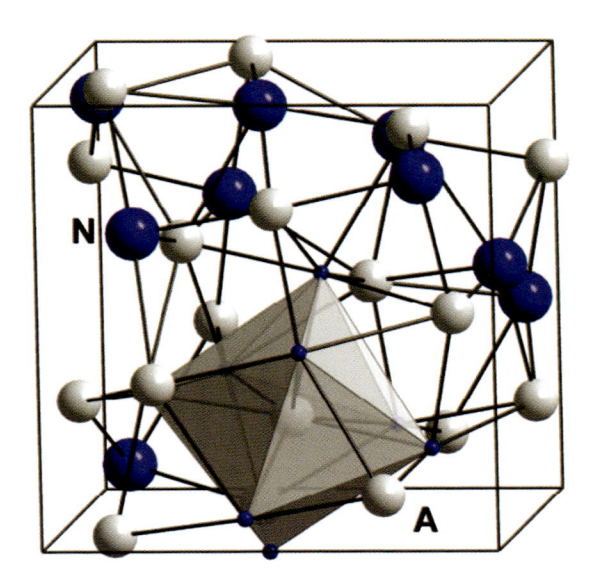

Figure 8.3 Ball-and-stick model of the Th$_3$P$_4$-type structure (space group $I\bar{4}3d$, no. 220) of c-A$_3$N$_4$, (A = Zr or Hf).

from the nanocrystalline Hf_3N_4, available via a few metastable reaction paths involving ammonolysis of organometallic precursors [148,149], Salamat et al. observed formation of tetragonal Hf nitride (space group $I4/m$) with unknown stoichiometry at 12 GPa after heating to 1500 K for ~90 s [150]. But the phase appears to be difficult to recover to ambient conditions. The heating of the nanocrystalline Hf_3N_4 to 2000 K at higher pressures of 19 GPa was reported to lead to an orthorhombic hafnium nitride or oxynitride recoverable to ambient conditions [150]. While the chemical composition was confirmed by careful electron probe microanalysis (EPMA) for c-A_3N_4 [146,147], the role of oxygen in stabilizing the tetragonal and orthorhombic Hf nitrides, with structures analogous to known HP phases of HfO_2 [150], needs further clarification.

Thin films of well-crystallized c-Zr_3N_4 were deposited on a variety of substrates such as silicon, quartz, sintered carbide, or even salt using an industrially viable modified filtered cathodic arc (FCA) method [151]. The films were found to have a high hardness of ~37 GPa (measured at an indentation load of 5 mN) and an excellent adhesion to sintered carbide substrate. These properties result in a 10 times better wear resistance compared to that of titanium mononitride, the basic material for advanced coatings presently used in industry. Films of nanocrystalline c-A_3N_4 were also deposited on silicon or glass using radiofrequency or direct-current magnetron sputtering [152–154].

Experimental values of B_0 for c-A_3N_4 were derived from HP-EoS [155,156] and from LU measurements [157]. In the latter work, bulk and shear moduli were measured for a porous sample and the B_0 and G_0 values for the dense material, given in Table 8.1, were derived applying the porosity analysis based on the Hashin–Shtrickman approach [157]. The presence of a contaminating Hf—N—O phase in the macroscopic samples of c-Hf_3N_4 [147] explains the lack of experimental G_0 values in the literature. Vickers indentation testing of a porous sample (25–30% of porosity) of c-Zr_3N_4 gave $H_V = 12$ GPa, similar to that of the single crystal of ZrN, and a high indentation fracture toughness of $K_{Ic\text{-}if} = 3.2$ MPa·m$^{1/2}$ [147]. Experimental data on H_V of pore-free c-A_3N_4 samples, expected to exceed 30 GPa, are not available because a method of densification of these ceramic materials is still not developed. The c-Zr_3N_4 phase persists upon heating in air up to $T = 773$ K [147] (Table 8.1), similar to the mononitrides of group 4 elements. Thus, c-A_3N_4 might find applications as hard wear-resistant materials. But they could also be suitable for optoelectronic or photovoltaic applications. Soft X-ray spectroscopic measurements combined with theoretical predictions show that c-Zr_3N_4 and c-Hf_3N_4 are low bandgap semiconductors with direct and nearly direct bandgaps of 1.6 and 1.8 eV, respectively. Accordingly, c-A_3N_4 can be considered as alternatives for GaAs, used in demanding technical applications such as near-IR LEDs and lasers for optoelectronics, since they have similar bandgap sizes and their structure appears to tolerate p- and n-doping on both cation and anion sites.

In the case of tantalum, a dense binary nitride having orthorhombic U_2S_3-type structure, η-Ta_2N_3, (Figure 8.4) forms at $P > 9$ GPa and high temperatures and can be recovered to ambient conditions [158,159]. As follows from the chemical

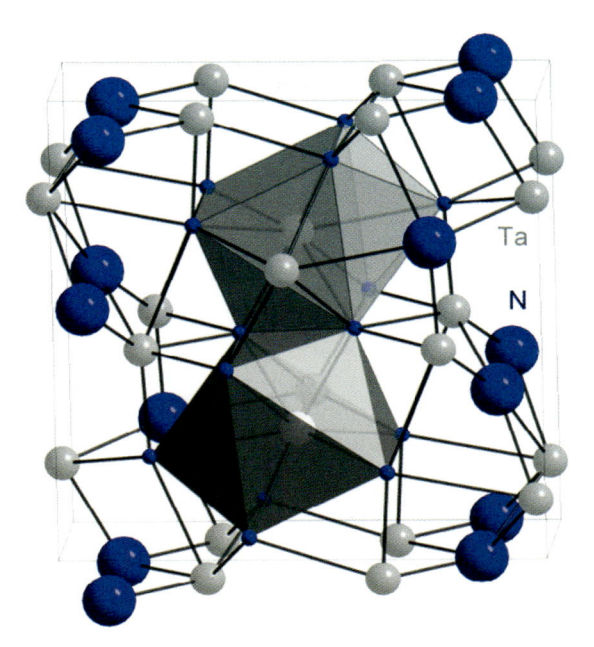

Figure 8.4 Ball-and-stick model of the U_2S_3-type structure (space group *Pbnm*, no. 62) of η-Ta_2N_3.

formula, confirmed by a careful EPMA analysis, 50% of the cations have the maximum possible oxidation state of +5 and the remaining 50% thereof have the oxidation state +4. Furthermore, 50% of the cations are coordinated by 8 N-anions and the remaining ones by 7 N-anions. As in the case of c-A_3N_4, macroscopic samples of η-Ta_2N_3, needed for mechanical properties measurements, were obtained in a multianvil HP-HT apparatus and exhibited a significant porosity of ~16% [160,161]. However, the surfaces of the samples polished mechanically show a much lower porosity of ~4% due to a surface self-healing behavior (surface densification) that is unique for ceramic materials. This surface self-healing manifests itself by the closing of pores having sizes similar to or smaller than that of the abrasive grains (0.25 µm in our case) used at the last step of polishing [160,161]. The effect was explained by a relatively high malleability (ductility) of η-Ta_2N_3 comparable to that of Fe, Nb, or Mo because η-Ta_2N_3, on one hand, and the elemental metals, on the other hand, have similarly high values of the B_0/G_0 ratio (suggested to be a reliable indicator of malleability [22,162]) exceeding two. The elastic moduli of η-Ta_2N_3 were measured by two independent methods, namely, laser ultrasonics [161] and nanoindentation testing combined with EoS data [160]. Applying the porosity analysis mentioned before, the first technique gave $B_0 = 281$ GPa and $G_0 = 123$ GPa that are consistent with the values obtained using the second method. The Vickers hardness of the porous η-Ta_2N_3 sample (~16%) was determined to be $H_V = 16$ GPa and H_V of the dense material estimated to exceed 30 GPa [163]. More important is a very high fracture toughness of η-Ta_2N_3 measured to be $K_{Ic-if} = 4.6$ MPa·m$^{1/2}$, comparable to the values reported for densified samples of hexagonal Si_3N_4 and SiC known as tough structural materials [163]. We can thus conclude that η-Ta_2N_3 is a promising ceramic material exhibiting self-healing behavior by mechanical polishing in addition to a high hardness and fracture toughness.

8.4
Chances, Expectations, and Prospects of Computational Simulations

Predicting hardness of materials poses a significant challenge. As outlined previously in this chapter, hardness is an engineering quantity, which characterizes the resistance of a material to plastic deformations such as scratching or indentation of its surface. Hence, the process of hardness determination involves various mechanisms, among which are volume compression, shear deformation, bond breaking, and creation and motion of dislocations. Consequently, hardness is a multiscale and multiphysics phenomenon and the ultimate goal of hardness prediction is to model experiments and measurements comprehensively. Nevertheless, there are several "simple" materials properties and structural signatures that serve very well to project hardness of a material.

The phenomenological correlations between hardness and elastic properties of a material serve excellently to estimate hardness of a single crystal. Formulas of Chen, $H_V = 2 (G \cdot k^2)^{0.585} - 3$ [24], and Tian, $H_V = 0.92 \cdot k^{1.137} G^{0.708}$ [4], use bulk modulus B and shear modulus G, together with their combination to Pugh's modulus ratio, $k = G/B$, and provide virtually identical values for hardness exceeding 5 GPa. Estimated hardness and experimental data agree better than 2 GPa [4], which is a great success.

Based on this strong correlation, exploration of potential superhard single crystalline compounds has its merits – and the availability of first-principle methods together with smart search algorithms has provided many examples of potential candidates for superhardness. With growing computer resources, we can expect with certainty more reports of possible "new superhard all sp^3-carbon" phases, since there are infinite many 4-connected networks that will provide candidate structures of new carbon polymorphs. Likewise, new boron carbide structures that combine packing of icosahedra and extended networks are waiting to be discovered together with various combinations of metal atoms with boron or boron–carbon networks. Since first-principle methods provide reliable data of thermodynamic properties, calculations will also help to estimate the viability of these compounds.

Innovative directions of hardness prediction target the possible enhancement of hardness through size effects in nanocrystalline materials. It is proposed that a combination of the Hall–Petch effect and quantum confinement effects in nanocomposite material yields hardness values exceeding single-crystal data of the constituting phases. At present, direct simulations of this phenomenon, for example, by computing elastic moduli of a nanocrystalline model, are limited by the necessary size of models and the resources needed for accurate first-principle methods. Further advances in electronic structure calculations will, however, shed some light on this prospect to generate novel superhard materials in the near future.

Yet another direction of simulations will involve multiscale modeling of the indentation (or scratching) process to model the experiment of hardness determination. Complete accurate atomistic simulation of an indentation process will not be feasible for a long time due to length and timescales involved. With understanding of atomistic structure and dynamics, properties of collections of atoms, and materials properties, the atomistic problem can – in principle – be coarse-grained and, further, homogenized. It still remains a challenge to do this multiscale, hierarchical modeling from nano- via meso- to macroscale without loss of intrinsic structural information. Advances in materials modeling will show whether the development of effective methods that include molecular dynamics, coarse grain simulation, micromechanics, and finite element methods will be successful.

8.5 Conclusions

In general, the production of technical materials and devices heavily involves various machining steps on different levels of the processing procedure in order to meet the performance specifications related to the individual application. Therefore, the search for novel advanced super- and/or ultrahard materials that outreach the performance and properties of the well-known diamond and c-BN still remains a challenging and interdisciplinary research topic in the near and far future.

The need for alternative hard materials is required for a great variety of technical applications in the fields of materials production and processing as well as surface machining or finishing of products and devices. The availability of hard materials with properties precisely adapted to the specific application is an important economic factor in order to lower production time and costs. Moreover, environment-friendly materials will contribute to save energy and to reduce waste during materials and device manufacturing.

In this chapter, we summarized the latest developments in modeling and synthesis as well as potential properties of advanced materials with super- and ultrahard mechanical properties. In this context, most promising compounds are based on light elements (B, C, N, and O), eventually containing silicon or transition metals as it has been shown by the amount of literature published in recent times. Despite the open questions that still remain open with regard to the existence and stability of carbon nitrides and boron carbonitrides, which are predicted as ultrahard materials, we can expect a significant development in this field in near future due to knowledge-based materials design approaches and novel synthesis strategies involving high-temperature and high-pressure techniques.

References

1 Haines, J., Leger, J.M., and Bocquillon, G. (2001) Synthesis and design of superhard materials. *Ann. Rev. Mater. Res.*, **31**, 1–23.
2 Vogelgesang, R. et al. (1996) Brillouin and Raman scattering in natural and isotopically controlled diamond. *Phys. Rev. B*, **54** (6), 3989–3999.
3 Datchi, F. et al. (2007) Equation of state of cubic boron nitride at high pressures and temperatures. *Phys. Rev. B*, **75** (21), 214104.
4 Tian, Y., Xu, B., and Zhao, Z. (2012) Microscopic theory of hardness and design of novel superhard crystals. *Int. J. Refract. Met. Hard Mater.*, **33**, 93–106.
5 Veprek-Heijman, M.G.J. et al. (2009) Non-linear finite element constitutive modeling of indentation into super- and ultrahard materials: the plastic deformation of the diamond tip and the ratio of hardness to tensile yield strength of super- and ultrahard nanocomposites. *Surf. Coat. Tech.*, **203** (22), 3385–3391.
6 Veprek, S. (2000) Nanostructured superhard materials, in *Handbook of Ceramic Hard Materials* (ed. R. Riedel), Wiley-VCH Verlag GmbH, Weinheim.
7 Dubrovinskaia, N. et al. (2006) High-pressure/high-temperature synthesis and characterization of boron-doped diamond. *Z. Naturforsch. B*, **61**, 1561.
8 Dubrovinskaia, N. et al. (2007) Superhard nanocomposite of dense polymorphs of boron nitride: noncarbon material has reached diamond hardness. *Appl. Phys. Lett.*, **90** (10), 3.
9 Nye, J.F. (1979) *Physical Properties of Crystals*, Oxford University Press, Oxford.
10 Levy, M., Bass, H., and Stern, R. (eds) (2000) *Handbook of Elastic Properties of Solids, Liquids, and Gases*, Academic Press.
11 Voigt, W. (1928) *Lehrbuch der Kristallphysik*, Leipzig, Teubner.
12 Reuss, A. (1929) *Berechnung der Berechnung der Fließgrenze von Mischkristallen auf Grund der Plastizitätsbedingung für Einkristalle*. *J. Appl. Math. Mec.*, **9** (1), 49–58.
13 Hill, R. (1952) The elastic behaviour of a crystalline aggregate. *Proc. Phys. Soc. London A*, **65** (389), 349–355.
14 Riedel, R. (ed.) (2000) *Handbook of Ceramic Hard Materials*, vol. 1, Wiley-VCH Verlag GmbH, Weinheim.
15 Li, X.D. and Bhushan, B. (2002) A review of nanoindentation continuous stiffness measurement technique and its applications. *Mater. Charact.*, **48** (1), 11–36.
16 Brazhkin, V. et al. (2004) What does 'harder than diamond' mean? *Nat. Mater.*, **3** (9), 576–577.
17 Oliver, W.C. and Pharr, G.M. (1992) An improved technique for determining hardness and elastic-modulus using load and displacement sensing indentation experiments. *J. Mater. Res.*, **7** (6), 1564–1583.
18 Oliver, W.C. and Pharr, G.M. (2004) Measurement of hardness and elastic modulus by instrumented indentation: advances in understanding and refinements to methodology. *J. Mater. Res.*, **19** (1), 3–20.
19 Liu, A.Y. and Cohen, M.L. (1989) Prediction of new low compressibility solids. *Science*, **245**, 841–842.
20 Cynn, H. et al. (2002) Osmium has the lowest experimentally determined compressibility. *Phys. Rev. Lett.*, **88** (13), 135701.
21 Zhang, R.F. et al. (2011) Superhard materials with low elastic moduli: three-dimensional covalent bonding as the origin of superhardness in B_6O. *Phys. Rev. B*, **83** (9), 092101.
22 Pugh, S.F. (1954) Relations between the elastic moduli and the plastic properties of polycrystalline pure metals. *Philos. Mag.*, **45** (367), 823–843.
23 Teter, D.M. (1998) Computational alchemy: the search for new superhard materials. *MRS Bull.*, **23** (1), 22–27.
24 Chen, X.Q. et al. (2011) Modeling hardness of polycrystalline materials and bulk metallic glasses. *Intermetallics*, **19** (9), 1275–1281.
25 Gao, F. et al. (2003) Hardness of covalent crystals. *Phys. Rev. Lett.*, **91** (1), 015502.

26 Simunek, A. and Vackar, J. (2006) Hardness of covalent and ionic crystals: first-principle calculations. *Phys. Rev. Lett.*, **96** (8), 085501.

27 Li, K.Y. *et al.* (2008) Electronegativity identification of novel superhard materials. *Phys. Rev. Lett.*, **100** (23), 235504.

28 Li, K.Y. and Xue, D.F. (2009) Hardness of materials: studies at levels from atoms to crystals. *Chin. Sci. Bull.*, **54** (1), 131–136.

29 Hall, E.O. (1951) The deformation and ageing of mild steel: III. discussion of results. *Proc. Phys. Soc. London A*, **64** (381), 747–753.

30 Petch, N.J. (1953) The cleavage strength of polycrystals. *J. Iron Steel Inst.*, **174** (1), 25–28.

31 Wollmershauser, J.A. *et al.* (2014) An extended hardness limit in bulk nanoceramics. *Acta Mater.*, **69**, 9–16.

32 Huang, X. *et al.* (2013) Strengthening metals by narrowing grain size distributions in nickel–titanium thin films. *J. Mater. Res.*, **28** (10), 1289–1294.

33 Veprek, S. (2013) Recent search for new superhard materials: go nano!. *J. Vac. Sci. Technol. A*, **31** (5), 050822.

34 Veprek, S. (2011) Recent attempts to design new super- and ultrahard solids leads to nano-sized and nano-structured materials and coatings. *J. Nanosci. Nanotechnol.*, **11** (1), 14–35.

35 Veprek, S., Prilliman, S.G., and Clark, S.M. (2010) Elastic moduli of nc-TiN/a-Si_3N_4 nanocomposites: compressible, yet superhard. *J. Phys. Chem. Solids*, **71** (8), 1175–1178.

36 Kroke, E. and Schwarz, M. (2004) Novel group 14 nitrides. *Coord. Chem. Rev.*, **248** (5–6), 493–532.

37 Guo, Y.J. and Goddard, W.A. (1995) Is carbon nitride harder than diamond: no, but its girth increases when stretched (negative Poisson ratio). *Chem. Phys. Lett.*, **237** (1–2), 72–76.

38 Liu, A.Y. and Wentzcovitch, R.M. (1994) Stability of carbon nitride solids. *Phys. Rev. B*, **50** (14), 10362–10365.

39 Teter, D.M. and Hemley, R.J. (1996) Low-compressibility carbon nitrides. *Science*, **271**, 53–55.

40 Mo, S.-D. *et al.* (1999) Interesting physical properties of the new spinel phase of Si_3N_4 and C_3N_4. *Phys. Rev. Lett.*, **83**, 5046–5049.

41 Kroll, P. and Hoffmann, R. (1999) Theoretical tracing of a novel route from molecular precursors through polymers to dense, hard C_3N_4 solids. *J. Am. Chem. Soc.*, **121** (19), 4696–4703.

42 Wang, J., Lei, J., and Wang, R. (1998) Diffraction-pattern calculation and phase identification of crystalline C_3N_4. *Phys. Rev. B*, **58** (18), 11890–11896.

43 Matsumoto, S., Xie, E.Q., and Izumi, F. (1999) On the validity of the formation of crystalline carbon nitrides, C_3N_4. *Diamond Relat. Mater.*, **8** (7), 1175–1182.

44 Malkow, T. (2000) Critical observations in the research of carbon nitride. *Mater. Sci. Eng. A*, **292** (1), 112–124.

45 Goglio, G., Foy, D., and Demazeau, G. (2008) State of Art and recent trends in bulk carbon nitrides synthesis. *Mater. Sci. Eng. R.*, **58** (6), 195–227.

46 Cote, M. and Cohen, M.L. (1997) Carbon nitride compounds with 1:1 stoichiometry. *Phys. Rev. B*, **55** (9), 5684–5688.

47 Kim, E. *et al.* (2001) Tetragonal crystalline carbon nitrides: theoretical predictions. *Phys. Rev. Lett.*, **86** (4), 652–655.

48 Kim, E. *et al.* (2001) Theoretical study of a body-centered-tetragonal phase of carbon nitride. *Phys. Rev. B*, **64** (9), 6.

49 Hao, J. *et al.* (2015) Prediction of a superhard carbon-rich C–N compound comparable to diamond. *J. Phys. Chem. C*, **119** (51), 28614–28619.

50 Wang, X.L. (2012) Polymorphic phases of sp^3-hybridized superhard CN. *J. Chem. Phys.*, **137** (18), 4.

51 Tang, X., Hao, J., and Li, Y.W. (2015) A first-principles study of orthorhombic CN as a potential superhard material. *Phys. Chem. Chem. Phys.*, **17** (41), 27821–27825.

52 Zhang, M.G. *et al.* (2014) A novel superhard tetragonal carbon mononitride. *J. Phys. Chem. C*, **118** (6), 3202–3208.

53 Dong, H.F. *et al.* (2015) The phase diagram and hardness of carbon nitrides. *Sci. Rep.*, **5**, 5.

54 Li, Q. et al. (2012) A novel low compressible and superhard carbon nitride: body-centered tetragonal CN_2. *Phys. Chem. Chem. Phys.*, **14** (37), 13081–13087.

55 Horvath-Bordon, E. et al. (2007) High-pressure synthesis of crystalline carbon nitride imide, $C_2N_2(NH)$. *Angew. Chem. Int. Ed.*, **46** (9), 1476–1480.

56 Salamat, A. et al. (2009) Tetrahedrally bonded dense C_2N_3H with a defective wurtzite structure: X-ray diffraction and Raman scattering results at high pressure and ambient conditions. *Phys. Rev. B*, **80** (10), 104106.

57 Sougawa, M. et al. (2011) Crystal Structure of new carbon–nitride-related material $C_2N_2(CH_2)$. *Jpn. J. Appl. Phys.*, **50** (9). doi: 10.1143/JJAP.50.095503

58 Sougawa, M. et al. (2013) Bulk modulus and structural changes of carbon nitride $C_2N_2(CH_2)$ under pressure: the strength of C–N single bond. *J. Appl. Phys.*, **113** (5). doi: 10.1063/1.4789020

59 Kojima, Y. and Ohfuji, H. (2013) Structure and stability of carbon nitride under high pressure and high temperature up to 125 GPa and 3000 K. *Diamond Relat. Mater.*, **39**, 1–7.

60 Lowther, J.E. (2005) Potential super-hard phases and the stability of diamond-like boron–carbon structures. *J. Phys.: Condens. Matter*, **17** (21), 3221–3229.

61 Nkambule, S.M. and Lowther, J.E. (2010) Crystalline and random "diamond-like" boron–carbon structures. *Solid State Commun.*, **150** (1–2), 133–136.

62 Xu, L.F. et al. (2010) Prediction of a three dimensional conductive superhard material: diamond-like BC_2. *J. Phys. Chem. C*, **114** (51), 22688–22690.

63 Yang, J. et al. (2007) Diamond-like BC_3 as a superhard conductor identified by ideal strength calculations. *J. Phys. Condens. Matter*, **19** (34), 346223.

64 Liu, Z.Y. et al. (2006) Prediction of a sandwichlike conducting superhard boron carbide: first-principles calculations. *Phys. Rev. B*, **73** (17), 172101.

65 Liu, H.Y. et al. (2011) Superhard and superconductive polymorphs of diamond-like BC_3. *Phys. Lett. A*, **375** (3), 771–774.

66 Liang, Y.C. et al. (2009) Superhardness, stability, and metallicity of diamondlike BC_5: density functional calculations. *Phys. Rev. B*, **80** (11), 113401.

67 Calandra, M. and Mauri, F. (2008) High-T_c superconductivity in superhard diamondlike BC_5. *Phys. Rev. Lett.*, **101** (1), 016401.

68 Li, Q.A. et al. (2010) Superhard and superconducting structures of BC_5. *J. Appl. Phys.*, **108** (2), 023507.

69 Liu, H.Y. et al. (2011) Superhard polymorphs of diamond-like BC_7. *Solid State Commun.*, **151** (9), 716–719.

70 Dong, B.W. et al. (2011) Superconductive superhard phase of BC_7: predicted via *ab initio* calculations. *Diamond Relat. Mater.*, **20** (3), 454–457.

71 Xu, L.F. et al. (2011) Prediction of a superconductive superhard material: diamond-like BC_7. *J. Appl. Phys.*, **110** (1), 013501.

72 Lowther, J.E. (2005) Potential super-hard phases and the stability of diamond-like boron–carbon structures. *J. Phys. Condes. Matter*, **17** (21), 3221–3229.

73 Moussa, J.E. and Cohen, M.L. (2006) Two bounds on the maximum phonon-mediated superconducting transition temperature. *Phys. Rev. B*, **74** (9), 094520.

74 Moussa, J.E. and Cohen, M.L. (2008) Constraints on T_c for superconductivity in heavily boron-doped diamond. *Phys. Rev. B*, **77** (6), 064518.

75 Zhang, M. et al. (2015) Superhard BC_3 in cubic diamond structure. *Phys. Rev. Lett.*, **114** (1), 5.

76 Mikhaylushkin, A.S., Zhang, X.W., and Zunger, A. (2013) Crystal structures and metastability of carbon–boron compounds C_3B and C_5B. *Phys. Rev. B*, **87** (9), 094103.

77 Guo, X.J. et al. (2007) First-principles investigation of dense B_4C_3. *J. Phys. Chem. C*, **111** (37), 13679–13683.

78 Zhang, Y.F. and Wang, L. (2013) Prediction of a superhard material: B_4C_4. *J. Phys. Soc. Jpn.*, **82** (7), 073702.

79 Zinin, P.V. et al. (2006) Phase transformation in the $BC_{1.6}$ phase under high pressure and high temperature. *J. Appl. Phys.*, **100** (1), 013516.

80 Solozhenko, V.L. et al. (2009) Ultimate metastable solubility of boron in diamond: synthesis of superhard diamondlike BC$_5$. *Phys. Rev. Lett.*, **102** (1), 015506.

81 Solozhenko, V.L. et al. (2009) Erratum: Ultimate metastable solubility of boron in diamond: synthesis of superhard diamondlike BC$_5$ [Phys. Rev. Lett. 102, 015506 (2009)]. *Phys. Rev. Lett.*, **102** (17), 179901.

82 Zinin, P.V. et al. (2012) Synthesis of cubic dense BC$_3$ nanostructured phase under high-pressure and high-temperature. *J. Appl. Phys.*, **111** (11), 114905.

83 Zhao, Y. et al. (2002) Superhard B–C–N materials synthesized in nano-structured bulks. *J. Mater. Res.*, **17** (12), 3139–3145.

84 Langenhorst, F. and Solozhenko, V.L. (2002) ATEM-EELS study of new diamond-like phases in the B–C–N system. *Phys. Chem. Chem. Phys.*, **4** (20), 5183–5188.

85 Sikora, A. et al. (2009) Effect of boron incorporation on the structure and electrical properties of diamond-like carbon films deposited by femtosecond and nanosecond pulsed laser ablation. *Thin Solid Films*, **518** (5), 1470–1474.

86 Shirasaki, T. et al. (2000) Synthesis and characterization of boron-substituted carbons. *Carbon*, **38** (10), 1461–1467.

87 Liu, Y.S. et al. (2009) Effect of deposition temperature on boron-doped carbon coatings deposited a BCl$_3$–C$_3$H$_6$–H$_2$ mixture using low pressure chemical vapor deposition. *Appl. Surf. Sci.*, **255** (21), 8761–8768.

88 Zinin, P.V. et al. (2012) Phase transition in BC$_x$ system under high-pressure and high-temperature: synthesis of cubic dense BC$_3$ nanostructured phase. *J. Appl. Phys.*, **111** (11), 114905.

89 Zinin, P.V. et al. (2009) Ultraviolet Raman spectroscopy of the graphitic BC$_x$ phases. *Diamond Relat. Mater.*, **18** (9), 1123–1128.

90 Odake, S. et al. (2013) Formation of the high pressure graphite and BC$_8$ phases in a cold compression experiment by Raman scattering. *J. Raman Spectrosc.*, **44** (11), 1596–1602.

91 Zinin, P.V. et al. (2014) Elastic properties, sp^3 fraction, and Raman scattering in low and high pressure synthesized diamond-like boron rich carbides. *J. Appl. Phys.*, **116** (13), 9.

92 Lowther, J.E. (2011) The role played by computation in understanding hard materials. *Mech. Compos. Mater.*, **4**, 1104–1116.

93 Bundy, F.P. (1967) Hexagonal diamond: a new form of carbon. *J. Chem. Phys.*, **46** (9), 3437.

94 Kraus, D. et al. (2016) Nanosecond formation of diamond and lonsdaleite by shock compression of graphite. *Nat. Commun.*, 7, 10970.

95 Taniguchi, T. et al. (1997) *In-situ* X-ray observation of phase transformation of rhombohedral boron nitride under static high pressure and high temperature. *Diamond Relat. Mater.*, **6** (12), 1806–1815.

96 Liu, A.Y., Wentzcovitch, R.M., and Cohen, M.L. (1989) Atomic arrangement and electronic structure of BC$_2$N. *Phys. Rev. B*, **39** (3), 1760–1765.

97 Tateyama, Y. et al. (1997) Proposed synthesis path for heterodiamond BC$_2$N. *Phys. Rev. B*, **55** (16), R10161–R10164.

98 Yuge, K. et al. (2008) First-principles-based phase diagram of the cubic BNC ternary system. *Phys. Rev. B*, **77** (9), 094121.

99 Zhou, X.-F. et al. (2009) A tetragonal phase of superhard BC$_2$N. *J. Appl. Phys.*, **105** (9), 093521 (4 pp.).

100 Li, Q. et al. (2009) Rhombohedral superhard structure of BC$_2$N. *J. Appl. Phys.*, **105** (5), 053514.

101 Luo, X. et al. (2008) Refined crystal structure and mechanical properties of superhard BC$_4$N crystal: first-principles calculations. *J. Phys. Chem. C*, **112**, 9516–9519.

102 Bhat, S. (2016) Studies on boron–carbon–nitrides (B–C–N) leading to the discovery of the novel boron oxynitride B$_6$N$_4$O$_3$. PhD thesis, TU Darmstadt. http://tuprints.ulb.tu-darmstadt.de/5540/

103 Jiang, X. et al. (2011) Mapping distributions of mechanical properties and formation ability on the ternary B–C–N phase diagram. *Diamond Relat. Mater.*, **20** (7), 891–895.

104 Bill, J., Riedel, R., and Passing, G. (1992) Amin-Borane als Precursoren für Borcarbidnitrid. *Z. Anorg. Allg. Chem.*, **610** (4), 83–90.

105 Riedel, R., Bill, J., and Passing, G. (1991) A novel carbon material derived from pyridine-borane. *Adv. Mater.*, **3** (11), 551–552.

106 Hubacek, M. and Sato, T. (1995) Preparation and properties of a compound in the BCN system. *J. Solid State Chem.*, **114**, 258.

107 Badzian, A.R. (1981) Cubic boron nitride - diamond mixed crystals. *Mater. Res. Bull.*, **16** (11), 1385–1393.

108 Sasaki, T. et al. (1993) Simultaneous crystallization of diamond and cubic boron nitride from the graphite relative boron carbide nitride (BC_2N) under high pressure/high temperature conditions. *Chem. Mater.*, **5** (5), 695–699.

109 Nakano, S. et al. (1994) Segregative crystallization of several diamond-like phases from the graphitic BC_2N without an additive at 7.7 GPa. *Chem. Mater.*, **6** (12), 2246–2251.

110 Knittle, E. et al. (1995) High-pressure synthesis, characterization, and equation of state of cubic C-BN solid solutions. *Phys. Rev. B*, **51** (18), 12149–12156.

111 Komatsu, T. et al. (1996) Synthesis and characterization of a shock-synthesized cubic B–C–N solid solution of composition $BC_{2.5}N$. *J. Mater. Chem.*, **6** (11), 1799.

112 Komatsu, T. et al. (1999) Creation of superhard B–C–N heterodiamond using an advanced shock wave compression technology. *J. Mater. Process. Technol.*, **85** (1–3), 69–73.

113 Komatsu, T. (2004) Bulk synthesis and characterization of graphite-like B-C-N and B-C-N heterodiamond compounds. *J. Mater. Chem.*, **14** (2), 221–227.

114 Solozhenko, V.L. et al. (2001) Synthesis of superhard cubic BC_2N. *Appl. Phys. Lett.*, **78** (10), 1385.

115 Langenhorst, F. and Solozhenko, V.L. (2007) ATEM-EELS study of diamond-like phases in the B-C-N system. *Geochim. Cosmochim. Acta*, **71** (15), A542–A542.

116 Tang, M. et al. (2012) Superhard solid solutions of diamond and cubic boron nitride. *Scr. Mater.*, **66** (10), 781–784.

117 Leinenweber, K. et al. (1999) Synthesis and structure refinement of the spinel, γ-Ge_3N_4. *Chem. Eur. J*, **5**, 3076–3078.

118 Schwarz, M. et al. (2000) Spinel-Si_3N_4: Multi-anvil press synthesis and structural refinement. *Adv. Mater.*, **12**, 883–887.

119 Serghiou, G. et al. (1999) Synthesis of a cubic Ge_3N_4 phase at high pressures and temperatures. *J. Chem. Phys.*, **111**, 4659–4662.

120 Zerr, A. et al. (1999) Synthesis of cubic silicon nitride. *Nature*, **400**, 340–342.

121 Jiang, J.Z. et al. (2000) Structural characterisation of cubic silicon nitride. *Europhys. Lett.*, **51**, 62–67.

122 Scotti, N. et al. (1999) Sn_3N_4, a tin(IV) nitride: syntheses and the first crystal structure determination of a binary tin-nitrogen compound. *Z. Anorg. Allg. Chem.*, **625**, 1435–1439.

123 Lima, R.S. et al. (1991) Magnetron sputtered tin nitride. *Solid State Commun.*, **79** (5), 395–398.

124 Maruyama, T. and Morishita, T. (1995) Tin nitride thin films prepared by radio-frequency reactive sputtering. *J. Appl. Phys.*, **77**, 6641–6645.

125 Caskey, C.M. et al. (2015) Semiconducting properties of spinel tin nitride and other IV_3N_4 polymorphs. *J. Mater. Chemistry C*, **3** (6), 1389–1396.

126 Shemkunas, M.P. et al. (2004) Hardness, elasticity, and fracture toughness of polycrystalline spinel germanium nitride and tin nitride. *J. Mater. Res.*, **19**, 1392–1399.

127 Soignard, E., McMillan, P.F., and Leinenweber, K. (2004) Solid solutions and ternary compound formation among Ge_3N_4-Si_3N_4 nitride spinels synthesized at high pressure and high temperature. *Chem. Mater.*, **16**, 5344–5349.

128 Schwarz, M. et al. (2002) Spinel sialons. *Angew. Chem. Int. Ed.*, **41**, 789–793.

129 Wang, W.D. et al. (2012) Superhard composites of cubic silicon nitride and diamond. *Diamond Relat. Mater.*, **27–28**, 49–53.

130 Zerr, A. et al. (2006) Recent advances in new hard high-pressure nitrides. *Adv. Mater.*, **18**, 2933–2948.

131 Sekine, T. et al. (2000) Shock-induced transformtion of β-Si_3N_4 to a high-pressure cubic-spinel phase. *Appl. Phys. Lett.*, **76**, 3706–3708.

132 Kohler, A. et al. (2015) The role of oxygen in shockwave-synthesized γ-Si_3N_4 material. *J. Eur. Ceram. Soc.*, **35** (12), 3283–3288.

133 Yakushev, V.V. et al. (2015) Formation of cubic silicon nitride from the low-pressure phase by high-temperature shock compression. *Combust. Explos.*, **51** (5), 603–610.

134 Yunoshev, A.S. (2004) Shock-wave synthesis of cubic silicon nitride. *Combust. Explos.*, **40** (3), 370–373.

135 Horvath-Bordon, E. et al. (2006) High-pressure chemistry of nitride-based materials. *Chem. Soc. Rev.*, **35**, 987–1014.

136 Jiang, J.Z. et al. (2002) Compressibility and thermal expansion of cubic silicon nitride. *Phys. Rev. B*, **65**, 161202.

137 Zerr, A. et al. (2002) Elastic moduli and hardness of cubic silicon nitride. *J. Am. Ceram. Soc.*, **85**, 86–90.

138 Somayazulu, M.S. et al. (2000) High pressure–high temperature synthesis of spinel Ge_3N_4, in *Science and Technology of High Pressure, Proceedings of AIRAPT-17* (eds M. Manghnani, W.J. Nellis, and M. Nicol), Universities Press, Hyderabad, India, pp. 663–666.

139 Pradhan, G.K. et al. (2010) Elastic and structural instability of cubic Sn_3N_4 and C_3N_4 under pressure. *Phys. Rev. B*, **82** (14), 144112.

140 Jiang, J.Z. et al. (2001) Hardness and thermal stability of cubic silicon nitride. *J. Phys. Condens. Matter*, **13**, L515–L520.

141 He, H. et al. (2001) Phase transformation of germanium nitride (Ge_3N_4) under shock wave compression. *J. Appl. Phys.*, **90**, 4403–4406.

142 Museur, L., Zerr, A., and Kanaev, A. (2016) Photoluminescence and electronic transitions in cubic silicon nitride. *Sci. Rep.*, **6**, 18523.

143 Boyko, T.D. et al. (2010) Class of tunable wide band gap semiconductors γ-$(Ge_xSi_{1-x})_3N_4$. *Phys. Rev. B*, **81** (15), 155207.

144 Boyko, T.D. et al. (2013) Electronic structure of spinel-type nitride compounds Si_3N_4, Ge_3N_4, and Sn_3N_4 with tuneable band gaps: application to light emitting diodes. *Phys. Rev. Lett.*, **111** (9), 097402.

145 Zerr, A., Miehe, G., and Riedel, R. (2003) Synthesis of cubic zirconium and hafnium nitride having Th_3P_4-structure. *Nat. Mater.*, **2**, 185–189.

146 Dzivenko, D.A. et al. (2007) High-pressure multianvil synthesis and structure refinement of oxygen-bearing cubic zirconium(IV) nitride. *Adv. Mater.*, **19**, 1869–1873.

147 Dzivenko, D. et al. (2009) Synthesis and properties of oxygen bearing c-Zr_3N_4 and c-Hf_3N_4. *J. Alloys Comp.*, **480**, 46–49.

148 Fix, R., Gordon, R.G., and Hoffman, D.M. (1991) Chemical vapor deposition of titanium, zirconium, and hafnium nitride thin films. *Chem. Mater.*, **3**, 1138–1148.

149 Li, J. et al. (2005) Synthesis of nanocrystalline Zr_3N_4 and Hf_3N_4 powders from metal dialkylamides. *Z. Anorg. Allg. Chem.*, **631**, 1449–1455.

150 Salamat, A. et al. (2013) Synthesis of tetragonal and orthorhombic polymorphs of Hf_3N_4 by high-pressure annealing of a prestructured nanocrystalline precursor. *J. Am. Chem. Soc.*, **135** (25), 9503–9511.

151 Chhowalla, M. and Unalan, H.E. (2005) Thin films of hard cubic Zr_3N_4 stabilized by stress. *Nat. Mater.*, **4**, 317–322.

152 Sui, Y.R. et al. (2009) Preparation, characterization and properties of N-rich Zr–N thin film with Th_3P_4 structure. *Appl. Surf. Sci.*, **255**, 6355–6358.

153 Meng, Q.N. et al. (2011) Preferred orientation, phase transition and hardness for sputtered zirconium nitride films grown at different substrate biases. *Surf. Coat. Technol.*, **205**, 2865–2870.

154 Gu, Z.Q. et al. (2015) Identification and thermodynamic mechanism of the phase transition in hafnium nitride films. *Acta Mater.*, **90**, 59–68.

155 Dzivenko, D.A. et al. (2006) Equation of state of cubic hafnium(IV) nitride having Th_3P_4-type structure. *Solid State Commun.*, **139**, 255–258.

156 Dzivenko, D.A. et al. (2007) Elastic moduli and hardness of c-

157 Zerr, A. et al. (2010) Elastic moduli of hard c-Zr_3N_4 from laser ultrasonic measurements. *Phys. Status Solidi Rapid Res. Lett.*, **4** (12), 353–355.

Previous entry continuation: $Zr_{2.86}(N_{0.88}O_{0.12})_4$ having Th_3P_4-type structure. *Appl. Phys. Lett.*, **90**, 191910.

158 Friedrich, A. et al. (2010) *In situ* observation of the reaction of tantalum with nitrogen in a laser heated diamond anvil cell. *J. Alloys Comp.*, **502** (1), 5–12.

159 Zerr, A. et al. (2009) High-pressure synthesis of tantalum nitride having orthorhombic U_2S_3 structure. *Adv. Funct. Mater.*, **19**, 2282–2288.

160 Bourguille, J. et al. (2015) Elastic moduli and hardness of η-Ta_2N_3 from nanoindentation measurements. *EPL*, **111** (1), 18006.

161 Zerr, A. et al. (2012) Elastic moduli of η-Ta_2N_3, a tough self-healing material, via laser ultrasonics. *Phys. Status Solidi Rapid Res. Lett.*, **6**, 484–486.

162 Gandhi, C. and Ashby, M.F. (1979) Fracture-mechanism maps for materials which cleave: fcc, bcc and hcp metals and ceramics. *Acta Metal.*, **27** (10), 1565–1602.

163 Bourguille, J., Brinza, O., and Zerr, A. (2016) Vickers microhardness and indentation fracture toughness of tantalum sesquinitride, η-Ta_2N_3. *Ceram. Int.*, **42** (1), 982–985.

9
Self-healing Materials

Martin D. Hager[1,2]

[1]*Friedrich Schiller University Jena, Laboratory for Organic and Macromolecular Chemistry, Humboldtstr. 10, 07743 Jena, Germany*
[2]*Friedrich Schiller University Jena, Jena Center for Soft Matter (JCSM), Philosophenweg 7, 07743 Jena, Germany*

9.1
Introduction

The grand dream of indestructible materials has inspired mankind all times. For instance, in Greek mythology adamantine (derived from the Greek word *adamas* = invincible) was said to be an indestructible material. In Homer's *Iliad*, the hero Achilles wore an armor forged by Hephaestus that was said to be impenetrable [1]. This short list could go on demonstrating the perpetual quest of mankind for superior materials. Both materials science and engineering provided novel materials, which could outperform the by then common materials. Consequently, impressing materials with fascinating properties could be achieved: "Unbreakable glass" and ultrastrong glass [2] were developed, ultrastrong polymer composites have been fabricated [3,4], ultrahigh performance concrete [5] was designed, and high-strength steels were developed [6]. With these materials in hand, previously unthinkable applications became possible, for example, airplanes consisting of over 50% advanced composites, skyscrapers reaching a height of 1000 m, and so on. However, despite all these efforts to prevent any damage, the occurrence of damage will happen sooner or later. Within this context not necessarily a catastrophic failure is meant. Damage on the molecular level, as well as microcracks, will accumulate over time leading finally to a failure of the material. Damage is inevitable!

Therefore, another concept was developed in materials science: *damage management* [7]. In contrast to the above-described concept of *damage prevention*, the materials are no longer designed to be stronger, tougher, harder, and so on in order to circumvent any damage (at least for some time), the novel materials should be able to deal ("live") with the damage (see Figure 9.1). That means, an occurring damage can be healed prolonging the lifetime of the material. Noteworthy, the initial material properties do not have to be superior to the classical

---

Handbook of Solid State Chemistry, First Edition. Edited by Richard Dronskowski, Shinichi Kikkawa, and Andreas Stein.
© 2017 Wiley-VCH Verlag GmbH & Co. KGaA. Published 2017 by Wiley-VCH Verlag GmbH & Co. KGaA.

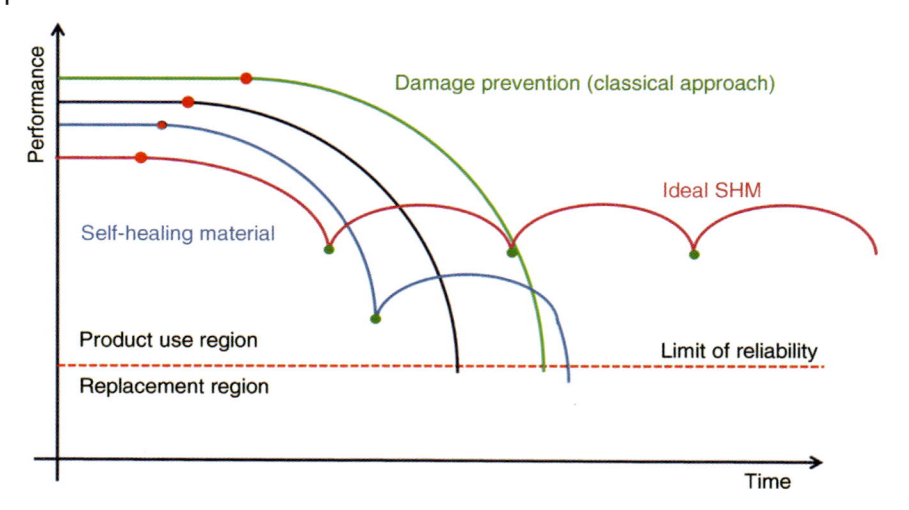

Figure 9.1 Schematic representation of the design approaches for materials: dark blue – material; green – damage prevention; light blue – one time self-healing material; violet – ideal self-healing material (according to Ref. [8]); red dot – starting of damage; and green dot – healing event.

material as long as the performance fulfills the requirements of the corresponding application. Considering an ideal self-healing material, the lifetime will be in theory infinite due to multiple healing events, which restore (partially) the original properties/performance of the material.

Nature shows impressively that self-healing is not utopia [8]. We almost take it for granted that a small incision in the finger and even a broken bone in the leg will heal [9]. Furthermore, our genetic information in the DNA is healed on the molecular level. Interestingly, it is not only the "living material" that shows the ability for self-healing. The byssus threads of a certain mussel can restore their original properties after damage [10]. Consequently, nature has always been a source of inspiration for self-healing materials [11–14].

9.2
Definitions and General Principles

A self-healing material can restore (partially) its original properties/function [15]. Noteworthy, the properties and functions include, but are not limited to mechanical properties. First, self-healing materials were limited to the healing of cracks and scratches after a mechanical damage. However, in recent years self-healing functional materials gained significant interest [16].

In recent years, the following distinctions have been established in the scientific literature:

- *Extrinsic and intrinsic* self-healing materials, the first mentioned require the introduction of additional healing agents (e.g., by encapsulation). The material

itself does not possess the capability for self-healing. In contrast, intrinsic self-healing materials feature the desired ability itself. The first approach is more general, allowing the achievement of self-healing for many materials, which will serve a matrix for the encapsulated healing agent. No development of novel materials/building blocks is required. In contrast, the second approach often includes the design of a novel material due to the introduction of the capability for self-healing into the material. Therefore, also the fabrication of tailor-made materials is possible.
- *Autonomic and nonautonomic self-healing materials*: This classification is a controversial one. In autonomic self-healing materials, the damage itself is the trigger for the self-healing process. Thus, nonautonomic self-healing materials require an additional external trigger (e.g., light or heat) to start the healing process. One might consider these materials rather healable/remendable than really self-healing. However, in the current literature also these materials are classified as self-healing materials. That is legitimate, particularly considering a self-healing under real-life conditions. A material might be exposed to conditions, for example, very low temperatures, which do not allow the autonomic process. In contrast, materials utilized under temperature cycles might feature the required temperature for the healing process under usage. For instance, a self-healing car coating in the midday sun.

In 2010, we defined a general common principle of self-healing materials (see Figure 9.2) [15]. Already a closer look at the above-mentioned example of a cut in the finger reveals this principle. The small incision results in a bleeding providing a "mobile phase." In a self-healing material (a) the crack (b) can be closed by the generation of a mobile phase (c). This process is induced by the damage itself or requires an external trigger. Now back to our bleeding finger – the blood clotting starts resulting in a closure of the wound. The original properties as well as functions are restored (i.e., the blood circulation is sealed, pathogens are blocked, etc.) and the scratch is closed. Later the scab will fall off, revealing the newly formed skin. In materials, the mobile phase will close the scratch (d) and

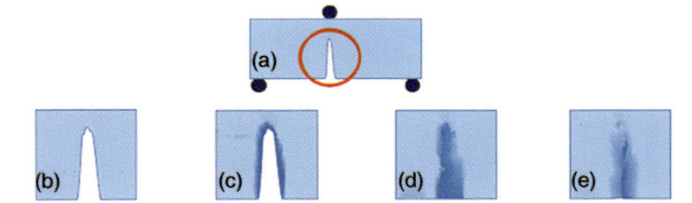

Figure 9.2 Schematic representation of the general mechanism of the self-healing process of a damaged specimen (a). After the damage event (b) a mobile phase (c) has to be induced, either by the damage itself of by an external trigger. The generated mobility allows the closing of the crack (d). Subsequently, the mobile phase is immobilized again (e) resulting in the restoration of the original properties. (Reproduced with permission from Ref. [16])

after the subsequent immobilization (e) the material properties are (partially) restored. The natural analog shows also that the mobile phase and the subsequent immobilized materials are not necessarily the same as for the prior material.

Taking the general principle into account, the required conditions for the healing process will strongly depend on the material. Both metals and ceramics will require higher temperatures, that is, >600 and >800 °C, respectively, in order to induce enough mobility. In contrast, in concrete the required processes can already occur at room temperature. Moreover, rather low temperatures (mostly below 120 °C) will be required both for self-healing asphalt and for self-healing polymers. Furthermore, due to the differences of the mobile species (e.g., single atoms in metals up to large polymer chains), the proportions of the healable damage will be different.

In principle, several general mechanisms can be considered, which will enable a healing process [15]. Table 9.1 summarizes the suitability of different strategies for the different material classes. For instance, the encapsulation of healing agents is a rather general approach, which is suitable for most of the different materials. In contrast reversible cross-links are by far less suited as a general approach. This strategy is strongly linked to polymers. Furthermore, it is quite obvious that polymeric materials are suited for many approaches due to their simple functionalization. In contrast, self-healing metals can be achieved only by limited approaches.

Returning back to the biomimetic self-healing, some of the above-described general mechanisms can be related to biological processes. The encapsulation and the channel transport mimic bleeding [13]. In addition, reversible interactions are the basis for adaptive materials in nature. The following paragraphs will provide an overview on the different self-healing material classes providing selected examples: from metals followed by ceramics and concrete to asphalt and polymers.

Table 9.1 Summary of potential healing mechanisms for the presented material classes.

Mechanism	Encapsulation	Channel transport	Expanding phases	Temperature increase (higher mobility)	Reversible cross-links	Biological processes	Electro-chemical processes	Phase separation	Shape-memory-assisted healing
Metals	░	░	▓				▓	░	▓
Ceramics	▓	▓	▓				▓	░	
Asphalt	▓	▓		▓	░				
Concrete	▓					▓			
Polymer and polymer composites	▓	▓		▓	▓			▓	▓

Source: Table modified from Ref. [15].
The color code indicates the potential or already demonstrated level of the corresponding approach (from dark gray – positive over light gray – potential to white – very unlikely).

9.3
Self-healing Metals

Metals are by far the least investigated class of self-healing materials. In recent years, some approaches have been developed to obtain self-healing metals [17,18]. The mobile phase required for the healing process is constituted of (single) atoms. Moreover, the mobility is induced at rather high temperatures making the design of self-healing metals challenging. Recently, Grabowski and Tasan introduced a classification of self-healing metals, which distinguishes the size of the healed damage: from healing of nanoscale voids to healing of macroscale cracks (see Figure 9.3). The most common approach is the high-temperature precipitation (Figure 9.3a). A solute (virtually the healing agent) is embedded in the metal/metal alloy. If a damage is generated, that is, occurring of a nanovoid at the grain boundary, the solute will be transported by diffusion to the nanovoid resulting in precipitation within the nanovoid. As a consequence, the nanovoid cannot progress to form larger damages (i.e., microcracks). Examples based on this strategy were described utilizing boron atoms in steel [19] as well as boron together with nitrogen [20] resulting in the formation of elementary boron and boron nitride, respectively. Similar processes have been observed in Fe-Au-B-N alloys, leading to the segregation of gold [21]. In comparison, the low-temperature precipitation (Figure 9.3b) is quite similar to the above described approach. The transport of the solute is already possible at lower temperatures, for example, copper in Al-alloys [22]. A novel concept – introduced by Grabowski and Tasan – is the incorporation of nanoscale shape-memory nanoalloys (Figure 9.3c). Although a practical demonstration has not been done up to now, the concept is promising.

In contrast to the above-described approaches for damages in the nanometer-range, the following four strategies allow the healing of macroscale damages. In Figure 9.3d, shape-memory alloy wires are introduced, which will bring the crack surfaces after damage in contact. The damage leads to a crack in the weaker solder material. Moreover, the stress induces a transition of the shape-memory alloy (SMA) wires. Subsequent heating causing the melting of the matrix will close the crack completely [23]. Recently, NiTi SMA wires were incorporated in a zinc ZA-8 die-casting alloy. Thermal heating resulted in a healing of up to 30% of the original ultimate tensile strength and ductility [24]. Another approach, which is more closely related to the well-known encapsulation of healing agent in polymers (see below) is the incorporation of solder tubes/capsules (Figure 9.3e) [25]. Thus, a damage, that is, a crack, can be closed by thermal treatment. The lower melting point of the solder allows the selective melting of the encapsulated healing agent, which results in a crack closure. For instance, aluminum was utilized as matrix. Aluminum oxide tubes were used as the "capsules" for the healing agent. Latter component was a Sn_3Pb_2 solder, which should close the crack after melting. However, due to the poor interaction of the matrix and the tubes the solder could not sufficiently close the crack [26]. Comparable to the latter approach, the solder can also be applied as coating (Figure 9.3f). The last strategy is the electrohealing (Figure 9.3g). In contrast to the previous approaches, no additional healing agent is

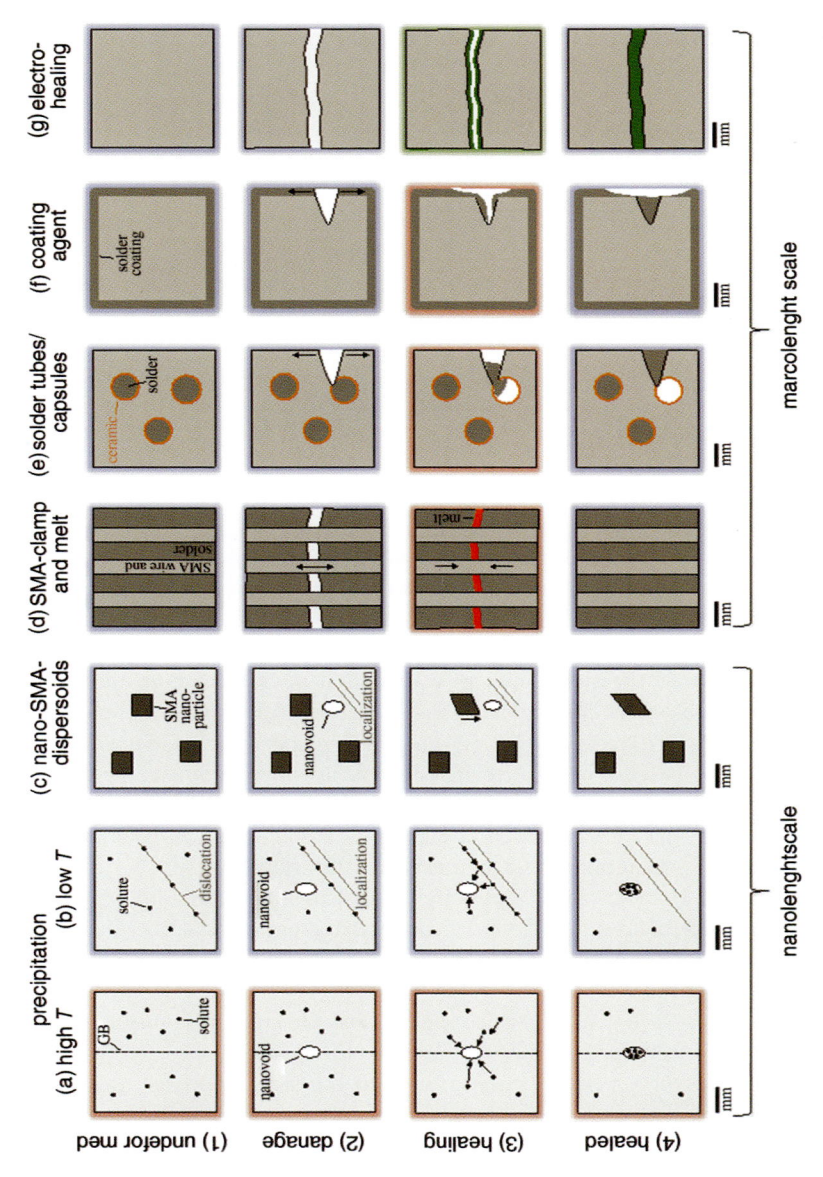

Figure 9.3 Overview on the different strategies to obtain self-healing metals. (Reproduced with permission from Ref. [9].)

required. The cracks can be closed by the application of a voltage in the presence of a corresponding electrolyte. Subsequently, material is deposited in the crack. By this method, cracks in nickel could be healed [27].

9.4
Self-healing Ceramics

The usage of ceramic materials has a long history. The use of clay as building material by humans dates back to 24 000 BC. Pottery of vessels was started later. Chinese porcelain was a first highlight in the usage in these materials. In recent years, research in ceramics covered many important topics – only to mention a few: thermal barrier coatings [28], zirconia for dental applications [29], superconductors, and membranes for various applications [30]. Consequently, self-healing ceramics are also of great interest [31]. The design of self-healing ceramics is challenging, in particular if the healing process should occur at temperatures below 1000 °C. The required activation energies for solid-state diffusion are quite high [31]. In addition, the bonds feature both ionic and covalent characters resulting in not beneficial relaxation phenomenon [32]. Due to these intrinsic properties, the available healing mechanisms are limited. Diffusional sintering can lead to crack regression; cracks can be filled by a formed oxidation product. The latter approach seems to be the most promising strategy and was investigated with filler materials as well as in MAX phases. The repair filler forms an oxide at high temperatures closing the cracks. For instance, SiC was utilized as the healing agent leading to the formation of gaseous CO_2 and SiO_2, which will close the crack (see Figure 9.4). Alumina with SiC nanoparticles (20–50 nm) could recover its strength completely. However, smaller particles did not result in an efficient healing.

This extrinsic approach can also be transferred to other ceramic materials. SiC/spinel nanocomposites showed healing at 1545 °C within 1 min [33]. Comparable with the SiC filler also, a TiC filler was tested as healing agent for thermal barrier coatings [34]. Other potential repair fillers are Si and $CrSi_2$ or $TiSi_2$. The latter two materials feature even higher volume expansion factors (1.2 and 1.67, respectively) [31]. In contrast to the above-described strategy, also intrinsic self-healing ceramics have been investigated. The so-called MAX phases are ternary compounds consisting of a transition metal ion M, an element A from the group IIIA

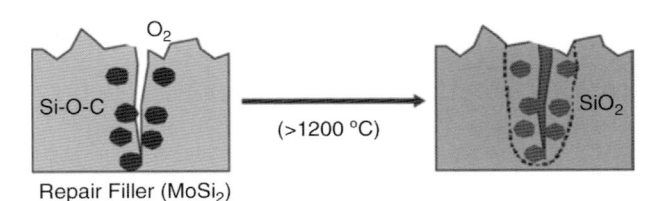

Figure 9.4 Schematic representation of the self-healing of ceramic materials based on a repair filler. (Reproduced with permission from Ref. [16].)

to VIA, and an element X (C or N) with the total composition of $M_{n+1}AX_n$ (with $n = 1/4$, 1, 2, or 3). The materials are also capable of forming metal oxides upon thermal treatment, that is, the ceramic material itself is the healing agent. For instance, Ti_2AlC ceramics have been investigated [35,36]. Upon oxidation Al_2O_3 and TiO_2 are formed, which can both potentially close the crack. Interestingly, small cracks are filled only by alumina. The healing process can also be initiated multiple times (seven cycles have been tested, see Figure 9.5) [36]. Due to their

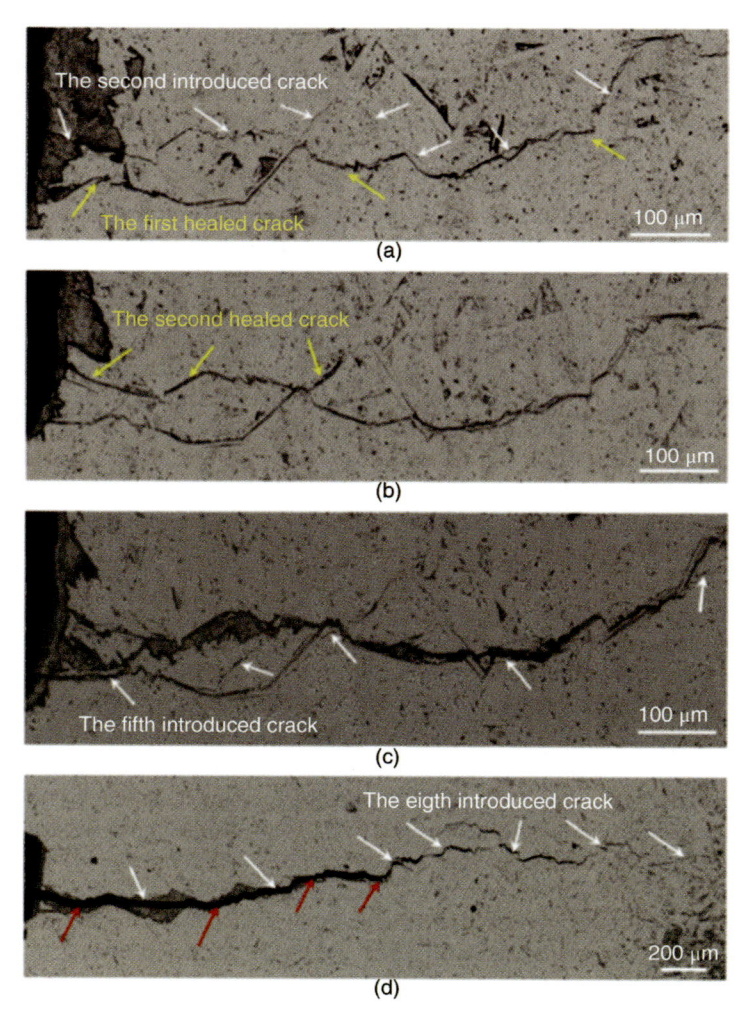

Figure 9.5 Back-scattered scanning electron micrographs of the Ti_2AlC ceramic. (a) Crack within the ceramic (white), which was induced after the first healing cycle (the yellow arrows indicate the healed crack), (b) two complete healing cycles, (c) crack, which was induced after the fourth completed cycle, and (d) crack, which was induced after the seventh cycle; red arrows indicate not-healed parts of the cracks. (Reproduced with permission from Ref. [37].)

structural diversity, many different MAX phases are possible (9 M elements, 12 A elements, 2 X elements with many different combinations and compositions; 75 are known). Therefore, it is of interest to identify interesting candidates besides the known self-healing ceramics Ti_3AlC_2 and Ti_2AlC. A conceptual study was performed to identify further promising materials [37]. Keeping in mind the general healing mechanism (see above), four main criteria were identified for the creation of a healable ceramic material: Preferential oxidation of the A element, fast diffusion of A element, volume expansion upon oxidation, and adhesion of the oxidation product to the matrix [37]. Thus, the most promising materials are V_2AlC, V_3AlC_2, Nb_2AlC, Zr_2AlC, Zr_2AlN, and Ti_3SiC_2.

Within this paragraph, self-healing glass is also presented in brief. Utilizing the general approach of encapsulation, vanadium boride was embedded within a glassy material [38]. After a crack, oxidation of the healing agent will form vanadium oxide and boroxide. Both materials can close the crack.

9.5
Self-healing Concrete

Concrete is by far the most used construction material. In 2014, more than a billion tons have been produced worldwide [39]. Due to the importance of this material, self-healing concrete is of great interest. Fortunately, concrete – depending on the composition – has a natural ability for self-healing. Not without reason, Roman buildings still stand after over 2000 years. This is an impressive time span considering the present condition of some buildings (e.g., ailing bridges) nowadays. Concrete is prone to crack formation. As a consequence, steel reinforcements are utilized. However, if the concrete cracks, the reinforcement might corrode resulting in the failure of the construction. Therefore, in recent years, several approaches to design self-healing concrete have been investigated [39–41].

In intrinsic self-healing concrete, the autonomous healing is achieved by two possible processes. As mentioned above, concrete can feature the natural ability of self-healing, which is based on the hydration of unhydrated particles and the dissolution and carbonization of calcium hydroxide [42]. Based on these processes, smaller cracks can be healed. This natural ability can be improved by three factors (see Figure 9.6). By the restriction of the crack width, the occurring

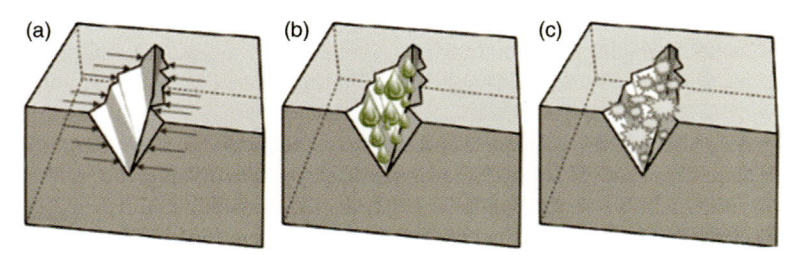

Figure 9.6 Schematic representation of the factors promoting the autonomous self-healing in concrete: (a) restriction of the crack width, (b) water supply, and (c) crystallization [42].

damage is within the limit, which can be healed (Figure 9.6a). Furthermore, water is required for the hydration and the dissolution of calcium hydroxide (Figure 9.6b), which should crystallize after the reaction with CO_2 (Figure 9.6c).

Over 25 years ago, Li *et al.* took advantage of this natural ability and designed engineered cementitious composites (ECCs), which were reinforced by PE fibers limiting the maximum crack width occurring (typically below 60 μm) [43]. This approach was investigated intensively in the following years [40,44,45]. Different polymeric materials have been studied as the fiber material, for instance, poly(vinyl alcohol) (PVA) [46], steel cord, and polypropylene (PP) [47]. The most promising fibers were the PVA fibers. In contrast to the polyolefines such as polyethylene and PP, the PVA features hydroxyl groups, which can presumably interact with the present ions. Steel cord was not suitable due to the corrosion. In the future, natural fibers will also play a role due to their low price as well as the presence of functional groups within these materials [45]. In addition, superabsorbent polymers can provide the required water, because not every crack will be filled by water from the outside [45].

Moreover, the two general approaches encapsulation (i.e., microcapsules) and hollow fibers have been utilized [40,41]. The latter approach was investigated by Dry *et al.* for the delivery of healing agents (e.g., polymerizable monomers) [48]. Different healing agents have been encapsulated in microcapsules. The utilized healing agents range from calcium hydroxide, reactive monomers (e.g., cyanoacrylates, methacrylates, and epoxy) [49], and polymeric precursors [50].

One of the most fascinating approaches in self-healing materials is the biobased self-healing concrete [51]. While for other materials biological processes have been mimicked, the self-healing of this concrete is based on bacteria. In short, bacteria or, to be more precise, bacteria spores are encapsulated in the concrete. In addition, a nutrition source (e.g., urea, calcium lactate, etc.) is utilized. Upon damage, water will penetrate into the crack and the bacteria "awake" from their long Sleeping Beauty slumber. The nutrition source will be converted and these special bacteria precipitate calcium carbonate, which can close the crack (see Figure 9.7). Interestingly, this approach is already far beyond the lab scale. First field tests have been successfully performed [51]. Thus, cracks in the concrete of a car park could be healed.

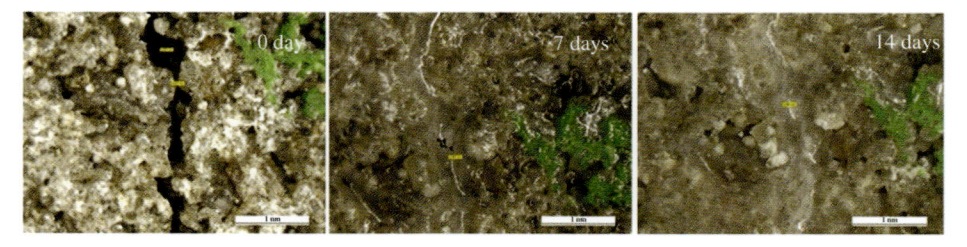

Figure 9.7 Self-healing of concrete based on bacteria within 14 days. (Reproduced with permission from Ref. [52].)

9.6
Self-healing Asphalt

Considering the importance of road networks, self-healing asphalt would provide many benefits. The continuous load caused by traffic combined with the environmental conditions (e.g., very hot in summer and freezing water in winter) will result in damage sooner or later. Ultimately, potholes – well-known and feared by every car driver – will occur, requiring the repair of the road. In 2013, the EU countries spent more than € 10 billion on the maintenance of the road infrastructure [53]. In recent years, significant effort was spent on the development of self-healing asphalt [52]. Commonly, stone mastic asphalts are utilized as road surface. These systems include, among others, coarse aggregates, and/or sand, and/or limestone fine aggregate, and/or filler and bitumen. The last constituent is the key component for a potential self-healing process. At current, there are three main strategies for the self-healing in asphalt: induction heating and rejuvenation [52]. Different nanoparticles have been incorporated into the asphalt in order to improve the resulting properties. For instance, nanoclay was utilized for this purpose [54–56]. The nanoscale filler has another beneficial effect, it will move toward the crack tip preventing further crack growth [57]. The same authors also investigated nanorubbers as additional filler materials resulting in self-healing asphalt.

The induction heating is a more promising approach. Bitumen, the key component, gets softer with increasing temperature and the mobility required for the healing is provided. Moreover, the required conductivity can be introduced by the incorporation of conductive fibers (i.e., steel wool) into the asphalt mixture [58]. The optimal temperature for the healing process was found to be 85 °C [59]. Too high temperatures cause a higher expansion of the binder leading to negative side effects. Interestingly, this approach is far beyond the lab scale. Since 2010, a part of the A 58 in the Netherlands is paved with self-healing asphalt [60]. The test is still ongoing and first investigations indicate that this asphalt will last 30 years!

One reason for the aging process of asphalt is the loss of volatile components of both the bitumen and the polymerization or oxidation of the present constituents [52]. Consequently, the bitumen will become more viscous under the age-hardening. Encapsulated rejuvenators can prevent this process by reducing the viscosity. Different encapsulated healing agents (i.e., rejuvenators) have been tested, *inter alia* sun flower oil [61], aromatic oils [62], or dimethylphenol [63].

9.7
Self-healing Polymers

By far the most investigated material class in the context of self-healing materials are polymers and polymer composites. The origin of self-healing polymers can be traced back to the 1960s when the healing of cracks in polymers (e.g., poly (vinyl acetate)) was investigated [64]. However, it took until the publication [65] of S.R. White *et al.* in 2001 whereby the research field of self-healing materials

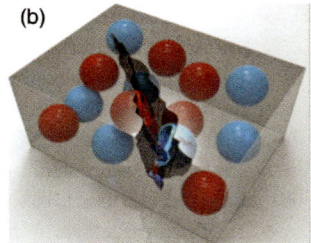

Figure 9.8 Schematic representation of self-healing polymers based on encapsulated healing agents: (a) healing agent in capsule and catalyst and (b) two different healing agents. (Reproduced with permission from Ref. [76].)

started to grow significantly. Noteworthy, this particular publication is by now the most cited in the field [66]. In recent years, many different approaches have been investigated resulting in self-healing polymers [67–72]. Extrinsic self-healing polymers have been obtained by the encapsulation of healing agents, for instance, like in the previously mentioned publication, dicyclopentadiene was encapsulated in microcapsules. Upon damage, the capsules are ruptured by the crack and the healing agent is released. The additionally present catalyst (i.e., a ruthenium complex) initiates the polymerization resulting in a closure of the crack (see Figure 9.8). Many different healing agents have been evaluated [73,74]. Recently, Du Prez and coworker summarized the fifteen most promising chemistries for these systems [75]. Besides the already mentioned ROMP-based healing agents, epoxy-based systems – also in combination with amines or thiols [76] – are worth mentioning. In general, all applied healing agents are liquids, which can flow into the crack. Subsequently, the healing agent is immobilized by, for example, a polymerization. This process can be initiated by a catalyst or a two-component system is used (i.e., reaction of healing agents A and B). The best systems result in healing efficiencies higher than 80% (healing efficiency defined as the ratio of the critical load of the healed and the original sample). Besides these "reactive" systems, pure solvents have been utilized as healing agents [77]. Interestingly, urea-formaldehyde capsules filled with chlorobenzene showed healing efficiencies of 80%.

Impediments for a wide use of the capsules are their stability during the processing of the polymer matrix and their thermal stability. Latter constraint was improved by the usage of core–shell capsules featuring an outer shell of polydopamine improving the thermal stability significantly [78]. However, the major disadvantage of the capsule-based healing is the one-time healing. Upon the self-healing process, the healing agent is consumed. A second crack in the previously damaged region will not be healed. Therefore, an additional mass transport (i.e., transport of the healing agent) is required. Hollow fibers can provide the required supply of healing agent. For instance, hollow glass fibers have been integrated into polymer composites [79,80]. Moreover, a microvascular network was utilized for the transport of dicyclopentadiene allowing multiple (i.e., seven) healing cycles with healing efficiencies between 30 and 70% [81]. The concept was extended to a dual system based on an epoxy resin and an amine hardener

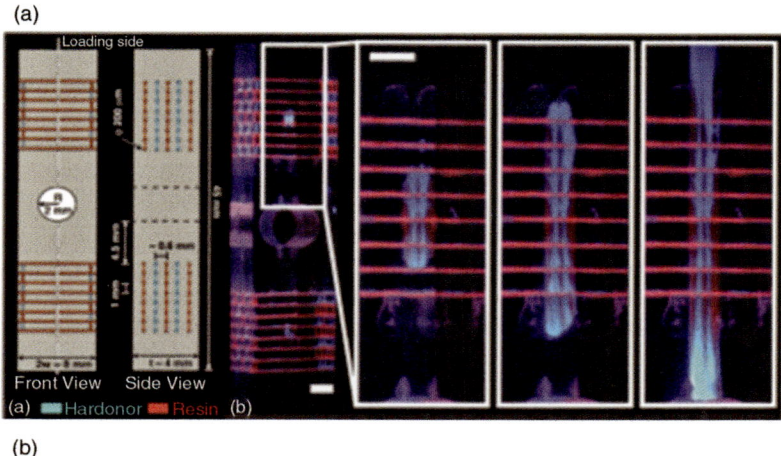

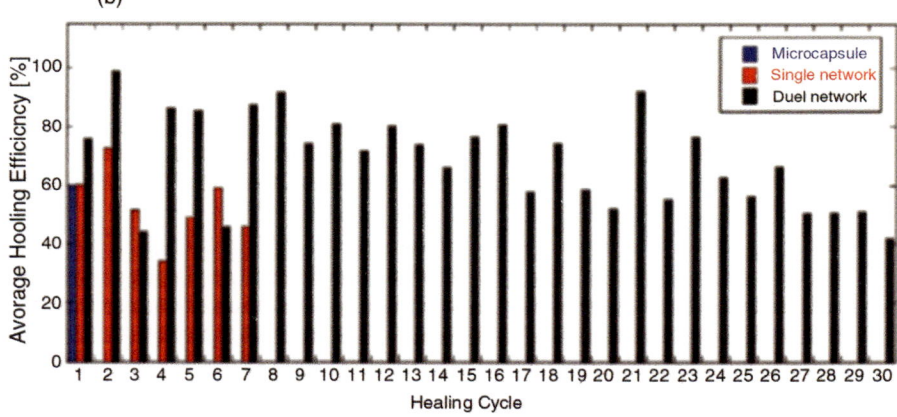

Figure 9.9 (a) Test specimen with microvascular network (healing agents resin and hardener); scalebar 2 mm. (b) Healing efficiencies of capsules (Ref. [67]), single network (Ref. [83]) and a dual interpenetrating network (Ref. [84]). (Reproduced with permission from Refs. [84,85].)

as healing agents (Figure 9.9) [82]. After the occurrence of a crack, both healing agents are released and mixed leading to the curing of the epoxy resin. Optimization of the architecture of the microvascular network, the healing efficiencies could be improved, allowing 30 healing cycles [83].

The concept was already transferred to fiber composites [85]. A full recovery of the mechanical properties was achieved. Furthermore, a ternary microvascular system was utilized to shorten the required healing time [84]. The third channel provides a liquid for heating of the specimen leading to a faster curing of the epoxy–resin.

A critical aspect of the self-healing process is the maximum healable damage. This size is often in the range of several micrometers. In contrast, recently an example of a self-healing polymer allowing the healing of a damage in the size of 3 cm was reported [86]. Microchannels are capable of delivering the required amount of healing agent to close such large damage volumes. For the transport, a rather low viscosity of the healing agent is required; however, the low viscosity is counterproductive for the healing (e.g., dripping of the healing agent). In order to circumvent this problem, a two-stage healing process was applied. First, a gelator system (aldehyde and acylhydrazine resulting in the formation of acylhydrazone function) is utilized, which increases the viscosity by gelation. Second, a monomer is polymerized within the gel, leading to a complete closure of the damage after immobilization.

Besides the extrinsic self-healing polymers, intrinsic self-healing polymers have been investigated intensively in the past few years. One prerequisite for these materials is a kind of reversibility within the material, that is, the presence of reversible binding units and cross-linkers. This behavior can be achieved in supramolecular polymers [87–90] as well as polymers with reversible covalent bonds [91,92]. In the latter case, different reversible moieties have been utilized to generate mobility (often triggered by heat). The prime example is the Diels–Alder reaction (DA) [93]. This thermally reversible [4+2] cycloaddition between a diene and a dienophile allows for the generation of the required mobility. In 2002, F. Wudl and coworkers described the first example of a thermally remendable polymer [94]. A monomer with four furan moieties was combined with a monomer with three maleimide groups resulting in a polymer network. By heating, the retro Diels–Alder reaction occurs, leading to a healing of a crack (healing efficiency of approx. 50%) (see Figure 9.10). Tuning of the monomers improved the achievable healing efficiency to 80% [95]. This standard system, that is, the furan maleimide pair, has been utilized since then intensively: thermosetting polyketones [96], polymer coatings [97,98], and epoxy resins [99,100].

In addition, DA-based self-healing polymers have been combined with shape-memory polymers [101]. The Diels–Alder shape-memory-assisted self-healing was obtained by the fabrication of polyurethane. PCL segments were introduced in order to obtain the shape-memory effect, which enhances the self-healing behavior by bringing the damaged surfaces closer together. In contrast to the furan-maleimide pair, other reversible systems have been investigated less. The anthracene maleimide system requires much higher healing temperatures of over 160 °C; however, the thermal stability is also improved [102]. The self-addition of cyclopentadiene allows the fabrication of polymer networks with only one monomer species. Healing was achieved at temperatures between 80 and 100 °C [103].

In contrast to the Diels–Alder systems, acylhydrazones offer further opportunities. Besides the hydrolytic cleavage resulting in an aldehyde and an acylhydrazide, different exchange reactions are possible. This reversible interaction was applied both in dynamers (i.e., dynamic polymers) and in self-healing polymers [91,104]. Comparable exchange reactions between reversible moieties have been utilized for the self-healing based on disulfide bonds [105] and dynamic boronic esters [106]. Recently, a room-temperature dynamic polydimethylsiloxane network was described [107]. Anionic silanol end groups within the PDMS network enable a dynamic exchange. Finally, as the last example of reversible covalent polymers dynamic urea-based polymers are described. Hindered ureas (i.e., isocyanate and a sterically hindered amine) allow the fabrication of reversible polymers [108]. Healing efficiencies of up to 87% have been achieved.

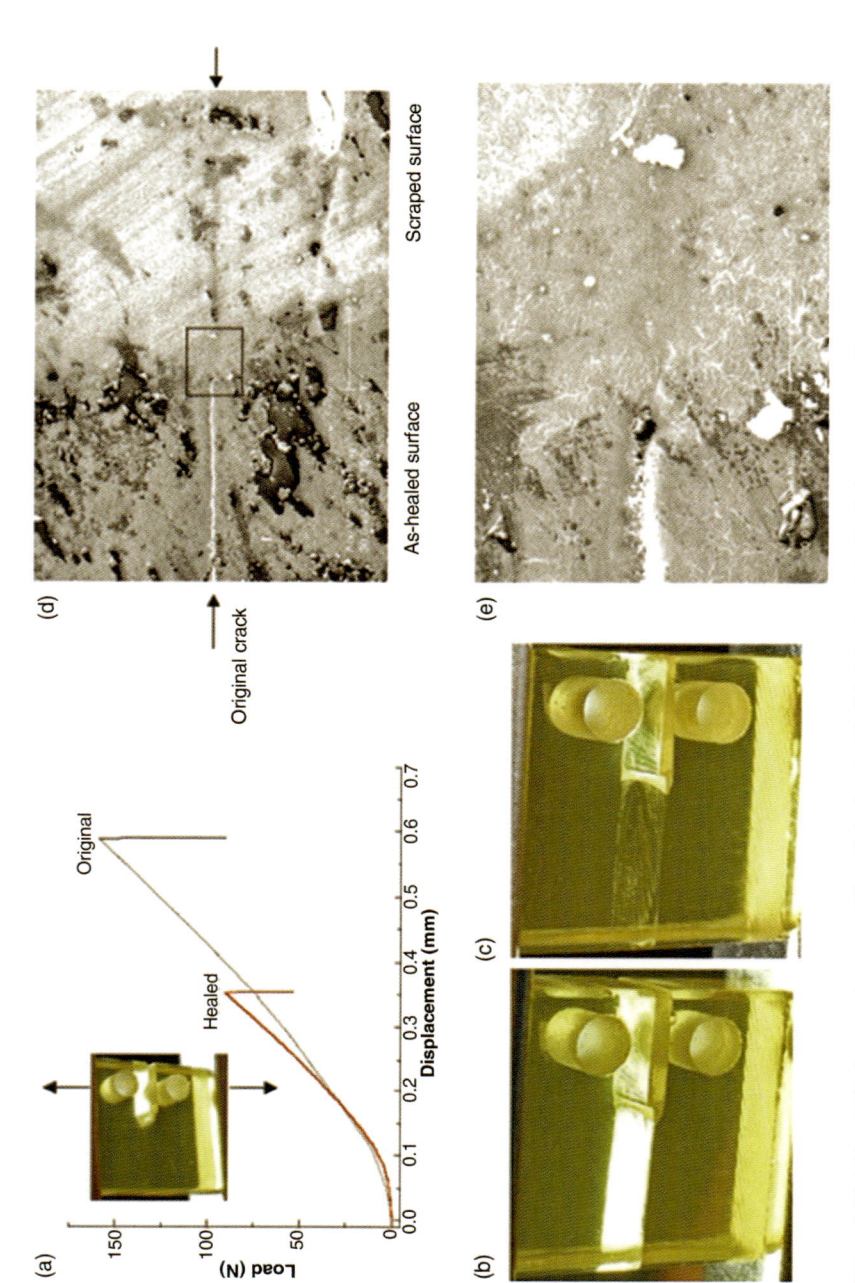

Figure 9.10 (a) Load displacement curves of the original materials (black line) and the healed specimen (red line). Images of the broken specimen before thermal treatment (b) and the healed specimen (c). (d) SEM image of the healed specimen and (e) zoom of the boxed area in (d). (Reproduced with permission from Ref. [96].)

In conclusion, the design of self-healing polymers based on reversible covalent interactions is often a balancing act between dynamics and mechanical properties. Polymers, which are highly dynamic at lower temperatures or at room temperature, often feature weaker mechanical properties. In contrast, polymers with very good mechanics require elevated temperatures for the healing process.

Supramolecular polymers and supramolecular polymer networks [109] provide reversible noncovalent interactions. Consequently, these materials are of particular interest for the design of self-healing materials [87–90]. Besides π–π interactions [110,111] and host–guest interactions [112], mainly ionic interactions, hydrogen bonds and metal–ligand interactions have been utilized for self-healing polymers. In the following, the three main approaches will be presented.

Ionomers, that is, polymers with ionic groups (up to 15%), have been investigated intensively [113]. In particular, ethylene-*co*-methacrylic acid ionomers have been studied after ballistic impact [114]. The projectile puncture is closed again. Due to the presence of ionic clusters within the material, reversible interactions are available, which can contribute to the healing of the material. Moreover, copolymers of *n*-butyl acrylate and acrylic acid were neutralized with different metal ions (Na^+, Zn^{2+}, Co^{2+}; ion content between 2 and 10%) [115]. The resulting materials were studied via rheology providing information about the supramolecular bond lifetime. Bose *et al.* were able to determine an optimal region (lifetime between 10 and 100 s) in which the materials featured good mechanical properties while still showing the ability for healing.

Hydrogen bonds are a consistently recurring motif in supramolecular polymers [116]. Due to the high reversibility and dynamic nature, these supramolecular moieties are promising candidates for self-healing polymers. The most prominent example is self-healing supramolecular rubber described by Leibler *et al.* in 2008 [117,118]. The rubber, based on fatty acids, diethylene triamine, and urea, showed excellent self-healing behavior. Depending on the passed time after the damage, excellent restoration of the mechanical properties was achieved. This polymer was also utilized as binder for silicone particle anodes [119]. During charging, the huge volume expansion of the silicone particles causes large stresses within the electrode resulting in mechanical damage, which consequently leads to a loss in capacity. Electrodes based on the self-healing binder showed excellent capacity retention. Furthermore, hydrogen bonds have also been utilized in more complex polymer architectures, for instance, in block copolymers [120] and V- and H-shaped polymers [121].

As the last example, self-healing metallopolymers [122] will be presented. One of the first examples was described in 2011 [123]. An end-functionalized polymer (poly(ethylene-*co*-butylene)) bearing two ligands (2,6-bis(1′-methylbenzimidazolyl) pyridine) was complexed with Zn(II) and La(III) ions resulting in main-chain metallopolymers and networks, respectively. Due to the large difference between the metal complexes and the polymer chain, complex-rich domains are formed within the material. Healing was achieved by irradiation with UV-light (also leading to temperatures >200 °C). In several studies, terpyridine-containing polymethacrylates have been studied in detail. Cross-linking with Fe(II) resulted in self-healing polymer coatings [124]. The required healing temperature (100 °C) for these systems could be successfully reduced to 80 and 60 °C by complexation with either Cd-acetate [125] or Mn(II) chloride, respectively [126]. Both can also act as bridging ligands providing a more dynamic cross-linking. In principle, these metallopolymers can feature ionic clusters (compare above ionomers) as well as reversible bonds. Both interactions can contribute to the healing process. The presence of clusters was revealed by SAXS [124] and the contribution of the exchange/decomplexation was corroborated by theoretical calculations [127].

In summary, self-healing supramolecular polymers often show high reversibility and the required dynamics for self-healing. However, these materials can often be considered as rather soft materials limiting potential applications. In addition, creep – the major obstacle for supramolecular polymers – has to be overcome. Nevertheless, supramolecular polymers are very promising for the design of further self-healing polymers.

9.8
Commercial Applications

The above-described examples illustrate the huge variety of different approaches for self-healing materials developed in fundamental research. The respective applications of these materials is discussed hear. According to market reports, the market of self-healing materials will strongly grow in the next years [128]. In 2020, the market volume estimated to be close to $ 3 billion and it is expected to cross $ 7 billion in 2022.

Considering the different material classes, self-healing metals and to a lesser extent self-healing ceramics seem to be still far away from a commercial application. Self-healing asphalt was utilized on a highway in the Netherlands. A precommercial status is already reached. A similar trend can be seen for self-healing concrete, which also reached a precommercial status. Interestingly, the bacteria-based systems also reached this level.

Self-healing polymers are the material class with the most representatives in commercial systems. Polyurethane-based coatings systems, which show a temperature-enabled reflow have been utilized as car coatings [129] and other coating systems [130]. Interestingly, these systems have also been applied on mobile phones and in the interior of automobiles [131]. Capsule-based self-healing coatings will enter the market [132]. In addition, the self-healing ionomer Surlyn® is commercially available. Even self-healing supramolecular polymers – in particular the hydrogen bond-based systems – are commercial products [133,134].

9.9
Conclusions and Outlook

Many different types of self-healing materials have been investigated in the past few years. The research on self-healing metals is still in its infancy. Some promising approaches have been developed; however, due to the intrinsic properties of metals, a self-healing ability is difficult to achieve. High temperatures are required and only few atoms can be mobilized. In contrast, self-healing ceramics have been investigated to a greater extent. A closer look reveals that mainly two approaches have been utilized. The encapsulation of healing agents, which form in the heat oxides, was studied. Furthermore, MAX phases, that is, ceramics capable of forming oxides have been utilized as self-healing ceramics/ceramic coatings. In contrast to the previous two material classes, self-healing concrete was designed by many different approaches, resulting in promising approaches for commercial systems. Both ECCs and biobased healing of concrete are commercial and precommercial materials, respectively. Rather similarly, self-healing asphalt has reached this status. Self-healing asphalt based on induction heating was utilized for a highway. The largest variety of different design strategies has been employed for self-healing polymers. Many different possible mechanisms have been utilized up to now resulting in different materials. Thus, the required healing conditions (e.g., temperature) can be tuned for this material class depending on the later application. Considering the different general mechanisms, encapsulation and channel transport (e.g., in vascular systems) have proven to be the most universal approaches. Virtually any material can be transferred into a self-healing material, if the right healing agent can be embedded via microcapsules or microchannels within the material. However, this approach is strongly limited by the processability of the materials (i.e., stability of the capsules during processing). The design of intrinsic self-healing materials – reversible cross-links in polymers, expanding phases in ceramics or concrete – indeed requires the synthesis/fabrication of new materials. However, the properties can be tuned in a wider range. Already the short summary of commercial applications reveals that self-healing materials have left the lab behind and start to become part of our everyday life in many different products and applications. The grand dream seems to become true.

In recent years self-healing is not limited anymore to the restoration of mechanical properties or the closure of cracks after a mechanical damage event. Self-healing functional materials emerge into the spotlight [16]. The function restored can include not only the mechanical properties (like for most of the presented examples in this chapter) but also the optical properties and electric conductivity. For instance, the last property was restored in polymer composites by the encapsulation of liquid conductive additives (e.g., liquid metals or charge transfer salts) resulting in the restoration of the conductivity in broken conductive tracks [135,136]. Considering the increasing complexity of future materials – maybe even material systems – self-healing materials will also have to address the huge variety of required properties and functions.

Acknowledgments

S. Bode is acknowledged for critical reading of the manuscript. MDH thanks US Schubert for continuous support. The DFG is acknowledged for financial support within the framework of SPP 1568 "Design and Generic Principles of Self-healing Materials."

References

1. Paton, W.R. (1912) The armour of Achilles. *Classical Rev.*, **26**, 1–4.
2. Wondraczek, L., Mauro, J.C., Eckert, J., Kühn, U., Horbach, J., Deubener, J., and Rouxel, T. (2011) Towards ultrastrong glasses. *Adv. Mater.*, **23**, 4578–4586.
3. Podsiadlo, P., Kaushik, A.K., Arruda, E.M., Waas, A.M., Shim, B.S., Xu, J., Nandivada, H., Pumplin, B.G., Lahann, J., Ramamoorthy, A., and Kotov, N.A. (2007) Ultrastrong and stiff layered polymer nanocomposites. *Science*, **318**, 80–83.
4. Zhu, W.-K., Cong, H.-P., Yao, H.-B., Mao, L.-B., Asiri, A.M., Alamry, K.A., Marwani, H.M., and Yu, S.-H. (2015) Bioinspired, ultrastrong, highly biocompatible, and bioactive natural polymer/graphene oxide nanocomposite films. *Small*, **11**, 4298–4302.
5. Zadeh, D.B., Bahari, A., and Tirandaz, F. (2009) Ultra-high performance concrete, in *Excellence in Concrete Construction Through Innovation* (eds M.C. Limbachiya and H.Y. Kew), Taylor Francis, London, pp. 275–280.
6. Senuma, T. (2001) Physical metallurgy of modern high strength steel sheets. *ISIJ International*, **41**, 520–532.
7. van Der Zwaag, S. (2007) An introduction to material design principles: damage prevention versus damage management, in *Self Healing Materials: An Alternative Approach to 20 Centuries Materials Science* (ed. S. van der Zwaag), Springer, Dordrecht, pp. 1–18.
8. Harrington, M.J., Speck, O., Speck, T., Wagner, S., and Weinkamer, R. (2016) Biological archetypes for self-healing materials. *Adv. Polym. Sci.* doi: 10.1007/12_2015_334
9. Taylor, D., Hazenberg, J.G., and Lee, T.C. (2007) Living with cracks: damage and repair in human bone. *Nat. Mater.*, **6**, 263–268.
10. Schmitt, C.N.Z., Politi, Y., Reinecke, A., and Harrington, M.J. (2015) Role of sacrificial protein–metal bond exchange in mussel byssal thread self-healing. *Biomacromolecules*, **16**, 2852–2861.
11. Diesendruck, C.E., Sottos, N.R., Moore, J.S., and White, S.R. (2015) Biomimetic self-healing. *Angew. Chem. Int. Ed.*, **54**, 10428–10447.
12. Trask, R.S. and Bond, I.P. (2006) Biomimetic self-healing of advanced composite structures using hollow glass fibres. *Smart Mater. Struct.*, **15**, 704–710.
13. Trask, R.S., Williams, H.R., and Bond, I.P. (2007) Self-healing polymer composites: mimicking nature to enhance performance. *Bioinspir. Biomim.*, **2**, P1–P9.
14. van der Zwaag, S., van Dijk, N.H., Jonkers, H.M., Mookhoek, S.D., and Sloof, W.G. (2009) Self-healing behaviour in man-made engineering materials: bioinspired but taking into account their intrinsic character. *Phil. Trans. R. Soc. A*, **367**, 1689–1704.
15. Hager, M.D., Greil, P., Leyens, C., van der Zwaag, S., and Schubert, U.S. (2010) Self-healing materials. *Adv. Mater.*, **22**, 5424–5430.
16. Ahner, J., Bode, S., Micheel, M., Dietzek, B., and Hager, M.D. (2016) Self-healing functional polymeric materials. *Adv. Polym. Sci.* **273**, 247–283.
17. Grabowski, B. and Tasan, C.C. (2016) Self-healing metals. *Adv. Polym. Sci.* **273**, 387–407.
18. Ferguson, J.B., Schultz, B.F., and Rohatgi, P.K. (2014) Self-healing metals and metal matrix composites. *JOM*, **66**, 866–871.
19. Laha, K., Kyono, J., Kishimoto, S., and Shinya, N. (2005) Beneficial effect of B segregation on creep cavitation in a type 347 austenitic stainless steel. *Scr. Mater.*, **52**, 675–678.
20. Shinya, N., Kyono, J., and Laha, K. (2006) Self-healing effect of boron nitride

precipitation on creep cavitation in austenitic stainless steel. *J. Intell. Mater. Sys. Struct.*, **17**, 1127–1133.

21 Zhang, S., Kwakernaak, C., Tichelaar, F.D., Sloof, W.G., Kuzmina, M., Herbig, M., Raabe, D., Brück, E., van der Zwaag, S., and van Dijk, N.H. (2015) Autonomous repair mechanism of creep damage in Fe-Au and Fe-Au-B-N alloys. *Metallurgical Mater. Trans. A*, **46**, 5656–5669.

22 Lumley, R.N. (2007) Self healing in aluminium alloys, in *Self Healing Materials: An Alternative Approach to 20 Centuries of Materials Science* (ed. S. van der Zwaag), Springer, Dordrecht, pp. 219–254.

23 Manuel, M.V. (2009) Principles of self-healing in metals and alloys: an introduction, in *Self-Healing Materials: Fundamentals, Design Strategies, and Applications* (ed. G. Ghosh), Wiley-VCH Verlag GmbH, Weinheim, pp. 251–265.

24 Ferguson, J.B., Schultz, B.F., and Rohatgi, P.K. (2015) Zinc alloy ZA-8 shape memory alloy self-healing metal matrix composite. *Mater. Sci. Eng. A*, **620**, 85–88.

25 Nosonovsky, M. and Rohatgi, P.K. (2012) *Biomimetics in Materials Science: Self-Healing, Self-Lubricating, and Self-Cleaning Materials*, Springer.

26 Lucci, J.M., Rohatgi, P.K., and Schultz, B. (2008) Experiment and computational analysis of self-healing in an aluminum alloy. In: Proceedings of the ASME International Mechanical Engineering Congress and Exposition, Boston, USA.

27 Zheng, X.G., Shi, Y.N., and Lu, K. (2013) Electro-healing cracks in nickel. *Mat. Sci. Eng. A*, **561**, 52–59.

28 Cao, X.Q., Vassen, R., and Stoever, D. (2004) Ceramic materials for thermal barrier coatings. *J. Euro. Ceram. Soc.*, **24**, 1–10.

29 Denry, I. and Kelly, J.R. (2008) State of the art of zirconia for dental applications. *Dental Mater.*, **24**, 299–307.

30 Deville, S. and Stevenson, A.J. (2015) Mapping ceramics research and its evolution. *J. Am. Ceram. Soc.*, **98**, 2324–2332.

31 Greil, P. (2012) Generic principles of crack-healing ceramics. *J. Adv. Ceramics*, **1**, 249–267.

32 Jarvis, E.A. and Carter, E.A. (2006) A nanoscale mechanism of fatigue in ionic solids. *Nano Lett.*, **6**, 505–509.

33 Tavangarian, F. and Lin, G. (2015) Crack healing and strength recovery in SiC/spinel nanocomposite. *Ceramics Int.*, **41**, 2828–2835.

34 Ouyang, T., Wu, J., Yasir, M., Zhou, T., Fang, X., Wang, Y., Liu, D., and Suo, J. (2016) Effect of TiC self-healing coatings on the cyclic oxidation resistance and lifetime of thermal barrier coatings. *J. Alloys Comp.*, **656**, 992–1003.

35 Yang, H.J., Pei, Y.T., Rao, J.C., De Hosson, J., and Th., M. (2012) Self-healing performance of Ti_2AlC ceramic. *J. Mater. Chem.*, **22**, 8304–8313.

36 Li, S., Song, G., Kwakernaak, K., van der Zwaag, S., and Sloof, W.G. (2012) Multiple crack healing of a Ti_2AlC ceramic. *J. Euro. Ceram. Soc.*, **32**, 1813–1820.

37 Farle, A.-S., Kwakernaak, C., van der Zwaag, S., and Sloof, W.G. (2015) A conceptual study into the potential of $M_{n+1}AX_n$-phase ceramics for self-healing of crack damage. *J. Euro. Ceram. Soc.*, **35**, 37–45.

38 Coillot, D., Méar, F.O., Podor, R., and Montagne, L. (2010) Autonomic self-repairing glassy materials. *Adv. Funct. Mater.*, **20**, 4371–4374.

39 Patel, P. (2015) Helping concrete heal itself. *ACS Cent. Sci.*, **1**, 470–472.

40 Van Tittelboom, K. and De Belie, N. (2013) Self-healing in cementitious materials: a review. *Materials*, **6**, 2182–2217.

41 Wu, M., Johannesson, B., and Geiker, M. (2012) A review: self-healing in cementitious materials and engineered cementitious composite as a self-healing material. *Construct. Build. Mater.*, **28**, 571–583.

42 Edvardsen, C. (1999) Water permeability and autogenous healing of cracks in concrete. *ACI Mater. J.*, **96**, 448–454.

43 Li, V.C., Lim, Y.M., and Chan, Y.-W. (1998) Feasibility study of a passive smart self-healing cementitious composite. *Composites B*, **29**, 819–827.

44 Yıldırım, G., Keskin, Ö.K., Keskin, S.B., Sahmaran, M., and Lachemi, M. (2015) A

review of intrinsic self-healing capability of engineered cementitious composites: recovery of transport and mechanical properties. *Constr. Build. Mater.*, **101**, 10–21.
45 Snoeck, D. and De Belie, N. (2015) From straw in bricks to modern use of microfibres in cementitious composites for improved autogenous healing: a review. *Constr. Build. Mater.*, **95**, 774–787.
46 Li, V.C., Wu, C., Wang, S., Ogawa, A., and Saito, T. (2002) Interface tailoring for strain-hardening PVA-ECC. *ACI Mater. J.*, **99**, 463–472.
47 Homma, D., Mihashi, H., and Nishiwaki, T. (2009) Self-healing capability of fiber reinforced cementitious composites. *J. Adv. Concr. Technol.*, **7**, 217–228.
48 Dry, C.M. (1994) Matrix cracking repair and filling using active and passive modes for smart timed release of chemicals from fibers into cement matrices. *Smart Mater. Struct.*, **3**, 118–123.
49 Dry, C.M. (1999) Repair and prevention of damage due to transverse shrinkage cracks in bridge decks. *Smart Struct. Mater.*, **3671**, 253–256.
50 Feiteira, J., Gruyaert, E., and De Belie, N. (2016) Self-healing of moving cracks in concrete by means of encapsulated polymer precursors. *Construc. Building Mater.*, **102**, 671–678.
51 Tziviloglou, E., Van Tittelboom, K., Palin, D., Wang, J., Sierra-Beltrán, M.G., Erşan, Y.Ç., Mors, R., Wiktor, V., Jonkers, H.M., Schlangen, E., and De Belie, N. (2016) Bio-based self-healing concrete: from research to field application. *Adv. Polym. Sci.* **273**, 345–385.
52 Tabaković, A. and Schlangen, E. (2016) Self-healing technology for asphalt pavements. *Adv. Polym. Sci.* **273**, 286–306.
53 European Union Road Federation, Yearbook 2014–2015, European Union Road Federation, Brussels.
54 Fang, C., Yu, R., Liu, S., and Li, Y. (2013) Nanomaterials applied in asphalt modification: a review. *J. Mater. Sci. Technol.*, **29**, 589–594.
55 You, Z., Mills-Beale, J., Foley, J.M., Roy, S., Odegard, G.M., Dai, Q., and Goh, S.W. (2011) Nanoclay-modified asphalt materials: preparation and characterization. *Constr. Building Mater.*, **25**, 1072–1078.
56 Santagata, E., Baglieri, O., Tsantilis, L., and Chiappinelli, G. (2015) Fatigue properties of bituminous binders reinforced with carbon nanotubes. *Inter. J. Fatigue*, **80**, 30–39.
57 Qiu, J., van de Ven, M.F.C., Wu, S., Yu, J., and Molenaar, A.A.A. (2009) Investigating the self healing capability of bituminous binders. *Road Mater. Pavement Des.*, **10**, 81–94.
58 García, Á., Schlangen, E., van de Ven, M., and Liu, Q. (2009) Electrical conductivity of asphalt mortar containing conductive fibers and fillers. *Construct. Build. Mater.*, **23**, 3175–3181.
59 Liu, Q., Schlangen, E., García, Á., and van de Ven, M. (2010) Induction heating of electrically conductive porous asphalt concrete. *Construct. Build. Mater.*, **24**, 1207–1213.
60 Schlangen, E. (2015) selfhealingasphalt.blogspot.de/ (accessed February 15, 2016).
61 Garcia, A., Jelfs, J., and Austin, C.J. (2015) Internal asphalt mixture rejuvenation using capsules. *Construct. Build. Mater.*, **101**, 309–316.
62 Su, J.-F., Schlangen, E., and Wang, Y.-Y. (2015) Investigation the self-healing mechanism of aged bitumen using microcapsules containing rejuvenator. *Construct. Build. Mater.*, **85**, 49–56.
63 Chung, K., Lee, S., Park, M., Yoo, P., and Hong, Y. (2015) Preparation and characterization of microcapsule-containing self-healing asphalt. *J. Ind. Eng. Chem.*, **29**, 330–337.
64 Malinskii, Y.M., Prokopenko, V.V., Ivanova, N.A., and Kargin, V.A. (1970) Investigation of self-healing of cracks in polymers. *Polym. Mechan.*, **6**, 240–244.
65 White, S.R., Sottos, N.R., Geubelle, P.H., Moore, J.S., Kessler, M.R., Sriram, S.R., Brown, E.N., and Viswanathan, S. (2001) Autonomic healing of polymer composites. *Nature*, **409**, 794–817.
66 Web of Science, Search: self healing materials; 1305 citations (accessed February 25, 2016).

67 Yang, Y. and Urban, M.W. (2013) Self-healing polymeric materials. *Chem. Soc. Rev.*, **42**, 7446–7467.

68 Binder, W.H. (2013) *Self-Healing Polymers: From Principles to Applications*, Wiley-VCH Verlag GmbH, Weinheim, Germany.

69 Blaiszik, B.J., Kramer, S.L.B., Olugebefola, S.C., Moore, J.S., Sottos, N.R., and White, S.R. (2010) Self-healing polymers and composites. *Annu. Rev. Mater. Res.*, **40**, 179–211.

70 Yang, Y., Ding, X., and Urban, M.W. (2015) Chemical and physical aspects of self-healing materials. *Progr. Polym. Sci.*, **49–50**, 34–59.

71 Syrett, J.A., Becer, C.R., and Haddleton, D.M. (2010) Self-healing and self-mendable polymers. *Polym. Chem.*, **1**, 978–987.

72 Murphy, E.B. and Wudl, F. (2010) The world of smart healable materials. *Progr. Polym. Sci.*, **35**, 223–251.

73 Billiet, S., Hillewaere, X.K.D., Teixeira, R.F.A., and Du Prez, F.E. (2013) Chemistry of crosslinking processes for self-healing polymers. *Macromol. Rapid Commun.*, **34**, 290–309.

74 Zhu, D.Y., Rong, M.Z., and Zhang, M.Q. (2015) Self-healing polymeric materials based on microencapsulated healing agents: from design to preparation. *Progr. Polym. Sci.*, **49–50**, 175–220.

75 Hillewaere, X.K.D. and Du Prez, F.E. (2015) Fifteen chemistries for autonomous external self-healing polymers and composites. *Progr. Polym. Sci.*, **49–50**, 121–153.

76 Yuan, Y.C., Rong, M.Z., Zhang, M.Q., Chen, J., Yang, G.C., and Li, X.M. (2008) Self-healing polymeric materials using epoxy/mercaptan as the healant. *Macromolecules*, **41**, 5197–5202.

77 Caruso, M.M., Delafuente, D.A., Ho, V., Sottos, N.R., Moore, J.S., and White, S.R. (2007) Solvent-promoted self-healing materials. *Macromolecules*, **40**, 8830–8832.

78 Kang, S., Baginska, M., White, S.R., and Sottos, N.R. (2015) Core–shell polymeric microcapsules with superior thermal and solvent stability. *ACS Appl. Mater. Interfaces*, **7**, 10952–10956.

79 Trask, R.S., Williams, G.J., and Bond, I.P. (2007) Bioinspired self-healing of advanced composite structures using hollow glass fibres. *JR Soc. Interface*, **4**, 363–371.

80 Trask, R.S. and Bond, I.P. (2006) Biomimetic self-healing of advanced composite structures using hollow glass fibres. *Smart Mater. Struct.*, **15**, 704–710.

81 Toohey, K.S., Sottos, N.R., Lewis, J.A., Moore, J.M., and White, S.R. (2007) Self-healing materials with microvascular networks. *Nat. Mater.*, **6**, 581–585.

82 Hamilton, A.R., Sottos, N.R., and White, S.R. (2010) Self-healing of internal damage in synthetic vascular materials. *Adv. Mater.*, **22**, 5159–5163.

83 Hansen, C.J., Wu, W., Toohey, K.S., Sottos, N.R., White, S.R., and Lewis, J.A. (2009) Self-healing materials with interpenetrating microvascular networks. *Adv. Mater.*, **21**, 4143–4147.

84 Hansen, C.J., White, S.R., Sottos, N.R., and Lewis, J.A. (2011) Accelerated self-healing via ternary interpenetrating microvascular networks. *Adv. Funct. Mater.*, **21**, 4320–4326.

85 Patrick, J.F., Hart, K.R., Krull, B.P., Diesendruck, C.E., Moore, J.S., White, S.R., and Sottos, N.R. (2014) Continuous self-healing life cycle in vascularized structural composites. *Adv. Mater.*, **26**, 4302–4308.

86 White, S.R., Moore, J.S., Sottos, N.R., Krull, B.P., Santa Cruz, W.A., and Gergely, R.C.R. (2014) Restoration of large damage volumes in polymers. *Science*, **344**, 620–623.

87 de Espinosa, L.M., Fiore, G.-L., Weder, C., Foster, E.J., and Simon, Y.C. (2015) Healable supramolecular polymer solids. *Progr. Polym. Sci.*, **49–50**, 60–78.

88 Hart, L.R., Harries, J.L., Greenland, B.W., Colquhoun, H.M., and Hayes, W. (2013) Healable supramolecular polymers. *Polym. Chem.*, **4**, 4860–4870.

89 Herbst, F., Döhler, D., Michael, P., and Binder, W.H. (2013) Self-healing polymers via supramolecular forces. *Macromol. Rapid Commun.*, **34**, 203–220.

90 Enke, M., Döhler, D., Bode, S., Binder, W.H., Hager, M.D., and Schubert, U.S.

(2016) Intrinsic self-healing polymers based on supramolecular interactions: state of the art and future directions. *Adv. Polym. Sci.* **273**, 59–112.
91 Roy, N., Bruchmann, B., and Lehn, J.-M. (2015) DYNAMERS: dynamic polymers as self-healing materials. *Chem. Soc. Rev.*, **44**, 3786–3807.
92 Kuhl, N., Bode, S., Hager, M.D., and Schubert, U.S. (2016) Self-healing polymers based on reversible covalent bonds. *Adv. Polym. Sci.* **273**, 1–58.
93 Liu, Y.-L. and Chuo, T.-W. (2013) Self-healing polymers based on thermally reversible Diels–Alder chemistry. *Polym. Chem.*, **4**, 2194–2205.
94 Chen, X., Dam, M.A., Ono, K., Mal, A., Shen, H., Nut, S.R., Sheran, K., and Wudl, F. (2002) A thermally re-mendable cross-linked polymeric material. *Science*, **295**, 1698–1702.
95 Chen, X., Wudl, F., Mal, A.K., Shen, H., and Nutt, S.R. (2003) New thermally remendable highly cross-linked polymeric materials. *Macromolecules*, **36**, 1802–1807.
96 Zhang, Y., Broekhuis, A.A., and Picchioni, F. (2009) Thermally self-healing polymeric materials: the next step to recycling thermoset polymers? *Macromolecules*, **42**, 1906–1912.
97 Kötteritzsch, J., Stumpf, S., Hoeppener, S., Vitz, J., Hager, M.D., and Schubert, U.S. (2013) One-component intrinsic self-healing coatings based on reversible crosslinking by Diels–Alder cycloadditions. *Macromol. Chem. Phys.*, **214**, 1636–1649.
98 Wouters, M., Craenmehr, E., Tempelaars, K., Fischer, H., Stroeks, N., and van Zanten, J. (2009) Preparation and properties of a novel remendable coating concept. *Prog. Org. Coat.*, **64**, 156–162.
99 Tian, Q., Yuan, Y.C., Rong, M.Z., and Zhang, M.Q. (2009) A thermally remendable epoxy resin. *J. Mater. Chem.*, **19**, 1289–1296.
100 Turkenburg, D.H. and Fischer, H.R. (2015) Diels–Alder based thermo-reversible cross-linked epoxies for use in self-healing composites. *Polymer*, **79**, 187–194.

101 Rivero, G., Nguyen, L.T.T., Hillewaere, X.K.D., and Du Prez, F.E. (2014) One-pot thermo-remendable shape memory polyurethanes. *Macromolecules*, **47**, 2010–2018.
102 Yoshie, N., Saito, S., and Oya, N. (2011) A thermally-stable self-mending polymer networked by Diels–Alder cycloaddition. *Polymer*, **52**, 6074–6079.
103 Murphy, E.B., Bolanos, E., Schaffner-Hamann, C., Wudl, F., Nutt, S.R., and Auad, M.L. (2008) Synthesis and characterization of a single-component thermally remendable polymer network: Staudinger and Stille revisited. *Macromolecules*, **41**, 5203–5209.
104 Kuhl, N., Bode, S., Bose, R.K., Vitz, J., Seifert, A., Hoeppener, S., Garcia, S.J., Spange, S., van der Zwaag, S., Hager, M.D., and Schubert, U.S. (2015) Acylhydrazones as reversible covalent crosslinkers for self-healing polymers. *Adv. Funct. Mater.*, **25**, 3295–3301.
105 Canadell, J., Goossens, H., and Klumperman, B. (2011) Self-healing materials based on disulfide links. *Macromolecules*, **44**, 2536–2541.
106 Cromwell, O.R., Chung, J., and Guan, Z. (2015) Malleable and self-healing covalent polymer networks through tunable dynamic boronic ester bonds. *J. Am. Chem. Soc.*, **137**, 6492–6495.
107 Schmolke, W., Perner, N., and Seiffert, S. (2015) Dynamically cross-linked polydimethylsiloxane networks with ambient-temperature self-healing. *Macromolecules*, **48**, 8781–8788.
108 Ying, H., Zhang, Y., and Cheng, J. (2014) Dynamic urea bond for the design of reversible and self-healing polymers. *Nat. Commun.*, **5**, 3218.
109 Rossow, T. and Seiffert, S. (2015) Supramolecular polymer networks: preparations, properties, and potentials. *Adv. Polym. Sci.*, **268**, 1–46.
110 Burattini, S., Colquhoun, H.M., Fox, J.D., Friedmann, D., Greenland, B.W., Harris, P.J.F., Hayes, W., Mackay, M.E., and Rowan, S.J. (2009) A self-repairing, supramolecular polymer system: healability as a consequence of donor–acceptor π–π stacking interactions. *Chem. Commun.*, 6717–6719.

111 Hart, L.R., Nguyen, N.A., Harries, J.L., Mackay, M.E., Colquhoun, H.M., and Hayes, W. (2015) Perylene as an electron-rich moiety in healable, complementary π–π stacked, supramolecular polymer systems. *Polymer*, **69**, 293–300.

112 Nakahata, M., Takashima, Y., and Harada, A. (2016) Highly flexible, tough, and self-healing supramolecular polymeric materials using host–guest interaction. *Macromol. Rapid Commun.*, **37**, 86–92.

113 Kalista, S.J., Jr., Pflug, J.R., and Varley, R.J. (2013) Effect of ionic content on ballistic self-healing in EMAA copolymers and ionomers. *Polym. Chem.*, **4**, 4910–4926.

114 Grande, A.M., Castelnovo, L., Di Landro, L., Giacomuzzo, C., Francesconi, A., and Rahman, M.A. (2013) Rate-dependent self-healing behavior of an ethylene-*co*-methacrylic acid ionomer under high-energy impact conditions. *J. Appl. Polym. Sci.*, **130**, 1949–1958.

115 Bose, R.K., Hohlbein, N., Garcia, S.J., Schmidt, A.M., and van der Zwaag, S. (2015) Connecting supramolecular bond lifetime and network mobility for scratch healing in poly(butyl acrylate) ionomers containing sodium, zinc and cobalt. *Phys. Chem. Chem. Phys.*, **17**, 1697–1704.

116 Binder, W.H. and Zirbs, R. (2007) Hydrogen bonded polymers. *Adv. Polym. Sci.*, **207**, 1–78.

117 Cordier, P., Tournilhac, F., Soulié-Ziakovic, C., and Leibler, L. (2008) Self-healing and thermoreversible rubber from supramolecular assembly. *Nature*, **451**, 977–980.

118 Montarnal, D., Tournilhac, F., Hidalgo, M., Couturier, J.-L., and Leibler, L. (2009) Versatile one-pot synthesis of supramolecular plastics and self-healing rubbers. *J. Am. Chem. Soc.*, **131**, 7966–7967.

119 Lopez, J., Chen, Z., Wang, C., Andrews, S.C., Cui, Y., and Bao, Z. (2016) The effects of cross-linking in a supramolecular binder on cycle life in silicon microparticle anodes. *ACS Appl. Mater. Interfaces*, **8**, 2318–2324.

120 Hentschel, J., Kushner, A.M., Ziller, J., and Guan, Z. (2012) Self-healing supramolecular block copolymers. *Angew. Chem. Int. Ed.*, **51**, 10561–10565.

121 Chen, S., Mahmood, N., Beiner, M., and Binder, W.H. (2015) Self-healing materials from V- and H-shaped supramolecular architectures. *Angew. Chem. Int. Ed.*, **54**, 10188–10192.

122 Sandmann, B., Bode, S., Hager, M.D., and Schubert, U.S. (2013) Metallopolymers as an emerging class of self-healing materials. *Adv. Polym. Sci.*, **262**, 239–258.

123 Burnworth, M., Tang, L., Kumpfer, J.R., Duncan, A.J., Beyer, F.L., Fiore, G.L., Rowan, S.J., and Weder, C. (2011) Optically healable supramolecular polymers. *Nature*, **472**, 334–337.

124 Bode, S., Zedler, L., Schacher, F.H., Dietzek, B., Schmitt, M., Popp, J., Hager, M.D., and Schubert, U.S. (2013) Self-healing polymer coatings based on crosslinked metallosupramolecular copolymers. *Adv. Mater.*, **25**, 1634–1638.

125 Bode, S., Bose, R.K., Matthes, S., Ehrhardt, M., Seifert, A., Schacher, F.H., Paulus, R.M., Stumpf, S., Sandmann, B., Vitz, J., Winter, A., Hoeppener, S., Garcia, S.J., Spange, S., van der Zwaag, S., Hager, M.D., and Schubert, U.S. (2013) Self-healing metallopolymers based on cadmium bis (terpyridine) complex containing polymer networks. *Polym. Chem.*, **4** 4966–4973.

126 Bode, S., Enke, M., Bose, R.K., Schacher, F.H., Garcia, S.J., van der Zwaag, S., Hager, M.D., and Schubert, U.S. (2015) Correlation between scratch healing and rheological behavior for terpyridine complex based metallopolymers. *J. Mater. Chem. A*, **3**, 22145–22153.

127 Kupfer, S., Zedler, L., Guthmuller, J., Bode, S., Hager, M.D., Schubert, U.S., Popp, J., Gräfe, S., and Dietzek, B. (2014) Self-healing mechanism of metallopolymers investigated by QM/MM simulations and Raman spectroscopy. *Phys. Chem. Chem. Phys.*, **16**, 12422–12432.

128 Markets for Self-Healing Materials: 2015–2022 (2015). Available at http://ntechresearch.com/market_reports/markets-for-self-healing-materials-2015-2022 (accessed February 21, 2016).

129 Scratch Shield. Available at http://www.nissan-global.com/EN/TECHNOLOGY/

OVERVIEW/scratch.html (accessed February 21, 2016).
130 PU self-healing coatings utilize polymer memory to repair scratches. Available at http://www.coatings.covestro.com/en/Technologies/Functional-Coatings/Self-Healing-Coatings (accessed February 21, 2016).
131 Selbstheilende Polymere (2015). Available at www.ruehl-ag.de/index.php?id=160 (accessed February 21, 2016).
132 www.autonomicmaterials.com/ (accessed February 21, 2016).
133 http://www.suprapolix.com/pages/polymers (accessed February 21, 2016).
134 Self-healing elastomer enters industrial production (2009). Available at http://www.arkema.com/en/media/news/news-details/Self-healing-elastomer-enters-industrial-production/ (accessed February 21, 2016).
135 Odom, S.A., Caruso, M.M., Finke, A.D., Prokup, A.M., Ritchey, J.A., Leonard, J.H., White, S.R., Sottos, N.R., and Moore, J.S. (2010) Restoration of conductivity with TTF-TCNQ charge-transfer salts. *Adv. Funct. Mater.*, **20**, 1721–1727.
136 Odom, S.A., Chayanupatkul, S., Blaiszik, B.J., Zhao, O., Jackson, A.C., Braun, P.V., Sottos, N.R., White, S.R., and Moore, J.S. (2012) A self-healing conductive ink. *Adv. Mater.*, **24**, 2578–2581.

10
Functional Surfaces for Biomaterials

Akiko Nagai, Naohiro Horiuchi, Miho Nakamura, Norio Wada, and Kimihiro Yamashita

Tokyo Medical and Dental University, Institute of Biomaterials and Bioengineering, 2-3-10 Kanda-Surugadai, Chiyoda, Tokyo 101-0062, Japan

10.1
Introduction

Solid surfaces have different properties from the bulk solid due to the difference in the energy of surface atom. Therefore, the interface between biomaterials and biological systems, which is the solid–liquid interface, is metastable. Because the properties can change depending on circumstances such as pH, osmotic pressure, and temperature, the surface chemical composition can be altered when the molecular mobility is enhanced. In addition, at this local microenvironment, other factors, such as blood flow or immune systems, will participate in this dynamic field. To develop appropriate material designs for medical devises, we need to understand the dynamic force and various biocomponents that create physicochemical interactions at the interface and to control the surface properties using appropriate bulk materials and the application of surface modification (Figure 10.1).

The most important functions of the biomaterial surface are biocompatibility. D.F. Williams defines biocompatibility as follows: the ability of a biomaterial to perform its desired function with respect to a medical therapy, without eliciting any undesirable local or systemic effects on the recipient or beneficiary of that therapy, but generating the most appropriate beneficial cellular or tissue response in that specific situation, and optimizing the clinically relevant performance of that therapy [1]. Therefore, the "functions" of biomaterials vary widely depending on the intended use. For bone prostheses, bioactive materials, which can directly bond to adjacent tissues, are needed, and for vascular grafts, bioinert materials are favored to suppress inflammation and thrombosis that occur in response to the materials [2–5].

Key factors involved in the surface properties of an implanted material include surface chemistry, surface topology, and mechanical properties [6,7]. Chemical compositions of material surfaces are often different from the bulk and can be

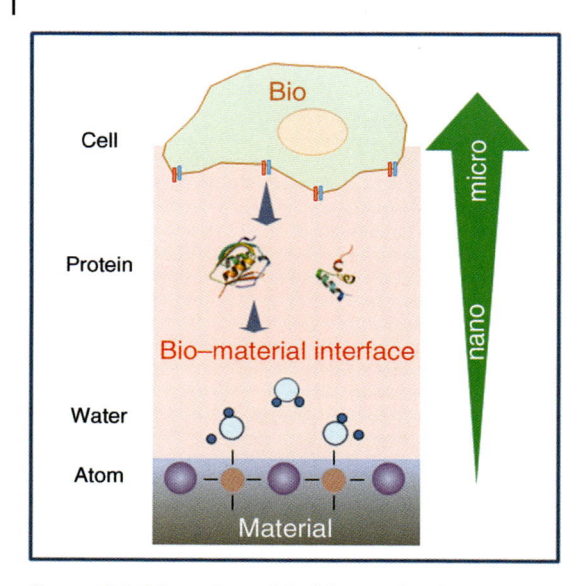

Figure 10.1 Schematic model of the interface between a material and biocomponents.

modified adding functional groups such as self-assembled monolayers (SAMs) [8,9] or adding polymers like 2-methacryloyloxyethyl phosphorylcholine (MPC) [5] or oxidizing metal substrata [10,11]. SAMs influence the amount of adsorbed protein and the protein conformation, and then, the protein layer may subsequently affect cell responses. MPC polymer suppressed protein adsorption and endothelial cell adhesion. The titania film on titanium substrate alters its surface free energy, making the surface hydrophilic [12–15]. Surface topology has been subdivided into macro-, micro-, and nanoscale roughness [16,17]. The response of tissue and cells to surface topology is not the same fashion. Micro- and nanostructure of cardiovascular devices influence endothelialization and antithrombosis of the luminal surface [18,19]. In the case of dental implants, macroscale roughness relates to mechanical interlocking between the implant and the bone early fixation and stability. Microscale roughness contributes enhancements to osteoblast attachment and differentiation as well as interlocking [20–22]. Protein adsorption and cell adhesion of the material surface are affected by nanoscale roughness [23,24]. For neuronal progenitor cells, microtopographical patterns influence the neuronal subtype differentiation [25,26]. The mechanical properties are linked to the potential surface stresses to the hardness of the material, or to mechanical tensile strains from surface elasticity of the material [27,28]. Cells sense and respond to the stress force through mechanical receptors on the cell membrane. The stiffness was reported to direct stem cell lineage specification and cell proliferation [29,30], and the mechanical strain regulated cell proliferation and differentiation [31,32].

Because many factors like those mentioned above constitute surface properties of materials, it is generally difficult to evaluate surface properties. For the systematic analysis of one surface property, there should be no differences in the rest of the factors. However, some surface modification techniques alter several surface properties at a time, or limited substrata must be chosen to obtain an appropriate surface. Numerous attempts have been made in previous studies to solve these problems such as replicas of original surfaces prepared for topology analysis or polyacrylamide gels used as cell culture substrata for evaluating the effects of elastic stiffness [29,33].

Here, we will introduce hydroxyapatite (HAp) electrets and several studies evaluating their surface properties. In our previous studies, electrified apatite was proved to become polarized hydroxyapatite (see Section 10.2). We focus on the wettability, surface free energy of the polarized hydroxyapatite (Section 10.3), and the interaction with cultured cells, and also introduce the effects on the crystallization (Section 10.5).

10.2
Polarized Hydroxyapatite

Hydroxyapatite is a calcium apatite with the general formula $Ca_{10}(PO_4)_6X_2$ where the X is, in most cases, a halogen ion (F, Cl, Br) or a hydroxide (OH^-) ion. The most noteworthy feature is that hydroxyapatite contains OH^- ions in its structure. The symmetry is, in general, expressed by $P6_3/m$ (space group 176) [34]; therefore, the unit cell is depicted with a hexagonal prism. The structure consists of PO_4 tetrahedra, Ca^{2+} ions, and columns of OH^- ions oriented along the c-axis in the hexagonal prism. The unit cell formula of the hexagonal HAp is expressed as $Ca_{10}(PO_4)_6(OH)_2$. The Ca^{2+} ions are divided into two types: four Ca_1 and six Ca_2 ions. The Ca_1 is called columnar calcium ion and is arranged in columns that repeatedly crosses the lattice. The Ca_2 calcium ions form triangles on the ab plane, and one triangle is located at $\pi/3$ angle with respect to the neighboring triangles. The stacking of the Ca triangles forms a column that encloses the OH^- ion column. The OH column is important for the discussion and understanding of the properties of HAp [35]. The OH columns cause interesting electric properties, such as proton conductivity [36–40], ferroelectric-like properties [41–45], and electret behavior [46–52].

Acceleration and deceleration of bone-like crystal growth on polarized ceramic HAp, depending on the electric polarization conditions, was found [46]. Sintered ceramic HAp disks that were polarized by electrical process were immersed in simulated body fluid. The growth rate of the bone-like crystal demonstrated acceleration and deceleration effects of polarization. The polarization (poling) procedure is as follows. HAp disk samples, with a size of 10 mm diameter and 1 mm thickness, were sandwiched between platinum plates, heated to 300 °C in air, then subjected to a DC field of 1–1000 V for 2 min or 1 h, and thereafter cooled to room temperature under the electric field. In such a polarization treatment, the polarization remains after release of the electric field, and the metastable state is maintained for a certain time. In this situation, we have to

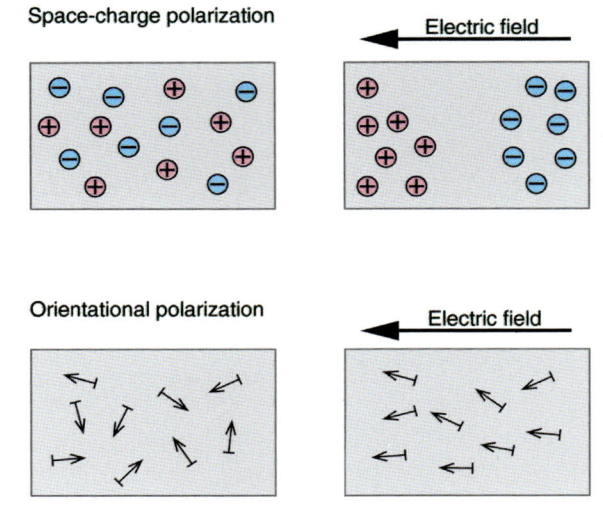

Figure 10.2 Schematic illustrations for space-charge polarization and orientational (dipole) polarization.

consider two types of polarization: space-charge polarization and dipole (orientation) polarization. Schematic explanations for the two polarizations are presented in Figure 10.2. The space-charge polarization takes place as a result of transport of electric charges. The electric charges, for example, ions and electrons, are moved by the applied electric field. The motions are relatively long-range order and generate a distribution of charges in the materials. In the case of HAp, the electric charges are protons of OH^- ions. The dipole (orientation) polarization arises from orientations of molecules that have electric dipole moments as shown in Figure 10.2. When an external electric field is applied to the material, the electric field changes the orientations of dipoles. In HAp, OH^- ions and some kind of defect pairs work as the electric dipoles.

Polarizations generated by poling processes can be characterized by measuring of thermally stimulated depolarization current (TSDC) ([53], Chapter 3). The polarized materials are heated at a constant rate and the current generated by the heating is measured using an ammeter. The measured depolarization current can be represented as a function of temperature. Figure 10.3 presents the TSDC curves for HA polarized under different poling conditions, while the electrical field and time were 1.0 kV/cm and 1 h, respectively. Each curve has one peak and the peak height increases as the poling temperature increases. In addition, the peak temperature that gives maximum depolarization currents increases with increasing poling temperature. This tendency implies that the induced polarization is caused by the space-charge polarization that is due to migration of protons in HAp structure. Activation energies for the depolarization processes can be determined by TSDC curves. The calculated values for the TSDC curves in Figure 10.3 were 0.72–0.82 eV [47], which were comparable to the reported

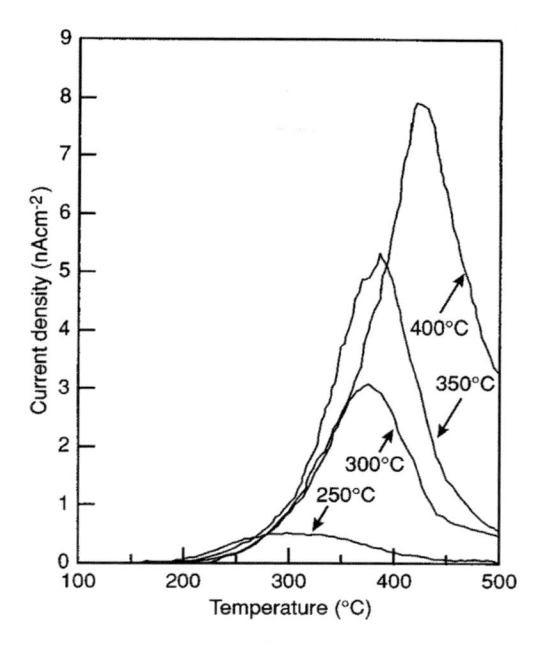

Figure 10.3 TSDC curves of polarized HAp ceramics. The poling conditions are DC field of 1.0 kV/cm for 1 h at 250, 300, 350, and 400 °C, and the heating rate was 5.0 °C/min. (Reproduced with permission from Ref. [47].)

value of the activation energy of proton migration (e.g., 0.73 eV [40]). The space-charge polarization can be illustrated as shown in Figure 10.4 [48]. Tanaka *et al.* reported that the TSDC curves had strong dependence on the microstructure, which implies that the protons cannot migrate across grain boundaries as shown in the figure. The TSDC curves can be deconvoluted into several elements of polarization [48,49]. The polarization elements can be roughly classified into two groups. One is the space charge polarization due to the long-range migration of protons. The other one inclueds polarizations that originate from defects [49]. The defects are considered to be displaced protons around PO_4^{3-} tetrahedral [48] or defect associate dipoles that are formed through dehydration, which was formed by coupling of vacancies of OH^- ion and O^{2-} ions at the OH^- lattice sites. The Kelvin probe is another method to characterize polarization of HAp and allows noncontact measurements of the surface charge. The charge on bare surfaces can be evaluated using a sensing electrode that is placed over a sample. Surface charge induced by the polarizations produces a potential difference (surface potential: ϕ_{sp}) between the sample surface and the sensing electrode that is placed over the sample. The potential difference can be changed to surface charge density (σ_s) using the following equation: $\sigma_s = \epsilon_0 \epsilon_s \varphi_{sp}/d$, where ε_s is the relative permittivity of the sample (a value of 20 was used for the relative permittivity of HAp) and d is the sample thickness. Figure 10.5 presents the surface potential and surface charge density at the positive and negative electrode

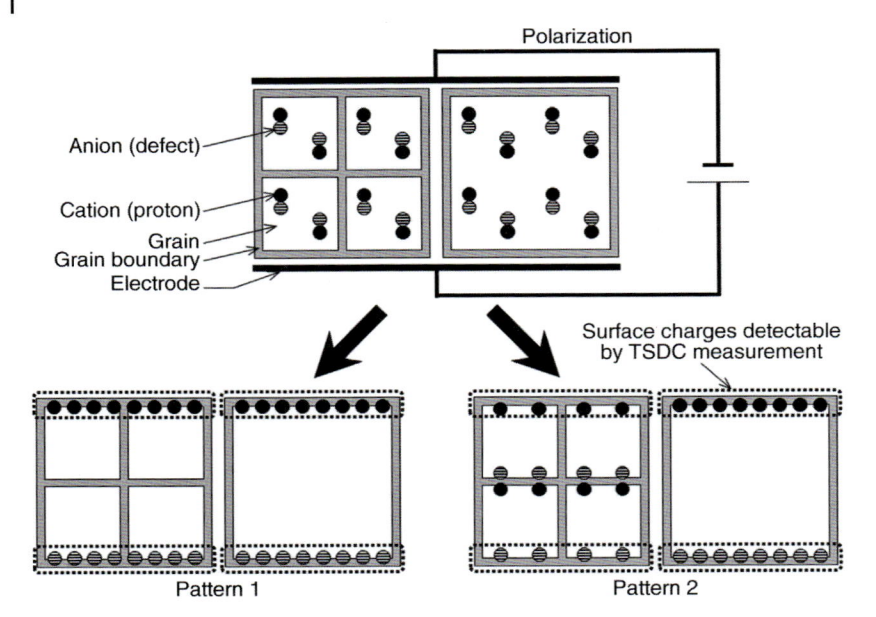

Figure 10.4 Schematic illustration of two models assumed for poling via solid-state ionic migration in HAp ceramics with different grain sizes. Pattern 1 shows the poling state provided by ionic migration through the grain boundaries. Pattern 2 shows the state in which the ions are completely trapped at the grain boundary. Pattern 2 can well explain the observed TSDC behavior. The ions seldom migrate beyond the grain boundaries. (Reproduced with permission from Ref. [48].)

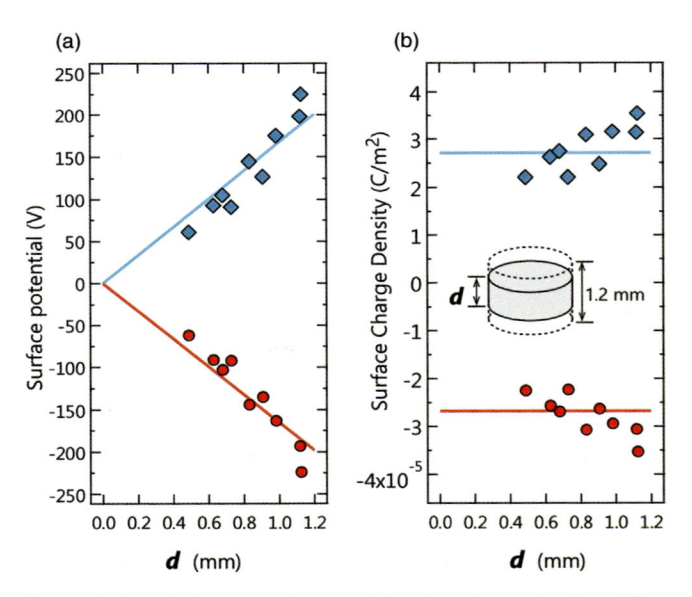

Figure 10.5 (a) Surface potentials and (b) surface charge density of HAp samples with different thicknesses. The horizontal lines in (b) are calculated from the gradients of the lines in (a). Reproduced with permission from Ref. [51].)

sides of the HAp samples with different thicknesses [51]. The measured surface potential and thickness showed a linear relationship as shown in Figure 10.5a, and the potentials were converted to the surface charge densities as shown in Figure 10.5b. The surface charge is independent of the thickness, suggesting that the surface charge does not originate from local charge that is inhomogeneously distributed (e.g., the interfacial charge near the electrode). The surface charge is believed to be induced by polarization that is uniformly distributed in the HAp samples.

10.3
Wettability and Surface Free Energy

Polarization of ceramic biomaterials improves the surface wettability by increasing the surface free energies compared to the conventional ceramic surfaces [54]. The surface free energy was investigated using the two liquid-phase method [55,56]. In this method, contact angle values of water on the specimens were measured in hydrocarbon oils such as hexane (18.4 mJ/m^2), heptane (20.1 mJ/m^2), octane (21.7 mJ/m^2), decane (23.8 mJ/m^2), and hexadecane (27.5 mJ/m^2). The surface free energy was calculated using Jouany's Eq. (10.1):

$$\gamma_W - \gamma_H + \gamma_{HW} \cos\theta = 2\sqrt{\gamma_S^d}\left(\sqrt{\gamma_W^d} - \sqrt{\gamma_H}\right) + I_{SW}^p, \tag{10.1}$$

where the subscripts W, H, and S refer to water, hydrocarbon, and the solid, respectively, γ_S^d represents the dispersion components, and γ_S^p represents the polar components [57].

The values of the contact angles in hydrocarbon oils were plotted for HA and β-TCP (tricalcium phosphate) as shown in Figure 10.6a and b, respectively. The dispersion and polar components of the surface free energy were calculated using the values of the slope and y-intercept from the plots of the contact angles in the hydrocarbon oils. In addition, the surface free energy was calculated by summing the dispersion and polar components. The surface energies were calculated to be 31.0 mJ/m^2 for HA and 27.6 mJ/m^2 for β-TCP. The values of the surface energy increased by approximately 6 and 15 times for the polarized specimens compared to the conventional HA and β-TCP, respectively. Changes to the surface free energy affect polar molecules such as water and proteins.

To investigate the surface wettability, contact angle measurements were performed on the specimens in air water using distilled and deionized water. The contact angles were calculated using Young's Eq. (10.2):

$$\gamma_{SV} = \gamma_{LV} \cos\theta + \gamma_{SL}, \tag{10.2}$$

where the subscripts S, L, and V refer to the solid, liquid, and vapor, respectively.

The contact angle values significantly decreased for the ceramic biomaterials exposed to electrical polarization, as illustrated in Figure 10.6c and d. The surfaces of the conventional ceramic biomaterials without polarization exhibited

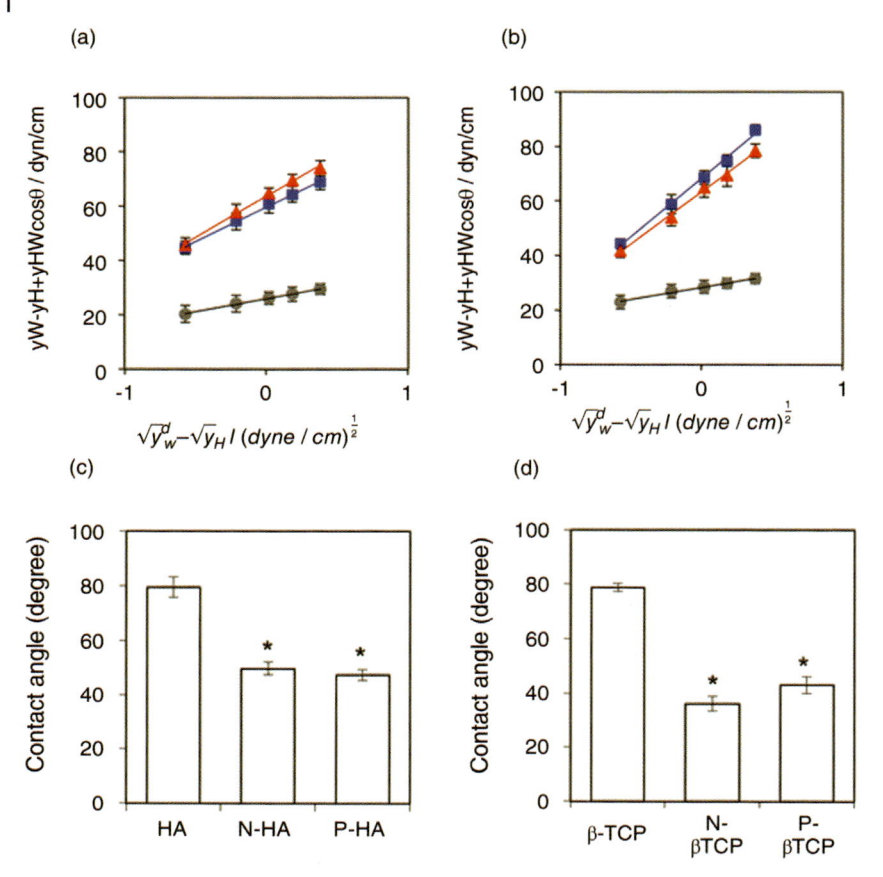

Figure 10.6 Contact angle measurements using the two liquid-phase method. The contact angles in hydrocarbon oils including hexane (18.4 mJ/m^2), heptane (20.1 mJ/m^2), octane (21.7 mJ/m^2), decane (23.8 mJ/m^2), and hexadecane (27.5 mJ/m^2) using distilled and deionized water are plotted for (a) HA and (b) βTCP. Black circles, blue squares and red triangles show the conventional, negatively charged and positively charged surfaces, respectively. Contact angle values of (c) HA and (d) βTCP using distilled and deionized water in air. (*$p < 0.005$ compared with the conventional surfaces.)

contact angles of 70–80°, while the polarized surfaces of HA and β-TCP exhibited contact angles of 50 and 40°, respectively. The ceramic material surfaces with polarization treatment exhibited lower angles, thus being more hydrophilic. The contact angle of water decreases as the surface free energy increases on the polarized ceramic surfaces, indicating that the wettability of the polarized ceramic surfaces improved.

In general, surface characteristics are determined by the surface roughness, surface crystallinity, constituent elements at the surface [58–60], and the incorporation of ions such as carbonate or fluorine [61,62]. In addition, electron-induced surface energy modifications such as photoluminescence and surface photovoltage spectroscopy can enhance surface characteristics. The electrical polarization produced no changes in the surface roughness, crystallinity, crystal phase, morphology of crystal grains, or constituent elements [63]. The present study employed ceramic biomaterials with or without polarization with an almost equal surface roughness using the same types of materials. A possible explanation for the improved wettability due to polarization involves the changes in the surface free energy.

10.4 Initial Cell Behaviors

Generally, adhesion of the cells to ceramic biomaterials plays a predominant role in the regulation of the subsequent differentiation and formation of the extracellular matrix following spreading and motility. Adhesion of the osteoblasts is primarily the result of two subjects, namely, the surface characteristics of the biomaterials involved and stimulation from outside the cells. The former subject, surface characteristics, including the topography, constituent elements, functional group, and wettability of biomaterials affects osteoblast attachment and adhesion [58,59,64,65]. The surface characteristics were reportedly affected by the surface roughness, the surface crystallinity [58,59,64], the constituent elements at the surface, and the incorporation of ions such as carbonate or fluorine [61,62]. In addition, electron-induced surface energy modifications such as photoluminescence and surface photovoltage spectroscopy were also effective in the improvement of the surface characteristics [66]. The latter subject, stimulation from outside the cells, which includes electrical stimulation such as capacitive coupling, inductive coupling, and combined electromagnetic fields, affect osteoblast attachment, adhesion, and motility [66].

These behaviors of osteoblasts can be attributed to the differences in surface characteristics because the differences affect the signal transduction pathways. The signal transduction pathways involved in the adhesion of osteoblasts on biomaterials were confirmed by the subsequential expression of $\alpha_v\beta_1$-integrins [60]. Cell shapes on substrates are dependent on the integrin-mediated cytoskeletal and signal transduction molecules, such as actin filaments and vinculin [67,68], and are important during cell–substrate adhesion for subsequent cell behaviors such as proliferation and differentiation [62,69].

Morphological observation and quantitative analysis of osteoblasts revealed that the typical adhered cells had a round or spherical shape on the HA but had a spindle- or fanlike spreading configuration on the N-HA and P-HA surfaces of polarized HA 1 h after seeding (Figure 10.7) [63]. However, the cell areas positively stained for actin, which showed the degree of cell spreading on the HA specimens, were distinctly larger on the N-HA and P-HA than on the HA 1 h after seeding. These results demonstrated that the charges induced on the HA surfaces affected the percentages of the number of initial cells spreading against the total number of adhered cells and on the degree of cell spreading on each cell.

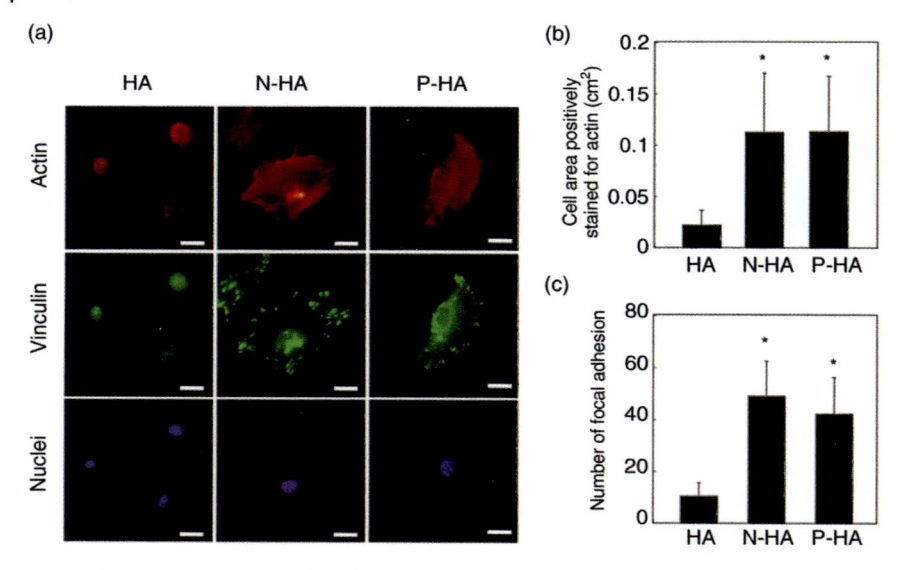

Figure 10.7 (a) Morphology of the adhered osteoblasts on HA, N-HA, and P-HA 1 h after seeding. Actin (cell cytoskeleton), vinculin (focal adhesion), and nuclei fluorescence images are shown. Actin staining showed spherical shapes and colocalization with vinculin on HA. The cells already showed a spread shape on N-HA and P-HA, as observed from the actin images. Focal adhesions positively stained for vinculin were found in the peripheral regions of the cells on N-HA and P-HA. Bar = 20 μm. (b) To quantify the degree of cell spreading, the cell areas positively stained for actin were measured. The cell area was significantly larger on the N-HA and P-HA, compared to the HA ($^*<0.005$ compared with HA). (c) To quantify the formation of vinculin-positive focal adhesions, the number of focal adhesions per cell was measured. The number of vinculin-positive focal adhesions was significantly larger on N-HA and P-HA, compared to HA ($^*<0.002$ compared with HA).

One of the important reasons for the enhancement of cell spreading on the N-HA and P-HA is certainly the improved surface wettability. Changes in surface wettability of HA resulted from structural transformation. For example, A-type carbonate apatite and fluoridated HA affected cellular behavior [61]. A-type carbonate apatite, corresponding to a location of carbonate ions at monovalent anionic OH$^-$ sites, had lower wettability, lower cell spreading, and similar cell proliferation compared to HA. Incorporation of fluorine in HA caused an insignificant increase in surface roughness and a slight increase in contact angle [62]. The fluoridated HA had no effect on initial cell attachment but did have an effect on the cell-spreading process. Surface wettability on several biomaterials other than apatite affects cell adhesion on polymer [70], glass microscope slides [71], yttria partially stabilized zirconia [72], poly(methyl methacrylate) [58], and titanium [59,73]. Despite the differences in substrates or methods for the improvement of surface wettability, changes in the surface wettability of biomaterials brought out advantages in cellular behaviors such as attachment and spreading. Therefore, the improvement of surface wettability by polarization contributed to the acceleration of cell spreading on the N-HA and P-HA.

The mechanism of cell adhesion on biomaterials varied according to the type of substrate and depended on the kinds and amount of proteins previously adsorbed on the biomaterials [74,75]. Human osteoblast-like cells initially attached and spread more quickly on the HA surface than they did on titanium [60] since the molecules regulating cell spreading on HA, such as vitronectin and fibronectin, were apparently different from those on titanium [76]. Furthermore, HA and titanium surfaces influence the gene expression of adhesion proteins early both during adhesion and during the proliferation and differentiation phases [64]. The signal transduction pathways involved in the adhesion of osteoblastic cells on HA and titanium were confirmed by the sequential expression of $\alpha_v\beta_1$-integrins, focal adhesion kinase (FAK), and extracellular signal-regulated kinase genes followed by the expression of the c-jun and c-fos genes for proliferation and the ALP gene for differentiation [60]. Mitogen-activated protein kinase (MAPK) phosphorylation, spreading, and focal adhesion formation were enhanced on $Ti_6 Al_4 V$ and FN-coated glass compared to poly-l-lysine (PLL)-coated glass [77]. These effects on the signal transduction pathway through the adsorbed proteins on the various types of substrates resembled the effects detected on the polarized HA surfaces. The N-HA and P-HA accelerated the adsorption of one of the cell adhesion proteins such as fibrinogen *in vivo* [78]. In comparison with other biomaterials, the surface wettability of HA had advantages for protein adsorption [76]. One of the reasons for the enhancement of the initial cell attachment on the N-HA and P-HA was thought to be accelerated protein adsorption. In addition, there was a possibility that the polarized HA affected the kinds of signal pathways in the osteoblastic cells.

10.5
Crystallization of Calcium Carbonate on Polarized Hydroxyapatite Substrates

10.5.1
Theoretical Approaches to Heterogeneous Nucleation in the Presence of Electric Field and the Property of the Electric Field Generated by the Polarized Substrate

By assuming a homogeneous nucleation of a parent phase in the presence of a uniformly applied electric field E under constant pressure and temperature, the isothermal reversible work required for the homogeneous formation of a spherical nucleus of n atoms, $\Delta G_{\text{hom}}(n)$, is presented in the following equation:

$$\Delta G_{\text{hom}}(n) = \Delta G_{\text{hom } 0}(n) + \Delta W_E(n), \tag{10.3}$$

where $\Delta G_{\text{hom } 0}(n)$ is the work of the formation in the absence of an electric field and $\Delta W_E(n)$, the free energy change owing to the electric field [79,80]. According to the classic theory for nucleation, the value of $\Delta G_{\text{hom } 0}(n)$ is given by $-nkT \ln \alpha + \beta \sigma n^{2/3}$, where k is the Boltzmann constant, T is the absolute temperature of the parent phase, α is the supersaturation ratio of the parent phase, σ is the specific surface energy, and β is a constant depending on the geometrical shape of the cluster.

To calculate $\Delta W_E(n)$ in Eq. (10.3) according to electromagnetism, we consider that v_m is the mean atom volume. Then, the radius, r_n, of the spherical nucleus is connected with n by the relation, $(3v_m/4\pi)^{1/3}n^{1/3}$, we obtain $\Delta W_E(n) = -kTaE^2n$, where $a = v_m\varepsilon_p(\varepsilon_p - \varepsilon_c)/2kT(2\varepsilon_p + \varepsilon_c)$, ε_c and ε_p being the permittivity of crystalline and parent phase, respectively, is a constant. Hence, the work required to form a spherical nucleus in the presence of electric field is presented in the following equation:

$$\Delta G_{hom}(n) = -nkT(\ln \alpha + aE^2) + \beta\sigma n^{2/3}, \tag{10.4}$$

where β is a geometrical factor represented by an equation, $(4\pi)^{1/3}(3v_m)^{2/3}$. From differentiation of Eq. (10.4), the activation energy for homogeneous nucleation of s nucleus of critical size in the presence of electric field, ΔG^*_{hom} is presented in the following equation [79,80]:

$$\Delta G^*_{hom} = 16\pi v_m^2 \sigma^3 / \{3[kT(\ln \alpha + aE^2)]^2\}. \tag{10.5}$$

The equation of ΔG^*_{hom} indicates that the activation energy decreases as the electric field strength and the supersaturation ratio increase.

Classical nucleation theory points out that the contact angle is important for heterogeneous nucleation. The Young's equation, $\gamma_{SA} = \gamma_{LA} \cos \theta + \gamma_{SL}$, relates the cosine of equilibrium contact angle θ to the three interfacial tensions, γ_{LA}, γ_{SA}, and γ_{SL}, where γ_{LA} is the interfacial tension of a nucleus, γ_{SA} that of a substrate, and γ_{SL} that between a nucleus and a substrate [81]. When a cup-shaped nucleus with the contact angle θ nucleates on the substrates according to the classic theory for nucleation, the activation energy for heterogeneous nucleation, ΔG^*_{het}, is given by $\Delta G^*_{het} = \Delta G^*_{hom} f(\theta)$, where $f(\theta)$ is $(2 - 3\cos\theta + \cos^3\theta)/4$, whose function changes monotonically as θ increases in the range, $0 \leq f(\theta) \leq 1$. Hence, ΔG^*_{het} is represented in the following equation:

$$\Delta G^*_{het} = [(2 - 3\cos\theta + \cos^3\theta)/4]16\pi v_m^2 \sigma^3 / \{3[kT(\ln \alpha + aE^2)]^2\}. \tag{10.6}$$

The equation of ΔG^*_{het} indicates that (a) the increase in α and E decreases ΔG^*_{het}; (b) the decrease in the contact angle θ decreases ΔG^*_{het} [82]. The activation energy decreases as the contact angle θ becomes smaller as shown in Figure 10.8, where normalized activation energy ($\Delta G^*_{het}/\Delta G^*_{hom}$) is plotted against contact angle. The rate of heterogeneous nucleation at any given supersaturation level, J_{het}, is given by $A \exp(-\Delta G^*_{het}/kT)$, where A is the preexponential factor, suggesting that the rate of heterogeneous nucleation increases with α and E, while it decreases with θ.

According to electrostatics, the electric potential V at a place near the polarized substrate is estimated to be $(t\rho/4\pi\varepsilon)\omega$, where t is the thickness of a polarized substrate, ρ the surface charge density, ε the permittivity of a medium, and ω a solid angle. Assuming a disk with radius r and thickness t, the electric potential

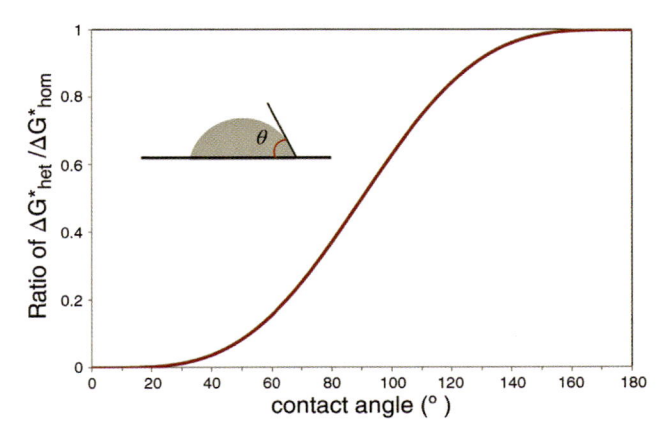

10.5 Crystallization of Calcium Carbonate on Polarized Hydroxyapatite Substrates

Figure 10.8 A theoretical graph of normalized activation energy, $\Delta G^*_{het}/\Delta G^*_{hom}$, of heterogeneous nucleation plotted against contact angle.

V at a distance x from the center of the polarized substrates is calculated to be $(t\rho/2\varepsilon)\{1 - x/(r^2 + x^2)^{1/2}\}$. The profile of the electric potential, normalized by V_0 (V at $x=0$), as a function of the distance normalized by t, is shown in Figure 10.9 [82]. As the electric field strength E is $|-dV/dx|$, it is calculated to be $|(t\rho/2\varepsilon)\{r^2/(r^2 + x^2)^{3/2}\}|$. The profile of the electric field normalized by E_0 with $x=0$ was plotted against distance normalized by t as shown in Figure 10.9 [82]. This profile shows that the magnitude of electric field at a normalized distance of 0.05 is 0.99 of the value of E_0, suggesting that a local electric field uniformly exists in the polarized substrate vicinity. The effect of the electrostatic force density induced by electric fields on an ion moving in an electrolyte solution is discussed as follows. The electrostatic force density f is defined as the force per unit volume. An electrostatic force density is given by qE, where q is the charge

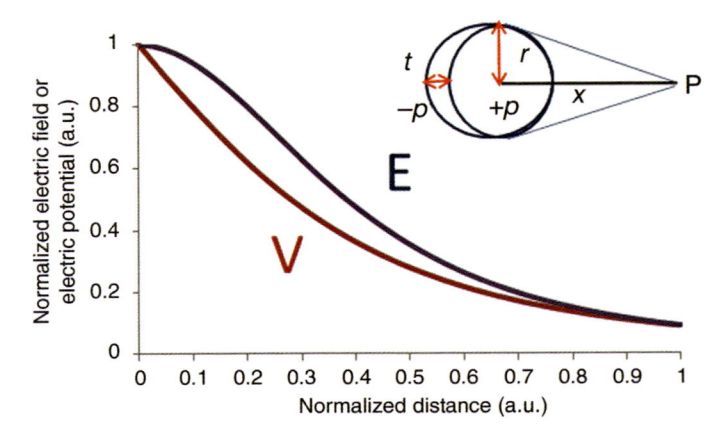

Figure 10.9 Theoretical graphs of normalized electric potential (V) and normalized electric field (E) plotted against normalized distance.

density of the ion (including its sign, either positive or negative). The direction of f acting on an ion located in a given electric field depends on the direction of the electric field. Namely, the direction with the positive q is opposite to the direction with negative q. Ca^{2+} ions play a key role in the crystallization of $CaCO_3$. Thus, the electric field controls a concentration of Ca^{2+} ions in the vicinity of the polarized surface due to the electrostatic force. In contrast, electrostatic interactions between CO_3^{2-} ions and the polarized surface may be negligible due to the relatively low charge density of the CO_3^{2-} ions.

For measurements of contact angles using the sessile drop method with a water droplet, it was indicated that significant decreases in the contact angles were observed on the polarized surfaces, compared to nonpolarized surfaces [54,63]. The results also suggest that the Young's equation should be modified to accommodate the effect of surface electric potential due to the polarized substrates. An electrowetting equation, which accommodates the change of contact angle caused by an externally applied electric potential, has been formulated from classical electrostatics as follows:

$$\gamma_{LA}\cos\theta' = \gamma_{SA} - \gamma_{SL} + \epsilon V^2/2d, \qquad (10.7)$$

where θ', the contact angle under the externally applied electric potential of V, ε the permittivity constant of a dielectric layer beneath a water droplet, and d a thickness of the dielectric layer [83]. The above equation can be rewritten as follows:

$$\cos\theta' = \cos\theta_0 + \epsilon V^2/2\gamma_{LA}d, \qquad (10.8)$$

where θ_0 is the contact angle without the externally applied electric potential. This equation indicates that the contact angle, θ', is lowered by applied electric potential [84,85]. This equation thus supports our results of the contact angle measurements on the polarized substrates.

From the about discussion, we concluded the effects surface electric field due to the polarized materials on crystallization as follows: (1) decrease in the activation energy for nucleation, (2) control of the mass transport of reactant ions and regulation of a local supersaturation, (3) adsorption and conformation change of macromolecules with functional groups as an additive, and (4) decrease in contact angle.

10.5.2
$CaCO_3$ Crystallization on Polarized HAp Substrates

Reports on the control of crystallization by magnetic field so far have been many [86–91], but fewer have been on the control by an electric field [92,93]. The electric fields generated by electric polarized materials have been used for crystallization in a few cases [82,94–98]. We here report the crystallization of calcium carbonate in the presence of the surface electric fields due to polarized hydroxyapatite (HAp) ceramics with and without polyacrylic sodium (PAA) as soluble additive through the diffusion of CO_2 vapor from solid $(NH_4)_2CO_3$ to $CaCl_2$ solution (a

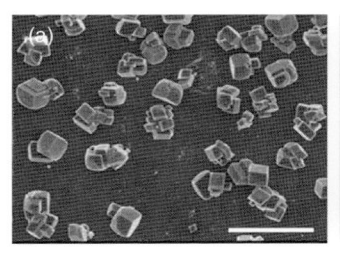

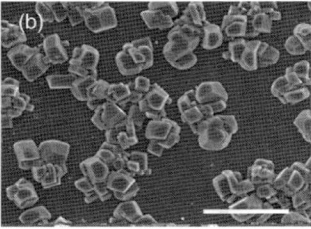

Figure 10.10 Scanning electron micrographs of calcite crystals formed on the nonpolarized and polarized HAp substrates. (a) Calcite crystals on the nonpolarized HAp substrates. (b) Calcite crystals on the polarized HAp substrates. Scale bars in insets = 50 μm.

slow diffusion method) [94,95]. PAA (MW 2000) as a soluble additive is a weak polyacid having a single structure with carboxylic acid groups without hydrophobic side chains and plays a significant role in the crystallization of calcium carbonate, namely, as a template of epitaxial growth, inhibition of growth, and promotion of an aggregation of crystals due to its adsorption [98–105].

X-ray diffraction (XRD) analysis indicated that the only polymorph of $CaCO_3$ crystals formed under all the experimental conditions were calcite. Figure 10.10 presents typical SEM microphotographs of calcite crystals formed on nonpolarized and polarized HAp substrates without PAA. The calcite crystals with a well-established rhombohedra equilibrium morphology formed on the substrates. The crystal density on the polarized surface was larger than that on a nonpolarized surface, indicating that the surface electric field due to the electrets favors calcite nucleation. This result is explained by the nucleation theory in the presence of an external electric field as described already. On the other hand, the addition of PAA dramatically changed the surface morphology and fracture of the precipitates as shown in Figure 10.11, namely, the hemispheric aggregates or flat island-shaped aggregates. These aggregates were made of calcite needles that consisted of the connected submicron crystals. The individual needles were in close contact with each other, which formed a macroscopically coherent deposit layer. The top of these needles was capped with a triangular or rhombohedral end face. Only hemispheric aggregates of calcite formed on the nonpolarized substrates and became an isolated partial film. The calcite aggregates formed on the polarized HAp substrates developed into a continuous film through growth and coalescence. The cooperation of PAA and the surface electric field due to the polarized HAp substrates favored the formation of calcite thin films. Calcite thin films with a fan-like structure in the cross section and with a columnar structure were made of the hemispheric and flat island-shaped aggregates, respectively (Figure 10.11). Calcite continuous films with a columnar structure ranged from 4 to 10 μm, while those films with a fan-like structure ranged from 2 to 7 μm [94]. The XRD pattern of the as-precipitated continuous films with a columnar structure formed on the polarized HAp substrates in the presence of PAA is shown in Figure 10.12. This pattern shows two strong peaks assigned to

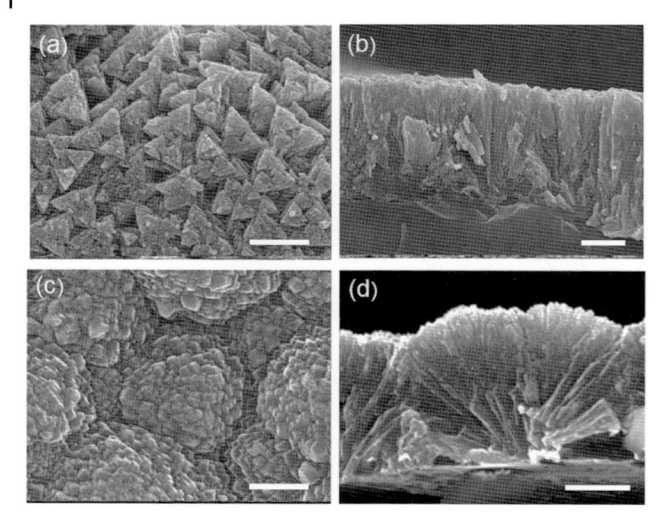

Figure 10.11 Scanning electron micrographs of the surface morphology and fracture of calcite films formed on the polarized HAp substrates. (a, b) The surface morphology and fracture of calcite film with a columnar structure in the cross section. (c, d) The surface morphology and fracture of calcite film with a fan-like structure in the cross section. Scale bars in insets = 1 μm.

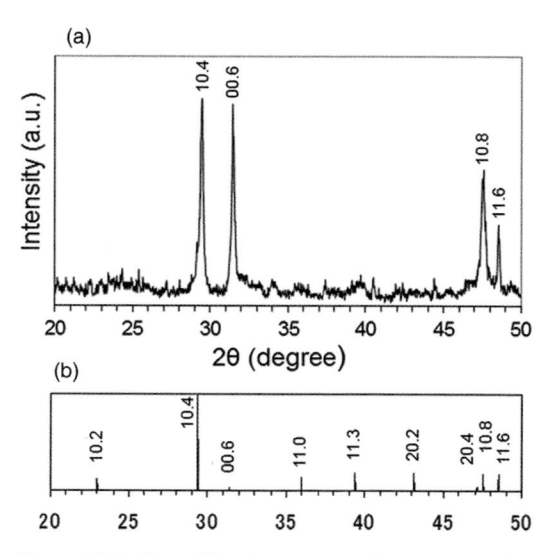

Figure 10.12 X-ray diffraction pattern of continuous calcite film. (a) XRD pattern of calcite film formed on the polarized HAp substrate in the presence of PAA. (b) The bar chart of the standard pattern of calcite (ICDD No. 5–0586).

the (00.6) and (10.4) peaks of calcite. For a quantitative analysis of the orientational uniformity, we normalized the measured intensities of the peaks in these spectra by the standard intensities of the peaks for randomly oriented calcite powder. The percentage of calcite crystals in the different orientations (hk.l) obtained in the range 20–50° was estimated by the following equation: $\%_{hk.l} = (I_{hk.l}/I^*_{hk.l})/\Sigma(I_{hk.l}/I^*_{hk.l}) \times 100$, where $I_{hk.l}$ is the measured intensity of the peaks in the XRD profile of the calcite prisms grown under our experimental conditions and $I^*_{hk.l}$ is the intensity of the peaks of calcite listed in ICDD No. 5–0586 [106]. The percentages of the (00.1), (10.4), (10.8), and (11.6) peaks of as-precipitated calcite continuous film with a columnar structure formed on the polarized HAp surface are 86, 4, 8, and 2%, respectively. Combining the XRD data with the SEM observations of the surface morphology and fracture of calcite aggregates, it is confirmed that the hemispheric- or flat island-shaped aggregates were made of the calcite crystals that elongated along the c-axis. The morphology of the PAA-Ca^{2+} complex assemblies adsorbing on the surfaces prior to the calcite nucleation was an important factor to control the structure of calcite aggregates formed. The conformation of the PAA-Ca^{2+} complex assembly, which adsorbed onto the polarized HAp substrates, was controlled by the balance of the spatial charge distribution in its structure and the properties of the surface electric field due to the polarized HAp, which led to the different morphologies of the calcite thin films [94,95]. These results indicate that the addition of PAA and the polarization treatment had the greatest impact on the crystallization of calcite. This oriented growth was explained by epitaxy on the PAA-Ca^{2+} complexes that adsorbed onto the nonpolarized or polarized HAp substrate and was improved by the surface electric field [94,95].

References

1 Williams, D.F. (2008) On the mechanisms of biocompatibility. *Biomaterials*, **29** (20), 2941–2953.

2 Wang, G., Moya, S., Lu, Z. *et al.* (2015) Enhancing orthopedic implant bioactivity: refining the nanotopography. *Nanomedicine (Lond.)*, **10** (8), 1327–1341.

3 Ban, J., Kang, S., Kim, J. *et al.* (2015) MicroCT analysis of micro-nano titanium implant surface on the osseointegration. *J. Nanosci. Nanotechnol*, **15** (1), 172–175.

4 Solouk, A., Cousins, B.G., Mirzadeh, H., and Seifalian, A.M. (2011) Application of plasma surface modification techniques to improve hemocompatibility of vascular grafts: a review. *Biotechnol. Appl. Biochem.*, **58** (5), 311–327.

5 Liu, Y., Inoue, Y., Mahara, A. *et al.* (2014) Durable modification of segmented polyurethane for elastic blood-contacting devices by graft-type 2-methacryloyloxyethyl phosphorylcholine copolymer. *J. Biomater. Sci. Polym. Ed.*, **25** (14–15), 1514–1529.

6 Albrektsson, T. and Wennerberg, A. (2004) Oral implant surfaces: Part 1: review focusing on topographic and chemical properties of different surfaces and *in vivo* responses to them. *Int. J. Prosthodont.*, **17** (5), 536–543.

7 Roach, P., Eglin, D., Rohde, K., and Perry, C.C. (2007) Modern biomaterials: a review – bulk properties and implications of surface modifications. *J. Mater. Sci. Mater. Med.*, **18** (7), 1263–1277.

8 Murphy, W.L., Mercurius, K.O., Koide, S., and Mrksich, M. (2004) Substrates for cell adhesion prepared via active site-directed immobilization of a protein domain. *Langmuir*, **20** (4), 1026–1030.

9 Keselowsky, B.G., Collard, D.M., and Garcia, A.J. (2005) Integrin binding specificity regulates biomaterial surface chemistry effects on cell differentiation. *Proc. Natl. Acad. Sci. UAA*, **102** (17), 5953–5957.

10 Nagai, A., Ma, C., Kishi, S. et al. (2012) Surface properties of Al_2O_3-YSZ ceramic composites modified by a combination of biomimetic coatings and electric polarization. *Appl. Surf. Sci.*, **262**, 45–50.

11 Tsutsumi, Y., Nishimura, D., Doi, H. et al. (2009) Difference in surface reactions between titanium and zirconium in Hanks' solution to elucidate mechanism of calcium phosphate formation on titanium using XPS and cathodic polarization. *Mater. Sci. Eng. C*, **29** (5), 1702–1708.

12 Att, W., Hori, N., Iwasa, F. et al. (2009) The effect of UV-photofunctionalization on the time-related bioactivity of titanium and chromium-cobalt alloys. *Biomaterials*, **30** (26), 4268–4276.

13 Shibata, Y., Suzuki, D., Omori, S. et al. (2010) The characteristics of *in vitro* biological activity of titanium surfaces anodically oxidized in chloride solutions. *Biomaterials*, **31** (33), 8546–8555.

14 Ma, C., Nagai, A., Yamazaki, Y. et al. (2012) Electrically polarized micro-arc oxidized TiO_2 coatings with enhanced surface hydrophilicity. *Acta Biomater.*, **8** (2), 860–865.

15 Rupp, F., Gittens, R.A., Scheideler, L. et al. (2014) A review on the wettability of dental implant surfaces. I: theoretical and experimental aspects. *Acta Biomater.*, **10** (7), 2894–2906.

16 Junker, R., Dimakis, A., Thoneick, M., and Jansen, J.A. (2009) Effects of implant surface coatings and composition on bone integration: a systematic review. *Clin. Oral Implants Res.*, **20** (Suppl 4), 185–206.

17 Anselme, K. and Bigerelle, M. (2011) Role of materials surface topography on mammalian cell response. *Int. Mater. Rev.*, **56** (4), 243–266.

18 Robotti, F., Franco, D., Banninger, L. et al. (2014) The influence of surface micro-structure on endothelialization under supraphysiological wall shear stress. *Biomaterials*, **35** (30), 8479–8486.

19 Nozaki, K., Shinonaga, T., Ebe, N. et al. (2015) Hierarchical periodic micro/nano-structures on nitinol and their influence on oriented endothelialization and anti-thrombosis. *Mater. Sci. Eng. C*, **57**, 1–6.

20 Zhao, G., Schwartz, Z., Wieland, M. et al. (2005) High surface energy enhances cell response to titanium substrate microstructure. *J. Biomed. Mater. Res. A*, **74** (1), 49–58.

21 Nagai, A., Yamazaki, Y., Ma, C.F. et al. (2012) Response of osteoblast-like MG63 cells to TiO_2 layer prepared by micro-arc oxidation and electric polarization. *J. Eur. Ceram. Soc.*, **32** (11), 2647–2652.

22 Olivares-Navarrete, R., Hyzy, S.L., Haithcock, D.A. et al. (2015) Coordinated regulation of mesenchymal stem cell differentiation on microstructured titanium surfaces by endogenous bone morphogenetic proteins. *Bone*, **73**, 208–216.

23 Lima, A.C. and Mano, J.F. (2015) Micro-/nano-structured superhydrophobic surfaces in the biomedical field: part I: basic concepts and biomimetic approaches. *Nanomedicine (Lond.)*, **10** (1), 103–119.

24 Khang, D., Choi, J., Im, Y.M. et al. (2012) Role of subnano-, nano- and submicron-surface features on osteoblast differentiation of bone marrow mesenchymal stem cells. *Biomaterials*, **33** (26), 5997–6007.

25 Qi, L., Li, N., Huang, R. et al. (2013) The effects of topographical patterns and sizes on neural stem cell behavior. *PLoS One*, **8** (3), e59022.

26 Tan, K.K., Tann, J.Y., Sathe, S.R. et al. (2015) Enhanced differentiation of neural progenitor cells into neurons of the mesencephalic dopaminergic subtype on topographical patterns. *Biomaterials*, **43**, 32–43.

27 Xuereb, M., Camilleri, J., and Attard, N.J. (2015) Systematic review of current

dental implant coating materials and novel coating techniques. *Int. J. Prosthodont.*, **28** (1), 51–59.

28 Abazari, A.M., Safavi, S.M., Rezazadeh, G., and Villanueva, L.G. (2015) Modelling the size effects on the mechanical properties of micro/nano structures. *Sensors (Basel)*, **15** (11), 28543–28562.

29 Engler, A.J., Sen, S., Sweeney, H.L., and Discher, D.E. (2006) Matrix elasticity directs stem cell lineage specification. *Cell*, **126** (4), 677–689.

30 Bae, Y.H., Mui, K.L., Hsu, B.Y. et al. (2014) A FAK-Cas-Rac-lamellipodin signaling module transduces extracellular matrix stiffness into mechanosensitive cell cycling. *Sci. Signal.*, **7** (330), ra57.

31 Kapur, S., Baylink, D.J., and Lau, K.H. (2003) Fluid flow shear stress stimulates human osteoblast proliferation and differentiation through multiple interacting and competing signal transduction pathways. *Bone*, **32** (3), 241–251.

32 Yan, Y.X., Gong, Y.W., Guo, Y. et al. (2012) Mechanical strain regulates osteoblast proliferation through integrin-mediated ERK activation. *PLoS One*, **7** (4). doi. org/10.1371/journal. pone.0035709

33 Schuler, M., Kunzler, T.P., M., deWild. et al. (2009) Fabrication of TiO_2-coated epoxy replicas with identical dual-type surface topographies used in cell culture assays. *J. Biomed. Mater. Res. A*, **88** (1), 12–22.

34 Kay, M.I., Young, R.A., and Posner, A.S. (1964) Crystal structure of hydroxyapatite. *Nature*, **204** (4963), 1050–1052.

35 Uskoković, V. (2015) The role of hydroxyl channel in defining selected physicochemical peculiarities exhibited by hydroxyapatite. *RSC Adv.*, **5** (46), 36614–36633.

36 Takahashi, T., Tanase, S., and Yamamoto, O. (1978) Electrical conductivity of some hydroxyapatites. *Electrochim. Acta*, **23** (4), 369–373.

37 Yamashita, K., Kitagaki, K., and Umegaki, T. (1995) Thermal instability and proton conductivity of ceramic hydroxyapatite at high temperatures. *J. Am. Ceram. Soc.*, **78** (5), 1191–1197.

38 Yashima, M., Yonehara, Y., and Fujimori, H. (2011) Experimental visualization of chemical bonding and structural disorder in hydroxyapatite through charge and nuclear-density analysis. *J. Phys. Chem. C*, **115** (50), 25077–25087.

39 Yashima, M., Kubo, N., Omoto, K. et al. (2014) Diffusion path and conduction mechanism of protons in hydroxyapatite. *J. Phys. Chem. C*, **118** (10), 5180–5187.

40 Wei, X. and Yates, M.Z. (2012) Yttrium-doped hydroxyapatite membranes with high proton conductivity. *Chem. Mater.*, **24** (10), 1738–1743.

41 Tofail, S.A.M., Haverty, D., Stanton, K.T., and McMonagle, J.B. (2005) Structural order and dielectric behaviour of hydroxyapatite. *Ferroelectrics*, **319** (1), 117–123.

42 Haverty, D., Tofail, S.A.M., Stanton, K.T., and McMonagle, J.B. (2005) Structure and stability of hydroxyapatite: density functional calculation and Rietveld analysis. *Phys. Rev. B*, **71** (9), 094103.

43 Lang, S.B., Tofail, S.A.M., Kholkin, A.L. et al. (2013) Ferroelectric polarization in nanocrystalline hydroxyapatite thin films on silicon. *Sci. Rep.*, **3**, 2215.

44 Gandhi, A.A., Wojtas, M., Lang, S.B. et al. (2014) Piezoelectricity in poled hydroxyapatite ceramics. *J. Am. Ceram. Soc.*, **97** (9), 2867–2872.

45 Horiuchi, N., Wada, N., Nozaki, K. et al. (2016) Dielectric relaxation in monoclinic hydroxyapatite: observation of hydroxide ion dipoles. *J. Appl. Phys.*, **119** (8), 084903.

46 Yamashita, K., Oikawa, N., and Umegaki, T. (1996) Acceleration and deceleration of bone-like crystal growth on ceramic hydroxyapatite by electric poling. *Chem. Mater.*, **8** (12), 2697–2700.

47 Nakamura, S., Takeda, H., and Yamashita, K. (2001) Proton transport polarization and depolarization of hydroxyapatite ceramics. *J. Appl. Phys.*, **89** (10), 5386–5392.

48 Tanaka, Y., Iwasaki, T., Nakamura, M. et al. (2010) Polarization and microstructural effects of ceramic

hydroxyapatite electrets. *J. Appl. Phys.*, **107** (1), 014107.
49 Horiuchi, N., Nakamura, M., Nagai, A. et al. (2012) Proton conduction related electrical dipole and space charge polarization in hydroxyapatite. *J. Appl. Phys.*, **112** (7), 074901–074906.
50 Wada, N., Mukougawa, K., Horiuchi, N. et al. (2013) Fundamental electrical properties of ceramic electrets. *Mater. Res. Bull.*, **48** (10), 3854–3859.
51 Horiuchi, N., Nakaguki, S., Wada, N. et al. (2014) Polarization-induced surface charges in hydroxyapatite ceramics. *J. Appl. Phys.*, **116** (1), 014902.
52 Wada, N., Horiuchi, N., Mukougawa, K. et al. (2016) Electrostatic induction power generator using hydroxyapatite ceramic electrets. *Mater. Res. Bull.*, **74**, 50–56.
53 Braunlich, P. (1979) *Thermally Stimulated Relaxation in Solids*, Springer-Verlag, Berlin.
54 Nakamura, M., Hori, N., Namba, S. et al. (2015) Wettability and surface free energy of polarised ceramic biomaterials. *Biomed. Mater. Bristol Engl.*, **10** (1), 011001.
55 Schultz, J., Tsutsumi, K., and Donnet, J.-B. (1977) Surface properties of high-energy solids. *J. Colloid Interface Sci.*, **59** (2), 272–276.
56 Schultz, J., Tsutsumi, K., and Donnet, J.-B. (1977) Surface properties of high-energy solids. *J. Colloid Interface Sci.*, **59** (2), 277–282.
57 Chassin, P., Jounay, C., and Quiquampoix, H. (1986) Measurement of the surface free-energy of calcium-montmorillonite. *Clay Miner.*, **21**, 899–907.
58 Lampin, M., Warocquier-Clérout, N., Legris, C. et al. (1997) Correlation between substratum roughness and wettability, cell adhesion, and cell migration. *J. Biomed. Mater. Res.*, **36** (1), 99–108.
59 Advincula, M.C., Rahemtulla, F.G., Advincula, R.C. et al. (2006) Osteoblast adhesion and matrix mineralization on sol–gel-derived titanium oxide. *Biomaterials*, **27** (10), 2201–2212.
60 Okumura, A., Goto, M., Goto, T. et al. (2001) Substrate affects the initial attachment and subsequent behavior of human osteoblastic cells (Saos-2) *Biomaterials*, **22** (16), 2263–2271.
61 Redey, S.A., Nardin, M., Bernache-Assolant, D. et al. (2000) Behavior of human osteoblastic cells on stoichiometric hydroxyapatite and type A carbonate apatite: role of surface energy. *J. Biomed. Mater. Res.*, **50** (3), 353–364.
62 Wang, Y., Zhang, S., Zeng, X. et al. (2008) Initial attachment of osteoblastic cells onto sol–gel derived fluoridated hydroxyapatite coatings. *J. Biomed. Mater. Res. A*, **84** (3), 769–776.
63 Nakamura, M., Nagai, A., Hentunen, T. et al. (2009) Surface electric fields increase osteoblast adhesion through improved wettability on hydroxyapatite electret. *ACS Appl. Mater. Interfaces*, **1** (10), 2181–2189.
64 Rouahi, M., Champion, E., Hardouin, P., and Anselme, K. (2006) Quantitative kinetic analysis of gene expression during human osteoblastic adhesion on orthopaedic materials. *Biomaterials*, **27** (14), 2829–2844.
65 Schwartz, Z. and Boyan, B.D. (1994) Underlying mechanisms at the bone-biomaterial interface. *J. Cell. Biochem.*, **56** (3), 340–347.
66 Aronov, D., Karlov, A., and Rosenman, G. (2007) Hydroxyapatite nanoceramics: basic physical properties and biointerface modification. *J. Eur. Ceram. Soc.*, **27** (13–15), 4181–4186.
67 Clark, E.A. and Brugge, J.S. (1995) Integrins and signal transduction pathways: the road taken. *Supramol. Sci.*, **268** (5208), 233–239.
68 Miyamoto, S., Teramoto, H., Coso, O.A. et al. (1995) Integrin function: molecular hierarchies of cytoskeletal and signaling molecules. *J. Cell Biol.*, **131** (3), 791–805.
69 Ben-Ze'ev, A., Farmer, S.R., and Penman, S. (1980) Protein synthesis requires cell-surface contact while nuclear events respond to cell shape in anchorage-dependent fibroblasts. *Cell*, **21** (2), 365–372.
70 Xu, L.-C. and Siedlecki, C.A. (2007) Effects of surface wettability and contact time on protein adhesion to biomaterial

surfaces. *Biomaterials*, **28** (22), 3273–3283.

71 Webb, K., Hlady, V., and Tresco, P.A. (1998) Relative importance of surface wettability and charged functional groups on NIH 3T3 fibroblast attachment, spreading, and cytoskeletal organization. *J. Biomed. Mater. Res.*, **41** (3), 422–430.

72 Hao, L., Lawrence, J., and Chian, K.S. (2005) Osteoblast cell adhesion on a laser modified zirconia based bioceramic. *J. Mater. Sci. Mater. Med.*, **16** (8), 719–726.

73 Das, K., Bose, S., and Bandyopadhyay, A. (2007) Surface modifications and cell–materials interactions with anodized Ti. *Acta Biomater.*, **3** (4), 573–585.

74 Kilpadi, K.L., Sawyer, A.A., Prince, C.W. *et al.* (2004) Primary human marrow stromal cells and Saos-2 osteosarcoma cells use different mechanisms to adhere to hydroxylapatite. *J. Biomed. Mater. Res. A*, **68** (2), 273–285.

75 Masiello, L.M., Fotos, J.S., Galileo, D.S., and Karin, N.J. (2006) Lysophosphatidic acid induces chemotaxis in MC3T3-E1 osteoblastic cells. *Bone*, **39** (1), 72–82.

76 Matsumura, H., Kawasaki, K., Okumura, N. *et al.* (2003) Characterization of the surface of protein-adsorbed dental materials by wetting and streaming potential measurements. *Colloids Surf. B*, **32** (2), 97–103.

77 Krause, A., Cowles, E.A., and Gronowicz, G. (2000) Integrin-mediated signaling in osteoblasts on titanium implant materials. *J. Biomed. Mater. Res.*, **52** (4), 738–747.

78 Nakamura, M., Sekijima, Y., Nakamura, S. *et al.* (2006) Role of blood coagulation components as intermediators of high osteoconductivity of electrically polarized hydroxyapatite. *J. Biomed. Mater. Res. A*, **79** (3), 627–634.

79 Kashchiev, D. (1972) Nucleation in external electric field. *J. Cryst. Growth*, **13–14**, 128–130.

80 Dhanasekaran, R. and Ramasamy, P. (1986) Two-dimensional nucleation in the presence of an electric field. *J. Cryst. Growth*, **79** (1–3, Part 2), 993–996.

81 Young, T. (1805) An essay on the cohesion of fluids. *Philos. Trans. R. Soc. Lond.*, **95**, 65–87.

82 Wada, N., Horiuchi, N., Nakamura, M. *et al.* (2015) Controlled calcite nucleation on polarized calcite single crystal substrates in the presence of polyacrylic acid. *J. Cryst. Growth*, **415**, 7–14.

83 Kang, K.H. (2002) How electrostatic fields change contact angle in electrowetting. *Langmuir*, **18** (26), 10318–10322.

84 Welters, W.J.J. and Fokkink, L.G.J. (1998) Fast electrically switchable capillary effects. *Langmuir*, **14** (7), 1535–1538.

85 Ricks-Laskoski, H.L. and Snow, A.W. (2006) Synthesis and electric field actuation of an ionic liquid polymer. *J. Am. Chem. Soc.*, **128** (38), 12402–12403.

86 Higashitani, K., Kage, A., Katamura, S. *et al.* (1993) Effects of a magnetic field on the formation of $CaCO_3$ particles. *J. Colloid Interface Sci.*, **156** (1), 90–95.

87 Madsen, H.E.L. (1995) Influence of magnetic field on the precipitation of some inorganic salts. *J. Cryst. Growth*, **152** (1–2), 94–100.

88 Wakayama, N.I., Ataka, M., and Abe, H. (1997) Effect of a magnetic field gradient on the crystallization of hen lysozyme. *J. Cryst. Growth*, **178** (4), 653–656.

89 Yanagiya, S., Sazaki, G., Durbin, S.D. *et al.* (2000) Effects of a magnetic field on the growth rate of tetragonal lysozyme crystals. *J. Cryst. Growth*, **208** (1–4), 645–650.

90 Chibowski, E., Szczes, A., and Holysz, L. (2005) Influence of sodium dodecyl sulfate and static magnetic field on the properties of freshly precipitated calcium carbonate. *Langmuir*, **21** (18), 8114–8122.

91 Knez, S. and Pohar, C. (2005) The magnetic field influence on the polymorph composition of $CaCO_3$ precipitated from carbonized aqueous solutions. *J. Colloid Interface Sci.*, **281** (2), 377–388.

92 Anderson, B.J. and Hallett, J. (1979) Influence of environmental saturation and electric field on growth and evaporation of epitaxial ice crystals. *J. Cryst. Growth*, **46** (3), 427–444.

93 Shichiri, T. and Araki, Y. (1986) Nucleation mechanism of ice crystals

under electrical effect. *J. Cryst. Growth*, **78** (3), 502–508.

94 Wada, N., Nakamura, M., Tanaka, Y. et al. (2009) Formation of calcite thin films by cooperation of polyacrylic acid and self-generating electric field due to aligned dipoles of polarized substrates. *J. Colloid Interface Sci.*, **330** (2), 374–379.

95 Wada, N., Tanaka, Y., Nakamura, M. et al. (2009) Controlled crystallization of calcite under surface electric field due to polarized hydroxyapatite ceramics. *J. Am. Ceram. Soc.*, **92** (7), 1586–1591.

96 Wada, N., Nakamura, M., Wang, W. et al. (2011) Controlled deposition of calcite crystals on yttria-stabilized zirconia ceramic electrets. *Cryst. Growth Des.*, **11** (1), 166–174.

97 Wada, N., Horiuchi, N., Nakamura, M. et al. (2013) Effect of poly(acrylic acid) and polarization on the controlled crystallization of calcium carbonate on single-phase calcite substrates. *Cryst. Growth Des.*, **13** (7), 2928–2937.

98 Wada, N., Horiuchi, N., Nakamura, M. et al. (2014) Cooperative effects of polarization and polyaspartic acid on formation of calcium carbonate films with a multiple phase structure on oriented calcite substrates. *J. Cryst. Growth*, **402**, 179–186.

99 Zhang, S. and Gonsalves, K.E. (1998) Influence of the chitosan surface profile on the nucleation and growth of calcium carbonate films. *Langmuir*, **14** (23), 6761–6766.

100 Hosoda, N., Sugawara, A., and Kato, T. (2003) Template effect of crystalline poly (vinyl alcohol) for selective formation of aragonite and vaterite $CaCO_3$ thin films. *Macromolecules*, **36** (17), 6449–6452.

101 Wada, N., Suda, S., Kanamura, K., and Umegaki, T. (2004) Formation of thin calcium carbonate films with aragonite and vaterite forms coexisting with polyacrylic acids and chitosan membranes. *J. Colloid Interface Sci.*, **279** (1), 167–174.

102 Miura, T., Kotachi, A., Oaki, Y., and Imai, H. (2006) Emergence of acute morphologies consisting of iso-oriented calcite nanobricks in a binary poly(acrylic acid) system. *Cryst. Growth Des.*, **6** (2), 612–615.

103 Aschauer, U., Ebert, J., Aimable, A., and Bowen, P. (2010) Growth modification of seeded calcite by carboxylic acid oligomers and polymers: toward an understanding of complex growth mechanisms. *Cryst. Growth Des.*, **10** (9), 3956–3963.

104 Lee, S., Lee, S.G., Sim, M. et al. (2011) Control over the vertical growth of single calcitic crystals in biomineralized structures. *Cryst. Growth Des.*, **11** (11), 4920–4926.

105 Sarkar, A., Dutta, K., and Mahapatra, S. (2013) Polymorph control of calcium carbonate using insoluble layered double hydroxide. *Cryst. Growth Des.*, **13** (1), 204–211.

106 Aizenberg, J., Black, A.J., and Whitesides, G.M. (1999) Oriented growth of calcite controlled by self-assembled monolayers of functionalized alkanethiols supported on gold and silver. *J. Am. Chem. Soc.*, **121** (18), 4500–4509.

11
Functional Materials for Gas Storage. Part I: Carbon Dioxide and Toxic Compounds

L. Reguera[1,2] and E. Reguera[1]

[1]National Polytechnic Institute, Center for Applied Science and Advanced Technology, Legaria Unit, Legaria 694, Col. Irrigación, Del. M. Hidalgo, 11500 Mexico City, Mexico
[2]University of Havana, Faculty of Chemistry, Zapata s/n e/G and C. Aguirre, Vedado, 10400 Havana City, Cuba

11.1 Introduction

Because of the combustion of coal, oil, and natural gas, the emission of carbon dioxide to atmosphere has sharply risen. About 80% of global CO_2 emissions result from the combustion of fossil fuels. According to the United Nations Intergovernmental Panel on Climate Change (IPCC), the estimated average temperature rise for the year 2050 is 2–2.5 °C, which will have dramatic effects on the sea level increase, above the allowed level for many coastal regions and insular countries. The rising level in CO_2 concentration in the atmosphere has many other consequences, among them, an increase in acidity of the oceans, changes in the wind circulation and rain patterns, and so on. This explains the urgency for reversing the growing tendency of CO_2 emissions. A route toward this goal is CO_2 capture from exhaust gases of industrial installations that have a large contribution to the increasing emissions. The uses of CO_2 are limited, at least compared to its generation by fossil fuels combustion, and from here, there is a need of its capture followed by compression in high-pressure vessels to allow transport to secure storage sites in the ground or oceanic geological reservoirs to avoid reemission.

The fossil fuel exploitation and the increase in the industrial activity are responsible for the emission of large amounts of gaseous pollutants, such as volatile organic compounds (VOCs), inorganic gases, and vapors (e.g., NH_3, CO, NO_x, H_2S, SO_2, Cl_2, BCl_3, etc.). These pollutants may also be present in the air due to accidents in industrial plants or by inappropriate prevention and control in industrial processes. Adsorption in porous solids is by far the simplest and most economical technology applied to remove gaseous pollutants from air. This chapter provides a summary on the main materials nowadays available for

Handbook of Solid State Chemistry, First Edition. Edited by Richard Dronskowski, Shinichi Kikkawa, and Andreas Stein.
© 2017 Wiley-VCH Verlag GmbH & Co. KGaA. Published 2017 by Wiley-VCH Verlag GmbH & Co. KGaA.

adsorption of toxic chemicals. In addition, from the First World War up to 1993, when the Chemical Weapons Convention demilitarized the use of toxic chemicals, a large amount of sophisticated chemical warfare agents were created and protection devices based on the adsorption of these toxic agents were developed. This subject is briefly reviewed.

11.2
Evaluation of Gases' Adsorption

11.2.1
Physical Adsorption of Gases and Vapors

Adsorption is an exothermic and reversible process, if it takes place through physical interactions. Under the hypothesis of reversibility, to liberate the adsorbed species, an energy amount equal to the adsorption energy is required. The attractive forces involved in the adsorption process, with the exception of weak guest–host coordination interaction, are of electrostatic nature. From this fact, the finding of suitable adsorbents for gas storage implies the optimization of the involved adsorption heat (ΔH°_{ads}) and of parameters regarding the material under consideration [1]:

- The heat of adsorption (ΔH°_{ads}) of the gas on the solid surface, since it is the energy released in the adsorption process, which determines the molecule stability as adsorbed species at a given temperature. This parameter determines the operating temperature and pressure in systems based on the gas adsorption and desorption.
- The pore system features, which determine the storage density at high pressures, in particular, the accessible pore volume.
- The material density, which is a determining factor in the size and weight of the storage units.

11.2.2
Gas Adsorption Measurement Techniques

Obtaining accurate measurements of gas adsorption in porous materials involves optimization of a group of experimental factors. Some of the first reports on gas adsorption in porous materials did not include enough information to confirm the validity of the reported data and for some cases, the results are unreproducible. The amount of gas adsorbed is typically measured using volumetric (frequently called manometric) or gravimetric techniques. In the volumetric case, known doses of gas are taken and introduced into the sample cell *step by step*, following the adsorption process through measurements of pressure, temperature, and volume, and then using an equation of state for the gas in order to calculate the adsorbed amount. At each point, the amount of adsorbate molecules in the gas phase is calculated and any missing molecule is attributed to the adsorbed phase. In order to determine the number of moles in the gas phase, however, the available volume must be a known parameter. The dead volume is the difference between the volume of the empty sample cell and the volume occupied by the sample and the adsorbate. Determination of the combined sample and adsorbate volume is intrinsically problematic [2] and can be further complicated by the sample swelling.

The gravimetric technique, on the other hand, directly determines the amount adsorbed from the weight measurement using a microbalance. In this case, the amount of gas introduced does not need to be calculated, and the error accumulation associated with the volumetric method is also avoided [3]. However, the microbalance reading must be corrected for the buoyancy effect of the sample in the surrounding fluid (gas), for which knowledge of the volume of the sample and the adsorbate is required, in an analogous way to the volumetric method [4].

The problems encountered in the absorption measurements of gases have been usually associated with three factors: (1) the existence of small leaks in the measuring equipment, (2) adsorption of impurities – both present in the gas used as in the vacuum system, (3) instrumental errors related to insufficient sensitivity of the measuring equipment initially used. The first source of measurement errors is very important in high-pressure volumetric determinations.

To obtain reliable data, measurement systems must meet a set of basic requirements given as follows:

- Operation of the high-vacuum systems under extremely clean conditions, use of diaphragm type metal seals, and oil-free vacuum pumps, usually turbomolecular pumps with evacuation capacity up to 10^{-10} bar.
- If possible, use a purification system for gases, even when ultrapure gases are used. This is particularly important for measurements at high pressures, when the adsorption impurities become much larger than when we are working at atmospheric pressure [1b,5].

The resolution of the recorded isotherms should be sufficient to avoid unambiguous definition of its shape. In that sense, certain experimental protocols for the data collection must be considered; some of them are as follows:

- Confirm the absence of significant amounts of adsorbed impurities by checking the reproducibility for the collected data, particularly when the equilibrium is reached slowly.
- Both the adsorption and the desorption isotherms must be recorded in order to ensure that data correspond to equilibrium states (free of kinetic effects).
- The determination of adsorption enthalpies by isosteric method must be carried out under low coverage conditions of the adsorption sites in order to minimize the contribution from adsorbate–adsorbate interactions.
- Introduce the required corrections to the recorded data-related flotation or dead volume for gravimetric and volumetric instruments, respectively. This is particularly important for the adsorption at high pressures [1b].

- In commercially available volumetric equipment, the dead volume is usually determined introducing He, which supposes that it is not adsorbed. This hypothesis is invalid for microporous solids, and large evacuation periods are then required to remove the He atoms that were adsorbed before the adsorption isotherm recording. This is particularly important for the study of H_2 adsorption [6].

11.2.3
Adsorption Data Evaluation

The requirements for adsorbent materials to be used in gas storage are closely related to the physical properties of the gas molecule. The storage of vapors with critical temperature (T_c) well above the working temperature (T_w) is possible using mesoporous solids (2–50 nm pore size) and even macroporous materials (pore diameter > 50 nm). For noncondensable gases ($T_w < T_c$) and solids of weak adsorption potential, the use of microporous solids (pore size < 2 nm) is a requirement in order to favor a high molecule confinement within the pores.

The features of the pore system, the adsorbate nature and concentration of the adsorbate molecules in the gas phase determine the adsorption model to be used to describe the adsorption data. The Henry isotherm is the simplest available adsorption model, where the amount of adsorbed molecules is proportional to the applied pressure, which determines the adsorbate gravimetric density in the gas phase:

$$n_{ad} = K_H P, \tag{11.1}$$

where n_{ad} is the amount of adsorbed molecules (adsorption *in excess*) and K_H is the Henry constant and senses the strength for the adsorbate–adsorbent interaction. This model is applicable for very low concentration (high dilution) of adsorbate molecules in the gas phase and negligible lateral interactions between adsorbed molecules. Such condition is satisfied, for instance, in the separation of mixtures of gases by differential adsorption of the involved molecules where the concentration of the injected gaseous phase is very low (high dilution, $K_H P \ll 1$).

The model that follows, in order of complexity, is the Langmuir isotherm. It is useful to describe the chemical sorption and the physical adsorption in monolayers. This model supposes that the adsorbent solid is composed by n_s equivalent and independent adsorption sites. At certain values of pressure and temperature, only a fraction (n_{ad}) of the available sites remains occupied. The equation for this model is as follows [7]:

$$n_{ad} = \frac{n_p K P}{1 + KP}, \tag{11.2}$$

where K is the Langmuir constant or adsorption coefficient, and it is related to the absolute value of the adsorption energy (E) and measurement temperature (T) according to the following:

$$K = k \exp\left(\frac{E}{RT}\right), \tag{11.3}$$

where R is the universal gas constant and k is a pre-exponential factor. For low surface coverage ($KP \ll 1$), Eq. (11.2) reduces to Eq. (11.1).

For condensable molecules (vapors), where the adsorbed phase could be found forming multilayers, the most convenient adsorption model is given by the Brunauer–Emmett–Teller (BET) isotherm. The BET isotherm can be considered an extension of the Langmuir model to the adsorption in multilayers, when the adsorbed phase behaves as a new adsorption site to allow the adsorption of additional molecules to form a system of adsorbed layers, for adsorbate pressure below its vapor pressure (p^o). The equation for the BET isotherm is as follows [7]:

$$\frac{p}{n_{ad}(p^o - p)} = \frac{1}{n_m C} + \frac{C-1}{n_m C} \cdot \frac{p}{p^o}, \tag{11.4}$$

where $C \approx \exp\left(\frac{E_1 - E_L}{RT}\right)$.

E_1 is the positive value of the so-called "energy of adsorption of the first layer," under the hypothesis of the absence of lateral interactions between adsorbed molecules in that first layer; and E_L has the same value as the adsorbate liquefaction energy. Common VOCs, except methane, have critical temperature above 300 K and under ambient conditions are vapors, whose adsorption isotherms can be described by the BET model.

These three adsorption models suppose that the adsorption process takes place because of the interaction of the adsorbent surface with the adsorbate molecules. For microporous solids, such hypothesis is invalid since the adsorption potential affects the entire micropore volume.

The adsorption forces that determine the adsorption processes are of relatively short range, usually with a distance dependence of r^{-3}–r^{-6} (discussed below). From this fact, adsorption is favored for molecules confined to small volume, particularly in microporous materials. In these materials, the adsorption has a volumetric character; the guest molecule senses an adsorption field in the solid volume as a whole, through the porous network. In such porous solids, the surface available to interact with the adsorbate molecules includes the pore surface, which is the main contribution to the solid specific surface area. Under such confinement conditions, the adsorption field usually favors the pore filling, and the micropore volume (W_o) appears as the geometrical parameter that determines the limit storage capacity.

For the adsorption of molecules below their critical temperature (T_c), the pore filling takes place through the formation of a condensed phase within the accessible free volume. The adsorption isotherms recorded under such conditions are described by the Dubinin–Astakhov equation, derived from the *Dubinin model for the volumetric filling of micropores* [8]. The amount of adsorbed molecules that occupies the solid micropore volume W_o or limit storage capacity is given by

$$n_p = \frac{W_o \rho_x}{M(x)}, \tag{11.5}$$

where n_p is expressed in moles, $M(x)$ is the adsorbate molar mass, and $\rho(x)$ is the adsorbate density in its liquid state at the measurement temperature. The number of adsorbate molecules accommodated within the micropore volume (n_{ad}) and micropore coverage (θ) depends on the experiment temperature:

$$\theta = \frac{n_{ad}}{n_p}. \quad (11.6)$$

The Dubinin's theory for the volumetric filling of micropores is based on two postulates: the first one is taken from the adsorption potential of Polanyi and it establishes that for a given gas and an adsorbed amount, the adsorption potential (A) –also called molar adsorption work – is temperature independent. This adsorption potential is equivalent to the free energy required to transfer a mole of adsorbate from its liquid state to the adsorbed phase [9]. The temperature independence of A supposes that $\left(\frac{\partial A}{\partial T}\right)_{n_{ad}} = 0$, where

$$A = -\Delta G = RT \ln\left(\frac{P_v}{P}\right). \quad (11.7)$$

In Eq. (11.7), R is the universal gas constant, T is the measurement temperature, P_v is the adsorbate vapor pressure, and P is the equilibrium pressure. The theory of volumetric filling of micropores takes the adsorbate liquid state as the reference state to calculate the thermodynamic properties (enthalpy, entropy, and free energy). Because of this fact, the model is applicable only for the adsorption below the adsorbate critical temperature (for vapors).

The second postulate results from an experimental basis and it establishes that the dependence between the volumetric filling (θ) (Eq. (11.6)) and the adsorption potential is given by a Weibull-type distribution:

$$\theta = \exp\left[-\left(\frac{A}{E_o}\right)^n\right]. \quad (11.8)$$

If Eqs (11.6) and (6.7) are now substituted in Eq. (11.8), the expression for Dubinin–Astakhov (DA) isotherm for the volumetric filling of micropores results as follows[10]:

$$n_{ad} = n_p \cdot \exp\left[-\left(\frac{RT}{E_o}\ln\left(\frac{P_v}{P}\right)\right)^n\right]. \quad (11.9)$$

In this equation, E_o is known as characteristic energy and its value senses the average adsorption energy, and n is called heterogeneity parameter and is related to the heterogeneity of the adsorption sites. Because in this model the energy follows a Weibull-type distribution, to a larger n value corresponds a narrower adsorption energy distribution [11].

The evaluation of adsorption isotherms corresponding to noncondensable adsorbates, for example, H_2 above its critical temperature (32.7 K), is usually carried out by the Langmuir–Freundlich model derived from the *osmotic theory of the adsorption*. This last theory supposes that an osmotic equilibrium between two solutions (composed of vacancies and guest molecules) of different concentration is established. One of these solutions is present inside the micropores and the other one in the gaseous phase. In this model, the role of dissolvent corresponds to the vacancies, for example, the available adsorption sites (or vacuum). By this reason, this model is also known as the *theory of vacancy solution* [12]. This theory not only allows reaching a more general equation than the one proposed by Dubinin but also relies on the volumetric filling of the micropores.

These solutions remain in equilibrium when only one of them is immersed in an external field. The model assumes that the corresponding potential field can be virtually represented by an osmotic pressure, π. For example, inside the cavities and channels of a porous solids, an adsorption field is present, which may be virtually represented by a osmotic pressure between the gas and the adsorbed phase. Therefore, considering the adsorption space as an inert volume, the effect of adsorption would be caused by a virtual π pressure applied to compress the adsorbed phase into that volume. That is, the effect of adsorption field is replaced by an external pressure π.

According to this theory, during the adsorption in the micropores the formation of a dissolution of molecules and vacancies takes place, within the micropores. The vacancies correspond to the micropore volume not occupied by the adsorbate molecules. Then, if we add the volume occupied by the adsorbate molecules (V_a) to the free volume (V^x), the micropore volume (as all) will be given by

$$V_a + V^x = V_p. \tag{11.10}$$

If we consider the volume occupied by an adsorbate molecule "b" equals the volume of a vacancy (hypothesis), the following dependences result:

$$\frac{V_a}{b} + \frac{V^x}{b} = n_{ad} + N^x = \frac{V_p}{b} = n_p, \tag{11.11}$$

where n_p is the maximum number of adsorbate molecules that can be accommodated in the micropore volume. If now this equation is expressed according to the molar fraction of adsorbate molecules, $X_{ad} = n_{ad}/n_p$ and the molar fraction of vacancies, $X^x = N^x/n_p$, the sum of these two fractions must be equal to the unit:

$$X_{ad} + X^x = 1. \tag{11.12}$$

If we consider that the adsorption in the micropores' system can be described as an osmotic process, where the vacuum (vacancies) is the solvent and molecules adsorbed are the solute, and applying the thermodynamics of the process of osmosis to the model described above, the following expression for the adsorption isotherm results as follows [13]:

$$n_{ad} = \frac{n_p K P^B}{1 + K P^B}. \tag{11.13}$$

This equation is commonly referred to as Sips or Bradley isotherm, or simply osmotic adsorption isotherm, and for $B = 1$ it reduces to the Langmuir isotherm (Eq. (11.2)).

However, because of its greater practical utility, the model known as Langmuir–Freundlich isotherm is more frequently used. It is obtained replacing in Eq. (11.13) n_{ad} by $n_p/2$, and P by $P_{0.5}$, results as follows:

$$\frac{n_p}{2} = \frac{n_p K P_{0.5}^B}{1 + K P_{0.5}^B}. \tag{11.14}$$

Then, simplifying,

$$K = \frac{1}{P_{0.5}^B} = \left(\frac{1}{P_{0.5}}\right)^{-\frac{1}{g}}, \tag{11.15}$$

where g is known as osmotic coefficient ($B = 1/g$). Now, combining Eqs. (11.13) and (6.15), the expression for the Langmuir– Freundlich isotherm is obtained as follows:

$$P = P_{0.5}\left(\frac{n_{ad}}{n_p - n_{ad}}\right)^g. \tag{11.16}$$

A value of $g=1$ indicates that the solution of the adsorbate in the vacancies behaves as ideal (the adsorbate–adsorbent interactions are negligible), while $g>1$ indicates that the (solute–solvent) adsorbate–adsorbent interactions are favored. The value of g increases parallel to the adsorbate–adsorbent energy of interaction [14]. Many other adsorption models are available in the specialized literature [13,15], but the above-mentioned ones are sufficient to evaluate the most common data of gas storage in materials.

In the literature reports on gas storage by adsorption in solids, it is common to find information on the storage capacity at certain values of temperature and pressure, as only data and without further clarification. It is important to note that usually the information on the adsorption capacity arises from direct values of adsorption data using gravimetric or volumetric equipment. Such data acquisition methods provide the "adsorption in excess," which is indeed the magnitude that must be reported. The "adsorption in excess" is different from the "absolute" or "total adsorption capacity." Nevertheless, the latter are parameters that must be considered since they are of interest for comparative studies using different materials, under the hypothesis that the primary adsorption data were recorded under the same experimental conditions. Gross *et al.* illustrate the meaning of these terms through the adsorption of H_2 whose isotherms are usually recorded under adsorbate supercritical state [16]. In the context of supercritical gas adsorption under high pressures, these parameters and concepts deserve to be reviewed in some detail.

11.3
Nature of Adsorbate–Adsorbent Interactions

The adsorption potential and corresponding adsorption heat for a given adsorbate depends on both material surface and adsorbate properties. First, we are going to refer to the properties of the gases under consideration, which determine the type of interactions in which they can participate and their scope in terms of interaction energy dependence on the adsorbent–adsorbate electron cloud distance. The interaction energy is the key parameter to be considered for adsorbent materials design.

As already mentioned, the gases of interest for energy and environmental applications, and to be stored in solids, are H_2, CH_4, and CO_2. In adsorption studies, other gases, such as He, Xe, and N_2, must be considered. He is used to evaluate the dead volume in volumetric measurements. Xe is an excellent probe adsorbate to explore the geometry and local properties of cavities in porous solids combining adsorption and ^{129,131}Xe NMR spectroscopy [17]. N_2 is the adsorbate commonly used to evaluate the specific surface area using the BET method [18]. For materials of narrow channels and pore window sizes below the N_2 kinetic diameter, Ar is used as a probe to evaluate the material textural properties. Light hydrocarbons and vapors of other organic molecules are frequently used as probes to evaluate the surface properties of solids, particularly, the presence of acid and basic sites, and of local electric field and its anisotropy [19]. In this and the following chapter, the adsorption process is the main subject to be considered and not as structural technique. The adsorption in porous solids is the main mechanism to remove toxic chemicals from air, which is relevant for the personnel protection against chemical accidents and chemical weapons. For instance, filters based on porous solids, usually activated carbons impregnated with other active species, are the main component of gas masks used for personnel protection against toxic chemicals.

Any adsorbate molecule (or atom) is susceptible to be adsorbed in solids through dispersive-type interactions related to the instantaneous fluctuation of their electron clouds, which induces dipole and quadrupole moments. Such induced dipole and quadrupole moments are responsible for attractive interactions between the adsorbate molecule and the adsorbent surface. This short distance interaction has an r^{-6} dependence, where r is the distance between the involved charge distributions. This interaction depends on electron density on both the adsorbent and the adsorbate, through their polarizability (α). The role of the electron density in the attractive forces through dispersive interactions is appreciated when comparing the critical temperature (T_c) for the noble gases, in K: He (5.19) < Ne (44.4) < Ar (150.87) < Xe (289.77) [20]. For Xe, with 157 electrons, a T_c value of 16.5 °C is observed, close to room temperature. In the absence of charge centers in the solids, able to polarize the guest molecule, the dispersive forces are the only attractive interactions to allow the adsorption of He, Xe, and CH_4 in materials, for instance. For methane (10 electrons), similar to Ar, the critical temperature is 190.56 K [21], which is related to the molecule polarizability; 2.593×10^{-24} cm^3, versus 1.6411×10^{-24} cm^3 for Ar ($T_c = 150.87$ K) [20].

Carbon dioxide is a molecule with quadrupole moment ($\mathbf{Q} = -14.25 \times 10^{-40}$ cm^2) [22] and, as a consequence, is able to have an attractive interaction with a local electric field gradient ($\nabla \mathbf{E}$). The energy (E) involved in this interaction has an r^{-3} dependence, $E = (1/3)\,\mathbf{Q} \cdot \nabla \mathbf{E}$. Neighboring molecules with quadrupole

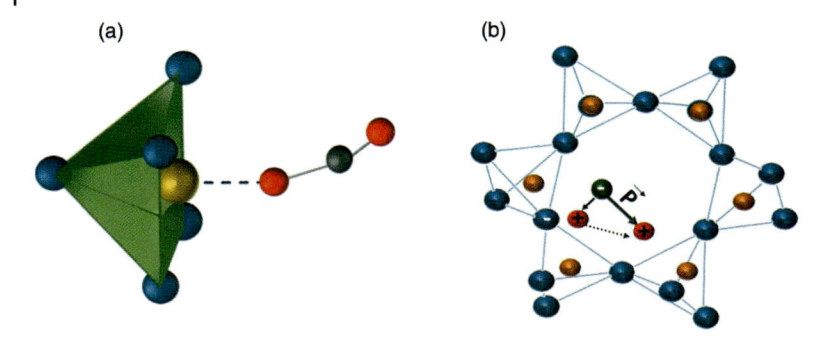

Figure 11.1 Illustration for (a) CO_2 coordination interaction with a partially naked transition metal at a solid surface; (b) charge–dipole interaction during the adsorption of polar molecules in porous solids.

moment interact among themselves through a quadrupole–quadrupole coupling, which has an r^{-5} dependence. This is relevant, for instance, in the volumetric filling of micropores and for the adsorption in multilayers.

In the presence of charge centers, for example, exchangeable ions in zeolites and metal–organic frameworks (MOFs), the polarization of the electron cloud of guest molecules, with an r^{-4} dependence, contributes to the adsorption of the guest molecules. Similar to the dispersive interactions, it depends on the molecule polarizability (α): $E = \alpha E^2/2$, where $\mathbf{E}$ is the local electric field sensed by the molecule. The oxygen ends of carbon dioxide have basic character and are able to interact with acidic sites in the adsorbent, while the C atom has acid features and can interact with basic sites. The partially naked transition metal sites in porous solids behave as Lewis acidic sites and are able to contribute to the adsorption of CO_2 molecules through the formation of a coordination bond (Figure 11.1) [23]. The interaction through the C atom leads to the formation of relatively highly stable chemical compounds, for example, carbonates, when it reacts with CaO, and carbamate for the reaction with primary amines. The relatively high polarizability of CO_2 (2.911), in 10^{-24} cm^3 units [20], is a favorable condition for its adsorption through dispersive and polarization interactions.

The nature of the adsorbate–surface interaction involved in gas adsorption determines the formation of a layer of adsorbed phase on the solid surface. In the presence of interactions of relatively large scope, for example, r^{-3} and r^{-4}, additional layers of adsorbed molecules can be accumulated on the surface. This results in an effect equivalent to the adsorbate compression, substantially increasing the density of the gas adsorbed at the solid surface [24]. In microporous solids, such interactions favor the micropore filling, even for supercritical gases.

The adsorption forces mentioned above are present for practically all pollutant molecules. For those with dipole moments, two additional attractive forces could be present, related to the charge–dipole and dipole–dipole interactions, with r^{-2} and r^{-3} energy dependence, respectively.

In addition to the involved adsorption heat, the criteria for materials design for a given molecule adsorption must also include the highest possible surface area, large pore volume, thermal stability well above the activation temperature of the material, and low weight in order to increase the gravimetric storage density in storage devices.

11.4
Adsorption Heats and Gas Adsorption–Desorption

Adsorption is an exothermic process and it liberates energy. For simplicity, in the following text, the adsorption heat will be given as its absolute value. To estimate the optimal heat of adsorption (ΔH^o_{ads}) required to maintain a given molecule as adsorbed species at a certain temperature and pressure, a model to describe the gas molecule adsorption must be assumed. As already mentioned, if the adsorption takes place on localized sites and without lateral interactions between the adsorbed species, an appropriate model to describe the adsorption data is the Langmuir isotherm (Eq. (11.2)). This model is an acceptable approximation for the adsorption of molecules with weak guest–guest interactions, where a single layer of adsorbed species is formed [25]. Based on this model, Bhatia and Myers [25a] have illustrated that from simple considerations, it is possible to estimate the optimal adsorption heat value required to maintain a molecule as adsorbed species between two given pressure values, P_1 and P_2 ($P_1 > P_2$), under isothermal conditions. Then, the amount of molecules desorbed (released), D, when the pressure changes from P_1 to P_2, is given by

$$D(K, P_1 P_2) = \frac{KP_1 n_p}{(1 + KP_1)} - \frac{KP_2 n_p}{(1 + KP_2)}. \tag{11.17}$$

For fixed values of P_1 and P_2, and taking the derivative of D with respect to K, the maximum value of D is obtained for K given by

$$K = \frac{1}{\sqrt{P_1 P_2}}. \tag{11.18}$$

The value of K depends on the entropy change, ΔS^o, relative to the standard pressure, P_o (1 bar), on the enthalpy change in the adsorption process, ΔH^o, and on the temperature (T), according to

$$K = \frac{\left(e^{\Delta S^o/R}\right)\left(e^{-\Delta H^o/RT}\right)}{P_o}, \tag{11.19}$$

where R is the ideal gas constant, ΔH^o value represents the average or integrated heat adsorption between P_1 and P_2 and its absolute value is equal to the isosteric heat. From these considerations, the integral heat of adsorption for the condition of maximum desorption is given by

$$\Delta H^o_{opt} = T\Delta S^o + \frac{RT}{2} \ln\left(\frac{P_1 P_2}{P_o^2}\right). \tag{11.20}$$

The values of ΔH^o are usually obtained through the isosteric method, recording two isotherms at close temperatures (T_1 and T_2), for example, 77 and 87 K, and using the Clausius–Clapeyron equation [7,13,26]. Another method to get the value of ΔH^o, and at the same time the entropic contribution, ΔS^o, is taking the linear form of Eq. (11.20) and combining it with Eq. (11.2) to obtain:

$$\ln\left(\frac{n_{ad}}{(n_p - n_{ad})P}\right) = \frac{\Delta S^o}{R} - \frac{\Delta H^o}{RT}. \tag{11.21}$$

Instead of the amount adsorbed "n_{ad}" at the pressure P, one can take any other magnitude proportional to "n_{ad}", such as the integrated intensity "A" for the ν (H–H) vibration, in the case of H_2 molecule [27]. From the measured "A" values, the intercept and the slope of the linear fitting of $\ln(A/(A_p - A)P)$ versus $1/T$ give us directly the value of $\Delta S^o/R$ and $\Delta H^o/R$, respectively. In principle, this method is applicable for any adsorbed species at low adsorption coverage in order to have experimental conditions where the Langmuir model remains valid.

For practical applications, when the value for ΔH^o and ΔS^o can be estimated or obtained from other techniques, it seems convenient to use an inverse reasoning: use these values to estimate the corresponding optimal temperature (T_{opt}) for the adsorption and desorption processes, according to

$$T_{opt} = \frac{\Delta H^o}{\left[\Delta S^o + (R/2)\ln(P_1 P_2/P_o^2)\right]}. \tag{11.22}$$

Carbon dioxide, for instance, must be captured (adsorbed) from gaseous streams to be desorbed and accumulated in pressed vessels for the transport and storage in secure sites to avoid its release into the atmosphere. If ΔH^o is known, for operation pressure values (P_1 and P_2), Eq. (11.22) provides an estimated value for the optimal operation temperature [25a].

11.5
Carbon Dioxide Capture

The main processes from which CO_2 must be captured are (a) postcombustion, which corresponds to the CO_2 contained in flue gas from energy generation plants using coal, natural gas, or oil as energy sources; (b) precombustion, which concerns with removal of CO_2 from natural gas and landfill gas where the main components are CH_4, CO_2, and N_2, for example, gasification of coal for hydrogen production, fertilizer production plants, and so on; (c) oxy-fuel combustion, where the oxidizer is oxygen previously separated from nitrogen in order to have a higher combustion efficiency, resulting a flue gas composed of practically pure CO_2. This last process usually involves a recycle of flue to optimize the oxygen consumption.

The flue gas composition of fossil fuel combustion is very varied, in v/v: CO_2: 2–15%; N_2 up 78%, if air is used as oxidizer; water vapor, up to 7%; and minor fractions of O_2, CO, SO_2, SO_3, NO_x, and volatile hydrocarbons [28]. This explains that CO_2 capture from flue gas necessarily must be accompanied by a

11.5.1
Carbon Dioxide Capture by Chemisorption

Only a few CO_2 capture processes are available at technological level, among them, adsorption by aqueous amine solutions (20–40 wt%), such as monoethanol amine (MEA) where the amine participates in a nucleophilic attack of the acidic C atom of CO_2 to form a C—N bond, resulting in the formation of carbamate [29]. This reaction involves energy in the 50–100 kJ/mol range, and it can be classified as a chemisorption process [30]. MEA is a primary amine, which forms stable complexes with CO_2 enabling its facile capture but limiting the regeneration process. Tertiary amines, for example, triethanolamine, shows low ability for CO_2 capture and its reactivation involves relatively low energy consumption. Secondary amines, for example, diethanolamine, are intermediate in behavior. This explains that these three amines are commonly used together or through the formation of composites [31]. This is a mature technology, as a result of about 60 years of research and development. Its main drawbacks are as follows: (1) the amine solution usually has low thermal stability on heating, limiting the available regeneration temperature; (2) its corrosive behavior and the need of adding anticorrosive inhibitors or reduction of the amine concentration below 40 wt%; (3) large energy consumption for the absorbent regeneration, depending on the amine concentration, up to 40 wt%; (4) and negative environmental impact of amine loss [32]. For this type of capture, other compounds containing amine groups have been considered, including 2-amino-2-methyl-1-propanol, N-methyldiethanolamine, and piperazine [28b,33]. The use of ammonia solutions has also been explored, because it has the advantage that other acidic gases such a NO_x and SO_x can also be absorbed in a one-step process [34].

Other CO_2 capture systems, many of them evaluated at pilot-plant levels, are based on chemical reactions. This group includes metal oxides of high reactivity for CO_2, for example, CaO and MgO; alkali and alkali earth hydroxides, such as KOH and $Ca(OH)_2$; and alkali metal based ceramics, for example, Li_2ZrO_3 and Na_2ZrO_3, among others [35]. The involved chemical reaction of CO_2 with these materials is highly exothermic, as it takes place at above 650 °C for CaO, for instance, and as a result, the endothermic regeneration process consumes a large amount of energy; the $CaCO_3$ decomposition reaction is observed at about 700 °C [36].

11.5.2
Carbon Dioxide Capture by Adsorption in Porous Solids

This section concerns the use of solid materials for CO_2 adsorption. Several families of materials with high specific surface are available and their interactions with the CO_2 molecule will be discussed here. These materials include (a) zeolites, aluminas, silicas, and related porous solids; (b) carbon-based materials, for example, activated carbons, activated carbon nanofibers, and graphene nanostructures; (c) coordination polymers; (d) metal–organic frameworks. As already mentioned, the carbon dioxide molecule is able to participate in three types of adsorptive interactions: (1) dispersive forces (van der Waals type), (2) electrostatic forces through polarization of its electron cloud and involving the molecule quadrupole moment, and (3) coordination through its O-end with partially naked transition metals, which is a relatively weak interaction and is reversible without large energy consumption. These three adsorption mechanisms have the advantage of their high reversibility. Furthermore, since the CO_2 capture involves its selective adsorption from a gaseous stream at a pressure around 1 bar and with a relatively low partial pressure of CO_2, the adsorption potential and selectivity for this molecule at low pressure must be optimized. The gaseous stream usually contains accompanying species that compete for the available adsorption sites with CO_2, and this imposes limitations for the available materials. An acid–base reaction through its acidic carbon atom is possible in porous solids functionalized with amine groups.

11.5.3
Zeolites, Aluminas, and Silicas

Zeolites are porous aluminosilicate materials, of natural or synthetic origin, with a robust crystalline structure, high thermal stability, and a network of interconnected pores and channels with size in the 0.5–1.2 nm range. The Si/Al ratio determinates the framework negative charge and the amount of charge-balancing cations located inside the pores and cavities. The charge-balancing cations commonly are alkaline or alkaline earth metals. They can polarize the electron cloud of the CO_2 molecule to allow its adsorption. Because of their textural properties, zeolites 13X, 5A, and chabazite are among the most studied porous solids for CO_2 capture. By far, zeolite 13X is the most studied adsorbent for CO_2 capture, due to its high selectivity for this molecule. The adsorption capacity of zeolites is in the 3–25 wt% range at room temperature for pure CO_2 and up to 12 wt% at a partial CO_2 pressure of 15% [37]. The largest adsorption capacities have been reported for Li^+ as exchangeable metal, in correspondence with a strong interaction with the molecule quadrupole moment [38]. LTA-type zeolites have been studied for natural gas upgrading, tuning the selectivity for CO_2 through the Si/Al ratio; for a ratio of 1, a selectivity of 1271 toward CO_2 over CH_4 at low coverage has been reported [39]. The removal of CO_2 from natural gas is relevant for its transport; it is responsible for corrosion of pipelines in the presence of moisture in the gas streams. Zeolites show a fast adsorption at relatively low partial pressure of CO_2, which is highly attractive for their selectivity toward this molecule. Unfortunately, the exchangeable metals responsible for such fast adsorption and selectivity also favor the uptake of water molecules, which compete with carbon dioxide for the adsorption sites. The corresponding adsorption heat is found in the 30–42 kJ/mol range, where the exact value depends on the cation found inside the pores and channels [39].

Porous silicas and aluminas are mesoporous materials free of charge centers, where the adsorption forces for the carbon dioxide molecule are of dispersive nature. Because their pores and channels are relatively large compared to the molecule size, the contribution of these interactions to the adsorption forces is insufficient to support a high CO_2 adsorption capacity in these materials. Porous silicas as functionalized materials have shown an interesting performance for CO_2 capture (discussed next).

11.5.4
Carbon-based Materials

In the absence of functionalization, the adsorption forces for CO_2 in activated carbon are due dispersive interactions, and from this fact, the involved adsorption heats are significantly lower (about 20 kJ/mol) than the values found for zeolites (close to 40 kJ/mol for zeolite 13X) [40]. This suggests that, compared to zeolites, activated carbons have a smaller selectivity for CO_2 adsorption, which negatively impacts its capture from complex mixtures of gases. Nevertheless, with respect to zeolites, activated carbon has important advantages for CO_2 capture: (1) its hydrophobic character is favorable for the CO_2 adsorption from a wet gas stream, where zeolites deactivate; (2) activated carbons can be prepared with surface areas close to 3600 m^2/g, well above the typical values for zeolites (SA_{BET} = 726 m^2/g for zeolite 13 X); (3) such a large surface area favors higher adsorption capacities at high pressure; (4) relatively low adsorption heat values that result in cost reduction for their regeneration; (5) a superior thermal conductivity that is a convenient feature for the desorption process through moderate heating; (6) a good chemical resistance for aggressive species contained in flue gas, for example, SO_x, NO_x, and CO; (7) related to their chemical composition and large porosity, the density of activated carbons is significantly smaller than that of zeolites; and (8) their porosity is easily controlled through the conditions used in the activation process. These features of activated carbons are favorable for their applications in carbon dioxide capture from both precombustion and postcombustion processes. Activated carbons have the additional advantage that they are prepared using well-known physical and chemical routes, starting with readily available low-cost raw materials, for example, carbonaceous minerals, vegetal wastes, carbon-containing resins, and plastics. Activated carbon can be prepared with a fiber morphology, which is advantageous for handling, compared to granular or powdered adsorbents [41].

11.5.5
Prussian Blue and Related Materials

Prussian blue analogs and transition metal nitroprussides will be considered to shed light on the role of the different factors that contribute to the adsorption potential for the carbon dioxide molecule in these types of porous materials.

Prussian blue and its analogs are formed by the assembling of octahedral hexacyanometallate anions, $[T_i^{n+}(CN)_6]^{6-n}$, through a second transition metal (T_e) linked to the N end of the CN bridge. The solid framework appears as a 3D array of $\cdots N\equiv C\text{-}T_i\text{-}C\equiv N\text{-}T_e\text{-}N\equiv C \cdots$ chains. The inner metal (T_i) is a transition metal with a maximum of six nd electrons, the number of electrons that can be accommodated in the t_{2g} orbitals for the metal with an octahedral coordination. This condition is satisfied for Ti(III), V(III), Cr(III), Mn(III), Fe(II), Fe(III), Ru(II), Os(II), Co(III), Rh(III), Ir(III), Pd(IV), and Pt(IV), for instance. The T_i metal is always found with a low spin electronic configuration. The metal linked at the N adopts a high spin state, except for Co(III), and it is found with a wide variety of valence states, usually from +1 to +4. Rigorously, the term Prussian blue analogs must be used for those compositions where both, T_i and T_e metals, have octahedral coordination, and the solids crystallize with a cubic structure. The most notorious deviation of that regularity is found for Zn, which under certain conditions, adopts a tetrahedral coordination to the N ends of the CN bridges, forming solids of hexagonal structure. The coordination chemistry of transition metal hexacyanometallates has been the subject of textbooks and review articles [42].

The charge balance (stoichiometry) in the formed coordination complex salt determines the porosity of Prussian blue analogs. This will be illustrated with some examples. For the combination of T_i(III) and T_e(II), solids of formula unit $(T_e)_3[T_i(CN)_6]_2 \cdot xH_2O$ are formed. In a cubic array of $\cdots N\equiv C\text{-}T_i\text{-}C\equiv N\text{-}T_e\text{-}N\equiv C \cdots$ chains, the structure of this composition has one-third of vacancies for the $[T_i(CN)_6]^{3-}$ building block. This implies that, on average, one-third of the coordination for the metal T_e remains free or is occupied by water molecules in the as-synthesized solids. If these vacancies are randomly distributed, the anhydrous material has one-third (33%) of its volume unoccupied, but if the vacancies are ordered, the free volume is 50% [43]. For T_i(II) and T_e(III), the material formula unit is $(T_e)_4[T_i(CN)_6]_3 \cdot xH_2O$. This is the case of $(Fe)_4[Fe(CN)_6]_3 \cdot xH_2O$, known as insoluble Prussian blue, which in its anhydrous state and with a random vacancy distribution has 25% of its volume free. Many other combinations of T_i and T_e metals are possible to form porous solids [44]. For T_i(IV) and T_e(II), the resulting solids are free of vacancies. In the framework of Prussian blue analogs, the cavities generated by the presence of vacancies remain in communication by the interstitial free spaces of about 4 Å diameter. For T_i(III) with T_e = Zn(II), the anhydrous solids crystallize with a hexagonal structure related to a tetrahedral coordination for the Zn atom. The typical framework of the Prussian blue analog, $(T_e)_3[T_i(CN)_6]_2$, and of the related zeolite-like hexagonal solid, $Zn_3A_2[T_i(CN)_6]_2$, with T_i(II) and Zn (T_e), are also porous solids with the hexagonal structure, with the following formula unit, $Zn_3A_2[T_i(CN)_6]_2 \cdot xH_2O$ where A is a charge-balancing cation, usually an alkali metal. The water molecules form the solvation sphere of these exchangeable metals. The accessible free volume of this phase is formed by ellipsoidal shape cavities of about $15.5 \times 11.1 \times 7.9$ Å. Six elliptical windows of about 6.8×8 Å communicate neighbor cavities [45].

For the isostructural series, $Zn_3[Co(CN)_6]_2$ (hexagonal phase) and $Zn_3A_2[Fe(CN)_6]_2$ with A = Na, K, Rb, Cs, the strength of adsorption forces for CO_2 follows

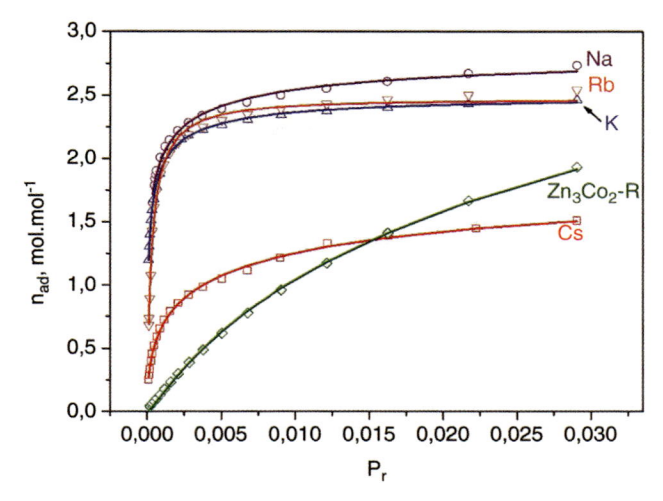

Figure 11.2 CO_2 adsorption isotherms for $Zn_3[Co(CN)_6]_2$ (hexagonal phase) and $Zn_3A_2[Fe(CN)_6]_2$ with A = Na, K, Rb, Cs [46].

the order Na > K > Rb > Cs > $Zn_3[Co(CN)_6]_2$ (Figure 11.2) [46]. An analogous result is obtained when Fe is replaced by Ru and Os [26b]. The obtained values for the characteristic energies (E_o) in the Dubinin–Astakhov isotherm (Eq. (11.9)) in kilojoule per mole, as a sensor for the strength of the guest–host interaction in that series are Na(23.1), K (22), Rb (19.5) > 16.81 (Cs) and 8.2 for $Zn_3[Co(CN)_6]_2$. The electrostatic interactions dominate the adsorption forces for carbon dioxide molecules in these materials. For $Zn_3[Co(CN)_6]_2$, all the interactions are of dispersive type because that composition is free of exchangeable metal ions and both Fe and Zn have saturated their coordination environment. The presence of exchangeable metals within their cavity has a minor effect on the number of CO_2 molecules accommodated in the cavity volume: Na (6.74) > K (6.1) > Rb (5.9) > Cs (5.6) < 8.2 in $Zn_3[Co(CN)_6]_2$. These values are equivalent to limit storage capacity (in excess) of 22.3, 19.2, 15.5, 14.7, and 28 wt %, respectively. In the $Zn_3A_2[Fe(CN)_6]_2$ series, the exchangeable metal is found close to the N atom of the CN group, the most negative site in the host framework. The CO_2 molecules are accommodated in the center of the cavity, maximizing the electrostatic interactions with neighboring alkali cations. In $Zn_3[Co(CN)_6]_2$ (hexagonal phase), all the cavity volume is available to accommodate CO_2 molecules and a limit adsorption capacity (estimated from isotherms recorded at 0 °C and up to 1 bar) of 7.9 molecules per cavity was observed [46].

In Prussian blue analog series, for example, $(T_e)_3[Co(CN)_6]_2$, with T_e = Mn, Co, Ni, Cu, Zn, and Cd, as anhydrous phase, the T_e atom is always found at the cavity surface with a partially naked coordination sphere. The values for the characteristic energy (E_o), in kilojoule per mole, follow the order Ni(23.7) > Cu (12.35) > Co (10.94) > Mn (10.64) > Cd (10.22) > Zn (9.35). This is the order of the metal polarizing power except for Zn, suggesting that the adsorption forces for CO_2 are dominated by the electrostatic interaction. The deviation for Zn from the expected value is ascribed to the pseudotetrahedral coordination of the Zn atom in the anhydrous phase of Prussian blue analogs, even within a cubic unit cell, limiting the metal interaction with the guest CO_2 molecule [43c]. The limit adsorption in excess in that series, calculated from adsorption isotherms recorded at 0 °C and up to 1 bar, and using the Dubinin–Astakhov model, is close to 4 CO_2/cavity, below the number of metal sites available to interact with the guest molecules (six sites per cavity) [43c]. It seems, steric repulsions of adsorbed molecules within the cavity prevent more dense packing in the available free volume.

Transition metal nitroprussides represent another typical series of porous coordination polymers. For nitroprussides with narrow cavities, Mn, Zn, and Cd, the value for the characteristic energy (E_o), in kilojoule per mole, is 16.4, 12.8 and 15.1, respectively, quite different from those of the cubic phase with cavities of about 9 Å diameter (Fe, Co, Ni) [47]. For these last compositions, the value of E_o is significantly lower, Fe (8.6), Co (8.2) and Ni (9.9) in kilojoule per mole. A greater confinement for the carbon dioxide molecule favors a stronger guest-host interaction, and for relatively large available cavity volume, the dominant factor in the adsorption forces is the metal ability to stabilize the molecule as adsorbed species through electrostatic interactions. This explains that within the cubic phase the higher value of E_o corresponds to Ni and those solid solutions containing this metal [47].

11.5.6
Metal–Organic Frameworks

MOFs Are hybrid materials resulting from molecular blocks, of organic nature, linked through transition metal ions to form 3D porous solids. The large variety of available organic ligands (blocks) results in a wide diversity of MOF-type solids. This also facilitates their rational design with tailored pore shape, size and volume, and a flexible framework [48]. Above 20,000 MOF type materials have been reported up to date, with BET surface in the range of 1,000 to 10,000 m^2/g [49]. Figure 11.3 shows three typical pore topologies of MOFs. In such a universe

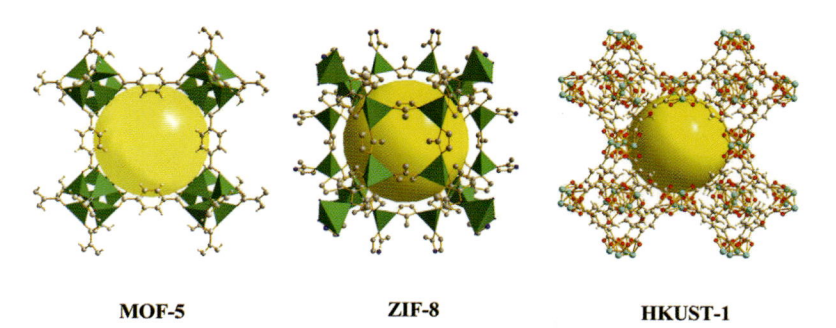

MOF-5   **ZIF-8**   **HKUST-1**

Figure 11.3 Porous framework of three typical MOFs. These images were prepared from the CIF files available at Cambridge Crystallographic Data Centre for these materials (CCDC 256965, CCDC 973356, and CCDC 943008, respectively).

of solids with a porous framework, the rationalization of knowledge is only possible by summarizing the reported results according to the involved adsorption interaction. The potential of MOFs for gas adsorption has been known for over two decades [48a]. Since then, the study of these materials has mainly concentrated on crystal engineering and development of new synthesis methods [50]. MOFs have received special attention for gas storage, catalysis, and as ion exchangers due to their versatile open framework [50a,51], applications in which zeolites have been the traditionaly used materials. The synthesis under design and wide structural diversity are probably the main advantages of MOFs respect to zeolites, while their lower thermal stability appears as a disadvantage [51b]. MOFs are composed a light elements, usually H, C, N and O, with a minor fraction of metal ions as linkers, which is an attractive feature for gases storage applications because, in principle, high gravimetric storage density could be obtained.

In MOFs, carbon dioxide adsorption is possible through three types of interactions: (1) dispersive forces; (2) electrostatic interactions; (3) coordination to a partially naked metal centers, through CO_2 lone-pair orbitals. In MOFs free of polarizing charge centers and without partially naked transition metals, for example, ZIFs series without polar substituents and MOF-5 and its isoreticular analogs, the adsorption forces for carbon dioxide molecule are of dispersive nature. The adsorption isotherms for these materials have sigmoidal shape, related to a weak adsorption interaction at low pressures, and an increasing storage capacity at high pressure [52]. For ZIF-8 (zinc 2-methy-imidazolate), the isosteric adsorption heat is 15.93 kJ/mol [53], a value typical of dispersive interactions for CO_2 in a nonconfined space. For activated carbons with only dispersive interactions contributing to the adsorption forces, the isosteric heat of adsorption is close to 20 kJ/mol [40], because the guest molecule remains confined to a small volume. The porous frameworks for the ZIFs series have relatively large pores, with diameters above 7 Å [54], higher than the kinetic diameter of CO_2, 0.33 nm, and from this fact, within the porous framework it is found interacting with only the cavity wall.

The presence of polar groups as substituents in MOFs favors the occurrence of electrostatic interactions with the carbon dioxide molecule enabling its adsorption. Within the ZIFs series, the higher CO_2 adsorption capacities are reported for compositions with polar substituents, and the uptake values at 298 K and 1 bar follow the order $-NO_2$ (ZIF-78) > -CN, -Br > -Cl (ZIF-82, -81, -69) > C_6H_6, -Me (ZIF-68, -79) > -H (ZIF-70) [54]. This is the order expected for electrostatic interactions (dipole–quadrupole) between the substituent groups and CO_2.

Metal centers located at the cavity surface of MOFs behave as charge centers able to polarize the electron cloud of the guest CO_2 molecules within the cavities and to interact with its quadrupole moment. This explains that in the design and preparation of MOFs for CO_2 separation and capture, special attention is given to frameworks containing metals with an unsaturated coordination environment. In such porous frameworks, the adsorption forces are notably enhanced making possible the adsorption at low pressures and the filling of cavities with CO_2 molecules at high pressures [55]. The enhancement of the adsorption forces at low pressure is relevant for carbon dioxide uptake from a complex mixture of gases, as is the case of flue gas from postcombustion. In materials containing metal centers with an unsaturated coordination environment, instead the sigmoid type isotherms observed in the absence of such charge centers, the recorded isotherms are of type I, according to the IUPAC classification [18,28b]. The reported adsorption heats for CO_2 in MOFs with such features are found in the 26 to 80 kJ/mol range [55]. This large range of adsorption heats is related to the nature of the framework ligands. Molecules with high donor character reduce the metal polarizing power and its ability to stabilize the carbon dioxide molecule as adsorbed species.

When the metal at the cavity surface has empty orbitals of relatively low energy to accept electrons from the carbon dioxide molecule, through the p orbitals of its O atom, a coordination could be formed (Figure 11.1b). This is the limiting case for electrostatic interactions. The formation of a coordination bond between carbon dioxide and metals with an unsaturated coordination sphere has effectively been confirmed for Ni_2(dhtp) using X-ray diffraction and IR spectroscopy data [50]. On the charge donation to the metal center, the O-C-O conformation deviates from its typical linear structure. This modifies the molecule's vibrational pattern, which is easily appreciated in the IR spectra for its coordinated state. IR spectroscopy is the simplest technique to follow the coordination interaction for CO_2 during its adsorption in porous solids with partially naked transition metals [56]. The end-on coordination of CO_2 to a metal with an unsaturated coordination environment within a relatively large cavity, has certain rotational freedom, and such behavior can be followed through NMR spectra of adsorbed $^{13}CO_2$ [57].

Metal-organic frameworks form a large family of porous solids, with a widespread diversity of structural features, BET surface areas in the 1,000 to 10,000 m^2/g range, and with tunable adsorption potential for a given adsorbate. Notwithstanding such properties of MOFs, their practical applications for CO_2 adsorption are limited. Except for ZIF frameworks without hydrolysable substituents, practically all the remaining MOFs compositions are highly susceptible to hydrolysis and this is a limitation to operate in the presence of moisture. The relatively low mechanical, thermal and chemical stability of many MOFs are also drawbacks to overcome in the finding of new MOF type materials for carbon dioxide separation and storage.

11.5.7
Functionalized Porous Solids

The obvious functional groups to incorporate in high surface area solids for applications in CO_2 capture are amine-containing molecules. Two routes are commonly used to functionalize porous solids with amines: a) impregnation (the simplest way) with small amine molecules, for example, MEA; 2) incorporation of ligands containing amine groups during the material preparation. The first option reduces the material pore volume because the available space is, at least, partially occupied by the impregnated amine molecule. For zeolite 13X impregnated with MEA, the adsorption capacity at 120 °C increases from 4 to 14 mL/g and the reacted solid shows presence of carbamate groups [58]. The immobilization has been widely considered in silica-based materials, for example, MCM-41 and SBA-15, incorporating the amines into the reactants to obtain a solid framework containing accessible amine functionalities [59]. The amine-containing porous silicas show a high affinity for CO_2.

The impregnation route of MOFs with amines results in a dramatic increase of their adsorption capacity. For instance, the treatment of MIL-101 with ethanol and then with NH_4F results in an increase for CO_2 storage from 79.2 to 176 wt% at 304 K and 50 bar [60]. As already mentioned, MOFs free of charge centers show a high CO_2 storage capacity at high pressures but not at low pressures. The incorporation of amine and other nitrogen–containing functionalities, with basic properties, in their structure allow the carbon dioxide adsorption also at low pressure. For instance, the presence of organic linkers containing oxazole and tetrazole groups favors a significant increase in the material adsorption potential for CO_2, through the interaction of the negatively charged framework region with the guest molecule [61]. An analogous behavior results from the incorporation of ligands with aromatic groups in the framework [55]. The availability of N-containing molecules with basic character to form a part of the MOF's framework with high affinity for the carbon dioxide capture, is practically infinite. From this point of view, the materials availability with such potentiality is not a problem; the key issue is their performance in practical applications. As already mentioned, MOFs are usually unstable in the presence of moisture, and up to now their large scale production involves large costs due to the used chemicals and the preparation conditions.

11.5.8
Porous Organic Polymers

As mentioned above, MOFs are porous solids with extremely large surface area and pore volume and tunable functionalities, features that are highly attractive for carbon dioxide separation and capture but, unfortunately these materials usually are unstable in high-temperature, moisture and other severe conditions and from this fact, their potential for industrial applications is low. An alternative is the design of more robust materials in terms of stability in the presence of water vapor, a higher thermal stability and mechanical resistance, appropriate chemical stability, and at the same time incorporating a high affinity and selectivity for CO_2, with large surface area and accessible pore volume. An option under study is the class of organic porous polymers incorporating amine functionalities, including aromatic amines to tune the required regeneration temperature. Aromatic amines are less basic than aliphatic ones, and this makes it possible to prepare materials with tunable selectivity for carbon dioxide and using relatively low temperature to liberate the chemically linked CO_2 molecule. Several types of porous organic polymers are now under study, among them those that involve the formation of strong covalent bonds [62], for example, covalent organic frameworks (COFs) [63], conjugated microporous polymers (CMPs) [64], covalent triazine-based frameworks (CTFs) [65], and porous aromatic frameworks (PAFs) [66]. Porous organic polymers can be prepared with extremely large surface area, above 4000 m^2/g for COFs, for example [67]. Among the figures of merit for design and preparation of this type of materials, are the involved adsorption heats and their selectivity for CO_2 adsorption in mixtures of CO_2/N_2 and CO_2/CH_4 to allow the applications in postcombustion carbon dioxide capture and in methane purification (CO_2 removal from natural gas streams). The adsorption heats in porous organic polymers can be tuned to values around 30 kJ/mol, to maintain the carbon dioxide molecule as adsorbed species slightly above room temperature and at the same avoiding the use of high temperatures in the regeneration process. In addition to carbon dioxide affinity for amine groups, it has a relatively large quadrupole moment, -14.25×10^{-40} cm^2 [22], which favors its adsorption through electrostatic interactions. The design of organic porous polymers for CO_2 capture usually incorporates the presence of polar groups to enable framework interactions with guest molecules by electrostatic adsorption forces.

Some porous organic polymers suffer from the same limitations mentioned for MOFs. For instance, the perspective of COFs' applications is limited by poor stability of the boronate or boroxine linkages against hydrolysis [68]. At present, the preparation of this type of material appears complex, involving nonusual monomers, expensive catalysts, and complicated synthesis conditions. In the near future, the study of this type of porous solids is probably limited to an academic context because of current limitations to scale up to a level of industrial interest.

11.6
Carbon Dioxide Capture by Clathrates

Clathrates are inclusion compounds where the guest species remain trapped within a host framework, without chemical interactions among them. Under this concept, the above-discussed guest–host systems resulting from the physical adsorption of gases and vapors in porous solids could be considered as clathrates. In a context of materials for gas storage, those clathrates formed by a crystalline array of water molecules with the gas molecules as guest in the framework free spaces are receiving increasing attention [28a,69]. Such crystalline arrays of water molecules result from hydrogen bonding interactions between them. In the absence of guest molecules, the structure of the host framework collapses. The guest molecule must be gaseous or liquid species of low molecular weight, small size, and nonpolar character, for example, O_2, N_2, H_2, CO_2, CH_4, H_2S, Ar,

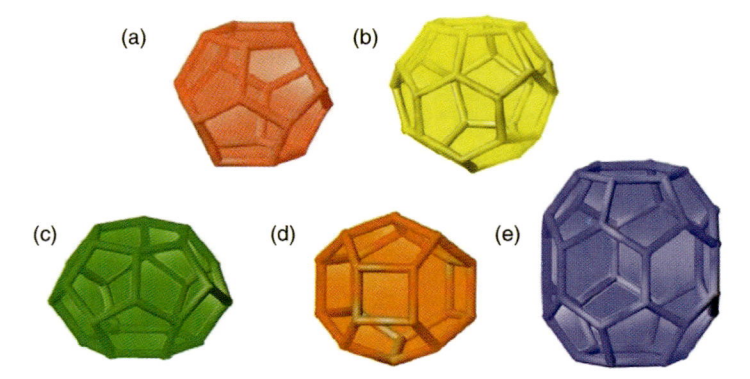

Figure 11.4 Illustration for the types of cages reported for *sI*, *sII*, and *sH* of hydrates. (a) Pentagonal dodecahedron (5^{12}), (b) tetrakaidecahedron ($5^{12}6^2$), (c) hexakaidecahedron ($5^{12}6^4$), (d) irregular dodecahedron ($4^35^66^3$), and (e) icosahedron ($5^{12}6^8$) [70a].

Kr, and Xe. The formation of clathrates with light hydrocarbons and polar molecules containing highly hydrophobic fragments is also possible. The cages can also be formed by ammonia, methane, heavy noble gases, and diverse molecular species. For host framework of noble gases and methane, the clathrate formation requires high pressures and low temperatures, because the stabilizing forces are of dispersive nature [28a,70]. Clathrates are usually formed under high pressure. For this reason, in the following the pressure will be expressed in MPa or GPa units (1 MPa = 10 bar).

Clathrates (hydrates) are crystalline solids with certain structural regularities, independent of the guest molecules. This is related to the different topologies that result from the tetrahedral coordination and hydrogen bonding network formed between them ("ice rules") [71]. Such structures may be represented as a set of close-packed polyhedra (Figure 11.4) built from the water molecules, with neighboring polyhedra sharing common faces, to form three types of structures, namely, *sI*, *sII*, and *sH*. The first one crystallizes with a cubic unit cell in the $Pm\overline{3}n$ space group, with two types of cavities: a small pentagonal dodecahedral one denoted 5^{12}, which comprises 12 pentagons as faces, and a large tetrakaidecahedral cavity, denoted $5^{12}6^2$, which comprises 12 pentagons and 2 hexagons in its faces. The unit cell of this phase contains two small and six large cages, and its stoichiometric formula is $G_8 H_{46}$ where G is the guest and H is the host. The *sII* phase crystallizes in an fcc unit cell ($Fd\overline{3}m$ space group): each unit cell contains 16 small 5^{12} cages and 8 hexakaidecahedral cages, of larger size, denoted $5^{12}6^4$ delimited by 12 pentagons and 4 hexagons. The stoichiometry of this phase is $G_{24} H_{136}$ [69].

The third phase, *sH*, crystallizes with a hexagonal symmetry (*P6/mmm* space group), with three types of cages: 5^{12} (smaller cavity), a mid-sized $4^35^66^3$ one delimited by 3 square, 6 pentagonal, and 3 hexagonal faces, and a large icosahedral cage, denoted $5^{12}6^8$, with 12 pentagonal and 8 hexagonal faces [72]. The formed clathrate does not always have a definite stoichiometry, because the cages could be partially occupied. For this reason, it is convenient to use a parameter known as the hydration degree that indicates the ratio of water molecules to guest molecules in the formed structure. For the structure sI, with all the cages occupied by guest molecule, the hydration degree is 5.75, and an alternative formula unit is Gas·5.75H_2O [28a].

Hydrates are promising materials for CO_2 separation and capture from precombustion fuel gas postcombustion flue and natural gas streams. The potential of hydrates for such applications is determined by their ability to stabilize structures with high affinity for CO_2 at economic and technological viable values of pressure and temperature. CO_2, without accompanying species, forms nonstoichiometric sI hydrates; the cages remain partially occupied by the guest molecule, with storage capacity in the 3.1–4.3 wt%, corresponding to hydration degree of 5.7 and 7.7, respectively [28a]. Its formation requires a relatively high pressures: up to 0.7 GPa in the 260–280 K range [73]. By using promoting molecules, the hydrate stability can be significantly increased. Such promoting molecules include THF and ammonium salts, for example, TBAF, TBACl, TBABr, and $TBANO_3$. The formed semiclathrates are stable under ambient conditions. The cost of this stability improvement is a marked reduction in storage capacity.

The most attractive and intensively studied promotor of hydrate formation and stability is THF, because it favors a decrease in the formation pressure at relatively high temperatures and, at the same time, this molecule has relatively small size leaving enough space to accommodate the accompanying guest species. For an aqueous solution containing 3 mol% of THF, the pressure of formation is 1.5 and 4.5 MPa at 11 and 17 °C, respectively. An increase in the THF concentration results in a structural transition in the formed hydrate, from the sI to the sII phase [28a].

For selective retention of CO_2 molecules by hydrate formation in the presence of accompanying gaseous species, of N_2 in flue gas, H_2 in precombustion fuel, and CH_4 in natural gas, the formation conditions and the stability of their hydrates must be quite different. The hydrates corresponding to these molecules are formed under high pressures and at low temperatures. Since the forces that determine the stability of clathrates are of dispersive nature and these molecules, individually or as clusters, fit the available space in the clathrates cages, the selectivity of the clathrates for a given molecule is not too high. For instance, the size of CO_2 and CH_4 is very similar, and the separation capability would be determined by the difference in electron density. For an equimolar mixture of CO_2 and CH_4, in the formed hydrate, their ratio is $CH_4/CO_2 = 0.25$. The same behavior has been reported when a promoter is used [74].

The study of clathrates for gas storage is an incipient and active research area. To date, the main efforts have been oriented toward reducing the formation conditions, pressure, and temperature, to bring them to technologically viable ranges. In the case of carbon dioxide separation and capture, the main efforts will probably be focused on the clathrate selectivity.

11.7
Air Purification by Adsorption of Toxic Chemicals

The removal of toxic chemicals from air is relevant in three areas: environmental protection, prevention of occupational diseases, and protection against chemical weapons. In chemical, petrochemical, food processing, printing, and semiconductors industries, among others, emissions of harmful gases and vapors are common, which must be removed for obvious reasons of personnel protection and odor control. These harmful chemicals include volatile organic compounds, and volatile species of inorganic nature such as NO_x, SO_2, CO, HCN, Cl_2, NH_3, CNCl, and H_2S. For a simple application, like the cigarette filters, the use of an appropriate filter material becomes a health-care subject. In the same sense, the removal of VOCs and other harmful chemicals from indoors is another area of attention for air purification as a result of increasing environmental pollution [75]. Regarding protection against chemical weapons, even though the Chemical Weapons Convention of 1993 demilitarized the use of toxic chemicals, research and development activities in these areas continue [76]. Some of these toxic chemicals have been used after 1993 in terrorist attacks, for example, sarin nerve gas liberated by the Aum Shinrikyo cult in the Tokyo subway in 1995 was responsible for the killing of several civilians. In the following, from the available public information, which is limited, brief mention to the most important known toxic chemicals used in chemical weapons is included, with special attention to those materials employed for the protection of people.

The most studied and applied materials for the above-mentioned three application areas are activated carbons. In these applications, the material remains in contact with water vapor from the environment and activated carbons have hydrophobic character, which reduces the possibility of deactivation by water adsorption. By their nature, activated carbons have high affinity for the adsorption of gases and vapors of organic nature. Such affinity depends on the molecular weight of the toxic compound and on its boiling point. A high molecular weight (M_w) is equivalent to a larger polarizability value and to stronger adsorption forces through dispersive interactions with the host solids. The higher the boiling point (BP), the greater is the chance for molecules to condense within the porous matrix. The reported activated carbons adsorption affinity (AAf) for VOCs follows the order of both molecular weight and boiling point, for example, in the order of M_w, BP in °C, and AAf in %: nitrobenzene (121; 211; 51); tetrachloroethane (166; 47; 40); tetrachloroethylene (165; 121; 35); styrene (104; 145; 25); xylene (106; 138; 21); toluene (92; 11; 20); benzene (78; 80; 12); hexane (86; 68; 7); methylene chloride (84; 40; 2); acetone (58; 56; 0.8); and vinyl chloride (62; −14; 0.7) [77].

Regarding the adsorption of chemical warfare agents, activated carbon appears to be the most effective and most used material. It is effective in absorbing both nerve agents and mustard vesicants, for example, nerve agent VX (*O*-ethyl *Z*-[2-diisopropylamino)ethyl] methylphosphonothioate), nerve agent sarin (isopropyl methylphosphonofluoridate), and mustard agent (Bis(2-chloroethyl)sulfide) [76].

Activated carbon-based masks for air purification, including those of military use, could include a wide variety of impregnants that react with VOCs, at low partial pressure, and with toxic chemicals of inorganic nature. These impregnants contain metal salts, and acidic and basic functionalities, for example, copper, zinc, silver, molybdenum and triethylenediamine [78]. The presence of acidic species in the adsorbent material favors the trapping of basic pollutants such as ammonia (NH_3), while the basic impregnants allow the adsorption of acidic chemicals. The impregnation with metal salts provides protection against highly toxic species, among them, hydrogen cyanide, cyanogen chloride, and carbon monoxide (blood agents). These last toxic chemicals strongly bind to the iron atom of hemoglobin blocking its ability to transport oxygen causing tissue hypoxia. During the impregnation of activated carbon with both acidic and basic functionalities, a fraction of these species could react reducing the material ability for the removal of toxic compounds. This is, probably, one of the main disadvantage of activated carbon-based composites for the adsorption and removal of a wide variety of pollutants from air.

MOFs and porous organic polymers (POPs) are porous materials that have been evaluated for air purification. For instance, POPs containing metal-catecholate groups show ability to remove ammonia, cyanogen chloride, sulfur dioxide, and octane; for ammonia even under humid conditions [79]. The ammonia capture is relevant because of the use of this chemical in many industrial technologies, such as pharmaceutical, cleaning products, and fertilizers industries. Ammonia is a toxic species frequently associated with accidents in chemical plants. MOFs show a high flexibility to be functionalized with acidic and basic functionalities at separated sites to avoid neutralization reactions, and with accessible partially naked metal centers. All these features appear ideal for their use in air purification. Such possibility appears documented from studies conducted along the past two decades [80]. The accessible metal centers at the surface of cavities have a double role: (1) they are adsorption sites for a wide variety of adsorbates, including blood agents that could form stable complexes with transition metals; (2) organic species adsorbed by polarization of their electron cloud, could then be decomposed by hydrolysis reactions catalyzed by these metal centers in the presence of water, which is always present in air. This second possibility appears well documented from experimental and simulation studies for nerve agents, such as sarin, soman ($C_7H_{16}FO_2P$), and VX, and vesicants, such as sulfur mustard and lewisite ($C_2H_2AsCl_3$) in $Cu_3(BTC)_2$ [80,81]. The presence of water vapor in air, which makes possible the hydrolysis of many toxics of organic nature, appears to be one of the major handicaps of MOFs for air purification, where the material to be used in the filtering devices must be stable under ambient conditions. As discussed above, MOFs based on carboxylate building blocks, for example, MOF-5 and its isostructural analogs, are susceptible to hydrolysis in the presence of moisture. Fortunately, there are MOF series, such as ZIFs, which are stable in aqueous media, but many of them have limitations to incorporate the appropriate functionalities to prepare filtering materials with abilities for trapping a wide diversity of toxic chemicals. In principle, the presence of transition metal centers at the cavity surfaces with unsaturated coordination sphere could be adsorption sites for hydrogen cyanide, cyanogen chloride, and carbon monoxide by their ability to form stable complexes with these metals. The reported studies for the trapping of these species in MOFs are still limited, except for CO [80]. The study of MOFs for air purification is an active research area and important progress could be obtained in the coming years.

11.8
Outlook on Materials for CO_2 Capture and Air Purification

Both the CO_2 capture from gaseous streams and the air purification are processes that must operate under low partial pressure of the adsorbate, and usually in the presence of a complex mixture of gases. By this fact, materials to be used for such applications should be characterized by a high selectivity toward the molecules to be adsorbed. The presence of water vapor in such mixtures of gases represents an additional handicap, particularly for CO_2 capture because the water molecule competes by the adsorption sites for carbon dioxide. An attractive alternative under consideration to overcome such handicap is the preparation of microporous organic polymers (MOPs), containing hydrophobic groups, with large surface area and pore size below 2 nm [82]. Such functionality could also be implemented in MOFs, but their low stability in the presence of water is a limitation. Microporous organic polymer-type porous solids also represent an attractive option for VOC's removal in air purification tasks, so does incorporating functional groups capable of adsorption of inorganic pollutants. Hydrates (clathrates) with CO_2 as guest molecules are the basis of a very active research area on materials for carbon dioxide capture. The formed guest–host system could be used to maintain the CO_2 molecules trapped for their storage in secure sites in the ground or in oceanic geological reservoirs. The main global reserves of methane are found below the ocean floor as hydrates and, in principle, CO_2 hydrates could be stored in such geological conditions. Nevertheless, at least in the short term, composites of activated carbon-containing metal salts and basic substances to form filters with broad functionalities for adsorption of VOCs, acidic oxides, ammonia, and other pollutants will continue to play an important role both in personnel protection in chemical industries and chemical laboratories and in dealing with chemical accidents.

Acknowledgment

The preparation of this overview on functional materials for gases storage was partially supported by the CONACyT project 2013–231461, particularly financing the stay of Dr. L. Reguera as Visiting Researcher at CICATA-IPN (Mexico). The authors thank Dr. Velasco Arias for the figures prepartion.

References

1 (a) Benard, P., Chahine, R., Chandonia, P.A., Cossement, D., Dorval-Douville, G., Lafi, L., Lachance, P., Paggiaro, R., and Poirier, E. (2007) *J. Alloys Compd.*, **446–447**, 380–384; (b) Thomas, K.M. (2007) *Catal. Today*, **120**, 389–398.
2 (a) Blach, T.P. and Gray, E.M. (2007) *J. Alloys Compd.*, **446–447**, 692–697; (b) Webb, C.J. and Gray, E.M. (2014) *Int. J. Hydrogen Energy*, **39**, 366–375.
3 Webb, C.J. and Gray, E.M. (2014) *Int. J. Hydrogen Energy*, **39**, 7158–7164.
4 (a) Broom, D.P. and Moretto, P. (2007) *J. Alloys Compd.*, **446–447**, 687–691; (b) Broom, D.P., Webb, C.J., Hurst, K.E., Parilla, P.A., Gennett, T., Brown, C.M., Zacharia, R., Tylianakis, E., Klontzas, E., Froudakis, G.E., Steriotis, T.A., Trikalitis, P.N., Anton, D.L., Hardy, B., Tamburello, D., Corgnale, C., Hassel, B.A., Cossement, D., Chahine, R., and Hirscher, M. (2016) *Appl. Phys. A*, **122** 1–21.
5 Wang, L., Lachawiec, J.A.J., and Yang, R.T. (2013) *RSC Adv.*, **3**, 23935–23952.
6 Micromeritics (2005), Application Note No. 136: Using the ASAP 2020 for determining the hydrogen adsorption capacity of powders and porous materials.
7 Rouquerol, F., Rouquerol, J., and Sing, K. (1999) *Adsorption by Powders and Solids: Principles, Methodology and Applications*, Academic Press, London.
8 Stoeckli, F. (2001) *Russ. Chem. B*, **50**, 2265–2272.
9 Simonot Grange, M.H. (1987) *J. Chim. Phys.*, **84**, 1161–1174.
10 Dubinin, M.M. and Stoeckli, H.F. (1980) *J. Colloid Interface Sci.*, **75**, 34–42.
11 Stoeckli, H.F. (1981) *Carbon*, **19**, 325–326.
12 Bering, B.P. and Serpinskii, V.V. (1974) *Izv. Akad. Nauk SSSR, Ser. Khim.*, **11**, 2427–2440.
13 Roque-Malherbe, R.M.A. (2007) *Adsorption and Diffusion in Nanoporous Materials*, Taylor & Francis Group, Boca Raton.
14 (a) Yakubov, T.S., Bering, B.P., Dubinin, M.M., and Serpinskii, V.V. (1977) *Izv. Akad. Nauk SSSR, Ser. Khim.*, **2**, 463–466; (b) Zhou, L. (2002) *Adsorption: Theory, Modeling and Analysis* (ed. J. Tóth), Marcel Dekker, Inc., New York.
15 (a) Bhattacharya, S. and Gubbins, K.E. (2006) *Langmuir*, **22**, 7726–7731; (b) Birkett, G. and Do, D.D. (2006) *Langmuir*, **22**, 7622–7630; (c) Chilev, C., Weinberger, B., Passarello, J.P., Darkrim-Lamari, F., and Pentchev, I. (2009) *AIChE J.*, **55**.
16 (a) Gross, K.J., Carrington, K.R., Barcelo, S., Karkamkar, A., Purewal, J., Ma, S., Kevin Ott, D.P., Burrell, T., Semeslberger, T., Pivak, Y., Dam, B., and Chandra, D. (2013) Recommended best practices for the characterization of stor-age properties of hydrogen storage materials, https://www1.eere.energy.gov/hydrogenandfuelcells/pdfs/best_practices_hydrogen_storage.pdf 579; (b) Langmi, H.W., Ren, J., North, B., Mathe, M., and Bessarabov, D. (2014) *Electrochim. Acta*, **128**, 368–392.
17 Lima, E., Balmaseda, J., and Reguera, E. (2007) *Langmuir*, **23**, 5752–5756.
18 Sing, K.S. (1985) *Pure Appl. Chem*, **57**, 603–619.
19 Autié-Castro, G., Autié, M.A., Rodríguez-Castellón, E., Santamaría-González, J., and Reguera, E. (2014) *J. Surf. Interfac. Mater.*, **2**, 220–226.
20 Aman, A. (2004) *Handbook of Chemistry and Physics*, CRC Press, Florida, USA.
21 (a) Ambrose, D. (1987) *Handbook of the Thermodynamics of Organic Compounds* (eds. R.M. Stevenson and S. Malanowski), Elsevier, New York. (b) Ambrose, D. and Tsonopoulos, C. (1995) *J. Chem. Eng. Data*, **40**, 531.
22 Watson, J.N., Craven, I.E., and Ritchie, G.L.D. (1997) *Chem. Phys. Lett.*, **274**, 1–6.
23 Dietzel, P.D.C., Johnsen, R.E., Fjellvag, H., Bordiga, S., Groppo, E., Chavan, S., and Blom, R. (2008) *Chem. Commun.*, 5125–5127.
24 Reguera, L., Roque, J., Hernández, J., and Reguera, E. (2010) *Int. J. Hydrogen Energy*, **35**, 12864–12869.
25 (a) Bhatia, S.K. and Myers, A.L. (2006) *Langmuir*, **22**, 1688–1700; (b) Matranga, K.R., Myers, A.L., and Glandt, E.D. (1992) *Chem. Eng. Sci.*, **47**, 1569–1579.

26 (a) Natesakhawat, S., Culp, J.T., Matranga, C., and Bockrath, B. (2007) *J. Phys. Chem. C*, **111**, 1055–1060; (b) Reguera, L., Balmaseda, J., Krap, C.P., Avila, M., and Reguera, E. (2008) *J. Phys. Chem. C*, **112**, 17443–17449.

27 Areán, C.O., Manoilova, O., Bonelli, B., Palomino, G.T., and Garrone, E. (2003) *Chem. Phys. Lett.*, **370**, 631–635.

28 (a) Ma, Z.W., Zhang, P., Bao, H.S., and Deng, S. (2016) *Renew. Sust. Energ. Rev.*, **53**, 1273–1302; (b) Sumida, K., Rogow, D.L., Mason, J.A., McDonald, T.M., Bloch, E.D., Herm, Z.R., Bae, T.-H., and Long, J.R. (2012) *Chem. Rev.*, **112**, 724–781.

29 Lee, S.-Y. and Park, S.-J. (2015) *J. Ind. Eng. Chem.*, **23**, 1–11.

30 Blanchon le Bouhelec, E., Mougin, P., Barreau, A., and Solimando, R. (2007) *Energy Fuels*, **21**, 2044–2055.

31 Rochelle, G.T. (2009) *Supramol. Sci.*, **325**, 1652–1654.

32 E. F., daSilva. and Svendsen, H.F. (2007) *Int. J. Greenh. Gas Control*, **1**, 151–157.

33 Freeman, S.A., Dugas, R., Van Wagener, D.H., Nguyen, T., and Rochelle, G.T. (2010) *Int. J. Greenh. Gas Control*, **4**, 119–124.

34 Shakerian, F., Kim, K.-H., Szulejko, J.E., and Park, J.-W. (2015) *Appl. Energy*, **148**, 10–22.

35 (a) Li, L., Wen, X., Fu, X., Wang, F., Zhao, N., Xiao, F., Wei, W., and Sun, Y. (2010) *Energy Fuels*, **24**, 5773–5780; (b) Ochoa-Fernández, E., Rønning, M., Grande, T., and Chen, D. (2006) *Chem. Mater.*, **18**, 6037–6046; (c) Sanna, A., Uibu, M., Caramanna, G., Kuusik, R., and Maroto-Valer, M.M. (2014) *Chem. Soc. Rev.*, **43**, 8049–8080; (d) Boot-Handford, M.E., Abanades, J.C., Anthony, E.J., Blunt, M.J., Brandani, S., Dowell, N.Mac., Fernandez, J.R., Ferrari, M.-C., Gross, R., Hallett, J.P., Haszeldine, R.S., Heptonstall, P., Lyngfelt, A., Makuch, Z., Mangano, E., Porter, R.T.J., Pourkashanian, M., Rochelle, G.T., Shah, N., Yao, J.G., and Fennell, P.S. (2014) *Energy Environ. Sci.*, **7**, 130–189; (e) Jo, H.G., Yoon, H.J., Lee, C.H., and Lee, K.B. (2016) *Ind. Eng. Chem. Res*, **55**, 3833–3839.

36 Donat, F., Florin, N.H., Anthony, E.J., and Fennell, P.S. (2012) *Environ. Sci. Technol.*, **46**, 1262–1269.

37 Dunne, J.A., Mariwala, R., Rao, M., Sircar, S., Gorte, R.J., and Myers, A.L. (1996) *Langmuir*, **12**, 5888–5895.

38 Zhang, J., Singh, R., and Webley, P.A. (2008) *Micropor. Mesopor. Mat.*, **111**, 478–487.

39 Palomino, M., Corma, A., Rey, F., and Valencia, S. (2010) *Langmuir*, **26**, 1910–1917.

40 Plaza, M.G., García, S., Rubiera, F., Pis, J.J., and Pevida, C. (2010) *Chem. Eng. J.*, **163**, 41–47.

41 Yoon, S.-H., Lim, S., Song, Y., Ota, Y., Qiao, W., Tanaka, A., and Mochida, I. (2004) *Carbon*, **42**, 1723–1729.

42 Sharpe, A.G. (1976) *Chemistry of Cyano Complexes of the Transition Metals*, Academic Press.

43 (a) Jiménez-Gallegos, J., Rodríguez-Hernández, J., Yee-Madeira, H., and Reguera, E. (2010) *J. Phys. Chem. C*, **114**, 5043–5048; (b) Balmaseda, J., Reguera, E., Rodríguez-Hernández, J., Reguera, L., and Autie, M. (2006) *Micropor. Mesopor. Mat.*, **96**, 222–236; (c) Roque, J., Reguera, E., Balmaseda, J., Rodríguez-Hernández, J., Reguera, L., and del Castillo, L.F. (2007) *Micropor. Mesopor. Mat.*, **103**, 57–71.

44 Reguera, E., Fernández-Bertrán, J., Dago, A., and Díaz, C. (1992) *Hyperfine Interact.*, **73**, 295–308.

45 (a) Avila, M., Reguera, L., Vargas, C., and Reguera, E. (2009) *J. Phys. Chem. Solids*, **70**, 477–482; (b) Krap, C.P., Zamora, B., Reguera, L., and Reguera, E. (2009) *Micropor. Mesopor. Mat.*, **120**, 414–420; (c) Rodríguez-Hernández, J., Reguera, E., Lima, E., Balmaseda, J., Martínez-García, R., and Yee-Madeira, H. (2007) *J. Phys. Chem. Solids*, **68**, 1630–1642.

46 Reguera, L., Balmaseda, J., del Castillo, L.F., and Reguera, E. (2008) *J. Phys. Chem. C*, **112**, 5589–5597.

47 Reguera, L., Balmaseda, J., Krap, C.P., and Reguera, E. (2008) *J. Phys. Chem. C*, **112**, 10490–10501.

48 (a) Yaghi, O.M., Li, G., and Li, H. (1995) *Nature*, **378**, 703–706; (b) Rosseinsky, M.J. (2004) *Micropor. Mesopor. Mat.*, **73**, 15–30.

49 Furukawa, H., Cordova, K.E., O'Keeffe, M., and Yaghi, O.M. (2013) *Supramol. Sci.*, **341**, 1230444.

50 (a) Forster, P.M. and Cheetham, A.K. (2004) *Micropor. Mesopor. Mat.*, **73**, 57–64; (b) Gardner, G.B., Venkataraman, D., Moore, J.S., and Lee, S. (1995) *Nature*, **374**, 792–795; (c) MacGillivray, L.R., Subramanian, S., and Zaworotko, M.J. (1994) *J. Chem. Soc. Chem. Commun.*, 1325–1326; (d) Yaghi, O.M. and Li, G. (1995) *Angew. Chem. Int. Ed.*, **34**, 207–209.

51 (a) Rosi, N.L., Eckert, J., Eddaoudi, M., Vodak, D.T., Kim, J., O'Keeffe, M., and Yaghi, O.M. (2003) *Supramol. Sci.*, **300**, 1127–1129; (b) Kaskel, S., Schuth, F., and Stocker, M. (2004) *Micropor. Mesopor. Mat.*, **73**, 1; (c) Forster, P.M., Eckert, J., Chang, J.-S., Park, S.-E., Férey, G., and Cheetham, A.K. (2003) *J. Am. Chem. Soc.*, **125**, 1309–1312; (d) Hwang, Y.K., Hong, D.Y., Chang, J.S., Jhung, S.H., Seo, Y.K., Kim, J., Vimont, A., Daturi, M., Serre, C., and Férey, G. (2008) *Angew. Chem. Int. Ed.*, **47**, 4144–4148; (e) Prasad, T.K., Hong, D.H., and Suh, M.P. (2010) *Chem. Eur. J.*, **16**, 14043–14050.

52 Millward, A.R. and Yaghi, O.M. (2005) *J. Am. Chem. Soc.*, **127**, 17998–17999.

53 Pérez-Pellitero, J., Amrouche, H., Siperstein, F.R., Pirngruber, G., Nieto-Draghi, C., Chaplais, G., Simon-Masseron, A., Bazer-Bachi, D., Peralta, D., and Bats, N. (2010) *Chem. Eur. J.*, **16**, 1560–1571.

54 Banerjee, R., Furukawa, H., Britt, D., Knobler, C., O'Keeffe, M., and Yaghi, O.M. (2009) *J. Am. Chem. Soc.*, **131**, 3875–3877.

55 Andirova, D., Cogswell, C.F., Lei, Y., and Choi, S. (2016) *Micropor. Mesopor. Mat.*, **219**, 276–305.

56 Bordiga, S., Regli, L., Bonino, F., Groppo, E., Lamberti, C., Xiao, B., Wheatley, P.S., Morris, R.E., and Zecchina, A. (2007) *Phys. Chem. Chem. Phys.*, **9**, 2676–2685.

57 Kong, X., Scott, E., Ding, W., Mason, J.A., Long, J.R., and Reimer, J.A. (2012) *J. Am. Chem. Soc.*, **134**, 14341–14344.

58 Jadhav, P.D., Chatti, R.V., Biniwale, R.B., Labhsetwar, N.K., Devotta, S., and Rayalu, S.S. (2007) *Energy Fuels*, **21**, 3555–3559.

59 Le, Y., Guo, D., Cheng, B., and Yu, J. (2013) *J. Colloid Interface Sci.*, **408**, 173–180.

60 Llewellyn, P.L., Bourrelly, S., Serre, C., Vimont, A., Daturi, M., Hamon, L., De Weireld, G., Chang, J.-S., Hong, D.-Y., Hwang, Y. Kyu, Jhung, S. Hwa, and Férey, G. (2008) *Langmuir*, **24**, 7245–7250.

61 Poloni, R., Smit, B., and Neaton, J.B. (2012) *J. Am. Chem. Soc.*, **134**, 6714–6719.

62 Huang, J., Zhou, X., Lamprou, A., Maya, F., Svec, F., and Turner, S.R. (2015) *Chem. Mater.*, **27**, 7388–7394.

63 Feng, X., Ding, X., and Jiang, D. (2012) *Chem. Soc. Rev.*, **41**, 6010–6022.

64 Jiang, J.-X., Su, F., Trewin, A., Wood, C.D., Campbell, N.L., Niu, H., Dickinson, C., Ganin, A.Y., Rosseinsky, M.J., Khimyak, Y.Z., and Cooper, A.I. (2007) *Angew. Chem.*, **119**, 8728–8732.

65 Bhunia, A., Boldog, I., Moller, A., and Janiak, C. (2013) *J. Mater. Chem. A*, **1**, 14990–14999.

66 (a) Ben, T., Pei, C., Zhang, D., Xu, J., Deng, F., Jing, X., and Qiu, S. (2011) *Energy Environ. Sci.*, **4**, 3991–3999; (b) Ben, T., Ren, H., Ma, S., Cao, D., Lan, J., Jing, X., Wang, W., Xu, J., Deng, F., Simmons, J.M., Qiu, S., and Zhu, G. (2009) *Angew. Chem.*, **121**, 9621–9624.

67 Cote, A.P., Benin, A.I., Ockwig, N.W., O'Keeffe, M., Matzger, A.J., and Yaghi, O.M. (2005) *Supramol. Sci.*, **310**, 1166–1170.

68 Lanni, L.M., Tilford, R.W., Bharathy, M., and Lavigne, J.J. (2011) *J. Am. Chem. Soc.*, **133**, 13975–13983.

69 Krishna, L. and Koh, C.A. (2015) *MRS Energy Sust.*, **2**, 1–17.

70 (a) Struzhkin, V.V., Militzer, B., Mao, W.L., Mao, H.k., and Hemley, R.J. (2007) *Chem. Rev.*, **107**, 4133–4151; (b) Veluswamy, H.P., Wong, A.J.H., Babu, P., Kumar, R., Kulprathipanja, S., Rangsunvigit, P., and Linga, P. (2016) *Chem. Eng. J.*, **290**, 161–173; (c) Song, Y. (2013) *Phys. Chem. Chem. Phys.*, **15**, 14524–14547; (d) Chidester, B.A. and Strobel, T.A. (2011) *J. Phys. Chem. A*, **115**, 10433–10437.

71 (a) Bernal, J.D. and Fowler, R.H. (1933) *J. Chem. Phys.*, **1**, 515–548; (b) Pauling, L. (1948) *The Nature of the Chemical Bond*, Cornell University Press, Ithaca, NY.

72 (a) Koh, C.A. (2002) *Chem. Soc. Rev.*, **31**, 157–167; (b) Ripmeester, J.A., John, S.T., Ratcliffe, C.I., and Powell, B.M. (1987) *Nature*, **325**, 135–136.

73 Bollengier, O., Choukroun, M., Grasset, O., Le Menn, E., Bellino, G., Morizet, Y., Bezacier, L., Oancea, A., Taffin, C., and Tobie, G. (2013) *Geochim. Cosmochim. Acta*, **119**, 322–339.

74 Tang, J., Zeng, D., Wang, C., Chen, Y., He, L., and Cai, N. (2013) *Chem. Eng. Res. Des.*, **91**, 1777–1782.

75 (a) Raso, R.A., Zeltner, M., and Stark, W.J. (2014) *Ind. Eng. Chem. Res.*, **53**, 19304–19312; (b) Simoni, M., Jaakkola, M., Carrozzi, L., Baldacci, S., Di Pede, F., and Viegi, G. (2003) *Eur. Respir. J.*, **21**, 15s–20s.

76 Osovsky, R., Kaplan, D., Nir, I., Rotter, H., Elisha, S., and Columbus, I. (2014) *Environ. Sci. Technol.*, **48**, 10912–10918.

77 Sheherd, A. (May 2001) 13th Annual EnviroExpo, Boston, Massachusetts.

78 Smith, J., Romero, J., Dahn, T., Dunphy, K., Croll, L., and Dahn, J. (2012) *J. Hazard. Mater.*, **235**, 279–285.

79 Weston, M.H., Peterson, G.W., Browe, M.A., Jones, P., Farha, O.K., Hupp, J.T., and Nguyen, S.T. (2013) *Chem. Comm.*, **49**, 2995–2997.

80 DeCoste, J.B. and Peterson, G.W. (2014) *Chem. Rev.*, **114**, 5695–5727.

81 Roy, A., Srivastava, A.K., Singh, B., Mahato, T., Shah, D., and Halve, A. (2012) *Micropor. Mesopor. Mat.*, **162**, 207–212.

82 Li, G. and Wang, Z. (2013) *J. Phys. Chem. C*, **117**, 24428–24437.

12
Functional Materials for Gas Storage. Part II: Hydrogen and Methane

L. Reguera[1,2] *and E. Reguera*[1]

[1]*Center for Applied Science and Advanced Technology, Legaria Unit, National Polytechnic Institute, Mexico DF, Mexico*
[2]*Faculty of Chemistry, University of Havana, Cuba*

12.1
Introduction

The view with which the gas storage is considered has changed in the last two decades. Until the middle of the 1990s, gas storage was considered as the gas accumulation in the free volume of traditional porous solids commonly used as catalysts, as molecular sieves and ion exchangers. More recently, the materials for gases storage have resulted from a design process in order to optimize the guest–host interactions according to: (1) highest possible surface area [1], (2) large accessible free volume to accommodate guest molecules and, (3) optimal adsorption heats to maintain the molecules as adsorbed species at certain values of temperature and pressure [2]. The desirable materials features include the possibilities of obtaining solids with gate opening and flexible framework behavior to allow the gas storage at high pressure and its release at low pressure, particularly for the storage of molecules with relatively weak physical interactions with the host framework, for example, H_2, CH_4 [3]. Such a change of view for the adsorption in porous solids has been motivated by the need to find a suitable way to store these two energy vectors, H_2 and CH_4, at high gravimetric density for their use as fuel in mobile technologies. In this regard, hydrogen appears to be the most promising option because of its clean burning (it generates water as by-product) and the high calorific value (142.26 kJ/g) [4], about three times that of gasoline. Another alternative would be the storage of methane at high density. Methane, as other fossil fuels, is a nonrenewable energy source, but its global availability is very high. It represents about 80% of the natural gas content and is an appropriate fuel for mobile technologies. In the following, when we refer to retention of H_2 or CH_4, the term of "storage" will be used. In addition to porous solids, H_2 and CH_4 can also be stored as clathrates, which for methane are well known because the largest global reserves of methane are found in the form of water clathrates in the depth of

Handbook of Solid State Chemistry, First Edition. Edited by Richard Dronskowski, Shinichi Kikkawa, and Andreas Stein.
© 2017 Wiley-VCH Verlag GmbH & Co. KGaA. Published 2017 by Wiley-VCH Verlag GmbH & Co. KGaA.

oceans, below the ocean floor [5]. H_2 clathrates do not exist in nature but they have been prepared in order to evaluate their potentialities for H_2 storage [6]. The state of the art of CH_4 and H_2 clathrates will be briefly discussed.

12.2
On the Adsorption Forces for H_2 and CH_4 in Porous Solids

H_2 and CH_4 are supercritical gases at ordinary conditions, their critical temperatures are 32.97 and 190.56 K, respectively [7]. H_2 is a molecule with quadrupole moment (2.21×10^{-40} cm^2) while CH_4 is a spherical molecule without such electrical feature [8]. Their polarizabilities, which determine the strength of dispersive and polarization interactions, are also different, in units of 10^{-24} cm^3, 0.802 43 and 2.593, respectively [7]. In the adsorption of H_2, five attractive interactions could be participating: (1) dispersive interactions, which depend on the material surface electron cloud polarizability; (2) polarization of its electron cloud in the presence of a charge center; (3) the molecule interaction, through its quadrupole moment, with a local anisotropic electric field; (4) quadrupole–quadrupole interaction between neighboring H_2 molecules; (5) coordination to a partially naked transition metal to form so-called Kubas type complex, where the metal receives electron density from the molecule σ orbital and at the same time the metal donates charge to the H_2 molecule antibonding orbital (σ^*) (Figure 12.1a) [9]. The coordination interaction appears attractive because it involves stabilization energy values in the range of 10–80 kJ/mol [9], and in principle, modulating its strength the H_2 molecule could be maintained as "adsorbed" species at temperatures close to room conditions. However, it must be noted that this interaction allows the formation of a single monolayer of guest molecules and from this fact, the resulting gravimetric density of H_2 storage is low. An option to increase the H_2 storage is through combination with electrostatic type interactions (discussed below).

The nature of the adsorbate–surface interaction involved in gas adsorption determines the formation of a layer of guest molecules as adsorbed phase on the solid

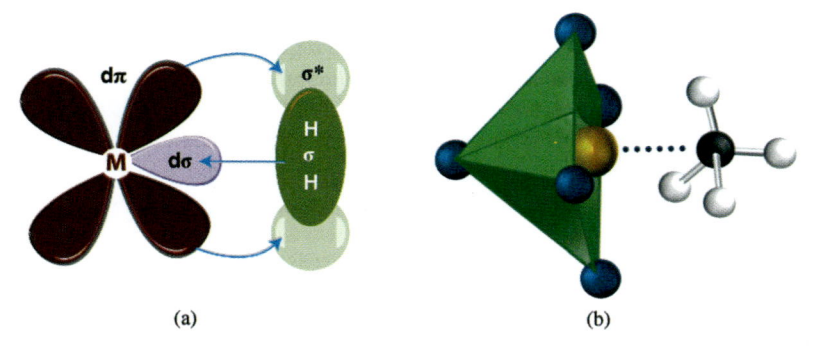

Figure 12.1 Illustration for: (a) H_2 coordination to a partially naked transition metal [2a]; (b) polarization of the electron cloud of CH_4 molecule by a charge center.

surface. In the presence of interactions of relatively large scope, for example, r^{-3} and r^{-4} (interaction of H_2 quadrupole moment with an electric field gradient, and polarization forces, respectively), additional layers of adsorbed molecules can be accumulated on the surface. This results in an effect equivalent to the adsorbate compression, substantially increasing the density of the gas adsorbed at the solid surface [10]. In microporous solids such interactions favor the micropore filling, even for supercritical gases. For H_2, the optimal pore size is probably about 1 nm in order to favor their stability within the pore through dispersive and quadrupole–quadrupole interactions. For H_2 adsorption in Cu-containing Prussian blue analogues, for instance, all the above mentioned guest–host interactions are possible, and in the recorded isotherms at 75 K and up to 10 bar, at relatively low pressure (3 bar), a maximum is observed (Figure 12.2), which suggests that the pore volume is completely occupied by H_2 molecules. This family of materials has a pore diameter of about 0.9 nm (discussed below) [11]. An analogous behavior is observed for Ni nitroprusside, with pores also 0.9 nm in diameter. At 75 K and 5 bar, the adsorption isotherm shows a maximum, where the adsorption in excess corresponds to the accommodation of about 10 H_2 in the pore volume [10]. This is equivalent to a local density of H_2 storage close to the density of solid H_2. In these two materials, the involved adsorption heat is close to 10 kJ/mol [10,11]. These two examples illustrate that through an appropriate materials design, optimizing all the possible guest–host interactions [2a], adsorbents for H_2 storage at high density could be prepared.

For CH_4, only the first three interactions are possible. Due to the relatively high polarizability for CH_4, its adsorption by solids containing charge

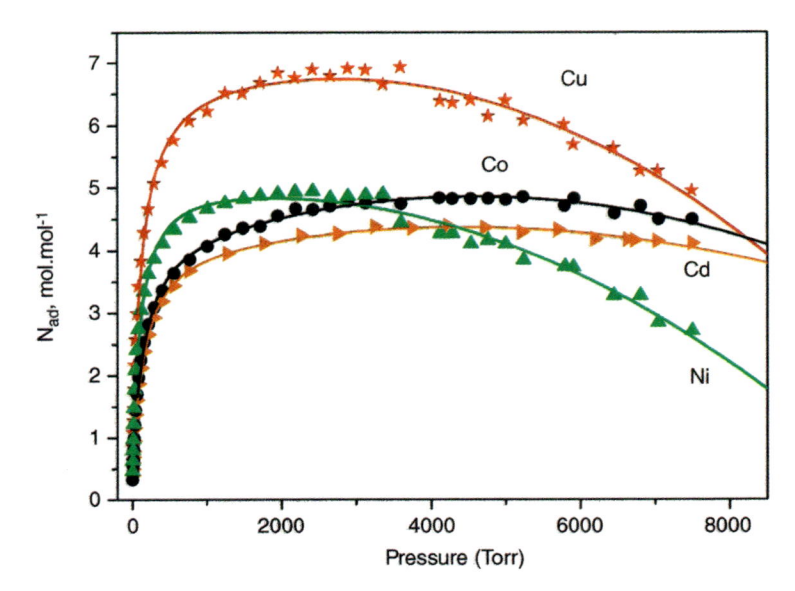

Figure 12.2 H_2 adsorption isotherms in $T_3[Co(CN)_6]_2$ recorded at 75 K. The observed maximum in these isotherms is typical for excess adsorption of supercritical adsorbates. Experimental details are available from Ref. [11].

centers is an attractive route to increase its gravimetric storage density (Figure 12.1b).

Due to the relatively weak guest–host interaction for H_2 and CH_4, to achieve high gravimetric storage densities, the adsorption process must be carried out under high pressures. The study of gases at high pressures involves special cautions (see Chapter 11 of this volume). The adsorption data evaluation from high-pressure experiments also requires of special considerations.

For supercritical adsorbate, the maximum adsorption amount should correspond to the adsorbate monolayer completion on the adsorbent surface. In theory, the same maximum should be reached at all temperatures in the limit of high pressure. However, this is not the case for micropore filling, which supposes that a pseudocondensed phase is formed inside the pores. Micropore filling appears as a dominant mechanism for adsorption in many porous materials of interest for H_2 and CH_4 storage. The maximum amount adsorbed in microporous materials corresponds to the entire filling of the adsorption space [12]. The adsorption isotherms for supercritical gases, recorded up to high pressures, show a maximum (Figure 12.2) at certain pressure value. The maximum occurs at the pressure value where the gas densities at the cavity and the bulk gas are increasing at the same rate, so that a pressure increase has no effect on the amount adsorbed. Above that point, the gas density in the cavity saturates, while the bulk gas density keeps increasing, resulting in the observed negative gain for the excess amount adsorbed. A temperature dependence of the adsorption capacity may occur as a result of volumetric expansion of the adsorbed phase. Zhou and others described an exponential decrease in the maximum adsorption amount with temperature for activated carbons [13]. In fact, the maximum excess adsorption amount decreases with temperature for most adsorbents. These authors have proposed the following empirical model for supercritical adsorbates to estimate the absolute limit adsorption capacity (n_o^s) from data of excess adsorption versus pressure:

$$n^\sigma = n_o^s \left[1 - \exp(-KP^B)\right] - \left(\sum_{i=1} c_i P^i\right)\left(\sum_{j=0} c_j P^j\right) \quad (12.1)$$

In Eq. (12.1), K is a parameter related to the adsorption energy (analogous to its meaning in the Langmuir model, see Part I of this chapter) and $B = 1/g$, contains information on both, the adsorption energy and heterogeneity of the guest–host interactions. The value of B decreases from unity, for a uniform surface, down to close to zero for a heterogeneous surface and stronger interactions. The empirical parameters c_i and c_j are related with the compressibility of the adsorbate in the gas and adsorbed phase, respectively, on the gas pressure [13a].

12.3
Adsorbent Materials for H_2 Storage

A large variety of porous solids have been prepared and evaluated along the last three decades for gas storage and a vast literature on this subject is available. To rationalize that huge volume of information, the following discussion on properties of gases storage materials is organized according to the above-discussed guest–host interactions involved in the adsorption process.

12.3.1
Carbon-Based Materials

Carbon based materials are highly attractive for H_2 storage for four reasons: (1) carbon is an element of high natural abundance in the Earth; (2) it has low weight so that high gravimetric storage density could be obtained; (3) the existence of a large variety of structures for solid carbon is known, which results in diverse types of materials; (4) their high chemical and thermal stability.

In the absence of functionalization and doping, the adsorption forces for carbon-based materials with hydrogen molecules are of dispersive nature. Carbon is an atom with only six electrons and with a low polarizability, for example, for a graphene layer it is 0.867, in units of 10^{-24} cm^3 [14]. This explains that the dispersive interactions between H_2 and carbon-based materials must be weak. The limit adsorption of supercritical gases in these materials is proportional to the available surface area, and it is reached at high pressure [15]. The value of this last parameter is determined by the synthesis conditions, and carbonization and activation processes employed during their preparation [16]. Within carbon-based materials we will highlight activated carbons, nanofibers, and carbon nanotubes, both of single- or multiple wall. In such materials, the existence of a porous structure is determined by the spatial arrangement of the graphene layers, which may be stacked or rolled, resulting in the development of porous structures.

Related to the weak adsorption forces generated by the dispersive interactions in carbon-based materials, the adsorption/desorption of H_2 are characterized by relatively fast kinetics, and the corresponding isotherms, therefore, have no hysteresis. This is an attractive feature for systems that require a high speed of H_2 loading and unloading. However, the applicability of these materials is limited because, as expected, the adsorption capacity is highly dependent on the temperature and pressure used [17]. As already discussed, the involved adsorption heat is close to 7 kJ/mol, and from this fact, the operating temperature is about 138 K, quite low for practical applications.

The literature shows large inconsistencies regarding the adsorption capacities of H_2 in carbon nanostructures. Such inconsistencies are related to the complex and diverse synthesis methodologies employed [18], small amounts of sample that are obtained, which hinders a reliable measurement process and makes it more imprecise [19], other instrumental factors, and an inappropriate purity for the H_2 used. For single-wall carbon nanotubes (SWCNT), at 80 K and 40 bar, 8 wt% of H_2 adsorption was initially reported [20]. Further studies on the same material resulted in ∼2.5 wt% (77 K) and ∼0.5 wt% (298 K) at 70 bar [18b]. For carbon nanofibers, reports on storage capacities in the 1–20 wt% (in materials doped with Li and K) range are found [21]. However, more recent studies for this latter type of materials report adsorption capacity of 0.7 wt% at 77 K and 104 atm [22]. For traditional activated carbons, similar to the AX-21 type, the adsorption enthalpies are about 6 kJ/mol, slightly above the values found for SWCNT, which are in the range 4.3–4.5 kJ/mol [22b]. The highest values of adsorption heats in carbon-based materials, close to 7.5 kJ/mol, are found for chemically activated carbons using KOH [23]. For activated carbon obtained by bituminous mineral activation with KOH, at 77 K and 10 bar, the H_2 adsorption capacity is 6.8 wt% [15]. However, the value falls well below 1 wt% at room temperature. At room temperature, the strength of such interactions is of the same order as the thermal energy for gaseous hydrogen [18b,24]. This explains that all the efforts to use carbon-based materials for H_2 storage have failed. Such studies continue but with the aim to understand the possible role of structural and compositional factors on the hydrogen adsorption capacity and related adsorption heats.

The above mentioned attractive features of carbon-based materials for H_2 storage explain the large effort devoted to the modification of carbon structures, through functionalization of their surface, in order to favor a stronger interaction with H_2 and to increase the available surface area and pore volume [15]. The functionalization routes used include the incorporation of metal ions, metal clusters, and heteroatoms. In some cases, when Pt or Pd metal particles are incorporated, a process called spillover takes place. This is an intermediate storage mechanism between the chemical hydride and the surface physical adsorption, as is going to be discussed below. This procedure has been seen as a viable alternative to increase the amount of H_2 stored at room temperature [25].

The doping with alkali and alkali earth ions has been used as an attempt to stabilize the H_2 molecule as adsorbed species through electrostatic interactions of polarization and using the molecule quadrupole moment. The results in that sense are limited, with adsorption heats below 10 kJ/mol, and adsorption capacity proportional to the surface area of the solid [26]. Another research effort with the same aim is the doping with transition metals to favor both, the electrostatic interactions and a possible metal coordination to the hydrogen molecule. The doping with transition metals results in the formation of metallic clusters, and this limits their interaction with the hydrogen molecules. This probably explains that no notable improvements in the hydrogen storage capacity have been reported in such studies [27]. For instance, after doping with Ti(4+), the reported storage capacity is 3 wt% at 77 K and 1 bar [28].

The doping of activated carbons with heteroatoms for H_2 storage is supported in an attempt to modify the adsorbent surface polarity. In such studies, doping with B, Si, N, O, and S has been considered [29]. From both, experimental and theoretical studies, contradictory results have been reported. Some authors report certain improvements for the storage capacity [29a–c] while others discard such a possibility from experimental results, arguing that the doping heteroatoms favor the appearance of repulsive interactions between the hydrogen molecule and the heteroatoms [17a,29d]. It seems, that the effect of doping with heteroatoms is negative or at least, it has a marginal effect on the material storage capacity.

12.3.2
Silica, Alumina, and Zeolites

A large group of silicas, aluminas, and zeolites has been evaluated for H_2 storage. These materials are composed of relatively heavy metals, at least compared with carbon-based solids, and this affects the gravimetric H_2 density that is possible in their porous framework [30]. This probably explains the relatively low gravimetric H_2 storage density in these materials, below 2.5 wt% at cryogenic conditions (77 K) [31]. Among them, zeolites have been the most attractive porous solids for studies related with H_2 storage. Zeolites have a regular network of channels and cavities of different size, quite different from the framework of activated carbons, for instance. Such features are attractive to evaluate the effect of structural factors on H_2 storage in porous solids. The substitution of Si^{4+} by Al^{3+} in the composition of these materials results in the need to have charge-compensating cations within their system of channels and cavities. Such cations are charge centers able to interact with the hydrogen molecule through polarization of its electron cloud. The electric field created by these charge centers is anisotropic, and this allows their attractive interaction with the H_2 quadrupole moment. Zeolites containing H^+, Li^+, Na^+, K^+, Mg^{2+}, Ca^{2+}, and other cations have been considered in studies on these interactions. The H_2 adsorption isosteric heats for such materials are in the 4–15 kJ/mol range [32]; an exception is found for Mg^{2+} with a value of 17.5 kJ/mol [33]. The nature of such low values of adsorption heats has been ascribed to the interaction of the metal with the negative host framework shielding the metal polarizing capacity [32d]. Zeolites exchanged with Cu^+ have served as a model to study the coordination of the H_2 molecule to the copper atom [2a,34], obtaining the values for ΔH_{ads} and ΔS from the intensity for the ν(H–H) stretching band, from IR spectra, and then applying Eq. (11.21) [33b].

Mesoporous silica has been used as a support for grafting benzyl Ti(III) for H_2 uptake through its coordination to the metal centers (Figure 12.2) [35]. Analogous studies have been carried out for Ti(III), V(III), Cr(II), and Mn (II) but supported in hydrazine gels [36]. For such systems, an experimental heat of adsorption of 39 kJ/mol has been reported, sufficient to maintain the hydrogen molecule as adsorbed species at room temperature. The H_2 adsorption in these materials has a practically linear dependence on the applied pressure within the considered pressure range, up to 140 bar, with an increase for the H_2 binding energy on the coverage rising, opposite to the one observed in pure physisorption [35b,36a,b,d]. The binding of a H_2 molecule modifies the energy for the frontier orbitals of the available adsorption sites, bringing the energy of the LUMO and HOMO of these last ones closer to the HOMO and LUMO for the free H_2 molecule [37]. This favors the adsorption of the following H_2 molecule, up to reach the maximum coordination capacity for the metal center. For an isolated metal, the coordination capacity has been estimated to be six molecules per metal [38]. For supported Ti(III), the coordination capacity, according to the experimental results, is 2.4 molecules per metal [35].

12.3.3
Metal-Organic Frameworks (MOFs)

MOFs offer the possibility to increase the H_2 storage capacity optimizing all the possible interactions that stabilize the molecule as adsorbed species. This is possible through a crystal engineering process where their high surface area and relatively lightweight can be combined with pore size below 2 nm, availability of partially naked transition metals at the pore surface, presence of an anisotropic electric field in the pore region, and large pore volume [39]. The discussion on H_2 storage MOFs type materials is organized according to the type of adsorbent–adsorbate interaction that prevails.

12.3.4
H_2 Storage in MOFs Through Dispersive Interactions

The simplest MOFs type porous solids correspond to the formation of a stoichiometric assembling of a metal oxide as secondary building unit linked by organic blocks containing carboxylate groups. In such a subset of MOFs the coordination sphere of the involved metal center remains saturated by the ligand group from the organic block. These porous solids are free of charge centers and of partially naked metals, and all the adsorption forces for H_2 result from dispersive interactions with the framework surface. This is the case for instance, of the isoreticular (IRMOF) series, IRMOF-1, -6, -11, and -20, formed by the linking of Zn acetate unit through linear organic ligands of different length, with two carboxylate groups at their ends [40]. Related to the ligand length, the formed porous solids have different pore volumes and Langmuir surface areas. For this series and other MOFs, a linear correlation between the maximum H_2 uptake and the surface area has been observed [39a]; this is evidenced that only the dispersive forces are participating in the H_2 adsorption. This behavior is similar to that discussed above for carbon-based porous solids. An illustrative example on the linear correlation of surface area and maximum H_2 uptake is found for MOF-210, with an extended microporosity and extraordinarily high surface areas, up to 10 400 m^2/g (Langmuir) [39a]. For this material, the H_2 capacity limit of excess adsorption is 7.9 wt% at 77 K and ~80 bar, that corresponds to 15 wt% of absolute adsorption. To the best of our knowledge, this is the record of H_2 storage in porous solids, of course, under cryogenic conditions.

Imidazole and its derivatives also form MOFs type frameworks with $Zn(2+)$, $Co(2+)$, and other transition metal ions. The angle between the pyridinic and pyrrolic N ends of these ligands, at 1,3 positions, is 145°, similar to the O—Si—O angle found in zeolites. For this reason, usually the resulting solids crystallize with structures and pore topologies similar to those found in many zeolites. This explains that this subset of MOFs is known as zeolite-like imidazolate frameworks (ZIFs) [41]. An attractive feature of ZIF type porous solids is their high thermal stability, above 400 °C [42]. These materials are obtained as anhydrous solids because their structure is free of charge centers. For ZIFs, the adsorption forces for H_2 are of dispersive nature and the maximum H_2 uptake is proportional to the framework surface area [42]. The reported adsorption in excess for H_2 at 77 K and 1 bar is 1.3, 1.4, and 1.1 wt % for ZIF-8 ($Zn(MIm)_2$; MIm = 2-methyl-imidazolate), ZIF-11 ($Zn(BIm)_2$; BIm = benzimidazolate) and ZIF-20 ($Zn(Pur)_2$: Pur = purine), respectively [41]. As expected, an increase in the pressure results in a higher H_2 uptake capacity, which for ZIF-8, is 3.1 wt% at 77 K and 55 bar [42]. As in other porous solids where the adsorption forces for the hydrogen molecule are dominated by dispersive interactions, the adsorption enthalpies in ZIFs remain below 10 kJ/mol. For ZIF-20, for instance, it is 8.5 kJ/mol [41]. Such values for ΔH_{ads} are slightly higher than the ones found for carbon-based materials and other subsets of MOFs. According to results from neutron diffraction, the adsorption sites for H_2 in ZIFs are located close to the π cloud of the organic ligands and to the N atoms [43]. These are ligand regions rich in electron density, which favors dispersive type interactions with the hydrogen molecules.

According to the dependence of dispersive forces on the electron cloud polarizability, attention has been paid to the effect of aromatic ligands inclusion on the MOFs ability for H_2 uptake. This has been evaluated for different MOFs structures, among them MOF-5 ($Zn_4O(BDC)_3$; BDC = 1,4-benzenedicarboxylate = and NOTT (from Nottigham) series [44]. For NOTT-110 ($[Cu_2(L)(H_2O)_2]$ $(DMF)_{7.5}(H_2O)_5$, where L = (2,7-phenanthrenediyl)diso-phthalate) and NOTT-111 (L = [2,7-(9.10-dihydrophenanthrenediyl)]diiso-pthalate). The inclusion of phenanthrene groups in the MOFs results in an enhancement for the H_2 adsorption [45]. As expected, the results for both, the amount of H_2 adsorbed and the increase for the adsorption enthalpy are marginal, at least considering the practical requirements [45].

12.3.5
Ionic MOFs: H_2 Adsorption Through Electrostatic Interactions

The preparation of anionic MOFs opens the possibility to enhance the electrostatic interactions with the hydrogen molecule by incorporation of exchangeable metals into their porous framework. As already discussed, these are interactions of relatively long range, r^{-3} and r^{-4}, which favors the pore filling even for supercritical gases. Two routes are used to obtain anionic MOFs: (a) the reduction of organic ligands in the framework accompanied by insertion of a charge balancing cation [46]; (b) direct synthesis of an anionic framework with metals in its free spaces [47]. When in the resulting material cations of organic nature are found inside the pores, for example, H_2ppz^{2+} or Me_2NH^{2+}, which usually are of large size and remain linked to the pore surface by hydrogen bonds, they can be exchanged by alkali ions of small size and large polarizing power, like Li^+ and Na^+ [48]. The interaction with hydrogen molecules is favored for small cations. Furthermore, the electrostatic interaction of the cation with the host framework modulates the electronic structure of the latter. Usually the exchangeable ions also modify the surface area and the pore volume.

The incorporation of alkali metal ions (Li^+, Na^+, and K^+) by ligand reduction has been carried out for several MOF-type frameworks, among them MOF-5 [49] and twofold inter-woven MOF, $Zn_2(NDC)_2(diPyNI)$ (NDC = 2,6-naphthalenedicarboxylate, diPyNI = N,N'-di-(4-pyridyl)-1,4,5,8-naphthalenetetracarboxydiimide) [46,50]. In these two MOFs, the presence of an alkali ion in the porous framework results in an increase of both, the amount of H_2 adsorbed and the related heat of adsorption, compared with their unreduced form. For the second type of MOFs, the heat of adsorption changes in the order $Li^+ > Na^+ > K^+$; with the higher value for Li^+, 5.96 kJ/mol versus 4.31 kJ/mol for the unreduced form [51]. An inverse order was observed for the adsorption capacity at 77 K and 1 bar, which is ascribed to changes in the solid framework on the ligand reduction process [50]. In the NOTT series, several compositions are obtained with bulky cations as charge balancing ions [48,52]. When these cations are exchanged by alkali cations, an increase in H_2 storage capacity is observed [52,53].

ZMOFs are a subset of MOFs type materials with zeolite type topology and with anionic character, and from this last feature, they contain exchangeable cations in their porous framework. ZMOFs can be prepared from a large variety of organic ligands, for example, 4,5-imidazoledicarboxylic acid (H_3ImDC) and 4,6-pyrimidinedicarboxylic acid. These materials appear as an excellent platform to evaluate the role of the electrostatic interactions to enhance the H_2 storage in MOFs with zeolite type structures. For ZMOFs with topologies analogous to zeolites rho, Y, or sod, where the imidazolium ion have been exchanged by Li^+, Na^+, K^+, and Cs^+, an H_2 adsorption capacity lower than that corresponding to the nonexchanged solid was observed [54]. For instance, in SodZMOF the adsorption capacity at 77 K and 10 bar was 1.6 wt%, while in the sample exchanged with Na^+, it was 1.2 wt% [54]. Such behavior was ascribed to an increase in the pore volume. Similar studies for other ZMOFs type solids have demonstrated that the improvement in the H_2 adsorption and the related adsorption heats are limited [55].

Surprisingly, for beryllium benzene tribenzoate (Be-BTB), a nonionic MOF, the highest value for the H_2 excess adsorption capacity at room temperature has been reported to be 2.3 wt.% at ~95 bar [56]. Such behavior is ascribed to the appearance of a large charge density anisotropy, which results in a strong electrostatic interaction with the hydrogen molecule. DFT calculations for the H_2 adsorption in that material reveal that Be 2p orbitals can interact with H_2 molecular orbitals in a similar way to Kubas interaction [57].

12.3.6
H_2 Adsorption in MOFs with Partially Naked Transition Metal Centers

The formation of Kubas type complexes of H_2 with transition metals [9] has stimulated the study of MOFs with unsaturated coordination environments. The same motivation explains the above discussed studies on H_2 adsorption on partially naked Ti(III), V(III), Cr(II), and Mn (II) supported in silica and hydrazine gels, with adsorption heats up to 39 kJ/mol [35,36,55c]. Such values of

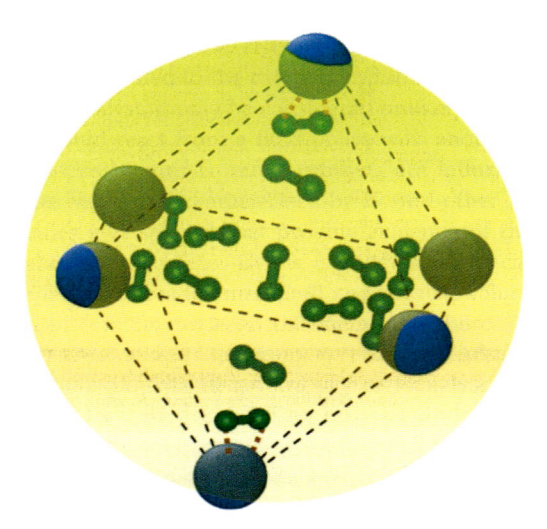

Figure 12.3 Model for pore filling with H_2 molecules combining coordination, electrostatic and dispersive interactions. The H_2 quadrupole moment determines the transversal disposition of neighboring molecules.

adsorption heats are highly attractive to achieve reversible H_2 storage at technologically viable conditions of pressure and temperature. The metal centers that belong to the solid framework but that are located at the cavity surface, usually have a mixed coordination sphere formed by framework ligands and molecules of the solvent used during the material synthesis. If these last molecules are removable by moderate heating, for instance, without the framework disrupting, the liberated coordination sites would be available to bind H_2 molecules through a nondissociative interaction. For metals of relatively high polarizing power, once H_2 molecules saturate their coordination sphere, the metal center could preserve certain polarizing capacity to allow the stabilization of additional adsorbate molecules through electrostatic interactions. For cavities about 1 nm in diameter, the remaining available volume could be occupied by molecules stabilized by dispersive and quadrupole–quadrupole interactions with molecules already adsorbed (Figure 12.3). This is the interaction model for H_2 storage in MOFs containing partially naked transition metals, where the five possible adsorption interactions for this adsorbate would be participating. This explains that the main efforts in MOFs preparation for H_2 storage have been aimed at obtaining frameworks with transition metals located at the surface of their cavities. In that sense, at least a thousand MOFs structures have been prepared, characterized, and evaluated for H_2 adsorption [39b–d,56,58]. In the following, the results reported for representative compositions will be discussed.

The H_2 adsorption by the isostructural M_2(dhtp) series with M = Mg, Mn, Co, Ni, and Zn, with dhtp = 2,5-dihydroxyterephthalate demonstrates that the heat of adsorption is in the 8.5–12.9 kJ/mol range and follows the order Zn < Mn < Mg < Co < Ni [59]. These values of adsorption heats are higher than the ones found for MOFs with exchangeable metals in the cavities and of course, of those where the dispersive interactions determine the adsorption process. The highest value of ΔH_{ads} corresponds to Ni^{2+}, the transition metal ion of higher polarizing power, with contributions of coordination interaction.

Spectroscopic and structural studies of H_2 adsorption in Cu^+-exchanged zeolites reveal that the H_2 molecule coordinates to the copper atom [2a,34]. For this reason, MOFs with partially naked Cu atom at the cavity surface have received certain attention. For $Cu_2(abtc)_3$ and $Cu_2((abtc)(dmf)_2)_3$ where abtc = 1,1′-azobenzene-3,3′,5,5′-tetyracarboxylate, with unsaturated and saturated coordination sphere for the copper atom, the *excess* adsorption at 77 K and 1 bar is 2.87 and 1.83 wt%, respectively. The corresponding adsorption heats at low coverage are 11.60 and 6.53 kJ/mol [60]. Both, the excess adsorption and the adsorption heats are conclusive clues about the presence of a relatively strong interaction of the hydrogen molecule with the copper atom. In the zinc/copper MOF, $Zn(BDC)_3[Cu(Pyen)].(DMF)_5(H_2O)_5$ where H_2BDC = 1,4-benzedicarboxylic acid and H_2Pyen = 5-methyl-4-oxo-1,4-dihydro-pyridine-3-carbaldehyde, after its dehydration, the copper atom appears with a planar coordination, with available coordination sites at both side of the plane. The H_2 adsorption enthalpy for this MOFs is unusually high, 12.3 kJ/mol [61], as evidence of a relatively strong interaction with the copper atom is present. The coordinative binding of H_2 to surface-exposed copper atoms in MOFs has been confirmed by neutron diffraction [62].

Hydrogen storage with partially naked metal centers has been studied for all the 3d series. Typically, a significant increase in the adsorption capacity and in the involved adsorption heats is reported [39b,d]. For $[Mn(DMF)_6][Mn_4Cl)_3(BTT)_8(H_2O)_{12}.42DMF.11H_2O.20CH_3OH$ where BTT = 1,3,5-benzetristetrazole, the DMF molecules can be removed without affecting the solid structure to obtain a material where the H_2 adsorption at 77 K and 90 bar is 6.9 wt% and the heat of adsorption is 10.1 kJ/mol. That value of excess adsorption is equivalent to 60 g H_2/L, about 85% the density of liquid hydrogen [39d].

12.4
Prussian Blue Analogues and Related Porous Solids

Figure 12.4 illustrates a typical framework of the porous Prussian blue analogue, $(T_e)_3[T_i(CN)_6]_2$, and the related zeolite-like hexagonal solids, $Zn_3A_2[T_i(CN)_6]_2$, where A is a charge-balancing cation [63], as was described in Chapter 11 of this volume.

The porous framework of $Zn_3[T_i(CN)_6]_2$ with T_i = Fe(III), Co(III), Rh(III) and Ir(III), is free of exchangeable metals and of partially naked transition metals. In these solids, the only possible guest–host interactions enabling adsorption of hydrogen molecules are dispersive forces. For $Zn_3[Co(CN)_6]_2$ the extrapolated limit capacity of H_2, from an isotherm recorded at 75 K and 1 bar and using the Langmuir–Freundlich model, is 1.82 wt%, equivalent to 11.6 H_2/cavity, with an adsorption heat of 6.2 kJ/mol [64].

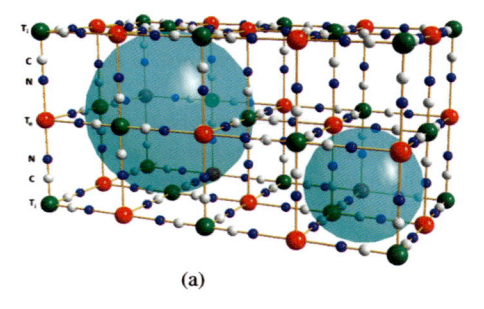

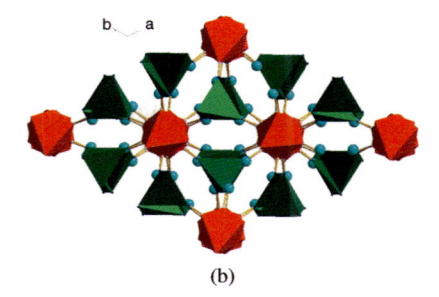

Figure 12.4 Typical porous framework for the anhydrous phase of: (a) Prussian blue analogues $(T_e)_3[T_i(CN)_6]_2$, (b) the related zeolitic-like solids $Zn_3A_2[T_i(CN)_6]_2$ with A = Na$^+$, K$^+$, Rb$^+$ Cs$^+$, (NH$_4$)$^+$ (small spheres close to the N atoms). These images were prepared form the corresponding ICSD files (416746 and 421892, respectively).

To study the role of the electrostatic interactions in the H$_2$ adsorption in this family of materials, the series $Zn_3A_2[T_i(CN)_6]_2$ with T_i = Fe, Ru, Os, and A = Na, K, Rb, Cs have been considered [64,65]. The extrapolated limit adsorption capacity, from isotherms recorded at 75 K and 1 bar, is in the 11.6–8.4 H$_2$/cavity (1.66–0.95 wt% range) and follows the order K > Rb > Cs. An inverse order is observed for the adsorption heats, in kJ/mol: K (8.4) > Rb (6.8) > Cs (6.4). This is the order of the metal polarizing power. The available volume determines the limiting amount of molecules accommodated within a cavity. For Na, the adsorption curves show a pronounced hysteresis, even for large equilibrium times, which has been ascribed to a strong interaction of the cations with hydrogen molecules, limiting their diffusion through the porous framework. In these materials, the cations occupy positions close to the cavity windows, near the N ends of the CN ligands, where the framework negative charge is concentrated. The relatively low values observed for the adsorption heats in these materials suggests that the metal polarizing power is at least partially shielded by the framework electron cloud [64,66].

Prussian blue analogues are prototypical porous solids to study the role of partially naked transition metals on the H$_2$ adsorption [67]. The first studies in that sense were carried out in the series $(T_e)_3[Co(CN)_6]_2$ with T_e = Mn, Fe, Co, Ni, Cu, Zn, and Cd [67a,68]. The highest adsorption capacity (1.24 wt %), at 75 K and 1 bar, was observed for Cu. The adsorption heats in the series are in the 5.3–6.9 kJ/mol range, with the lowest and highest values for Mn and Ni, respectively [15f,g]. For Cu, neutron diffraction studies found the hydrogen molecule adsorbed close to the copper atom with an H—H distance that suggests its coordination to the metal center [69]. The results of a systematic study for H$_2$ adsorption in copper Prussian blue analogues, $Cu_3[T_i(CN)_6]_2$ with T_i = Co, Rh, Ir, and the mixed compositions, $Cu_{3-x}Mn_x[Co(CN)_6]_2$, support the presence of a coordination type interaction between the copper atom and the hydrogen molecules [67d]. Copper Prussian blue analogues show a relatively high H$_2$ adsorption capacity, about 30% higher than the remaining compositions within this family of porous solids (Figure 12.2). This has been ascribed to an ordered distribution of vacancies for the copper analogues [11]. Prussian blue analogues can be prepared as substitutional solid solutions combining different metals in the T_i and T_e crystallographic positions. This enables the modulation of their adsorption potential for the hydrogen molecule within the cavities [67c,d].

12.5
Transition Metal Nitroprussides

The pentacyanonitrosylferrate anion, $[Fe(CN)_5NO]^{2-}$, forms porous solids with divalent transition metals (T_e), of formula unit $T_e[Fe(CN)_5NO]\cdot xH_2O$ [70]. These solids are usually known as nitroprussides. In the formed solids, the NO group remains unlinked at its O end and this determines the presence of free spaces within the material framework. When the NO groups from neighboring $[Fe(CN)_5NO]$ blocks are oriented to the same crystallographic position, corresponding to a T_e metal, a metal vacancy is generated, and to satisfy the charge neutrality, a vacancy for the building block appears at certain network positions. The absence of building blocks generates pores of about 9 Å diameter, which remain in communication through interstitial free spaces of about 4 Å diameter. This structural feature is found for T_e = Fe, Co, Ni, and the solids crystallize with a cubic unit cell. For the remaining metals, T_e = Mn, Cu, Zn, Cd, solids with different structures and a lower porosity are formed [71]. Similar to Prussian blue analogues, in cubic nitroprussides, the cavity surface has six metal centers with available coordination sites, which in the as-synthesized material are occupied by water molecules. These water molecules evolve at moderate heating without affecting the solid framework. This behavior and the structural features of cubic nitroprussides explain their interest for studies related to H_2 storage [10,71,72].

The recorded H_2 adsorption isotherms, at 75 K and up to 10 bar, for cubic nitroprussides, including their substitutional solid solutions of Ni and Co at the cavity surface, saturate (have a maximum) below that pressure. The pressure values, in bar, where the maximum is observed follow the order Ni(4.7) > $Ni_{0.5}Co_{0.5}$(4.3) > $Ni_{0.3}Co_{0.7}$(3.2) > Co(2.3). This is also the order of limit adsorption in excess. For Ni, the recorded isotherm saturates at higher pressure and the amount of hydrogen molecules accumulated within a cavity is larger within these compositions, close to 10 H_2/cavity, equivalent to approximately to a local H_2 density of about 77.6 g/l, higher than the density of liquid hydrogen (71 g/l). This is in correspondence with the value for the involved adsorption heat, which is 7.5 kJ/mol, above the sublimation enthalpy of H_2 (1.1 kJ/mol) [10]. Ni^{2+} is the most polarizing metal within the 3d series and this explains the H_2 adsorption behavior in Ni nitroprusside. It seems, the adsorption forces in this family of materials are dominated by the electrostatic interactions, with r^{-3} and r^{-4} dependences, which favors the pore filling with hydrogen molecules, according to the model of Figure 12.3.

The behavior of cubic nitroprussides for H_2 storage is different from that observed for Prussian blue analogues. This could be explained by the role of the NO group in the solid electronic structure. The nitrosyl ligand subtracts a large amount of electron density from the iron atom linked at its N atom by a π-back donation mechanism. This results in a minor availability of charge density on the iron atom to participate in the π-back donation with the CN groups. The charge removed by this way from the metal linked at the C end of the CN group is accumulated on its N atom and then partially donated to the metal bounded to this last atom, reducing its polarizing power. In summary, compared with Prussian blue analogues, in nitroprussides, the metal found at the cavity surface preserves a higher polarizing power and this favors the H_2 adsorption through electrostatic interactions. This example illustrates one of the many available routes to increase the adsorption potential for the H_2 molecule in porous solids.

12.6
Porous Organic Polymers

Common organic polymers have conformational and rotational freedom, and structures with efficient packaging. This implies that such materials do not normally have large surface areas. However, two small groups of light organic polymers with extended surface areas are known: hypercrosslinked organic polymers and polymers of intrinsic microporosity (PIMs) [73].

Hypercrosslinked polystyrene material has been obtained from suspension polymerization of vinylbenzyl chloride followed by a Friedel–Crafts-type post-crosslinking in dichloroethane at 80 °C using $FeCl_3$ as the catalyst. This material has shown BET and Langmuir surface areas of 1466 and 2138 m^2/g, respectively; and hydrogen adsorption capacity (excess) of 3.04 wt.% (the highest value reported up to the best of my knowledge for organic polymers), at 77 K and 15 bar [73b].

Another family of hypercrosslinked microporous polymers, with surface areas ranging from 600 to 1800 m^2/g has been prepared through template-free Friedel–Crafts alkylation reactions between benzene/biphenyl/1,3,5-triphenylbenzene as co-condensing rigid aromatic building blocks and 1,3,5-tris(bromomethyl)benzene or 1,3,5-tris(bromomethyl)-2,4,6-trimethylbenzene as cross-linkers under the catalysis of anhydrous $AlCl_3$ or $FeCl_3$. The highest hydrogen adsorption capacity (excess) achieved within this family was 1.9 wt. % at 77 K and 1 bar [74].

For organic polymers of intrinsic microporosity (PIMs) the hydrogen excess adsorption capacity is in the range 1.44–1.70 wt.%, at 77 K and 10 bar. When the pressure is reduced to 1 bar, the adsorption capacity was found to be lower, between 1.04 and 1.43%. These materials are composed wholly of fused ring subunits designed to provide highly rigid and contorted macromolecular structures that pack space inefficiently, forming solids with large amounts of interconnected free volume. The BET surface areas for these materials are in the 760–830 m^2/g range [73d].

In organic polymers, free of doping with species that favor electrostatic and coordination interactions, the adsorption forces for the hydrogen molecule are determined by dispersive interactions and this explains the low values obtained for H_2 storage.

12.7
Current State, Targets, and Perspectives on Materials for H_2 Storage

The globally accepted targets for H_2 storage are those established by the DOE (Department of Energy of the United States of America), which were readjusted in 2013: 7.5 wt% (0.055 kg H_2/kg system) and in volumetric terms, 70 g H_2/l system (Ultimate DOE targets) [75]. According to the required adsorption enthalpy to maintain the H_2 molecule as adsorbed species at room temperature, the target would be 20–30 kJ/mol. As already discussed, the H_2 adsorption takes place through three types of interactions, dispersive forces (van der Walls), of electrostatic nature (interaction of the molecule quadrupole moment with the local electric field gradient, polarization of the H_2 electron cloud by a charge center, and quadrupole–quadrupole), and by H_2 binding to a partially naked transition metal center (formation of Kubas H_2 complex, Figure 12.2a).

The H_2 Kubas complex is formed with extended metal 3d orbitals to allow their overlapping with the H_2 σ^* orbital. At the same time, a metal orbital is capable of accepting electron density from the hydrogen molecule through a lateral σ donation (Figure 12.1a). Copper is an atom with a large density of electrons in the 3d orbitals, but that feature favors the metal charge donation to the H_2 σ^* orbital and this induces the compensating lateral σ donation from the hydrogen molecule to the copper atom. The H_2 Kubas complexes were initially documented for bulky nd atoms, among them Os [9], but for obvious reasons, for H_2 storage only 3d metals are considered. As mentioned above, the binding of H_2 to 3d metal centers in zeolites, gels of hydrazine and silica, MOFs, and Prussian blue analogues is well supported from high-resolution neutron powder diffraction, neutron vibrational spectroscopy, solid-state ^{1}HNMR, and IR spectroscopy measurements. Both, the σ donation to the metal center, and the retrodonation of this last to the H_2 σ^* orbital induce an elongation in the H—H distances that is detected by these techniques [9].

According to the above-discussed summary on materials for H_2 storage, and on the involved adsorption heats, the dispersive type interaction provides a maximum of 7 kJ/mol [2a], it is 17.5 kJ/mol for electrostatic interactions [33], and up to 39 kJ/mol for the binding to Ti(III), V(III), Cr(II), and Mn(II) [35a,36]. Only the coordination interaction satisfies the target in terms of adsorption heat values (Figure 12.5). This type of interaction supports the adsorption of a single layer of H_2 molecules, and from this fact, the resulting storage density is low, below the DOE target. Theoretical calculations suggest that up to six H_2 molecules can be accommodated around an isolated metal [38]. For a metal that forms part of the surface of a cavity, two H_2 could be accommodated as

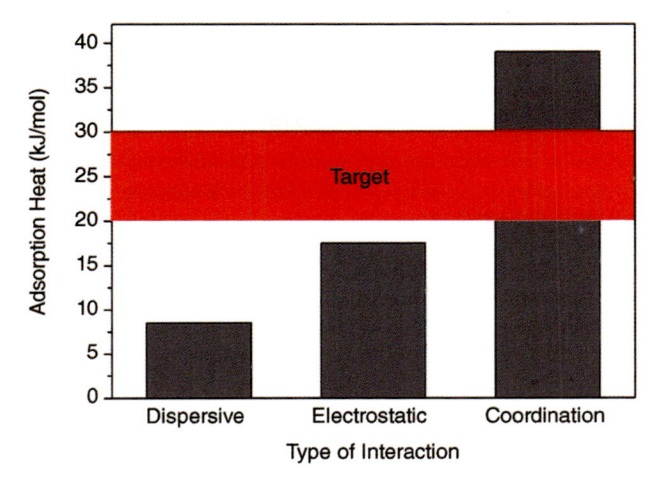

Figure 12.5 Guest–host interactions contribution to the H_2 adsorption heats and their relative position relative to the DOE target.

coordinated species, and four if the metal belongs to a molecular wire with the possibility of coordination around its axis. For Ti with an atomic weight of 47.867, these two configurations are equivalent to 8.3 and 16.3 wt% of H_2 storage. This suggests a possible research route to obtaining materials appropriate for H_2 storage, for example, the preparation of molecular wires containing metals able to form a Kubas H_2 complex (Figures 12.1a and 12.6). An ideal option would be the combination of the coordination interaction with electrostatic forces. Unfortunately, the metals with the highest polarizing power are unfavorable for the coordination interaction.

The above-mentioned results on H_2 storage in Ni nitroprusside [10] suggest that an appropriate design of porous solids with highly polarizing charge centers at the cavity surface, and highly anisotropic local electric field in the cavity

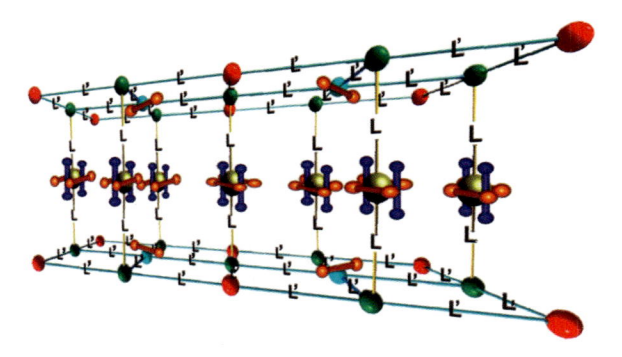

Figure 12.6 Model of porous solid to favor an optimal contribution of the coordination interaction to both, high H_2 storage capacity and large adsorption heat.

region, could be enough to produce local hydrogen density close to the ultimate DOE target (70 g/l system) and near to the density of solid hydrogen (86 g/l).

12.8
Materials for Methane Storage

The worldwide availability of natural gas, which is predominantly methane, has dramatically increased in the last two decades, because of development of new extraction technologies. This fact together with the high hydrogen to carbon ratio for methane (4: 1), the highest within fossil fuels, have stimulated an interest in the use of methane as a fuel for mobile technologies. As a gas, at standard temperature and pressure (STP), 298 K and 1 bar, its volumetric energy density is very low, 0.0378 MJ/l versus 34.2 MJ/l for gasoline, for instance. This explains that for massive introduction of methane as fuel in mobile technologies, its storage at high density is a requirement. Compression and liquefaction are the common routes to increase the methane density. In recent decades, the option of storage by adsorption in solids has gained interest by the appearance of lightweight materials with large surface area and pore volume to adsorb amounts of methane similar to those stored in vessels at high pressure. The properties of these materials and the methane adsorption in them are subjects to be discussed below. As in the case of hydrogen, the worldwide-accepted targets for methane storage are those established by DOE. These targets have changed along the last two decades at 35 bar and STP conditions, from 150 v/v in 1993, to 180 v/v nearly a decade later (2000), to the current value of 350 v/v and a gravimetric storage of 0.5 g (CH_4)/g (sorbent) [76]. Such an increase in the DOE target reveals the interest in the introduction of methane as fuel for mobile technologies.

As mentioned above, methane is a molecule free of a dipole or quadrupole moment, where the adsorption forces are determined by dispersive and polarization forces. The molecule's sensitivity to these interactions is determined by its polarizability (α), which is 2.60×10^{-24} cm^3, above three times the value for H_2 [7]. The value of α also determines its critical temperature (T_c), which is 190.56 K (−82.59 °C). This suggests that for temperatures above T_c the methane molecule must be adsorbed by forming a Langmuir type monolayer, and from this fact, the adsorption capacity could be determined by the material's accessible surface. Up to date, the materials of highest methane adsorption capacity are activated carbons and MOF-type porous solids with extremely high surface areas, particularly Langmuir surface areas. The amount of CH_4 storage in a given material depends of other factors, among them, the available free pore volume, the pore size and its distribution (MPSD), the pore geometry, and the guest–host interaction strength. Systematic studies on methane storage in activated carbons suggests that in the storage capacity, the MPSD is a relevant parameter. The highest adsorption capacities are observed for activated carbons with narrow pore size distributions, which not necessarily corresponds to the materials with the highest accessible surface area [77]. The presence of pores with a size that significantly exceeds the optimal value is equivalent to wasting free spaces in the solids. It seems, the better activated carbons for methane storage are those prepared by chemical routes under mild conditions, for example, using KOH as activation agent. Regarding the pore size and geometry, for slit type pores, an optimal value of 8 Å is reported, considering the adsorption of two adjacent layers of methane molecules, at the pore walls [78]. This value ignores the required mass transfer in the solid, which is limited by the molecule adsorption at the entrance of the porous framework channels. When this factor is considered, the optimal pore diameter is slightly higher. Experimental studies for methane adsorption in porous carbons show isotherms free of hysteresis for pore size in the 12–15 Å range [79]. Very narrow pores could affect the release of molecules from the porous framework, particularly because in practical applications, the discharge process takes place at a pressure close to atmospheric pressure. To the best of our knowledge, the highest adsorption capacity found for activated carbons (around 180 v/v) remains below the current DOE target (350 v/v) [76]. Regarding the adsorption enthalpy, in activated carbons it remains below 25 kJ/mol, which favors the discharge process at room conditions.

The presence of charge centers in the solid favors the adsorption of methane, through polarization of its electron cloud and the appearance of induced dipole and quadrupole moments in the molecule, which contribute to the adsorption forces. In materials designed for methane storage, this interaction must be tuned. If the porous framework contains a highly polarizing cation, for example, Na^+, K^+, the electrostatic interaction could be strong enough to prevent molecular diffusion through the solid free space. This is observed in the recorded adsorption isotherms through the appearance of pronounced hysteresis or very low adsorption, although the test with other adsorbates reveals the availability of a large free volume and pore windows of sufficient size to allow methane molecules access to the pores [80].

MOFs are the porous solids that have attracted greatest attention for methane storage in the last two decades. As already mentioned, the surface area, pore size and pore size distribution are critical parameters to obtain large methane adsorption capacity and to allow a fast charge–discharge kinetic in devices for methane storage. In activated carbons, these parameters are determined through the activation conditions without possibility of their fine control. Crystalline solids are highly attractive in that sense, because they have a regular structure (framework) with pores and access windows of well-defined shape and size. This means that MOFs with tailored structure and adsorption potential for methane storage can be prepared after a detailed design of their composition and structural features. Such an approach has proved to be effective [79]. In MOFs free of partially naked metal centers, the adsorption sites for methane are the pore surface regions rich in electron density, where the electron cloud is more polarizable and from this fact, stronger adsorption forces originate by dispersive interactions. For MOF-5, for instance, the adsorption site, according to structural studies, from high-resolution powder neutron diffraction data, is close to the metal oxide cluster [81]. In the ZIFs subset of MOFs, also free of unsaturated metal sites, the primary adsorption site is on top of the organic linkers [81]. This is a region rich in π-electron density that favors methane adsorption in the solid. The heat of adsorption for methane in this type of MOF remains below 14 kJ/mol [82], lower than the value reported for activated carbons. This is probably related to molecular confinement, which is greater in activated carbons of slit type pores. In MOFs, the pore size is usually larger than the optimal value for methane adsorption through dispersive interactions and at room temperature, a fraction of the available volume is wasted. Regarding storage capacity, for MOF-5 and ZIF-8, it is 110 and 99 v/v (STP) [83]. In standard conditions, methane hydrates (clathrates) have a storage capacity of 176 v/v. This fact has stimulated the study of methane adsorption in the presence of water molecules in the ZIF-8 cavities. The results correspond to an increase of 56% in the v/v (STP) storage capacity by formation of methane hydrates within the cavities [83b]. Unlike MOFs based on carboxylate groups, which are unstable in humid air, ZIFs are stable in aqueous media. The adsorption of methane in MOFs at temperatures below its condensation point, $\approx$111 K, sheds light on the state of methane as a confined condensed phase within the cavities. For MOF-5 and ZIF-8, a structural transition appears close to 60 K, related to interactions between neighboring molecules; this transition is not observed in bulk methane at low temperature [81]. The physical properties of condensed methane phases as small molecular clusters remains poorly studied. The use of porous solids to form these clusters in a confined region provides an opportunity for such studies.

In the study of MOFs for methane storage, compositions that contain metals with an unsaturated coordination sphere have received special attention, as they favor electrostatic interactions with guest molecules. For the MOFs series M_2(dhtp) with M = Mg, Mn, Co, Ni, Zn, the reported storage capacity, in v/v (STP), is 149, 158, 174, 190, 171, respectively [81], close to the theoretical value, in the 160–170 v/v (STP) range, supposing the adsorption of a methane molecule per metal center. The reported adsorption heats in that study are, in kJ/mol, 18.5 (Mg), 19.1 (Mn), 19.6 (Co), 20.2 (Ni), and 18.3 (Zn). These values of adsorption heats are significantly higher than those found for MOFs free of unsaturated metal sites, for example, <14 kJ/mol for MOF-5 and ZIF-8. Neutron diffraction studies corroborate that the primary adsorption site is the metal center [84]. Once these adsorption sites are occupied, the following molecules are positioned close to the organic linkers, particularly close to regions rich in electron density. The same behavior has been reported for many other MOFs with partially naked metal centers. For PCN-14, $Cu_2(H_2O)_2$(adip)·2DMF (adip = 5,5′-(9,10-anthracenediyl)di-isophthalate), with partially naked copper atoms at the cavity surface, the methane storage capacity is close to 230 cm^3 (STP)/cm^3 with adsorption heat of 30 kJ/mol [85].

In addition to large cavities, the framework of many MOFs has pocket-shaped pores with a size appropriate to accommodate methane molecules, which have a kinetic diameter of 0.38 nm [7]. In these small pores, the guest molecule confinement favor a strong dispersive interaction with the host solid, with adsorption energies similar to those found for activated carbons of slit type pores of about 0.8 nm. This is the case, for instance, for $Cu_3(BTC)_2$ (BTC = benzene-1,3,5-tricarboxylate), also known as HKUST-1. Its porous framework is formed by two types of large intersecting 3D channels, with channel diameters of about 0.9 nm. At the surface of one of these large channels, copper atoms with unsaturated coordination sphere are found. In addition to these large pores, the framework contains smaller pocket-shaped pores delimited by the $C_3H_6(O_2C)_3$ ligand moieties. These small pores are hydrophobic and remain in communication with the large channel through triangular windows of about 0.38 nm [86]. Both, the pore size and the hydrophobic nature of the framework are favorable features for methane adsorption in that free space. For methane storage in $Cu_3(BTC)_2$ probably a combination of factors, availability of unsaturated copper sites and the pocket-shaped pores contribute. The reported storage capacity of this material is 230 v (STP)/v, with an adsorption heat of 21 kJ/mol [87].

Methane is a gas with greenhouse effect, with undesirable impact on the environment; however, there is about 220 times less methane in the Earth's atmosphere than carbon dioxide so that the methane contribution to the greenhouse effect is of minor relevance compared with the role of carbon dioxide. For a hypothetical scenario of an increasing liberation of methane to the atmosphere, for example, by inappropriate operation of gas natural extraction technologies, the above mentioned materials, nowadays under study for methane storage, could be used for methane capture.

12.9
Methane and Hydrogen Clathrates

As discussed in Chapter 11 of this volume, the most interesting clathrates for gas storage are those where water molecules maintained together through hydrogen bond interactions form the host framework, called hydrates. Such interactions are relatively weak and by this reason for their formation and stability, high pressures and low temperatures are required.

12.9.1
Methane and Natural Gas Hydrates

The best known gas hydrates are those formed with methane and natural gas, for two reasons: (1) the largest global reserves of methane (above 90% of the natural gas composition) are found in the form of water clathrates in the deep of oceans, below the ocean floor, where the water temperature remains around 2 °C and the water–gas mixture remains under high pressure; (2) by its formation in routine operations of natural gas processing, when liquid water is condensed in the presence of methane at relatively high pressure, for example, in pipelines of natural gas conduction, where the formed methane or gas natural hydrates are responsible for the pipelines blocking.

Regarding clathrate formation, methane and natural gas show a different behavior, resulting from the presence of small amounts ($\approx$3%) of larger hydrocarbons in natural gas, like propane and isobutene, which determines the type of structure formed and its thermal stability. Methane molecules fit into both, small and large cavities of *sI*, and from this fact, the methane clathrate crystallizes with the *sI* structure. The presence of even small amounts of propane inhibits the formation of the *sI* form because this molecule can not be accommodated within its cages, and the clathrate is formed in the *sII* form. This explains that the majority of naturally occurring deposits of gas hydrates are found with *sI* structure because they are composed of methane from biogenic sources, which are free of heavier hydrocarbons. When the natural gas deposit has thermogenic nature, the presence of heavier hydrocarbons determines the formation of hydrates in the *sII* form [5,88]. The current studies on hydrates of methane and natural gas are oriented to: (1) development of technologies for methane recovery from the available clathrates under the ocean and permafrost; (2) have a tunable control for the clathrates formation kinetic that makes their applications for gas storage possible; (3) inhibit clathrate formation and/or reduce their stability when they can be formed in oil and natural gas transport pipelines. Concerning methane recovery from oceanic deposits, a successful production test was reported in March 2013 using depressurization at about 1 km depth and 270 m below the ocean floor [89].

From the study of the formation conditions for natural gas clathrates, valuable information for methane storage as clathrates is obtained. For this last application, the clathrate must be formed at technologically viable pressure and temperature values, that is, a low pressure and noncryogenic temperature. The addition of small amounts of propane (0.01 mol%), for instance, to the methane to be stored, favors storage conditions at lower pressures. At 277.6 K, the hydrate of pure propane (*sII* structure) is formed at 4.3 bar. For pure methane, at the same temperature (*sI* structure), the required pressure is 40.6 bar, and for a 50 : 50 mixture of methane and propane, it is 8.0 bar [5b]. Many other molecules are available to reduce the formation temperature for methane hydrates, among them, THF [90]. These accompanying molecules are known as promoters and determine the structure to be formed and the hydrate stability. The methane storage capacity of *sII* is 15.7 wt%, similar to the storage capacity for the *SI* form, 15.6 wt%.

The efforts on the study of natural gas clathrates have two objectives: (1) inhibit or retard their formation in natural gas transport pipelines. For this purpose, different additives are commonly used, among them, those that reduce the stability of the hydrogen bonds between water molecules, for example, additives that contains OH, NH_2, NHCO, and SO_3H groups, which compete by the hydrogen bonds with the water molecules [51]; (2) natural gas storage using pellets of hydrates to allow its transport and distribution in far sites or to facilitate certain applications of this energy vector [5a].

12.9.2
Hydrogen Storage in Clathrates

Although hydrogen clathrates do not exist in nature; they have been prepared and studied to evaluate their potentiality for H_2 storage [6,91]. A reduced number of molecular systems containing H_2 as guest species have resulted from these studies, for example, H_2–CH_4 [6c,92], H_2–H_2O [6c,91], H_2–Ar [93], H_2–Xe, H_2–SiH_4, H_2–GeH_4, and H_2–NH_3BH_3 [94]. Within these systems, the highest H_2 storage capacity corresponds to clathrates formed by methane as host structure, 33.4 wt%. Because the host framework is formed by dispersive interactions between methane molecules, the required pressure to obtain such clathrates is extremely high, above 5 GPa [95]. From a practical point of view, the most attractive systems are the hydrogen hydrates, for several reasons: (1) this is a friendly storage method because it involves a reversible use of water; (2) hydrogen is stored in its molecular form; (3) within the cages, a relatively high amount of H_2 can be accommodated, close to the DOE targets for some compositions; (4) the stored hydrogen is easily recovered by thermal stimulation and depressurizing the system, and with minimal risks of explosion; (5) using appropriate promotors, the storage is possible at moderate conditions of pressure and temperature. The main drawbacks of this storage method continue to be: (a) the relatively high pressures and low temperatures required to stabilize the clathrate structure, quite different from ambient conditions; (b) the slow kinetics that characterize the process of clathrate formation process, in some cases it takes days; (c) the relatively low hydrogen storage capacity when clathrate formation promoters are used to favor a faster formation kinetic under mild pressure and temperature values [94]. The promoters are accompanying guest species that favor the formation of a given clathrate structure and allow the tuning of their stability conditions. In this sense, a well-known promoter is tetrahydrofuran (THF). The role of THF is similar to that mentioned above for propane in the formation and stability of methane clathrates. THF promotes the clathrate crystallization with the *sII* structure, where the hydrogen molecules are found occupying the small cages while THF is accommodated in the largest ones, together with H_2 molecules, depending on the amount of THF as guest species. For hydrogen hydrates using THF as promotor, a pressure reduction of 60 times (from 300 to 5 MPa) at 279.6 K has been reported, relative to the same system free of THF [90a]. For hydrogen hydrates, the three above-mentioned structures (*sI*, *sII* and *sH*) have been reported. In addition, a distinct structure, named *sVI*, has been identified for hydrates formed with *tert*-butyl amine and hydrogen as guest species. This structure has two types of cages, 8-hedra (4^45^4) and 17-hedra ($4^35^96^27^3$) [96]. A set of hydrates stabilized with ammonium salts, for example, tetrabutyl ammonium fluoride, chloride, bromide, and nitrate has been reported [97]. These relatively large molecules behave as promotors, significantly reducing the pressure required to have stable hydrates at relatively high temperatures. The anion of these salts forms part of the hydrate host while the cation occupies the available space in the formed cages, particularly within a large cages. It seems that electrostatic anion–cation attractive interaction favor the stability of the formed hydrate. These hydrates are known as semiclatharates.

Theoretical studies have estimated up to 10 wt% of H_2 storage capacity in *sI* hydrate, with five molecules occupying the small cages (5^{12}) and seven accommodated in the $5^{12}6^2$ cages, although these predictions are not validated by experimental results. This discrepancy has been ascribed to the release of guest molecules through the hexagonal faces of the host framework. The blocking of these releasing windows with methane molecules results in an increase for H_2 storage capacity [98]. Hydrogen hydrates of *sII* type are probably the most studied clathrates for hydrogen storage, because of the possibility of tuning the pressure and temperature of formation combining hydrogen with accompanying molecules that stabilizes this structure, as already mentioned when THF participates as guest molecule. Cyclopentane, cyclohexanone, furan, 2,5-dihydrofuran, 1,3-dioxalane, tetrahydropyran, and other small molecules play a role similar to THF. The *sII* structure could also be stabilized combining H_2 with other gaseous molecules, among them, methane, ethane, and propane [99]. Hydrogen hydrates with *sH* structure are also obtaining for mixed guest molecules. This structure has three types of cages, of small, medium, and large size. H_2 is mainly accommodated in small and medium size cages while the stabilizing molecule, usually >7 Å, occupies the largest cages [100]. Hydrates with *sVI* structure appear particularly attractive for H_2 storage; up to 6 wt% of H_2 could be stored in this mixed hydrate [96].

12.10
Materials for Dissociative Hydrogen Storage

The moderate advances in hydrogen storage as molecular species have stimulated the search for alternative routes. Two of these routes will be briefly discussed in the following section, storage as hydrides and through spillover in metal doped porous solids, both involving the dissociation of hydrogen molecules. Three types of hydrides are usually considered: simple metal hydrides, complex metal hydrides, and chemical hydrides.

Simple metal hydrides, MH_x, result from a direct reaction between molecular hydrogen and the metal. For alkaline metals, alkaline earths and Al, the formed hydride has ionic nature. The most extensively studied metal hydrides include: LiH, NaH, MgH_2, CaH_2, and AlH_3, with hydrogen content from 4.2 (NaH) to 12.6 wt% (LiH). Their applications are limited by the large energy barrier to overcome for the dehydration reaction. Hydrogen can also be stored as transition metal hydrides, for example, $FeTiH_x$, where the hydrogen atom occupies interstitial positions in the metal lattice. In addition to the relatively high temperatures required for hydrogen release from these materials, the gravimetric storage capacity is low.

Complex hydrides include aluminum hydrides, amides, and borohydrides. Of these complexes, only aluminum hydrides have ionic nature. This last series comprise several compositions: $NaAlH_4$, $KAlH_4$, Na_2LiAlH_6, K_2LiAlH_6, K_2NaAlH_6, and $KAlH_6$. The dehydrogenation of these last complexes takes place through at least two steps. For instance, on heating, $NaAlH_4$ decomposes around 160 °C to form $NaH + Al$, releasing 4 wt% of H_2. This last mixture can then be rehydrogenated at moderate conditions [101]. The analogous behavior has been reported for the remaining complex aluminum hydrides [102]. The charge–discharge cycles are favored in the presence of titanium compounds as catalyst, for example, $TiCl_3$, $TiCl_4$, $TiBr_4$, TiF_3, and $Ti(OBu^n)_4$. Unfortunately, the cycling reversibility is limited even if a catalyst is used [103]. Furthermore, the hydrogenation enthalpy for metal hydrides is relatively high, >33 kJ/mol H_2, which negatively impacts the onboard refueling vehicles during the potential applications of these materials for hydrogen storage in mobile technologies.

Chemical hydrides have been explored as materials for hydrogen storage because these compounds show high storage capacity, sufficient to satisfy the DOE targets. Chemical hydrides that have received great attention for hydrogen storage are amides and borohydrides. Alkali-metal amides are well known reagents for their use in synthetic organic chemistry. $LiNH_2$ is a prototypical amide whose reaction with LiH liberates a hydrogen molecule according to: $LiNH_2 + 2LiH \rightarrow Li_2NH + LiH + H_2 \rightarrow Li_3N + 2H_2$ [104]. These reactions release 5.5 and 5.2 wt% of H_2, respectively. Analogous reactions are known for other alkali and alkali earth metal amides [103a]. For Mg, the reaction is $Mg(NH_2)_2 + 4LiH \rightarrow (1/3)Mg_3N_2 + (4/3)Li_3Ni + 4H_2$, which corresponds to an overall storage capacity of 9.1 wt%. The main drawback of amides for this application is their decreasing storage capacity on adsorption–desorption cycles, by decomposition with evolution of ammonia [103a].

Borohydrides (tetrahydroborates) can be prepared by direct reaction between the metal, boron and hydrogen on heating in the 550–700 °C and under an H_2 pressure of 3–15 MPa, according to: $M + B + 2H_2 \rightarrow MBH_4$ where M = Li, Na, K, (1/2)Mg, ... [105]. The inverse reaction liberates part of the hydrogen used in the preparation reaction. Some borohydrides release the hydrogen spontaneously at room temperature, for example, $Al(BH_4)_3$ [103a]. This compound is liquid at room temperature and has a hydrogen storage capacity of 17 wt%. One of the main limitations of borohydrides for hydrogen storage is their low formation kinetics and decomposition with evolution of diborane [103a].

In summary, the current state of the art concerning hydrogen storage as hydrides reveals that these materials have potential for hydrogen storage but their stability on prolonged cycles of charge–discharge together with low temperature operation and appropriate kinetics for these two processes are desirable features addressed by current research in this area. Probably, the best candidates correspond to alkali and alkali earth complex hydrides using appropriate catalysts.

Hydrogen storage in porous solids through spillover has received increasing attention because of its potential to meet the DOE targets. In the limiting case of H_2 coordination to a metal center, the charge donated by the metal to the H_2 σ^* orbital could be sufficient to induce its σ bond rupture, producing two protons and two electrons. Both, protons and electrons are reactive species able to diffuse through the solids using electronic and ionic conduction channels to the reaction sites; which are defects and reducible species for the electrons, and atoms (C, O, S, ...) with available orbitals to form a bond with a proton. In the presence of water molecules, the formation of hydronium ions, $(H_3O)^+$, favors the proton mobility in the solid. The physical mechanism involved in hydrogen spillover has been discussed in excellent review papers [106], although it is only partially understood. The rupture of the hydrogen molecule supposes that the host solid contains metal centers suitable for the catalytic reaction. Usually, these catalytic centers are Pt, Pd, or Rh nanoparticles. These are metals with extended nd orbitals, with large electron density in their metallic state, features that favor H_2 coordination interaction to the metals and H—H bond rupture (Figure 12.2a).

Activated carbon is the material that has received most attention for hydrogen storage through the spillover mechanism [107]. The doping of activated carbon with nanoparticles of reducible metals, for example, Pt, Pd, is relatively simple. For these reasons, composites between activated carbons and MOFs, exploiting the qualities of both, are also attractive materials for hydrogen storage by spillover [107]. For instance, the composite Pt-ACs-MOF-5 shows a hydrogen storage capacity of 2.3 wt% at 298 K and 100 bar [108]. In the analogous composite using IRMOF-8, the reported storage capacity is 4 wt% under similar conditions of temperature and pressure and with estimated adsorption heats in the 20–23 kJ/mol range [109]. Hydrogen storage by this mechanism is also possible for MOFs without mixing with activated carbons, for zeolites and other solids of extended surface [107b]. The spillover phenomenon is used to explain the behavior of many catalytic reactions [106].

12.11
Outlook on Materials for H_2 and CH_4 Storage

Two parameters determine the ability of a given material for gas storage, its adsorption capacity at a given values of pressure and temperature, and the corresponding heat of adsorption. For H_2, practically all the materials evaluated to date remain short with respect to the DOE ultimate targets. In that sense, four research areas appear to be the most promising to achieve the required targets in the short and medium term: (1) design, preparation, and evaluation of porous solids that combine large pore volume, light weight, stability under operation conditions, and adsorption heats in the 20–30 kJ/mol range; (2) availability of hydrogen hydrates with fast formation kinetic at technologically viable values of pressure and temperature; (3) preparation of complex hydrides with low hydrogenation and dehydrogenation temperatures and large stability on cycling use; (4) promote the preparation and study of porous solids able for the hydrogen storage at room conditions through the hydrogen spillover.

The storage of methane at technologically viable conditions is possible by adsorption in porous solids and as methane hydrates. In the near future, the main efforts will probably be aimed to optimize the performance of these materials in order to have larger storage capacity and attractive economical operation conditions, at least compared with the use of oil derivatives as energy vectors. The potential use of methane as energy vector for mobile technologies is stimulating the study and technological development for its storage at high gravimetric density. The use of methane as raw material in the chemical industry, instead of oil, probably favors a progressive development of technologies, and of course, of materials for its storage.

Acknowledgment

The preparation of this overview on functional materials for gases storage was partially supported by the CONACyT project 2013-231461, particularly financing the stay of Dr. L. Reguera as visiting researcher at CICATA-IPN (Mexico).

References

1 Furukawa, H., Cordova, K.E., O'Keeffe, M., and Yaghi, O.M. (2013) *Science*, **341**, 974–986.
2 (a) Reguera, E. (2009) *Int. J. Hydrogen Energy*, **34**, 9163–9167. (b) Bhatia, S.K. and Myers, A.L. (2006) *Langmuir*, **22**, 1688–1700.
3 (a) Hyun, S.-M., Lee, J.H., Jung, G.Y., Kim, Y.K., Kim, T.K., Jeoung, S., Kwak, S.K., Moon, D., and Moon, H.R. (2016) *Inorg. Chem.*, **55**, 1920–1925. (b) Casco, M.E., Cheng, Y.Q., Daemen, L.L., Fairen-Jimenez, D., Ramos-Fernandez, E.V., Ramirez-Cuesta, A.J., and Silvestre-Albero, J. (2016) *Chem. Commun.*, **52**, 3639–3642. (c) Choi, H.J., Dincă, M., and Long, J.R. (2008) *J. Am. Chem. Soc.*, **130**, 7848–7850. (d) Kitaura, R., Seki, K., Akiyama, G., and Kitagawa, S. (2003) *Angew. Chem., Int. Ed.*, **42** 428–431.
4 Zhang, Y.-h., Jia, Z.-C., Yuan, Z.-M., Yang, T., Qi, Y., and Zhao, D.-L. (2015) *J. Iron Steel Res. Int.*, **22**, 757–770.
5 (a) Koh, C.A., Sloan, E.D., Sum, A.K., and Wu, D.T. (2011) *Annu. Rev. Chem. Biomol. Eng.*, **2**, 237–257. (b) Sloan, E.D. and Koh, C.A., (2008) *Clathrate Hydrates of Natural Gases*, 3rd edn, CRC Press, Boca Raton.
6 (a) Struzhkin, V.V., Militzer, B., Mao, W.L., Mao, H.K., and Hemley, R.J. (2007) *Chem. Rev.*, **107**, 4133–4151. (b) Dyadin, Y.A., Larionov, E.G., Manakov, A.Y., Zhurko, F.V., Aladko, E.Y., Mikina, T.V., and Komarov, V.Y. (1999) *Mendeleev Commun.*, **9**, 209–210. (c) Mao, W.L. and Mao, H.-K. (2004) *Proc. Natl. Acad. Sci. USA*, **101**, 708–710.
7 Lide, D.R. (2004) *Handbook of Chemistry and Physics*, CRC Press, Fl., USA.
8 Lochan, R.C. and Head-Gordon, M. (2006) *Phys. Chem. Chem. Phys.*, **8**, 1357–1370.
9 Kubas, G.J. (2007) *Chem. Rev.*, **107**, 4152–4205.
10 Reguera, L., Roque, J., Hernández, J., and Reguera, E. (2010) *Int. J. Hydrogen Energy*, **35**, 12864–12869.
11 Jiménez-Gallegos, J., Rodríguez-Hernández, J., Yee-Madeira, H., and Reguera, E. (2010) *J. Phys. Chem. C*, **114**, 5043–5048.
12 Gross, K.J., Carrington, K.R., Barcelo, S., Karkamkar, A., Purewal, J., Ma, S., Dantzer, P., Ott, K., Burrell, T., Semeslberger, T., Pivak, Y., Dam, B., and Chandra, D. (2013) Recommended Best Practices for the Characterization of Storage Properties of Hydrogen Storage Materials. Available at https://www1.eere.energy.gov/hydrogenandfuelcells/pdfs/best_practices_hydrogen_storage.pdf, 579.

13 (a) Zhou, L. (2002) *Adsorption: Theory, Modeling and Analysis* (ed. J. Tóth), Marcel Dekker, Inc., NY. (b) Amankwah, K.A.G. and Schwarz, J.A. (1995) *Carbon*, **33**, 1313–1319.
14 Yu, E.K., Stewart, D.A., and Tiwari, S. (2008) *Phys. Rev. B*, **77**, 195406.
15 Tellez-Juárez, M.C., Fierro, V., Zhao, W., Fernández-Huerta, N., Izquierdo, M.T., Reguera, E., and Celzard, A. (2014) *Int. J. Hydrogen Energy*, **39**, 4996–5002.
16 Thomas, K.M. (2007) *Catal. Today*, **120**, 389–398.
17 (a) Zhao, X.B., Xiao, B., Fletcher, A.J., and Thomas, K.M. (2005) *J. Phys. Chem. B*, **109**, 8880–8888. (b) Zhao, X., Villar-Rodil, S., Fletcher, A.J., and Thomas, K.M. (2006) *J. Phys. Chem. B*, **110**, 9947–9955.
18 (a) Kuznetsova, A., Mawhinney, D.B., Naumenko, V., Yates, J.T., Liu, J., and Smalley, R.E. (2000) *Chem. Phys. Lett.*, **321**, 292–296. (b) Panella, B., Hirscher, M., and Roth, S. (2005) *Carbon*, **43**, 2209–2214.
19 (a) Benard, P., Chahine, R., Chandonia, P.A., Cossement, D., Dorval-Douville, G., Lafi, L., Lachance, P., Paggiaro, R., and Poirier, E. (2007) *J. Alloys Compd.*, **446–447**, 380–384. (b) Poirier, E., Chahine, R., Tessier, A., and Bose, T.K. (2005) *Rev. Sci. Instrum.*, **76**, 055101–055106.
20 Ye, Y., Ahn, C.C., Witham, C., Fultz, B., Liu, J., Rinzler, A.G., Colbert, D., Smith, K.A., and Smalley, R.E. (1999) *Appl. Phys. Lett.*, **74**, 2307–2309.
21 (a) Yang, R.T. (2000) *Carbon*, **38**, 623–626. (b) Pinkerton, F.E., Wicke, B.G., Olk, C.H., Tibbetts, G.G., Meisner, G.P., Meyer, M.S., and Herbst, J.F. (2000) *J. Phys. Chem. B*, **104**, 9460–9467.
22 (a) Poirier, E., Chahine, R., and Bose, T.K. (2001) *Int. J. Hydrogen Energy*, **26**, 831–835. (b) Benard, P. and Chahine, R. (2007) *Scr. Mater.*, **56**, 803–808.
23 Anson, A., Jagiello, J., Parra, J.B., Sanjuan, M.L., Benito, A.M., Maser, W.K., and Martinez, M.T. (2004) *J. Phys. Chem. B*, **108**, 15820–15826.
24 Züttel, A., Sudan, P., Mauron, P., and Wenger, P. (2004) *Appl. Phys. A Mater. Sci. Process.*, **78**, 941–946.

25 Park, N., Choi, K., Hwang, J., Kim, D.W., Kim, D.O., and Ihm, J. (2012) *Proc. Natl. Acad. Sci. USA*, **109**, 19893–19899.
26 Xia, Y., Yang, Z., and Zhu, Y. (2013) *J. Mater. Chem. A*, **1**, 9365–9381.
27 Bhat, V.V., Contescu, C.I., and Gallego, N.C. (2010) *Carbon*, **48**, 2361–2364.
28 (a) Dash, R., Chmiola, J., Yushin, G., Gogotsi, Y., Laudisio, G., Singer, J., Fischer, J., and Kucheyev, S. (2006) *Carbon*, **44**, 2489–2497. (b) Gogotsi, Y., Dash, R.K., Yushin, G., Yildirim, T., Laudisio, G., and Fischer, J.E. (2005) *J. Am. Chem. Soc.*, **127**, 16006–16007.
29 (a) Cho, J.H., Yang, S.J., Lee, K., and Park, C.R. (2011) *Int. J. Hydrogen Energy*, **36**, 12286–12295. (b) Sankaran, M., Viswanathan, B., and Srinivasa Murthy, S. (2008) *Int. J. Hydrogen Energy*, **33**, 393–403. (c) Sevilla, M., Fuertes, A.B., and Mokaya, R. (2011) *Int. J. Hydrogen Energy*, **36**, 15658–15663. (d) Zhao, W., Fierro, V., Fernández-Huerta, N., Izquierdo, M.T., and Celzard, A. (2013) *Int. J. Hydrogen Energy*, **38**, 10453–10460.
30 Nijkamp, M.G., Raaymakers, J.E.M.J., van Dillen, A.J., and de Jong, K.P. (2001) *Appl. Phys. A Mater. Sci. Process.*, **72**, 619–623.
31 Langmi, H.W., Book, D., Walton, A., Johnson, S.R., Al-Mamouri, M.M., Speight, J.D., Edwards, P.P., Harris, I.R., and Anderson, P.A. (2005) *J. Alloys Compd.*, **404–406**, 637–642.
32 (a) Areán, C.O., Palomino, G.T., Carayol, M.R.L., Pulido, A., Rubeš, M., Bludský, O., and Nachtigall, P. (2009) *Chem. Phys. Lett.*, **477**, 139–143. (b) Jhung, S.H., Sun Lee, J., Woong Yoon, J., Pyo Kim, D., and Chang, J.-S. (2007) *Int. J. Hydrogen Energy*, **32**, 4233–4237. (c) Palomino, G.T., Bonelli, B., Areán, C.O., Parra, J.B., Carayol, M.R.L., Armandi, M., Ania, C.O., and Garrone, E. (2009) *Int. J. Hydrogen Energy*, **34**, 4371–4378. (d) Torres, F.J., Vitillo, J.G., Civalleri, B., Ricchiardi, G., and Zecchina, A. (2007) *J. Phys. Chem. C*, **111** 2505–2513.
33 (a) Torres, F.J., Civalleri, B., Terentyev, A., Ugliengo, P., and Pisani, C. (2007) *J. Phys. Chem. C*, **111**, 1871–1873. (b) Otero Areán, C., Turnes Palomino, G., and Llop Carayol, M.R. (2007) *Appl.*

Surf. Sci., **253**, 5701–5704. (c) Turnes Palomino, G., Llop Carayol, M.R., and Otero Arean, C. (2006) *J. Mater. Chem*., **16**, 2884–2885.

34 (a) Georgiev, P.A., Albinati, A., Mojet, B.L., Ollivier, J., and Eckert, J. (2007) *J. Am. Chem. Soc.*, **129**, 8086–8087. (b) Serykh, A.I. and Kazansky, V.B. (2004) *Phys. Chem. Chem. Phys.*, **6**, 5250–5255.

35 (a) Hamaed, A., Hoang, T.K.A., Trudeau, M., and Antonelli, D.M. (2009) *J. Organomet. Chem.*, **694**, 2793–2800. (b) Hamaed, A., Trudeau, M., and Antonelli, D.M. (2008) *J. Am. Chem. Soc.*, **130**, 6992–6999.

36 (a) Hamaed, A., Hoang, T.K.A., Moula, G., Aroca, R., Trudeau, M.L., and Antonelli, D.M. (2011) *J. Am. Chem. Soc.*, **133**, 15434–15443. (b) Hoang, T.K.A., Morris, L., Rawson, J.M., Trudeau, M.L. and Antonelli, D.M. (2012) *Chem. Mater.*, **24**, 1629–1638. (c) Hoang, T.K.A., Morris, L., Sun, J., Trudeau, M.L., and Antonelli, D.M. (2013) *J. Mater. Chem. A*, **1**, 1947–1951. (d) Hoang, T.K.A., Webb, M.I., Mai, H.V., Hamaed, A., Walsby, C.J., Trudeau, M., and Antonelli, D.M. (2010) *J. Am. Chem. Soc.*, **132** 11792–11798.

37 (a) Skipper, C.V., Antonelli, D.M., and Kaltsoyannis, N. (2012) *J. Phys. Chem. C* **116**, 19134–19144. (b) Skipper, C.V., Hamaed, A., Antonelli, D.M., and Kaltsoyannis, N. (2010) *J. Am. Chem. Soc.*, **132**, 17296–17305.

38 Gagliardi, L. and Pyykkö, P. (2004) *J. Am. Chem. Soc.*, **126**, 15014–15015.

39 (a) Furukawa, H., Ko, N., Go, Y.B., Aratani, N., Choi, S.B., Choi, E., Yazaydir, A.Ö., Snurr, R.Q., O'Keeffe, M., Kim, J., and Yaghi, O.M. (2010) *Science*, **329**, 424–428. (b) Langmi, H.W., Ren, J., North, B., Mathe, M., and Bessarabov, D. (2014) *Electrochim. Acta*, **128**, 368–392. (c) Dincă, M. and Long, J.R. (2008) *Angew. Chem., Int. Ed.*, **47**, 6766–6779. (d) Dinca, M., Dailly, A., Liu, Y., Brown, C.M., Neumann, D.A., and Long, J.R. (2006) *J. Am. Chem. Soc.*, **128** 16876–16883.

40 Wong-Foy, A.G., Matzger, A.J., and Yaghi, O.M. (2006) *J. Am. Chem. Soc.*, **128**, 3494–3495.

41 Hayashi, H., Cote, A.P., Furukawa, H., O'Keeffe, M., and Yaghi, O.M. (2007) *Nat. Mater.*, **6**, 501–506.

42 Park, K.S., Ni, Z., Coté, A.P., Choi, J.Y., Huang, R., Uribe-Romo, F.J., Chae, H.K., O'Keeffe, M., and Yaghi, O.M. (2006) *Proc. Natl. Acad. Sci. USA*, **103**, 10186–10191.

43 Wu, H., Zhou, W., and Yildirim, T. (2007) *J. Am. Chem. Soc.*, **129**, 5314–5315.

44 Rosi, N.L., Eckert, J., Eddaoudi, M., Vodak, D.T., Kim, J., O'Keeffe, M., and Yaghi, O.M. (2003) *Science*, **300**, 1127–1129.

45 Yang, S., Lin, X., Dailly, A., Blake, A.J., Hubberstey, P., Champness, N.R., and Schröder, M. (2009) *Chem. Eur. J.*, **15**, 4829–4835.

46 Mulfort, K.L. and Hupp, J.T. (2007) *J. Am. Chem. Soc.*, **129**, 9604–9605.

47 Dincă, M. and Long, J.R. (2007) *J. Am. Chem. Soc.*, **129**, 11172–11176.

48 Yang, S., Martin, G.S.B., Titman, J.J., Blake, A.J., Allan, D.R., Champness, N.R., and Schröder, M. (2011) *Inorg. Chem.*, **50**, 9374–9384.

49 Chu, C.-L., Chen, J.-R., and Lee, T.-Y. (2012) *Int. J. Hydrogen Energy*, **37**, 6721–6726.

50 Mulfort, K.L. and Hupp, J.T. (2008) *Inorg. Chem.*, **47**, 7936–7938.

51 Tariq, M., Rooney, D., Othman, E., Aparicio, S., Atilhan, M., and Khraisheh, M. (2014) *Ind. Eng. Chem. Res.*, **53**, 17855–17868.

52 Yang, S., Lin, X., Blake, A.J., Thomas, K.M., Hubberstey, P., Champness, N.R., and Schroder, M. (2008) *Chem. Comm.*, 6108–6110.

53 Yang, S., Callear, S.K., Ramirez-Cuesta, A.J., David, W.I.F., Sun, J., Blake, A.J., Champness, N.R., and Schroder, M. (2011) *Faraday Discuss.*, **151**, 19–36.

54 Calleja, G., Botas, J.A., Sánchez-Sánchez, M., and Orcajo, M.G. (2010) *Int. J. Hydrogen Energy*, **35**, 9916–9923.

55 (a) Nouar, F., Eckert, J., Eubank, J.F., Forster, P., and Eddaoudi, M. (2009) *J. Am. Chem. Soc.*, **131**, 2864–2870. (b) Sumida, K., Horike, S., Kaye, S.S., Herm, Z.R., Queen, W.L., Brown, C.M., Grandjean, F., Long, G.J., Dailly, A., and

Long, J.R. (2010) *Chem. Sci.*, **1**, 184–191. (c) Sumida, K., Stück, D., Mino, L., Chai, J.-D., Bloch, E.D., Zavorotynska, O., Murray, L.J., Dincă, M., Chavan, S., Bordiga, S., Head-Gordon, M., and Long, J.R. (2013) *J. Am. Chem. Soc.*, **135** 1083–1091.

56 Lim, W.-X., Thornton, A.W., Hill, A.J., Cox, B.J., Hill, J.M., and Hill, M.R. (2013) *Langmuir*, **29**, 8524–8533.

57 Lee, H., Huang, B., Duan, W., and Ihm, J. (2010) *Appl. Phys. Lett.*, **96**, 143120.

58 Suh, M.P., Park, H.J., Prasad, T.K., and Lim, D.-W. (2012) *Chem. Rev.*, **112**, 782–835.

59 Zhou, W., Wu, H., and Yildirim, T. (2008) *J. Am. Chem. Soc.*, **130**, 15268–15269.

60 Lee, Y.-G., Moon, H.R., Cheon, Y.E., and Suh, M.P. (2008) *Angew. Chem., Int. Ed.*, **47**, 7741–7745.

61 Chen, B., Zhao, X., Putkham, A., Hong, K., Lobkovsky, E.B., Hurtado, E.J., Fletcher, A.J., and Thomas, K.M. (2008) *J. Am. Chem. Soc.*, **130**, 6411–6423.

62 Yan, Y., Telepeni, I., Yang, S., Lin, X., Kockelmann, W., Dailly, A., Blake, A.J., Lewis, W., Walker, G.S., Allan, D.R., Barnett, S.A., Champness, N.R., and Schröder, M. (2010) *J. Am. Chem. Soc.*, **132**, 4092–4094.

63 (a) Avila, M., Reguera, L., Vargas, C., and Reguera, E. (2009) *J. Phys. Chem. Solids*, **70**, 477–482. (b) Krap, C.P., Zamora, B., Reguera, L., and Reguera, E. (2009) *Microporous. Mesoporous. Mater.*, **120**, 414–420. (c) Rodríguez-Hernández, J., Reguera, E., Lima, E., Balmaseda, J., Martínez-García, R., and Yee-Madeira, H. (2007) *J. Phys. Chem. Solids*, **68**, 1630–1642.

64 Reguera, L., Balmaseda, J., del Castillo, L.F., and Reguera, E. (2008) *J. Phys. Chem. C*, **112**, 5589–5597.

65 Reguera, L., Balmaseda, J., Krap, C.P., Avila, M., and Reguera, E. (2008) *J. Phys. Chem. C*, **112**, 17443–17449.

66 Rodríguez, C., Reguera, E., and Avila, M. (2010) *J. Phys. Chem. C*, **114**, 9322–9327.

67 (a) Chapman, K.W., Southon, P.D., Weeks, C.L., and Kepert, C.J. (2005) *Chem. Commun.*, 3322–3324. (b) Kaye, S.S., Dailly, A., Yaghi, O.M., and Long, J.R. (2007) *J. Am. Chem. Soc.*, **129**, 14176–14177. (c) Krap, C.P., Balmaseda, J., del Castillo, L.F., Zamora, B., and Reguera, E. (2010) *Energy Fuels*, **24**, 581–589. (d) Reguera, L., Krap, C.P., Balmaseda, J., and Reguera, E. (2008) *J. Phys. Chem. C*, **112**, 15893–15899.

68 Kaye, S.S. and Long, J.R. (2005) *J. Am. Chem. Soc.*, **127**, 6506–6507.

69 Hartman, M.R., Peterson, V.K., Liu, Y., Kaye, S.S., and Long, J.R. (2006) *Chem. Mater.*, **18**, 3221–3224.

70 Reguera, E., Dago, A., Gómez, A., and Bertrán, J.F. (1996) *Polyhedron*, **15**, 3139–3145.

71 Reguera, L., Balmaseda, J., Krap, C.P., and Reguera, E. (2008) *J. Phys. Chem. C*, **112**, 10490–10501.

72 Culp, J.T., Matranga, C., Smith, M., Bittner, E.W., and Bockrath, B. (2006) *J. Phys. Chem. B*, **110**, 8325–8328.

73 (a) Germain, J., Hradil, J., Fréchet, J.M.J., and Svec, F. (2006) *Chem. Mater.*, **18**, 4430–4435. (b) Lee, J.-Y., Wood, C.D., Bradshaw, D., Rosseinsky, M.J., and Cooper, A.I. (2006) *Chem. Commun.*, 2670–2672. (c) McKeown, N.B. and Budd, P.M. (2006) *Chem. Soc. Rev.*, **35**, 675–683. (d) McKeown, N.B., Gahnem, B., Msayib, K.J., Budd, P.M., Tattershall, C.E., Mahmood, K., Tan, S., Book, D., Langmi, H.W., and Walton, A. (2006) *Angew. Chem., Int. Ed.*, **45** 1804–1807.

74 Liu, G., Wang, Y., Shen, C., Ju, Z., and Yuan, D. (2015) *J. Mater. Chem. A*, **3**, 3051–3058.

75 O'Malley, K., Ordaz, G., Adams, J., Randolph, K., Ahn, C.C., and Stetson, N.T. (2015) *J. Alloys Compd.*, **645** (Supplement 1), S419–S422.

76 Rana, M.K., Koh, H.S., Zuberi, H., and Siegel, D.J. (2014) *J. Phys. Chem. C*, **118**, 2929–2942.

77 Lozano-Castelló, D., Cazorla-Amorós, D., and Linares-Solano, A. (2002) *Energy Fuels*, **16**, 1321–1328.

78 Cracknell, R.F., Gordon, P., and Gubbins, K.E. (1993) *J. Phys. Chem.*, **97**, 494–499.

79 Düren, T., Sarkisov, L., Yaghi, O.M., and Snurr, R.Q. (2004) *Langmuir*, **20**, 2683–2689.

80 Zamora, B., Roque, J., Balmaseda, J., and Reguera, E. (2010) *Z. Anorg. Allg. Chem.*, **636**, 2574–2578.
81 Wu, H., Zhou, W., and Yildirim, T. (2009) *J. Phys. Chem. C*, **113**, 3029–3035.
82 Guo, H.-C., Shi, F., Ma, Z.-F., and Liu, X.-Q. (2010) *J. Phys. Chem. C*, **114**, 12158–12165.
83 (a) Zhou, W., Wu, H., Hartman, M.R., and Yildirim, T. (2007) *J. Phys. Chem. C*, **111**, 16131–16137. (b) Mu, L., Liu, B., Liu, H., Yang, Y., Sun, C., and Chen, G. (2012) *J. Mater. Chem.*, **22**, 12246–12252.
84 Wu, H., Zhou, W., and Yildirim, T. (2009) *J. Am. Chem. Soc.*, **131**, 4995–5000.
85 Ma, S., Sun, D., Simmons, J.M., Collier, C.D., Yuan, D., and Zhou, H.-C. (2008) *J. Am. Chem. Soc.*, **130**, 1012–1016.
86 Autie-Castro, G., Autie, M.A., Rodríguez-Castellón, E., Aguirre, C., and Reguera, E. (2015) *Colloids Surf. A*, **481**, 351–357.
87 Peng, Y., Krungleviciute, V., Eryazici, I., Hupp, J.T., Farha, O.K., and Yildirim, T. (2013) *J. Am. Chem. Soc.*, **135**, 11887–11894.
88 Lu, H., Seo, Y.-T., Lee, J.-W., Moudrakovski, I., Ripmeester, J.A., Chapman, N.R., Coffin, R.B., Gardner, G., and Pohlman, J. (2007) *Nature*, **445**, 303–306.
89 Krishna, L. and Koh, C.A. (2015) *MRS Energy Sustain.*, **2**, 1–17.
90 (a) Florusse, L.J., Peters, C.J., Schoonman, J., Hester, K.C., Koh, C.A., Dec, S.F., Marsh, K.N., and Sloan, E.D. (2004) *Science*, **306**, 469–471. (b) Strobel, T.A., Hester, K.C., Koh, C.A., Sum, A.K., and Sloan, E.D. (2009) *Chem. Phys. Lett.*, **478**, 97–109.
91 Mao, W.L., Mao, H.-K., Goncharov, A.F., Struzhkin, V.V., Guo, Q., Hu, J., Shu, J., Hemley, R.J., Somayazulu, M., and Zhao, Y. (2002) *Science*, **297**, 2247–2249.
92 Somayazulu, M.S., Finger, L.W., Hemley, R.J., and Mao, H.K. (1996) *Science*, **271**, 1400–1402.
93 Loubeyre, P., Letoullec, R., and Pinceaux, J. (1994) *Phys. Rev. Lett.*, **72**, 1360–1363.
94 Veluswamy, H.P., Wong, A.J.H., Babu, P., Kumar, R., Kulprathipanja, S., Rangsunvigit, P., and Linga, P. (2016) *Chem. Eng. J.*, **290**, 161–173.
95 Mao, W.L. and Mao, H.-K. (2003) *Proc. Natl. Acad. Sci. USA*, **101**, 708–710.
96 Prasad, P.S., Sugahara, T., Sum, A.K., Sloan, E.D., and Koh, C.A. (2009) *J. Phys. Chem. A*, **113**, 6540–6543.
97 Fowler, D., Loebenstein, W., Pall, D., and Kraus, C.A. (1940) *J. Am. Chem. Soc.*, **62**, 1140–1142.
98 Willow, S.Y. and Xantheas, S.S. (2012) *Chem. Phys. Lett.*, **525**, 13–18.
99 Zhang, S.-X., Chen, G.-J., Ma, C.-F., Yang, L.-Y., and Guo, T.-M. (2000) *J. Chem. Eng. Data*, **45**, 908–911.
100 Valdés, A.L. and Kroes, G.-J. (2012) *J. Phys. Chem. C*, **116**, 21664–21672.
101 Bogdanović, B. and Schwickardi, M. (1997) *J. Alloys Compd.*, **253**, 1–9.
102 Graetz, J. (2012) *ISRN Mater. Sci.*, **2012**, 1–18.
103 (a) Orimo, S.-I., Nakamori, Y., Eliseo, J.R., Züttel, A., and Jensen, C.M. (2007) *Chem. Rev.*, **107**, 4111–4132. (b) Srinivasan, S.S., Brinks, H.W., Hauback, B.C., Sun, D., and Jensen, C.M. (2004) *J. Alloys Compd.*, **377**, 283–289.
104 Bergstrom, F. and Fernelius, W.C. (1933) *Chem. Rev.*, **12**, 43–179.
105 Goerrig, D. (1958) Verfahren zur herstellung von boranaten. German Patent F27373 IVa/12i.
106 Prins, R. (2012) *Chem. Rev.*, **112**, 2714–2738.
107 (a) Wang, L., Lachawiec, J.A.J. and Yang, R.T. (2013) *RSC Adv.*, **3**, 23935–23952. (b) Wang, L. and Yang, R.T. (2008) *Energy Environ. Sci.*, **1**, 268–279.
108 Lee, S.-Y. and Park, S.-J. (2011) *Int. J. Hydrogen Energy*, **36**, 8381–8387.
109 Li, Y. and Yang, R.T. (2006) *J. Am. Chem. Soc.*, **128**, 8136–8137.

13
Supported Catalysts

Isao Ogino,[1] Pedro Serna,[2] and Bruce C. Gates[3]

[1]Hokkaido University, Graduate School of Engineering, Division of Applied Chemistry, Kita 13, Nishi 8, Kita-ku, 060-8628 Sapporo, Japan
[2]ExxonMobil Research and Engineering Co., Corporate Strategic Research, 1545 Route 22 E, Annandale, NJ 08801, USA
[3]University of California, Department of Chemical Engineering, One Shields Avenue, Davis, CA 95616, USA

13.1
Introduction

Catalysts make chemical reactions go faster, often by many orders of magnitude in comparison with purely thermal processes. The great majority of biological and industrial reactions are catalytic. The dominant technological catalysts are solids, which offer the important practical benefit of being readily separated from fluid-phase products – held in place in reactors such as tubes, with reactants continuously flowing in one end and products out the other, as in petroleum refineries and chemical plants. Some of these reactors are large, but small ones are important too, such as the catalytic converters that clean up automobile exhaust. Most of these solid catalysts are robust and applied at high temperatures – a great benefit because reaction rates typically increase exponentially with increasing temperature, and the higher the rate the smaller and more economical the reactor, for a given production rate. Solid catalysts are usually porous, having high internal surface areas (even hundreds of square meters per gram) and therefore high numbers of catalytically active species per unit of reactor volume to facilitate high rates of catalytic reaction.

Common industrial catalysts include metals (e.g., platinum, palladium, nickel, rhodium, and copper) [1], metal oxides (e.g., γ-Al_2O_3, mixed oxides of bismuth and molybdenum), zeolites [2], and metal sulfides [3,4]. Others are metal carbides, metal phosphides [5,6], and metal nitrides. These materials are often finely dispersed on the surfaces of other materials called supports. Supports are usually robust, inexpensive porous materials such as metal oxides, zeolites, or carbon. Supports such as γ-Al_2O_3 offer the advantages of wide ranges of tailorable

Handbook of Solid State Chemistry, First Edition. Edited by Richard Dronskowski, Shinichi Kikkawa, and Andreas Stein.
© 2017 Wiley-VCH Verlag GmbH & Co. KGaA. Published 2017 by Wiley-VCH Verlag GmbH & Co. KGaA.

physical properties, such as internal surface area and pore size distribution that enable control of diffusion in the pores. Catalytically active components on supports, such as metals, are often expensive, and it is economical to disperse them as nanosized particles on the internal support surface; the smaller the catalytic particle, the greater the fraction of atoms available for catalysis at its surface.

13.2
Classes of Solid Catalysts

13.2.1
Bulk Solid Catalysts

Catalysts may also be applied in the form of bulk materials. Bulk catalysts in large-scale processes – when economics is important – are almost always porous, such as Raney nickel, which is used in hydrogenation of vegetable oils.

Nonporous bulk materials have also been investigated as catalysts but rarely for application. Investigations of single crystals of metals, and, to a degree, metal oxides and metal sulfides, have provided fundamental insight into the workings of supported catalysts. The most fruitful investigations of single-crystal catalysts have been carried out with metal crystals exposing known faces and mounted in ultrahigh vacuum (UHV) chambers where they are interrogated with beams of unimpeded photons, electrons, or small particles. Thus, characterizations of single crystals by low-energy electron diffraction and X-ray photoelectron spectroscopy, for example, have provided essential details of the surface compositions and structures and bonding of reactive species on the surfaces. The techniques work best with samples that are good electrical conductors, such as metals, and they are more challenging for metal oxides, metal sulfides, and so on. A limitation of the UHV approach is that many species are too weakly bound to the surfaces to be present at the extremely low pressures of the experiments.

Nonetheless, a fundamental understanding of catalysis on well-defined single-crystal surfaces of many metals is a strong foundation for understanding catalysis on surfaces of small particles of metals that present facets with structures matching the faces of single crystals; such structures include various planes of the metal, which typically include step, edge, and corner sites. Many practical catalysts incorporate supported particles that are large enough to be in this category, and understanding of them rests on research with single crystals. The catalytic properties typically vary from one metal face to another and are different from those of steps, edges, and corners; defect sites may be different still.

13.2.2
Supported Catalysts

The most thoroughly investigated supported catalysts incorporate particles of supported species that have the character of the same material in the bulk, for example, particles of metal that consist of thousands or more metal atoms and present crystal faces and other structural features matching those of single metal crystals. Supported metal particles in this class are commonly used as catalysts, but supported metal oxides, metal sulfides, and so on in this class are much less typical and considered in only little depth in this review.

When supported particles become smaller, their properties increasingly deviate from those of the bulk. If they are small enough, their properties may be influenced by quantum size effects and interactions with the support, and they may incorporate structures that do not reflect structural elements characteristic of bulk single crystals. This category of supported catalyst includes not only small particles of metal, metal sulfide, and so on but also layers only a few atoms thick, essentially two-dimensional structures that may incorporate many atoms; an important example is layers of MoS_2 on Al_2O_3.

As the supported species become smaller, reaching the subnanosize range, they increasingly take on the character of molecules. Such species are often called clusters (but the cluster/particle distinction is vague). A cluster may consist of only a few (say, <10) metal atoms. Then, its properties are strongly influenced by its surroundings, including the support and other groups on the cluster; groups chemically bonded to metals are called ligands, including the support. The effects of supports as ligands are only partially understood, as summarized below.

In the limiting case of the smallest supported species, they are single atoms. Isolated metal atoms in supported catalysts are almost always present as cations, typically anchored through metal–oxygen bonds; they usually have other ligands bonded to them as well. We call these species supported metal complexes (or "single-site catalysts" or "site isolated catalysts"), because they are analogs of soluble molecular species.

Supported metal catalysts often incorporate more than one kind of metal. Supported bimetallics are alloy-like when the metals are in particles with bulk-like properties, but as the particles become smaller, deviations from bulk structure and behavior become significant. As bimetallic particles become smaller, the distributions of atoms between the interior and the surface change. In some supported catalysts, the metals are even segregated into separate components on the support, sometimes with a noble metal in the zero-valent state in particles and an oxophilic metal present as cationic species bonded to the support. Some bimetallic particles, however, incorporate unsegregated noble and oxophilic metals; the latter may bond to the support and at the same time to other oxophilic metal atoms, with noble metal atoms bonded both to oxophilic and to noble metal atoms but not to the support.

Supported catalysts are typically highly nonuniform in character and challenging to characterize structurally; often, the structural information leads only to imprecise models of average structures. Some literature fails to make clear the degree of nonuniformity of the material.

13.3
Preparation of Supported Catalysts

13.3.1
Industrial Catalyst Preparation Methods

The preparation of common supported catalysts involves support synthesis, deposition of catalyst precursor components, and treatment to convert the supported species into catalytically active species. Industrial catalysts are also subjected to additional processing to control the sizes and shapes of the final product. Often, catalysts or supports are extruded as pastes to give cylinders, typically with diameters of about 1/8 inch. Some are made as pellets resembling pharmaceutical pills, and some are nearly spherical particles, formed by spray drying. Catalyst particle sizes are chosen to minimize pressure drop in fixed-bed reactors and to facilitate mixing of the reactants flowing through the bed of particles. Some catalysts incorporate binders (e.g., vitrified clay) to "glue" support microparticles together for stability, and inert components may be included to facilitate heat transfer and prevent thermal instability in exothermic or endothermic reactions. Other common catalyst components are promoters that improve catalytic activity, selectivity, or stability; *chemical promoters* are present in minor amounts, whereas *textural promoters* are present in much greater amounts and typically provide stability such as resistance to sintering.

Supports offer high surface areas per unit mass (specific surface areas) to accommodate highly dispersed catalytic species. Common preparation routes used in industry to prepare supported catalysts include the following [7,8]:

- Precipitation of a metal oxide from precursors initially dissolved in a solution, usually upon changes in the temperature and/or pH to reach supersaturation, nucleation, and particle growth that determine the particles size and morphology.
- Gelation/flocculation, from colloidal solutions that use micelles separated by repulsive electrostatic interactions. Neutralization of the micellar charges leads to the formation of solids with sizes that depend on the micelle sizes, ripening rate, and drying conditions.
- Hydrothermal synthesis, whereby crystallization of a solid phase is achieved from an aqueous gel at relatively high temperatures, as in the synthesis of zeolites. The concentration of the gel components, pH, temperature, and time of crystallization determine the final structures [2].

Metals are almost exclusively deposited onto supports, for example, by direct contact between two types of solid precursors (e.g. ball milling) and by "dry" and "wet" methods to afford catalysts with high dispersions of the active phase. Aqueous solutions of inorganic metal salts are used. In the "dry" processes, the volume of precursor solution nearly matches the support pore volume, resulting in a gel-like mixture with essentially all the liquid in the pores. In the "wet" technique, an excess of liquid is used, leading to suspensions of the support. In the former method, the precursor is fully deposited onto the support as the solvent is evaporated. In the latter, some precursor often remains in the supernatant solution, requiring expensive recovery and recycling.

Typical industrial catalyst precursors are water-soluble inorganic salts, and the species in contact with the support are ionic. The deposition of metal-containing species on metal oxide supports is largely controlled by polar interactions; the pH strongly influences the charge of the support surface species. The isoelectric point of the support (the pH at which a specific support has a neutral surface) allows prediction of whether the support surface is positively or negatively charged and thus of the species that are deposited.

Ion-exchange occurs when a metal ion from the solution replaces an ion on the support. Alternatively, because most metal oxides offer sites with associated electrostatic charges at surface defects, these sites may initiate nucleation and growth of the metal-containing species. Coprecipitation of the support precursors with the active metal-phase precursors is also an industrial option, used, for example, to prepare Cu-ZnO/Al_2O_3 catalysts for the synthesis of methanol from CO and H_2 (according to this designation, Al_2O_3 is a support) [9,10].

Subsequent steps involve drying of powders, calcination in air, and (if the catalyst is a supported metal) reduction in H_2. The goals include removal of undesired species that could cause instability or deactivation of the catalyst. Drying and calcination remove water and anion residues introduced with the precursors (e.g., Cl^-, NO_3^-), generating oxidized species on the support; subsequent reduction with H_2 gives zero-valent metal particles.

Selection of industrial catalyst preparation methods is mainly driven by cost and scalability. A consequence of the above-mentioned methods is that the interactions between precursors and supports are via electrical attraction/repulsion forces, and the methods are not effective for discriminating among various anchoring sites on the support, so that the structures of the supported catalysts are nonuniform and ill-defined. Thus, these catalysts do not lend themselves to precise structure characterizations or structure–performance correlations.

13.3.2
Synthesis of Supported Metals with Well-Defined Structures

In contrast, supported catalysts that are essentially molecular and structurally well-defined allow incisive characterization and facilitate fundamental understanding. Synthesis of such catalysts on high-area supports often involves organometallic precursors with reactivities that match the synthetic goals [11–16].

The organometallic precursors may incorporate ligands that react cleanly with supports while other ligands remain on the metal. Ligands include acetylacetonate (acac) in complexes such as $Rh(C_2H_4)_2(acac)$, $Ir(C_2H_4)_2(acac)$, and $Au(CH_3)_2(acac)$ [17–19]. The acac ligands react with OH groups on metal oxides, zeolites, and metal organic frameworks (MOFs), with the metal complexes becoming anchored through metal–oxygen bonds and liberating Hacac. Furthermore, some mononuclear metal complexes can be converted to well-defined metal clusters. Alternatively, supported metal clusters can be made from molecular precursors such as $Os_3(CO)_{12}$ and $Ir_4(CO)_{12}$, which can be decarbonylated after adsorption [20].

Zeolites, being crystalline, allow precise synthesis of supported metal complexes and clusters at specific bonding sites, OH groups bonded near framework Al^{3+} ions. Because of their structural uniformity, such supported catalysts are represented well by theory [21–24]. Other crystalline porous materials that offer similar advantages are MOFs. There are only a few examples of thermally stable MOFs, and their use as catalyst supports is rare. Catalysts can be bonded to either the organic linkers or the nodes of MOFs, and some nodes are essentially small pieces of metal oxide, exemplified by the $[Zr_6(\mu_3\text{-}O)_4(\mu_3\text{-}OH)_4(OH)_4(OH_2)_4]^{8+}$ and $[Zr_6(\mu_3\text{-}O)_4(\mu_3\text{-}OH)_4]^{12+}$ nodes of NU-1000 and UiO-66, respectively. The node OH and OH_2 groups are anchoring sites for metals. MOF-node-supported $Ir(C_2H_4)_2$ was made from $Ir(C_2H_4)_2(acac)$ on UiO-66 or NU-1000 [25,26].

13.4
Characterization of Supported Catalysts

Supported catalysts are characterized by combinations of microscopic and spectroscopic methods summarized below, complemented with equilibrium chemisorption measurements and temperature-programmed methods such as reduction, oxidation, and decomposition [27]. Recall that most supported catalysts are highly nonuniform materials, so that spectra represent averages over sample space, and transmission electron microscopy (TEM) is often needed to provide a measure of sample heterogeneity.

13.4.1
Transmission Electron Microscopy

TEM provides images of supported species and information about their sizes and degrees of uniformity. Imaging of metal particles is accomplished with conventional TEM, provided that the metals are heavy. It is much more challenging to image individual metal atoms because of the smallness of the dispersed species and the changes induced by the electron beam [28], which typically causes sintering of metal and deterioration of supports such as MOFs and zeolites [29]. Images of supported metal particles are shown in Figure 13.1 [30,31]. Advances in TEM have led to many atomic-resolution images of metals in supported catalysts. Aberration-corrected scanning transmission electron microscopy (STEM) allows visualization of single atoms [32–36] and even determines details of their bonding to supports [37–39]. Images of some catalysts resembling industrial catalysts show that they consist largely of atomically dispersed metals and clusters of only several metal atoms each [40].

Application of high-dose rates with short exposure times may allow high-resolution imaging of beam-sensitive samples [41–43]. The images shown in Figure 13.2 show the locations of Ir atoms in the pores of zeolite HSSZ-53 and

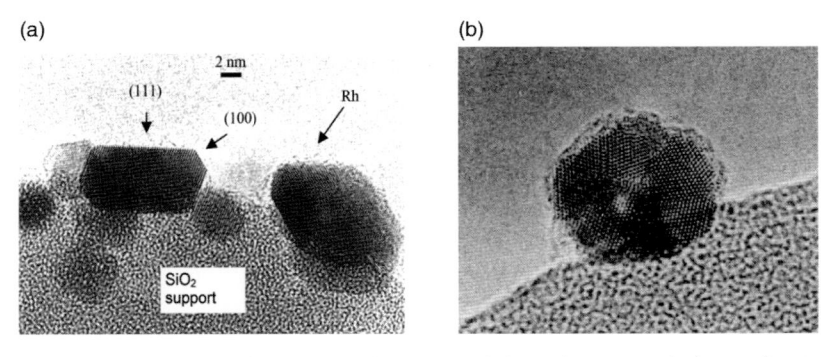

Figure 13.1 (a) Image of rhodium particles on model silica spheres annealed in H_2 showing well-defined surface facets and (b) after oxidation–reduction treatment showing a rougher surface that has a much different reactivity (courtesy of Prof. A. Datye) [30].

how they migrate and form small clusters upon exposure to the electron beam [41]. First, Ir_2 forms, and then larger clusters, ultimately leading to the destruction of the zeolite framework.

13.4.2
X-ray Absorption Spectroscopy

X-ray absorption fine structure (XAFS) spectroscopy is an element-specific technique that provides local structural information about supported metals even when they do not have long-range order [44]. Extended XAFS (EXAFS) spectra recorded near a metal X-ray absorption edge provide information about the average numbers of nearby backscattering atoms of various types and average distances between the absorbing metal atom and the nearby atoms. EXAFS spectroscopy is one of the few techniques that provide information about the metal–support interface. Typical metal–oxygen distances in metal oxide- and zeolite-supported metal complexes and clusters of Group 7 and Group 8 metals are 2.1 ± 0.1 Å [45], a bonding distance. Because it is hardly possible to distinguish low-Z backscatterers such as carbon and oxygen from each other in the analysis of EXAFS data, identification of these ligands requires complementary information, such as IR spectra. When supported metal complexes or clusters bear carbonyl ligands, analysis of the EXAFS data is facilitated because of the multiple scattering of linear M–C–O moieties. Structures are determined from EXAFS data by fitting, and theoretical calculations facilitate the analyses, narrowing the list of candidate models [46,47]. X-ray absorption near edge spectra (XANES, measured simultaneously with EXAFS spectra) provide information about the oxidation states of metals and the symmetry of supported metal complexes.

EXAFS spectroscopy and STEM were used to elucidate the activation of NaY zeolite-supported mononuclear gold complexes prepared by adsorption of $Au(CH_3)_2(acac)$ [48]. Images show that the precursors were adsorbed at T5 sites while still incorporating acac ligands. Exposure to $CO + O_2$ removed acac

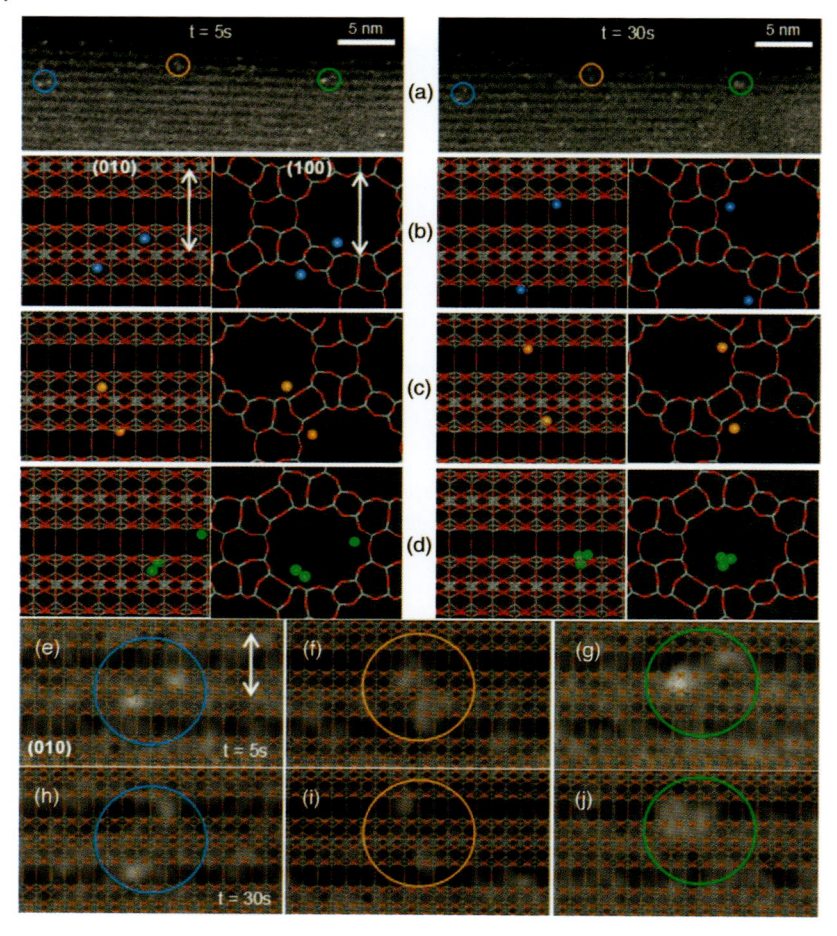

Figure 13.2 Examples illustrating the movement of Ir atoms in SSZ-53 channels. (a) Experimental image acquired in the (010) projection of a zeolite crystallite after an exposure to the electron beam for $t = 5$ s (left) and 30 s (right). (b–d) Crystallographically simulated framework of the zeolite in the (100) and (010) projections with the locations of Ir atoms or clusters corresponding to the STEM images acquired at $t = 5$ s (left) and $t = 30$ s (right); the colors of the Ir atoms correspond to the colors of the circles in (a). (e–g) Crystallographically simulated model of SSZ-53 in the (010) projection superimposed on the encircled areas of the experimental image at $t = 5$ s. (h–j) Crystallographically simulated model of SSZ-53 in the (010) projection superimposed on the zoomed areas containing the corresponding encircled region of the experimental image taken at $t = 30$ s. The double-headed white arrows indicate the diameter of a SSZ-53 channel in each projection [41].

ligands and oxidized the gold, which had chemisorbed CH_3 ligands that were replaced by CO as some of the gold complexes migrated to T6 sites.

This combination of techniques was also used to elucidate the 3D structure of the metal–support interface in MgO-supported triosmium clusters [39]. The sample was prepared from $Os_3(CO)_{12}$, forming supported $[Os_3(CO)_{11}]^{2-}$. The images show that each Os atom on the Mg(110) surface was located on top of a column of Mg atoms, with the Os atoms in each image of a triosmium cluster defining a triangle well approximated as isosceles (but not equilateral). When it was assumed that the Os–$O_{support}$ bonding distance was 2.1 Å (the observed EXAFS value for the sample as a whole [with the clusters on various MgO faces]), it followed that the triosmium frames were tilted at an angle of 38° with respect to the MgO(110) surface.

13.4.3
Chemisorption Equilibrium Measurements

Numbers of exposed metal atoms of supported noble metal particles are commonly measured by equilibrium adsorption of H_2, CO, or NO. In some cases, supported metals are first oxidized by O_2, followed by titration of chemisorbed surface oxygen atoms by H_2 [49]. Such data are used to estimate the metal surface areas and determine the metal dispersion, defined as the fraction of metal atoms exposed at a surface.

Measurements of surface areas of supported metals require information about the stoichiometry of the species adsorbed on the metal, such as H or CO; it is commonly assumed that there is one H atom or one CO molecule per exposed atom of noble metal. Furthermore, it is generally assumed that the probes do not change the structures of the supported metals. These assumptions are not generally valid, especially when the particles are quite small; for example, γ-Al_2O_3-supported platinum clusters with an average diameter of about 1.1 nm when exposed to H_2 were shown by EXAFS spectroscopy to have Pt–Pt and Pt–support oxygen distances substantially greater than those of the sample under vacuum [50]. And when the same sample was exposed to O_2, fragmentation of the platinum clusters was observed as platinum oxide species formed.

Because of these complications, it is essential to use combinations of techniques to verify structural conclusions. TEM provides cluster/particle size distributions. IR spectra distinguish terminal from bridging CO ligands and may thereby give evidence of isolated and neighboring metal centers, respectively. EXAFS spectra provide information about the structural changes associated with the chemisorption of probe molecules and about metal dispersions. Determination of metal dispersions by EXAFS spectroscopy requires a model providing a relationship between the observed metal–metal coordination number and the average metal particle size, which is a function of the particle geometry [51]. The shape of the average metal particle was calculated as a function of the relative magnitude of the metal–metal and metal–support interaction energy by minimizing the total energy; the shape of a metal particle with $n+1$ atoms was obtained from that of the particle with n atoms by placing the extra metal atom at the position of minimum energy.

Table 13.1 Comparison of dispersions of platinum determined from chemisorption, TPD, and EXAFS data for Pt/γ-Al$_2$O$_3$ reduced with H$_2$ at 673 K [52].

Chemisorption	Chemisorption	TPD	EXAFS
From hydrogen chemisorption data	From data characterizing hydrogen titration of adsorbed oxygen	From H$_2$ desorption data at temperatures <673 K	From the Pt–Pt first-shell coordination number
0.58	0.57	0.58	0.58

13.4.4
Case Study Illustrating Application of Complementary Characterization Methods [52]

A prototypical supported metal catalyst characterized by complementary techniques is platinum clusters/particles supported on porous γ-Al$_2$O$_3$ (Pt/γ-Al$_2$O$_3$). The sample was synthesized by reduction in H$_2$ of adsorbed [PtCl$_2$(PhCN)$_2$] and characterized by chemisorption measurements to count the platinum surface sites. The adsorbing gases were H$_2$ or CO, or O$_2$ followed by titration of the chemisorbed O atoms with H$_2$. Temperature-programmed desorption (TPD) of hydrogen and EXAFS spectroscopy provided complementary data characterizing the platinum particles under vacuum after reduction with H$_2$. The EXAFS data were interpreted on the basis of a model giving a relationship between the average platinum cluster diameter and the EXAFS Pt–Pt first-shell coordination number [51] based on the simplifying assumption of an epitaxial arrangement of Pt atoms on the γ-Al$_2$O$_3$ (111) surface. Accordingly, the average platinum particle diameter was found to be about 2 nm, and the number of Pt atoms per average particle about 280. The results obtained with the several techniques are summarized in Table 13.1; they demonstrate excellent internal consistency.

EXAFS characterization indicates morphology changes of the supported platinum clusters that are not evidenced by chemisorption and TPD data. For example, when the catalyst was treated sequentially with O$_2$ at 473–673 K and H$_2$ at 673 K, the data show a decrease in the coordination number of the Pt–Pt second shell and an increase in the coordination number of the third shell as the treatment temperature increased, consistent with the inference that the treatment caused the platinum particles to become raft-like.

13.5
Reactivities of Supported Catalysts

13.5.1
Adsorption

The reactivity of a supported metal – if it is small – is strongly influenced by its electronic structure and environment. Reactivity is tested with a reactive gas such as CO, H_2, O_2, H_2O, NO, or C_2H_4. In a catalytic reaction, at least one reactant is bonded to the catalyst, and so if any of the reactive gases is not bonded to the catalyst, it may be considered unlikely to be converted catalytically (although it could react from a fluid phase with another reactant bonded to the catalyst). Species bonded to metal catalysts are influenced by the electronic structure of the metal and characterized by IR and other spectroscopies. Good probe molecules give spectra that provide evidence of the bonding and thus information about the surface. CO is a good probe because IR spectra of the adsorbed species are readily measured and because values of ν_{CO} are sensitive to the electron transfer between the metal and the carbon atoms, which influences the C—O bonding [53]. Strong ν_{CO} bands of adsorbed CO ligands appear in uncluttered regions of IR spectra of many catalysts, and the small size of CO allows it access to many metal sites. In some cases, the sharpness of the ν_{CO} bands provides information about the structural uniformity of the supported species – broadbands characterize CO bonded to metal species that are nonuniform in structure and in their bonding to the support [54].

Reactions of supported metal species with probe molecules may change the structures of the supported metal species, and such changes are characterized by techniques such as XAFS spectroscopy and STEM. Some of these techniques are performed with the sample in a reactive environment to give real-time evidence of the changes [27,55–57]. Experiments characterizing the replacement of one adsorbed species with another can be carried out advantageously with flow reactors because equilibrium limitations are overcome as a flowing carrier gas acts as a sweep gas to remove even tightly bound species – all that is needed is for a minute amount to leave the surface for it to be continually removed and replaced by another species in the flowing gas.

Subtle changes of adsorbates on the most highly dispersed metals significantly affect their reactivities. For example, zeolite HY-supported $M(CO)_2$ (M = Rh, Ir) undergoes facile ligand exchange of one of the two CO ligands with C_2H_4, forming supported $M(CO)(C_2H_4)$ and $M(CO)(C_2H_4)_2$ [21,22]. However, exchange of the second CO ligand is much more difficult. Hydrogen can also be incorporated into these supported metal complexes, giving, for example, $Rh(CO)(H)_2$ and $Rh(CO)(H)$ from supported $Rh(CO)(C_2H_4)$, but not directly from supported $Rh(CO)_2$ [58]. Similar data have been obtained for metal oxide-supported Ir_4 and Ir_6 clusters, although the compositions and structures of the adsorbed species are less than fully defined [59]. These clusters incorporate π-bonded ethylene, ethylidene, di-σ-bonded ethylene, and ethyl ligands.

Adsorption on bulk-like metal particles is similar to that on metal clusters and single metal crystals, but for the most part less precisely defined, because different species form on various surface sites such as facets, corners, edges, and so on.

Water adsorbs dissociatively on the acidic sites of supports including zeolites and metal oxides, giving monatomic hydrogen species, and the hydrogen migrates to nearby supported metals and forms metal hydrides in a process called reverse spillover. Vayssilov *et al.* used theory to characterize Rh_6 clusters at six-rings of a faujasite zeolite, showing that reverse spillover of hydrogen from zeolite OH groups onto the Rh_6 clusters led to oxidation of the Rh atoms bonded to the zeolite [60]. The reaction forms supported Rh_6 clusters each incorporating three H atoms, with the clusters bonded to the support via three Rh—O bonds; EXAFS data are consistent with this predicted bonding. The calculations showed that the positively charged Rh atoms at the metal–support interface are approximately Rh(I), similar to that in $Rh(CO)_2$ bonded to the zeolite [47].

13.5.2
Restructuring of Supported Catalysts

A traditional view of active sites on catalyst surfaces is based on the naive assumption that their structures remain unchanged even when reactive atmospheres are present and catalysis occurs. But recent work shows that supported metals (and catalysts generally) have a dynamical nature under operating conditions, particularly when the catalytic species consist of no more than a few atoms.

Interconversions of supported metal species were investigated with samples present initially as HY zeolite-supported $Ir(C_2H_4)_2$ complexes; these were converted into Ir_4 clusters (with various hydrocarbon ligands) by treatment with H_2 or H_2 and ethylene [56,61]. H_2 reduced the Ir(I) complexes to give the tetranuclear iridium clusters, and the process was reversed when the sample was exposed to ethylene alone, aided by OH groups of the zeolite, which oxidatively fragmented the clusters. The transformation was found to be stoichiometric as indicated by isosbestic points in XANES spectra recorded during the transformations. The ratio of mononuclear to tetranuclear iridium species was controlled by the gas-phase H_2 to ethylene ratio, and the structure ultimately controlled the catalytic activity of the iridium for ethylene hydrogenation. With rhodium, the interconversion of rhodium complexes and clusters caused alteration of the catalyst selectivity to form n-butenes selectively on the former and ethane on the latter [57].

13.5.3
Oxidation of Supported Catalysts

Supported clusters of oxophilic metals are easily oxidized and give various structures that are broadly classified as supported metal oxides; some are mononuclear metal species and some are approximately described as metal oxide clusters; some are monolayers or nearly so. Supported noble metal clusters have been oxidatively fragmented in treatments with gases including O_2, CO, or NO [62], and supported noble metal particles or clusters may be oxidized at their surfaces and retain (nearly) metallic cores, allowing titration with H_2 to determine metal surface areas.

13.5.4
Sulfidation of Supported Catalysts

Sulfidation treatments in atmospheres such as $H_2S + H_2$ are used to form supported metal sulfides from various metal-containing structures on supports; the metal-containing structures are typically metal oxide-like complexes. Sulfided catalysts containing MoS_2 are widely used in technology to remove sulfur-containing compounds in fossil fuels (e.g., mercaptans, thiophenes, etc.) by reactions with H_2 to form H_2S (the reactions are called hydrodesulfurization) [4]. MoS_2 is made more active by promotion with nickel or cobalt [3].

The catalytically active species consist of few-atom thick layers of MoS_2 with the layer edges providing the catalytically active sites and providing locations for nickel or cobalt promoters, bonded as atomically dispersed cations. MoS_2 and WS_2 act similarly. On single crystals of gold acting as supports, small triangular clusters of MoS_2 have been formed and they have been investigated with various techniques and imaged with scanning tunneling microscopy (STM) [63–65]. The relative free energies of the cluster edges have been inferred to determine the nanocluster shape: triangular if the Mo edges have a lower energy than the S edges and hexagonal if the two edges have similar energies. MoS_2 cluster shape depends on the sulfidation conditions [64,66]; MoS_2 clusters on Au(111) have a triangular shape with the particles having Mo edges with full sulfur coverage in an H_2S-rich H_2S/H_2 mixture, but they have a truncated triangular shape in an H_2-rich mixture. Supports also influence the shapes of MoS_2 clusters; when supported on Ti(110), the morphology becomes highly elongated because of the epitaxy between the MoS_2 and the oxide surface. Elongated clusters (layers) have been observed by TEM on γ-Al_2O_3 supports and are inferred to be present in industrial catalysts.

13.6
Catalyst Performance: Structure–Property Relationships

13.6.1
Structure Sensitivity and Structure Insensitivity in Catalysis by Supported Metal Clusters and Particles

Reactions that occur on a catalyst surface at a rate (per active site) that depends on the arrangement of the surface atoms are referred as *structure sensitive*. *Structure-insensitive* reactions, in contrast, are unaffected by the site architecture. These terms were coined [67] after the observation that some catalytic reactions, such as alkane hydrogenolysis [68] and isomerization [69], depend strongly on the average metal particle size, whereas others, such as cyclopropane ring opening [70] or olefin hydrogenation [71], are virtually independent of it. It was suggested [72] that the term should more rigorously describe the relationship between the specific reaction rate and the structures of the surface planes and other features, because changes in metal particle size also affect the abundance of steps, corners, defects, and metal/support interphase sites.

13.6.2
Dependence of Catalytic Activity on Crystal Faces

The synthesis of ammonia from N_2 and H_2 is an industrial example of a structure-sensitive reaction. The process has been investigated extensively with single-crystal as well as supported metal catalysts, and the reaction rates can be largely understood by considering changes in the number of nearest metal neighbors for various crystallographic orientations of, for example, iron, and the related changes in the binding energy of adsorbed nitrogen species [73]. Proposals of specific geometric arrangements of Fe atoms on (111) surfaces as the most active catalytic sites are consistent with the observations [73] and theory [74].

The effect of the structure of a crystal face on catalysis is typically diagnosed in experiments with single crystals exposing various faces and often rationalized on the basis of relationships between the crystallographic orientations and the adsorbate–adsorbent (reactant–catalyst) binding energies [75]. Structure-insensitive reactions, however, may result when repulsive adsorbate–adsorbate interactions at high surface coverages lead to weak adsorbate–adsorbent interactions.

13.6.3
Roles of Structural Features such as Steps and Corners

Discontinuities such as steps and vacancies on metal single crystals influence the catalytic events by offering sites of low coordination compared to sites on well-defined planes (terraces). For example, rough surfaces of platinum are more active for hydrocarbon hydrogenolysis than terraces, with the rates of the C–C scission correlating well with the number of defect sites and steps [76]. Steps, corners, and defects are common in metal particles in industrial supported metal catalysts, but the synthetic challenge of making stable catalysts that have these properties optimized and stabilized is formidable. A common approach is to try to control the structural properties by control of the average size of the supported species, and geometries of small particles provide guidance [77]. Variables such as the nature and specific surface area of the support, the metal loading, and the conditions of preparation and activation of the materials are often considered in attempts to control the metal particle size range, but the methods are empirical and inexact.

Hydrogenation reactions catalyzed by supported gold particles illustrates the importance of low coordination sites. When gold is the catalyst and H_2 the reductant, the rate-determining step in the catalytic reaction may often be the scission of the H—H bond. In contrast to other noble metals, gold facilitates the dissociation of H_2 at significant rates only at edges, steps, and corners of nanoparticles [78]. This inference is supported by correlations between the catalytic activity for H/D exchange in the reaction of H_2 with D_2 to form HD and the abundance of this type of site in supported gold catalysts, determined by adsorption of CO at low temperatures and IR spectroscopy [79]; support for the interpretation is provided by theory [78]. Nanosized gold opened the door to selective catalysis that is difficult to achieve with other noble metals, as in the reduction of substituted nitroaromatic compounds [80] or the partial reduction of C≡C bonds [81].

13.6.4
Effects of Supports as Ligands and Modifiers of Metals

Practical metal catalysts present highly dispersed species that must survive the often harsh conditions (high temperature and pressure) common in processing, and the support can be important in helping to slow metal coalescence and aggregation leading to particle growth and loss of surface area. The nature of the support can significantly influence the size and morphology of the metal species on it [82], and the support can affect the electron density of the metal via charge transfer [17]. An extreme example is the so-called strong metal–support interaction (SMSI) phenomenon [83]: when the support is a partially reducible metal oxide such as TiO_2, typical procedures to activate noble metals in H_2 atmospheres may result in SMSI effects, such as when partially reduced metal oxide species from the support cover part of the metal surface. SMSI effects may lead to modified catalyst activity and selectivity, enhanced thermal stability, and significant transfer of electron density to the metal.

An example illustrating the importance of the support is the selective hydrogenation of nitroaromatic compounds in the presence of —C=C, —C≡C, or —CHO groups. Selective hydrogenation of nitro compounds was reported with gold nanoparticles supported on TiO_2 [80], whereby the metal–support boundary is responsible for selective activation of the NO_2 group, and gold is necessary for activation of H_2, as inferred from kinetics of the catalytic reaction, spectroscopy, and theory [84]. To achieve similar results, previous alternatives required the use of toxic metal salts or additives with a supported platinum catalyst. SMSI concepts were later applied to design selective platinum hydrogenation catalysts without those additives. Pt/TiO_2 reduced at 723 K in H_2 resulted in platinum nanoparticles partially covered by TiO_x species that are highly selective for the hydrogenation of substituted nitroaromatic compounds [85], because (a) the metal surface lacks terraces (now blocked by TiO_x patches) that are responsible for the undesired side reactions [85] and (b) the metal/support interface where selective NO_2 activation occurs [86] is augmented.

Roles of the support are maximized for atomically dispersed supported metals because of the direct metal–support bonding. Effects of the support were elucidated with a series of supported isostructural molecular catalysts synthesized precisely on a variety of supports, as described in Section 13.3.2. For example, investigations of $Ir(C_2H_4)_2$ on various supports (each bonded through two Ir—O bonds), synthesized as described above, gave evidence of increasing hydrogenation activity with increasing electron-deficiency of the metal caused by more electron-withdrawing supports. At 298 K and ~ 1 bar (333 mbar of C_2H_4 and 666 mbar of H_2), a turnover frequency (reaction rate per metal atom, TOF) of $0.71 \, s^{-1}$ was observed when the support was zeolite HY, whereas the value was only $0.03 \, s^{-1}$ when the support was the electron-donating MgO [87]; ethane was the main product [88]. Additional data have been reported with MOFs as the supports [26]. The data show how the electron-donor properties of a family of metal oxide, zeolite, and MOF-supported iridium complexes influence the catalytic properties of these supported complexes for the ethylene hydrogenation and dimerization [26,35].

$Ir(C_2H_4)CO$, $Rh(C_2H_4)_2CO$, and $Rh(C_2H_4)CO$ were found when the support was the zeolite with the sample in an atmosphere of ethylene and CO, but they were absent when the support was MgO (and then stable iridium *gem*-dicarbonyl species were observed) [89].

The support can also play a direct role in catalysis by acting in concert with the metal. The support may even play a key role in determining catalyst selectivity. For example, $Ru(C_2H_4)_2(acac)$ [90,91] and $Rh(C_2H_4)_2$ [57,92] on zeolite HY are precursors of catalysts for ethylene dimerization in the presence of H_2, whereas the same species are 100% selective for ethylene hydrogenation when the support is MgO [93]. These results are intriguing because rhodium (and ruthenium) is typically regarded as inactive for C—C bond formation unless labile ligands such as chloride are present and H_2 is absent (otherwise, hydrogenation dominates).

13.6.5
Effects of Metal Nuclearity

Modern catalysis science reflects an increasing awareness of the relevance and singularities of subnanometer-sized metal species as (a) potential new catalysts with unanticipated properties and (b) excellent models to understand catalysis more fundamentally – at a molecular level.

As the metal—metal bonds become discrete covalent bonds in small metal clusters, and not part of the continuum characteristic of metallic particles, new classes of structures arise that bond to and activate reactants differently. For example, consider the hydrogenation of light alkenes by small iridium clusters (with less than six atoms per catalytic unit) [59,94]. The reaction is considered structure-insensitive according to classical surface science models and, thus, independent of the particle size for nanoparticles greater than about 1 nm in diameter. In contrast, a marked dependence of the reaction rates on the nuclearity of metal species was observed with supported metal clusters that are smaller – small enough to have molecular character. For example, Ir_4 and Ir_6 clusters on γ-Al_2O_3 show a 10-fold rate difference from each other in the hydrogenation of ethylene at 288 K [94].

Mononuclear metal complexes on supports are also molecular in nature, but they lack the M—M bonds characteristic of metal clusters and surfaces. The existence of neighboring metal centers is important for a number of processes, and their role in H—H bond scission is illustrative [95]. Comparisons between $M(C_2H_4)_2$ species and $M_x(C_2H_5)_y$ clusters ($x<6$; $0<y<12$) on a particular support facilitated quantification of the effect of the metal structure without significant complications associated with the nature of the support and the ligands bonded to the metal center. For example, rhodium dimers on MgO are ~60 times more active than mononuclear rhodium complexes on the same support for the hydrogenation of ethylene at 298 K and a H_2:C_2H_4 molar ratio of 4 [96]. Similarly, iridium complexes on MgO catalyze ethylene hydrogenation with a TOF of $0.03\,s^{-1}$ at 298 K, 333 mbar of C_2H_4 and 666 mbar of H_2, whereas Ir_4 clusters on the same support are characterized by a TOF of $0.18\,s^{-1}$ under identical conditions [87]. These differences in catalytic activity are reflected in similar differences in the activity of the clusters versus the complexes for H–D exchange [87], consistent with the observation that H_2 dissociation is the rate-determining reaction step when these catalytic units lack metal—metal bonds [87].

13.6.6
Effects of Ligands Other Than Supports

Metals bonded to solid surfaces must interact with reactants from the fluid phase to initiate catalysis. Molecules adsorbed at various positions on a metal nanoparticle can be considered ill-defined ligands relative to their molecular analogs in organometallic complexes; their role in the latter is much better understood than that on surfaces, and fundamental understanding can be applied to the design of catalytically active sites. Supported molecular metal complexes and clusters fill the gap between the common supported metal particles and the molecular organometallic catalysts in solution, and they have been used to investigate the effects of small ligands on the metal centers during catalysis. The sources of these ligands are diverse, from elements present in the metal precursors used in the synthesis to the components of the reaction mixture, including the reactants, products, and impurities.

CO is a common ligand bonded to metals in industrial chemical processes. It bonds strongly to many metals. It is commonly encountered in industrial processing as a result of the partial oxidation of hydrocarbons or as a product of processes such as steam reforming of feedstocks such as natural gas. Because of the high strengths of many metal—CO bonds, CO is a poison in many catalytic processes; it causes catalyst deactivation by blocking catalytic sites, hindering the interactions with reactants that are necessary for catalysis. However, CO can also be used advantageously for subtle control of selectivity in, for example, partial hydrogenation processes. It has been shown that whereas σ-bonded terminal CO ligands bond irreversibly to rhodium dimers in the presence of 1,3-butadiene and H_2 at 333 K, bridging CO ligands can be replaced by the reactants, leading to the exclusive formation of *n*-butenes (>99%) at almost complete conversion of the diene in catalysis [96]. The result is of interest because (a) diene impurities are responsible for a loss of quality of polymers formed from 1-butene [97] and (b) rhodium is typically regarded as unselective for the partial hydrogenation of 1,3-butadiene [98,99].

13.6.7
Metal Oxidation State

In typical supported metal catalysts, the metal oxidation state (and more generally the metal electron density) is influenced by the support and other ligands bonded to the metal centers, by the structure of the metal particle, and by the composition of the metal frame in multimetallic catalysts. As different species usually exist in a variety of chemical environments, common catalysts typically include metal atoms in various sites and in various oxidation states – and these can be altered during catalysis as a result of interactions with reactants. The role of the metal oxidation state is not easily separated from effects of metal particle size and morphology.

However, the issues may be resolvable when the catalysts are essentially molecular, as shown by results characterizing gold catalysts containing clusters of virtually the same size but variable fractions of reduced gold versus gold atoms, as determined by XANES and EXAFS spectra [100]. Exposure of a catalyst initially consisting of Au(III) species on MgO to $CO + O_2$ with variable ratios led to the formation of small gold clusters that incorporated Au(I) and Au(0) species in a ratio depending on the composition of the gas. The steady-state catalytic activity for CO oxidation increased as the fraction of partially oxidized gold species increased with decreasing CO/O_2 ratios, indicating that Au(I) species, presumably providing low-coordination sites on the gold clusters or sites at the gold–support interface that play a role in the CO oxidation reaction.

13.6.8
Catalyst Deactivation by Sintering and Poisoning by Compounds Containing P, N, S, or As

Supported metal catalysts lose activity in operation at high temperatures in the presence of feed streams with various components, including minor impurities. Among multiple deactivation mechanisms of supported catalysts, we highlight metal sintering, coke deposition, and chemical poisoning as the most significant in industrial processes; these may all occur simultaneously.

Metal sintering results from the exposure of a supported metal catalyst to reactants (e.g., H_2 and others) at high temperatures that lead to weakening of the metal–metal and metal–support bonding and enable the species to migrate and recombine on the support surface to form larger (and thermodynamically more stable) structures. Metals may enter the gas phase and be transported to nearby particles, where they are deposited to form larger particles [101], or metal particles may migrate on support surfaces and coalesce [102]. Both processes lead to losses in the number of sites available for catalysis as the metal dispersion decreases.

In processes with hydrocarbon feedstocks, catalyst deactivation takes place as a result of the essentially irreversible deposition of heavy carbonaceous residues called coke (typically containing polyaromatic rings). Coke formation is catalyzed by acidic centers in zeolites, for example, and by metal nanoparticles that break and rearrange C—C bonds of hydrocarbons. The deposits form on catalytic sites and prevent reactants from bonding to them. Catalyst deactivation also arises from blockage of catalyst pores by coke, limiting the transport of reactants and products. Coke can also block the interstices between particles of catalyst in a reactor and plug it.

Chemical poisoning, like coke deposition, results in the deactivation of catalytic sites as the poison bonds strongly to them and prevents bonding of the reactants. Sulfur-, nitrogen-, and phosphorus-containing compounds, which are often present as impurities in industrial feed streams to catalytic reactors, are common poisons of metal and acid sites. For example, sulfur-containing compounds such as H_2S or mercaptans bond so strongly to metals that they poison them almost irreversibly. Economical processing often requires purification units upstream of reactors to remove poisons; alternatively, poison-resistant catalysts are sought. Thus, supported metal sulfides are for the most part much less susceptible to poisoning by sulfur-containing compounds than metals, and they are used in a number of hydrogenation processes such as hydrodesulfurization and hydrodenitrogenation. But metal sulfides are usually much less active for hydrogenation reactions than noble metals.

13.6.9
Catalyst Regeneration

Because almost all catalysts are deactivated during operation, they need to be periodically regenerated or replaced. Regeneration may refer to a process whereby an undesired deposit such as coke is removed from the catalyst surface upon application of a suitable gas and temperature. Coke is burned in air or diluted O_2 and removed as CO_2 and water (and CO) at temperatures that depend on the nature of the carbonaceous residue (typically >773 K) and do not take the catalyst to so high a temperature as to destroy it. A challenge arises in the regeneration of poisoned supported metal catalysts, because the harsh conditions needed for removal of the poison often lead to sintering of the metal species and deactivation attributed to decreased dispersion. Supported metal particles that have sintered (e.g., platinum) may be redispersed by treatment in reactive gases (e.g., Cl_2) that form volatile compounds that redeposit on the support and are treated (e.g., in H_2) to form smaller metal particles.

Acknowledgment

The work at the University of California was supported by the US Department of Energy, Office of Science, Basic Energy Sciences, Grant DEFG02-04ER15513.

References

1 Anderson, J.A. and García, M.F. (eds) (2005) *Supported Metals in Catalysis*, vol. **5** in Catalytic Science Series, Imperial College Press, London.
2 Corma, A. (1997) From microporous to mesoporous molecular sieve materials and their use in catalysis. *Chem. Rev.*, **97** (6), 2373–2420.
3 Topsøe, H. and Clausen, B.S. (1986) Active-sites and support effects in hydrodesulfurization catalysts. *Appl. Catal.*, **25** (1–2), 273–293.
4 Schuit, G.C.A. and Gates, B.C. (1973) Chemistry and engineering of catalytic hydrodesulfurization. *AIChE J.*, **19** (3), 417–438.
5 Prins, R. and Bussell, M. (2012) Metal phosphides: preparation, characterization and catalytic reactivity. *Catal. Lett.*, **142** (12), 1413–1436.
6 Oyama, S.T., Gott, T., Zhao, H., and Lee, Y.-K. (2009) Transition metal phosphide hydroprocessing catalysts: a review. *Catal. Today*, **143** (1–2), 94–107.
7 Perego, C. and Villa, P. (1997) Catalyst preparation methods. *Catal. Today*, **34** (3–4), 281–305.
8 Campanati, M., Fornasari, G., and Vaccari, A. (2003) Fundamentals in the preparation of heterogeneous catalysts. *Catal. Today*, **77** (4), 299–314.
9 Pinna, P. (1998) Supported metal catalysts preparation. *Catal. Today*, **41** (1–3), 129–137.
10 Behrens, M., Studt, F., Kasatkin, I., Kühl, S., Hävecker, M., Abild-Pedersen, F., Zander, S., Girgsdies, F., Kurr, P., Kniep, B.-L., Tovar, M., Fischer, R.W., Nørskov, J.K., and Schlögl, R. (2012) The active site of methanol synthesis over Cu/ZnO/Al$_2$O$_3$ industrial catalysts. *Science*, **336** (6083), 893–897.
11 Basset, J.-M., Gates, B.C., Candy, J.P., Choplin, A., Leconte, M., Quignard, F., and Santini, C. (eds) (2012) *Surface Organometallic Chemistry: Molecular Approaches to Surface Catalysis*, Kluwer, Dordrecht.
12 Copéret, C., Chabanas, M., Petroff Saint-Arroman, R., and Basset, J.-M. (2003) Homogeneous and heterogeneous catalysis: bridging the gap through surface organometallic chemistry. *Angew. Chem. Int. Ed.*, **42** (2), 156–181.
13 Wegener, S.L., Marks, T.J., and Stair, P.C. (2012) Design strategies of the molecular level synthesis of supported catalysts. *Acc. Chem. Res.*, **45** (2), 206–214.
14 Tada, M. and Iwasawa, Y. (2007) Advanced design of catalytically active reaction space at surfaces for selective catalysis. *Coord. Chem. Rev.*, **251** (21–24), 2702–2716.
15 Conley, M. and Copéret, C. (2014) State of the art and perspectives in the "molecular approach" towards well-defined heterogeneous catalysts. *Top. Catal.*, **57** (10–13), 843–851.
16 Lefebvre, F. (2015) Synthesis of well-defined solid catalysts by surface organometallic chemistry, in *Atomically-Precise Methods for Synthesis of Solid Catalysts* (eds S. Hermans and T.V. de Bocarme), The Royal Society of chemistry, Cambridge, pp. 1–26.
17 Serna, P. and Gates, B.C. (2014) Molecular metal catalysts on supports: organometallic chemistry meets surface science. *Acc. Chem. Res.*, **47** (8), 2612–2620.
18 Guzman, J. and Gates, B.C. (2003) Supported molecular catalysts: metal complexes and clusters on oxides and zeolites. *Dalton Trans*, (17), 3303–3318.
19 Ogino, I., Chen, C.Y., and Gates, B.C. (2010) Zeolite-supported metal complexes of rhodium and of ruthenium:

a general synthesis method influenced by molecular sieving effects. *Dalton Trans.*, **39** (36), 8423–8431.

20 Kulkarni, A., Lobo-Lapidus, R.J., and Gates, B.C. (2010) Metal clusters on supports: synthesis, structure, reactivity, and catalytic properties. *Chem. Commun.*, **46** (33), 5997–6015.

21 Liang, A.J., Craciun, R., Chen, M., Kelly, T.G., Kletnieks, P.W., Haw, J.F., Dixon, D.A., and Gates, B.C. (2009) Zeolite-supported organorhodium fragments: essentially molecular surface chemistry elucidated with spectroscopy and theory. *J. Am. Chem. Soc.*, **131** (24), 8460–8473.

22 Martinez-Macias, C., Chen, M., Dixon, D.A., and Gates, B.C. (2015) Single-site zeolite-anchored organoiridium carbonyl complexes: characterization of structure and reactivity by spectroscopy and computational chemistry. *Chem. Eur. J.*, **21** (33), 11825–11835.

23 Chen, M., Serna, P., Lu, J., Gates, B.C., and Dixon, D.A. (2015) Molecular models of site-isolated cobalt, rhodium, and iridium catalysts supported on zeolites: ligand bond dissociation energies. *Comput. Theor. Chem.*, **1074**, 58–72.

24 Ogino, I. (2015) Zeolite-supported molecular metal complex catalysts, in *Atomically-Precise Methods for Synthesis of Solid Catalysts* (eds S. Hermans and T.V. de Bocarme), The Royal Society of chemistry, Cambridge, pp. 27–54.

25 Yang, D., Odoh, S.O., Wang, T.C., Farha, O.K., Hupp, J.T., Cramer, C.J., Gagliardi, L., and Gates, B.C. (2015) Metal-organic framework nodes as nearly ideal supports for molecular catalysts: NU-1000- and UiO-66-supported iridium complexes. *J. Am. Chem. Soc.*, **137** (23), 7391–7396.

26 Yang, D., Odoh, S.O., Borycz, J., Wang, T.C., Farha, O.K., Hupp, J.T., Cramer, C.J., Gagliardi, L., and Gates, B.C. (2016) Tuning Zr_6 MOF nodes as catalyst supports: site densities and electron-donor properties influence molecular iridium complexes as ethylene conversion catalysts. *ACS Catal.*, **6**, 235–247.

27 Fierro-Gonzalez, J.C., Kuba, S., Hao, Y., and Gates, B.C. (2006) Oxide- and zeolite-supported molecular metal complexes and clusters: physical characterization and determination of structure, bonding, and metal oxidation state. *J. Phys. Chem. B*, **110** (27), 13326–13351.

28 Pan, M. and Crozier, P.A. (1993) Low-dose high-resolution electron microscopy of zeolite materials with a slow-scan CCD camera. *Ultramicroscopy*, **48** (3), 332–340.

29 Bursill, L.A., Thomas, J.M., and Rao, K.J. (1981) Stability of zeolites under electron-irradiation and imaging of heavy cations in silicates. *Nature*, **289** (5794), 157–158.

30 Logan, A.D., Sharoudi, K., and Datye, A.K. (1991) Oxidative restructuring of rhodium metal surfaces: correlations between single crystals and small metal particles. *J. Phys. Chem.*, **95** (14), 5568–5574.

31 Kalakkad, D., Anderson, S.L., Logan, A.D., Peña, J., Braunschweig, E.J., Peden, C.H.F., and Datye, A.K. (1993) *n*-Butane hydrogenolysis as a probe of surface sites in rhodium metal particles: correlation with single crystals. *J. Phys. Chem.*, **97** (7), 1437–1444.

32 Nellist, P.D., Chisholm, M.F., Dellby, N., Krivanek, O.L., Murfitt, M.F., Szilagyi, Z.S., Lupini, A.R., Borisevich, A., Sides, W.H., and Pennycook, S.J. (2004) Direct sub-Angstrom imaging of a crystal lattice. *Science*, **305** (5691), 1741.

33 Pennycook, S.J., Varela, M., Hetherington, C.J.D., and Kirkland, A.I. (2006) Materials advances through aberration-corrected electron microscopy. *MRS Bull.*, **31** (1), 36–43.

34 Hansen, L.P., Ramasse, Q.M., Kisielowski, C., Brorson, M., Johnson, E., Topsøe, H., and Helveg, S. (2011) Atomic-scale edge structures on industrial-style MoS_2 nanocatalysts. *Angew. Chem. Int. Ed.*, **50** (43), 10153–10156.

35 Yang, M., Liu, J., Lee, S., Zugic, B., Huang, J., Allard, L.F., and Flytzani-Stephanopoulos, M. (2015) A common single-site Pt(II)−O(OH)$_x^-$ species stabilized by sodium on "active" and "inert" supports catalyzes the water–gas

shift reaction. *J. Am. Chem. Soc.*, **137** (10), 3470–3473.

36 Qiao, B., Wang, A., Yang, X., Allard, L.F., Jiang, Z., Cui, Y., Liu, J., Li, J., and Zhang, T. (2011) Single-atom catalysis of CO oxidation using Pt_1/FeO_x. *Nat. Chem.*, **3** (8), 634–641.

37 Browning, N.D., Aydin, C., Lu, J., Kulkarni, A., Okamoto, N.L., Ortalan, V., Reed, B.W., Uzun, A., and Gates, B.C. (2013) Quantitative Z-contrast imaging of supported metal complexes and clusters: a gateway to understanding catalysis on the atomic scale. *ChemCatChem*, **5** (9), 2673–2683.

38 Ortalan, V., Uzun, A., Gates, B.C., and Browning, N.D. (2010) Direct imaging of single metal atoms and clusters in the pores of dealuminated HY zeolite. *Nat. Nanotechnol.*, **5** (7), 506–510.

39 Kulkarni, A., Chi, M.F., Ortalan, V., Browning, N.D., and Gates, B.C. (2010) Atomic resolution of the structure of a metal–support interface: triosmium clusters on MgO(110). *Angew. Chem. Int. Ed.*, **49** (52), 10089–10092.

40 Bradley, S., Sinkler, W., Blom, B., Bigelow, W., Voyles, P., and Allard, L. (2012) Behavior of Pt atoms on oxide supports during reduction treatments at elevated temperatures, characterized by aberration corrected STEM imaging. *Catal. Lett.*, **142** (2), 176–182.

41 Aydin, C., Lu, J., Liang, A.J., Chen, C.-Y., Browning, N.D., and Gates, B.C. (2011) Tracking iridium atoms with electron microscopy: first steps of metal nanocluster formation in one-dimensional zeolite channels. *Nano Lett.*, **11** (12), 5537–5541.

42 Martinez-Macias, C., Xu, P., Hwang, S.-J., Lu, J., Chen, C.-Y., Browning, N.D., and Gates, B.C. (2014) Iridium complexes and clusters in dealuminated zeolite HY: distribution between crystalline and impurity amorphous regions. *ACS Catal.*, **4** (8), 2662–2666.

43 Kistler, J.D., Chotigkrai, N., Xu, P., Enderle, B., Praserthdam, P., Chen, C.Y., Browning, N.D., and Gates, B.C. (2014) A single-site platinum CO oxidation catalyst in zeolite KLTL: microscopic and spectroscopic determination of the locations of the platinum atoms. *Angew. Chem. Int. Ed.*, **53** (34), 8904–8907.

44 Koningsberger, D.C., Mojet, B.L., van Dorssen, G.E., and Ramaker, D.E. (2000) XAFS spectroscopy: fundamental principles and data analysis. *Top. Catal.*, **10** (3–4), 143–155.

45 Alexeev, O. and Gates, B.C. (2000) EXAFS characterization of supported metal-complex and metal-cluster catalysts made from organometallic precursors. *Top. Catal.*, **10** (3–4), 273–293.

46 Ferrari, A.M., Neyman, K.M., Mayer, M., Staufer, M., Gates, B.C., and Rösch, N. (2000) Faujasite-supported Ir_4 clusters: a density functional model study of metal-zeolite interactions. *J. Phys. Chem. B*, **103** (25), 5311–5319.

47 Goellner, J.F., **Gates, B.C.**, Vayssilov, G.N., and Rösch, N. (2000) Structure and bonding of a site-isolated transition metal complex: rhodium dicarbonyl in highly dealuminated zeolite Y. *J. Am. Chem. Soc.*, **122** (33), 8056–8066.

48 Lu, J., Aydin, C., Browning, N.D., and Gates, B.C. (2012) Imaging isolated gold atom catalytic sites in zeolite NaY. *Angew. Chem. Int. Ed.*, **51** (24), 5842–5846.

49 Benson, J.E., Hwang, H.S., and Boudart, M. (1973) Hydrogen–oxygen titration method for the measurement of supported palladium surface areas. *J. Catal.*, **30** (1), 146–153.

50 Alexeev, O.S., Li, F., Amiridis, M.D., and Gates, B.C. (2005) Effects of adsorbates on supported platinum and iridium clusters: characterization in reactive atmospheres and during alkene hydrogenation catalysis by X-ray absorption spectroscopy. *J. Phys. Chem. B*, **109** (6), 2338–2349.

51 Kip, B.J., Duivenvoorden, F.B.M., Koningsberger, D.C., and Prins, R. (1987) Determination of metal particle size of highly dispersed Rh, Ir, and Pt catalysts by hydrogen chemisorption and EXAFS. *J. Catal.*, **105** (1), 26–38.

52 Alexeev, O., Kim, D.-W., Graham, G.W., Shelef, M., and Gates, B.C. (1999) Temperature-programmed desorption of hydrogen from platinum particles on

γ-Al$_2$O$_3$: evidence of platinum-catalyzed dehydroxylation of γ-Al$_2$O$_3$. *J. Catal.*, **185** (1), 170–181.

53 Hadjiivanov, K.I. and Vayssilov, G.N. (2002) Characterization of oxide surfaces and zeolites by carbon monoxide as an IR probe molecule. *Adv. Catal.*, **47**, 308–511.

54 Miessner, H., Burkhardt, I., Gutschick, D., Zecchina, A., Morterra, C., and Spoto, G. (1989) The formation of a well defined rhodium dicarbonyl in highly dealuminated rhodium-exchanged zeolite-Y by interaction with CO. *J. Chem. Soc. Faraday Trans. 1*, **85**, 2113–2126.

55 Odzak, J.F., Argo, A.M., Lai, F.S., Gates, B.C., Pandya, K., and Feraria, L. (2001) A flow-through X-ray absorption spectroscopy cell for characterization of powder catalysts in the working state. *Rev. Sci. Instrum.*, **72** (10), 3943–3945.

56 Uzun, A. and Gates, B.C. (2008) Real-time characterization of formation and breakup of iridium clusters in highly dealuminated zeolite Y. *Angew. Chem. Int. Ed.*, **47** (48), 9245–9248.

57 Serna, P. and Gates, B.C. (2011) Supported molecular iridium catalysts: resolving effects of metal nuclearity and supports as ligands. *J. Am. Chem. Soc.*, **133** (40), 4714–4717.

58 Khivantsev, K., Vityuk, A., Aleksandrov, H.A., Vayssilov, G.N., Alexeev, O.S., and Amiridis, M.D. (2015) Effect of Si/Al ratio and Rh precursor used on the synthesis of Hy zeolite-supported rhodium carbonyl hydride complexes. *J. Phys. Chem. C*, **119** (30), 17166–17181.

59 Argo, A.M., Odzak, J.F., Lai, F.S., and Gates, B.C. (2002) Observation of ligand effects during alkene hydrogenation catalysed by supported metal clusters. *Nature*, **415** (6872), 623–626.

60 Vayssilov, G.N., Gates, B.C., and Rösch, N. (2003) Oxidation of supported rhodium clusters by support hydroxy groups. *Angew. Chem. Int. Ed.*, **42** (12), 1391–1394.

61 Uzun, A. and Gates, B.C. (2009) Dynamic structural changes in a molecular zeolite-supported iridium catalyst for ethene hydrogenation. *J. Am. Chem. Soc.*, **131** (43), 15887–15894.

62 Asakura, K., Chun, W.-J., Shirai, M., Tomishige, K., and Iwasawa, Y. (1997) In-situ polarization-dependent total-reflection fluorescence XAFS studies on the structure transformation of Pt clusters on α-Al$_2$O$_3$(0001) *J. Phys. Chem. B*, **101** (28), 5549–5556.

63 Lauritsen, J.V. and Besenbacher, F. (2006) Model catalyst surfaces investigated by scanning tunneling microscopy. *Adv. Catal.*, **50**, 97–147.

64 Lauritsen, J.V., Bollinger, M.V., Lægsgaard, E., Jacobsen, K.W., Nørskov, J.K., Clausen, B.S., Topsøe, H., and Besenbacher, F. (2004) Atomic-scale insight into structure and morphology changes of MoS$_2$ nanoclusters in hydrotreating catalysts. *J. Catal.*, **221** (2), 510–522.

65 Walton, A.S., Lauritsen, J.V., Topsøe, H., and Besenbacher, F. (2013) MoS$_2$ nanoparticle morphologies in hydrodesulfurization catalysis studied by scanning tunneling microscopy. *J. Catal.*, **308**, 306–318.

66 Füchtbauer, H., Tuxen, A., Li, Z., Topsøe, H., Lauritsen, J., and Besenbacher, F. (2014) Morphology and atomic-scale structure of MoS$_2$ nanoclusters synthesized with different sulfiding agents. *Top. Catal.*, **57** (1–4), 207–214.

67 Boudart, M. (1969) Catalysis by supported metals. *Adv. Catal.*, **20**, 153–166.

68 Anderson, J.R. and Shimoyama, Y. (1972) Proceedings International Congress on Catalysis, p. 965.

69 Boudart, M., Aldag, A.W., Ptak, L.D., and Benson, J.E. (1968) On selectivity of platinum catalysts. *J. Catal.*, **11** (1), 35–45.

70 Boudart, M., Aldag, A., Benson, J.E., Doughart, Na., and Harkins, C.G. (1966) On specific activity of platinum catalysts. *J. Catal.*, **6** (1), 92.

71 Dorling, T.A. and Moss, R.L. (1966) Structure and activity of supported metal catalyst: 1. Crystallite size and specific activity for benzene hydrogenation of platinum/silica catalysts. *J. Catal.*, **5** (1), 111–115.

72 Bond, G.C. (1993) Strategy of research on supported metal-catalysts: problems of structure-sensitive reactions in the gas-phase. *Acc. Chem. Res.*, **26** (9), 490–495.

73 Strongin, D.R., Carrazza, J., Bare, S.R., and Somorjai, G.A. (1987) The importance of C7 sites and surface roughness in the ammonia synthesis reaction over iron. *J. Catal.*, **103** (1), 213–215.

74 Lang, N.D., Holloway, S., and Nørskov, J.K. (1985) Electrostatic adsorbate–adsorbate interactions: the poisoning and promotion of the molecular adsorption reaction. *Surf. Sci.*, **150** (1), 24–38.

75 Somorjai, G.A. and Carrazza, J. (1986) Structure sensitivity of catalytic reactions. *Ind. Eng. Chem. Fund.*, **25** (1), 63–69.

76 Davis, S.M., Zaera, F., and Somorjai, G.A. (1984) Surface-structure and temperature-dependence of n-hexane skeletal rearrangement reactions catalyzed over platinum single-crystal surfaces: marked structure sensitivity of aromatization. *J. Catal.*, **85** (1), 206–223.

77 Cleveland, C.L. and Landman, U. (1991) The energetics and structure of nickel clusters: size dependence. *J. Chem. Phys.*, **94** (11), 7376–7396.

78 Corma, A., Boronat, M., González, S., and Illas, F. (2007) On the activation of molecular hydrogen by gold: a theoretical approximation to the nature of potential active sites. *Chem. Commun.*, **32**, 3371–3373.

79 Serna, P., Boronat, M., and Corma, A. (2011) Tuning the behavior of Au and Pt catalysts for the chemoselective hydrogenation of nitroaromatic compounds. *Top. Catal.*, **54** (5), 439–446.

80 Corma, A. and Serna, P. (2006) Chemoselective hydrogenation of nitro compounds with supported gold catalysts. *Science*, **313** (5785), 332–334.

81 Jia, J., Haraki, K., Kondo, J.N., Domen, K., and Tamaru, K. (2000) Selective hydrogenation of acetylene over Au/Al_2O_3. *J. Phys. Chem. B*, **104** (47), 11153–11156.

82 Wu, Z., Schwartz, V., Li, M., Rondinone, A.J., and Overbury, S.H. (2012) Support shape effect in metal oxide catalysis: ceria-nanoshape-supported vanadia catalysts for oxidative dehydrogenation of isobutene. *J. Phys. Chem. Lett.*, **3** (11), 1517–1522.

83 Tauster, S.J., Fung, S.C., and Garten, R.L. (1978) Strong metal-support interactions: group-8 noble-metals supported on TiO_2. *J. Am. Chem. Soc.*, **100** (1), 170–175.

84 Serna, P., Concepción, P., and Corma, A. (2009) Design of highly active and chemoselective bimetallic gold–platinum hydrogenation catalysts through kinetic and isotopic studies. *J. Catal.*, **265** (1), 19–25.

85 Corma, A., Serna, P., Concepción, P., and Calvino, J.J. (2008) Transforming nonselective into chemoselective metal catalysts for the hydrogenation of substituted nitroaromatics. *J. Am. Chem. Soc.*, **130** (27), 8748–8753.

86 Boronat, M., Concepción, P., Corma, A., González, S., Illas, F., and Serna, P. (2007) A molecular mechanism for the chemoselective hydrogenation of substituted nitroaromatics with nanoparticles of gold on TiO_2 catalysts: a cooperative effect between gold and the support. *J. Am. Chem. Soc.*, **129** (51), 16230–16237.

87 Lu, J., Serna, P., Aydin, C., Browning, N.D., and Gates, B.C. (2011) Supported molecular iridium catalysts: resolving effects of metal nuclearity and supports as ligands. *J. Am. Chem. Soc.*, **133** (40), 16186–16195.

88 Martinez-Macias, C., Serna, P., and Gates, B.C. (2015) Isostructural zeolite-supported rhodium and iridium complexes: tuning catalytic activity and selectivity by ligand modification. *ACS Catal.*, **5** (10), 5647–5656.

89 Lu, J., Serna, P., and Gates, B.C. (2011) Zeolite- and MgO-supported molecular iridium complexes: support and ligand effects in catalysis of ethene hydrogenation and H–D exchange in the conversion of H_2+D_2. *ACS Catal.*, **1** (11), 1549–1561.

90 Ogino, I. and Gates, B.C. (2008) Molecular chemistry in a zeolite: genesis of a zeolite Y-supported ruthenium complex catalyst. *J. Am. Chem. Soc.*, **130** (40), 13338–13346.

91 Ogino, I. and Gates, B.C. (2009) Role of the support in catalysis: activation of a mononuclear ruthenium complex for ethene dimerization by chemisorption on dealuminated zeolite Y. *Chem. Eur. J.*, **15** (28), 6827–6837.

92 Serna, P. and Gates, B.C. (2011) A bifunctional mechanism for ethene dimerization: catalysis by rhodium complexes on zeolite HY in the absence of halides. *Angew. Chem., Int. Ed.*, **50** (24), 5528–5531.

93 Serna, P. and Gates, B.C. (2013) Zeolit - and MgO-supported rhodium complexes and rhodium clusters: tuning catalytic properties to control carbon–carbon vs. carbon–hydrogen bond formation reactions of ethene in the presence of H_2. *J. Catal.*, **308**, 201–212.

94 Argo, A.M., Odzak, J.F., and Gates, B.C. (2003) Role of cluster size in catalysis: spectroscopic investigation of gamma-Al_2O_3-supported Ir_4 and Ir_6 during ethene hydrogenation. *J. Am. Chem. Soc.*, **125** (23), 7107–7115.

95 Conner, W.C. and Falconer, J.L. (1995) Spillover in heterogeneous catalysis. *Chem. Rev.*, **95**, 759–788.

96 Yardimci, D., Serna, P., and Gates, B.C. (2013) Surface-mediated synthesis of dimeric rhodium catalysts on MgO: tracking changes in the nuclearity and ligand environment of the catalytically active sites by X-ray absorption and infrared spectroscopies. *Chem. Eur. J.*, **19** (4), 2035–2045.

97 Seth, D., Sarkar, A., Ng, F.T.T., and Rempel, G.L. (2007) Selective hydrogenation of 1,3-butadiene in mixture with isobutene on a Pd/alpha-alumina catalyst in a serni-batch reactor. *Chem. Eng. Sci.*, **62** (17), 4544–4557.

98 Boitiaux, J.P., Cosyns, J., and Robert, E. (1987) Hydrogenation of unsaturated hydrocarbons in liquid phase on palladium, platinum and rhodium catalysts: III. Quantitative selectivity ranking of platinum, palladium and rhodium in the hydrogenation of 1-butene, 1,3-butadiene and 1-butyne using a single reaction scheme. *Appl. Catal.*, **35** (2), 193–209.

99 Bond, G.C., Webb, G., Wells, P.B., and Winterbottom, J.M. (1965) The hydrogenation of alkadienes. Part I. The hydrogenation of buta-1,3-diene catalysed by the noble group VIII metals. *J. Chem. Soc.*, **1965**, 3218–3227.

100 Guzman, J. and Gates, B.C. (2004) Catalysis by supported gold: correlation between catalytic activity for CO oxidation and oxidation states of gold. *J. Am. Chem. Soc.*, **126** (9), 2672–2673.

101 Campbell, C.T. (2013) The energetics of supported metal nanoparticles: relationships to sintering rates and catalytic activity. *Acc. Chem. Res*, **46** (8), 1712–1719.

102 Hansen, T.W., DeLaRiva, A.T., Challa, S.R., and Datye, A.K. (2013) Sintering of catalytic nanoparticles: particle migration or Ostwald ripening? *Acc. Chem. Res.*, **46** (8), 1720–1730.

14
Hydrogenation by Metals

Xin Jin and Raghunath V. Chaudhari

University of Kansas, Center for Environmentally Beneficial Catalysis, Chemical & Petroleum Engineering Department, Lawrence, KS 66047, USA

14.1
Introduction

Hydrogenations by heterogeneous metal catalysts are extensively used in industrial processes in petroleum, coal processing, and purification of monomers, pharmaceuticals, fine and specialty chemicals, agrochemicals, and perfumery as well as food products. It is also considered as one of the powerful tools in multi-step synthesis of complex organic products, though in most cases it is one of the several steps in the synthesis of target products. Catalytic hydrogenation has been known for more than 100 years, when Paul Sabatier showed that olefins and benzene can be efficiently hydrogenated using Ni metal catalysts. Today, it is known as one of the most versatile and widely used processes with applications in petroleum, petrochemical, pharmaceuticals, and fine chemical industry. Catalytic hydrogenation has a very broad scope since several metals are known to catalyze the reactions in conversion of cheaper and abundantly available feedstocks to value added chemicals with high activity and selectivity. Similarly, a wide range of multifunctional molecules can be reduced by catalytic hydrogenation to produce value added products, very often with chemo-, regio-, and/or stereoselectivity. The availability of hydrogen in large volumes at reasonable cost and progress in the development of catalytic reactors facilitated rapid expansion of hydrogenation technologies for the production of a range of products. While for the large volume processes like ammonia synthesis, Fischer–Tropsch (F–T), methanol synthesis, purification of monomers, manufacture of H_2O_2, and hydroprocessing of petroleum crude (HDS, HDN, HCC, etc.). Catalytic hydrogenation became an indispensable technology; its application has also been growing rapidly in the last few decades to a large variety of intermediates and specialty products (for pharmaceuticals, perfumery, and fine chemicals) on small to medium scale. Catalytic hydrogenation is also recognized as an important alternative to the stoichiometric reagent based organic synthesis to develop "green"

Handbook of Solid State Chemistry, First Edition. Edited by Richard Dronskowski, Shinichi Kikkawa, and Andreas Stein.
© 2017 Wiley-VCH Verlag GmbH & Co. KGaA. Published 2017 by Wiley-VCH Verlag GmbH & Co. KGaA.

technologies with minimum waste, high-atom efficiency, safety, and environmental compatibility. It is therefore not surprising that recent years have witnessed significant advances in not only industrial applications of hydrogenation technologies but also in fundamental understanding of catalysis, surface science, and reaction engineering of this important class of chemical transformation. Hydrogenation reactions are catalyzed by heterogeneous as well as homogeneous catalysts, though due to practical considerations and economic usage, heterogeneous catalysts have been more widely applied with a few exceptions in the area of asymmetric catalysis for which homogeneous catalysts are preferred to achieve stereospecificity. In most practical examples, hydrogenation is carried out either in gas–solid catalytic or multiphase (gas–liquid–solid or gas–liquid–liquid–solid) catalytic reactors depending on volatility of the substrates, products, and solvents. In this case, the effective utilization of the catalysts also depends on engineering aspects involving interphase and intraparticle mass transfer and mixing of the fluid phases. Hydrogenation reactions are highly exothermic and hence selection of a suitable reactor type to enable safe and optimum operation is also an important requirement. For liquid phase hydrogenation, usually higher pressures are required since hydrogen is a sparingly soluble gas in most liquids. In addition to catalyst system consisting of metal, support, promoters, and cocatalysts, selection of a solvent is also equally important. Thus, developing catalysts and processes for hydrogenation requires multidisciplinary skills in surface science, catalysis, organic chemistry, reaction engineering, and process design. Several aspects of hydrogenation reactions covering catalysis [1–5] and reaction engineering [4] have been reviewed previously, which may be referred for details. In this chapter, a review of hydrogenation by metals is presented with recent advances in new catalytic materials with a focus on the synthesis, characterization, and application of mono- and multimetallic catalysts.

14.2
Classification of Hydrogenation Reactions

Catalytic "hydrogenation" reactions can be broadly classified into two categories: (a) reactions in which H_2 is added to the molecule across a C—C, C—N, or C—O bonds and (b) reactions involving H_2 addition with simultaneous deoxygenation, desulfurization, or denitrogenation reactions. Reactions in case (b) are also commonly referred as "Hydrogenolysis" which essentially involve cleavage of a C—C, C—O, C—N, C—S, or C—X (carbon–heteroatom) single bond. Typical examples are: hydroprocessing (hydrodesulphurization, hydrocracking and hydrodenitrogenation, HDN, and HC) of petroleum feedstock, coal liquefaction, and F–T synthesis. Similarly, hydrogenation of nitroaromatics to amines, and hydrodeoxygenation of biomass feedstocks (sugars, polyols, and bio-oils) will also fall in this category. However, both categories can also be generally referred as hydrogenation reactions. In a few cases, hydrogenation is used in combination with amination or alkylation to drive an equilibrium reaction to produce a final value-added product. For example, reductive alkylation of amines or reductive amination of ketones or aldehydes [6,7]. The classification of various types of hydrogenation reactions useful in industrial processes as well as organic synthesis is summarized in Table 14.1 for a wide range of substrates with different functional groups. Some illustrative examples are discussed with their characteristic features along with recent advances in the heterogeneous metal catalysts for these reactions.

14.2.1
Hydrogenation of Alkenes

Alkenes with a single double bond are readily hydrogenated to alkanes with several catalysts consisting of Raney Ni, supported Ni, Pd, Pt, Ru, and Rh at mild conditions and are often used as model reactions to study catalysis of hydrogenation kinetics. However, hydrogenation of polyunsaturated compounds with other substituent groups can be challenging to convert selectively to desired products. Hydrogenation of fatty acids and fatty acid esters is one of the largest scale applications of alkene hydrogenation in industry in which it is required to convert polyunsaturated oils to a controlled degree of unsaturation for food applications [8]. The oils consist of linolenic (triunsaturated), linoleic (diunsaturated), and oleic acids (monounsaturated) compounds, which are converted by catalytic hydrogenation to partially hydrogenated products. The olefin groups in these polyunsaturated compounds react at different rates and selectivity depending on the type of catalysts. Catalysts containing Ni powder, Cu-chromite, Raney Ni, alloys consisting of Ni–Cu, Cu–Al, Pd–Al, Pt, and Pd on charcoal have also been proposed [9]; however, Ni-based catalysts have been most commonly used due to lower cost and easy availability compared to other catalysts [10]. Hydrogenation of oils is carried out in agitated slurry, bubble column slurry, or multistage agitated reactors at 1–5 atm H_2 pressure and 110–220 °C. In spite of the high viscosity of the oils and products, no solvent is preferable to avoid contamination at the cost of lower rates of reaction. Other applications are: (a) selective hydrogenation of dienes (e.g., butadiene) to olefins in petroleum feedstocks, which are associated with simultaneous *cis–trans* isomerization and (b) selective hydrogenation/isomerization of cyclododecatriene to cyclododecene using Pd catalysts [11–13]. Several examples of olefin hydrogenation in organic synthesis are reviewed by Nishimura [4] with details of reaction mechanism and catalytic chemistry.

14.2.2
Partial Hydrogenation of Alkynes to Alkenes

Hydrogenation of carbon–carbon triple bond to olefinic products has been a subject of major interest in petrochemical and fine chemical industry. Since alkynes such as acetylene are easily saturated to corresponding alkanes with two

Table 14.1 Hydrogenation reactions in petrochemical and fine chemical processes.

Substrate	Product	Chemical bond	Examples
Hydrogenation			
Alkynes	Alkenes	C≡C	
Aldehydes, ketones	Alcohols, alkanes	C=O	
Carboxylic acids	Aldehydes	O=C–OH	
Carboxylic acids	Alcohols	O=C–OH	
Carbohydrates	Polyols	C=O	
Aromatic rings	Cycloalkane rings	(benzene ring)	

Substrate	Bond	Reaction	Product
Imines	C=N	PhO-C(=N-)-NH₂ $\xrightarrow{Pd}$ PhO-CH(NH-)-NH₂	Amines
Alkyl halides Hydrogenolysis	C–Cl, Br, I	Ph–Cl $\xrightarrow{Pd}$ PhH + HCl	Alkanes
Aromatic nitro compounds		Ph–NO₂ $\xrightarrow{Pd}$ Ph–N=O $\xrightarrow{Pd}$ Ph–NH₂	Aromatic amines
Polyols	C–O	HO–CH₂–CH(OH)–CH₂–OH $\xrightarrow{Ru}$ HO–CH(CH₃)–CH₂–OH or HO–CH₂–CH₂–CH₂–OH	Glycols, alcohols
Ester	C–O	Ph–C(=O)–O–CH₂–NH₂ $\xrightarrow{Ru}$ Ph–CH₃ + CO₂ + NH₃	Alcohols, alkanes, CO₂
Amines	C–N	Ph–CH₂–NH–C₆H₁₁ $\xrightarrow{Pd}$ C₆H₁₁ + H₂N–Ph	Alkanes, amines
Thiophene	C–S	dibenzothiophene $\xrightarrow{MoNi}$ biphenyl + H₂S	Alkanes, H₂S
CO	C–C	$n\,\text{CO} + (2n+1)\text{H}_2 \rightarrow \text{C}_n\text{H}_{(2n+2)} + n\,\text{H}_2\text{O}$	Alkanes, alkenes

Table 14.2 Reaction heats of partial hydrogenation of alkynes to alkenes at 82 °C.

Reactions	$-\Delta H$ (kJ/mol)
≡≡ →(H₂)→ ==	177
≡≡ →(2 H₂)→ —	313
≡— →(H₂)→ \\=	166
≡— →(2 H₂)→ \\—	291
≡— →(H₂)→ \\=/	155
≡— →(2 H₂)→ _/	274

molecules of hydrogen in consecutive steps, minimization of the second step hydrogenation to produce olefins is a major challenge. Table 14.2 presents the general scheme of selective hydrogenation of several common alkyne compounds. It is generally accepted that activity trend for different catalysts follows: Pd > Pt > Ni, Rh > Co > Fe > Cu > Ir > Ru > Au, while selectivity trend follows: Rh > Pt > Ru > Ir. Promoters such as Cr, V, K, Na, Pb are known to enhance the selectivity in partial hydrogenation of alkynes.

Palladium, nickel, and iron are among the most investigated catalysts for selective hydrogenation of alkynes. Palladium catalysts usually exhibit much higher hydrogenation activity compared with nickel and iron ones. Supported palladium catalysts including Pd/C, Pd/CaCO$_3$, Pd/BaSO$_4$ are widely used in fine chemical synthesis. Pyridine, quinolone, Pb(OAc)$_2$, and ammonia are used to poison palladium catalysts, to reduce the activity for second step hydrogenation and obtain higher selectivity to olefins. Most of the semi-hydrogenation reactions can be easily carried out at room temperature and 1–20 atm H$_2$ pressure, examples of which are shown in Table 14.3.

14.2.3
Hydrogenation of Aldehydes and Ketones

Platinum group metal catalysts have been shown to be active and selective for aliphatic and alicyclic ketone hydrogenation. Hydrogenation of C—O groups is involved in converting substrates consisting of aldehydes, ketones, carboxylic acids/esters, and ether groups. Isobutyl methyl ketone, cyclohexanone, cyclophentanone can be hydrogenated to corresponding alcohols in the presence of Pt/C, Pd/C, Rh/C, and Ru/C catalysts using acetic acid, methanol, ethyl acetate, water, aqueous NaOH, and aqueous HCl. In addition, 2-propanone, 2-butanone, 2-pentanone, 2-heptanone, 4-methyl-2-pentanone, 3-pentanone, 3-heptanone, 2,4-dimethyl-3-pentanone, cyclohexanone, 4-methylcyclohexanone, 2-

14.2 Classification of Hydrogenation Reactions

Table 14.3 Semi-hydrogenation of alkynes in the presence of solid metal catalysts.

Reactions	T, P, solvent	Catalyst	Reference
(PhC≡CH → cis-PhCH=CHPh 97.4% + PhCH₂CH₃ 1.7%)	25 °C, 1 atm, cyclohexane	PdPb/CaCO₃	[4]
(diene-diol → allylic alcohol)	25 °C, 1 atm, EtOAc	PdPb/CaCO₃	[14]
(vinylcyclohexene alkyne → alkene 86% + ethylbenzene 8%)	25°, 1 atm, petroleum ether	PdPb/CaCO₃	[15]
(diester alkyne → cis-diester alkene)	25 °C, 1 atm, methanol	Pd/BaSO₄	[16]
(cyclohexyl alkyne → cyclohexyl alkene)	25 °C, 1 atm, methanol	Pd/SrCO₃	[17]
(tertiary alcohol alkynes → tertiary alcohol alkenes)	25–75 °C, 1 atm or ultrasound assisted	Pd/polystyrene, PdPb/CaCO₃	[18–20]
(diol dialkyne → diol dialkene)	25 °C, 5 atm, methanol	Pd/ZnO,	[21]

(continued)

Table 14.3 (Continued)

Reactions	T, P, solvent	Catalyst	Reference
(isoprene → 2-methyl-2-butene)	25 °C, 1 atm, toluene	Pd/TiSi	[22]
(2-methyl-3-buten-2-ol derivative)			
(hexynol → cis-hexenol)	25–40 °C, 2 atm, methanol/ethanol	Pd/monolith, Pd/TiO$_2$	[23,24]
(butynediol → cis-butenediol)			
(Ph-propargyl alcohol → Ph-allyl alcohol)			
(geraniol-type alkyne → alkene)	23–27 °C, 1.6–2 MPa, no solvent	Pd/Al$_2$O$_3$	[25,26]
(2-methyl-3-butyn-2-ol → 2-methyl-3-buten-2-ol)	25 °C, 1 atm, ethanol	PdCu/SiO$_2$	[27,28]
(ethynylcyclohexene → vinylcyclohexene)	30 °C, 1 atm, cyclohexane	Raney Ni-B-Cu^{2+}	[29]
(long-chain alkyne → cis-alkene)	25 °C, 1 atm, THF	Ni/graphite	[30]
(butynediol → cis-butenediol)	50 °C, 10 MPa	Fe	[31]

TiSi: Titanium silicate.

methylcyclohexanone as well as other for aliphatic and alicyclic ketones show good reactivity in the presence of Raney nickel catalysts.

Pd/C and Pt/C catalysts can also effectively hydrogenate aromatic ketones to corresponding alcohols. In particular, acetophenone (in ethanol), propiophenone (in methanol, N,N-diethylnicotinamide), and benzhydrols with substituents (in methanol, nicotinamide solvents) can be hydrogenated to alcohols at room temperature and low hydrogen pressure. Modified nickel catalysts with copper and iron as promoters also show good performances for hydrogenation of benzophenone to benzhydrol but at 135 °C and 5.9 MPa pressure.

Rhodium catalysts show better selectivity for hydrogenation of cyclohexanone derivatives to axial alcohols compared with supported nickel and cobalt catalysts at room temperature in THF, propanol, and methanol as solvents. For equatorial alcohols from cyclohexanone derivatives, Pd/C, Raney Ni, Raney Co catalysts show good performance. Examples of hydrogenation reactions and catalysts are summarized for aldehydes (Table 14.4) and unsaturated aldehydes (Table 14.5). Hydrogenation of unsaturated aldehydes/esters without affecting the olefin group is highly challenging. For example, hydrogenation of cinnamaldehyde to cinnamyl alcohol [32] can be achieved with nanosize Pt catalysts or with appropriate promoters/modifiers.

Table 14.4 Hydrogenation of aldehydes.

Reactions	T, P, solvent	Catalyst	Reference
	140–150 °C, 10–18 MPa, water	Ru/C, Ni/C	[35]
	25 °C, 0.2–0.3 MPa, ethanol	Pd/C	[3]
	80–100 °C, 10 MPa	Raney Ni	[36]
	100 °C, 8 MPa, water	Raney Ni	[37]
	25 °C, 1 atm, ethanol	Raney Ni	[38]

Table 14.5 Hydrogenation of unsaturated aldehydes.

Reactions	T, P, solvent	Catalyst	Reference
PhCH=CH-CHO → PhCH=CH-CH$_2$OH	100–110 °C, 5.2–17.5 MPa, propanol	Os/C, Re$_2$O$_7$	[33]
Cyclohexane-CHO → Cyclohexane-CH$_2$OH	125 °C, 10 MPa, THF	CuCr	[34]

14.2.4
Hydrogenation of Carboxylic Acids, Esters, and Ethers

Generally, carboxylic acids groups in compounds with 4–12 carbons can be hydrogenated (Table 14.6) to alcohols in the presence of copper (promoted by chromite, barium) and nickel catalysts under relatively harsh conditions (300–350 °C, ~25 MPa). Platinum group metals such as ruthenium and rhenium display excellent activity under relatively milder conditions (150–170 °C, ~14 MPa) for hydrogenation of aliphatic carboxylic acids. For some carboxylic acids, such as succinic acid and acetic acid, severe reaction conditions are required (72–95 MPa).

Selective hydrogenation of carboxylic acids to aldehydes is difficult and limited research work exists for this type of reaction. Hydrogenation of sugar-derived carboxylic acids only leads to poor yield (14–45%) of corresponding aldehydes and alcohols. For example, hydrogenation of gluconic acid on platinum catalysts at room temperature and atmospheric pressure gives 14–28% yield of glucose.

14.2.5
Hydrogenation of Nitroaromaticso

The nitro group in aromatic compounds is most easily hydrogenated to amines compared to other functional groups and hence is widely used for the production of aromatic amines, with applications in dyes, pharmaceuticals, and antioxidants rubber and petroleum industries. The nitro group is so reactive that it is hydrogenated with high selectivity even in the presence of other reducible groups such as aldehydes, olefins, and acetylenes. Hydrogenation of nitroaromatics is a highly exothermic reaction but requires low pressure and temperature and hence used in not only commercial manufacture of amines but also as a synthetic tool. Some well-known examples are: hydrogenation of nitrobenzene to aniline, an important commodity and intermediate for large volume polyurethanes (methylene diphenyl diisocyanate, MDI), dinitrotoluene to toluene diamine, an intermediate for another large volume polyurethane (toluene diisocyanate, TDI), *p*-nitrocumene to *p*-cumidine, a herbicide, and nitrobenzene to *p*-aminophenol, a pharmaceutical intermediate. Generally, hydrogenation of nitro group proceeds through two steps via a nitroso (R-NO) and hydroxyl amine

Table 14.6 Summary of metal catalysts for hydrogenation of carboxylic acids, esters, and ethers.

Reactions	T, P, solvent	Catalyst	Reference
CH₃CHOHCH₂CO₂Et → CH₃CHOHCH₂CH₂OH CH₃CH₂CH(CO₂Et)₂ → CH₃CH₂CH(CH₃)CH₂OH EtOCH₂CO₂Et → EtOCH₂CH₂OH	160–250 °C, 17.5–44 MPa, methanol	CuBaCr, CuCr	[39]
PhCO₂Et → PhCH₂OH + PhCH₃	108–250 °C, 20–30 MPa	CuMgCr, CuBaCr	[40,41]
PhCH₂CO₂Et → PhCH₂CH₂OH + PhCH₂CH₃			
Naphthyl-CO₂Me → Naphthyl-CH₂OH + Naphthyl-CH₃			
Naphthyl-CO₂Me → Naphthyl-CH₂OH + Naphthyl-CH₃			
Ph–C≡C–CO₂Me → PhCH=CHCO₂Me	25 °C, 2 atm, methanol	Pd/monolith	[23]
γ-valerolactone → 1,4-pentanediol	250 °C, 20–30 MPa	CuCr, CuBaCr	[42]
γ-butyrolactone (substituted) → diol			
coumarin → 3-(2-hydroxyphenyl)propanol			

(continued)

Table 14.6 (Continued)

Reactions	T, P, solvent	Catalyst	Reference
octahydrochromene → decalin-type ether	250 °C, 10–20 MPa	Raney Ni	[43]
succinic anhydride → γ-butyrolactone	30–95 °C, 7.3–10.8 MPa, EtOAc	Pd/C, Pd/Al$_2$O$_3$	[44]
Ph-CH=CH-CH$_2$-O-Si(tBu)Me$_2$ → Ph-CH=CH-CH$_2$-OH; Ph-CH$_2$-CH$_2$-CH$_2$-O-Si(tBu)Me$_2$	20 °C, methanol, toluene	Pd/C	[45,46]
lactic acid → 1,2-propanediol	120 °C, 3.5–8 MPa	Ru/C, RuRe/C, RuW/C, RuMo/C	[47,48]
ethyl lactate → 1,2-propanediol + ethanol			
levulinic acid → 1,4-pentanediol	200 °C, 3 atm	Cu/SiO$_2$	[49]
succinic acid → 1,4-butanediol + γ-butyrolactone	150–200 °C, 0.35–8 MPa	Ru/C, RuRe/C, Ru/TiO$_2$	[50,51]
itaconic acid → 3-methyl-γ-butyrolactone + 2-methyl-γ-butyrolactone			

Table 14.7 Hydrogenation of nitroaromatics.

Substrate	Reaction condition	Catalyst	Reference
Halogenated nitrobenzene	60 °C, 0.2–1 MPa	Pd/C, Pd/Fe$_2$O$_3$, Ni/TiO$_2$	[52–54]
Nitrobenzene	Room temperature, 1 atm	Pd/Fe$_3$O$_4$, FeMn, AuCu	[55,56]
2,4-dinitrotoluene	100–130 °C	Ni-based catalysts	[57]

(R-NHOH) intermediate but the subsequent steps of hydrogenation of nitroso and hydroxyl amine groups are too fast and hence it is convenient to conduct hydrogenation of nitro group to amines in a single step. Another important application of this class of reaction is to produce halogenated amines such as hydrogenation of *o-*, *m-*, or *p*-nitrochloro benzene to corresponding chloroanilines, key intermediates in dyes, drugs, and pesticides. In this case, catalysts are suitably modified (poisoned) to protect the halogenated group and typically with sulfided Pt catalysts, high selectivity to chloroanilines can be achieved. Examples of efficient catalysts for hydrogenation of nitrocompounds are: Raney Ni, supported Ni, Pd/C, Pt/C, Ru/support, and Cu-chromite. Some important examples with recent development of catalysts are summarized in Table 14.7.

14.2.6
Reductive Amination and Alkylation Reactions

Reductive alkylation and amination are important reactions for the synthesis of value added higher alkylated amine derivatives (secondary and tertiary amines) with applications in dyestuff, rubber, and petroleum industries (as antioxidants). The unique feature of this class of reaction is it is a tandem synthesis of two steps: (a) condensation of an amine with aldehyde or a ketone to imine intermediate, a thermodynamically limiting noncatalytic reaction and (b) catalytic hydrogenation of imine intermediate to alkylated amine. This combination allows one step synthesis with high conversion and selectivity of the secondary and tertiary amine products in which the product of an equilibrium reaction is scavenged by an irreversible hydrogenation step. Typical conditions of hydrogenation are 100–150 °C and 50–100 atm pressure. The examples with type of catalysts used are presented in Tables 14.8 and 14.9.

14.2.7
Hydrogenolysis of C—O Bond

Hydrodeoxygenation of biomass derived molecules such as xylitol, sorbitol to fuel range compounds and value-added chemicals has gained considerable interest in recent years as an approach to synthesize renewable chemicals. Generally, these polyols tend to undergo dehydration and hydrogenation reactions to form

Table 14.8 Reductive alkylation of aromatic amines.

Substrate	Reaction condition	Catalyst	Reference
Aniline and methanol	350–450 °C	LiY, NaY	[61]
2-methyl-6-ethylaniline and methoxy-2-propanol	170 °C, 1 atm	Ca^{2+} treated $PtSn/SiO_2$	[62]
2-butoxy-5-tertoctylnitrobenzene and butanal	80 °C, ethanol	Pd/C	[63]
3-aminobenzoic acid and aqueous HCHO	50 °C, 1.5 MPa, methanol	Pd/C	[7]
Aromatic amines and aldehydes	30–70 °C, 1.5 MPa, methanol	Pt/C	[64]
Aromatic amines and aldehydes	30–70 °C, 1.5 MPa, methanol	Pt/C	[65]
$4\text{-}H_2NC_6H_4NHPh$ and Me_2CHCH_2COMe	160 °C	Ni/kieselguhr	[59]
$3,4\text{-}Me_2C_6H_3NH_2$ and Et_2CO	60 °C, 3 atm	Pt/C	[66]
$NH(C_6H_4NO_2)_2$ and methyl isobutyl ketone	80–110 °C, 1 MPa	Pt/C	[67]
Aniline and acetophenone	110 °C, 3–6 MPa, aniline	PtS/C	[68]
Aniline and acetone	100–140 °C, 3–4 MPa	Pt, Pd, Rh, Ru/C	[68]

alkanes in the presence of Pt and Pd catalysts on acidic supports, while C—C cleavage and hydrogenolysis of C—O bond easily occur in the presence of Ru, Ni, and Cu catalysts when basic promoters are used, to give glycols, linear alcohols and methane as the main products (Table 14.10).

Hydrogenolysis of glycerol in the presence of metal catalysts has been discussed in several reviews [69,78–80]. Hydrogenolysis of glycerol gives lactic acid, 1,3-propanediol (1,3-PDO), EG, and methanol as products with methane and CO_2 depending on the type of catalysts used. Several reports described possible reaction pathways for hydrogenolysis of glycerol, the overall scheme for which is shown in Figure 14.1 [81–85]. It is generally believed that glycerol conversion is initiated with dehydrogenation or dehydration steps on catalyst surface. The two possible pathways give glyceraldehyde and acetol as the key intermediates, respectively, which are instantaneously converted to alcoholic products in the presence of hydrogen.

Catalytic conversion of xylitol, sorbitol, and mannitol is of increasing interest because these can be converted to various products such as furfurals, furans, alcohols, and hydrocarbons as fuel additives [86,87] as well as other high value chemicals such as 1,2-PDO, EG, and glycerol [88–90]. For fuel production, Pt-based catalysts on acidic silica–alumina (SiAl) and zirconium phosphate (ZrP) supports were evaluated for catalytic upgrading to C_5 and C_6 polyols [87,91]. In

Table 14.9 Reductive alkylation of aromatic diamines.

Substrate	Reaction condition	Catalyst	Reference
p-Phenylenediamine and 2-octanone	165 °C, 5.2–6.9 MPa	Ni/kiselguhr	[58]
p-Phenylenediamine and 2-octanone	165 °C, 5.2–6.9 MPa	Ni/kiselguhr	[59]
p-Nitroaniline methyl ethyl ketone	160 °C, 6 MPa	$CrO_3+CuO+BaO$	[58]
p-Nitroaniline and acetone	100 °C, 6.2 MPa	Pt/Al_2O_3	[58]
N-Phenyl- p-phenylenediamine, methyl iso-butyl ketone	180 °C, 4.1 MPa	PtS/C	[58]
p-Nitroaniline, ethyl amyl ketone	135 °C, 8.3 MPa	PtS/Al_2O_3	[58]
p-Nitroaniline, methyl ethyl ketone	148 °C	Pt/Al_2O_3 containing fluorine	[58]
p-Nitroaniline, methyl ethyl ketone	160 °C, 10 MPa	Pt/Al_2O_3	[60]
4-Nitrodiphenylamine, methyl isoamyl ketone, methyl isobutyl ketone	150 °C, 2.8–3.5 MPa	Pt/C	[60]

comparison, PtRe/C catalysts displayed much lower yield of gasoline products ($Y = 40\%$) but improved RON = 89 under similar reaction conditions [91]. A recent report on Ni/ZSM-5+MCM-41 catalyst showed excellent C_6 alkane selectivity ($S = 90\%$) [92].

Hydrogenolysis of sorbitol and xylitol can also give glycols and alcohols. There is a lack of systematic study in this area, but several catalysts have been found to be effective, including Ru/SiO_2 [88], Ru/C [93], Ru/carbon nanofiber (CNF) and graphite-based composite materials [94], and RuRe/C catalysts [95]. Nonnoble metal catalysts such as kieselguhr supported Ni [96], $NiCe/Al_2O_3$ [97], Ni/NaY [98], supported Cu catalysts on solid base $CaO–Al_2O_3$ [75] are also known to give good performances in hydrogenolysis of these sugar derived polyols. Base promoters including NaOH, $Ca(OH)_2$, MgO are necessary for selective C—C and C—O cleavage to produce glycols and alcohols. The reaction conditions for hydrogenolysis of sugar polyols are in the range of 180–245 °C and 2–8 MPa pressure.

14.2.8
Hydrodesulphurization of Petroleum Crude

Hydrodesulphurization (HDS) of petroleum crude involves hydrogenolysis of C—S bonds to give hydrocarbon fuel and H_2S. HDS is widely used for the

Table 14.10 Hydrogenolysis of biomass derived feedstocks to fuels and chemicals.

Reactions	T, P, solvent	Catalyst	Reference
Glycerol → 1,2-propanediol + 1,3-propanediol	>200 °C, 2–8 MPa, water	Supported Ru, Pd, Pt, Cu, Ni, Co catalysts	[69]
Glycerol → 1,2-propanediol + 1,3-propanediol	220 °C, under N_2, water	Supported Pt, Pd, RuPt, CuPd,	[70,71]
Glycerol → 1,3-propanediol	100–130 °C, 2–8 MPa, water, ionic liquids	Supported $PtWO_3$, IrRe, RhRe	[72–74]
Sorbitol → 1,2-propanediol + ethylene glycol	>200 °C, 6–12 MPa, water	Raney Ni, supported Ru, Ni, Cu	[75]
Sorbitol → 1,2-propanediol + ethylene glycol			
Sorbitol → hexane + hexanediols	240 °C, >8 MPa, water	Pt/P-ZrO_2, Pt/SiAl, Pd/SiAl	[76,77]

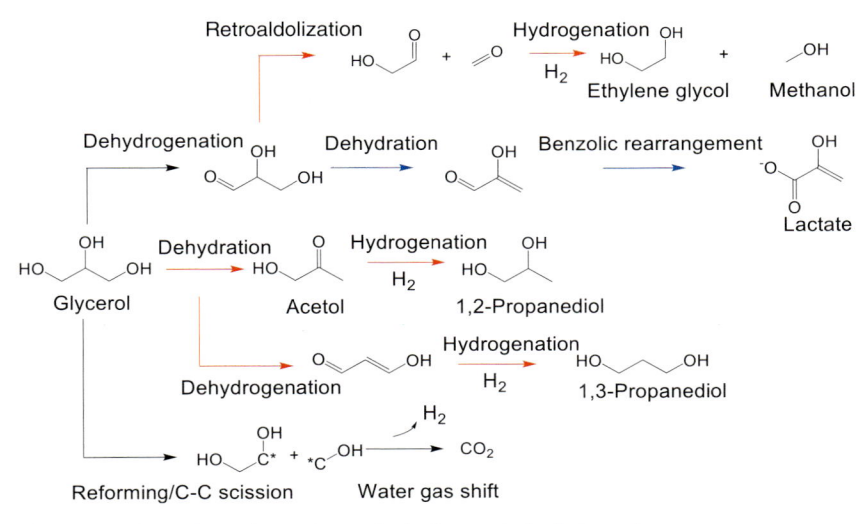

14.2 Classification of Hydrogenation Reactions

Figure 14.1 General reaction pathways for hydrogenolysis of glycerol.

removal of organic sulfur compounds from hydrocarbon mixtures. Raney Ni was proposed for desulfurization of aliphatic thioalcohols and disulfides in neutral and alkane conditions. This is particularly important for organic synthesis and purification of solvents. HDS technology has large scale industrial applications and is an indispensable process in any refinery. The most commonly used catalysts for HDS consist of supported nickel, cobalt, tungsten, and molybdenum as well as their binary, ternary, and quaternary mixtures [99]. The as-prepared materials need to be sulfided before use at 300–400 °C and 1–7 MPa H_2 conditions. While the catalytic chemistry and kinetics of HDS of petroleum have been reviewed previously, several advanced techniques in preparing novel-structured metal catalysts have shown promising features in enhancing the efficiency of HDS catalysts under relatively milder conditions (<370 °C, 3–5 MPa) [100–102].

The basic reactions involved in HDS processes are presented in Figure 14.2. Generally, sulfur compounds are converted to hydrocarbons and hydrogen sulfide (removed by washing) in order to prevent poisoning of downstream reforming process catalysts such as platinum. In addition, burning of sulfur compounds in fuel products such as gasoline and diesel can lead to the formation of sulfur oxide, which is a major pollutant for acid rain. HDS is thus very important for ultraclean fuel production. Methane thiol, dimethyl sulfide, thiophene, and dibenzothiophene are the main organosulfur compounds in petroleum crude. Under the reaction conditions (300–500 °C and 3–5 MPa), there is no thermodynamic limitation for hydrogenation of the organosulfur compounds. The reaction heat generated during HDS reactions is in the range of 11–70 kcal/mol [101].

Mo, W, MoCo, MoNi, and NiW catalysts supported on SiO_2–Al_2O_3, TiO_2–Al_2O_3, and ZrO_2–Al_2O_3 have been extensively studied by different research

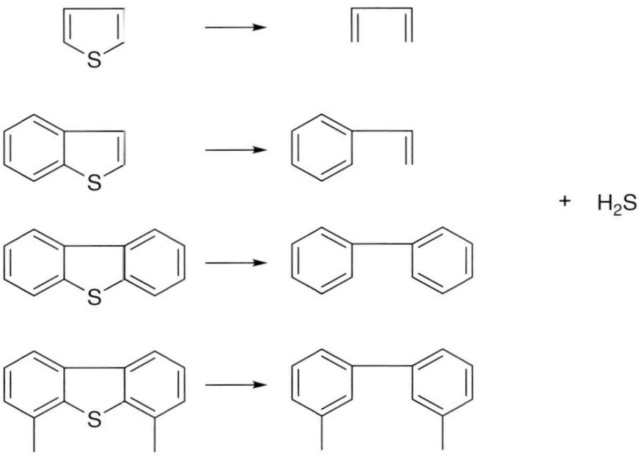

Figure 14.2 Reaction schemes for HDS of various organosulfur compounds.

groups [102]. The metal support interaction and promotional effects by addition of a second metal have been the main focus of recent reports. Obviously, due to the complexity of metal–metal as well as metal-support interactions it is impossible to exclude one factor while studying another. General findings in HDS performance studies are summarized here.

Metal Dispersion

It was found that active "OH" groups on support surface strongly influence Mo dispersion affecting HDS activity. With increasing SiO_2 content in the catalyst supports, HDS can be enhanced. In contrast, unfavorable side reactions such as hydrocracking can be prevented due to low protonic acidity of SiO_2–Al_2O_3 support [102]. Other promoters such as B_2O_3 are found to favorably promote the formation of octahedral Co while decreasing the amounts of tetrahedral Co in catalysts. This observation was later confirmed by another study in HDS of dibenzothiophene [103].

Metal Support Interaction

The nature of solid supports including acidity/basicity, and porosity [104–106] is critical for the performance for HDS processes. Klicpera and Zdrazil [107–109] proposed a basic MgO supported Mo catalyst prepared by a nonaqueous impregnation method, where damage of support due to interaction with water can be minimized. The HDS rate of thiophene on CoMo, NiMo catalysts prepared by such a method can be enhanced by 1.5–2.3 times compared to those supported on Al_2O_3. CoMo and Ni catalysts supported on zeolites (e.g., HY) were also studied. While their counterparts on acidic Al_2O_3 supports easily undergo deactivation due to coke formation [110–117], zeolite surface with strong Lewis acid centers favors the dispersion of MoS_2 such that catalytic activity can be enhanced due to the electron deficient character of sulfide particles [118].

Mesoporous supports such as MCM-41 and SBA-15 have also been investigated [119,120]. The unique mesoporosity for molecular size selectivity and more surface Lewis acid sites for sulfide formation are attributed to enhanced HDS performances.

Surface Characterization

A variety of characterization techniques have been used to reveal the possible reaction mechanism on metal catalyst surface and structural evolution during HDS reactions [121–128]. Xe-NMR studies on Mo-based catalysts reveal that, when the Mo loading increased, a single peak shifted further to low magnetic field and became gradually broader. This result is caused by a growth of inhibition and irregularity in the motion of Xe. This result suggests that Xe electronically interacted strongly with a coordinatively unsaturated site at the stacked MoS_2 crystallite [124–128]. ^{95}Mo NMR studies of Mo/Al_2O_3 catalysts show two important results. Spikelet echo spectra identified static adsorbed species (speculated to be predominantly adsorbed tetrahedral/octahedral Mo, and perhaps $Al_2(MoO_4)_3$), and a dynamic species (speculated to be disordered, surface interactive tetrahedral, and octahedral Mo), in the uncalcined catalyst samples [121]. ^{35}S studies revealed possible surface reaction mechanism on MoNi catalysts. The addition of Ni increases the total number of active sites of Mo/Al_2O_3 catalysts and mobility of sulfur compound species thus HDS activity is enhanced on bimetallic MoNi catalysts.

14.3
Types of Hydrogenation Catalysts

14.3.1
Nickel, Copper, Iron, and Cobalt

Nickel, cobalt, copper, and iron metals are commonly used nonnoble metal catalysts for hydrogenation reactions [129,130]. Properties and applications of these metals are summarized in Table 14.11 [131]. Nickel metal catalysts are widely

Table 14.11 Properties and hydrogenation applications of common nonnoble metal catalysts.

Metal	Ni	Co	Fe	Cu
Atomic radius (nm)	0.149	0.152	0.156	0.145
Crystallinity	fcc	sh	bcc	Fcc
Lattice constant (nm)	0.352	0.251, 0.407	0.287	0.361
Lattice angle	$\pi/2$	$\pi/2, \pi/3$	$\pi/2$	$\pi/2$

fcc: face centered cubic, sh: simple hexagonal, bcc: body centered cubic.

used as effective hydrogenation catalysts. For example, Raney Ni and its varied forms have been widely used for hydrogenation of nitriles, aldehydes, and ketones. Supported Ni catalysts on Al_2O_3, SiO_2, TiO_2, ZrO_2, and so on combined with other metals such as Mo, W, and Co are known to be good catalysts for hydrodesulphurization (HDS) of petroleum for the production of clean transportation fuels, as well as deoxygenation of biomass derived feedstocks. Although not as widely used as nickel catalysts, cobalt catalysts display promising performance in hydrogenation of aromatics, amines, and most importantly Fischer–Tropsch (F–T) synthesis [132–135]. Most commonly used cobalt catalysts include Raney Co and supported cobalt metal catalysts. In particular, cobalt metal is extensively used with molybdenum as effective HDS catalysts in petrochemical industry. Similar cobalt metal, iron is also an effective metal for F–T synthesis. Supported iron on Al_2O_3, SiO_2 are known to be effective catalysts for ammonia synthesis. Copper catalysts have been used for hydrogenation of aldehydes, ketones, esters, unsaturated alcohols, and deoxygenation of polyols to glycols and alcohols.

14.3.2
Platinum Group Catalysts

Platinum group metals, ruthenium, rhodium, palladium, osmium iridium, and platinum have been widely used as hydrogenation catalysts in industry. Platinum catalysts are used in petrochemical industries since early stages. It has been recognized that Ru, Rh, Pd groups often behave differently from Os, Ir, and Pt group metals. For example, it has been found that the former group displays strong isomerization effect during hydrogenation of aromatic compounds. The properties and corresponding hydrogenation applications of platinum group metals have been summarized in Table 14.12. Platinum catalysts display excellent activity even under mild conditions, for benzene hydrogenation, naphtha reforming, as well as acetylenic, olefinic, and nitro compounds. Supported palladium catalysts are used for a variety of industrial reactions, examples of which include selective hydrogenation of alkynes (Pd/$BaSO_4$), nitroaromatics (Pd/C), and cyclohexene (Pd/$CaCO_3$). Ruthenium metal is found to be much more active compared with nickel and cobalt catalysts. Thus, hydrogenation of amines, aromatics, and nitrile compounds can be carried out under much milder

Table 14.12 Properties and hydrogenation applications of platinum group metal catalysts.

Metal	Ru	Rh	Pd	Os	Ir	Pt
Atomic radius (nm)	0.178	0.173	0.169	0.185	0.180	0.177
Crystallinity	sh	fcc	fcc	sh	fcc	fcc
Lattice constant (nm)	0.271, 0.428	0.380	0.389	0.273, 0.432	0.384	0.392
Lattice angle	$\pi/2, \pi/3$	$\pi/2$	$\pi/2$	$\pi/2, \pi/3$	$\pi/2, \pi/3$	$\pi/2$

fcc: face centered cubic, *sh*: simple hexagonal.

conditions in the presence of ruthenium catalysts. Ruthenium metal is found to be very efficient in hydrogenolysis of C—O bond. This feature has been exploited extensively in recent years, particularly in biomass conversion. Similarly, rhodium and iridium metals combined with catalytic promoters also show good performances in selective C—O hydrogenolysis.

14.3.3
Preparation Methods-Conventional

Several books and journal publications have reviewed the catalyst preparation methods [136–140]. Conventionally, metal precursors are mixed with solid supports (or precursors) followed by sequential treatment such as calcination (decomposition of metal precursors to metal oxide forms and crystallization of support) and activation (reduced to metallic state). The general description of each method is summarized below.

14.3.3.1 Impregnation Method

Impregnation is one of the most commonly used methods in preparing supported metal catalysts [141]. In the incipient wetness impregnation method (IWI), solid support (e.g., activation carbon, zeolite) is blended with known (small amounts) amounts of metal precursor solution (e.g., $Ni(NO_3)_2$ in aqueous solution). The volume of liquid solution is predetermined so it will just fill in the whole pore volume in catalyst supports. Due to capillary effect, metal precursor solution will be captured in the pores (and some on the surface) of the solid support. The sample can be dried in oven before calcination. During calcination, metal precursor salts decompose to metal oxides (e.g., M_xO_y), after which solid catalysts can be further activated to the metallic form under a reducing environment such as under hydrogen. In an alternative method, large amounts of metal precursor solution and solid support can also be blended, after which solvent (e.g., H_2O, alcohols) can be removed to deposit the metal on the solid support. Following procedures such as drying, calcination, and activation, supported metal catalysts can be achieved. In the impregnation method, the types of metal precursors, solvent, preparation temperature, calcination, and activation conditions such as temperature, pressure influence the final morphology of the catalysts, such as metal particle size, shape, dispersion, as well as metal-support interaction.

14.3.3.2 Precipitation Method

In general, the precipitation method involves the use of metal precursors salts and a precipitation agent, which can be excluded in a calcination step. Precipitation can be induced by cooling and evaporation of solvent to reach supersaturation of precursor solution and also by addition of a precipitation agent. Metal precursors include nitrates, carbonates, or ammonium forms of metal salts. Hydroxides, oxohydrates, oxides, and carbonates are common products from the precipitation method. Several factors, such as solution composition, pH value, precipitation agent, aging, temperature, and solvent influencing the final properties of final catalysts, have been reviewed in details [138]. After calcination, catalytic materials can be further activated under a reducing environment to obtain metallic catalysts.

Sol–Gel Method [142]
Conventional impregnation and precipitation methods often have several issues such as catalyst sintering. Alternative methods such as the sol–gel method were thus developed to potentially solve such problems [143]. This technique involves mixing of metal precursors with metalorganic precursors, where homogeneous solution can be formed. A metalorganic precursor is hydrolyzed by simply adding water with precise control of pH and temperature. As hydrolysis proceeds, framework and colloidal particles are formed, which continue to grow until metal oxide gel formation.

14.3.4
Monometallic Nanocatalysts

Platinum is one of the most extensively investigated catalytic materials [144–146]. The most commonly used methodology in platinum nanoparticles synthesis is called *in situ* reduction, where platinum precursors are directly reduced to metallic state by reducing agents in the presence of polymers. Generally, the strategy for preparing well-defined platinum nanoparticles is to control the preparation temperature, the types of polymer (stabilizing formed particles from agglomeration), and reducing agent (e.g., $NaBH_4$). Table 14.13 shows several case studies for platinum nanoparticle synthesis, the morphologies of as prepared materials and potential applications in hydrogenation reactions. In addition, experimental work showed that modified precipitation methods, such as coprecipitation of metal precursors with support precursors (e.g., magnesium nitrate) in one pot also lead to uniformly distributed nanoparticles on solid supports, but other morphologies such as surface facets, shapes cannot be controlled by such methods. Due to the high cost and extreme scarcity of platinum element, researchers are exploring other cost-effective metals as catalysts for hydrogenation. Examples include palladium, ruthenium, gold, nickel, and cobalt nanocatalysts. While size control is not the major challenge in nanocatalysts preparation, facet formation and shape control are still the main tasks to fabricate well-defined nanocatalysts. Representative examples of palladium, ruthenium, gold, nickel, and cobalt nanocatalysts preparation and their applications are summarized in Tables 14.14 and 14.15.

14.3.5
Bimetallic Nanocatalysts

Bimetallic and multimetallic nanoparticle catalyst design is one of the hottest areas in recent years. The main reason that most bimetallic and multimetallic

14.3 Types of Hydrogenation Catalysts | 361

Table 14.13 Synthesis of monometallic Pt nanocatalysts and their applications in hydrogenation reactions.

Precursor	Stablizer	Reductant	Solvent	Morphological feature	Hydrogenation application	Reference
K_2PtCl_4	SPA	H_2	Water	4–18 nm, cubic, truncated octahedral, Al_2O_3 support	Propene	[178]
$Pt(acac)_2$	OLA	H_2	Toluene	10–20 nm, cubic, octapod	F–T	[179]
K_2PtCl_4	CTAB	Ascorbic, $NaBH_4$	Water	Cuboctahedra, cubic	Ethylene	[180,181]
H_2PtCl_6	PVP, Ag^+	EG	Water	Cubic, octahedral, SBA-15 support	Nitrobenzene	[182]
H_2PtCl_6	PVP	Ethanol	Water	2 nm, mesoporous carbon support	Olefins	[183]
K_2PtBr_4	APTS	H_2	Water	Spherical, $Fe_3O_4@SiO_2$ support	Alkenes and ketones	[184]
K_2PtCl_4	TTAB	$NaBH_4$	Water	5–10 nm, cubic, CeO_2–SiO_2 support	Hydroformylation	[185]
H_2PtCl_4	OLA	OLA	Toluene	3 nm, graphene support	Alkyene, alkene and nitroaromatics	[186]
$Pt(NH_3)_4]^{2+}$	NH_2- groups	H_2	Water	Amino functionalized NaY support	Nitrobenzene	[187]
$Pt(acac)_2$	OLA	OLA, carbonyl	Octadecene	4.3 nm, cubic	o-Halogenated aniline	[188]
H_2PtCl_6	NaOH	H_2	Water	~2 nm in TiO_2 nanotube	Enantioselective hydrogenation of 1-phenyl-1,2-propanedione	[189]
H_2PtCl_6	CNT	Sodium formate	Dry form	3.7–4.5 nm, encapsulated in nanotube	Asymmetric hydrogenation of α-ketoesters	[190]

acac: acetylacetonate; APTS: 3-(2-aminoethylamino) propyltrimethoxysilane; CNT: carbon nanotube; CTAB: hexadecyltrimethylammonium bromide; EG: ethylene glycol; NIPA: poly(N-isopropylacrylamide); OLA: oleylamine; PVP: poly(vinyl pyrrolidone); SPA: sodium polyacrylate; TTAB: tetradecyltrimethylammonium bromide.

Table 14.14 Synthesis of monometallic Pd nanocatalysts and their applications in hydrogenation reactions.

Precursor	Stabilizer	Reductant	Solvent	Morphological feature	Hydrogenation application	Reference
Pd(OAc)$_2$	NaOH, APTS	NaBH$_4$	Water	Embedded in SiO$_2$ support	Phenol to cyclohexanone	[160]
Pd$_2$(dba)$_3$	–	H$_2$	THF	Embedded in SiO$_2$ support	Alkenes, aromatics, nitriles, ketones	[161]
Pd(NO$_3$)$_2$	PVP, NaOH	Formaldehyde	Water	Embedded in SiO$_2$ support	Carboxybenzaldehyde to p-toluic acid	[162]
H$_2$PdCl$_4$	APTS, PVP	H$_2$	Water	Polymer coated on Pd particle surface	2,4-dimethyl-1,3-pentadiene	[163]
Pd(OAc)$_2$	–	Propanol	Propanol	Deposited on Fe$_3$O$_4$	Mono- and di-substituted alkenes	[164]
PdCl$_2$	CO	CO	Ethanol	Pd deposited on SiO$_2$-Al$_2$O$_3$ support surface	Phenylacetylene to styrene	[165]
Pd$_2$(dba)$_3$	TPP, TOP	H$_2$	THF	1.8–2.8 nm, immobilized on carbon support	Furfural to alcohols	[166]
Pd(NH$_3$)$_4$]$^{2+}$	NH$_2$- groups	H$_2$	Water	Amino functionalized carbon support	Nitrobenzene	[167]
Pd(acac)$_2$	PVP	NaBH$_4$	Ethanol, water	2–6 nm, spherical,	Alkynols	[168]
PdCl$_2$	APTS	NaBH$_4$	Water	>10 nm, magnetic, NiFe$_2$O$_4$ support	4-nitro aniline, 1,3-dinitrobenzene	[169]
K$_2$PdCl$_4$	thiosulfate	NaBH$_4$	Water, toluene	2–5 nm	Tandem semi-hydrogenation-isomerization of alkynes	[170]
Pd$_2$(dba)$_3$	NH$_2$- groups	H$_2$	Toluene	3.5 nm	Benzyl propargylamines	[171]
Pd(OAc)$_2$	Mg(OMe)$_2$, HF	H$_2$	Methanol	6–10 nm, MgF$_2$ support	Unsaturated olefins	[172]

H$_2$PdCl$_4$	—	H$_2$	NaOH, water	9 nm particles on layered double hydroxide support	Citral to citronellal	[173]
K$_2$PdCl$_4$	APTS	H$_2$	Water, urea	Ultrasmall particles on SiO$_2$ support	Succinic acid to γ-butyrolactone	[174]
Pd(NO$_3$)$_2$	—	DMF	DMF	3 nm, encapsulated in metal organic framework	Alkynes to alkenes	[175]
PdCl$_2$	—	NaBH$_4$	Water	4.1–5.9 nm, rGO support	Phenols to alcohols	[176]
[Pd(NH$_3$)$_4$]$^{2+}$	—	N$_2$H$_4$	Water	6 nm, spherical shape	2-methyl-3-butyn-2-ol	[177]

acac: acetylacetonate; DMF: dimethylformamide; HHDMA: adecyl-2-hydroxyethyl-dimethyl ammonium dihydrogen phosphate; PVP: poly(vinyl pyrrolidone); rGO: reduced graphene oxide; TCAB: tetraoctylammonium bromide; TOP: isooctylphosphine; TTAB: Tetradecyl trimethyl ammonium bromide.

Table 14.15 Synthesis of other monometallic nanocatalysts and their applications in hydrogenation reactions.

Precursor	Stabilizer	Reductant	Solvent	Morphological feature	Hydrogenation application	Reference
$RuCl_3$	1,5-cyclooctadiene	H_2	Dry form	3–5 nm in metal organic framework	Phenol to cyclohexanone and cyclohexanol	[147]
$Ru(acac)_3$	—	—	Physically blended with graphene	<1 nm	Ketones to alcohols	[148]
$Ru_3(CO)_{12}$	Sodium dodecylsulfate	H_2	acetonitrile	3–6 nm	Aliphatic and aromatic nitriles to amines	[149]
$RuCl_3$	EG, propanol	EG	EG	3–6 nm, supported on clay	Transfer hydrogenation of substituted nitrobenzenes to anilines	[150]
$RhCl_3$	HEA16X	$NaBH_4$	Water	3 nm, Fe_3O_4–SiO_2 support	Olefins and aromatic derivatives	[151]
$RhCl_3$	CO_2	H_2	Ethanol	>6 hollow polymer nanospheres support	Allyl alcohol	[152]
$Rh(NBD)_2BF_4$	—	$NaBH_4$	Methanol	3 nm, cation-exchange resin support	Substituted arenes	[153]
$(Ir(cod)Cl)_2$	BMI.PF6	H_2	BMI.PF6	2.7 nm, embedded in polymer membrane	C_6 alkenes	[154]
$IrCl_3$	1,10-phenanthroline	H_2	Ethanol	2–4 nm, N-doped carbon support	Catalytic transfer hydrogenation of aliphatic and cyclo ketones	[155]
$Ni(NO3)2$	NaOH, Na_2CO_3	H_2	Water	15–20 nm, TiO_2 support	o-chloronitrobenzene to o-chloroaniline	[53]
$NiCp_2$	—	H_2	Gas deposition	5 nm, embedded in Al_2O_3 tube	Cinnamaldehyde and nitrobenzene	[156]
$Ni(NO3)2$	NaOH, Na_2CO_3	H_2	Water	10 nm, ZnO-Al_2O_3 support	Citral	[157]
$Co(OAc)_2$	1,10-phenanthroline	H_2	Water	20–80 nm, carbon support	Catalytic transfer hydrogenation of nitroarenes	[158]
$CoCl_2$	—	NH_3BH_3	Water	Mesoporous carbon nitride	Catalytic transfer hydrogenation of nitroarenes	[159]

BMI.PF6: 1-n-butyl-3-methylimidazolium hexafluorophosphate; ((cod)Cl)₂: bis(1,5-cyclooctadiene)diiridium(I)dichloride; EG: ethylene glycol; HAP: Hydroxyapatite; HEA16X: N,N-dimethyl-N-cetyl-N-(2-hydroxyethyl) ammonium chloride and tetrafluoroborate salts, X = Cl or BF_4; Rh(NBD)₂BF₄: bis(norbornadiene)rhodium(I) tetrafluoroborate.

materials outperform monometallic catalysts is that the addition of a second metal often induces surface electro reconfiguration, leading to modified interaction with hydrogen and reactants during turnover cycles, thus enhancing catalytic activity. Preparation of such materials still follows the same strategy, where metal precursors mixed with polymer stabilizers are reduced by chemical agents. In order to obtain different surface morphologies, selection of reducing agents, such as H_2, $NaBH_4$, amines, diols is important. Applications of bimetallic nanocatalysts in hydrogenation reactions are summarized in Tables 14.16 and 14.17.

Table 14.16 Recent applications in hydrogenation using bimetallic catalysts.

Catalyst	Hydrogenation application
PtSn	Aromatic ketones to alcohols and alkanes [212], glycerol to 1,2-propanediol [213], ring open of cyclopentane [214], propane to alkanes [215]
PtZn	Halonitrobenzenes to haloanilines [216], dehydrolinalool to linalool [217]
PtCo	2,5-furandimethanol and 2,5-dimethylfuran [218], butane to alkanes [219]
PtRe	Levulinic acid to γ-valerolactone and 1,4-pentanediol [220], amides to amines [221], glycerol to 1,3-propanediol [222], cyclopentane to alkanes [223]
PtAu	Dehydrolinalool to linalool [217]
PtAg	Hydrogenation of 1-epoxy-3-butene [224]
PtPd	Dehydrolinalool to linalool [217], cellulose, hemicellulose, lignin to fuels [225,226], decane to alkanes [227]
PtRu	CO_2 to alcohols [228], benzene to cyclohexane [229], hydrogenation of arene [230], citral to geraniol and nerol [231], ring open of methylcyclopentane [232]
PtRh	Hydrogenation of arene [230], ring open of methylcyclopentane [232], chloronitrobenzene to chloroaniline [233], butyronitrile to butylamine [234], propane to alkanes [235], cyclohexane to hexane [236]
PtIr	Hydrogenation of arene [230], propane to alkanes [215]
PtFe, PtNi	Glycerol to 1,2-propanediol [237–239]
PtMo	2-(hydroxymethyl)tetrahydropyran to 1,6-hexanediol [240]
PtGa, PtIn	Citral to geraniol and nerol [241]
PtGe	Ring open of methylcyclopentane [242], ethylcyclopentane [243]
PdCo	F–T synthesis [244]
PdAu	Hydrogenation of alkenes and nitro compounds [245],
PdBi	Dichlorodifluoromethane to difluoromethane [246]
PdRu	Dimethyl terephthalate to dimethyl cyclohexane-1,4-dicarboxylate [247,248], naphthalene and cyclic polyenes to alkanes [249], dichlorodifluoromethane to difluoromethane [246]
PdRe	Succinic acid to γ-butyrolactone and 1,4-butanediol [220,250,251]
PdPb	Hydrogenation of 3-methylbutyne and isoprene [252]
PdCo	Ethylene glycol to alkanes and ethanol [253]

(*continued*)

Table 14.16 (*Continued*)

Catalyst	Hydrogenation application
PdCu	Imines to amines [254], glycerol to glycols [255–258], sorbitol to glycols [259]
PdFe, PdNi, PdZn	Glycerol to glycols [256], ethylene glycol to alkanes and ethanol [260], lignin to aromatics [261], propane to alkanes [262]
PdRh	Benzene to cyclohexa-1,3-diene [263]
PdMo	1,3-butadiene to butene [264]
RuRe	Succinic acid to γ-butyrolactone [220], levulinic acid to γ-valerolactone and 1,4-pentanediol [220], cyclohexylcarboxamide to cyclohexanemethylamine [265], sorbitol to glycols [95], glycerol to glycols [84,266]
RuNi	CO_2 to methane [267], benzene to cyclohexane [268], lignin to aromatics [261], butane to alkanes [269]
RuCo	Glycerol to 1,2-propanediol [270], butane to alkanes [269]
RuCu	Citral to geraniol and nerol [271], glycerol to 1,2-propanediol [272–274], paracetamol to 4-(acetamido)cyclohexanol [275], hexane to alkanes [275]
RuMo	Cyclohexylcarboxamide to cyclohexanemethylamine [276]
RuAu	Glycerol to glycols [277], succinic acid to γ-butyrolactone and 1,4-butanediol [277]
RuOs	Ring opening of tetralin [278]
IrRe	Glycerol to 1,3-propanediol [279,280]
IrRh	Hydrogenation of arene [230]
IrAu	Ring open of methylcyclopentane [281]
RhRe	Cyclohexylcarboxamide to cyclohexanemethylamine [265], glycerol to 1,3-propanediol [72], 2-(hydroxymethyl)tetrahydropyran to 1,6-hexanediol [282]
RhMo	2-(hydroxymethyl)tetrahydropyran to 1,6-hexanediol [240,283]
RhFe	Ethylene glycol to alkanes and ethanol [260]
RuNi, NiAu	Lignin to aromatics [261] and phenolics [284], propane to alkanes [285,286]
NiCu	Hydrogenation of 3-methylbutyne and isoprene [252], glycerol to 1,2-propanediol [287,288], propane to alkanes [289], cyclohexane to methane [290], levulinic acid to γ-valerolactone and 1,4-pentanediol [291], paracetamol to 4-(acetamido)cyclohexanol [275], hexane to alkanes [275]
NiFe	2,5-furandimethanol and 2,5-dimethylfuran [292]
NiW	Cellulose to polyols [293], phenol to benzene
CoFe	F–T synthesis [244,294], glycerol to glycols [295]
CuAg	Glycerol to 1,2-propanediol [296,297]

In classic catalysis theory, metal particle sizes, surface morphologies, metal compositions, and metal-support interactions are often the major parameters considered for catalytic (hydrogenation) performance. It is generally believed that, for metal catalysts, metal atoms located at defects, edges, and corners are often significantly more active catalytically than those on surfaces. Recent advances in nanoscience studies show that the fashion of metal atoms assembly and metal–metal interaction are also critical factors that determine the catalytic

Table 14.17 Preparation of novel structured bimetallic nanocatalysts and their applications in hydrogenation reactions.

Catalyst	Preparation method	Morphological feature	Hydrogenation application	Reference
Pt/CoFe oxide	Use APTS as ligand to grow Pt on CoFe oxide surface	6 nm	Cyclohexene	[191]
PtFe/Fe$_2$O$_3$	Use ODL and OLA as solvent and reductants	Nanoporous structure	Acetone to iso-propanol	[192]
PtNi/C	Use galvanic displacement to remove Ni with Pt	Wide size distribution	Tetralin to decalin	[193]
PtRu/SnO$_2$	Use PVP as stabilizer and ethanol as reductant	2.3 nm	o-chloronitrobenzene to o-chloroaniline	[194]
PtCo	Use oleic acid, 1,2-hexadecanediol to reduce metal precursors	Size control	Cinnamaldehyde to cinnamyl alcohol	[195]
PdCu/rGO	Direct reduce metal precursors on graphene oxide by NaBH$_4$	Dominant Cu (111) facet	Glycerol, xylitol, sorbitol to glycols	[196]
PdCu/ diatomite	Use NaBH$_4$ to reduce metal precursors	3–7 nm	Long-chain aliphatic esters	[197]
PdCu, PdRu	Solvothermal synthesis using SDS, PEG	Encapsulated in polymer	Styrene oxide	[198]
PdCu/ZnO	Use PVP and Ag$^+$ to fabricate PdCu core-shell	3.4 nm	Semi-hydrogenation of ehydroisophytol	[199]
PdAu/C	Use PVP as stabilizer, sodium formate as reductant in aqueous solution	4–15 nm	2-chloronitrobenze, H$_2$O$_2$ synthesis	[200]
PdRh/SiO$_2$	Solvothermal synthesis using OLA, TOP as solvent and stabilizer	Spherical, ~6 nm	Alkynes to alkenes	[201]
PdAg/SiO$_2$	PdSO$_4$ and Ag$_2$SO$_4$ impregnated followed NaBH$_4$ reduction	3 nm on dendrimer surface	Alkynes to alkenes	[202]

(continued)

Table 14.17 (Continued)

Catalyst	Preparation method	Morphological feature	Hydrogenation application	Reference
PdAg/MOF	Use Ag$^+$ to tune the growth of Pd particles on MOFs	3 nm	Alkynes to alkenes	[175]
RuNi/C	Use PVP and ethanol to fabricate Ni, then galvanic displacement to immobilize Ru	4.4 nm	Benzene to cyclohexane	[203]
RuCo/TiO$_2$	Add metal precursors to alkoxide solution	Nanocluster embedded in TiO$_x$ structures	Nitriles	[204]
PdCu/polymer	Use PEG and amines to form cross-linked polymer	—	Styrene oxide to 2-phenyl ethanol	[205]
NiFe/Fe$_3$O$_4$-SiO$_2$	Coprecipitation method to dope Fe and Ni on SiO$_2$	Mixed metal oxide particles	Furfural to furfuryl alcohol	[206]
CoNi/Al$_2$O$_3$	Sol–gel method to immobilize Co and Ni on support	Mixed metal oxide with layer doubled morphology	F–T synthesis	[207]
NiAs	Use TEG as stabilizer, hydrothermal method	Mixed metal oxide particles	p-Nitro phenol	[208]
NiB	Use to NaBH$_4$ or KBH$_4$ reduce Ni precursor	Mixed metal oxide particles	Fatty acid methyl esters, o-chloronitrobenzene	[209,210]
CoMn, CoFe	Hydrolysis method	Mixed metal oxide particles	F–T synthesis	[211]

MOF: metal organic framework; ODL: octadecylamine; OLA: oleylamine; rGO: reduced graphene oxide; TEG: tri ethylene glycol; TOP: trioctylphosphine.

activity for specific reactions. Basic theory of nanocatalysts, synthetic approaches, and relevant catalytic applications have been discussed in recent reviews [144–146,298–300]. Therefore, material scientists are developing various recipes to obtain metal nanoparticles with specific morphologies. In this context, novel structured catalysts, such as ultrasmall metal clusters, particles with uniform but specific surface morphologies (e.g., octahedral, tetrahedral, cubic, rhombic) as well as bimetallic or multimetallic nanoparticles with controlled structures (e.g., alloy, core–shell, cluster-in-cluster, layer-by-layer) are often found to show significantly enhanced performances compared to conventional catalysts.

In recent years, experimental studies have shown that classic cubic, octahedral, and rhombic morphologies often contain low-indexed surfaces such as (111), (100), and (110), which have limited numbers of defects and low coordinated metal atoms. Therefore, more recent efforts have been paid to fabricate nanocatalysts with high surface indexes, such as (411) and (221) [298,300]. High surface indexed nanocatalysts often exhibit novel shapes such as stars, concave shapes, flowers, and nanodendrites, examples of which are shown in Figure 14.3 [301]. The shapes of nanoparticle catalysts can also be tuned by the use of different types of polymer and good control of kinetic reduction rate of metal cations in the solution. "Nanocatalysts" have been extensively studied in the past decade and used for various other applications such as oxygen reduction reaction and CO oxidation, while there are only a few fundamental studies on hydrogenation using nanocatalysts. Currently, novel structured catalysts are mainly focused on platinum, ruthenium, gold, palladium, and iron metals for catalytic hydrogenation of olefins, furan, nitrobenzene, and biomass derived polyols and carboxylic acids.

14.3.6
Catalyst Support

Although unsupported metal catalysts such as Raney Ni, Raney Co, and platinum black are widely used for hydrogenation reactions, supported metal catalysts display several obvious advantages over unsupported one, including higher surface area of metal atoms exposed for catalytic reactions and tunable metal-support interaction for enhanced performance (see Section 14.3.7 for details). Examples of common solid catalyst supports include active carbon [71], graphite [306], $CaCO_3$ [307], Al_2O_3 [308], ZSM-5 [309], SiO_2 [310], and ZrO_2 [311]. To further enhance the catalytic performances, two or more metal oxides are combined, examples of which include TiO_2–SiO_2 [312], Nb_2O_5–SiO_2 [313], CeO_2–ZrO_2 [314], Al_2O_3–CeO_2 [315], Al_2O_3–ZrO_2 [316], La_2CO_3–Al_2O_3 [317], TiO_2–WO_3 [318], CaO–MgO [319], and CeO_2–TiO_2 [320], mesoporous support such as SBA-15 [321], MCM-41 [322], KIT-6 [323], and TUD-1 [324]. In recent years, increasing attention has been paid to novel structured catalyst supports, such as silicon carbide [325], carbon nanotube [326], mesoporous carbon materials [327], graphene [196], graphite fibers [328], and monoliths.

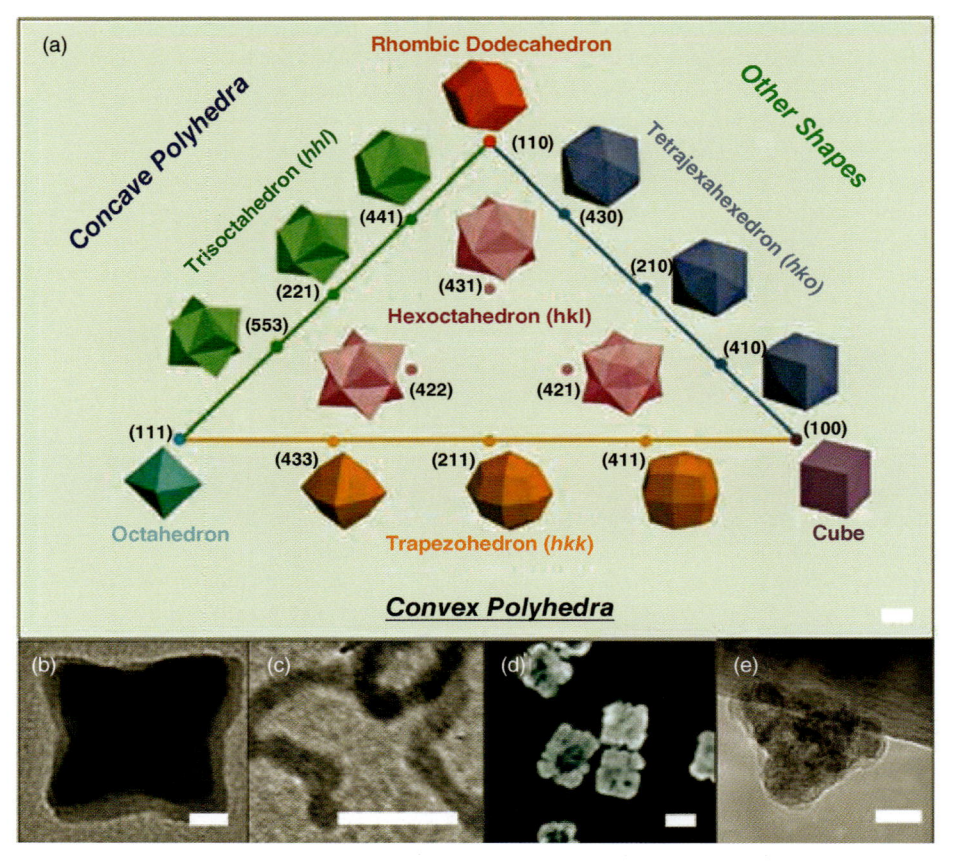

Figure 14.3 Examples of high surface-indexed nanocatalysts. (a) Schematic description of polyhedral crystals with high surface indexes. (Adapted from Ref. [301].) (b) Terahexahedron-shaped Pt nanoparticles. (Adapted from Ref. [302].). (c) PtAg nanowires. (Adapted from Ref. [303].) (d) PtPd nanocages (Adapted from Ref. [304].) (e) PtFe nanopyramid clusters (Adapted from Ref. [305].).

14.3.7
Catalyst Characterization

It is important to characterize the surface and structural properties of solid catalysts. The main purpose for characterization is to understand how surface physical and chemical properties influence the catalytic performances such as activity, selectivity, and stability. With the advancement of material science and instrument technologies, researchers are able to characterize the properties of catalyst surface on atomic levels. Examples of catalyst characterization techniques and their applications are summarized in Table 14.18 and Figure 14.4. Each of the following techniques has been revised and discussed

Table 14.18 Common catalyst characterization techniques.

Characterization technique	Information	Application examples
Volumetric adsorption	Amount of adsorbate as a function of pressure	Measure pore size, size distribution, pore volume of micro- [336], meso- [337,338], and macroporous [75] materials
IR-spectroscopy	Surface functional groups based on the vibrational excitation of covalently bonded atoms and groups use probe molecules such as CO, NO, NH_3, H_2O, and pyridine to obtain interaction information	Identify the presence of surface functional groups such as -C-O, -C=O, -O-H, -COOH, $-NH_2$ on catalyst surface [339]
Raman spectroscopy	Laser beam interaction with molecular vibrations, phonons, or other excitations in the system	Measure reaction kinetics by detecting the bond formation on catalyst surface [340,341]
UV-vis spectroscopy	The absorption or reflectance in the visible range influences the color of the chemicals involved. It deals with transitions from the ground state to the excited state	Measure kinetics of a chemical reaction where color change/shift is involved. Detect surface crystal structures, metal-ligand interaction, and band gap shift [342,343]
Temperature-programmed desorption (TPD)	Amounts of adsorbate as a function of temperature	Identify the bonding strength and amounts of H_2 [344], CO [345], CO_2 [346], and NH_3 [337] adsorbed on active sites.
Temperature-programmed reduction (TPR)	Reduction of catalyst surface by H_2 as a function of temperature	Measure metal oxide-support and metal oxide-H_2 interaction [347].
Temperature-programmed oxidation (TPD)	Oxidation of catalyst surface by O_2 as a function of temperature	Measure metal-support and metal-O_2 interaction [348].
Scanning electron microscopy (SEM)	Surface morphology in the range of 0.003–10 000 μm	Detect surface morphologies such as crystallinity, particle shape, and element composition [75,343].
Transmission electron microscopy (TEM)	Metal particle crystal lattice, surface indexes, element compositions	Measure lattice spacing between atomic layers [257], surface indexes [349], and element distribution within selected region [343]
X-ray diffraction (XRD)	Crystal species, crystalline degree, crystal size	Detect metallic crystals, mixed metal oxide structures on the surface of solid catalysts [350]
Extended X-ray absorption fine structure (EXAFS)	Local environment including coordination numbers, atomic distances	Identify the local environment for bimetallic nanoparticles [351]
X-ray absorption near edge structure (XANES)	Local coordination environment such as chemical state, unoccupied band structures	Probe the chemical state of local atomic environment.

Figure 14.4 Examples of characterization techniques used for metal catalysts (a) N_2 adsorption/desorption isotherm [75], (b) UV-vis spectroscopy [333], (c) IR spectroscopy [333], (d) temperature-programmed reduction [334], (e) transmission electron microscopy [257], (f) scanning electron microscopy and phase analysis [75], and (g) X-ray photoelectron spectroscopy [335].

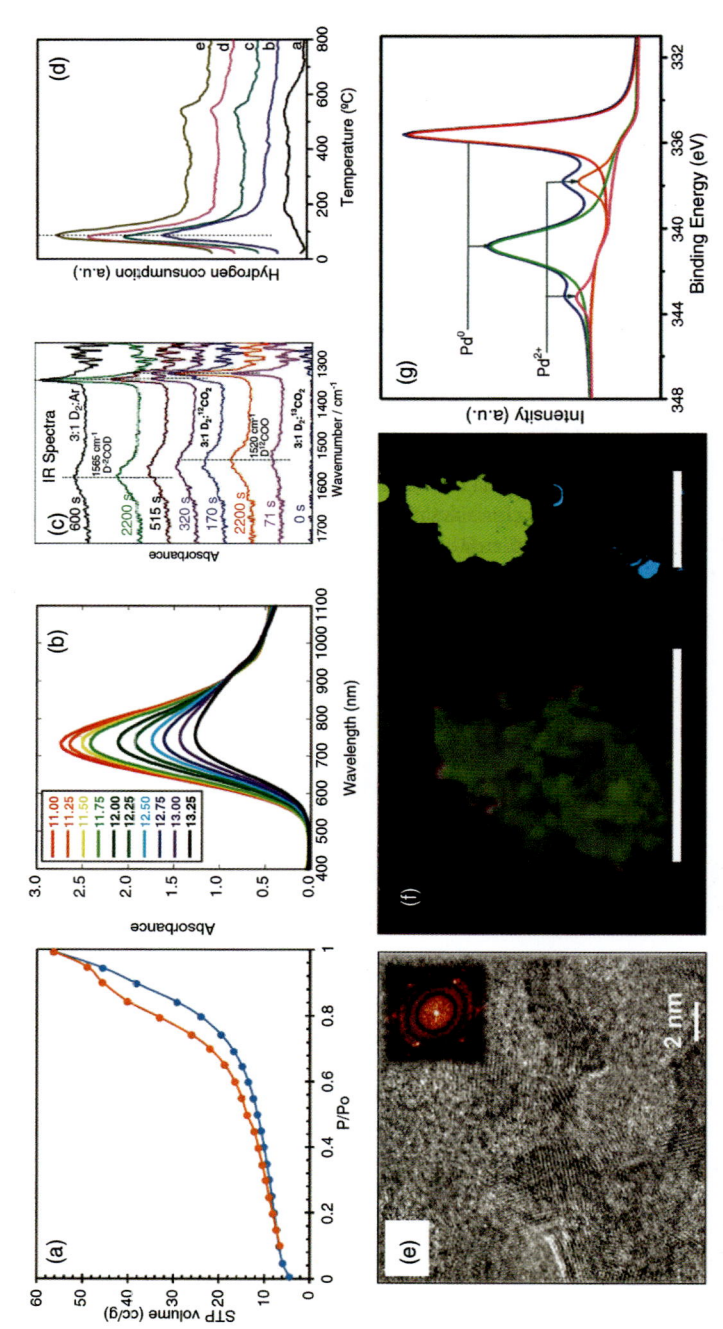

in recent publications [329–331]. In particular, volumetric adsorption technique is widely used to measure the pore size and volume of catalyst supports. This is important because these parameters determine the dynamics of molecular movement and diffusion on catalyst surface. In order to characterize the surface functional groups and possible reaction intermediates, IR and Raman spectroscopies are the most commonly used methods. Various groups on the catalyst surface as well as in active intermediates can be detected by these techniques, which provide insightful information about plausible reaction mechanism. UV-vis spectroscopy technique is often used to reflect any color change for metal-ligand interaction. Temperature programmed techniques are usually used to quantitatively measure the catalyst surface-reactant as well as metal-support interaction. Scanning electron microscopy coupled with X-ray diffraction technique is used to detect surface crystal structures, while transmission electron microscopy is often used to distinguished lattice structures and element dispersion within the range of 1–20 nm. Other X-ray techniques are often used to identify fine structures at atomic levels.

14.4
Summary and Conclusions

In this chapter, a brief review has been presented on applications of solid-state metal catalysts in hydrogenation reactions. In particular, classification of important hydrogenation reactions in petrochemicals, fine chemicals/pharmaceuticals synthesis, and biomass conversion, have been reviewed. Specific examples of each category with a focus on types of catalysts used, preparation methods, range of conditions used, and critical aspects controlling catalytic activity and selectivity are discussed. Emerging trends in the design of bimetallic catalysts for hydrogenation reactions have also been reviewed. Particularly, the nanosize mono- and bimetallic catalysts with novel surface configurations are believed to play a key role in enhancing catalytic performances in hydrogenation reactions.

References

1 Sheldon, R.A., Arends, I., and Hanefeld, U. (2007) *Green Chemistry and Catalysis*, Wiley-VCH Verlag GmbH.
2 Cerveny, L. (1986) *Catalytic Hydrogenation*, Elsevier.
3 Augustine, R.L. (1997) Selective heterogeneously catalyzed hydrogenations. *Catal. Today*, **37**, 419–440.
4 Nishimura, S. (2001) *Handbook of Heterogeneous Catalytic Hydrogenation for Organic Synthesis*, Wiley-VCH Verlag GmbH, New York, etc.
5 Rylander, P.N. (1985) *Hydrogenation methods*, Academic Press, London.
6 Xu, W., Ni, J., Zhang, Q., Feng, F., Xiang, Y., and Li, X. (2013) Tailoring supported palladium sulfide catalysts through H_2-

assisted sulfidation with H$_2$S. *J. Mater. Chem. A*, **1**, 12811–12817.

7 Roy, D., Jaganathan, R., and Chaudhari, R.V. (2005) Kinetic modeling of reductive alkylation of aniline with acetone using Pd/Al$_2$O$_3$ catalyst in a batch slurry reactor. *Ind. Eng. Chem. Res.*, **44**, 5388–5396.

8 Fahlbusch, K., Hammerschmidt, F., Panten, J., Pickenhagen, W., Schatkowski, D., Bauer, K., Garbe, D., and Surburg, H. (2005) *Ullmann's Encyclopedia of Industrial Chemistry*, Wiley-VCH Verlag GmbH, Weinheim.

9 Ramachandran, P. and Chaudhari, R. (1983) *Three-Phase Catalytic Reactors*, Gordon & Breach Science Pub.

10 Gray, J. and Russell, L. (1979) Hydrogenation catalysts–their effect on selectivity. *J. Am. Oil Chem. Soc.*, **56**, 36–44.

11 Benaissa, M.h., Carillo Le Roux, G., Joulia, X., Chaudhari, R.V., and Delmas, H. (1996) Kinetic modeling of the hydrogenation of 1, 5, 9-cyclododecatriene on Pd/Al$_2$O$_3$ catalyst including isomerization. *Ind. Eng. Chem. Res.*, **35**, 2091–2095.

12 Julcour, C., Jaganathan, R., Chaudhari, R.V., Wilhelm, A.-M., and Delmas, H. (2001) Selective hydrogenation of 1, 5, 9-cyclododecatriene in up-and down-flow fixed-bed reactors: experimental observations and modeling. *Chem. Eng. Sci.*, **56**, 557–564.

13 Chaudhari, R.V., Jaganathan, R., Mathew, S.P., Julcour, C., and Delmas, H. (2002) Hydrogenation of 1, 5, 9-cyclododecatriene in fixed-bed reactors: down-vs. upflow modes. *AIChE J.*, **48**, 110–125.

14 Mildner, P. and Weedon, B. (1953) 652. Carotenoids and related compounds. Part II. C 10 Intermediates for the synthesis of carotenoids. *J. Chem. Soc. (Resumed)*, 3294–3298. doi: 10.1039/JR9530003294.

15 Marvell, E.N. and Tashiro, J. (1965) Catalyst selectivity in semihydrogenation of some conjugated acetylenes1. *J. Org. Chem.*, **30**, 3991–3993.

16 Cram, D.J. and Allinger, N.L. (1956) Mold metabolites. VIII. Contribution to the elucidation of the structure of helvolic acid 1. *J Am Chem Soc*, **78**, 5275–5284.

17 Robins, P. and Walker, J. (1952) 117. Studies of the Diels–Alder reaction. Part I. The reaction between 1-vinylcyclohex-1-ene and benzoquinone, and the reduction of Δ2: 9 (14)-decahydro-1: 4-diketophenanthrene. *J. Chem. Soc. (Resumed)*, 642–649. doi: 10.1039/JR9520000642.

18 Sulman, E.M., Nikoshvili, L.Z., Matveeva, V.G., Tyamina, I.Y., Sidorov, A.I., Bykov, A.V., Demidenko, G.N., Stein, B.D., and Bronstein, L.M. (2012) Palladium containing catalysts based on hypercrosslinked polystyrene for selective hydrogenation of acetylene alcohols. *Top. Catal.*, **55**, 492–497.

19 Sulman, E., Sulman, M., Tyamina, I., Doluda, V., Nikoshvili, L., Lyubimova, N., Sidorov, A., and Matveeva, V. (2015) Adsorption processes for the synthesis of catalytically active metal nanoparticles in polymeric matrices. *Chem. Eng. Technol.*, **38**, 683–689.

20 Tripathi, B., Paniwnyk, L., Cherkasov, N., Ibhadon, A.O., Lana-Villarreal, T., and Gomez, R. (2015) Ultrasound-assisted selective hydrogenation of C-5 acetylene alcohols with Lindlar catalysts. *Ultrason. Sonochem.*, **26**, 445–451.

21 Protasova, L.N., Rebrov, E.V., Choy, K.L., Pung, S.Y., Engels, V., Cabaj, M., Wheatley, A.E.H., and Schouten, J.C. (2011) ZnO based nanowires grown by chemical vapour deposition for selective hydrogenation of acetylene alcohols. *Catal. Sci. Technol.*, **1**, 768–777.

22 Vile, G., Almora-Barrios, N., Mitchell, S., Lopez, N., and Perez-Ramirez, J. (2014) From the lindlar catalyst to supported ligand-modified palladium nanoparticles: selectivity patterns and accessibility constraints in the continuous-flow three-phase hydrogenation of acetylenic compounds. *Chem. Eur. J.*, **20**, 5926–5937.

23 Liguori, F. and Barbaro, P. (2014) Green semi-hydrogenation of alkynes by Pd@borate monolith catalysts under continuous flow. *J. Catal.*, **311**, 212–220.

24 Sittikun, J., Boonyongmaneerat, Y., Weerachawanasak, P., Praserthdam, P.,

and Panpranot, J. (2014) Pd/TiO$_2$ catalysts prepared by electroless deposition with and without SnCl$_2$ sensitization for the liquid-phase hydrogenation of 3-hexyn-1-ol. *React. Kinet. Mech. Cat.*, **111**, 123–135.

25 Kislyi, V.P., Tolkacheva, L.N., and Semenov, V.V. (2002) Hydrogenation on granular palladium-containing catalysts: II. Hydrogenation of nitroheterocyclic compounds. *Russ. J. Org. Chem.*, **38**, 269–271.

26 Tolkacheva, L.N., Kislyi, V.P., Taits, S.Z., and Semenov, V.V. (2002) Hydrogenation on granular palladium-containing catalysts: I. Hydrogenation of tertiary acetylene alcohols. *Russ. J. Org. Chem.*, **38**, 150–152.

27 Spee, M.P.R., Boersma, J., Meijer, M.D., Slagt, M.Q., van Koten, G., and Geus, J.W. (2001) Selective liquid-phase semihydrogenation of functionalized acetylenes and propargylic alcohols with silica-supported bimetallic palladium-copper catalysts. *J. Org. Chem.*, **66**, 1647–1656.

28 Spee, M.P.R., Grove, D.M., van Koten, G., and Geus, J.W. (1997) Simple preparation of bimetallic palladium-copper catalysts for selective liquid phase semihydrogenation of functionalized acetylenes and propargylic alcohols. *Stud. Surf. Sci. Catal.*, **108**, 313–319.

29 Nitta, Y., Sekine, F., Imanaka, T., and Teranishi, S. (1981) The effect of crystallite size of nickel on the enantioselectivity of modified nickel catalysts. *B Chem. Soc. Jpn.*, **54**, 980–984.

30 Savoia, D., Tagliavini, E., Trombini, C., and Umani-Ronchi, A. (1981) Active metals from potassium-graphite. Highly dispersed nickel on graphite as a new catalyst for the stereospecific semihydrogenation of alkynes. *J. Org. Chem.*, **46**, 5340–5343.

31 Reppe, W. (1955) Äthinylierung. *Justus Liebigs Ann. Chem.*, **596**, 1–4.

32 Durndell, L.J., Parlett, C.M., Hondow, N.S., Isaacs, M.A., Wilson, K., and Lee, A.F. (2015) Selectivity control in Pt-catalyzed cinnamaldehyde hydrogenation. *Sci. Rep. Uk*, 5. doi: 10.1038/srep09425.

33 Rylander, P.N. and Steele, D.R. (1969) Process for the selective hydrogenation of phenylphenols. Google Patents US3423472 A.

34 Falbe, J. and Korte, F. (1965) Synthesen mit kohlenmonoxyd, VI: lactame durch Ringschlußreaktion ungesättigter amine mit kohlenmonoxyd. *Chem. Ber.*, **98**, 1928–1937.

35 Kusserow, B., Schimpf, S., and Claus, P. (2003) Hydrogenation of glucose to sorbitol over nickel and ruthenium catalysts. *Adv. Synth. Catal.*, **345**, 289–299.

36 Elderfield, R.C., Pitt, B.M., and Wempen, I. (1950) Synthesis of the branched chain 5-Isopropylaminoamyl and 4-Isopropylaminobutyl ethers and of the bromides derived from them1, 2. *J. Am. Chem. Soc.*, **72**, 1334–1345.

37 Matsushima, Y. (1951) Studies on Amino-hexoses. II. The mechanism of the formation of so-called chitose from glucosamine. *B Chem. Soc. Jpn.*, **24**, 144–147.

38 Yeddanapalli, L. and Francis, D.J. (1962) Kinetics and mechanism of the alkali catalysed condensation of o-and p-methylol phenols by themselves and with phenol. *Die Makromol. Chem.*, **55**, 74–86.

39 Connor, R., Folkers, K., and Adkins, H. (1932) The preparation of copper-chromium oxide catalysts for hydrogenation. *J. Am. Chem. Soc.*, **54**, 1138–1145.

40 Mozingo, R. and Folkers, K. (1948) Hydrogenolysis of aromatic esters to alcohols. *J. Am. Chem. Soc.*, **70**, 229–231.

41 Adkins, H. and Burgoyne, E.E. (1949) Selective hydrogenation of esters containing a naphthalene nucleus. *J. Am. Chem. Soc.*, **71**, 3528–3531.

42 Cason, J., Brewer, P.B., and Pippen, E.L. (1948) Branched-chain fatty acids. VII. Simplified methods for preparing pure branched-chain alcohols and halides. *J. Org. Chem.*, **13**, 239–248.

43 Benneville, P.L.d. and Connor, R. (1940) The hydrogenation of coumarin and related compounds. *J. Am. Chem. Soc.*, **62**, 283–287.

44 Burks, J.R.E., Franko-Filipasic, B.R., and Kolyer, J.M. (1963) Preparation of

butyrolactone. Google Patents US3312718.
45 Sajiki, H., Ikawa, T., Hattori, K., and Hirota, K. (2003) A remarkable solvent effect toward the Pd/C-catalyzed cleavage of silyl ethers (pg 654, 2003). *Chem. Commun.*, 1106–1106. doi: 10.1039/b211313a.
46 Sajiki, H., Ikawa, T., Hattori, K., and Hirota, K. (2003) A remarkable solvent effect toward the Pd/C-catalyzed cleavage of silyl ethers. *Chem. Commun.*, 654–655.
47 Iqbal, S., Kondrat, S.A., Jones, D.R., Schoenmakers, D.C., Edwards, J.K., Lu, L., Yeo, B.R., Wells, P.P., Gibson, E.K., Morgan, D.J., Kiely, C.J., and Hutchings, G.J. (2015) Ruthenium nanoparticles supported on carbon: an active catalyst for the hydrogenation of lactic acid to 1,2-propanediol. *ACS Catal.*, **5**, 5047–5059.
48 Takeda, Y., Shoji, T., Watanabe, H., Tamura, M., Nakagawa, Y., Okumura, K., and Tomishige, K. (2015) Selective hydrogenation of lactic acid to 1,2-propanediol over highly active ruthenium-molybdenum oxide catalysts. *ChemSusChem*, **8**, 1170–1178.
49 Kasinathan, P., Yoon, J.W., Hwang, D.W., Lee, U.H., Hwang, J.S., Hwang, Y.K., and Chang, J.S. (2013) Vapor-phase hydrogenation of ethyl lactate over copper-silica nanocomposites. *Appl. Catal. A Gen.*, **451**, 236–242.
50 Primo, A., Concepcion, P., and Corma, A. (2011) Synergy between the metal nanoparticles and the support for the hydrogenation of functionalized carboxylic acids to diols on Ru/TiO$_2$. *Chem. Commun.*, **47**, 3613–3615.
51 Kang, K.H., Hong, U.G., Bang, Y., Choi, J.H., Kim, J.K., Lee, J.K., Han, S.J., and Song, I.K. (2015) Hydrogenation of succinic acid to 1,4-butanediol over Re-Ru bimetallic catalysts supported on mesoporous carbon. *Appl. Catal. A Gen.*, **490**, 153–162.
52 Lyu, J., Wang, J., Lu, C., Ma, L., Zhang, Q., He, X., and Li, X. (2014) Size-dependent halogenated nitrobenzene hydrogenation selectivity of Pd nanoparticles. *J. Phys. Chem. C*, **118**, 2594–2601.
53 Li, Y., Yu, J., Li, W., Fan, G., Yang, L., and Li, F. (2016) The promotional effect of surface defects on the catalytic performance of supported nickel-based catalysts. *Phys. Chem. Chem. Phys.*, **18**, 6548–6558.
54 Pelisson, C.H., Denicourt-Nowicki, A., Meriadec, C., Greneche, J.M., and Roucoux, A. (2015) Magnetically recoverable palladium(0) nanocomposite catalyst for hydrogenation reactions in water. *ChemCatChem*, **7**, 309–315.
55 Zhang, D.H., Chen, L., and Ge, G.L. (2015) A green approach for efficient p-nitrophenol hydrogenation catalyzed by a Pd-based nanocatalyst. *Catal. Commun.*, **66**, 95–99.
56 Gong, M., Jin, X., Sakidja, R., and Ren, S. (2015) Synergistic strain engineering effect of hybrid plasmonic, catalytic, and magnetic core–shell nanocrystals. *Nano. Lett.*, **15**, 8347–8353.
57 Raichle, A., Tsou, J.C., Neto, S., Penzel, U., Oehlenschlaeger, S., Zoellinger, M., Braunsberg, H., Haase, S., and Buettner, J. (2013) Process for preparing toluenediamine by hydrogenation of dinitrotoluene. US Patent 7511176.
58 Patil, N.G., Roy, D., Chaudhari, A.S., and Chaudhari, R.V. (2007) Kinetics of reductive alkylation of p-phenylenediamine with methyl ethyl ketone using 3% Pt/Al2O3 catalyst in a slurry reactor. *Ind. Eng. Chem. Res.*, **46**, 3243–3254.
59 Wilson, F.H. Jr. (1977) Modified nickel catalyst systems and their use in reductive alkylation reactions. US Patent 3739026.
60 Hoatson, J.R. and Rosenwald, R.H. (1957) Reductive alkylation process, in, Google Patents US2779789 A.
61 Su, B. and Barthomeuf, D. (1995) Effect of reaction temperature on the alkylation of aniline by methanolover almost neutral zeolites LiY and NaY. *Stud. Surf. Sci. Catal.*, **94**, 598–605.
62 Rusek, M. (1991) Effect of promoters on Pt/SiO$_2$ catalysts for the N-alkylation of sterically hindered anilines in the vapor phase. *Stud. Surf. Sci. Catal.*, **59**, 359–365.

63 Kaneko, M. and Tanaka, S. (1997) Method of manufacturing alkylaniline compounds, Eur. Patent 760360.

64 Lehtonen, J., Salmi, T., Vuori, A., and Tirronen, E. (1998) On the principles of modelling of homogeneous-heterogeneous reactions in the production of fine chemicals. A case study: reductive alkylation of aromatic amines. *Org. Process Res. Dev.*, **2**, 78–85.

65 Salmi, T., Lehtonen, J., Kaplin, J., Vuori, A., Tirronen, E., and Haario, H. (1999) A homogeneous-heterogeneously catalysed reaction system in a loop reactor. *Catal. Today*, **48**, 139–145.

66 Ross, L.J. and Levy, S.D. (1981) Reductive alkylation of substituted anilines. US Patent 4261926.

67 Watanabe, T. (1998) Preparation of di-(substituted amino) diphenylamines as antioxidants and antiozonants for polymers, Jpn. Patent 10168038.

68 Malz, R., Jancis, E., Reynolds, M., and O'Leary, S. (1995) *Reductive Alkylation of Acetophenone with Aniline*, Chemical Industries-Marcel-Dekker, New York, pp. 263–263.

69 Dam, J.ten. and Hanefeld, U. (2011) Renewable chemicals: dehydroxylation of glycerol and polyols. *ChemSusChem*, **4**, 1017–1034.

70 Roy, D., Subramaniam, B., and Chaudhari, R.V. (2010) Aqueous phase hydrogenolysis of glycerol to 1, 2-propanediol without external hydrogen addition. *Catal. Today*, **156**, 31–37.

71 Jin, X., Roy, D., Thapa, P.S., Subramaniam, B., and Chaudhari, R.V. (2013) Atom economical aqueous-phase conversion (APC) of biopolyols to lactic acid, glycols, and linear alcohols using supported metal catalysts. *ACS Sustain Chem. Eng.*, **1**, 1453–1462.

72 Nakagawa, Y., Shinmi, Y., Koso, S., and Tomishige, K. (2010) Direct hydrogenolysis of glycerol into 1, 3-propanediol over rhenium-modified iridium catalyst. *J. Catal.*, **272**, 191–194.

73 Kurosaka, T., Maruyama, H., Naribayashi, I., and Sasaki, Y. (2008) Production of 1, 3-propanediol by hydrogenolysis of glycerol catalyzed by Pt/WO3/ZrO$_2$. *Catal. Commun.*, **9**, 1360–1363.

74 Shimao, A., Koso, S., Ueda, N., Shinmi, Y., Furikado, I., and Tomishige, K. (2009) Promoting effect of Re addition to Rh/SiO$_2$ on glycerol hydrogenolysis. *Chem. Lett. (Jpn.)*, **38**, 540–541.

75 Jin, X., Shen, J., Yan, W.J., Zhao, M., Thapa, P.S., Subramaniam, B., and Chaudhari, R.V. (2015) Sorbitol hydrogenolysis over hybrid Cu/CaO-Al$_2$O$_3$ catalysts: tunable activity and selectivity with solid base incorporation. *ACS Catal.*, **5**, 6545–6558.

76 Kim, Y.T., Dumesic, J.A., and Huber, G.W. (2013) Aqueous-phase hydrodeoxygenation of sorbitol: A comparative study of Pt/Zr phosphate and Pt ReOx/C. *J. Catal.*, **304**, 72–85.

77 Li, N. and Huber, G.W. (2010) Aqueous-phase hydrodeoxygenation of sorbitol with Pt/SiO$_2$–Al$_2$O$_3$: Identification of reaction intermediates. *J. Catal.*, **270**, 48–59.

78 Zhou, C.-H.C., Beltramini, J.N., Fan, Y.-X., and Lu, G.M. (2008) Chemoselective catalytic conversion of glycerol as a biorenewable source to valuable commodity chemicals. *Chem. Soc. Rev.*, **37**, 527–549.

79 Pagliaro, M., Ciriminna, R., Kimura, H., Rossi, M., and Pina, C.Della. (2007) From glycerol to value-added products. *Angew. Chem., Int. Ed.*, **46**, 4434–4440.

80 Gallezot, P. (2012) Conversion of biomass to selected chemical products. *Chem. Soc. Rev.*, **41**, 1538–1558.

81 Miyazawa, T., Kunimori, K., and Tomishige, K. (2006) FUEL 108-Glycerol hydrogenolysis in the aqueous solution under hydrogen over Ru/C plus an ion-exchange resin and its reaction mechanism. *Abstr. Pap. Am. Chem. Soc.*, 232.

82 Wang, S.A., Zhang, Y.C., and Liu, H.C. (2010) Selective hydrogenolysis of glycerol to propylene glycol on Cu-ZnO composite catalysts: structural requirements and reaction mechanism. *Chem. Asian J.*, **5**, 1100–1111.

83 Zhao, B.B., Li, C.C., and Xu, C.L. (2012) Insight into the catalytic mechanism of glycerol hydrogenolysis using basal spacing of hydrotalcite as a tool. *Catal. Sci. Technol.*, **2**, 1985–1994.

84 Torres, A., Roy, D., Subramaniam, B., and Chaudhari, R.V. (2010) Kinetic modeling of aqueous-phase glycerol hydrogenolysis in a batch slurry reactor. *Ind. Eng. Chem. Res.*, **49**, 10826–10835.

85 Zhou, Z.M., Li, X., Zeng, T.Y., Hong, W.B., Cheng, Z.M., and Yuan, W.K. (2010) Kinetics of hydrogenolysis of glycerol to propylene glycol over Cu-ZnO-Al_2O_3 catalysts. *Chin. J. Chem. Eng.*, **18**, 384–390.

86 Huber, G.W., Cortright, R.D., and Dumesic, J.A. (2004) Renewable alkanes by aqueous-phase reforming of biomass-derived oxygenates. *Angew. Chem., Int. Edit.*, **43**, 1549–1551.

87 Li, N. and Huber, G.W. (2010) Aqueous-phase hydrodeoxygenation of sorbitol with Pt/SiO_2-Al_2O_3: Identification of reaction intermediates. *J. Catal.*, **270**, 48–59.

88 Sohounloue, D.K., Montassier, C., and Barbier, J. (1983) Catalytic hydrogenolysis of sorbitol. *React. Kinet. Catal. L.*, **22**, 391–397.

89 Sun, J.Y. and Liu, H.C. (2011) Selective hydrogenolysis of biomass-derived xylitol to ethylene glycol and propylene glycol on supported Ru catalysts. *Green Chem.*, **13**, 135–142.

90 Zhao, L., Zhou, J.H., Sui, Z.J., and Zhou, X.G. (2010) Hydrogenolysis of sorbitol to glycols over carbon nanofiber supported ruthenium catalyst. *Chem. Eng. Sci.*, **65**, 30–35.

91 Kim, Y.T., Dumesic, J.A., and Huber, G.W. (2013) Aqueous-phase hydrodeoxygenation of sorbitol: A comparative study of Pt/Zr phosphate and Pt-ReOx/C. *J. Catal.*, **304**, 72–85.

92 Zhang, Q., Jiang, T., Li, B., Wang, T.J., Zhang, X.H., Zhang, Q., and Ma, L.L. (2012) Highly selective sorbitol hydrogenolysis to liquid alkanes over Ni/HZSM-5 catalysts modified with pure silica MCM-41. *ChemCatChem*, **4**, 1084–1087.

93 Tronconi, E., Ferlazzo, N., Forzatti, P., Pasquon, I., Casale, B., and Marini, L. (1992) A mathematical-model for the catalytic hydrogenolysis of carbohydrates. *Chem. Eng. Sci.*, **47**, 2451–2456.

94 Zhao, L., Zhou, J., Sui, Z., and Zhou, X. (2010) Hydrogenolysis of sorbitol to glycols over carbon nanofiber supported ruthenium catalyst. *Chem. Eng. Sci.*, **65**, 30–35.

95 Jin, X., Subramaniam, B., and Chaudhari, R. (2013) Activity and selectivity of base promoted mono and bimetallic catalysts for hydrogenolysis of xylitol and sorbitol, ACS Symposium Series, pp. 273–285.

96 Clark, I.T. (1958) Hydrogenolysis of sorbitol. *Ind. Eng. Chem.*, **50**, 1125–1126.

97 Ye, L.M., Duan, X.P., Lin, H.Q., and Yuan, Y.Z. (2012) Improved performance of magnetically recoverable Ce-promoted Ni/Al2O3 catalysts for aqueous-phase hydrogenolysis of sorbitol to glycols. *Catal. Today*, **183**, 65–71.

98 Banu, M., Sivasanker, S., Sankaranarayanan, T.M., and Venuvanalingam, P. (2011) Hydrogenolysis of sorbitol over Ni and Pt loaded on NaY. *Catal. Commun.*, **12**, 673–677.

99 Nishimura, S. *Handbook of Heterogeneous Catalytic Hydrogenation for Organic Synthesis*, Wiley-VCH Verlag GmbH.

100 Brunet, S., Mey, D., Perot, G., Bouchy, C., and Diehl, F. (2005) On the hydrodesulfurization of FCC gasoline: a review. *Appl. Catal. A Gen.*, **278**, 143–172.

101 Vrinat, M.L. (1983) The Kinetics of the hydrodesulfurization process – a review. *Appl. Catal.*, **6**, 137–158.

102 Dhar, G.M., Srinivas, B.N., Rana, M.S., Kumar, M., and Maity, S.K. (2003) Mixed oxide supported hydrodesulfurization catalysts – a review. *Catal. Today*, **86**, 45–60.

103 Li, D., Sato, T., Imamura, M., Shimada, H., and Nishijima, A. (1998) The effect of boron on HYD, HC and HDS activities of model compounds over Ni-Mo/gamma-Al_2O_3-B_2O_3 catalysts. *Appl. Catal. B Environ.*, **16**, 255–260.

104 Dong, K.M., Wu, X.M., Lin, G.D., and Zhang, H.B. (2005) Mo-co catalyst supported on multiwalled carbon nanotubes for hydrodesulfurization of thiophene. *Chin. J. Catal.*, **26**, 550–556.

105 Guo, Y.A., Zeng, P.H., Ji, S.F., Wei, N., Liu, H., and Li, C.Y. (2010) Effect of Mo

promoter content on performance of Mo-Ni2P/SBA-15/cordierite monolithic catalyst for hydrodesulfurization. *Chin. J. Catal.*, **31**, 329–334.
106. Gutierrez, O.Y. and Klimova, T. (2011) Effect of the support on the high activity of the (Ni)Mo/ZrO$_2$-SBA-15 catalyst in the simultaneous hydrodesulfurization of DBT and 4,6-DMDBT. *J. Catal.*, **281**, 50–62.
107. Klicpera, T. and Zdrazil, M. (2001) High surface area MoO$_3$/MgO: preparation by reaction of MoO$_3$ and MgO in methanol or ethanol slurry and activity in hydrodesulfurization of benzothiophene. *Appl. Catal. A Gen.*, **216**, 41–50.
108. Klicpera, T. and Zdrazil, M. (2000) Synthesis of a high surface area monolayer MoO$_3$/MgO catalyst in a (NH4)(6)Mo7O24/MgO/methanol slurry, and its hydrodesulfurization activity. *J. Mater. Chem.*, **10**, 1603–1608.
109. Klicpera, T. and Zdrazil, M. (1999) High surface area MoO3/MgO: preparation by the new slurry impregnation method and activity in sulphided state in hydrodesulphurization of benzothiophene. *Catal. Lett.*, **58**, 47–51.
110. Silvy, R.P., Delannay, F., Grange, P., and Delmon, B. (1986) Effect of the activation conditions on the structure and catalytic activity of a Co-Mo gamma-Al$_2$O$_3$ hydrodesulfurization catalyst. *Polyhedron*, **5**, 195–198.
111. Zahradnikova, H., Karnik, V., and Beranek, L. (1985) Relation between reduction and sulfidation process during the activation of a Co-Mo/Al$_2$O$_3$ hydrodesulfurization catalyst. *Collect Czech. Chem. C*, **50**, 1573–1581.
112. Mitchell, P.C.H. and Valero, J.A. (1982) Hydro-demetallation and the promoter effect of vanadium compounds on a Co-Mo/Al$_2$O$_3$ hydrodesulfurization catalyst. *React. Kinet. Catal. L.*, **20**, 219–225.
113. Vrinat, M. and Demourques, L. (1981) High-pressure kinetic-study of dibenzothiophene hydrodesulfurization on Co-Mo-Gamma-Al$_2$O$_3$ catalyst. *Cr. Acad. Sci. II*, **292**, 581–584.
114. Levache, D., Guida, A., and Geneste, P. (1981) Hydrodesulfurization of sulfided polycyclic aromatic-compounds over Ni-Mo Gamma-Al$_2$O$_3$ catalyst. *B Soc. Chim. Belg.*, **90**, 1285–1292.
115. Gissy, H., Bartsch, R., and Tanielian, C. (1980) Hydrodesulfurization. 4. The effects of the feed components on Co-Mo-Al$_2$O$_3$ catalyst activation. *J. Catal.*, **65**, 158–165.
116. Dhainaut, E., Gachet, C., and Mourgues, L.D. (1979) Kinetics of dibenzothiophene hydrodesulfurization in presence of sulfided Co-Mo-Al$_2$O$_3$ catalyst – influence of hydrogen partial-pressure. *Cr. Acad. Sci. C Chim.*, **288**, 339–342.
117. Walton, R.A. (1976) Some remarks concerning x-ray photoelectron-spectra of Co-Mo-Al$_2$O$_3$ hydrodesulfurization catalyst system. *J. Catal.*, **44**, 335–337.
118. Bataille, F., Lemberton, J.L., Perot, G., Leyrit, P., Cseri, T., Marchal, N., and Kasztelan, S. (2001) Sulfided Mo and CoMo supported on zeolite as hydrodesulfurization catalysts: transformation of dibenzothiophene and 4,6-dimethyldibenzothiophene. *Appl. Catal. A Gen.*, **220**, 191–205.
119. Zeng, S.Q., Blanchard, J., Breysse, M., Shi, Y.H., Su, X.T., Nie, H., and Li, D.D. (2006) Mesoporous materials from zeolite seeds as supports for nickel-tungsten sulfide active phases – Part 2. Catalytic properties for deep hydrodesulfurization reactions. *Appl. Catal. A Gen.*, **298**, 88–93.
120. Zeng, S.Q., Blanchard, J., Breysse, M., Shi, Y.H., Su, X.T., Nie, H., and Li, D.D. (2005) Mesoporous materials from zeolite seeds as supports for nickel tungsten sulfide active phases Part 1. Characterization and catalytic properties in hydrocracking reactions. *Appl. Catal. A Gen.*, **294**, 59–67.
121. Edwards, J.C., Adams, R.D., and Ellis, P.D. (1990) A Mo-95 solid-state nmr-study of hydrodesulfurization catalysts. 1. Formation of fresh hds catalyst precursors by adsorption of polyoxomolybdates onto gamma-alumina. *J. Am. Chem. Soc.*, **112**, 8349–8364.
122. Kabe, T., Qian, W.H., and Ishihara, A. (1994) Study of hydrodesulfurization by the use of S-35 labeled dibenzothiophene. 2. Behavior of sulfur in Hds, Hdo, and

Hdn on sulfided Mo/Al$_2$O$_3$ catalyst. *J. Phys. Chem.*, **98**, 912–916.

123 Wang, D.H., Qian, W.H., Ishihara, A., and Kabe, T. (2002) Elucidation of sulfidation state and hydrodesulfurization mechanism on Mo/TiO$_2$ catalyst using S-35 radioisotope tracer methods. *Appl. Catal. A Gen.*, **224**, 191–199.

124 Hagiwara, K., Ebihara, T., Urasato, N., and Fujikawa, T. (2005) Application of Xe-129 NMR to structural analysis of MoS2 crystallites on Mo/Al$_2$O$_3$ hydrodesulfurization catalyst. *Appl. Catal. A Gen.*, **285**, 132–138.

125 Hagiwara, K., Ebihara, T., Urasato, N., and Fujikawa, T. (2005) Xe-129 NMR analysis of Mo/Al$_2$O$_3$ hydrodesulfurization catalyst – (Part 2) – Effect of cobalt addition. *J. Jpn. Petrol. Inst.*, **48**, 156–161.

126 Hagiwara, K., Ebihara, T., Urasato, N., and Fujikawa, T. (2005) Xe-129 NMR analysis of Mo/Al$_2$O$_3$ hydrodesulfurization catalyst. *J. Jpn. Petrol. Inst.*, **48**, 84–89.

127 Hagiwara, K., Ebihara, T., Urasato, N., and Fujikawa, T. (2007) Application of Xe-129 NMR spectroscopy to analysis of Co-Mo/Al$_2$O$_3$ hydrodesulfurization catalyst – Effect of sulfidation on Xe-129 NMR spectra. *J. Jpn. Petrol. Inst.*, **50**, 139–146.

128 Hagiwara, K. (2008) Characterization of Co-Mo/Al$_2$O(3) hydrodesulfurization catalyst by Xe-129 NMR Spectroscopy. *J. Jpn. Petrol. Inst.*, **51**, 32–41.

129 Singh, U.K. and Vannice, M.A. (2001) Kinetics of liquid-phase hydrogenation reactions over supported metal catalysts – a review. *Appl. Catal. A*, **213**, 1–24.

130 Stanislaus, A. and Cooper, B.H. (1994) Aromatic hydrogenation catalysis: a review. *Catal. Rev.*, **36**, 75–123.

131 Kittel, C. (2005) *Introduction to Solid State Physics*, Wiley-VCH Verlag GmbH.

132 Dry, M.E. (2002) High quality diesel via the Fischer–Tropsch process–a review. *J. Chem. Technol. Biotechnol.*, **77**, 43–50.

133 Van Der Laan, G.P. and Beenackers, A. (1999) Kinetics and selectivity of the Fischer–Tropsch synthesis: a literature review. *Catal. Rev.*, **41**, 255–318.

134 Dalai, A. and Davis, B. (2008) Fischer–Tropsch synthesis: a review of water effects on the performances of unsupported and supported Co catalysts. *Appl. Catal., A*, **348**, 1–15.

135 Khodakov, A.Y., Chu, W., and Fongarland, P. (2007) Advances in the development of novel cobalt Fischer-Tropsch catalysts for synthesis of long-chain hydrocarbons and clean fuels. *Chem. Rev.*, **107**, 1692–1744.

136 Ertl, G., Knözinger, H., and Weitkamp, J. (2008) *Preparation of Solid Catalysts*, Wiley-VCH Verlag GmbH.

137 Ross, J.R. (2011) *Heterogeneous Catalysis: Fundamentals and Applications*, Elsevier.

138 Campanati, M., Fornasari, G., and Vaccari, A. (2003) Fundamentals in the preparation of heterogeneous catalysts. *Catal. Today*, **77**, 299–314.

139 Pinna, F. (1998) Supported metal catalysts preparation. *Catal. Today*, **41**, 129–137.

140 Liu, C.-J., Vissokov, G.P., and Jang, B.W.-L. (2002) Catalyst preparation using plasma technologies. *Catal. Today*, **72**, 173–184.

141 Yamada, Y., Akita, T., Ueda, A., Shioyama, H., and Kobayashi, T. (2005) Instruments for preparation of heterogeneous catalysts by an impregnation method. *Rev. Sci. Instrum.*, **76**, 062226.

142 Frenzer, G. and Maier, W.F. (2006) Amorphous porous mixed oxides: sol-gel ways to a highly versatile class of materials and catalysts. *Annu. Rev. Mater. Res.*, **36**, 281–331.

143 Gonzalez, R.D., Lopez, T., and Gomez, R. (1997) Sol—Gel preparation of supported metal catalysts. *Catal. Today*, **35**, 293–317.

144 Gu, J., Zhang, Y.W., and Tao, F. (2012) Shape control of bimetallic nanocatalysts through well-designed colloidal chemistry approaches. *Chem. Soc. Rev.*, **41**, 8050–8065.

145 Lai, J.P., Niu, W.X., Luque, R., and Xu, G.B. (2015) Solvothermal synthesis of metal nanocrystals and their applications. *Nano Today*, **10**, 240–267.

146 Quan, Z.W., Wang, Y.X., and Fang, J.Y. (2013) High-index faceted noble metal

147 Ertas, I.E., Gulcan, M., ABulut, A., Yurderi, M., and Zahmakiran, M. (2016) Metal-organic framework (MIL-101) stabilized ruthenium nanoparticles: Highly efficient catalytic material in the phenol hydrogenation. *Microporous Mesoporous Mat.*, 94–103. doi: 10.1016/j.micromeso.2015.12.048.

148 Gopiraman, M., Babu, S.G., Khatri, Z., Kai, W., Kim, Y.A., Endo, M., Karvembu, R., and Kim, I.S. (2013) Dry synthesis of easily tunable nano ruthenium supported on graphene: novel nanocatalysts for aerial oxidation of alcohols and transfer hydrogenation of ketones. *J. Phys. Chem. C*, **117**, 23582–23596.

149 Ortiz-Cervantes, C., Iyanez, I., and Garcia, J.J. (2012) Facile preparation of ruthenium nanoparticles with activity in hydrogenation of aliphatic and aromatic nitriles to amines. *J. Phys. Org. Chem.*, **25**, 902–907.

150 Sarmah, P.P. and Dutta, D.K. (2012) Chemoselective reduction of a nitro group through transfer hydrogenation catalysed by Ru-0-nanoparticles stabilized on modified Montmorillonite clay. *Green Chem.*, **14**, 1086–1093.

151 Pelisson, C.H., Vono, L.L.R., Hubert, C., Denicourt-Nowicki, A., Rossi, L.M., and Roucoux, A. (2012) Moving from surfactant-stabilized aqueous rhodium (0) colloidal suspension to heterogeneous magnetite-supported rhodium nanocatalysts: Synthesis, characterization and catalytic performance in hydrogenation reactions. *Catal. Today*, **183**, 124–129.

152 Miao, S.D., Zhang, C.L., Liu, Z.M., Han, B.X., Xie, Y., Ding, S.F., and Yang, Z.Z. (2008) Highly efficient nanocatalysts supported on hollow polymer nanospheres: Synthesis, characterization, and applications. *J. Phys. Chem. C*, **112**, 774–780.

153 Moreno-Marrodan, C., Liguori, F., Mercade, E., Godard, C., Claver, C., and Barbaro, P. (2015) A mild route to solid-supported rhodium nanoparticle catalysts and their application to the selective hydrogenation reaction of substituted arenes. *Catal. Sci. Technol.*, **5**, 3762–3772.

154 Faria, V.W., Brunelli, M.F., and Scheeren, C.W. (2015) Iridium nanoparticles supported in polymeric membranes: a new material for hydrogenation reactions. *RSC Adv.*, **5**, 84920–84926.

155 Wang, Z., Huang, L., Geng, L.F., Chen, R.Z., Xing, W.H., Wang, Y., and Huang, J. (2015) Chemoselective transfer hydrogenation of aldehydes and ketones with a heterogeneous iridium catalyst in water. *Catal. Lett.*, **145**, 1008–1013.

156 Gao, Z., Dong, M., Wang, G.Z., Sheng, P., Wu, Z.W., Yang, H.M., Zhang, B., Wang, G.F., Wang, J.G., and Qin, Y. (2015) Multiply confined nickel nanocatalysts produced by atomic layer deposition for hydrogenation reactions. *Angew. Chem., Int. Edit.*, **54**, 9006–9010.

157 Yang, L., Jiang, Z.S., Fan, G.L., and Li, F. (2014) The promotional effect of ZnO addition to supported Ni nanocatalysts from layered double hydroxide precursors on selective hydrogenation of citral. *Catal. Sci. Technol.*, **4**, 1123–1131.

158 Jagadeesh, R.V., Banerjee, D., Arockiam, P.B., Junge, H., Junge, K., Pohl, M.M., Radnik, J., Bruckner, A., and Beller, M. (2015) Highly selective transfer hydrogenation of functionalised nitroarenes using cobalt-based nanocatalysts. *Green Chem.*, **17**, 898–902.

159 Zhao, T.J., Zhang, Y.N., Wang, K.X., Su, J., Wei, X., and Li, X.H. (2015) General transfer hydrogenation by activating ammonia-borane over cobalt nanoparticles. *RSC Adv.*, **5**, 102736–102740.

160 Zhang, F.W. and Yang, H.Q. (2015) Multifunctional mesoporous silica-supported palladium nanoparticles for selective phenol hydrogenation in the aqueous phase. *Catal. Sci. Technol.*, **5**, 572–577.

161 Dominguez-Quintero, O., Martinez, S., Henriquez, Y., D'Ornelas, L., Krentzien, H., and Osuna, J. (2003) Silica-supported palladium nanoparticles show remarkable hydrogenation catalytic activity. *J. Mol. Catal. A Chem.*, **197**, 185–191.

162 Li, K.T., Hsu, M.h., and Wang, K. (2008) Palladium core-porous silica shell-

nanoparticles for catalyzing the hydrogenation of 4-carboxybenzaldehyde. *Catal. Commun.*, **9**, 2257–2260.

163 Wu, T.B., Jiang, T., Hu, B.J., Han, B.X., He, J.L., and Zhou, X.S. (2009) Cross-linked polymer coated Pd nanocatalysts on SiO2 support: very selective and stable catalysts for hydrogenation in supercritical CO_2. *Green Chem.*, **11**, 798–803.

164 Kim, Y. and Kim, M.J. (2010) Magnetically recoverable palladium nanocatalyst for chemoselective olefin hydrogenation. *B Korean Chem. Soc.*, **31**, 1368–1370.

165 Shao, Z.F., Li, C.A., Chen, X.A., Pang, M., Wang, X.K., and Liang, C.H. (2010) A facile and controlled route to prepare an eggshell Pd catalyst for selective hydrogenation of phenylacetylene. *ChemCatChem*, **2**, 1555–1558.

166 Garcia-Suarez, E.J., Balu, A.M., Tristany, M., Garcia, A.B., Philippot, K., and Luque, R. (2012) Versatile dual hydrogenation-oxidation nanocatalysts for the aqueous transformation of biomass-derived platform molecules. *Green Chem.*, **14**, 1434–1439.

167 Sun, Q., Guo, C.Z., Wang, G.H., Li, W.C., Bongard, H.J., and Lu, A.H. (2013) Fabrication of magnetic yolk-shell nanocatalysts with spatially resolved functionalities and high activity for nitrobenzene hydrogenation. *Chem. Eur. J.*, **19**, 6217–6220.

168 Yarulin, A., Yuranov, I., Cardenas-Lizana, F., Abdulkin, P., and Kiwi-Minsker, L. (2013) Size-effect of Pd-(Poly(N-vinyl-2-pyrrolidone)) nanocatalysts on selective hydrogenation of alkynols with different alkyl chains. *J. Phys. Chem. C*, **117**, 13424–13434.

169 Demirelli, M., Karaoglu, E., Baykal, A., Sozeri, H., Uysal, E., and Duygulu, O. (2013) Recyclable NiFe2O4-APTES/Pd magnetic nanocatalyst. *J. Inorg. Organomet. P.*, **23**, 937–943.

170 Gavia, D.J., Koeppen, J., Sadeghmoghaddam, E., and Shon, Y.S. (2013) Tandem semi-hydrogenation/isomerization of propargyl alcohols to saturated carbonyl analogues by dodecanethiolate-capped palladium nanoparticle catalysts. *RSC Adv.*, **3**, 13642–13645.

171 Uberman, P.M., Costa, N.J.S., Philippot, K., Carmona, R.C., Dos Santos, A.A., and Rossi, L.M. (2014) A recoverable Pd nanocatalyst for selective semi-hydrogenation of alkynes: hydrogenation of benzyl-propargylamines as a challenging model. *Green Chem.*, **16**, 4566–4574.

172 Acham, V.R., Biradar, A.V., Dongare, M.K., Kemnitz, E., and Umbarkar, S.B. (2014) Palladium nanoparticles supported on magnesium hydroxide fluorides: a selective catalyst for olefin hydrogenation. *ChemCatChem*, **6**, 3182–3191.

173 Han, R.R., Nan, C.S., Yang, L., Fan, G.L., and Li, F. (2015) Direct synthesis of hybrid layered double hydroxide-carbon composites supported Pd nanocatalysts efficient in selective hydrogenation of citral. *RSC Adv.*, **5**, 33199–33207.

174 You, C.J., Zhang, C., Chen, L.F., and Qi, Z.W. (2015) Highly dispersed palladium nanoclusters incorporated in amino-functionalized silica spheres for the selective hydrogenation of succinic acid to gamma-butyrolactone. *Appl. Organomet. Chem.*, **29**, 653–660.

175 Chen, L.Y., Huang, B.B., Qiu, X., Wang, X., Luque, R., and Li, Y.W. (2016) Seed-mediated growth of MOF-encapsulated Pd@Ag core-shell nanoparticles: toward advanced room temperature nanocatalysts. *Chem. Sci.*, **7**, 228–233.

176 Wei, Z.J., Pan, R.F., Hou, Y.X., Yang, Y., and Liu, Y.X. (2015) Graphene-supported Pd catalyst for highly selective hydrogenation of resorcinol to 1, 3-cyclohexanedione through giant pi-conjugate interactions. *Sci. Rep.*, 5. doi: 10.1038/srep15664.

177 Semagina, N., Renken, A., Laub, D., and Kiwi-Minsker, L. (2007) Synthesis of monodispersed palladium nanoparticles to study structure sensitivity of solvent-free selective hydrogenation of 2-methyl-3-butyn-2-ol. *J. Catal.*, **246**, 308–314.

178 Yoo, J.W., Hathcock, D.J., and El-Sayed, M.A. (2003) Propene hydrogenation over truncated octahedral Pt nanoparticles

supported on alumina. *J. Catal.*, **214**, 1–7.

179 Ren, J. and Tilley, R.D. (2007) Preparation, self-assembly, and mechanistic study of highly monodispersed nanocubes. *J. Am. Chem. Soc.*, **129**, 3287–3291.

180 Lee, H., Habas, S.E., Kweskin, S., Butcher, D., Somorjai, G.A., and Yang, P.D. (2006) Morphological control of catalytically active platinum nanocrystals. *Angew. Chem., Int. Edit.*, **45**, 7824–7828.

181 Wang, F., Li, C.H., Sun, L.D., Xu, C.H., Wang, J.F., Yu, J.C., and Yan, C.H. (2012) Porous single-crystalline palladium nanoparticles with high catalytic activities. *Angew. Chem., Int. Edit.*, **51**, 4872–4876.

182 Rioux, R., Song, H., Grass, M., Habas, S., Niesz, K., Hoefelmeyer, J., Yang, P., and Somorjai, G. (2006) Monodisperse platinum nanoparticles of well-defined shape: synthesis, characterization, catalytic properties and future prospects. *Top. Catal.*, **39**, 167–174.

183 Ikeda, S., Ishino, S., Harada, T., Okamoto, N., Sakata, T., Mori, H., Kuwabata, S., Torimoto, T., and Matsumura, M. (2006) Ligand-free platinum nanoparticles encapsulated in a hollow porous carbon shell as a highly active heterogeneous hydrogenation catalyst. *Angew. Chem., Int. Edit.*, **45**, 7063–7066.

184 Jacinto, M., Landers, R., and Rossi, L. (2009) Preparation of supported Pt (0) nanoparticles as efficient recyclable catalysts for hydrogenation of alkenes and ketones. *Catal. Commun.*, **10**, 1971–1974.

185 Yamada, Y., Tsung, C.-K., Huang, W., Huo, Z., Habas, S.E., Soejima, T., Aliaga, C.E., Somorjai, G.A., and Yang, P. (2011) Nanocrystal bilayer for tandem catalysis. *Nat. Chem.*, **3**, 372–376.

186 Sheng, B.Q., Hu, L., Yu, T.T., Cao, X.Q., and Gu, H.W. (2012) Highly-dispersed ultrafine Pt nanoparticles on graphene as effective hydrogenation catalysts. *RSC Adv.*, **2**, 5520–5523.

187 Mandal, S., Roy, D., Chaudhari, R.V., and Sastry, M. (2004) Pt and Pd nanoparticles immobilized on amine-functionalized zeolite: excellent catalysts for hydrogenation and heck reactions. *Chem. Mater.*, **16**, 3714–3724.

188 Xie, R.G., Cao, X.Q., Pan, Y., and Gu, H.W. (2015) Synthesis of Pt nanocatalysts for selective hydrogenation of ortho-halogenated nitrobenzene. *Sci. China Chem.*, **58**, 1051–1055.

189 Campos, C.H., Torres, C.C., Osorio-Vargas, P., Mella, C., Belmar, J., Ruiz, D., Fierro, J.L.G., and Reyes, P. (2015) Immobilised chiral inducer on Pt-based mesoporous titanate nanotubes as heterogeneous catalysts for enantioselective hydrogenation. *J. Mol. Catal. A Chem.*, **398**, 190–202.

190 Chen, Z.J., Guan, Z.H., Li, M.R., Yang, Q.H., and Li, C. (2011) Enhancement of the performance of a platinum nanocatalyst confined within carbon nanotubes for asymmetric hydrogenation. *Angew. Chem., Int. Edit.*, **50**, 4913–4917.

191 Kim, D.J., Li, Y., Kim, Y.J., Hur, N.H., and Seo, W.S. (2015) A highly stable and magnetically recyclable nanocatalyst system: mesoporous silica spheres embedded with FeCo/graphitic shell magnetic nanoparticles and Pt nanocatalysts. *Chem. Asian J.*, **10**, 2754–2760.

192 Ji, Y.J., Wu, Y.E., Zhao, G.F., Wang, D.S., Liu, L., He, W., and Li, Y.D. (2015) Porous bimetallic Pt-Fe nanocatalysts for highly efficient hydrogenation of acetone. *Nano Res.*, **8**, 2706–2713.

193 Huang, Y.Q., Ma, Y., Cheng, Y.W., Wang, L.J., and Li, X. (2015) Supported nanometric platinum-nickel catalysts for solvent-free hydrogenation of tetralin. *Catal. Commun.*, **69**, 55–58.

194 Liu, M.h., Bai, Q., Xiao, H.L., Liu, Y.Y., Zhao, J., and Yu, W.W. (2013) Selective hydrogenation of o-chloronitrobenzene over tin dioxide supported platinum-ruthenium bimetallic nanocatalysts without solvent. *Chem. Eng. J.*, **232**, 89–95.

195 Tsang, S.C., Cailuo, N., Oduro, W., Kong, A.T.S., Clifton, L., Yu, K.M.K., Thiebaut, B., Cookson, J., and Bishop, P. (2008) Engineering preformed cobalt-doped platinum nanocatalysts for ultraselective hydrogenation. *ACS Nano*, **2**, 2547–2553.

196 Jin, X., Dang, L.N., Lohrman, J., Subramaniam, B., Ren, S.Q., and Chaudhari, R.V. (2013) Lattice-matched bimetallic CuPd-graphene nanocatalysts for facile conversion of biomass-derived polyols to chemicals. *ACS Nano*, **7**, 1309–1316.

197 Huang, C.L., Zhang, H.Y., Zhao, Y.F., Chen, S., and Liu, Z.M. (2012) Diatomite-supported Pd-M (M = Cu, Co, Ni) bimetal nanocatalysts for selective hydrogenation of long-chain aliphatic esters. *J. Colloid. Interf. Sci.*, **386**, 60–65.

198 Yadav, G.D. and Lawate, Y.S. (2013) Hydrogenation of styrene oxide to 2-phenyl ethanol over polyurea microencapsulated mono- and bimetallic nanocatalysts: activity, selectivity, and kinetic modeling. *Ind. Eng. Chem. Res.*, **52**, 4027–4039.

199 Yarulin, A., Yuranov, I., Cardenas-Lizana, F., Alexander, D.T.L., and Kiwi-Minsker, L. (2014) How to increase the selectivity of Pd-based catalyst in alkynol hydrogenation: effect of second metal. *Appl. Catal. A Gen.*, **478**, 186–193.

200 Corbos, E.C., Ellis, P.R., Cookson, J., Briois, V., Hyde, T.I., Sankar, G., and Bishop, P.T. (2013) Tuning the properties of PdAu bimetallic nanocatalysts for selective hydrogenation reactions. *Catal. Sci. Technol.*, **3**, 2934–2943.

201 Ren, M.R., Li, C.M., Chen, J.L., Wei, M., and Shi, S.X. (2014) Preparation of a ternary Pd-Rh-P amorphous alloy and its catalytic performance in selective hydrogenation of alkynes. *Catal. Sci. Technol.*, **4**, 1920–1924.

202 Karakhanov, E.A., Maximov, A.L., Zolotukhina, A.V., Yatmanova, N., and Rosenberg, E. (2015) Alkyne hydrogenation using Pd-Ag hybrid nanocatalysts in surface-immobilized dendrimers. *Appl. Organomet. Chem.*, **29**, 777–784.

203 Zhu, L.H., Sun, H.L., Fu, H., Zheng, J.B., Zhang, N.W., Li, Y.H., and Chen, B.H. (2015) Effect of ruthenium nickel bimetallic composition on the catalytic performance for benzene hydrogenation to cyclohexane. *Appl. Catal. A Gen.*, **499**, 124–132.

204 Gai, P.L., Kourtakis, K., and Boyes, E.D. (2005) *In situ* nanoscale wet imaging of the heterogeneous catalyzation of nitriles in a solution phase: novel hydrogenation chemistry through nanocatalysts on nanosupports. *Catal. Lett.*, **102**, 1–7.

205 Yadav, G.D. and Lawate, Y.S. (2011) Selective hydrogenation of styrene oxide to 2-phenyl ethanol over polyurea supported Pd-Cu catalyst in supercritical carbon dioxide. *J. Supercrit Fluid*, **59**, 78–86.

206 Halilu, A., Ali, T.H., Atta, A.Y., Sudarsanam, P., Bhargava, S.K., and Hamid, S.B.A. (2016) Highly selective hydrogenation of biomass-derived furfural into furfuryl alcohol using a novel magnetic nanoparticles catalyst. *Energy Fuels*. doi: 10.1021/acs.energyfuels.5b02826.

207 Nikparsa, P., Mirzaei, A.A., and Rauch, R. (2016) Impact of Na promoter on structural properties and catalytic performance of CoNi/Al_2O_3 nanocatalysts for the CO hydrogenation process: fischer-tropsch technology. *Catal. Lett.*, **146**, 61–71.

208 Shanbogh, P.P. and Peter, S.C. (2013) Low cost nano materials crystallize in the NiAs structure type as an alternative to the noble metals in the hydrogenation process. *RSC Adv.*, **3**, 22887–22890.

209 Li, T.L., Zhang, W.N., Lee, R.Z., and Zhong, Q.X. (2009) Nickel-boron alloy catalysts reduce the formation of Trans fatty acids in hydrogenated soybean oil. *Food Chem.*, **114**, 447–452.

210 Chen, Y.C. and Tan, C.S. (2007) Hydrogenation of p-chloronitrobenzene by Ni-B nanocatalyst in CO_2-expanded methanol. *J. Supercrit Fluid*, **41**, 272–278.

211 Sedighi, B., Feyzi, M., and Joshaghani, M. (2015) Preparation and characterization of Co-Fe nano catalyst for Fischer-Tropsch synthesis: optimization using response surface methodology. *J. Taiwan Inst. Chem. E*, **50**, 108–114.

212 Vetere, V., Merlo, A.B., and Casella, M.L. (2015) Chemoselective hydrogenation of aromatic ketones with Pt-based heterogeneous catalysts. Substituent effects. *Appl. Catal., A*, **491**, 70–77.

213 Barbelli, M.L., Santori, G.F., and Nichio, N.N. (2012) Aqueous phase hydrogenolysis of glycerol to bio-propylene glycol over Pt–Sn catalysts. *Bioresour. Technol.*, **111**, 500–503.

214 Bocanegra, S.A., Miguel, S.R.de., Borbath, I., Margitfalvi, J.L., and Scelza, O.A. (2009) Behavior of bimetallic PtSn/Al$_2$O$_3$ catalysts prepared by controlled surface reactions in the selective dehydrogenation of butane. *J. Mol. Catal. A Chem.*, **301**, 52–60.

215 Rice, R.W. and Keptner, D.C. (2004) The effect of bimetallic catalyst preparation and treatment on behavior for propane hydrogenolysis. *Appl. Catal. A*, **262**, 233–239.

216 Iihama, S., Furukawa, S., and Komatsu, T. (2015) Efficient catalytic system for chemoselective hydrogenation of halonitrobenzene to haloaniline using PtZn intermetallic compound. *ACS Catal.*, **6**, 742–746.

217 Sulman, E.M., Matveeva, V.G., Sulman, M.G., Demidenko, G.N., Valetsky, P.M., Stein, B., Mates, T., and Bronstein, L.M. (2009) Influence of heterogenization on catalytic behavior of mono-and bimetallic nanoparticles formed in poly (styrene)-block-poly (4-vinylpyridine) micelles. *J. Catal.*, **262**, 150–158.

218 Wang, G.-H., Hilgert, J., Richter, F.H., Wang, F., Bongard, H.-J., Spliethoff, B., Weidenthaler, C., and Schüth, F. (2014) Platinum–cobalt bimetallic nanoparticles in hollow carbon nanospheres for hydrogenolysis of 5-hydroxymethylfurfural. *Nat. Mater.*, **13**, 293–300.

219 Khemthong, P., Klysubun, W., Prayoonpokarach, S., Roessner, F., and Wittayakun, J. (2010) Comparison between cobalt and cobalt–platinum supported on zeolite NaY: Cobalt reducibility and their catalytic performance for butane hydrogenolysis. *J. Ind. Eng. Chem.*, **16**, 531–538.

220 Corbel-Demailly, L., Ly, B.K., Minh, D.P., Tapin, B., Especel, C., Epron, F., Cabiac, A., Guillon, E., Besson, M., and Pinel, C. (2013) Heterogeneous catalytic hydrogenation of biobased levulinic and succinic acids in aqueous solutions. *ChemSusChem*, **6**, 2388–2395.

221 Coetzee, J., Manyar, H.G., Hardacre, C., and Cole-Hamilton, D.J. (2013) The first continuous flow hydrogenation of amides to amines. *ChemCatChem*, **5**, 2843–2847.

222 Deng, C., Duan, X., Zhou, J., Chen, D., Zhou, X., and Yuan, W. (2014) Size effects of Pt-Re bimetallic catalysts for glycerol hydrogenolysis. *Catal. Today*, **234**, 208–214.

223 Moser, M.D., Lawson, R.J., Wang, L., and Parulekar, V.N. (1990) Platinum, germanium halogen on refractory support, in, Google Patents US4929332 A.

224 Loh, A.S., Davis, S.W., and Medlin, J.W. (2008) Adsorption and reaction of 1-epoxy-3-butene on Pt (111): Implications for heterogeneous catalysis of unsaturated oxygenates. *J. Am. Chem. Soc.*, **130**, 5507–5514.

225 Alonso, D.M., Wettstein, S.G., and Dumesic, J.A. (2012) Bimetallic catalysts for upgrading of biomass to fuels and chemicals. *Chem. Soc. Rev.*, **41**, 8075–8098.

226 Sankar, M., Dimitratos, N., Miedziak, P.J., Wells, P.P., Kiely, C.J., and Hutchings, G.J. (2012) Designing bimetallic catalysts for a green and sustainable future. *Chem. Soc. Rev.*, **41**, 8099–8139.

227 Elangovan, S., Bischof, C., and Hartmann, M. (2002) Isomerization and hydrocracking of n-decane over bimetallic Pt–Pd clusters supported on mesoporous MCM-41 catalysts. *Catal. Lett.*, **80**, 35–40.

228 He, Z., Qian, Q., Zhang, Z., Meng, Q., Zhou, H., Jiang, Z., and Han, B. (2015) Synthesis of higher alcohols from CO_2 hydrogenation over a PtRu/Fe$_2$O$_3$ catalyst under supercritical condition. *Philos. Trans. A Math. Phys. Eng. Sci.*, **373**, 20150006.

229 Liu, H., Fang, R., Li, Z., and Li, Y. (2015) Solventless hydrogenation of benzene to cyclohexane over a heterogeneous Ru–Pt bimetallic catalyst. *Chem. Eng. Sci.*, **122**, 350–359.

230 Dehm, N.A., Zhang, X., and Buriak, J.M. (2010) Screening of bimetallic heterogeneous nanoparticle catalysts for arene hydrogenation activity under

ambient conditions. *Inorg. Chem.*, **49**, 2706–2714.

231 Chatterjee, M., Zhao, F., and Ikushima, Y. (2004) Hydrogenation of citral using monometallic Pt and bimetallic Pt-Ru catalysts on a mesoporous support in supercritical carbon dioxide medium. *Adv. Synth. Catal.*, **346**, 459–466.

232 Samoila, P., Boutzeloit, M., Especel, C., Epron, F., and Marecot, P. (2009) Selective ring-opening of methylcyclopentane on platinum-based bimetallic catalysts. *Appl. Catal. A*, **369**, 104–112.

233 Wang, Y., Zhang, J., Wang, X., Ren, J., Zuo, B., and Tang, Y. (2005) Metal nanoclusters stabilized with simple ions and solvents—promising building blocks for future catalysts. *Top. Catal.*, **35**, 35–41.

234 Siepen, K., Bönnemann, H., Brijoux, W., Rothe, J., and Hormes, J. (2000) EXAFS/XANES, chemisorption and IR investigations of colloidal Pt/Rh bimetallic catalysts. *Appl. Organomet. Chem.*, **14**, 549–556.

235 Khatib, F.H.El. and Mahmoud, S.S. (2008) Propane hydrogenolysis over alumina-supported Pt-Rh catalyst. *Asian J. Chem.*, **20**, 1179.

236 Vasina, T., Masloboishchikova, O., Khelkovskaya-Sergeeva, E., Kustov, L., and Zeuthen, P. (2002) Cyclohexane transformations over metal oxide catalysts. 2. Selective cyclohexane ring opening to form n-hexane over mono- and bimetallic rhodium catalysts. *Russ. Chem. B*, **51**, 255–258.

237 Soares, A.V.-H., Perez, G., and Passos, F.B. (2016) Alumina supported bimetallic Pt–Fe catalysts applied to glycerol hydrogenolysis and aqueous phase reforming. *Appl. Catal., B*, **185**, 77–87.

238 Barbelli, M.L., Mizrahi, M.D., Pompeo, F., Santori, G.F., Nichio, N.N., and Ramallo-Lopez, J.M. (2014) EXAFS characterization of PtNi bimetallic catalyst applied to glycerol liquid-phase conversion. *J. Phys. Chem. C*, **118**, 23645–23653.

239 Iriondo, A., Cambra, J., Barrio, V., Guemez, M., Arias, P., Sanchez-Sanchez, M., Navarro, R., and Fierro, J. (2011) Glycerol liquid phase conversion over monometallic and bimetallic catalysts: effect of metal, support type and reaction temperatures. *Appl. Catal., B*, **106**, 83–93.

240 Alba-Rubio, A.C., Sener, C., Hakim, S.H., Gostanian, T.M., and Dumesic, J.A. (2015) Synthesis of supported RhMo and PtMo bimetallic catalysts by controlled surface reactions. *ChemCatChem*, **7**, 3881–3886.

241 Stassi, J.P., Zgolicz, P.D., Rodríguez, V.I., de Miguel, S.R., and Scelza, O.A. (2015) Ga and In promoters in bimetallic Pt based catalysts to improve the performance in the selective hydrogenation of citral. *Appl. Catal. A*, **497**, 58–71.

242 Győrffy, N., Bakos, I., Szabó, S., Tóth, L., Wild, U., Schlögl, R., and Paál, Z. (2009) Preparation, characterization and catalytic testing of GePt catalysts. *J. Catal.*, **263**, 372–379.

243 Serge Ello, A., Pirault-Roy, L., Wootsch, A., and Paál, Z. (2009) Ethylcyclopentane ring opening reaction over PtGe/Al2O3 catalysts prepared by controlled surface reaction. *Phys. Chem. News*, 103–109. doi: 10.1007/s11144-008-5343-1.

244 Ba, R., Zhao, Y., Yu, L., Song, J., Huang, S., Zhong, L., Sun, Y., and Zhu, Y. (2015) Synthesis of Co-based bimetallic nanocrystals with one-dimensional structure for selective control on syngas conversion. *Nanoscale*, **7**, 12365–12371.

245 Ma, J., Huang, X., Liao, X., and Shi, B. (2013) Preparation of highly active heterogeneous Au@ Pd bimetallic catalyst using plant tannin grafted collagen fiber as the matrix. *J. Mol. Catal. A: Chem.*, **366**, 8–16.

246 Song, J.H., Seo, K.W., Mok, Y.I., Park, K.Y., and Ahn, B.S. (2002) Catalytic hydrogenolysis of CFC-12 (CCl2F2) over bimetallic palladium catalysts supported on activated carbon. *Korean J. Chem. Eng.*, **19**, 246–251.

247 Chen, J., Liu, X., and Zhang, F. (2015) Composition regulation of bimetallic RuPd catalysts supported on porous alumina spheres for selective hydrogenation. *Chem. Eng. J.*, **259**, 43–52.

248 Chen, J., Guo, L., and Zhang, F. (2014) The role of hydrotalcite-modified porous alumina spheres in bimetallic RuPd catalysts for selective hydrogenation. *Catal. Commun.*, **55**, 19–23.

249 Johnson, B.F., Raynor, S.A., Brown, D.B., Shephard, D.S., Mashmeyer, T., Thomas, J.M., Hermans, S., Raja, R., and Sankar, G. (2002) New catalysts for clean technology. *J. Mol. Catal. A Chem.*, **182**, 89–97.

250 Shao, Z., Li, C., Di, X., Xiao, Z., and Liang, C. (2014) Aqueous-phase hydrogenation of succinic acid to γ-butyrolactone and tetrahydrofuran over Pd/C, Re/C, and Pd–Re/C catalysts. *Ind. Eng. Chem. Res.*, **53**, 9638–9645.

251 Ly, B.K., Tapin, B., Aouine, M., Delichere, P., Epron, F., Pinel, C., Especel, C., and Besson, M. (2015) Insights into the oxidation state and location of rhenium in Re-Pd/TiO_2 catalysts for aqueous-phase selective hydrogenation of succinic acid to 1,4-butanediol as a function of palladium and rhenium deposition methods. *ChemCatChem*, **7**, 2161–2178.

252 Bridier, B. and Pérez-Ramírez, J. (2011) Selectivity patterns in heterogeneously catalyzed hydrogenation of conjugated ene-yne and diene compounds. *J. Catal.*, **284**, 165–175.

253 Liao, F., Lo, T.B., Qu, J., Kroner, A., Dent, A., and Tsang, S.E. (2015) Tunability of catalytic properties of Pd-based catalysts by rational control of strong metal and support interaction (SMSI) for selective hydrogenolyic C—C and C—O bond cleavage of ethylene glycol units in biomass molecules. *Catal. Sci. Technol.*, **5**, 3491–3495.

254 Müslehiddinoğlu, J., Li, J., Tummala, S., and Deshpande, R. (2010) Highly diastereoselective hydrogenation of imines by a bimetallic Pd—Cu heterogeneous catalyst. *Org. Process Res. Dev.*, **14**, 890–894.

255 Feng, Y.-S., Liu, C., Kang, Y.-M., Zhou, X.-M., Liu, L.-L., Deng, J., Xu, H.-J., and Fu, Y. (2015) Selective hydrogenolysis of glycerol to 1,2-propanediol catalyzed by supported bimetallic PdCu-KF/γ-Al_2O_3. *Chem. Eng. J.*, **281**, 96–101.

256 Jiang, T., Huai, Q., Geng, T., Ying, W., Xiao, T., and Cao, F. (2015) Catalytic performance of Pd—Ni bimetallic catalyst for glycerol hydrogenolysis. *Biomass Bioenergy*, **78**, 71–79.

257 Jin, X., Dang, L., Lohrman, J., Subramaniam, B., Ren, S., and Chaudhari, R.V. (2013) Lattice-matched bimetallic CuPd-graphene nanocatalysts for facile conversion of biomass-derived polyols to chemicals. *ACS Nano*, **7**, 1309–1316.

258 Xia, S., Yuan, Z., Wang, L., Chen, P., and Hou, Z. (2011) Hydrogenolysis of glycerol on bimetallic Pd-Cu/solid-base catalysts prepared via layered double hydroxides precursors. *Appl. Catal. A*, **403**, 173–182.

259 Jia, Y. and Liu, H. (2015) Selective hydrogenolysis of sorbitol to ethylene glycol and propylene glycol on ZrO_2-supported bimetallic Pd-Cu catalysts. *Chin. J. Catal.*, **36**, 1552–1559.

260 Liao, F., Lo, B.T., Sexton, D., Qu, J., Ma, C., Chan, R.C.T., Lu, Q., Che, R., Kwok, W.M., and He, H. (2015) A new class of tunable heterojunction by using two support materials for the synthesis of supported bimetallic catalysts. *ChemCatChem*, **7**, 230–235.

261 Zhang, J., Teo, J., Chen, X., Asakura, H., Tanaka, T., Teramura, K., and Yan, N. (2014) A series of NiM (M = Ru, Rh, and Pd) bimetallic catalysts for effective lignin hydrogenolysis in water. *ACS Catal.*, **4**, 1574–1583.

262 Childers, D.J., Schweitzer, N.M., Shahari, S.M.K., Rioux, R.M., Miller, J.T., and Meyer, R.J. (2014) Modifying structure-sensitive reactions by addition of Zn to Pd. *J. Catal.*, **318**, 75–84.

263 Barbaro, P., Bianchini, C., Dal Santo, V., Meli, A., Moneti, S., Psaro, R., Scaffidi, A., Sordelli, L., and Vizza, F. (2006) Hydrogenation of arenes over silica-supported catalysts that combine a grafted rhodium complex and palladium nanoparticles: evidence for substrate activation on Rhsingle-site-Pdmetal moieties. *J. Am. Chem. Soc.*, **128**, 7065–7076.

264 Zina, M.S. and Ghorbel, A. (2004) Pd-Mo Bimetallic catalysts supported on Y-Zeolite: Effect of molybdenum on structural and catalytic properties of

palladium in partial hydrogenation of 1, 3 butadiene. *Stud. Surf. Sci. Catal.*, **154**, 2364–2370.
265 Beamson, G., Papworth, A.J., Philipps, C., Smith, A.M., and Whyman, R. (2011) Selective hydrogenation of amides using bimetallic Ru/Re and Rh/Re catalysts. *J. Catal.*, **278**, 228–238.
266 Ma, L. and He, D. (2010) Influence of catalyst pretreatment on catalytic properties and performances of Ru–Re/SiO$_2$ in glycerol hydrogenolysis to propanediols. *Catal. Today*, **149**, 148–156.
267 Lange, F., Armbruster, U., and Martin, A. (2015) Heterogeneously-catalyzed hydrogenation of carbon dioxide to methane using RuNi bimetallic catalysts. *Energy Technol.*, **3**, 55–62.
268 Zhu, L., Cao, M., Li, L., Sun, H., Tang, Y., Zhang, N., Zheng, J., Zhou, H., Li, Y., and Yang, L. (2014) Synthesis of different ruthenium nickel bimetallic nanostructures and an investigation of the structure–activity relationship for benzene hydrogenation to cyclohexane. *ChemCatChem*, **6**, 2039–2046.
269 Cerro-Alarcón, M., Maroto-Valiente, A., Rodríguez-Ramos, I., and Guerrero-Ruiz, A. (2004) Surface study of graphite-supported Ru–Co and Ru–Ni bimetallic catalysts. *Appl. Catal. A*, **275**, 257–269.
270 Feng, J., Xu, B., Jiang, W.D., Xiong, W., and Wang, J.B. (2012) Hydrogenolysis of glycerol on supported Ru-Co bimetallic catalysts, in *Advanced Materials Research*, Trans Tech Publications, pp. 297–300.
271 Álvarez-Rodríguez, J., Guerrero-Ruiz, A., Rodríguez-Ramos, I., and Arcoya, A. (2008) Changes in the selective hydrogenation of citral induced by copper addition to Ru/KL catalysts. *Microporous Mesoporous Mater.*, **110**, 186–196.
272 Soares, A.V., Salazar, J.B., Falcone, D.D., Vasconcellos, F.A., Davis, R.J., and Passos, F.B. (2016) A study of glycerol hydrogenolysis over Ru–Cu/Al$_2$O$_3$ and Ru–Cu/ZrO$_2$ catalysts. *J. Mol. Catal. A: Chem.*, **415**, 27–36.
273 Salazar, J.B., Falcone, D.D., Pham, H.N., Datye, A.K., Passos, F.B., and Davis, R.J. (2014) Selective production of 1,2-propanediol by hydrogenolysis of glycerol over bimetallic Ru–Cu nanoparticles supported on TiO$_2$. *Appl. Catal., A*, **482**, 137–144.
274 Liu, H., Liang, S., Jiang, T., Han, B., and Zhou, Y. (2012) Hydrogenolysis of glycerol to 1, 2-propanediol over Ru–Cu bimetals supported on different supports. *CLEAN–Soil, Air, Water*, **40**, 318–324.
275 Asedegbega-Nieto, E., Bachiller-Baeza, B., Guerrero-Ruiz, A., and Rodríguez-Ramos, I. (2006) Modification of catalytic properties over carbon supported Ru–Cu and Ni–Cu bimetallics: I. Functional selectivities in citral and cinnamaldehyde hydrogenation. *Appl. Catal., A*, **300**, 120–129.
276 Beamson, G., Papworth, A.J., Philipps, C., Smith, A.M., and Whyman, R. (2010) Selective hydrogenation of amides using ruthenium/molybdenum catalysts. *Adv. Synth. Catal.*, **352**, 869–883.
277 Villa, A., Chan-Thaw, C.E., Campisi, S., Bianchi, C.L., Wang, D., Kotula, P.G., Kübel, C., and Prati, L. (2015) AuRu/AC as an effective catalyst for hydrogenation reactions. *Phys. Chem. Chem. Phys.*, **17**, 28171–28176.
278 Eliche-Quesada, D., Mérida-Robles, J., Rodríguez-Castellón, E., and Jiménez-López, A. (2005) Ru, Os and Ru–Os supported on mesoporous silica doped with zirconium as mild thio-tolerant catalysts in the hydrogenation and hydrogenolysis/hydrocracking of tetralin. *Appl. Catal. A*, **279**, 209–221.
279 Deng, C., Leng, L., Duan, X., Zhou, J., Zhou, X., and Yuan, W. (2015) Support effect on the bimetallic structure of Ir–Re catalysts and their performances in glycerol hydrogenolysis. *J. Mol. Catal. A Chem.*, **410**, 81–88.
280 Amada, Y., Shinmi, Y., Koso, S., Kubota, T., Nakagawa, Y., and Tomishige, K. (2011) Reaction mechanism of the glycerol hydrogenolysis to 1, 3-propanediol over Ir–ReO x/SiO$_2$ catalyst. *Appl. Catal. B*, **105**, 117–127.
281 Chimentao, R., Valença, G., Medina, F., and Pérez-Ramírez, J. (2007) Hydrogenolysis of methylcyclopentane over the bimetallic Ir–Au/γ-Al$_2$O$_3$

catalysts. *Appl. Surf. Sci.*, **253**, 5888–5893.

282 Chia, M., O'Neill, B.J., Alamillo, R., Dietrich, P.J., Ribeiro, F.H., Miller, J.T., and Dumesic, J.A. (2013) Bimetallic RhRe/C catalysts for the production of biomass-derived chemicals. *J. Catal.*, **308**, 226–236.

283 Hakim, S.H., Sener, C., Alba-Rubio, A.C., Gostanian, T.M., O'Neill, B.J., Ribeiro, F.H., Miller, J.T., and Dumesic, J.A. (2015) Synthesis of supported bimetallic nanoparticles with controlled size and composition distributions for active site elucidation. *J. Catal.*, **328**, 75–90.

284 Zhang, J., Asakura, H., van Rijn, J., Yang, J., Duchesne, P., Zhang, B., Chen, X., Zhang, P., Saeys, M., and Yan, N. (2014) Highly efficient, NiAu-catalyzed hydrogenolysis of lignin into phenolic chemicals. *Green Chem.*, **16**, 2432–2437.

285 Yan, Z., Yao, Y., and Goodman, D.W. (2012) Dehydrogenation of propane to propylene over supported model Ni–Au catalysts. *Catal. Lett.*, **142**, 714–717.

286 Yan, Z. and Goodman, D.W. (2012) Silica-Supported Au–Ni Catalysts for the Dehydrogenation of Propane. *Catal. Lett.*, **142**, 517–520.

287 Jiang, T., Kong, D., Xu, K., and Cao, F. (2015) Effect of ZnO incorporation on Cu–Ni/Al_2O_3 catalyst for glycerol hydrogenolysis in the absence of added hydrogen. *Appl. Petrochem. Res.*, **5**, 221–229.

288 Yun, Y.S., Park, D.S., and Yi, J. (2014) Effect of nickel on catalytic behaviour of bimetallic Cu–Ni catalyst supported on mesoporous alumina for the hydrogenolysis of glycerol to 1, 2-propanediol. *Catal. Sci. Technol.*, **4**, 3191–3202.

289 Yao, Y. and Goodman, D.W. (2015) New insights into structure–activity relationships for propane hydrogenolysis over Ni–Cu bimetallic catalysts. *RSC Adv.*, **5**, 43547–43551.

290 Patil, S.P., Pande, J.V., and Biniwale, R.B. (2013) Non-noble Ni–Cu/ACC bimetallic catalyst for dehydrogenation of liquid organic hydrides for hydrogen storage. *Int. J. Hydrogen. Energy*, **38**, 15233–15241.

291 Obregón, I., Corro, E., Izquierdo, U., Requies, J., and Arias, P.L. (2014) Levulinic acid hydrogenolysis on Al_2O_3-based Ni-Cu bimetallic catalysts. *Chin. J. Catal.*, **35**, 656–662.

292 Yu, L., He, L., Chen, J., Zheng, J., Ye, L., Lin, H., and Yuan, Y. (2015) Robust and recyclable nonprecious bimetallic nanoparticles on carbon nanotubes for the hydrogenation and hydrogenolysis of 5-hydroxymethylfurfural. *ChemCatChem*, **7**, 1701–1707.

293 Fabičovicová, K., Malter, O., Lucas, M., and Claus, P. (2014) Hydrogenolysis of cellulose to valuable chemicals over activated carbon supported mono-and bimetallic nickel/tungsten catalysts. *Green Chem.*, **16**, 3580–3588.

294 Lögdberg, S., Tristantini, D., Borg, Ø., Ilver, L., Gevert, B., Järås, S., Blekkan, E.A., and Holmen, A. (2009) Hydrocarbon production via Fischer–Tropsch synthesis from H 2-poor syngas over different Fe-Co/γ-Al_2O_3 bimetallic catalysts. *Appl. Catal. B*, **89**, 167–182.

295 Liu, Q.y., Qiu, S.B., Wang, T.J., and Ma, L.L. (2013) Urchin-like CoCu bimetallic nanocomposites for catalytic hydrogenolysis of glycerol to propanediols. *Chin. J. Chem. Phys.*, **26**, 347–354.

296 Samson, K., Żelazny, A., Grabowski, R., Ruggiero-Mikołajczyk, M., Śliwa, M., Pamin, K., Kornas, A., and Lachowska, M. (2015) Influence of the carrier and composition of active phase on physicochemical and catalytic properties of CuAg/oxide catalysts for selective hydrogenolysis of glycerol. *Res. Chem. Intermediat.*, **41**, 9295–9306.

297 Zhou, J., Guo, L., Guo, X., Mao, J., and Zhang, S. (2010) Selective hydrogenolysis of glycerol to propanediols on supported Cu-containing bimetallic catalysts. *Green Chem.*, **12**, 1835–1843.

298 Lu, X.M., Rycenga, M., Skrabalak, S.E., Wiley, B., and Xia, Y.N. (2009) Chemical synthesis of novel plasmonic nanoparticles. *Annu. Rev. Phys. Chem.*, **60**, 167–192.

299 Chaudhuri, R.G. and Paria, S. (2012) Core/shell nanoparticles: classes, properties, synthesis mechanisms,

characterization, and applications. *Chem. Rev.*, **112**, 2373–2433.

300 Porter, N.S., Wu, H., Quan, Z.W., and Fang, J.Y. (2013) Shape-control and electrocatalytic activity-enhancement of pt-based bimetallic nanocrystals. *Acc. Chem. Res.*, **46**, 1867–1877.

301 Quan, Z., Wang, Y., and Fang, J. (2012) High-index faceted noble metal nanocrystals. *Acc. Chem. Res.*, **46**, 191–202.

302 Jin, M., Zhang, H., Xie, Z., and Xia, Y. (2011) Palladium concave nanocubes with high-index facets and their enhanced catalytic properties. *Angew. Chem., Int. Ed.*, **50**, 7850–7854.

303 Peng, Z., You, H., and Yang, H. (2010) Composition-dependent formation of platinum silver nanowires. *ACS Nano*, **4**, 1501–1510.

304 Zhang, H., Jin, M., Liu, H., Wang, J., Kim, M.J., Yang, D., Xie, Z., Liu, J., and Xia, Y. (2011) Facile synthesis of Pd–Pt alloy nanocages and their enhanced performance for preferential oxidation of CO in excess hydrogen. *ACS Nano*, **5**, 8212–8222.

305 Jin, X., Zhao, M., Yan, W., Zeng, C., Bobba, P., Thapa, P.S., Subramaniam, B., and Chaudhari, R.V. (2016) Anisotropic growth of PtFe nanoclusters induced by lattice-mismatch: efficient catalysts for oxidation of biopolyols to carboxylic acid derivatives. *J. Catal.*, **337**, 272–283.

306 Stein, M. and Breit, B. (2013) Catalytic hydrogenation of amides to amines under mild conditions. *Angew. Chem., Int. Ed.*, **52**, 2231–2234.

307 Auneau, F., Arani, L.S., Besson, M., Djakovitch, L., Michel, C., Delbecq, F., Sautet, P., and Pinel, C. (2012) Heterogeneous transformation of glycerol to lactic acid. *Top. Catal.*, **55**, 474–479.

308 Natesakhawat, S., Oktar, O., and Ozkan, U.S. (2005) Effect of lanthanide promotion on catalytic performance of sol–gel Ni/Al_2O_3 catalysts in steam reforming of propane. *J. Mol. Catal. A Chem.*, **241**, 133–146.

309 Khatamian, M., Khandar, A., Haghighi, M., Ghadiri, M., and Darbandi, M. (2010) Synthesis, characterization and acidic properties of nanopowder ZSM-5 type ferrisilicates in the Na+/K+ alkali system. *Powder Technol.*, **203**, 503–509.

310 Robles-Dutenhefner, P.A., da Silva, M.J., Sales, L.S., Sousa, E.M., and Gusevskaya, E.V. (2004) Solvent-free liquid-phase autoxidation of monoterpenes catalyzed by sol–gel Co/SiO_2. *J. Mol. Catal. A Chem.*, **217**, 139–144.

311 Bergamaschi, V., Carvalho, F., Santos, W., and Rodrigues, C. (2006) Synthesis and characterization of Ni-Cu/ZrO_2 and Co-Cu/ZrO_2 catalysts used for ethanol steam reforming, in *Materials Science Forum*, Trans. Tech. Publ., pp. 619–624.

312 Pabón, E., Retuert, J., Quijada, R., and Zárate, A. (2004) TiO_2–SiO_2 mixed oxides prepared by a combined sol–gel and polymer inclusion method. *Micropor. Mesopor. Mat.*, **67**, 195–203.

313 Francisco, M.S.P. and Gushikem, Y. (2002) Synthesis and characterization of SiO_2–Nb_2O_5 systems prepared by the sol–gel method: structural stability studies. *J. Mater. Chem.*, **12**, 2552–2558.

314 Liotta, L., Longo, A., Macaluso, A., Martorana, A., Pantaleo, G., Venezia, A., and Deganello, G. (2004) Influence of the SMSI effect on the catalytic activity of a Pt (1%)/Ce 0.6 $Zr_{0.4}O_2$ catalyst: SAXS, XRD, XPS and TPR investigations. *Appl. Catal., B*, **48**, 133–149.

315 Wey, M.-Y., Tseng, H.-H., Liang, Y.-S., Chang, Y.-C., and Lu, C.-Y. (2006) Effects of metal precursor in the sol–gel synthesis on the physicochemical properties of Pd/Al_2O_3–CeO_2 catalyst: CO oxidation. *J. Non Cryst. Solids*, **352**, 2166–2172.

316 Sajjadi, S.M., Haghighi, M., and Rahmani, F. (2014) Dry reforming of greenhouse gases CH_4/CO_2 over MgO-promoted Ni–Co/Al_2O_3–ZrO_2 nanocatalyst: effect of MgO addition via sol–gel method on catalytic properties and hydrogen yield. *J. Sol–Gel Sci. Techn.*, **70**, 111–124.

317 Araujo, J., Zanchet, D., Rinaldi, R., Schuchardt, U., Hori, C., Fierro, J., and Bueno, J. (2008) The effects of La_2O_3 on the structural properties of La_2O_3–Al_2O_3 prepared by the sol–gel method and on the catalytic performance of Pt/La_2O_3–Al_2O_3 towards steam reforming and

partial oxidation of methane. *Appl. Catal., B*, **84**, 552–562.
318 Ramos-Delgado, N., Gracia-Pinilla, M., Maya-Trevino, L., Hinojosa-Reyes, L., Guzman-Mar, J., and Hernández-Ramírez, A. (2013) Solar photocatalytic activity of TiO_2 modified with WO_3 on the degradation of an organophosphorus pesticide. *J. Hazard. Mater.*, **263**, 36–44.
319 Jiménez, R., García, X., López, T., and Gordon, A. (2008) Catalytic combustion of soot. Effects of added alkali metals on CaO–MgO physical mixtures. *Fuel Process Technol.*, **89**, 1160–1168.
320 Francisco, M., Mastelaro, V., Florentino, A., and Bazin, D. (2002) Structural study of copper oxide supported on a ceria-modified titania catalyst system. *Top. Catal.*, **18**, 105–111.
321 Morey, M.S., O'Brien, S., Schwarz, S., and Stucky, G.D. (2000) Hydrothermal and postsynthesis surface modification of cubic, MCM-48, and ultralarge pore SBA-15 mesoporous silica with titanium. *Chem. Mater.*, **12**, 898–911.
322 Kumar, N., Lazuen, A., Kubicka, D., Heikkilä, T., Lehto, V.-P., Karhu, H., Salmi, T., and Murzin, D.Y. (2006) Synthesis of Pt-modified MCM-41 mesoporous molecular sieve catalysts: influence of methods of Pt introduction in MCM-41 on physico-chemical and catalytic properties for ring opening of decalin. *Stud. Surf. Sci. Catal.*, **162**, 401–408.
323 Soni, K., Rana, B., Sinha, A., Bhaumik, A., Nandi, M., Kumar, M., and Dhar, G. (2009) 3-D ordered mesoporous KIT-6 support for effective hydrodesulfurization catalysts. *Appl. Catal., B*, **90**, 55–63.
324 Simons, C., Hanefeld, U., Arends, I.W., Sheldon, R.A., and Maschmeyer, T. (2004) Noncovalent anchoring of asymmetric hydrogenation catalysts on a new mesoporous aluminosilicate: application and solvent effects. *Chem. Eur. J.*, **10**, 5829–5835.
325 Ledoux, M.J. and Pham-Huu, C. (2001) Silicon carbide: a novel catalyst support for heterogeneous catalysis. *Cattech*, **5**, 226–246.
326 Planeix, J., Coustel, N., Coq, B., Brotons, V., Kumbhar, P., Dutartre, R., Geneste, P.,
Bernier, P., and Ajayan, P. (1994) Application of carbon nanotubes as supports in heterogeneous catalysis. *J. Am. Chem. Soc.*, **116**, 7935–7936.
327 Mahata, N., Gonçalves, F., Pereira, M.F.R., and Figueiredo, J.L. (2008) Selective hydrogenation of cinnamaldehyde to cinnamyl alcohol over mesoporous carbon supported Fe and Zn promoted Pt catalyst. *Appl. Catal., A*, **339**, 159–168.
328 Park, C., Rodriguez, N., and Baker, R. (1996) Use of graphite nanofibers as a novel catalyst support medium for hydrogenation reactions, in *MRS Proceedings*, Cambridge Univ Press, pp. 21.
329 Somorjai, G.A. and Li, Y. (2010) *Introduction to Surface Chemistry and Catalysis*, John Wiley & Sons.
330 O'Connor, J., Sexton, B., and Smart, R.S. (2013) *Surface Analysis Methods in Materials Science*, Springer Science & Business Media.
331 Kane, P.F. and Larrabee, G.B. (2013) *Characterization of Solid Surfaces*, Springer Science & Business Media.
332 Inoue, F., Philipsen, H., Radisic, A., Armini, S., Civale, Y., Shingubara, S., and Leunissen, P. (2012) Electroless copper bath stability monitoring with UV-vis spectroscopy, pH, and mixed potential measurements. *J. Electrochem. Soc.*, **159**, D437–D441.
333 Zaera, F. (2014) New advances in the use of infrared absorption spectroscopy for the characterization of heterogeneous catalytic reactions. *Chem. Soc. Rev.*, **43**, 7624–7663.
334 Cheng, H., Lin, W., Li, X., Zhang, C., and Zhao, F. (2014) Selective hydrogenation of m-dinitrobenzene to m-nitroaniline over $Ru-SnO_x/Al_2O_3$ catalyst. *Catalysts*, **4**, 276–288.
335 Sun, J., Fu, Y., He, G., Sun, X., and Wang, X. (2014) Catalytic hydrogenation of nitrophenols and nitrotoluenes over a palladium/graphene nanocomposite. *Catal. Sci. Technol.*, **4**, 1742–1748.
336 Leofanti, G., Padovan, M., Tozzola, G., and Venturelli, B. (1998) Surface area and pore texture of catalysts. *Catal. Today*, **41**, 207–219.

337 Yan, W., Ramanathan, A., Patel, P.D., Maiti, S.K., Laird, B.B., Thompson, W.H., and Subramaniam, B. (2016) Mechanistic insights for enhancing activity and stability of Nb-incorporated silicates for selective ethylene epoxidation. *J. Catal.*, **336**, 75–84.

338 Yan, W., Ramanathan, A., Ghanta, M., and Subramaniam, B. (2014) Towards highly selective ethylene epoxidation catalysts using hydrogen peroxide and tungsten- or niobium-incorporated mesoporous silicate (KIT-6). *Catal. Sci. Technol.*, **4**, 4433–4439.

339 Imelik, B. and Vedrine, J.C. (2013) *Catalyst Characterization: Physical Techniques for Solid Materials*, Springer Science & Business Media.

340 Upare, P.P., Lee, M., Lee, S.-K., Yoon, J.W., Bae, J., Hwang, D.W., Lee, U.-H., Chang, J.-S., and Hwang, Y.K. (2015) Ru nanoparticles supported graphene oxide catalyst for hydrogenation of bio-based levulinic acid to cyclic ethers. *Catal. Today*, **265**, 174–183.

341 Huang, J., Zhu, Y., Lin, M., Wang, Q., Zhao, L., Yang, Y., Yao, K.X., and Han, Y. (2013) Site-specific growth of Au—Pd alloy horns on Au nanorods: a platform for highly sensitive monitoring of catalytic reactions by surface enhancement Raman spectroscopy. *J. Am. Chem. Soc.*, **135**, 8552–8561.

342 Torres, C., Campos, C., Fierro, J.L.G., Oportus, M., and Reyes, P. (2013) Nitrobenzene hydrogenation on Au/TiO$_2$ and Au/SiO$_2$ catalyst: synthesis, characterization and catalytic activity. *Catal. Lett.*, **143**, 763–771.

343 Jin, X., Zhao, M., Shen, J., Yan, W., He, L., Thapa, P.S., Ren, S., Subramaniam, B., and Chaudhari, R.V. (2015) Exceptional performance of bimetallic Pt$_1$Cu$_3$/TiO$_2$ nanocatalysts for oxidation of gluconic acid and glucose with O$_2$ to glucaric acid. *J. Catal.*, **330**, 323–329.

344 Yao, Y. and Goodman, D.W. (2014) Direct evidence of hydrogen spillover from Ni to Cu on Ni—Cu bimetallic catalysts. *J. Mol. Catal. A Chem.*, **383**, 239–242.

345 Du, B., Su, H., and Wang, S. (2015) Palladium supported on carbon nanofiber coated monoliths for three-phase nitrobenzene hydrogenation: influence of reduction temperature and oxidation pre-treatment. *J. Ind. Eng. Chem.*, **21**, 997–1004.

346 Sudhakar, M., Kumar, V.V., Naresh, G., Kantam, M.L., Bhargava, S., and Venugopal, A. (2016) Vapor phase hydrogenation of aqueous levulinic acid over hydroxyapatite supported metal (M = Pd, Pt, Ru, Cu, Ni) catalysts. *Appl. Catal. B*, **180**, 113–120.

347 Vilé, G., Dähler, P., Vecchietti, J., Baltanás, M., Collins, S., Calatayud, M., Bonivardi, A., and Pérez-Ramírez, J. (2015) Promoted ceria catalysts for alkyne semi-hydrogenation. *J. Catal.*, **324**, 69–78.

348 Serrano-Lotina, A. and Daza, L. (2014) Long-term stability test of Ni-based catalyst in carbon dioxide reforming of methane. *Appl. Catal. A*, **474**, 107–113.

349 Hong, C.L., Jin, X., Totleben, J., Lohrman, J., Harak, E., Subramaniam, B., Chaudhari, R.V., and Ren, S.Q. (2014) Graphene oxide stabilized Cu$_2$O for shape selective nanocatalysis. *J. Mater. Chem. A*, **2**, 7147–7151.

350 O'Neill, B.J., Jackson, D.H., Crisci, A.J., Farberow, C.A., Shi, F., Alba-Rubio, A.C., Lu, J., Dietrich, P.J., Gu, X., and Marshall, C.L. (2013) Stabilization of copper catalysts for liquid-phase reactions by atomic layer deposition. *Angew. Chem., Int. Ed.*, **52**, 13808–13812.

351 Nakagawa, Y., Tamura, M., and Tomishige, K. (2014) Catalytic materials for the hydrogenolysis of glycerol to 1,3-propanediol. *J. Mater. Chem. A*, **2**, 6688–6702.

15
Catalysis/Selective Oxidation by Metal Oxides

Wataru Ueda

Kanagawa University, Department of Material and Life Chemistry, 3-27-1, Rokkakubashi, Kanagawa-ku, 221-868 Yokohama, Japan

15.1
Introduction

Catalysts have been utilized in many industrial chemical processes [1] and will undoubtedly continue to grow in importance. The reason is that catalysis are necessary for environmental systems, new energy generation, and new chemicals production based on biomass and natural gasses. In addition to these new chemical processes, the application of catalytic methodology will expand to other fields like medical chemistry, nuclear power generation, cosmology, and marine science. In other words, catalysis is regarded as a key technology for developing sustainable society. This trend must be the right direction because in principle catalysis is a phenomenon with extremely low or practically no impact on nature.

To respond to the above demands promptly and correctly, however, there are still a lot of problems in the development of catalysis chemistry and technology. Although there are many examples of catalytic materials, catalytic reactions, and catalytic systems that have been developed under the tremendously elaborative research, the technological knowledge accumulated during the development of these already achieved catalysts is not straightforwardly applicable to developing new catalysts. This is simply due to the fact that new catalytic reactions need new catalytic functions based on new catalytic materials. In addition, new catalytic systems are usually more complex than ever achieved. Since catalysis has to be understood in terms of collaborative action of elements in atomic scale, catalyst preparation to realize this situation has been categorized in materials synthesis chemistry in nanometer scale. More recently the term of catalytic reaction field where many catalytic elements are involved at the same time has also become important. This causes catalyst complexity and high dimensionality which, however, become more indispensable factors for expected new catalysis application. The understanding of this term can then help developing new

Handbook of Solid State Chemistry, First Edition. Edited by Richard Dronskowski, Shinichi Kikkawa, and Andreas Stein.
© 2017 Wiley-VCH Verlag GmbH & Co. KGaA. Published 2017 by Wiley-VCH Verlag GmbH & Co. KGaA.

advanced synthetic methodologies of catalytic materials, although there are always many unpredictable and uncontrollable factors even in such advanced synthetic methodology.

By focusing on selective oxidation catalysts based on metal oxides, this chapter categorically summarizes the catalyst development of the typical metal oxide catalysts for selective oxidation in terms of catalyst complexity and high-dimensionality in order to deduce what the next oxide catalysts and selective oxidation catalysts will be.

15.2
Development of Selective Oxidation Catalysts Based on Complex Metal Oxides

Oxidation of organic compounds produces energy as well as chemicals, so that the oxidation reaction impacts the environment strongly and directly. The biosphere on the earth utilizes oxygen in air to generate energy for lives. On the other hand, photosynthesis, which is regarded as the reverse reaction of the oxidation of organic compounds, takes place at the same time in the biosphere to balance the two reactions: energy consumption and accumulation. Although artificial photosynthesis catalysts based on complex metal oxides are under development, there are at present practically no reactions compensating artificial oxidations, like combustion of organic compounds based on fossil chemical materials. Therefore, catalytic oxidation processes should be developed more and applied to many more cases for achieving highly energy efficient chemicals utilization processes. In this context, fuel cells are a good example. In the fuel cells catalysts are used while in an internal combustion engines no catalysts are used.

Concerning the selective oxidation catalysts and processes for chemical industries, the development was remarkable in the past. Representative developed industrial selective oxidation processes using metal oxide catalysts are listed in Table 15.1 along with catalytic oxidation processes under development. It should be mentioned that most of the important selective oxidation processes were developed and industrialized during the 1960s and 1980s [2,3]. On the other hand, the number of the industrialized processes developed in recent years is not remarkable although as listed in the table there are many new oxidation reactions that have been planed and researched but not yet fully developed. The reactions are alkane selective oxidations, which are long desired selective oxidations but have not been achieved practically yet, since they are mostly classified as a highly difficult reactions which require catalysts with multifunctional properties, so-called complexity.

The complexity found in the selective oxidation catalysts results from the principle of catalytic oxidation itself, that is, complex activation steps of molecular oxygen and variedness of the reaction of activated organic molecules and activated oxygen. For accelerating oxidation reactions to a desired direction, catalysts have to play a major role and control every elemental step precisely. This

Table 15.1 Catalytic processes for the selective oxidation of light alkenes and alkanes.

Reactant	Product	Catalyst	Development stage
Methane	Methanol	Zeolite, metal oxide	Research
Methane	Formaldehyde	Supported metal oxide	Research
Methane	Ethene	Oxide with solid-base	Pilot plant
Ethene	Acetic acid	HPC	Industrialized
Ethane	Acetic acid	Mo–V–M–O	Industrialized
Ethane	Ethene	Mo–V–M–O	Research
Propene	Acrylic acid	Multicomponent oxide, Mo–V–O	Industrialized
Propene	Acrylonitrile	Multicomponent oxide	Industrialized
Propane	Acrylic acid	Mo–V–M–O	Pilot plant
Propane	Acrylonitrile	Mo–V–M–O	Industrialized
Propane	Propene	Metal oxide	Research
1-butene	Maleic anhydride	V–P–O	Industrialized
1-butene	Butadiene	Multi-component oxide	Pilot plant
n-butane	Maleic anhydride	V–P–O	Industrialized
n-butane	Butadiene	Metal oxide	Research
i-butene	Methacrylic acid	Multicomponent oxide, HPC	Industrialized
i-butane	Methacrylic acid	HPC	Research
i-butane	i-butene	Metal oxide	Research

HPC: heteropoly compounds.

implies that catalysts with simple function like in the Mars-Van Krevelen mechanism would not be able to complete the reaction satisfactorily but catalysts with multifunctions that can collaboratively work at the same time would be suitable for selective oxidation. The fact that most of the successful selective oxidation catalysts are composed of a number of catalytic elements in the form of an oxide confirms the need for multifunctional oxide catalysts. In the line of the progress in selective oxidation catalysts along with the concept of multifunctionality, many new examples of materials with particular crystal structure composed of multielements have recently evolved for selective oxidation.

15.3
Distinct Types of Complex Metal Oxide Catalysts [2]

Representative complex metal oxide catalysts for selective oxidations are classified in Table 15.2 on the basis of catalyst structure, size of active sites, and its complexity. Table 15.2 also shows relevant catalytic reactions and the evolution of complexity of the metal oxide catalyst. Most of the listed oxide catalysts are based on single crystal structures that create catalytic activity for selective oxidations except multicomponent Bi–Mo catalysts [4] in the last line in the table. The multicomponent Bi–Mo catalysts assume a multiphase structure and work very efficiently as oxidation catalysts for olefin oxidation compared to single

Table 15.2 Five distinct complex metal oxide catalysts for selective oxidation and their catalytic characteristics [2].

Catalyst	Structure view	Catalytic reaction	Evolution of catalyst material	Catalytic characteristic
Ti-silicate		Oxidation with H_2O_2 Ammoximation	From amorphous TiO_2–SiO_2 $\rightarrow$ to metal-substituted crystal zeolite	1. Single site in zeolite framework 2. Microporosity
$(VO)_2P_2O_7$		Butane oxidation to maleic anhydride	From amorphous V–P–O $\rightarrow$ to crystalline material	1. Twin octahedra site isolated with phosphates 2. Redox property at isolated site 3. Crystal plane dependency
$Cs_xH_yPMo_{12}O_{40}$		Methacrolein oxidation Oxidation with H_2O_2	Synthetic crystalline material $\rightarrow$ synthetic crystalline material with pore	1. Discrete polyoxometalate site 2. Redox property in discrete site 3. Brønsted acidity 4. Microporosity
$Mo_3VO_{11.2}$		Acrolein oxidation Ethane oxidation Propane ammoxidation Propane oxidation	From amorphous Mo–V–O $\rightarrow$ to crystalline material and to crystalline material with multi-element	1. Isolate octahedra unit 2. Redox in discrete zone 3. Crystal plane dependency 4. Multi-element action in catalysis 5. Brønsted acidity 6. Microporosity
Multicomponent Bi–Mo–O		Olefin oxidation Olefin ammoxidation	From simple metal oxide $\rightarrow$ to mixed oxide $\rightarrow$ then multiphase metal oxide $\rightarrow$ and now surface zone oxide	1. Particle zone 2. Redox in zone 3. Multielement and multiphase action in catalysis

phasic Bi–Mo–O oxide catalyst. $Mo_3VO_{11.2}$ that has been recently synthesized by a hydrothermal process is a new member of the selective oxidation catalysts for alkane oxidation and acrolein oxidation in the gas phase [5]. Divanadyl pyrophosphate catalyst [6], which is single phasic, is the sole catalytic material representing high performance of n-butane oxidation to maleic anhydride. Heteropoly compounds [7] are widely applied as catalysts not only for gas phase oxidation but also for liquid phase oxidation using peroxides. Metal-substituted zeolite catalysts [8], typically titanosilicates, may not be simply categorized into complex metal oxides but shows unique catalytic oxidation ability in the liquid phase using peroxides as oxidants similar to homogeneous catalysts.

15.3.1
Titanosilicates

Titanosilicates, in which micropores are present in the structure and Ti species are isolated in the framework lattice, are well-known oxidation catalyst with hydrogen peroxide; they are used, for example, for propene oxidation to propene oxide [9] and ammoximation of cyclohexanone to cyclohexanone oxime [10]. Careful synthesis for avoiding the hydrolysis of titanium source can give titanosilicates in which titanium is perfectly isolated in the Si—O zeolite framework. Due to the porous Si—O framework, every titanium atom in the framework is accessible for the reaction with hydrogen peroxide to form peroxo species (Scheme 15.1) and enable the catalytic reaction [8]. Moreover, titanium isolation in the framework is highly profitable for stabilization against leaching during the hydrogen peroxide oxidation in the liquid phase. Before the discovery of the titanosilicates, amorphous type of TiO_2/SiO_2 catalysts was known for the epoxidation of propene to propene oxide. Therefore, it is clear that strategy of the introduction of crystal structure in metal oxide catalysts is highly useful in order to create the perfect element isolation and introduction of a secondary structure like a porous structure is also effective in terms of catalytic activity. The titanosilicates are prominent examples that illustrate how crystal structure generates multifunctions.

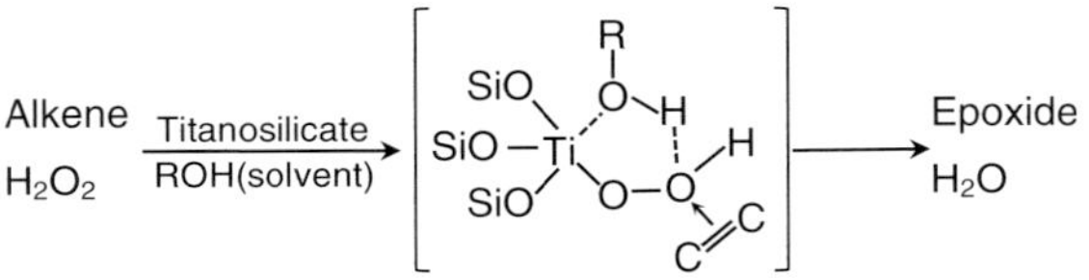

Alkene + H_2O_2 →[Titanosilicate, ROH(solvent)] [SiO–Ti intermediate] → Epoxide + H_2O

Scheme 15.1 Alkene epoxidation with hydrogen peroxide at single Ti site isolated in titanosilicate catalyst.

15.3.2
Divanadyl Pyrophosphate

The second example is a crystalline divanadyl pyrophosphate [6] catalyst which is active for the selective oxidation of *n*-butane to maleic anhydride.

The characteristic point in this catalyst is that twin octahedral vanadyl is structurally isolated by the six surrounding tetrahedral phosphate anions (structure, see Table 15.2). This is so called site-isolation [11]. The importance of crystal formation should be emphasized because only crystallized divanadyl pyrophosphate particles showed activity and selectivity for the reaction. Before the discovery of the crystallized divanadyl pyrophosphate catalyst, an amorphous type of V–P–O was known as the catalyst for 1-butene oxidation to malic anhydride. However, the amorphous type of V–P–O was not a good catalyst for the *n*-butane oxidation to maleic anhydride, suggesting that a V–P–O system needs a particular crystal structure for forming oxygen species more active for *n*-butane oxidation than for 1-butene oxidation which has much weaker C—H bond than *n*-butane has. Furthermore, the particular structure has to be uniform throughout the catalyst materials, otherwise various undesirable reactions will occur at the same time, causing lower product selectivity. In fact, similar to the titanosilicate catalyst, the progressive development of V–P–O catalysts by introduction of crystal structure has also been successfully done, leading to a highly active catalyst for the *n*-butane oxidation to maleic anhydride.

It is believed that the twin vanadyl sites surrounded by six tetrahedral phosphate anions can provide active oxygen species through dynamic structure deformation during the catalysis [12] and also transfer electrons to complete the reaction of *n*-butane to maleic anhydride. The twin structure exposed on the basal plane of the plate-type crystal of divanadyl pyrophosphate is also believed to be suitable for adsorption of *n*-butane [13] and for converting to maleic anhydride [14] (Scheme 15.2). In other words, the isolated twin vanadyl site can complete one catalytic cycle and tetrahedral phosphate anions play a role of framework structure formation.

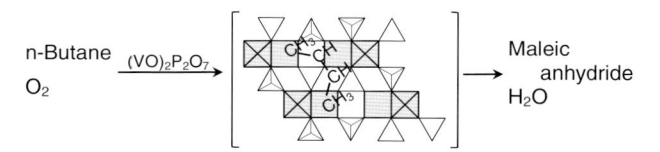

Scheme 15.2 *n*-Butane oxidation to maleic anhydride at twin V site on (020) basal plane of crystalline divanadyl pyrophosphate catalyst.

This oxidation process has been industrialized but the yield of maleic anhydride in the current process is still around 60% which is not as high as yields in alkene oxidation processes. Therefore, much more research for improving the catalytic performance based on crystalline V–P–O material has been done but all trials so far have apparently been not successful. The main reason seems to be that it is hard to introduce different metal elements beside V and P within the same pyrophosphate structure. In other words, crystal structure refuses modification, which has to be overcome by totally different catalyst design concepts.

15.3.3
Heteropoly Compounds

The third is polyoxometalate or heteropoly compounds [7,15] which are also a 3-dimensional crystalline materials constructed with multi (more than two)-nuclear octahedra anion(or cluster) with a discrete structure, typically Keggin anions (structure, see in Table 15.2). Catalytic properties in the compounds derive from the anions with various discrete structures, elemental composition diversity, and 3-dimensional structure of the anion arrangements. In addition, the uniformity and homogeneity of the anion in the 3-dimensional materials can also be an important factor for controlling product selectivity in catalytic oxidation. Therefore, heteropoly compound catalysts have been considered to resemble homogeneous catalysts. In fact, the heteropoly anions having their own catalytic properties form uniform catalytic active surface or 3-dimensional catalysis field (space) with 3-dimensional arrangement of the polyoxoanions. There are thus now high possibilities to create new solid-state oxide catalysts based on heteropoly compounds, if polyoxoanions with a discrete uniform structure can be arranged in high dimensional manner into crystalline structures.

The oxidizing ability of the solid-state heteropoly compounds is strongly affected by the kinds and substituted amount of counter cations as well as composition of the heteropolyanion itself. First, the oxidizing ability of the Keggin-type heteropolyanions is influenced largely by the kinds of addenda atoms. The oxidizing ability increases in general in the order of $W(VI) \ll Mo(VI) < V(V)$, which are addenda atoms [16]. This is one of the reasons why the mixed-addenda heteropolyanions $PMo_{12-n}V_nO_{40}^{(3+n)-}$ act as a good oxidation catalyst. As for heteroatoms, the oxidizing ability decreases linearly with an increase in the negative charge of the heteropolyanion. As for salts with non-reducible cations like alkali metal ions, the oxidizing abilities decrease monotonically with an increase in the substituted amount of counter cations [17]. On the other hand, when the salts form with reducible cations like Pd, Ag, and Cu, the reduction of counter cations precedes than that of heteropolyanion [18]. Thus, the reducibility of the counter cations reflects the oxidizing ability of the heteropoly compounds. Above all, the oxidizing ability can be tuned rationally by replacing structural element without largely changing the whole structure, which is the unique point of the heteropoly compound catalysts. Microporosity can be also introduced in this materials system without using organic linker, to form all-inorganic porous polyoxometalates [19]. Based on these chemistries of heteropoly compounds, one can rationally create catalysts that show high performance in the selective oxidation either with molecular oxygen or with hydrogen peroxide.

$H_{4+x}PMo_{12-x}V_xO_{40}$ and their salts (NH_4^+ and Cs^+ salts) show catalytic activities for oxidation by O_2; for instance, selective oxidations of methacrolein (Scheme 15.3), isobutene, and propane, and oxidative dehydrogenation of isobutyric acid to methacrylic acid, and oxidation of alcohols, and liquid-phase oxidation using peroxide like H_2O_2 and *tert*-butyl peroxide including hydroxylation of

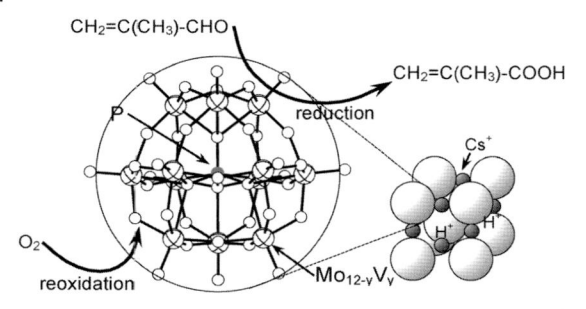

Scheme 15.3 Methacrolein oxidation to methacrylic acid at polyoxomolybdate cluster of crystalline $H_{3-x}Cs_xPMo_{12-y}V_yO40$ ($2 < x < 3$, $0 < y < 2$) catalyst.

aromatic compounds and oxidation of thioesters [7]. Substitution of Mo with V enhances dramatically the catalytic performance as stated above.

15.3.4
Crystalline $Mo_3VO_{11.2}$

The above three types of the crystalline catalysts have clear distinct structures that ultimately affect the catalytic oxidation performance. These materials are, however, rather classical. Now here is a new member of crystalline complex metal oxide catalyst; that is the crystalline $Mo_3VO_{11.2}$ catalysts [5]. This material (structure, see Table 15.2) has all the properties that three catalysts individually have, namely, a framework structure comprised of discrete heteropoly units, isolated octahedral oxide clusters, redox properties, acid properties, and porosity.

Interestingly, a Mo–V–O-based catalyst system in a noncrystalline form, so-called multicomponent Mo–V catalyst, has long been known as an efficient catalyst for gas phase acrolein oxidation to acrylic acid [3]. Later, it was found that Mo–V–O materials related to the acrolein oxidation catalysts also showed high activity for the oxidative dehydrogenation of ethane to ethene at relatively low reaction temperature. Therefore, this material attracted much attention as a catalyst having properties of strong oxidative C—H bond activation. Eventually, a new finding was made by Mitsubishi Chemicals who successfully carried out propane ammoxidation using Mo–V–O catalysts combined with other elements as the third and the fourth components to be active and selective for the reaction [20]. The new finding is based on the crystal formation in the presence of all the elements and it was revealed that only the crystal material showed the activity. Meanwhile, it was found that the same crystalline material was synthesized without the third and the fourth components by using a soft-synthetic method like the hydrothermal method [21]. This fundamental result reveals that only Mo and V elements can be arranged in the particular oxide structure (structure, see Table 15.2) and the resulting solid oxidatively activates the strong C—H bond of alkane molecules [22,23].

Crystalline Mo–V–O is now recognized as the real active phase in the non-crystalline multicomponent Mo–V catalyst for acrolein oxidation to acrylic acid [22] as well as for ethane oxidative dehydrogenation to ethene [23]. Both of the reactions only need catalytic activation of single strong C—H bonds by active oxygen species of the catalyst to form the products. To achieve such a simple reaction effectively and selectivity, surprisingly catalysts needed to have multicatalytic functionality based on a complicated crystal structure. This can be done over the crystalline $Mo_3VO_{11.2}$ catalyst and the details will be described later.

15.3.5
Multicomponent Bi–Mo–O

The multicomponent Mo–Bi oxide catalysts for propene selective oxidation are constructed with Mo as a main component and with nearly 10 kinds of constituent elements, including Bi, Fe, and Co [3,4]. When the multicomponent Mo–Bi oxide catalysts were compared with Mo–Bi–O oxide which is known as a basic active catalyst, the oxidation activity is increased by an order of magnitude, and shows activity even at 100 °C lower reaction temperature. The catalyst also shows high selectivity for acrolein formation. It is necessary for trivalent metal (usually Fe^{3+}) to coexist with divalent metal of more than one kind at least except for Mo, Bi for activity enhancement. Because the surface area is small in Mo–Bi–O, the increase of the surface area in the multicomponent material is one reason for the activity increase but the main reason is that the specific activity has increased. In order to understand this, the structure of the catalyst particles is first described and then the concept of function separation is considered [4].

In reflection of the complicated component, the multicomponent Mo–Bi oxide catalysts consist of various metal oxide and complex metal oxide phases either in crystalline or in amorphous states. However, the catalyst forms a particle with particular structures that are not the simple mixtures of the phases. M(II) and M(III) components, when the Mo–Bi–M(II)–M(III) ($Mo_{12}Bi_1Fe_3Co_8Ox$) system is taken as an example, are exclusively located in the bulk inside of the primary particle. On the other hand, Bi condenses on the surface region because the Bi component has a strong tendency to concentrate on the surface. As a result, a Mo–Bi–O phase forms fine particles in the surface region of the crystal particle of M(II) molybdate and M(III) molybdate solid-solution. The reason for this uneven distribution is that the ionic radius of Bi is bigger than that of the other metal ions (Figure 15.1). Actually, the preparation of the multicomponent Mo–Bi oxide catalysts is often divided into two part intentionally: preparation of M(II)–M(III) molybdate particles first and then impregnation of Mo–Bi–O on the molybdate surface.

Structurally, the catalyst particles are considered to be the simple three-dimensional molybdate crystals that are constructed with Mo oxoanions. The cationic site, on the other hand, is occupied with M(II) and M(III) ions from the particle center to surrounding surface while Bi ion occupies the sites in the surface region only. In addition, different metal ions with different valencies (M(II)

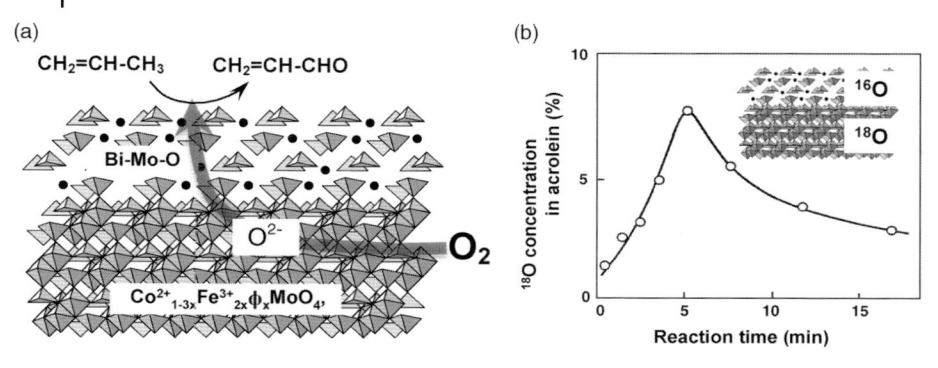

Figure 15.1 Structure model (a) of the multicomponent Bi–Mo–O catalysts and function separation demonstrated by ^{18}O tracer study (b) [4].

and M(III) ion) exist at the same time, so that a defect structure is introduced in the whole particle.

Next the elementary steps of the propene oxidation reaction are summarized and then a relationship between the catalyst structure and function separation and the principle of the synergy can be discussed. The oxidation of propene on the oxide catalysts occurs by the oxidation–reduction mechanism where the lattice oxygen ions of the catalyst particles participate in the reaction via an allyl intermediate. As elementary reaction steps, there are the dissociative activation of oxygen molecules, the abstraction of allyl hydrogen, the shift of an allyl intermediate on metal oxo species, the electron transfer in the catalyst through the lattice defect and a migration of the active oxygen in the bulk of the oxide catalysts. Only when the catalytic cycle of each reaction step proceeds smoothly, the oxidation reaction occurs. Although the catalytic oxidation activity can be obtained over the catalyst comprised of only Bi and Mo (Bi–Mo–O), there is the situation that many roles impose on Bi and Mo. Particularly, excessive roles impose on Bi. Bi directly participates in the allyl hydrogen abstraction which is the rate-determining step, but at the same time multiple Bi sites cooperate to activate oxygen molecules and must lead them to lattice oxygen while maintaining the structure without impeding the mobility of lattice oxygen. Therefore, the mean number of the Bi site that can be involved in the allyl hydrogen abstraction step under reaction conditions could be limited at the dynamic reaction condition. This is the reason for the low catalytic activity of the simple Bi–Mo–O catalyst.

In the case of multicomponent Mo–Bi oxide catalysts, other elements take over the role of Bi during activation of oxygen molecules. In this case, trivalent cations such as Fe activate the oxygen molecules. But as for the reaction, it is necessary for oxygen to be supplied to the Bi which is the catalytic main constituent for the rate-determining step of the propene oxidation. If the activated oxygen remains on Fe, no reaction proceeds. As described above, however, the lattice defect is formed in the crystal particle of M(II) molybdate and M(III) molybdate solid-solution and facilitates electronic transfer and lattice oxygen migration. Therefore, lattice oxygen can migrate from Fe to Bi in the bulk (Figure 15.1). Then Bi can devote itself to hydrogen abstraction from propene, more effectively, resulting in high oxidation activity. Furthermore, the concern that a useless oxidation reaction with oxygen species formed by partial reduction of molecular oxygen can take place causing lower selectivity will be diminished by the rapid migration of lattice oxygen and electron transfer. This situation has the additional benefit that reduction of oxygen with the Fe site occurs at lower temperature more. This situation remarkably increases the product selectivity. It may be said that this is a wonderful separation of function and collaboration supported by the catalyst structure.

The situation described above is totally different from those of the crystalline catalysts. Since lattice oxygen of the multicomponent Bi–Mo oxide catalyst is intrinsically active and involved in the reaction, dynamic behavior like reduction–oxidation not only on the surface region but also in the whole catalysts particles is a major concern in the catalysis. In other words, regional or zonal properties rather than local elemental arrangement determined by crystal structure is determinant. In the selective oxidation of n-butane, on the other hand, a particular local elemental arrangement, like the twin octahedral vanadyl in the case of the crystalline divanadyl pyrophosphate catalyst, is necessary for the formation of strong active oxygen species by particular element arrangement.

15.4
$Mo_3VO_{11.2}$ as an Unique Selective Oxidation Catalyst with High Dimensionally Structures

Effective utilization of alkanes as sources of alkenes and production of their derivatives has long been targeted in chemistry and technological researches and the interest is even now expanding because of the shale gas revolution and methane hydrate. The targets are, however, still challenging. The main problem is that light alkanes are poorly reactive due to the lack of lone pairs of electrons, absence of easily accessible empty orbitals, and low polarity of the C—H bonds, so that desired processes for the utilization will have to rely on technology, particularly catalytic oxidation. Complex metal oxides with high dimensional structures at the nanoscale appear to have high potentiality, since the complex metal oxide materials can attain multifunctional catalytic properties. The most prominent example is a crystalline Mo_3VO_x, which is a new type of complex metal oxide catalysts. This section will first focus on recent developments on complex metal oxide catalysts based on Mo_3VO_x materials for selective alkane oxidation and then more details on pure $Mo_3VO_{11.2}$ materials will be described in terms of crystal structure, structure formation, element incorporation, and alkane catalytic oxidation chemistry, particularly for ethane oxidation.

15.4.1
Oxide Catalysts for Oxidative Dehydrogenation of Ethane to Ethene

There are many types of complex metal oxide catalysts active for alkane oxidation. Here the catalytic oxidative dehydrogenation of ethane to ethene is taken as an example. The oxide catalysts for this reaction can be classified roughly into at least three groups depending on elements and their principal catalytic functions [24]. The first group consists of oxide catalysts constructed mainly with metal oxides possessing a solid-base property, such as alkaline, alkali-earth, and lanthanum oxides. For example, MgO-based mixed oxides [25], rare earth metal oxides [26], and oxyhalides [27] have been reported. These catalysts normally show oxidation activity at relatively high temperatures greater than 500 °C, because surface basic property become active for dissociating saturated C—H bonds of ethane only at such high temperature. Instead, these catalysts tend to show very high selectivity to ethene at high ethane conversion because the catalysts have poor capability of activating molecular oxygen to form active oxygen species which usually cause total oxidation to carbon oxides.

The second group consists of oxide catalysts containing group 8 elements such as Ni that exhibit high reaction rate at relatively low temperature around 400 °C. Ni—Nb—O [28] is a representative example of this group. The catalytic activity of this material is apparently controlled by the nature and state of group 8 elements [29] since binary oxides of group 8 elements are usually very active for nonselective oxidation of ethane to carbon oxides. In other words, metal oxides to be combined with Ni, Nb_2O_5 in the case of Ni—Nb—O catalyst, must play a crucial role of modification of the nature of NiO.

The third consists of catalysts of group 5 and 6 metal oxides, particularly vanadium and molybdenum oxides. The resulting metal oxides tend to have solid acidity. The catalysts in this group are quite varied, consisting mostly of complex metal oxides but in many cases contain V as a key element [24]. Among them $Mo_3VO_{11.2}$ complex oxide, which is a crystalline solid, has been recognized as the most active catalysts for the oxidative dehydrogenation of ethane [23]. Thorsteinson reported in 1978 [30] for the first time that amorphous type Mo–V–O showed high ethane oxidation activity even at temperatures lower than 300 °C. Even though the material was ill-crystallized oxide, the materials having strong a XRD peak at 22° (Cu–Kα) tend to show high activity and Nb is the most effective additive element to give material with this XRD peak. Therefore, a crystal structure in this material, Mo–V–Nb–O, appears to be highly responsible for the catalytic oxidation ability. Because of the great difficulty of introducing crystalline structures to this material system, no appreciable progress has occured. Meanwhile, oxide catalysts in which Te or Sb was added to Mo–V–Nb–O were developed for propane ammoxidation [20] and interestingly the catalyst had many distinct XRD peaks as well as a peak at 22°, indicating a crystalline material was formed in this chemical composition. It is a question whether or not Te, Sb, and Nb are necessary elements for the formation of the crystalline phase and also for the alkane oxidation. Eventually, Nb turned out to be an essential element for creating a basic crystalline phase and Te and Sb assisted the formation of the crystal. This is true as long as the materials are synthesized by a solid-state reaction at high temperature. However, this does not directly mean that these elements are indispensable for forming the crystal structure. In fact, well-crystallized single phasic $Mo_3VO_{11.2}$ materials with the same structure of Mo–V–Te(Sb)–Nb–O were synthesized by using a so-called soft-synthesis like the hydrothermal process without high temperature heat treatment [31]. This catalyst showed extremely high activity for the oxidative dehydrogenation of ethane, which is better than that of the Mo–V–Te–Nb–O catalysts. Therefore, the elements other than Mo and V are unessential for the structure formation and the catalytic oxidation of ethane.

15.4.2
Crystal Structures of $Mo_3VO_{11.2}$ Catalysts

By the hydrothermal reaction two new crystalline solids of $Mo_3VO_{11.2}$ with orthorhombic and trigonal structures can be synthesized [5]. Both the synthesized $Mo_3VO_{11.2}$ materials in the orthorhombic and the trigonal phases assume a layer-type structure with the network arrangement of MO_6 (M=Mo, V) octahedra forming slabs with the pentagonal $\{(Mo)Mo_5O_{21}\}$ building units, hexagonal rings, and heptagonal rings in (100) plane. The orthorhombic phase has lattice parameters ($a = 21.085(10)$ Å, $b = 26.556(13)$ Å, $c = 3.9974(17)$ Å, space group Pba2 (No.16)) and the trigonal phase lattice parameters ($a = 21.433(3)$ Å, $c = 4.0045(18)$ Å, space group P3 (No.143)). Prominent in these materials is that the pentagonal unit is constructed with molybdenum only and the linker between the pentagonal units is vanadium. The rest of the parts are either molybdenum or vanadium. These configurations are fully confirmed by the HAADF-STEM analysis images where the Mo pentagonal unit with bright z-contrast is clearly observed for the both samples (see Figure 15.1) [32]. From these structural analyses, the two $Mo_3VO_{11.2}$ solids are regarded as structure variants with the same structure units, that is, the pentagonal unit, the hexagonal channel, and the heptagonal channel. The difference is in the arrangements of them and the structures are determined by the direction of arrangement of the pentagonal units.

15.4.3
Formation Mechanism of the Crystal Structures [33]

Orthorhombic and the trigonal Mo_3VO_x can be synthesized with the precursor solution which is instantly obtained by mixing solutions of ammonium heptamolybdate and vanadyl sulfate. The purple-colored solution gave characteristic Raman bands at 1000–700 cm^{-1} with a main band at 870 cm^{-1}, which were similar to those of three different polyoxomolybdates, $Mo_{72}V_{30}$ $\{[K_{10}Mo_{72}V_{30}O_{282}(H_2O)_{56}(SO_4)_{12}]^{26-}\}$, Mo_{132} $\{[Mo_{132}O_{372}(H_2O)_{72}(CH_3CO_2)_{30}]^{42-}\}$, and $Mo_{57}V_6$ $\{[H_3Mo_{57}V_6(NO)_6O_{183}(H_2O)_{18}]^{21-}\}$ [34–37]. The common structural feature among three polyoxomolybdates is the pentagonal $\{(Mo)Mo_5O_{21}\}$ unit in the discrete polyoxo structures. Therefore, the pentagonal $\{(Mo)Mo_5O_{21}\}$ unit is already present in the precursor solution before the hydrothermal reaction and most probably the $Mo_{72}V_{30}$ compound is formed in the precursor solution because $Mo_{72}V_{30}$ gave almost the same Raman spectra to the precursor solution. This means that the first step in the material synthesis is the formation of the $Mo_{72}V_{30}$-type materials by the reaction of Mo and V compounds, followed by the transformation of this discrete $Mo_{72}V_{30}$-type compound into the layered solids, keeping the pentagonal units but with different connectivity because of their interaction with VO^{2+} cations as linkers under hydrothermal condition. Moreover, the structure of the layered solids formed after the hydrothermal reaction strongly depends on the pH of the precursor solution although the formed polyoxo species is the same. The controlling factors for the crystal formation are as follows:

1) The desired solid materials were not formed when the $Mo_{72}V_{30}$-type compound was not present in the solution.
2) The structure of the layered solids formed after the hydrothermal reaction strongly depends on the pH (2.2 for trigonal and 3.2 for orthorhombic) of the precursor solution although the formed polyoxo species is the same.
3) The pH range for the formation of the trigonal phase was quite narrow compared to that for the orthorhombic phase.
4) No solid was formed under pH conditions higher than 3.6 even though the precursor solution contains the $Mo_{72}V_{30}$-type compound in higher concentration.
5) Even at such high pH condition the desired solid materials were formed when corresponding structural solids were added to the precursor solution as seeds.
6) When the concentration of the precursor solution is high, an amorphous type of $Mo_3VO_{11.2}$ forms because of rapid nucleation under the reaction condition. The amorphous type materials have a similar layered structure to the other crystalline $Mo_3VO_{11.2}$ but the arrangement of the pentagonal $\{(Mo)Mo_5O_{21}\}$ unit in the slab is random.
7) Heat-treatment of the orthorhombic $Mo_3VO_{11.2}$ gives a tetragonal $Mo_3VO_{11.2}$ whose structure is still constructed with the pentagonal $\{(Mo)Mo_5O_{21}\}$ units but without heptagonal channels.

Based on these pieces of evidence, a crystal formation scheme is speculated as illustrated in Figure 15.2. The formation of the $Mo_{72}V_{30}$-type compound with the pentagonal $\{(Mo)Mo_5O_{21}\}$ unit is indispensable but a dynamic inter-conversion of this compound to less-condensed species comprised with a few pentagonal units with linkers is also important for the formation of the layered solids under the hydrothermal conditions. The formation of less-condensed species, dimers or trimers of pentagonal units, might be dependent on the pH and thus the pH can control the structure of the final solid. More importantly, the direction of pentagonal unit networking decides the structures and the pentagonal network forms open voids where octahedral pentamers or trimers can locate in the orthorhombic and trigonal structures, respectively.

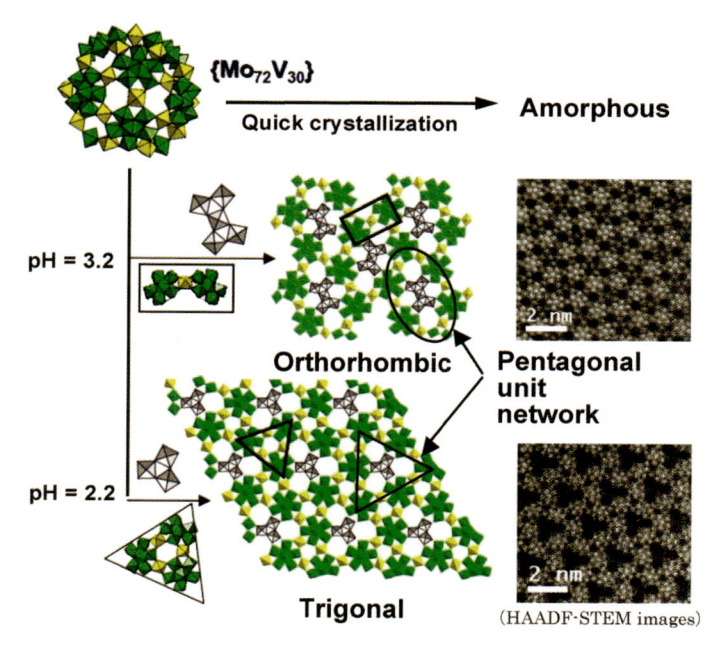

Figure 15.2 Formation of orthorhombic/trigonal $Mo_3VO_{11.2}$ catalysts through $Mo_{72}V_{30}$.

Apparently, the synthetic methodology for catalyst preparation is a key factor for creating the above high-dimensional structure because a simple high-temperature solid-state reaction never gives such meta-stable materials. It has to be emphasized that so-called soft-methods should be applied to this purpose. In the present soft-synthetic process of $Mo_3VO_{11.2}$, first, the pentagonal $\{(Mo)Mo_5O_{21}\}$ unit forms in the precursor solution before the hydrothermal reaction and interacts with VO^{2+} cations to form a polyoxo-type species consisting of $\{(Mo)Mo_5O_{21}\}$ unit in the solution, and then assembly of the polyoxo-type species occurs under hydrothermal conditions to form three-dimensional metal oxide solids. According to this formation mechanism unit assembly method under hydrothermal conditions can be employed in many other cases [38], to create new crystalline solid materials with high-dimensional networks of catalytic elements, and ultimately brings about catalysts with extremely high performance.

15.4.4
Catalysis Field of Crystalline $Mo_3VO_{11.2}$ for Selective Oxidation of Ethane

The pure and well-crystallized Mo_3VO_x catalyst has an outstanding catalytic property for the oxidation of ethane to ethene. Here in this section the genesis of such high oxidation activity of the crystalline $Mo_3VO_{11.2}$ catalyst is introduced. Since the above-mentioned structural features should relate to various

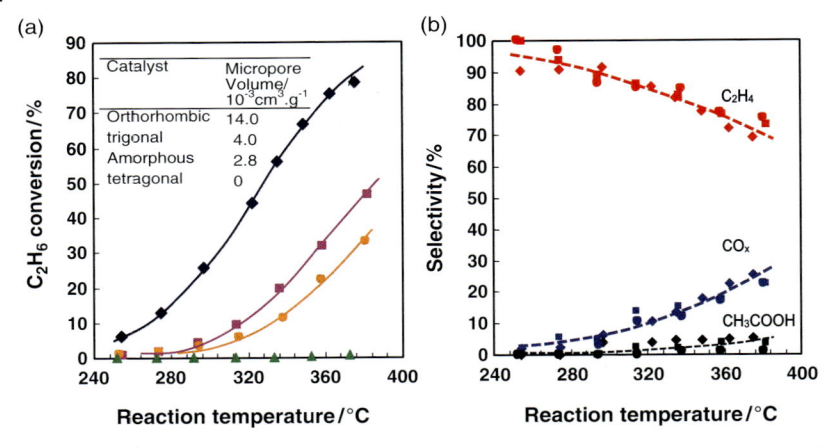

Figure 15.3 Ethane oxidative dehydrogenation to ethene over four $Mo_3VO_{11.2}$ catalyst with different crystal structures. (lozenges: orthorhombic, square: trigonal, circle: amorphous, and triangle: tetragonal) [23]. Ethane conversion (a), product selectivity (b), and micropore volume; inset.

properties, particularly oxidation catalysis, a comparison of four distinct types of Mo_3VO_x catalyst with different crystal structures is informative.

Figure 15.3 summarizes the catalytic activity and selectivity as the function of the reaction temperature. The catalyst samples are orthorhombic, trigonal, tetragonal, and amorphous types, all of which have almost the same chemical composition, Mo_3VO_x. Products detected were ethene, acetic acid, and COx over the catalysts. The highest ethane conversion was achieved on the crystalline orthorhombic Mo_3VO_x catalyst. The trigonal Mo_3VO_x catalyst showed less activity than that of the orthorhombic Mo_3VO_x catalyst in spite of the similar structure. The amorphous Mo_3VO_x catalyst showed much less activity and the tetragonal Mo_3VO_x catalyst was found almost inactive. The product selectivities were not dependent on the catalysts. The catalytic activity difference depending on the catalyst structure cannot be explained neither by the chemical composition nor the difference of the external surface area which are not much different among the catalysts. The crystal structure is obviously a determinant. Since all active or less active catalysts have the same pentagonal unit and the difference is only its arrangement, it can be considered that whether the materials have the heptagonal channel or not is determinative for the ethane oxidation activity.

The hexagonal channel structure was clearly observed as black parts in the HAADF-STEM images (Figure 15.2) and the pore size of the heptagonal channels is determined to be about 4.0 Å diameter. The heptagonal channels of the $Mo_3VO_{11.2}$ materials become empty after NH_4^+ ions occupying the heptagonal channels are removed as ammonia by the calcination at 673 K in air. This empty porous heptagonal straight channel can adsorb small molecules such as N_2, Ar, CO_2, Kr, CH_4, and C_2H_6 [39]. Moreover, the adsorption property can be tuned by the reduction-oxidation state of Mo_3VO_x without any fundamental changes in the crystal structure, which is a completely new phenomenon [40].

Because of this microporous property of the heptagonal channels, Figure 15.3 shows the values of the micropore volume of each catalyst. The value exhibits microporosity of the empty heptagonal channels in which ethane and oxygen molecules can enter. When this value is taken as a measure for evaluating the activity, a reasonable linear relationship between the micropore volume and the ethane conversion is obtained, indicating that the heptagonal channel plays a crucial role in the course of ethane oxidation.

The involvement of the heptagonal pore as a catalysis field for the ethane oxidation is further confirmed by the following results. The orthorhombic Mo_3VO_x catalysts with different external surface areas and the same micropore volume were tested for both selective oxidation of ethane which is small enough to enter the heptagonal channel and oxidation of acrolein to acrylic acid which are too large molecules to enter the channel. The results are summarized in Figure 15.4 [41]. The ethane oxidation rate was almost independent of the external surface area, while a usual dependency was observed in the case of acrolein oxidation. Clearly the ethane oxidation can take place inside the heptagonal channels, while the acrolein oxidation occurs over the external surface.

Above all, it is well demonstrated that the high-dimensional arrangement of the catalytic components, Mo, V, and O in the crystal structures is responsible for the observed catalytic activity of the crystalline orthorhombic Mo_3VO_x for the oxidative dehydrogenation of ethane to ethene. At the same time, the microporous property of the heptagonal channel primarily contributes to the activation of ethane molecules, because ethane is small enough to enter the pore of the heptagonal channels under the catalytic reaction conditions. Figure 15.4 (d) shows a reaction scheme for the selective oxidation of ethane to ethene in the micropore [42]. First, the empty heptagonal channel adsorbs ethane molecules effectively and provides them with a high probability of reaction with lattice oxygen in distorted octahedra of V and Mo near the heptagonal channel. There seems to be the distorted octahedra of V and Mo at the pentamer site near the heptagonal channel. The distorted octahedra can construct oxygen defective sites and provide active oxygen species by the reaction of molecular oxygen. This situation can explain the outstanding activity at low reaction temperature compared to other structurally related Mo–V–O catalysts. The high selectivity to ethene attained in the present catalyst system can be explained in such way that, because ethene is electron richer than ethane, ethene formed may not access back to the pore where ethane activation takes place.

15.4.5
Propane (amm)Oxidation Over Mo–V–Te(Sb)–Nb–O Catalysts

Since the discovery of the catalytic material Mo–V–(Nb)–(Te)-oxides for propane (amm)oxidation in the late twentieth century [20], there has been extensive researches for clarifying the origin of its remarkable activity for selective (amm)

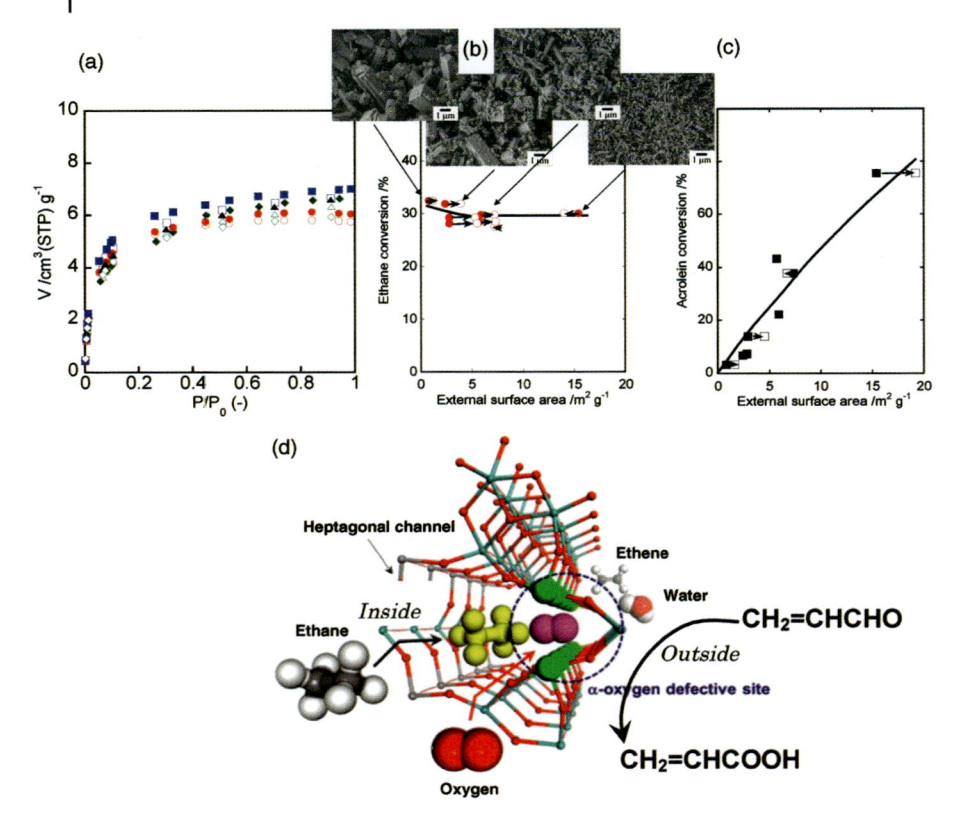

Figure 15.4 Adsorption isotherm (a) of ethane at room temperature on the orthorhombic $Mo_3VO_{11.2}$ catalysts with different particle size which are shown in SEM images (b), ethane oxidation (b) and acrolein oxidation (c) as a function of external surface area of the catalysts [41,42], and schematic presentation of reaction field (d), ethane oxidation inside pore and acrolein oxidation on the outside.

oxidation. Although the exact active centers based on the crystal structure has not been satisfactory elucidated yet, the orthorhombic phase, the so-called M1 phase, is undoubtedly responsible for the major activity. More important are the roles of each constituent of the Mo–V–(Nb)–(Te)-oxides.

The findings that the much simpler crystalline $Mo_3VO_{11.2}$ materials exhibited remarkable catalytic oxidation ability for propane oxidation as well as for the ethane selective oxidation as described above clearly support that Mo and V elements in the structure can provide an oxygen species having enough capability for alkane C—H bond activation. The role of the elements Te or Sb is also clear now because these elements can be introduced either in the hexagonal channels or in the heptagonal channels of the crystalline $Mo_3VO_{11.2}$ materials while keeping the original network structure [43]. The catalyst without these elements shows oxidation activity but gives very poor selectivity to the desired products, acrylic acid in the propane oxidation and acrylonitrile in the propane ammoxidation. On the other hand, the catalysts with these elements show appreciable selectivity to the products. Furthermore, Nb can be also incorporated in the pentagonal {(Mo)Mo$_5$} unit network and give rise a enhancement of the selectivity. Presumably propane is selectively oxidized to propene over the Mo–O–V sites and the formed propene immediately migrates to the Te element in the heptagonal channels to be oxidized via ally-intermediate process to the products. Nb might assist desorption of the desired products to prevent further deep oxidation to carbon oxide [44]. This fact strongly demonstrates that multicatalytic functions are constructed on the basis of crystal structure and catalytic oxidation can be understood by a molecular-base mechanism over multicatalytic functions [45]. This is totally different from the classical solid-state catalysis chemistry that has never given a true understandings of the molecular-based mechanism over a solid catalyst. It is hoped that this prominent example can promote further researches to create next generation catalysts with new levels of catalytic performance.

15.5
Oxide Catalysts as a New Material Category [2]

Table 15.3 lists the new catalyst materials created recently in various catalytic systems, particularly metal oxide-based catalysts. The classification in this table is not simply based on the complexity of elemental composition but on the new material state that exists only when various elements in various states interact intimately together in a form of structural arrangement. In Table 15.3, the materials are classified not only according to complexity as a common feature but also on the basis of structure units divided into main catalytic unit and subunit (structure-supporting unit), since it is generally seen in many examples that high-dimensional materials are basically constructed with these units. In addition, materials are classified into three large groups; one-dimensional materials like molecular catalysts, two-dimensional materials formed on surface and by strong contact between two different materials phases, and three-dimensional materials with crystallinity and high-surface area.

Molecular-type catalysts in one-dimension mostly need the co-existence of organic materials. Nevertheless, the coexisting organic material effectively leads to multifunctionality with much high-dimensionality. Complex metal oxide cluster systems are now attracting much attention, so that many more will be synthesized and applied extensively. The two-dimensional systems are rather classical, so that this most frequently appears in catalytic materials because catalysts having strong interaction between oxide-oxide and metal–metal oxide are quite popular. The two-dimensional systems are, however, most complicated among the listed materials because of the lack of detailed information of the interaction from the structural point of view, so that the structural characterization of the two-dimensional state needs to be done by various advanced

Table 15.3 Creation of metal oxide-based catalyst system with multifunctionality [2].

Classification	Catalytic material form	Principal catalyst structure unit	Counit for creating the principal catalyst	Brief description	Example
Molecular state	Multinuclear metal oxide cluster	Molecular assembly of elements with metal-oxygen bonds	Organic materials stabilizing metal oxide cluster, or in Zeolite framework	3-dimensional complex metal oxide cluster stabilized with surrounding organic moiety or with inorganic framework	$CaMn_4O_5$ cluster in oxygen evolving complex; Cu_3Ox/Zeolite
	High-dimensional polyoxo-metalates	Polyoxoanion of metals	Ionic linker	Orderly condensed oxoanions of different metals	Metal-substituted POM; Metal lacunary POM
Surface state, interface (2-dimensional)	Interface formed in metal on metal oxide surface	Nano-sized metal particles	Solid-state metal oxides	Interface state formed in the close contact of nano-sized metal particles with crystalline metal oxide surface	Au/TiO_2; Pd-$LaCoO_3$; Ru/hydrotalcite
	Surface material state formed in metal oxide on metal surface	2-Dimensional metal oxide	Metallic board; Metal particles	Surface material state formed either by covering metal particles with metal oxide or by interaction of metal oxide species with metallic board surface	FeOx/Pt; TiO_2/Pt; CrOx/Ru; NiOx/Au
3-dimensional solid	Interface formed in metal oxide on metal oxide surface	1,2,3-Dimensional metal oxides	Solid-state metal oxides	1-dimensional metal oxide like oxide clusters, 2-dimensional oxide like layered oxide formed by epitaxy, and 3-dimensional oxide formed on basal metal oxide	WO_3/ZrO_2; RuO_2/TiO_2; Re_2O_7/Al_2O_3; V_2O_5/TiO_2
	Multiordered porous metal oxide	Metal oxide	Crystalline porous materials, matrix	Organized porous materials, like high-dimensional porous metal oxides and/or added metal cation position-controlled zeolite	Metalosilicates; Organosilicates; Octahedral molecular sieve; Nanotube
	High-dimensional complex metal oxide	Polyoxoanion of metals	Linker	Polyoxoanion network with molecular polyoxoanions and linkers, forming high-dimensional complex metal oxides	$Mo_3V(M)O_{11.2}$; Organoployoxometalate; All-inorganic porous POM

methods. Nevertheless, much research is now undertaken to clarify this state structurally, for example by synthesizing new solid crystalline materials representing the interaction state between oxide-oxide or metal-oxide. Long-term research on these materials will ultimately end up with new two-dimensional materials existing only when the interaction occurs. The three-dimensional systems have already been introduced in the above sections by introducing the crystalline Mo–V–O catalysts. Other examples will be metal organic frameworks [46] and new zeolitic materials based on other elements like C beside Si and P. The ultimate goal in the three-dimensional systems, particularly of complex metal oxides, is to find synthetic methods where the structural main units and subunits can be freely combined with each other and also to create materials having a high ratio of surface to bulk which is most suitable in terms of catalysis.

15.6
Summary

Although at present, it is extremely difficult to design and synthesize solid-materials, particularly metal oxide catalysts, with high-dimensional structure and a unique catalysis field, a fundamental understanding that catalysts belong to a new material category will help great advancement in catalyst material synthesis. In order to respond to the current strong demand on catalysis and catalysts from various scientific and technology fields, this approach is inevitable. Along with advancing the hitherto methodology of catalyst preparation, there is a hope of arrival of new catalytic material concepts and synthetic methodology.

References

1 Centi, C., Cavani, F., and Trifiro, F. (eds) (2001) *Selective Oxidation by Heterogeneous Catalysis*, Kluwer Academic/Plenum Publishers, New York.
2 Ueda, W. (2012) Complication towards evolution in the history of oxide catalysts for selective oxidation, Annual Survey of Catalytic Science and Technologies, Special Edition for the 20th Anniversary, Catalysis Society of Japan, pp. 18–25.
3 Nojiri, N., Sakai, Y., and Watanabe, Y. (1995) Two catalytic technologies of much influence on progress in chemical process development in Japan. *Catal. Rev. Sci. Eng.*, **37**, 145–178.
4 Moro-oka, Y. and Ueda, W. (1994) Multicomponent bismuth molybdate catalyst – a highly functionalized catalyst system for the selective oxidation of olefin. *Adv. Catal.*, **40**, 233–273.
5 Sadakane, M., Watanabe, N., Katou, T., Nodasaka, Y., and Ueda, W. (2007) Crystalline Mo_3VO_x mixed metal oxide catalyst with trigonal symmetry. *Angew. Chem., Int. Ed.*, **46**, 1493–1496.
6 Bartley, J.K., Dummer, N.F., and Hutchings, G.J. (2009) Vanadium phosphate catalysts, in *Metal Oxide Catalysts* (eds S.D. Jackson and J.S. Hargreaves), Wiley-VCH Verlag GmbH & Co. KGaA, pp. 499–531.
7 Okuhara, T., Mizuno, N., and Misono, M. (1996) Catalytic chemistry of heteropoly compounds. *Adv. Catal.*, **41**, 113–252.

8 Notari, B. (1996) Microporous crystalline titanium silicates. *Adv. Catal.*, **41**, 253–334.

9 Celerici, M.G., Bellusi, G., and Romano, U. (1990) Synthesis of propene oxide from propylene and hydrogen peroxide catalyzed by titanium silicate. *J. Catal.*, **129**, 159–167.

10 Roffia, P., Leofanti, G., Cesana, A., Mantegazza, M.A., Padovan, M., Petrini, G., Tonti, S., and Gervasutti, P. (1990) Cyclohexanone ammoximation: a break through in the ε-caprolactam production process. *Stud. Surf. Sci. Catal.*, **55**, 43–52.

11 Grasselli, R.K. (1999) Advances and future trends in selective oxidation and ammoxidation catalysis. *Catal. Today*, **49**, 141–153.

12 Havecker, M., Mayer, R.W., K-Gericke, A., Bluhm, H., Kleimenov, E., Liskowski, A., Folloath, D.Su., Requejo, F.G., Ogletree, D.F., Salmeron, M., Sanchez, J.A.L., Bartly, J.K., Hutchings, G.H., and Schlogl, R. (2003) In situ investigation of the nature of the active surface of a vanadyl pyrophosphate catalyst during n-butane oxidation to maleic anhydride. *J. Phys. Chem. B*, **107**, 4587–4596.

13 Bordes, E. (1993) Nature of the active and selective sites in vanadyl pyrophosphate, catalyst of oxidation of n-butane, butane and pentane to maleic anhydride. *Catal. Today*, **16**, 27–38.

14 Kamiya, Y., Ueki, S., Hiyoshi, N., Yamamoto, N., and Okuhara, T. (2003) Preparation of catalyst precursors for selective oxidation of n-butane by exfoliation-reduction of $VOPO_4 \cdot 2H_2O$ in primary alcohol. *Catal. Today*, **78**, 281–290.

15 Kamiya, Y., Sadakane, M., and Ueda, W. (2013) Heteropoly compounds, in *Comprehensive Inorganic Chemistry II*, 2nd edn, Elsevier, pp. 185–204.

16 Kozhevnikov, I.V. and Matveev, K.I. (1983) Homogeneous catalysis based on heteropoly acid. *Appl. Catal.*, **5**, 135–150.

17 Komaya, T. and Misono, M. (1983) Activity patterns of $H_3PMo_{12}O_{40}$ and its alkali salts for oxidation reactions. *Chem. Lett.*, (8), 1177–1180.

18 Baba, T., Nomura, M., Ono, Y., and Ohno, Y. (1993) Solid-state proton MAS NMR study on the highly active protons in partially reduced trisilver dodecatungstophosphate ($Ag_3PW_{12}O_{40}$). *J. Phys. Chem.*, **97**, 12888–12893.

19 Zhang, Z.-X., Sadakane, M., Murayama, T., Izumi, S., Yasuda, N., Sakaguchi, N., and Ueda, W. (2014) Tetrahedral connection of ε-Keggin-type polyoxometalates to form an all-inorganic octahedral molecular sieve with an intrinsic 3D pore system. *Inorg. Chem.*, **53**, 903–911.

20 Ushikubo, T., Oshima, K., Kayou, A., Vaarkamp, M., and Hatano, M. (1997) Ammoxidation of propane over catalysts comprising mixed oxides of Mo and V. *J. Catal.*, **169**, 394–396.

21 Sadakane, M., Endo, K., Kodato, K., Ishikawa, S., Murayama, T., and Ueda, W. (2013) Assembly of a pentagonal polyoxomolybdate building block, $[Mo_6O_{21}]^{6-}$, into crystalline Mo-V oxides. *Eur. J. Inorg. Chem.*, **10–11**, 1731–1736.

22 Chen, C., Murayama, T., and Ueda, W. (2013) Single-crystalline phase Mo_3VOx: an efficient catalyst for partial oxidation of acrolein to acrylic acid. *ChemCatChem*, **5**, 2869–2873.

23 Konya, T., Murayama, T., Kato, T., Sadakane, M., Ishikawa, S., Battrey, D., and Ueda, W. (2013) An orthorhombic Mo_3VOx catalyst most active for oxidative dehydrogenation of ethane among related complex metal oxides. *Catal. Sci. Technol.*, **3**, 380–387.

24 Cavani, F., Ballarini, N., and Cericola, A. (2007) Oxidative dehydrogenation of ethene and propane: How far from commercial implementation? *Catal. Today*, **127**, 113–131.

25 Gaab, S., Machli, M., Find, J., Grasselli, R.K., and Lercher, J.A. (2003) Oxidative dehydrogenation of ethane over novel Li/Dy/Mg mixed oxides: Structure-activity study. *Top. Catal.*, **23**, 151–158.

26 Mulla, S.A.R., Buyevskaya, O.V., and Baerns, M. (2000) Ethylene and propene by oxidative dehydrogenation of ethane and propane: 'performance of rare-earth oxide-based catalysts and development of redox-type catalytic materials by combinatorial methods. *Catal. Today*, **62**, 91–99.

27 Ueda, W. and Lin, S.-W. (2003) Metal halide oxide catalysts active for alkane selective oxidation. *Catalysis*, **16**, 198–235.

28 Heracleous, E. and Lemonidou, A.A. (2006) Ni-Nb-O mixed oxides as highly active and selective catalysts for ethene production via ethane oxidative dehydrogenation. Part I: characterixation and catalytic performance. *J. Catal.*, **237**, 162–174.

29 Miller, J.E., Gonzales, M.M., Evans, L., Sault, A.G., Zhang, C., Rao, P., Whittwell, G., Maiti, A., and King-Smith, D. (2002) Oxidative dehydrogenation of ethane over iron phosphate catalysts. *Appl. Catal. A. General*, **231**, 281–292.

30 Thorsteinson, E.M., Wilson, T.P., Young, F.G., and Kasai, P.H. (1978) The oxidative dehydrogenation of ethane over catalysts containing mixed oxides of molybdenum and vanadium. *J. Catal.*, **52**, 116–132.

31 Watanabe, H. and Koyasu, Y. (2000) New synthesis route for Mo-V-Nb-Te mixed oxides catalyst for propane ammoxidation. *Appl. Catal. A. General*, **194–195**, 479–485.

32 Pyrz, W.D., Blom, D.A., Sadakane, M., Kodato, K., Ueda, W., Vogt, T., and Buttrey, D.J. (2010) Atomic-scale investigation of two-component MoVO complex oxide catalysts using aberration-corrected high-angle annular dark-field imaging. *Chem. Mater.*, **22**, 2033–2040.

33 Sadakane, M., Yamagata, K., Kodato, K., Endo, K., Toriumi, K., Ozawa, Y., Ozeki, T., Nagai, T., Matsui, Y., Sakaguchi, N., Pyrz, W.D., Buttrey, D.J., Bolm, D.A., Vogt, T., and Ueda, W. (2009) Synthesis of orthorhombic Mo-V-Sb oxide species by assembly of pentagonal Mo_6O_{21} polyoxometalate building blocks. *Angew. Chem., Int. Ed.*, **48**, 3782–3786.

34 Mueller, A., Todea, A.M., Van Slageren, J., Dressel, M., Boegge, H., Schmidtmann, M., Luban, M., Engelhardt, L., and Rusu, M. (2005) Triangular geometrical and magnetic motifs uniquely linked on a spherical capsule surface. *Angew. Chem., Int. Ed.*, **44**, 3857–3861.

35 Botar, B., Koegerler, P., and Hill, C.L. (2005) [{(Mo)$Mo_5O_{21}(H_2O)_3(SO_4)$}$_{12}(VO)_{30}(H_2O)_{20}]^{36-}$: a molecular quantum spin icosidodecahedron. *Chem. Commun.*, 3138–3140.

36 Mueller, A., Krichemeyer, E., Boegge, H., Schmidtmann, M., and Peters, F. (1998) Organizational forms of matter: An inorganic super fullerene and keplerate based on molybdenum oxide. *Angew. Chem., Int. Ed.*, **37**, 3359–3363.

37 Mueller, A., Krichemeyer, E., Dillinger, S., Boegge, H., Plass, W., Proust, A., Dloczik, L., Meyer, C., and Rohlfing, R. (1994) New perspectives in polyoxometalate chemistry by isolation of compounds containing very large moieties as transferable building blocks: $(NMe_4)_5[As_2Mo_8V_4AsO_{40}] \cdot 3H_2O$, $(NH_4)_{21}[H_3Mo_{57}V_6(NO)_6O_{183}(H_2O)_{18}] \cdot 65 H_2O$, $(NH_2Me_2)_{18}(NH_4)_6[Mo_{57}V_6(NO)_6O_{183}(H_2O)_{18}] \cdot 14 H_2O$, and $(NH_4)_{12}[Mo_{36}(NO)_4O_{108}(H_2O)_{16}] \cdot 33 H_2O$. *Z. Anorg. Allg. Chem.*, **620**, 599–619.

38 Long, D.-L., Tsunashima, R., and Cronin, L. (2010) Polyoxometalates: Building blocks for functional nanoscale systems. *Angew. Chem., Int. Ed.*, **49**, 1736–1758.

39 Sadakane, M., Kodato, K., Kuranishi, T., Nodasaka, Y., Sugawara, K., Sakaguchi, N., Nagai, T., Matsui, Y., and Ueda, W. (2008) Molybdenum-vanadium based molecular sieves with a microchannel of corner-sharing 7-membered metal-oxide-octahedron ring. *Angew. Chem., Int. Ed.*, **47**, 2493–2496.

40 Sadakane, M., Ohmura, S., Kodato, K., Fujisawa, T., Kato, K., Shimizu, K., Murayama, T., and Ueda, W. (2011) Redox tunable reversible molecular sieves: orthorhombic molybdenum vanadium oxide. *Chem. Commun.*, **47**, 10812–10814.

41 Ishikawa, S., Yi, X., Murayama, T., and Ueda, W. (2014) Heptagonal channel micropore of orthorhombic Mo_3VO_x as catalysis field for the selective oxidation of ethane. *Appl. Catal. A*, **474**, 10–17.

42 Ishikawa, S., Kobayashi, D., Konya, T., Ohmura, S., Murayama, T., Yasuda, N., Sadakane, M., and Ueda, W. (2015) Redox treatment of orthorhombic $Mo_{o29}V_{11}O_{112}$ and relationships between crystal structure, microporosity and catalytic performance for selective oxidation of ethane. *J. Phys. Chem. C*, **119**, 7195–7206.

43 Millet, J.M.M., Roussel, H., Pigamo, A., Dubois, J.L., and Jumas, J.C. (2002)

Characterization of tellurium in MoVTeNbO catalysts for propane oxidation or ammoxidation. *Appl. Catal., A*, **232**, 77–92.

44 Watanabe, N. and Ueda, W. (2006) Comparative study on the catalytic performance of single-phase Mo-V-O-based metal oxide catalysts in propane ammoxidation to acrylonitrile. *Ind. Eng. Chem. Res.*, **45**, 607–614.

45 Grasselli, R.K., Buttrey, D.J., DeSanto, P., Burrington, J.D., Lugmair, C.G., Volpe, A.F. Jr., and Weingand, T. (2004) Active centers in Mo-V-Nb-Te-Ox (amm) oxidation catalysts. *Catal. Today*, **91–92**, 251–258.

46 Kitagawa, S., Kitamura, R., and Noro, S. (2004) Functional porous coordination polymers. *Angew. Chem., Int. Ed.*, **43**, 2334–2375.

16
Activity of Zeolitic Catalysts

Xiangju Meng, Liang Wang, and Feng-Shou Xiao

Zhejiang University, Department of Chemistry, 38 Zheda Rd, Xihu Qu, Zhejiang 310027, China

16.1
Introduction

Approximately nine-out-of-ten chemical processes worldwide utilize heterogeneous catalysts currently, where selectivity for the shape/size of the reactant, product or intermediate together with stability are the vital aspects [1]. The combination of selectivity and stability for zeolites offers unique structural and chemical features that can be beneficial for many commercially important reactions/processes [2,3].

Looking back at the history of industrial chemistry, it can undoubtedly be declared that the impact of the introduction of zeolites into refinery operations has been revolutionary as it resulted in a significant increase in gasoline yield from catalytic cracking, leading to more efficient utilization of the petroleum feedstocks [4]. However, the global economy has experienced various spikes in crude oil prices in recent years. As a result, the economic developments emphasize the importance of considering change and flexibility of raw materials (e.g., coal, natural gas, biomass) [5,6].

Methanol to olefin (MTO) conversion is one of the most promising routes to produce useful chemicals from relatively abundant sources such as coal and natural gas [5]. The DMTO process (methanol-to-olefin developed from Dalian Insitute of Chemical Physics in China) has been successfully developed and applied in an industrial MTO plant for first time by Dalian Institute of Chemical Physics (DICP), Chinese Academy of Sciences in 2011. Notably, all reactions occurring in this process including alkylation, isomerization, homologation, oligomerization, cracking, cyclization, protonation, deprotonation, and aromatization involve zeolite catalysts.

Fischer–Tropsch synthesis (FTS) is also a key reaction in the utilization of nonpetroleum carbon resources (coal, natural gas, and shale gas) for sustainable production of clean liquid fuels from synthesis gas [7]. One of the most difficult

Handbook of Solid State Chemistry, First Edition. Edited by Richard Dronskowski, Shinichi Kikkawa, and Andreas Stein.
© 2017 Wiley-VCH Verlag GmbH & Co. KGaA. Published 2017 by Wiley-VCH Verlag GmbH & Co. KGaA.

challenges in the FTS is to control product selectivity, which generally follows the Anderson–Schulz–Flory (ASF) distribution. Recent studies indicate that zeolite-based bifunctional catalysts or catalytic systems combining the catalytic phases (sites) for CO hydrogenation and hydrocarbon hydrocracking/assembling can directly produce light olefins-, gasoline-, or diesel-range hydrocarbons with high selectivities.

The increasing demand for the production of chemicals and fuels from renewable carbon source such as biomass has focused much attention on the conversion of biomass (components) and derived intermediates to generate a valuable platform of chemicals and fuels. The distinct role of zeolite catalysts has been demonstrated in many of these biomass conversion routes [6].

With increasing environmental regulations to fight air pollution, which has reached alarming levels in certain parts of the world, new applications for zeolite catalysts in environmental protection have been opened [8]. A typical example is the selective catalytic reduction (SCR) process, which is currently the leading technology to revert the formed NO_x into harmless N_2 and H_2O by using a reducing agent such as urea or ammonia over a zeolite catalyst.

In this chapter, an overview of the current and potential applications of zeolites as catalysts in academic and commercially important chemical industry reactions/processes is provided.

16.2
Zeolites in Fluid Catalytic Cracking (FCC)

In 1942, the first commercial fluid catalytic cracker unit, PCLA No. 1, was built in Standard Oil of New Jersey's Baton Rouge, LA refinery, which opened a new chapter for oil refinery [4]. At the same time, a synthetic fluid cracking catalyst containing 13% alumina was developed to support the "powdered catalyst" operation in PCLA No. 1 [9]. A major breakthrough came in the 1960s when synthetic aluminosilicate zeolites were incorporated into fluid cracking catalysts to further enhance catalytic activity and selectivity for outstanding cracking properties [4]. In 1962, a new catalyst developed by Mobil, was prepared by adding a small amount of zeolite Y into the matrix of silica–alumina catalyst, which outperformed all existing catalysts in the market and resulted in a significant increase in gasoline production. This development was marked as the first commercial zeolite-based FCC catalysts and truly revolutionized the industry [4].

16.2.1
Y Zeolites in FCC

Zeolite Y is a large-pore zeolite with the same framework structure as natural zeolite called faujasite (framework code FAU). The three building units are the double 6-ring, the sodalite cage which is formed by connecting 4- and 6-rings in different arrangements and a very large cavity with four 12-ring windows known

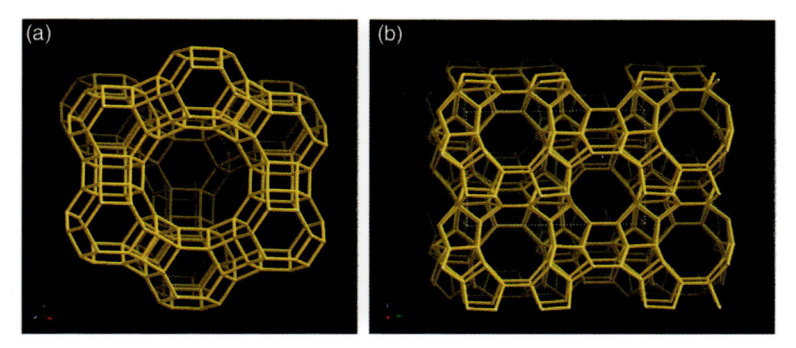

Figure 16.1 Skeletal diagram of (a) Y and (b) ZSM-5.

as the supercage. The connectivity of these cages allows molecules to diffuse in the three-dimensional lattice of identical pores, with an opening of around 7.4 Å in size (supercage opening, Figure 16.1).[1]

One important parameter controlling the cracking behavior of the zeolite Y is the framework composition that determines, in principle, the amount, strength, and distribution of the Brønsted acid sites associated with framework Al atoms. Notably, the zeolite is subjected to severe thermal and hydrothermal conditions in the riser/regenerator of FCC units, which requires ultrastability of the framework of zeolites. Therefore, ultrastable Y (USY) zeolites with low framework Al content prepared by dealumination under the severe conditions (temperatures above 700 °C in the presence of steam) are used in the regenerator of FCC units. However, it was found that the cracking behavior of dealuminated Y zeolite is also strongly contingent on the nature and amount of extraframework Al species as well as the mesoporosity generated during the dealumination treatments [3]. Therefore, great efforts have been devoted to the optimization of the cracking performance of Y zeolite through the manipulation of both extra framework Al and mesoporosity [2,3]. The requirement of the mesoporosity in the catalysts is due to the fact that the bulky molecules in commercial feedstocks can hardly access the acid sites located in the micropores and the first cracking events will take place, predominantly, on the external zeolite surface. Therefore, the accessibility of the zeolite within the catalyst particle plays a very important role. In this respect, the *in situ* synthesis of meso-/macroporous zeolites with more exposed external surface areas provide improved accessibility to the zeolite, which is very favorable for enhancement of catalytic activity [3].

16.2.2
MFI Zeolites in FCC

Zeolite ZSM-5 is an MFI type of zeolite, possessing an anisotropic framework with two intersecting 10-membered ring channels including straight channels

---

1) http://www.iza-structure.org/databases/.

(5.3 × 5.6 Å) parallel to the *b*-axis and zigzag channels (5.1 × 5.5 Å) parallel to the *a*-axis (Figure 16.1) [10].[1] The straight channels are favorable for diffusion and formation of relatively large products such as *o*-xylene (kinetic diameter about 5.8 Å), while the zigzag channels show difficulty for the adsorption of these molecules due to the shape selectivity [11].

ZSM-5 zeolite is well known as a highly shape selective zeolite in FCC, which could efficiently enhance gasoline octane number and increase the yield of light olefins (propylene and butenes) at the expense of gasoline [3,12]. ZSM-5 zeolite has been used as a separate FCC catalyst additive since 1986 [13], which significantly resolved the requirement for additional gasoline octane enhancers in the early 1980s as well as the considerably more favorable economics for producing light olefins especially in FCC units with propylene maximization as a main target (i.e., max-propylene units) in the later 1980s. In 1987, ZSM-5 additive manufacturers started to add phosphorus to their formulations, which dramatically enhanced the effectiveness of ZSM-5 and reduced the amount of additive required to achieve the desired effect in octane and light olefin yields from FCC operations [14,15], which was attributed to retarding aluminum from leaving the zeolite framework and reducing the initial acidity of the zeolite.

Notably, unlike Y zeolite as the main component in an FCC catalyst, ZSM-5 zeolite is only an additive. As a result, the amount of ZSM-5 is much lower than that of Y zeolites in FCC catalyst and can be varied in a wide range (typically 1–20%) depending on the objective of the refinery. In the initial stage there was a question of whether synergism occurred between zeolite Y and ZSM-5. Later, testing denied this hypothesis and the separated additive case was considered as more attractive, since it brings added flexibility [12].

16.2.3
Rare Earth (RE) in Zeolites in FCC

Rare earth (RE) elements such as lanthanum and cerium can efficiently stabilize aluminum atoms in the zeolite structure and enhance the hydrothermal stability of zeolites when they are introduced into an FCC catalyst via ion exchange procedures [16]. They prevent the aluminum atoms from leaving the zeolite lattice when the catalyst is exposed to harsh hydrothermal conditions in the FCC regenerator. A fully rare earth-exchanged zeolite equilibrates at a high unit cell size (UCS), whereas a nonrare earth zeolite equilibrates at a very low UCS. All intermediate levels of rare earth-exchanged zeolite can be produced. Rare earth cations therefore improve thermal and hydrothermal stability of the zeolites.

It is worth noting that the introduction of the rare earth ions in zeolites increases activity and gasoline selectivity with a loss in octane rating [16]. The octane loss is due to the promotion of hydrogen transfer reactions that increase also the coke production. The poorer gasoline quality is largely related to the transformation of olefins into corresponding paraffins to various degrees. Paraffins are more stable structures than olefins and so are less likely to undergo consecutive cracking themselves. The result is that lower gasoline quality usually

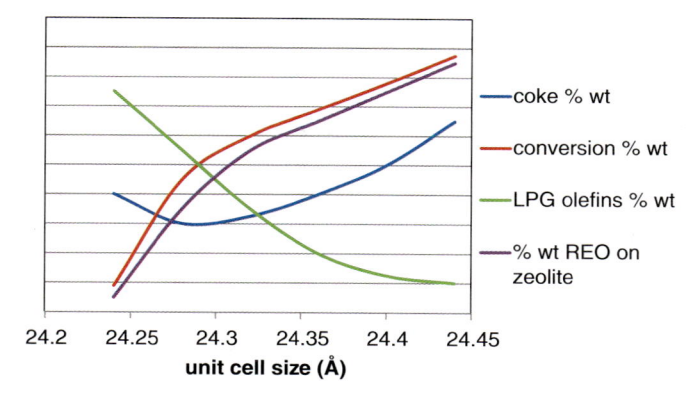

Figure 16.2 FCC yields and rare earth oxide (REO) correlation to unit cell size (UCS).

goes along with an improvement in gasoline yield. As shown in Figure 16.2, the decrease in rare earth content means a reduction in the UCS and therefore a decrease in the zeolite's equilibrium acidity. As a result, the catalyst manufacturer can adjust the acidity that the catalyst will develop once it is in equilibrium conditions in the unit by introducing more or fewer rare earth ions during preparation.

16.3
Zeolites in MTO

The discovery of MTO was first reported in the early 1970s by researchers at Mobil Chemical Research when they introduced a mixture of methanol and isobutene through HZSM-5, achieving products mainly consisting of liquid aromatics and gaseous products consisting of light olefins and alkanes [17,18]. In recent years, however, with increasing demand for light olefins from the polymer industry, various versions of methanol to olefin conversion processes were developed, especially MTP (methanol to propylene), became an attractive route to produce propylene. Particularly, China has built several commercial MTO or MTP plants, using MTO package (SAPO-34 based catalysts) by DICP [19,20]. Nearly at the same time, an unit of integrated MTO (based on SAPO-34 zeolite, UOP/Hydro) and olefin cracking process (OCP, Total/UOP) was established in the research and development center of Total at Feluy in Belgium since 2008 [21]. Lurgi Company developed a methanol-to-propylene (MTP) process based on a modified ZSM-5 catalyst with high propylene selectivity. The Demo Unit of MTP process was built by Lurgi in the Statoil methanol plant at Tjeldbergodden, Norway in 2001, and started in January 2002 [22]. Notably, the MTO process has also been a puzzling and still debating issue in several aspects of the mechanism. As a result, basic research on MTO, especially mechanism is still an ongoing subject [5].

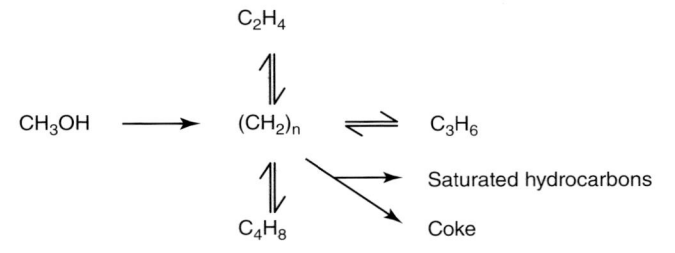

Figure 16.3 Representation of the carbon pool parallel mechanism originally proposed by Kolboe. (Reproduced with permission from Ref. [25]. Copyright 1996, Academic Press.)

16.3.1
Hydrocarbon Pool Mechanism (HCP)

For the sake of developing a new catalyst with high efficiency and better anticoking, a great deal of effort has been devoted in the past decades to elucidate the reaction mechanisms of MTO [5]. Among them, the hydrocarbon pool mechanism proposed by Dahl and Kolboe has received wide recognition (Figure 16.3) [23–25]. Kolboe *et al.* carried out systematic experiments to show that the organic species trapped inside the SAPO-34 cages (eight member ring opening of 3.8 Å) are vital for olefin formation and they found out that most of the olefin product was from ^{13}C-labeled methanol, and ethylene was almost inactive, when labeled methanol (^{13}C-MeOH) was used along unlabeled ethylene or propylene. The term "carbon pool" was a smart one as the exact structure of the $(CH_2)_n$ and "pool" species was not known at that time. However, it seems so accurate when one envisions a catalytic cycle, in which methanol or any C_1 species are fed into the pool on one side, and olefin products are produced on the nother side.

According to the type of hydrocarbon pool species, the mechanism can be further divided into an aromatic cycle and an alkene cycle. On the basis of the active intermediates involved, two routes have been proposed for the formation of light olefins via the polymethylbenzenes (polyMBs)-based aromatic cycle (Figure 16.4) [26,27], the side-chain route [28,29] and the paring route [30,31]. The polyMBs are considered as the most important hydrocarbon pool species for the aromatic cycle [26,27]. Although extensive experimental and theoretical investigations have been implemented to explore both the polyMB and the alkene cycles in MTO, the detailed contribution that each cycle makes to MTO and certain key issues involved in these two cycles are still rather ambiguous [32–35].

Recently, Dai *et al.* found that three-membered ring compounds, dienes, polymethylcyclopentenyl, and polymethylcyclohexenyl cations were formed before polymethylbenzenium cations in H-SAPO-34, which suggests that aromatics are produced via an olefin-based reaction cycle, and then, the aromatics-based cycle may be initiated to quickly dominate the steady process of the MTO reaction [37]. The structures of generated carbenium ions depend on the pore

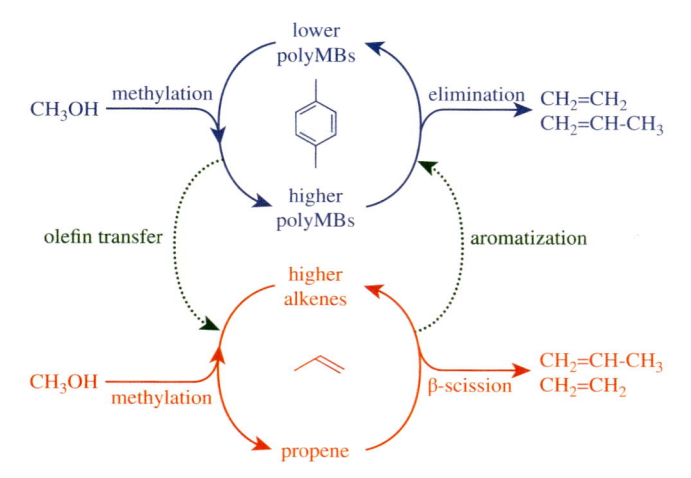

Figure 16.4 According to the sort of hydrocarbon pool species, the carbon pool parallel mechanism can be further divided into a polyMB cycle (or aromatic cycle) and an alkene cycle. (Reproduced with permission from Ref. [36]. Copyright 2015, American Chemical Society.)

systems of zeolite catalysts. Several types of carbenium ions, that is, polymethylcyclopentenyl, polymethylbenzenium, and dienylic carbenium, were recently observed in H-ZSM-5, H-Beta, H-SSZ-13, and H-DNL-6 with ^{13}C NMR and UV/Vis spectroscopies as well as confirmed by theoretical calculations [31,37–44]. These carbenium ions have been proved to be the active hydrocarbon species involved in the aromatic-based and olefin-based cycles.

Density functional theory considering dispersive interactions (DFT-D) has also been used to elucidate the catalytic roles that the polyMB and the alkene cycles may play in forming ethylene and propene from methanol in MTO over H-ZSM-5 [36]. The results demonstrate that ethylene and propene can be produced in nearly the same probability via the polyMB cycle, as they have a very close free energy height as well as a similar free energy barrier for the rate-determining steps. By the alkene cycle, however, propene is the dominant product, because the methylation and cracking steps to obtain propene have a much lower free energy barrier in comparison with those to form ethylene. As a result, ethylene is predominantly formed via the polyMB cycle, whereas propene is produced via both the polyMB and the alkene cycles. The contribution of the alkene cycle is probably larger than that of the polyMB cycle, resulting in a high fraction of propene in the MTO products. Meanwhile, both cycles are interdependent in MTO, as the aromatic species generated by aromatization via the alkene cycle can also serve as new active centers for the polyMB cycle, and vice versa. Moreover, the catalytic activity of H-ZSM-5 zeolite is directly related to its acid strength; weaker acid sites are unfavorable for the polyMB cycle and then enhance relatively the contribution of the alkene cycle to forming light olefins.

16.3.2
Direct Mechanism

Nevertheless, the debate on the direct mechanism in methanol conversion to hydrocarbons is still going on because the HCP mechanism cannot account for the origin of initial hydrocarbon pool species or the formation of the first C—C bond [5].

16.3.2.1 Carbene Mechanism

It is quite straightforward to consider carbene ($:CH_2$) as a key intermediate in MTO, especially for the initial C—C bond formation, as the following equation is so simple yet elegant:

$$:CH_2 +\ :CH_2 = CH_2CH_2$$

Chang et al. proposed the carbene mechanism in their earlier report on MTO [17,18]. Because carbene is an active species, direct spectroscopic observation of carbene was rare and most studies relied on indirect evidence that can be explained by a carbene intermediate. For example, Wang et al. observed the conversion of surface methoxy species (SMS) into hydrocarbons on zeolites, and claimed that the decomposition of SMS may occur via either the carbene or the oxonium ylide mechanism [45,46]. However, Sinclair and Lesthaeghe found that the energy barrier for the decomposition of SMS was very high and the rate coefficient was very low, which suggests that the possibility for the occurrence of the carbene mechanism is very small although recent IR spectroscopy studies identified the carbene species in the methylation of ethylene on H-ZSM-5 [47,48].

16.3.2.2 Oxonium Ylide Mechanism

The oxonium ylide mechanism involves the formation of two important intermediates, viz. trimethyl oxonium ion (TMO) and dimethyl oxonium methyl ylide (DOMY) species. TMO was not observed in the reaction of methanol and DME by FTIR and ^{13}C solid-state NMR spectroscopy on the working MTO catalyst [49–51]. The theoretical calculations show that TMO is stable in the H-ZSM-5 framework, but no possible route exists from TMO to DOMY, as the zeolite framework does not offer supplementary stabilization, making it a highly energetic species. Therefore, the oxonium ylide mechanism is unlikely to occur in the MTO process [52,53].

16.3.2.3 Others

Formaldehyde and methane were mainly observed when converting low pressure methanol (1–3 Pa) over H-ZSM-5 at 397 °C [54], resulting in the supposition of the methane-formaldehyde mechanism, which is also supported by the results obtained by Kubelková and Hutchings that methane was formed before C^{2+} hydrocarbons at low methanol coverage [55–58]. However, the computational results obtained with different models all confirm that the energy barrier of this step is high and the reaction rate constant is quite low [52,59]. Hence, the methane-formaldehyde mechanism is also difficult to occur, although the first step to release methane and formaldehyde is possible.

Haw and Song et al. proposed an unorthodox ideal with a so called "impurity mechanism" and they argued that there are significant amount of impurity compounds that already contain C—C bonds in any MTO reaction system, that is, acetone and ethanol in methanol (even the highest grade methanol is only 99.999%), and hydrocarbon residual species on zeolite after calcination [60]. These impurities could form hydrocarbon species that can be the starting materials to form the active organic centers for the hydrocarbon pool type mechanism. However, a contradictory result was obtained over surface methoxy species (SMS)-formed H-Y and H-SAPO-34 by the combined ^{13}C MAS NMR and diffuse reflectance UV-vis spectroscopy, which shows that the organic impurities (<30 ppm versus about 1000 ppm) in methanol do not have a significant influence on the product distribution [38]. Furthermore, it was found that aromatics and carbenium ions could not be detected on H-SAPO-34 in a similar way.

Recently, a combination of experimental and DFT calculation results led to a possible direct route for the formation of initial C—C bonds over H-SAPO-34 proposed by Li et al. [59]. This direct route involves in the formation of methoxymethyl cation ($CH_3OCH_2^+$) intermediates from SMS and DME despite the fact that methane is simultaneously formed. The methoxymethyl cation further couples with another DME or methanol molecule to form C—C bond-containing 1,2-dimethoxy ethane or 2-methoxyethanol, which gives propene as an initial alkene product through the formation of a series of oxonium cations by methylation, deprotonation, dealkylation, H-shift, and elimination. Validation requires kinetic isotope effect (KIE) studies and operando spectroscopy analysis, which are now in progress.

16.3.3
Catalyst Deactivation

Coking, on the other hand, is a universal problem in essentially all catalysts, but especially important for zeolite catalysts in MTO, where aromatics are important intermediates for olefin product, and yet are also precursors for coke species. H-ZSM-5 and H-SAPO-34, as typical commercial MTO catalysts, are by far the most extensively studied catalysts for the title reaction, also with regard to catalyst stability [5].

Schulz reported that catalytic tests performed using H-ZSM-5 at temperatures below 300 °C led to rapid activity loss and significant hydrocarbon deposition compared to tests employing the same catalyst at higher temperatures [61]. Schulz subjected the catalyst to temperature-programmed desorption after testing at 290 °C, and observed desorption of alkenes and lighter polymethylated benzenes at temperatures above 350 °C, suggesting that deactivation of H-ZSM-5 is caused by internal hydrocarbon deposits at temperatures below 350 °C, and by external coke at temperatures above 350 °C. Such mechanism has been confirmed by Weckhuysen et al. using UV-vis microspectroscopy and confocal fluorescence spectroscopy to monitor the evolution of hydrocarbon deposits in crystals of H-ZSM-5 at 390 °C [62]. They found that coke deposition proceeded faster on the triangular edges of the crystals, where straight channels are open to the surface. They also compared several H-ZSM-5 crystals with different Si/Al ratio at two temperatures (350 and 500 °C) and found that both higher acidic site density (Si/Al > 17) and temperature led to faster transformation of reactive species into polyaromatic coke molecules. Very recently, Lercher *et al.* intensively studied the relationship of the acidic sites and the deactivation of H-ZSM-5 at 450 °C by using many characterization methods [63], and they concluded that hydrocarbon deposits deactivated the catalyst by poisoning individual acidic sites in the catalyst, and not by hindering access to the catalyst pores by external coking, even at this high temperature.

In the case of H-SAPO-34, deactivation of the catalyst is accompanied by a gradual transformation of active polymethyl benzenes into less active polycyclic arenes with time on stream; the largest observed species being pyrene [64–66]. Recent results showed that deactivation was initiated by the formation of branched alkanes and alkenes which were trapped in the catalyst cavities, thereby hindering diffusion and enhancing polycyclic arene formation rather than polyaromatic species as the origin of the deactivation in the previous reports [64]. The temperature influence over H-SAPO-34 in MTO reaction is similar to that for H-ZSM-5. Very recently, Qian *et al.* applied a combination in operando UV/vis and IR spectroscopy, with online MS measurements during temperature-programmed MTO, finding that the rates of alkene (represented by propene) and deactivating polycyclic arene formation resembled each other at each temperature, the two competitive routes are based on the same reactants, that is, methoxy species and polymethyl benzenes [67].

16.4
Zeolites in FT Synthesis

A number of studies have pointed out that the catalytic behavior of a FT catalyst is a complicated interplay of many factors. Bifuncational catalysts have been designed to obtain controllable product distributions, particularly higher selectivity to middle-distillate hydrocarbons such as gasoline or diesel fuel and to light olefins such as ethylene and propene. The catalytic system contains both the active metal (Co, Fe carbide, or Ru nanoparticles) or oxides ($ZnCrO_X$) for the CO hydrogenation reaction (the primary reaction) and acidic zeolites for hydrocracking, isomerization, and/or carbine to light olefins.

16.4.1
Hybrid Bifunctional F-T Catalyst

A hybrid bifunctional FT catalyst containing both components (active sites and zeolites) can directly transform syngas into C_5–C_{11} hydrocarbons with good selectivity owing to the catalytic function of an acidic zeolite in hydrocracking/isomerization reactions. Several studies confirm that a hybrid catalyst with a suitable combination of different components might have been beneficial to catalyst stability because the coexistence of an acidic zeolite with the conventional FT catalyst might have inhibited deactivation of the F–T catalyst by simultaneous hydrocracking of waxes.

Zeolite-supported FT metal catalysts are simple bifunctional catalysts [7]. A few early studies already pointed out that the use of zeolites with strong acidity such as H-ZSM-5 as the supports could increase the selectivity to gasoline-range hydrocarbons with a large fraction of branched products [68–70]. Acidity and pore structure play key roles in controlling the product selectivity in these zeolite-supported catalysts. Compared with HY and H-mordenite, the use of H-ZSM-5 as a support for a cobalt catalyst showed relatively high selectivity to gasoline-range hydrocarbons owing to its strong acidity [68]. On the other hand, the pore structure of zeolites also plays a significant role in determining the product distribution. For example, among cobalt catalysts loaded on different zeolites including H-ZSM-5, H-ZSM-11, H-ZSM-12, and H-ZSM-34, the Co/H-ZSM-12 catalyst was more efficient for the formation of gasoline-range hydrocarbons, whereas a larger fraction of heavier n-paraffins remained over the Co/H-ZSM-34 catalyst, although the former catalyst exhibited a weaker acidity [69].

Notably, zeolites with hierarchical porous structures, particularly mesoporous zeolites containing both micro- and mesopores, have attracted considerable attention as a new type of promising catalytic materials in the past decade. Recently, mesoporous zeolites have also been employed for the construction of bifunctional FT catalysts. Particularly, the catalytic performance of mesoporous ZSM-5 and Beta zeolite supported FT catalysts have been intensively investigated by Wang's group [71,72]. They found that the use of meso-ZSM-5 instead of conventional H-ZSM-5 for loading Ru nanoparticles significantly decreased the selectivities to CH_4 and C_2–C_4 and increased that to C_5–C_{11} hydrocarbons, which should be attributed to the hydrocracking of primary hydrocarbons formed on Ru nanoparticles (Figure 16.5). Similar results have also been obtained for mesoporous Beta-supported Ru catalysts [72].

Sartipi *et al.* further compared three kinds of hierarchical zeolites, namely, meso-H-ZSM-5, delaminated MWW (ITQ-2), and meso-H-USY, as supports for cobalt catalysts (Co, 20 wt%) for F–T synthesis [73]. The CO conversion activities for Co/meso-H-USY and Co/meso-H-ZSM-5 were similar and higher than those for Co/ITQ-2 and Co/amorphous SiO_2. The selectivities are interesting in that C_5–C_{11} selectivity was the highest ($\approx$55%) over the Co/meso-H-ZSM-5 catalyst, much higher than those over Co/ITQ-2 and Co/meso-H-USY (about 40%). In contrast the C_{21+} selectivity decreased to 12% over Co/meso-H-USY and <5% over the Co/meso-H-ZSM-5 and Co/ITQ-2 catalysts. The difference in the selectivity is assigned to the acidity and density of the acid sites in the zeolites. Meso-H-USY exhibited a lower density of acid sites and its acidity was also weaker than that of the other two zeolites. On the other hand, meso-H-ZSM-5 had the strongest acidity. These results suggest that the acid density and strength

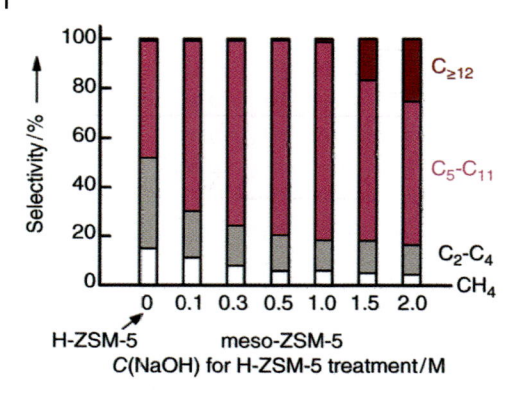

Figure 16.5 Product selectivity of Ru nanoparticles loaded on meso-ZSM-5 prepared by treating H-ZSM-5 with different concentrations of NaOH. (Reproduced with permission from Ref. [71]. Copyright 2011, Wiley-VCH Verlag GmbH.)

of the mesoporous zeolites are also essential factors to tune the product selectivities for F–T synthesis.

Very recently, Jiao et al. reported a surprising selectivity for lower olefins of about 80% of total hydrocarbons, produced directly from synthesis gas at 17% CO conversion under industrially attractive conditions at 400 °C, 25 bar, H_2/CO volume ratio of 1.5, and catalyst lifetimes exceeding 100 h (Figure 16.6) [74]. This bifunctional catalyst, OX-ZEO, consists of a metal oxide (Zn-CrO_x) and a zeolite (MSAPO). By using a metal oxide catalyst to activate CO, the authors prevent polymerization of CH_x species, allowing CO insertion followed by ketene formation, which could be the intermediate for olefin formation within the confined acidic pores of zeolites. This result should be of interest to both

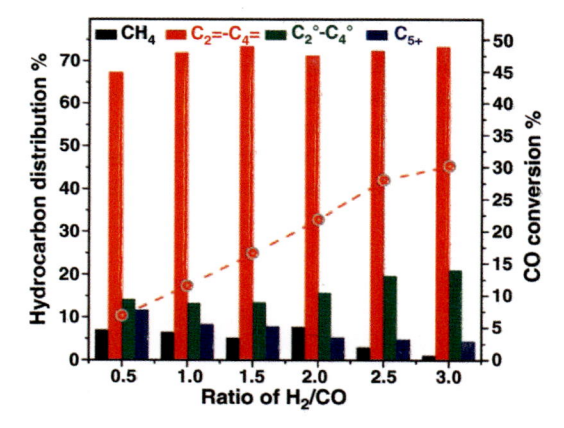

Figure 16.6 CO conversion and product distribution at different H_2/CO ratios in syngas over OX-ZEO catalyst. (Reproduced with permission from Ref. [74]. Copyright 2016, Science.)

academia and industry. It remains to be shown whether ketene or the thermodynamically more stable methanol is the key intermediate in the new process. The bifunctional nature of the catalyst invites structure performance studies including the effects of proximity of oxide and acid functions. The new process could become a serious competitor for industrial processes such as FTO and MTO.

Notably, several key issues should be considered for the development of efficient hybrid catalysts [7]: (1) the optimum reaction conditions for F–T synthesis and hydrocracking/isomerization reactions may be different; (2) intimate contact between the F–T active center and the acidic pores of zeolites is required to perform the two types of reactions smoothly; (3) the deactivation of zeolite may be faster because of carbon deposition on acid sites, whereas the frequent regeneration of the catalyst would be costly.

16.4.2
Core–Shell-Structured Catalyst Bifunctional F-T Catalyst

The spatial arrangement of the two components, which catalyze the two reactions in a tandem manner, can exert a significant effect on the process efficiency. To increase the efficiency of the tandem reactions, Tsubaki and coworkers successfully designed a kind of core–shell-structured bifunctional catalyst (Figure 16.7) [75–78]. They added F–T catalyst pellets into the zeolite precursor solution containing silicon and aluminum sources and a template for zeolite synthesis which was subjected to hydrothermal synthesis. Catalytic studies showed that CO conversions over Co/SiO_2 and the physically mixed hybrid catalyst were almost the same (94–98%), whereas they became slightly lower ($\approx$85%) over catalysts with the Co/SiO_2 core and H-ZSM-5 shell. The ratio of isoparaffins to n-paraffins (C_{iso}/C_n) increased from 0.49 to 1.88 upon increasing the thickness

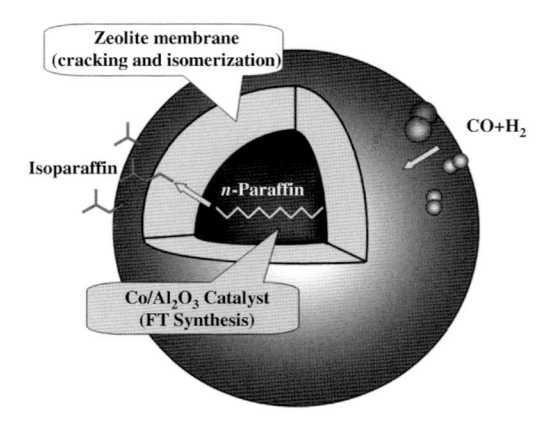

Figure 16.7 Representation of the core/shell ZSM-5 catalyst. (Reproduced with permission from Ref. [75]. Copyright 2008, Wiley-VCH Verlag GmbH.)

of the zeolite shell from 0 to about 24 mm [75]. Recently, Tsubaki and coworkers proposed new strategies to prepare core–shell-structured catalysts, by using zeolite silicalite-1 as the first intermediate layer coated on the Co/SiO$_2$ pellets and then the growth of acidic H-ZSM-5 shell on the intermediate layer [79]. The intermediate layer not only protects the core catalyst from damage in the subsequent hydrothermal synthesis but also facilitate the formation of the outside catalytic H-ZSM-5 shell. Catalytic studies showed CO conversions at 97.7% with C_{iso}/C_n ratio at 0.99 and the C_{11+} selectivity at only 0.4%. These results clearly support the role of the zeolite shell in the hydrocracking and isomerization of the long-chain hydrocarbons formed on the core catalyst.

ZSM-5 microspheres containing well-dispersed uniform Fe$_3$O$_4$ nanoparticles ($\approx$10 nm) inside the sphere were synthesized by an *in situ* crystallization route [80]. The C_5–C_{12} selectivity was 44.6% at a CO conversion of 87% after 110 h of reaction. The selectivity to C_{13}–C_{20} was low (3.2%), and no C_{21+} could be detected, indicating the role of the ZSM-5 zeolite in the hydrocracking of heavier hydrocarbons. However, this catalyst exhibited a higher CH$_4$ selectivity (16%).

16.5
Zeolites in Biomass Conversion

In the above sections, we have given a brief introduction of the widely application of zeolites in the fossil-based energy conversion, which is mainly attributed to the numerous positive features such as controllable acidity, shape-selectivity, and high stability. Notably, all these features also offer the wide-scale application in biomass conversion [6].

16.5.1
The Role of the Acidity

Bio-fats and oils with long linear hydrocarbon chains are structurally similar to petrochemical hydrocarbons. These molecules are very sensitive in acid catalyzed isomerization to yield monobranched-chain unsaturated fatty acids (MoBUFA), which are important as lubricants in cosmetics [81]. As the zeolite pores are already blocked at the beginning of the reaction due to coke formation, the isomerization takes place solely at the entrance of the micropores [82], making it a classic example of pore mouth catalysis. Based on the pore geometry, FER-based catalysts have proven to be some of the more efficient catalysts, especially for the formation of methyl MoBUFA (Figure 16.8) [83], while zeolites with larger pores (Y, USY, and Beta) mainly produce ethyl and propyl MoBUFA [84]. Further investigation showed that MoBUFA are only selectively produced when the Brønsted acid site speciation and density is optimized (up to 85% selectivity) [83].

At high temperature, the fats and oils would transform to hydrocarbon gases, gasoline, light naphtha fractions, and diesel as well as cokes via cracking over

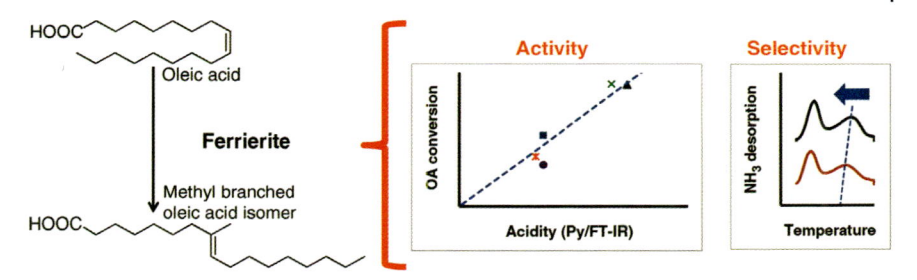

Figure 16.8 Isomerization of oleic acid over ferrierite. (Reproduced with permission from Ref. [75]. Copyright 2015, Elsevier.)

acidic zeolites [85–88]. In the case of lignocellulose, a pretreatment step (e.g., thermal degradation) is always required before cracking over zeolites since the lignocellulose molecules are much larger than fatty acids. Compared to cracking of fossil fuels, biomass cracking requires not only breaking of C—C bonds but also a considerable amount of C—O bonds. This change in substrate composition also translates into a switch of the optimal Si/Al needed, with higher Si/Al ratios for biomass cracking. For example, the highest activity is for Si/Al in the range of 5–8 in typical fossil fuel cracking, while the highest activity in biomass cracking can be found around 15 [89]. Such medium Si/Al zeolites show the best balance between availability of Brønsted acid sites on the one hand and maintaining enough space between the Brønsted acid sites to prevent side reactions on the other hand [89]. Not only the zeolite structure, but also the type and origin of the substrate influence the product distribution. For example, the presence of unsaturations in the alkyl chain affords more aromatic products than in case of saturated fats, while fats with only an alcohol function (fatty alcohols) produce more valuable products than fatty acids [90].

Notably, chemocatalytic conversion of biomass is mostly a liquid phase process which is a major difference with the gas phase processes in crude oil refining and upgrading of pyrolysis products. Zeolites have already shown potential in various Brønsted acidic reactions in the condensed phase. A well-known example hereof is the dehydration of sugars toward platform molecules like 5-hydroxymethylfurfural (5-HMF). However, the obtained 5-HMF yields in aqueous media over zeolites, (e.g., Y and ZSM-5) are still low (10–25%) [91]. The low yields are mainly rooted in the instability of the highly oxygenated molecules (both sugars, intermediates as end-products) at high temperatures and in acidic conditions, especially in aqueous media.

Lewis acid sites could be introduced into zeolites by the substitution of Si atoms by other metal atoms (Sn, Zr, V, Hf) [6,92–95]. In Lewis acid catalyzed reactions, protic solvents (e.g., water) always resulted in a significant decrease in catalyst activity. As a result, hydrophobic zeolite structures could efficiently limit the presence of water molecules within its structure and have been widely used as Lewis acid catalyst in presence of water. A typical example is Sn-Beta, a silicious beta framework containing small amounts of tetrahedral framework incorporated Sn [6].

Sn-Beta zeolite was discovered to be an efficient catalyst in aqueous hexose isomerization in 2010 [96,97], and later it has been used for conversion of sugar derived 5-HMF or trioses (glyceraldehyde, GLY, and 1,3-dihydroxyacetone, DHA) into lactic acid or alkyl lactates for biomass-derived platform molecules in aqueous or alcoholic media, respectively [6]. The catalytic performance of Sn-Beta, Sn-MFI, and Sn-MWW have been compared and Sn-MFI showed a high activity in water, yielding lactic acid, but significantly decreasing in activity when the reaction was performed in methanol, due to the pore-size restrictions in the MFI structure [98,99].

16.5.2
Shape-Selectivity of Zeolites in Biomass Conversion

ZSM-5 zeolite can be regarded as the typical example of a catalyst with excellent shape-selectivity in both oil refining and biomass production [100]. The original example of shape-selectivity in biomass conversion over H-ZSM-5 is the direct conversion of rubber latex and corn oil. Recently, the shape-selective properties of MFI induce a preferred hydrogenation of the slimmer *trans*-configured unsaturated fatty acids over *cis*-isomers, allowing the selective removal of undesired *trans*-isomers [101]. As a result, the obtained high-oleic and low-trans hydrogenated product is beneficial from a nutritional point of view, but also for biolubricant applications. The shape-selective properties of zeolites are also used to tune the product selectivity. For example, the product in the conversion of lignin-derived phenolic monomers to alkanes over metal-loaded MFI are mainly monocycloalkanes, while bicycloalkanes can selectively yield over zeolites with larger pores such as H-beta and H-Y [102–104].

Other examples of shape-selectivity in biomass conversions can be found in the upgrading of bioderived platform molecules. Notably, Sn-MFI and Ti-MFI are nearly inactive in hexose isomerization to fructose while Sn-Beta is very active, but Sn-MFI and Ti-MFI are very active in trioses isomerization, which are obviously related to the difference in their pore sizes [97,105].

16.5.3
Stability of Zeolites in Biomass Conversion

Normally, the stability of zeolites is measured with a gas phase under hydrothermal conditions [2,3]. However, many biomass conversions are performed with a liquid phase, which requires a different definition of the zeolite stability [6].

In the case of sugar transformations, the acidic products such as lactic acid and levulinic acid destroy the framework of zeolites. Especially in water, with formation of lactic acid, the USY zeolite is strongly affected by severe loss of pore volume, surface area, and acidity, which was attributed to the selective dealumination by the formed lactic acid [106]. Similar behavior was observed when the reaction was performed in neat levulinic acid [107].

Even in neutral super atmospheric liquid water (150–250 °C), tremendous changes in zeolite structure and properties occur [108,109], which are quite different from the phenomenon in steaming. For example, degradation of zeolite Y have been observed in neutral super atmospheric liquid water (150–250 °C) via hydrolysis of Si—O—Si bonds, initiated at silanol defects [108,109]. Notably, the stability of a zeolite in super atmospheric liquid water is strongly dependent on the zeolite topology. For example, there is a dependency between the framework stability of a zeolite and the framework density [110]. Zeolites with high framework densities, like MOR and MFI, are relatively stable up to 250 °C, while topologies with low density (BEA and especially FAU) undergo already extensive transformations at 150 °C and are largely transformed into amorphous silica-alumina after a treatment at 200 °C [109,110].

16.6
Zeolites in Selective Catalytic Reduction (SCR)

Currently, the catalysts used for SCR in industry are mainly based on TiO_2-supported V_2O_5, promoted with WO_3, which have also been used for heavy-duty diesel vehicles in Europe since 2005 [111]. However, problems persist due to the high activity for oxidation of SO_2 to SO_3, the rapid decrease in activity and selectivity at 550 °C, and the toxicity of the vanadia species, which begin to volatilize above 650 °C. Therefore, new catalyst should be developed to qualify more stringent legislation in the future [111].

16.6.1
Cu-Zeolites as SCR Catalyst

The exceptional NO decomposition activity over Cu-ZSM-5 was discovered by Iwamoto et al. [112]. Although other copper containing zeolite systems have been researched (e.g., Cu-MOR, Cu-BEA, and Cu-FAU), Cu-ZSM-5 has received the bulk of the attention due to its remarkable deNO$_x$ activity over a wide temperature range, 200–400 °C. Cu-ZSM-5 has shown superior conversion and selectivity in direct NO decomposition, hydrocarbon assisted SCR, and NH$_3$-SCR. However, the main drawback of Cu-ZSM-5 is its lack of hydrothermal stability. At higher temperatures, 700 °C and above, and in the presence of water vapor (steam), the catalyst deactivates, giving diminished SCR activity due to dealumination of the zeolite framework [111–113].

The pore size of zeolites was proposed to affect the hydrothermal stability of Cu-zeolite catalysts. Modenet et al. reported that the active Cu species in Cu-FER catalysts with smaller pore size (0.42 × 0.54 nm) showed higher stability than that in Cu-Beta catalysts with relatively larger pore size (0.66 × 0.67 nm) [114]. Later, Blakemanet et al. believed that the small pore structure in Cu-CHA (CHA =

chabazite) catalysts could prevent the degradation of the zeolite framework structure (0.38 × 0.38 nm) by Cu species as compared to that in Cu-Beta catalysts [115]. Fickelet et al. confirmed that the higher hydrothermal stability of the Cu-CHA catalyst was related to its small pore structure (0.38 × 0.38 nm) [116]. The detached Al(OH)$_3$ species in the zeolite framework with kinetic diameter of 0.503 nm could not exit the pores of the CHA structure, and they could reincorporate into the zeolite framework in the subsequent cooling stage. They concluded that the constricting dimension of the chabazite (CHA) structure was one of the most important factors for the high hydrothermal stability of Cu-CHA catalysts.

Because of the excellent hydrothermal stability of the CHA structure, the Cu-CHA catalysts (Cu-SSZ-13 and Cu-SAPO-34) in the NH$_3$-SCR reaction for the deNO$_x$ process from diesel exhaust have been widely investigated in recent years. Fickelet et al. confirmed that the Cu-SSZ-13 catalyst displayed superior SCR performance even after hydrothermal treatment at 750 °C for 16 h [116].

Notably, the structure-directing agents (SDA, N,N,N-trimethyl-1-adamantammonium hydroxide (TMAdaOH)) for the synthesis of SSZ-13 zeolite are very expensive, which would limit their wide applications [117]. To solve this problem, a new synthesis strategy, based on a one-pot synthesis, can greatly decrease the cost of Cu-CHA catalysts together with allowing controllable metal loadings. Ren et al. have theoretically compared the configuration of the CHA cage, a building unit of SSZ-13 zeolites, with a series of inexpensive inorganic or organic compounds. They found that a low-cost copper complex (Cu^{2+} coordinated with tetraethylenepentamine, Cu-TEPA) matches the CHA cage well (Figure 16.9). The copper complex was then used to synthesize the zeolite Cu-SSZ-13 (designated as Cu-ZJM-1,), showing excellent NH$_3$-SCR performance, high resistance to space velocity, and good hydrothermal stability simultaneously [118,119].

Although the superior performance of Cu-CHA catalysts in NH$_3$-SCR has been well recognized, the understanding of the structure–activity relationship is still ongoing. There is no doubt that isolated Cu^{2+} ions are the major active sites, but the location and the distribution of the active sites is still under debate. It is well known that there are four types of cationic sites in the chabazite structure [120,121]. Site I is displaced from the six-membered ring into the ellipsoidal cavity, site II is located near the center of the ellipsoidal cavity, site III is located in the center of the hexagonal prism, and site IV is located near the eight-membered-ring window. Because of the different locations of Cu^{2+} ions,

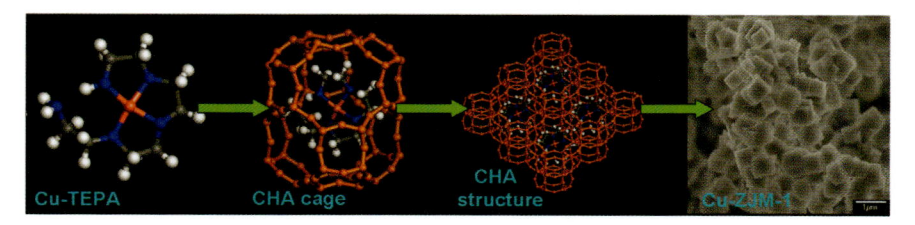

Figure 16.9 Mechanism on Cu-TEPA-templated Cu-SSZ-13 zeolites. (Reprinted with permission from Ref. [118]. Copyright 2011, Royal Society of Chemistry.)

different performance in NH$_3$-SCR was observed over various Cu-CHA catalysts even if they have the same type of CHA framework structure.

16.6.2
Fe-Zeolites as SCR Catalyst

Fe-exchanged zeolites have been proven to be the active catalysts for NH$_3$-SCR of NO$_x$. Among a number of Fe-based zeolite catalysts, the application of the MFI structure (Fe-ZSM-5) has attracted much attention since the end of the last century [122,123]. However, the hydrothermal stability of the Fe-ZSM-5 also hindered its application owing to the following reasons [124]: (i) the rapid dealumination of Al-OH-Si sites, (ii) the rapid depletion of dimeric iron species, and (iii) the slow migration of isolated Fe ions followed by the dealumination of Al sites. The remaining NH$_3$-SCR activity after hydrothermal aging under different conditions was mainly attributed to the residual monomeric Fe^{3+} species located at ion exchange sites, and the loss of active Fe sites with high redox ability, rather than the loss of Brønsted acidity, was responsible for the decreased activity [125]. In addition, hydrothermal deactivation could result in a slight increase in the high temperature NH$_3$-SCR activity of Fe-ZSM-5 catalysts, mainly due to a decrease in over-oxidation of the reducing agent NH$_3$ at high temperatures [124–126].

Recently, a kind of an Al-rich Beta zeolite with extra high hydrothermal stability was successfully synthesized in the absence of organic templates by the addition of calcined Beta crystals as seeds in the starting aluminosilicate gels [127], which also supply a good platform for the preparation of NH$_3$-SCR reaction catalysts. After ion-exchange with Fe, the relatively high concentration of Fe counter-ion species in Al-rich Fe-BEA caused greatly improved conversion of NO$_x$ over the low temperature range from 200 to 300 °C, which is very important for catalyst application in diesel exhaust gases. The rates for conversion of NO$_x$ to N$_2$ over Al-rich Fe-BEA are more than twice those of conventional Fe-BEA [128].

As the same in the case of Cu-zeolite, the active sites in Fe-zeolite catalysts for the NH$_3$-SCR are also under debate. A popular proposal is that all Fe species in Fe-ZSM-5 catalysts are active, although they exhibited different temperature dependencies in the SCR reaction and only monomeric Fe^{3+} species contribute to the SCR reaction below 300 °C [129]. As a result, Fe-zeolite catalysts with a maximum quantity of monomeric Fe^{3+} sites should be prepared, and the formation of Fe$_x$O$_y$ clusters should be avoided to obtain excellent NH$_3$-SCR activity at low temperatures.

16.7
Conclusions and Outlook

Owing to their applications in industrial processes, it can be concluded that zeolites are critical for modern society. Looking back at the industrial development, it is clear that the impact of the introduction of zeolites into refinery plants is revolutionary as it resulted in a significant increase in gasoline yield from catalytic cracking, giving more efficient utilization of the petroleum feedstocks. Later, many industrial applications of zeolites catalysts were found. In this century, applications of zeolite catalysts have novel perspectives broadened even more. For example, methanol to olefin conversion is one of the most promising routes to produce useful chemicals from relatively abundant carbon sources such as coal and methane; the application of zeolites for biomass conversions has also undergone a significant development. Obviously, zeolite catalysts are applied not only to traditional chemical processes but also to the emergent removal of environmental exhausts. For example, Cu- and Fe-based zeolite catalysts have been demonstrated to be efficient commercial SCR catalysts.

Although zeolites have been used industrially for a long time, the field of zeolite catalysis will continue to increase in relevance, building on the prospects for new structures and applications. It can be expected that zeolites will continue to play a vital role in chemical industry, as they can provide much-needed catalytic selectivity toward high-demand base chemicals as well as high-value fine chemicals and intermediates.

References

1 National Research Council (1992) *Catalysis Looks to the Future, (Panel on New Directions in Catalytic Science and Technology)*, National Academy Press, Washington D.C., p. 18.
2 Corma, A. (1995) Inorganic solid acids and their use in acid-catalyzed hydrocarbon reactions. *Chem. Rev.*, **95**, 559–614.
3 Corma, A. (1997) From microporous to mesoporous molecular sieve materials and their use in catalysis. *Chem. Rev.*, **97**, 2373–2419.
4 Hudec, P. (2011) FCC catalyst – key element in refinery technology. 45th International Petroleum Conference, Bratislava.
5 Olsbye, U., Svelle, S.K., Lillerud, P. *et al.* (2015) The formation and degradation of active species during methanol conversion over protonated zeotype catalysts. *Chem. Soc. Rev.*, **44**, 7155–7176.
6 Ennaert, T., Van Aelst, J., Dijkmans, J. *et al.* (2016) Potential and challenges of zeolite chemistry in the catalytic conversion of biomass. *Chem. Soc. Rev.*, **45**, 584–611.

7 Zhang, Q., Cheng, K., and Kang, J. (2014) Fischer-Tropsch catalysts for the production of hydrocarbon fuels with high selectivity. *ChemSusChem*, **7**, 1251–1264.
8 Brandenberger, S., Kröcher, O., Tissler, A. *et al.* (2008) The state of the art in selective catalytic reduction of NO_x by ammonia using metal-exchanged zeolite catalysts. *Catal. Rev.*, **50**, 492–531.
9 Reichle, A.D. (1992) Fluid catalytic cracking hits 50 year mark on the run. *Oil Gas J.*, **90**, 41–48.
10 Cundy, C.S. and Cox, P.A. (2003) The hydrothermal synthesis of zeolites: history and de-velopment from the earliest days to the present time. *Chem. Rev.*, **103**, 663–701.
11 Hwang, Y.K., Chang, J.-S., Park, S.E. *et al.* (2005) Microwave fabrication of MFI zeolite crystals with a fibrous morphology and their applications. *Angew. Chem., Int. Ed.*, **44**, 556–560.
12 Corma, A. and Martinez, A. (2005) Zeolitesin refining and petrochemistry. *Stud. Surf. Sci. Catal.*, **157**, 337–366.

13 Degnan, T.F., Chitnis, G.K., and Schipper, P.H. (2000) History of ZSM-5 fluid catalytic cracking additive development at Mobil. *Microporous Mesoporous Mater.*, **35–36**, 245–252.

14 Buchanan, J.S. (2000) The chemistry of olefins production by ZSM-5 addition to catalytic cracking units. *Catal. Today*, **55**, 207–212.

15 Kaeding, W.W. (1981) Selective alkylation toluene with methanol to produce para-xylene. *J. Catal.*, **67**, 159–174.

16 Vogt, E.T.C. and Weckhuysen, B.M. (2015) Fluid catalytic cracking: recent developments on the grand old lady of zeolite catalysis. *Chem. Soc. Rev.*, **44**, 7342–7370.

17 Chang, C.D. and Silvestri, A.J. (1977) The conversion of methanol and other O-compounds to hydrocarbons over zeolite catalysts. *J. Catal.*, **47**, 249–259.

18 Chang, C.D. (1983) Hydrocarbons from methanol. *Catal. Rev.*, **25**, 1–118.

19 Ondrey, G. (2011) Coal-to-chemicals. *Chem. Eng.*, **118**, 16–20.

20 OGJ editors (2010) Methanol-to-olefins unit starts up in China. *Oil Gas J.*, 108.

21 Chen, J.Q., Bozzano, A., Glover, B. *et al.* (2005) Recent advancements in ethylene and propylene production using the UOP/Hydro MTO process. *Catal. Today*, **106**, 103–107.

22 Koempel, H. and Liebner, W. (2007) Lurgi's methanol to propylene (MTP (R)) report on a successful commercialisation. *Stud. Surf. Sci. Catal.*, **167**, 261–267.

23 Dahl, I.M. and Kolboe, S. (1993) On the reaction mechanism for propene formation in the MTO reaction over SAPO-34. *Catal. Lett.*, **20**, 329–336.

24 Dahl, I.M. and Kolboe, S. (1994) On the reaction mechanism for hydrocarbon formation from methanol over SAPO-34: I. isotopic labeling studies of the Co-reaction of ethene and methanol. *J. Catal.*, **149**, 458–464.

25 Dahl, I.M. and Kolboe, S. (1996) On the reaction mechanism for hydrocarbon formation from methanol over SAPO-34 .2. Isotopic labeling studies of the co-reaction of propene and methanol. *J. Catal.*, **161**, 304–309.

26 Song, W.G., Haw, J.F., Nicholas, J.B. *et al.* (2000) Methylbenzenes are the organic reaction centers for methanol-to olefin catalysis on HSAPO-34. *J. Am. Chem. Soc.*, **122**, 10726–10727.

27 Mikkelsen, O., Ronning, P.O., and Kolboe, S. (2000) Use of isotopic labeling for mechanistic of the methanol-to-hydrocarbons reaction. Methylation of toluene with methanol over H-ZSM-5, H-mordenite and H-beta. *Microporous Mesoporous Mater.*, **40**, 95–113.

28 Mole, T., Bett, G., and Seddon, D. (1983) Conversion of methanol to hydrocarbons over ZSM-5 zeolite-anexamination of the role of aromatic-hydrocarbons using carbon-13-labeled and deuterium-labeled feeds. *J. Catal.*, **84**, 435–445.

29 Song, W.G., Nicholas, J.B., Sassi, A. *et al.* (2002) Synthesis of the heptamethylbenzenium cation in zeolite-beta: *in situ* NMR and theory. *Catal. Lett.*, **81**, 49–53.

30 Sullivan, R.F., Sieg, R.P., Langlois, G.E. *et al.* (1961) A new reaction that occurs in hydrocracking of certain aromatic hydrocarbons. *J. Am. Chem. Soc.*, **83**, 1156–1160.

31 Xu, T. and Haw, J.F. (1994) Cycleopentenyl carbenium ion formation in acidic zeolite-an *in-situ* NMR-study of cyclic precursors. *J. Am. Chem. Soc.*, **116**, 7753–7759.

32 Wang, C.M., Wang, Y.D., Liu, H.X. *et al.* (2012) Theoretical insight into the minor role of paring mechanism in the methanol-to-olefins conversion within HSAPO-34 catalyst. *Microporous Mesoporous Mater.*, **158**, 264–271.

33 Erichsen, M.W., Svelle, S., and Olsbye, U. (2013) H-SAPO-5 as nethanol to-olefins (MTO) model catalyst: towards elucidating the effect of acid strength. *J. Catal.*, **298**, 94–101.

34 Llias, S. and Bhan, A. (2014) The mechanism of aromatic dealkylation in methanol-to-hydrocarbons conversion on H-ZSM-5: what are the aromatic precursors to light olefins? *J. Catal.*, **311**, 6–16.

35 Sassi, A., Wildman, M.A., Ahn, H.J. *et al.* (2002) Methylbenzene chemistry on zeolite H Beta: multiple insights into

methanol-to-olefin catalysis. *J. Phys. Chem. B*, **106**, 2294–2303.

36 Wang, S., Chen, Y., Wei, Z. et al. (2015) Polymethylbenzene or alkene cycle? Theoretical study on their contribution to the process of methanol to olefins over H-ZSM-5 zeolite. *J. Phys. Chem. C*, **119**, 28482–28498.

37 Dai, W., Wang, C., Dyballa, M. et al. (2015) Understanding the early stages of the methanol-to-olefin conversion on H-SAPO-34. *ACS Catal.*, **5**, 317–326.

38 Jiang, Y., Wang, W., Reddy Marthala, V. et al. (2006) Effect of organic impurities on the hydrocarbon formation via the decomposition of surface methoxy groups on acidic zeolite catalysts. *J. Catal.*, **238**, 21–27.

39 Goguen, P.W., Xu, T., Barich, D.H. et al. (1998) Pulse-quench catalytic reactor studies reveal a carbon-pool mechanism in methanol-to-gasoline chemistry on zeolite HZSM-5. *J. Am. Chem. Soc.*, **120**, 2650–2651.

40 Haw, J.F., Nicholas, J.B., Song, W. et al. (2000) Roles for cyclopentenyl cations in the synthesis of hydrocarbons from methanol on zeolite catalyst HZSM-5. *J. Am. Chem. Soc.*, **122**, 4763–4775.

41 McCann, D.M., Lesthaeghe, D., Kletnieks, P.W. et al. (2008) A complete catalytic cycle for supramolecular methanol-to-olefins conversion by linking theory with experiment. *Angew. Chem., Int. Ed.*, **47**, 5179–5182.

42 Xu, S., Zheng, A., Wei, Y. et al. (2013) Direct observation of cyclic carbenium ions and their role in the catalytic cycle of the methanol-to-olefin reaction over chabazite zeolites. *Angew. Chem., Int. Ed.*, **125**, 11564–11568.

43 Haw, J.F., Richardson, B.R., Oshiro, I.S. et al. (1989) Reactions of propene on zeolite hy catalyst studied by insitu variable-temperature solid-state nuclear magnetic-resonance spectroscopy. *J. Am. Chem. Soc.*, **111**, 2052–2058.

44 Song, W., Nicholas, J.B., and Haw, J.F. (2001) A persistent carbenium ion on the methanol-to-olefin catalyst HSAPO-34: acetone shows the way. *J. Phys. Chem. B*, **105**, 4317–4323.

45 Wang, W., Seiler, M., and Hunger, M. (2001) Role of surface methoxy species in the conversion of methanol to dimethyl ether on acidic zeolites investigated by *in situ* stopped-flow MAS NMR spectroscopy. *J. Phys. Chem. B*, **105**, 12553–12558.

46 Wang, W. and Hunger, M. (2008) Reactivity of surface alkoxy species on acidic zeolite catalysts. *Acc. Chem. Res.*, **41**, 895–904.

47 Lesthaeghe, D., Van Speybroeck, V., Marin, G.B. et al. (2007) The rise and fall of direct mechanisms in methanol-to-olefin catalysis: an overview of theoretical contributions. *Ind. Eng. Chem. Res.*, **46**, 8832–8838.

48 Sinclair, P.E. and Catlow, C.R.A. (1997) Generation of carbenes during methanol conversion over Bronsted acidic aluminosilicates. A computational study. *J. Phys. Chem. B*, **101**, 295–298.

49 Forester, T. and Howe, R. (1987) *In situ* FTIR studies of methanol and dimethyl ether in ZMS-5. *J. Am. Chem. Soc.*, **109**, 5076–5082.

50 Anderson, M.W. and Klinowski, J. (1990) Solid-state NMR-studies of the shape-selective catalytic conversion of methanol into gasoline on zeolite ZSM-5. *J. Am. Chem. Soc.*, **112**, 10–16.

51 Munson, E.J., Lazo, N.D., Moellenhoff, M.E. et al. (1991) CO is neither an intermediate nor a catalyst in MTG chemistry on zeolite HZSM-5. *J. Am. Chem. Soc.*, **113**, 2783–2784.

52 Lesthaeghe, D., Van Speybroeck, V., Marin, G.B. et al. (2006) Understanding the failure of direct C-C coupling in the zeolite-catalyzed methanol-to-olefin process. *Angew. Chem., Int. Ed.*, **45**, 1714–1719.

53 Lesthaeghe, D., Van Speybroeck, V., Marin, G.B. et al. (2006) What role do oxonium ions and oxonium ylides play in the ZSM-5 catalysed methanol-to-olefin process? *Chem. Phys. Lett.*, **417**, 309–315.

54 Kubelková, L., Nováková, J., and Jíru, P. (1984) Properties of Y-type zeolites with various silicon aluminum ratios obtained by dealumination with silicon tetrachloride – distribution of aluminum and hydroxyl-groups and interaction with

ethanol. *Stud. Surf. Sci. Catal.*, **18**, 217–224.
55 Nováková, J., Kubelková, L., and Dolejšek, Z. (1986) Deuterium-labeled methanol in reactions with HZSM-5 zeolite. *J. Catal.*, **97**, 277–279.
56 Nováková, J., Kubelková, L., and Dolejšek, Z. (1987) Primary reaction steps in the methanol-to-olefin transformation on zeolites. *J. Catal.*, **108**, 208–213.
57 Hutchings, G.J., Gottschalk, F., Hall, M.V.M. *et al.* (1987) Hydrocarbon formation from methylating agents over the zeolite catalyst ZSM-5 – comments on the mechanism of carbon carbon bond and methane formation. *J. Chem. Soc. Faraday Trans. 1*, **83**, 571–583.
58 Hutchings, G.J., Gottschalk, F., and Hunter, R. (1987) Kinetic-model for methanol conversion to olefins with respect to methane formation at low conversion – comment. *Ind. Eng. Chem. Res.*, **26**, 635–637.
59 Li, J., Wei, Z., Chen, Y. *et al.* (2014) A route to form initial hydrocarbon pool species in methanol conversion to olefins over zeolites. *J. Catal.*, **317**, 277–283.
60 Song, W., Marcus, D.M., Fu, H. *et al.* (2002) An Oft-studied reaction that may never have been: direct catalytic conversion of methanol or dimethyl ether to hydrocarbons on the solid acids HZSM-5 or HSAPO-34. *J. Am. Chem. Soc.*, **124**, 3844–3845.
61 Schulz, H. (2010) "Coking" of zeolites during methanol conversion: basic reactions of the MTO-, MTP- and MTG processes. *Catal. Today*, **154**, 183–194.
62 Mores, D., Stavitski, E., Kox, M.H.F. *et al.* (2008) Space- and time-resolved in-situ spectroscopy on the coke formation in molecular sieves: methanol-to-olefin conversion over H-ZSM-5 and H-SAPO-34. *Chem. Eur. J.*, **14**, 11320–11327.
63 Müller, S., Liu, Y., Vishnuvarthan, M. *et al.* (2015) Coke formation and deactivation pathways on H-ZSM-5 in the conversion of methanol to olefins. *J. Catal.*, **325**, 48–59.
64 Hereijgers, B.P.C., Bleken, F., Nilsen, M.H. *et al.* (2009) Product shape selectivity dominates the Methanol-to-Olefins (MTO) reaction over H-SAPO-34 catalysts. *J. Catal.*, **264**, 77–87.
65 Arstad, B. and Kolboe, S. (2001) Methanol-to-hydrocarbons reaction over SAPO-34. Molecules confined in the catalyst cavities at short time on stream. *Catal. Lett.*, **71**, 209–212.
66 Fu, H., Song, W., Haw, J.F. *et al.* (2001) Polycyclic aromatics formation in HSAPO-34 during methanol-to-olefin catalysis: ex situ characterization after cryogenic grinding. *Catal. Lett.*, **76**, 89–94.
67 Qian, Q., Vogt, C., Mokhtar, M. *et al.* (2014) Combined operando UV/Vis/IR spectroscopy reveals the role of methoxy and aromatic species during the methanol-to-olefins reaction over H-SAPO-34. *ChemCatChem*, **6**, 3396–3408.
68 Bessell, S. (1993) Support effects in cobalt-based Fischer–Tropsch catalysis. *Appl. Catal. A*, **96**, 253–268.
69 Bessell, S. (1995) Investigation of bifunctional zeolite-supported cobalt Fischer–Tropsch catalysts. *Appl. Catal. A*, **126**, 235–244.
70 Jong, S.J. and Cheng, S. (1995) Reduction behavior and catalytic properties of cobalt containing ZSM-5 zeolites. *Appl. Catal. A*, **126**, 51–66.
71 Kang, J., Cheng, K., Zhang, L. *et al.* (2011) Mesoporous zeolite-supported ruthenium nanoparticles as highly selective Fischer–Tropsch catalysts for the production of C_5–C_{11} isoparaffins. *Angew. Chem., Int. Ed.*, **50**, 5200–5203.
72 Cheng, K., Kang, J., Huang, S. *et al.* (2012) Mesoporous Beta zeolite-supported ruthenium nanoparticles for selective conversion of synthesis gas to C_5–C_{11} isoparaffins. *ACS Catal.*, **2**, 441–449.
73 Sartipi, S., Alberts, M., Meijerink, M.J. *et al.* (2013) Toward liquid fuels from biosyngas: effect of zeolite structure in hierarchical-zeolite-supported cobalt catalysts. *ChemSusChem*, **6**, 1646–1650.
74 Jiao, F., Li, J., Pan, X. *et al.* (2016) Selective conversion of syngas to light olefins. *Science*, **351**, 1065–1068.
75 Bao, J., He, J., Zhang, Y. *et al.* (2008) A core/shell catalyst produces a spatially

confined effect and shape selectivity in a consecutive reaction. *Angew. Chem., Int. Ed.*, **47**, 353–356.

76 Yang, G., Tsubaki, N., Shamoto, J. *et al.* (2010) Confinement effect and synergistic function of H-ZSM-5/Cu-ZnO-Al_2O_3 capsule catalyst for one-step controlled synthesis. *J. Am. Chem. Soc.*, **132**, 8129–8136.

77 He, J., Liu, Z., Yoneyama, Y. *et al.* (2006) Multiple-functional capsule catalysts: a tailor-made confined reaction environment for the direct synthesis of middle isoparaffins from syngas. *Chem. Eur. J.*, **12**, 8296–8304.

78 Li, X., He, J., Meng, M. *et al.* (2009) One-step synthesis of H-beta zeolite-enwrapped Co/Al_2O_3 Fischer–Tropsch catalyst with high spatial selectivity. *J. Catal.*, **265**, 26–34.

79 Yang, G., Xing, C., Hirohama, W. *et al.* (2013) Tandem catalytic synthesis of light isoparaffin from syngas via Fischer–Tropsch synthesis by newly developed core–shell-like zeolite capsule catalysts. *Catal. Today*, **215**, 29–35.

80 Li, B., Sun, B., Qian, X. *et al.* (2013) *In-Situ* crystallization route to nanorod-aggregated functional ZSM-5 microspheres. *J. Am. Chem. Soc.*, **135**, 1181–1184.

81 Biermann, U. and Metzger, J.O. (2008) Synthesis of alkyl-branched fatty acids. *Eur. J. Lipid Sci. Technol.*, **110**, 805–811.

82 Wiedemann, S.C.C., Stewart, J.A., Soulimani, F. *et al.* (2014) Skeletal isomerisation of oleic acid over ferrierite in the presence and absence of triphenylphosphine: pore mouth catalysis and related deactivation mechanisms. *J. Catal.*, **316**, 24–35.

83 Wiedemann, S.C.C., Munõz-Murillo, A., Oord, R. *et al.* (2015) Skeletal isomerisation of oleic acid over ferrierite: influence of acid site number, accessibility and strength on activity and selectivity. *J. Catal.*, **329**, 195–205.

84 Zhang, Z., Dery, M., Zhang, S. *et al.* (2004) New process for the production of branched-chain fatty acids. *J. Surfactants Deterg.*, **7**, 211–215.

85 Katikaneni, S.P.R., Adjaye, J.D., and Bakhshi, N.N. (1995) Studies on the catalytic conversion of canola oil to hydrocarbons – influence of hybrid catalysts and steam. *Energy Fuels*, **9**, 599–609.

86 Milne, T.A., Evans, R.J., and Nagle, N. (1990) Catalytic conversion of microalgae and vegetable-oils to premium gasoline, with shape-selective zeolites. *Biomass*, **21**, 219–232.

87 Twaiq, F.A., Zabidi, N.A.M., and Bhatia, S. (1999) Catalytic conversion of palm oil to hydrocarbons: Performance of various zeolite catalysts. *Ind. Eng. Chem. Res.*, **38**, 3230–3237.

88 Twaiq, F.A.A., Mohamad, A.R., and Bhatia, S. (2004) Performance of composite catalysts in palm oil cracking for the production of liquid fuels and chemicals. *Fuel Process. Technol.*, **85**, 1283–1300.

89 Foster, A.J., Jae, J., Cheng, Y.-T. *et al.* (2012) Optimizing the aromatic yield and distribution from catalytic fast pyrolysis of biomass over ZSM-5. *Appl. Catal. A*, **423**, 154–161.

90 Cěrny, R., Kubu, M., and Kubička, D. (2013) The effect of oxygenates structure on their deoxygenation over USY zeolite. *Catal. Today*, **204**, 46–53.

91 Lima, S., Antunes, M.M., Fernandes, A. *et al.* (2010) Acid-catalysed conversion of saccharides into furanic aldehydes in the presence of three-dimensional mesoporous Al-TUD-1. *Molecules*, **15**, 3863–3877.

92 Mal, N.K. and Ramaswamy, A.V. (1997) Synthesis and catalytic properties of large-pore Sn-beta and Al-free Sn-beta molecular sieves. *Chem. Commun.*, 425–426.

93 Mal, N.K., Ramaswamy, V., Ganapathy, S. *et al.* (1994) Synthesis and characterization of crystalline, tin-silicate molecular-sieves with MFI structure. *J. Chem. Soc. Chem. Commun.*, 1933–1934.

94 Moliner, M. (2014) State of the art of Lewis acid-containing zeolites: lessons from fine chemistry to new biomass transformation processes. *Dalton Trans.*, **43**, 4197–4208.

95 Yang, G., Pidko, E.A., and Hensen, E.J.M. (2013) Structure, stability, and Lewis

acidity of mono and double Ti, Zr, and Sn framework substitutions in BEA zeolites: a periodic density functional theory study. *J. Phys. Chem. C*, **117**, 3976–3986.

96 Moliner, M., Román-Leshkov, Y., and Davis, M.E. (2010) Tin-containing zeolites are highly active catalysts for the isomerization of glucose in water. *Proc. Natl. Acad. Sci. USA*, **107**, 6164–6168.

97 Román-Leshkov, Y., Moliner, M., Labinger, J.A. et al. (2010) Mechanism of glucose isomerization using a solid Lewis acid catalyst in water. *Angew. Chem., Int. Ed.*, **49**, 8954–8957.

98 Osmundsen, C.M., Holm, M.S., Dahl, S. et al. (2012) Tin-containing silicates: structure-activity relations. *Proc. R. Soc. A Math. Phys. Eng. Sci.*, **468**, 2000–2016.

99 Taarning, E., Saravanamurugan, S., Holm, M.S. et al. (2009) Zeolite-catalyzed isomerization of triose sugars. *ChemSusChem*, **2**, 625–627.

100 Weisz, P.B., Haag, W.O., and Rodewald, P.G. (1979) Catalytic production of high-grade fuel (gasoline) from biomass compounds by shape-selective catalysis. *Science*, **206**, 57–58.

101 Philippaerts, A., Paulussen, S., Breesch, A. et al. (2011) Unprecedented shape selectivity in hydrogenation of triacylglycerol molecules with Pt/ZSM-5 zeolite. *Angew. Chem., Int. Ed.*, **50**, 3947–3949.

102 Hong, D.-Y., Miller, S.J., Agrawal, P.K. et al. (2010) Hydrodeoxygenation and coupling of aqueous phenolics over bifunctional zeolite-supported metal catalysts. *Chem. Commun.*, **46**, 1038–1040.

103 Zhao, C., Camaioni, D.M., and Lercher, J.A. (2012) Selective catalytic hydroalkylation and deoxygenation of substituted phenols to bicycloalkanes. *J. Catal.*, **288**, 92–103.

104 Zhao, C., Song, W., and Lercher, J.A. (2012) Aqueous phase hydroalkylation and hydrodeoxygenation of phenol by dual functional catalysts comprised of Pd/C and H/La-BEA. *ACS Catal.*, **2**, 2714–2723.

105 Lew, C.M., Rajabbeigi, N., and Tsapatsis, M. (2012) Tin-containing zeolite for the isomerization of cellulosic sugars. *Microporous Mesoporous Mater.*, **153**, 55–58.

106 West, R.M., Holm, M.S., Saravanamurugan, S. et al. (2010) Zeolite H-USY for the production of lactic acid and methyl lactate from C_3-sugars. *J. Catal.*, **269**, 122–130.

107 Luo, W., Deka, U., Beale, A.M. et al. (2013) Ruthenium-catalyzed hydrogenation of levulinic acid: Influence of the support and solvent on catalyst selectivity and stability. *J. Catal.*, **301**, 175–186.

108 Ennaert, T., Geboers, J., Gobechiya, E. et al. (2015) Conceptual frame rationalizing the self-stabilization of H-USY zeolites in hot liquid water. *ACS Catal.*, **5**, 754–768.

109 Ravenelle, R.M., Schüßler, F., D'Amico, A. et al. (2010) Stability of zeolites in hot liquid water. *J. Phys. Chem. C*, **114**, 19582–19595.

110 Lutz, W., Toufar, H., Kurzhals, R. et al. (2005) Investigation and modeling of the hydrothermal stability of technically relevant zeolites. *Adsorption*, **11**, 405–413.

111 Granger, P. and Parvulescu, V.I. (2011) Catalytic NOx abatement systems for mobile sources: From three-way to lean burn after-treatment technologies. *Chem. Rev.*, **111**, 3155–3207.

112 Iwamoto, M., Furukawa, H., Mine, Y. et al. (1986) Copper(ii) ion-exchanged zsm-5 zeolites as highly-active catalysts for direct and continuous decomposition of nitrogen monoxide. *J. Chem. Soc. Chem. Commun.*, 1272–1273.

113 Kwak, J.H., Tran, D., Burton, S.D. et al. (2012) Effects of hydrothermal aging on NH_3-SCR reaction over Cu/zeolites. *J. Catal.*, **287**, 203–209.

114 Moden, B., Donohue, J.M., Cormier, W.E. et al. (2008) Effect of Cu-loading and structure on the activity of Cu-exchanged zeolites for NH_3-SCR, in *Stud Surf Sc Catal*, Elsevier, pp. 1219–1222.

115 Blakeman, P.G., Burkholder, E.M., Chen, H.-Y. et al. (2014) The role of pore size on the thermal stability of zeolite supported Cu SCR catalysts. *Catal. Today*, **231**, 56–63.

116 Fickel, D.W., D'Addio, E., Lauterbach, J.A. et al. (2011) The ammonia selective catalytic reduction activity of copper-exchanged small-pore zeolites. *Appl. Catal. B*, **102**, 441–448.

117 Zones, S.I. (1991) Conversion of faujasites to high-silica chabazite SSZ-13 in the presence of N,N,N-trimethyl-1-adamantammonium iodide. *J. Chem. Soc., Faraday Trans.*, **87**, 3709–3716.

118 Ren, L., Zhu, L., Yang, C. et al. (2011) Designed copper-amine complex as an efficient template for one-pot synthesis of Cu-SSZ-13 zeolite with excellent activity for selective catalytic reduction of NO_x by NH_3. *Chem. Commun.*, **47**, 9789–9791.

119 Xie, L., Liu, F., Ren, L. et al. (2014) Excellent performance of one-pot synthesized Cu-SSZ-13 catalyst for the selective catalytic reduction of NO with NH_3. *Environ. Sci. Technol.*, **48**, 566–572.

120 Dědeček, J., Wichterlová, B., and Kubát, P. (1999) Siting of the Cu^+ ions in dehydrated ion exchanged synthetic and natural chabasites: a Cu^+ photoluminescence study. *Microporous Mesoporous Mater.*, **32**, 63–74.

121 Zamadies, M., Chen, X., and Kevan, L. (1992) Study of Cu(II) location and adsorbate interaction in CuH-SAPO-34 molecular sieve by electron Sp4 resoname and etectron spln echo modulath spectroscopies. *J. Phys. Chem.*, **96**, 2652–2657.

122 Ma, A.-Z. and Grünert, W. (1999) Selective catalytic reduction of NO by ammonia over Fe-ZSM-5 catalysts. *Chem. Commun.*, 71–72.

123 Long, R.Q. and Yang, R.T. (1999) Superior Fe-ZSM-5 catalyst for selective catalytic reduction of nitric oxide by ammonia. *J. Am. Chem. Soc.*, **121**, 5595–5596.

124 Brandenberger, S., Kröcher, O., Casapu, M., Tissler, A., and Althoff, R. (2011) Hydrothermal deactivation of Fe-ZSM-5 catalysts for the selective catalytic reduction of NO with NH_3. *Appl. Catal. B*, **101**, 649–659.

125 Brandenberger, S., Kröcher, O., Wokaun, A. et al. (2009) The role of Brønsted acidity in the selective catalytic reduction of NO with ammonia over Fe-ZSM-5. *J. Catal.*, **268**, 297–306.

126 Shi, X., Liu, F., Shan, W. et al. (2012) Hydrothermal deactivation of Fe-ZSM-5 prepared by different methods for the selective catalytic reduction of NO_x with NH_3. *Chin. J. Catal.*, **33**, 454–464.

127 Xie, B., Song, J. et al. (2008) An organotemplate-free and fast route for synthesizing Beta zeolite. *Chem. Mater.*, **20**, 4533–4535.

128 Sazama, P., Wichterlova, B., and Sklenak, S. (2014) Acid and redox activity of template-free Al-rich H-BEA∗ and Fe-BEA∗ zeolites. *J. Catal.*, **318**, 22–33.

129 Brandenberger, S., Kröcher, O., Tissler, A. et al. (2010) The determination of the activities of different iron species in Fe-ZSM-5 for SCR of NO by NH_3. *Appl. Catal. B*, **95**, 348–357.

17
Nanocatalysis: Catalysis with Nanoscale Materials

Tewodros Asefa[1,2] and Xiaoxi Huang[1]

[1]*Rutgers, The State University of New Jersey, Department of Chemistry and Chemical Biology, Piscataway, NJ 08854, USA*
[2]*Rutgers, The State University of New Jersey, Department of Chemical and Biochemical Engineering, Piscataway, NJ 08854, USA*

17.1
Introduction

17.1.1
Nanoscience and Nanotechnology

Nanoscience/nanotechnology is often defined as "the study, understanding, and control of matter at dimensions between approximately 1 and 100 nm, where unique phenomena enable novel applications." This definition applies to numerous materials ranging from metals, metal oxides, polymers, and dendrimers to porous materials such as aerogels, zeolites, mesoporous materials, and metal organic frameworks, which have nanoscale sizes and/or pores. Unquestionably, the terms ". . . unique phenomena enable novel applications . . ." in the above definition include some of the unique properties of nanomaterials that are relevant to catalysis.

Over the past two decades, nanoscience and nanotechnology have been generally among the most intensively explored fields of science and technology. This is attested by the large number of new literature being published pertaining to nanoscale materials, their properties, and potential applications in several areas. The fields of nanoscience and nanotechnology still remain among the most rapidly growing areas of science and technology today because nanomaterials are expected to bring more societal benefits, besides the ones they have already brought. The nanomaterials being investigated also allow researchers to learn a lot of fundamental science about materials in general.

Handbook of Solid State Chemistry, First Edition. Edited by Richard Dronskowski, Shinichi Kikkawa, and Andreas Stein.
© 2017 Wiley-VCH Verlag GmbH & Co. KGaA. Published 2017 by Wiley-VCH Verlag GmbH & Co. KGaA.

17.1.2
Catalysis

The term "catalyst" was first coined by Berzelius in 1836 to explain a new entity that was observed promoting a chemical reaction, or making the reaction give a higher yield of product. Since then, the field of "catalysis" has come a long way, and today over 90% of all the chemical products and synthetic materials we use "see" at least one catalyst, if not more, at some point during their synthesis/manufacturing. These catalysts literally enable better pathways for the transformation of various raw materials to the many valuable chemicals, pharmaceuticals, and synthetic products that we use everyday, and they make these possible in higher yield at less cost and with no/less amount of undesired byproducts. Catalysts also enable environmental remediation, energy conservation and energy storage.

Based on the relative phase of the catalyst with respect to that of the reactant (s), catalysts are most commonly or broadly divided into two groups: homogeneous catalysts and heterogeneous catalysts. The corresponding reactions are named homogeneous and heterogeneous catalysis. Homogenous catalysts are those that are in the same phase as the reactant(s) whereas heterogeneous catalysts are those that are in different phase with respect to the reactants, where typically the catalysts are in solid phase whereas the reactant(s) is/are often in gas or liquid phase, or both. Homogenous catalysts are generally soluble molecular or ionic compounds; thus, they have more easily accessible catalytic sites and often exhibit higher catalytic activity. Their structures and functional groups can also be relatively more easily tailored, and their structure–activity relationships are also relatively easy to study. However, they are harder to separate from reaction mixtures. Heterogeneous catalysts, on the other hand, can be easily separated, recovered, and reused after catalytic reactions. However, their structure–activity relationships are relatively harder to determine because their catalytic active species can inherently exist in many different conformations and states [1]. Their catalytic activity can also be relatively low because their ability to contact reactants can be limited due to mass diffusion.

17.1.3
Nanocatalysts and Nanocatalysts

Nanocatalysts can be defined as nanostructured materials that have catalytic properties or are used for catalysis. Although the early uses of nanomaterials in catalysis can be traced back to a century ago, or to the time when the Haber–Bosch process of ammonia synthesis based on iron-based particles was developed, research in nanocatalysts and nanocatalysis took off only after nanoscience and nanotechnology came to the stage and has begun making big strides since about five decades ago.

Nanocatalysts have some properties of both homogeneous and heterogeneous catalysts (i.e., they can give high catalytic activity (due to their large surface-to-volume) like many homogeneous and molecular catalysts while they are relatively easy to separate from reaction mixtures like many heterogeneous catalysts). But they additionally offer unique nanoscale related catalytic properties that cannot be found in traditional catalysts. This includes their size- and shape-dependent physical, electronic, and optical properties and quantum size effects, which can be useful to catalysis. Although their high surface area/volume ratio is generally beneficial to catalysis, as it allows the nanomaterials to contact reactants and achieve high activity, it can also make nanomaterials to aggregate/sinter, and lose their catalytic activity easily [2]. Hence, one of the major research efforts in nanocatalysis has focused on the development of efficient nanocatalysts that are robust and can remain stable [3].

17.2
Types of Nanocatalysts/Nanocatalysis Based on Activation

Chemical reaction pathways, in which bond breakages and bond reconstructions take place, involve some activation barriers. One of the jobs of a catalyst/nanocatalyst is to lower these barriers and make the reactions go faster/smoothly in the desired pathway. But, the (nano)catalyst would still need some energy in form of thermal, light, and electrical energy to overcome the remaining barriers. Accordingly, (nano)catalysts can be classified as thermocatalysts, photocatalysts, and electrocatalysts, and so on, and the corresponding reactions are called thermocatalysis, photocatalysis, and electrocatalysis, and so on. The energy may also enable additional processes, for example, generate the electron–hole pairs required for the reduction/oxidation reactions over the catalysts.

17.2.1
Nanophotocatalysis

Generally, for photocatalytic reactions over a given material to occur, the following four major processes must take place (Figure 17.1). (1) The material

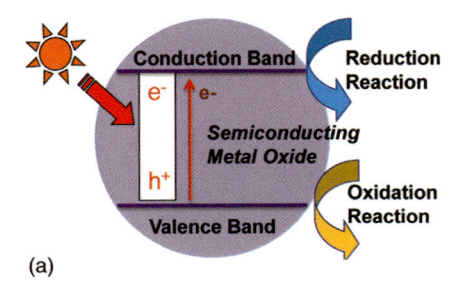

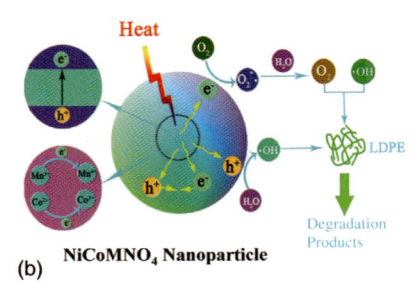

Figure 17.1 (a) Schematic illustration of photogeneration of charge carriers in a semiconducting nanophotocatalyst (e.g., TiO_2). (b) Schematic illustration of the generation of charge carriers in a temperature-sensitive nanocatalyst (e.g., $NiCoMnO_4$) by an indirect route (light → heat → charges). (Reproduced with permission from Ref. [4]. Copyright 2016, The Royal Society of Chemistry.)

should absorb radiation energy larger than its band gap and produce electron–hole pairs. (2) The electron–hole pairs should be separated. (3) The chemical potential of the electron–hole pairs must be big enough to promote the desired chemical reactions. (4) The electron–hole pairs should be able to diffuse to the surfaces of the material, where the reactions take place. Owing to their high surface area, ability to absorb light, and quantum size effects, many semiconducting nanomaterials meet these conditions to a good extent and exhibit photoinduced catalytic activity toward various reactions. It is also worth noting that the electron–hole pairs can be generated over some nanomaterials not directly by the radiation but rather indirectly by the thermal energy coming from the radiation. This is particularly true in materials that have good redox activity and high coefficient of temperature (e.g., NiCoMnO$_4$ nanomaterial, which produces electrons and holes hopping between its Mn^{3+} and Mn^{4+} as well as its Co^{2+} and Co^{3+} sites, and these electrons and holes are responsible for the catalytic reactions occurring over the surfaces of the nanomaterial, Figure 17.1,b) [4].

Unfortunately, because of recombination of electrons and holes, charge trapping, or low charge transport, many of the processes mentioned above take place only poorly over many nanomaterials, resulting in poor photocatalytic efficiency. For example, due to its relatively large band gap (3.1 eV), nanosized TiO$_2$, one of the most common nanophotocatalysts, absorbs mainly in the UV region of solar spectrum (<388 nm), which accounts for only a very small fraction (about 3%) of available solar energy [5–7]. As a result, nanosized TiO$_2$ works as a photocatalyst only in a very narrow window of Sun's light, or gives only a low efficiency in photocatalysis under direct solar irradiation.

So, improving the photoactivation response of nanophotocatalysts to a wide range (from UV to visible and infrared) remains one of the major goals in nanophotocatalysis research [8,9]. This has been attempted by: (1) tailoring the size, morphology, and structures of the nanomaterials [10,11]; (2) doping the nanomaterials with other elements such as N, C, and so on [9,12–19]; (3) finding inherently small band gap nanomaterials (e.g., g-C$_3$N$_4$, Bi$_2$WO$_6$, InNbO$_4$, etc.) [20–22]; (4) including visible-light or surface plasmon resonance (SPR) active nanoparticles (e.g., Au nanoparticles, Figure 17.2), and forming so-called plasmonic photocatalysts [6,23–25], and (5) finding nanomaterials responsive to low energy (infrared) light (e.g., Cu$_2$(OH)PO$_4$, Bi$_2$WO$_6$, H-doped WO$_3$, etc.) [26–30] or capable of converting low energy to high excitation energy for electrons (so-called upconversion photocatalysts) [31,32]. Many of these have been proven to photocatalyze a variety of reactions well [33,34].

When SPR active nanoparticles are combined with semiconducting nanomaterials, the former can serve as electron trapping sites and inhibit electron–hole recombinations, giving high quantum efficiency [35,36]. The former can also help generating energetic charge carriers, also called "hot" electrons and holes that can assist with photocatalysis [37–39]. These and other possible mecha-

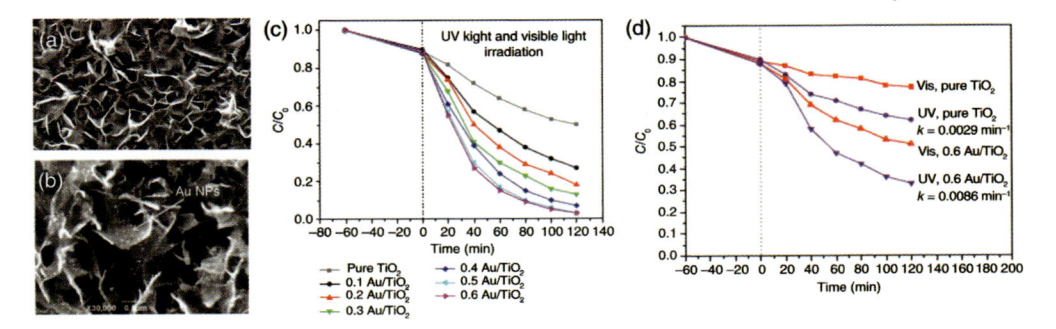

Figure 17.2 SEM images of pure nanostructured TiO_2 film before (a) and after (b) modification with Au nanoparticles (0.1 Au/TiO_2, in which the number indicates the wt.% of Au, i.e., 0.1% Au). (c) Normalized concentration of methylene blue (MB) over different Au/TiO_2 samples under UV and visible light irradiations (where C is the concentration of MB at the irradiation time t and C_0 is the concentration of MB at the adsorption equilibrium in the presence of the catalyst before irradiation). (d) Normalized concentration of MB over pure TiO_2 and 0.6 Au/TiO_2 films under UV and visible light irradiations. (Reproduced with permission from Ref. [25]. Copyright 2015, Yihan Chen et al.)

nisms by which plasmonic photocatalysts catalyze reactions more efficiently and their advantages are further illustrated in Figure 17.3 [40]. In the case of the upconversion photocatalysts, the nanomaterials convert multiple low energy (longer wavelength) photons to high energy (visible or even UV light) to excite their electrons to high enough bands, or to those that possess high enough chemical potential to promote the reactions [41–43].

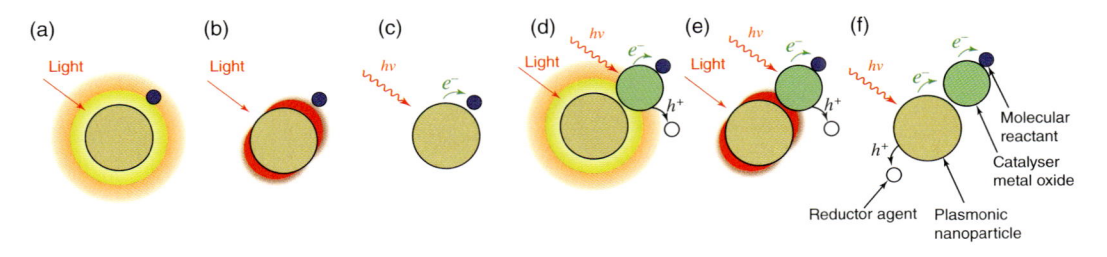

Figure 17.3 Main physical mechanisms involved in plasmon-assisted photochemistry. (a) The photoinduced heat generated by the nanoparticle provides heat to an adjacent reactant. (b) The enhancement of the optical near field at the vicinity of the nanoparticle increases the amount of photon seen by an adjacent reactant. (c) A photoinduced hot electron is transferred to a nearby reactant. (d) The rate of electron–hole (e^-–h^+) generated in a photocatalyst is enhanced by the heat generated by the nanoparticle. (e) The rate of (e^-–h^+) generated in a photocatalyst is enhanced by the strong optical near-field of the plasmonic nanoparticle. (f) The photocatalyst adjacent to the nanoparticle is activated by the hot electron transferred from the plasmonic nanoparticle. (Reprinted with permission from Ref. [40]. Copyright 2014, The Royal Society of Chemistry.)

17.2.2
Thermocatalysis

Heterogeneous thermocatalysts, which rely on heat, are most widely applied in many chemical processes [44]. But thermocatalysts have disadvantages too, especially compared with photocatalysts and electrocatalysts, because they generally consume more energy and operate at higher temperature, which are both less desirable catalytic conditions. Moreover, thermocatalysts need to be thermally durable, a condition that is relatively difficult to meet for many materials, especially nanostructured ones (because nanomaterials have inherent tendency to aggregate/grow/sinter).

So, unsurprisingly many conventional thermocatalysts are nonnanostructured materials; however, researchers have been trying to develop stable nanostructured materials that can serve as thermocatalysts because they can have improved catalytic activity, especially if they can work with the Sun's radiation. Among many materials studied for this purpose, nanostructured CeO_2 is a notable example. This nanomaterial exhibits good thermocatalytic property (e.g., photooxidizes air pollutants or volatile organic compounds (VOCs)) because its Ce^{4+} species undergo Ce^{4+}/Ce^{3+} redox processes and it has a large density of such exposed Ce^{4+} species due to its nanostructure. Another related material is a Ce ion-substituted cryptomelane-type octahedral molecular sieve, which can absorb a full solar spectrum and exclusively transform it to thermal energy, and then produce high enough temperature, above the light-off temperature to thermodecompose VOCs (Figure 17.4) [45].

Another major approach to achieve higher catalytic activity with nanomaterials is by making them to harvest as much energy as possible, preferably the full

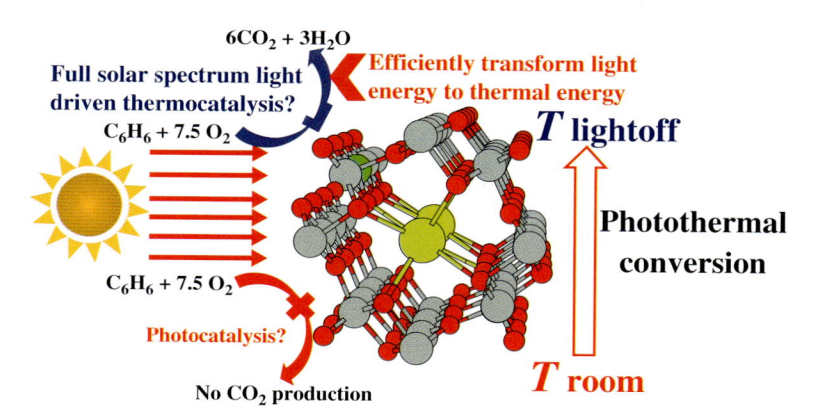

Figure 17.4 Schematic illustration of solar light driven thermocatalysis of a VOC by a Ce ion-substituted cryptomelane-type octahedral molecular sieve: Ce, gray sphere: Mn, red sphere: O, green sphere: Mn vacancy. (Reprinted with permission from Ref. [45]. Copyright 2015, The Royal Society of Chemistry.)

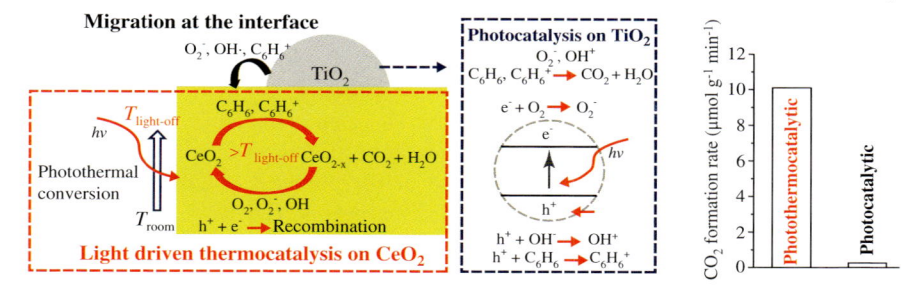

Figure 17.5 Schematic illustration of solar light-driven synergetic photothermocatalysis on TiO$_2$/CeO$_2$ nanomaterial. (Reprinted with permission from Ref. [47]. Copyright 2015, American Chemical Society.)

solar spectrum, and generate a large density of charge carriers for both photocatalysis and thermocatalysis, as in Figure 17.5 [46–48].

17.2.3
Electrocatalysis

Many reactions can also be catalyzed by electrochemical activation. Such reactions are generally called electrocatalysis, and the materials involved as catalyzing them are called electrocatalysts. Just like in photocatalysis, the rates of formation and transportation of charges are critical steps in electrocatalytic processes. Moreover, just like in many catalysts, higher surface area matters here as well, and this is why nanomaterials are interesting systems for electrocatalysis too. But for a nanomaterial to be called a good electrocatalyst, it should have the right catalytically active sites that can catalyze reactions at less electrochemical potential/activation energy and give the desired products.

However, most traditional solid-state electrocatalysts as well as nanoelectrocatalysts have involved noble metals such as Pt, Ir, Rh, and so on [49–51]. So, recently there has been a surge of interest and effort on the development of non-noble metal-based and metal-free nanostructured catalysts (sustainable catalysts) to replace those traditional ones based on less earth abundant elements, not only for electrocatalysis but also many other reactions [52,53]. The sustainable catalysts include layered materials such as graphene, graphene oxide, transition metal dichalogenides, and layered double hydroxides [53–55].

17.3
Synthetic Methods Used for Nanocatalysts

The catalytic activity and selectivity of nanocatalysts are heavily dependent on the size, shape, structure and composition of the nanocatalytic sites, and even the underlying support materials, if there are any [56]. So, several types of synthetic approaches to make nanomaterials with controlled size and structural and compositional features have been pursued to find materials that show high catalytic performance. Their compositions include metallic, metallic alloy, metal oxide, metal dichalcogenide, metal organic framework, covalent organic frameworks, heteroatom-doped carbon, metal nitride, metal phosphide, metal carbide, and so on. However, classifying nanocatalysts by summarizing all their possible compositions, the reactions they catalyze, or the possible synthetic procedures and methods used to make them would all be beyond the scope as well as the goal of this chapter. Besides, several excellent review papers have been written on these topics [57–63]. A broader yet reasonable way to classify nanocatalysts, which we use here, is based on their structures, namely, 0D, 1D, 2D, and 3D nanocatalysts, which is actually in line with the general classification of nanomaterials. Still, we wish to emphasize that these discussions would not be exhaustive, but rather exemplary and less on details.

The notable synthetic approaches that are commonly used to make nanocatalysts irrespective of their structure and compositions include: (1) soft or hard templating to generate nanoscale sizes and nanoporous structures; (2) colloidal synthesis under various conditions (room temperature, hydrothermal, chemical vapor, electrochemical- and electroless deposition, etc.) to nucleate and grow nanocatalysts in solution or on surfaces; (3) physical methods, such as exfoliation, doping, and so on; and (4) nanopatterning synthesis on substrates as thin films using chemical or physical methods or both.

17.3.1
Zero-Dimensional (0D) Nanocatalysts

These include catalytic active nanoparticles in solution; so, in many ways, these catalysts mimic homogeneous catalytic systems. The most common way to synthesize 0D nanocatalysts is by performing reactions that lead to stable dispersions of nanoparticles with compositions including metal, metallic alloy, metal oxide, metal chalcogenide, and so on. Their synthesis often requires reagents that produce the required composition of the nanoparticles and surface capping groups that arrest the particles when they are at the required nanoscale sizes. The surface capping agents also help the 0D nanocatalysts to remain stable before, during, and after catalysis. For example, the reduction of Au(III) or Ag(I) ions with $NaBH_4$ in the presence of capping agents such as 1-dodecanethiol can form catalytic active, stable Au or Ag nanoparticles in solution, and the reaction between Cd^{2+} and S^{2-} in the presence of 1-dodecanethiol can form photoactive thiolate-capped stable CdS nanoparticles.

Since the seminal independent works by Haruta and Hutchings showing the catalytic properties of nanoparticles [64,65], numerous 0D nanocatalysts have been reported (Figure 17.6) [66–71]. However, many of them have one common problem: they need capping agents to be stable while, at the same time, they need naked surfaces to catalyze reactions. Recently, Astruc and coworkers reported a synthetic approach that addresses this issue. They synthesized highly stable, active transition metal nanoparticle (TMNP) catalysts composed of Fe,

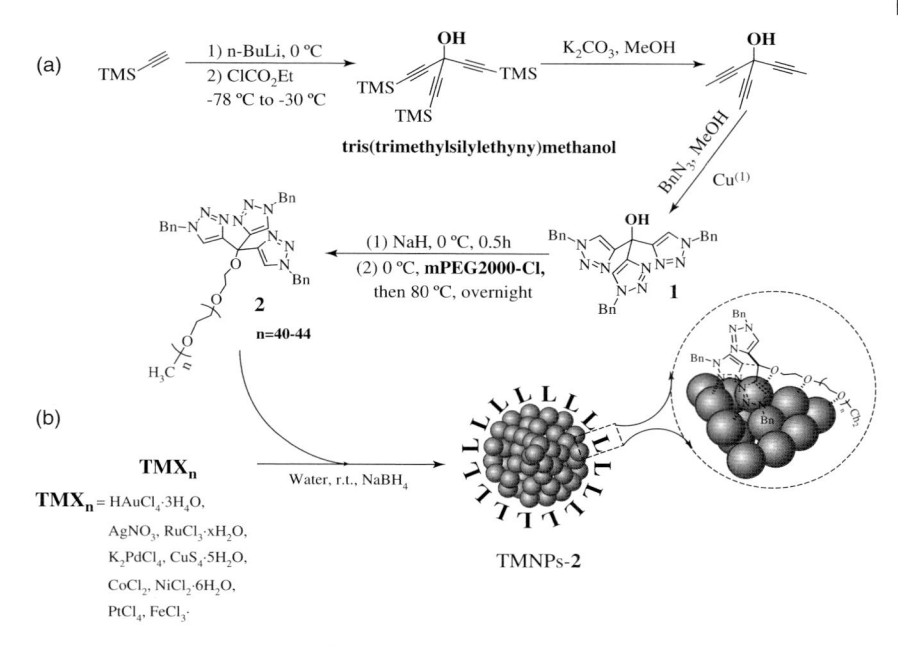

Figure 17.6 (a) Synthesis of amphiphilic bidentate ligand (*tris*(triazolyl)-polyethylene glycol (*tris*-trz-PEG) (**2**)) (b) and TMNP catalysts capped with **2** (TMNP-**2**). (Reproduced with permission from Ref. [66]. Copyright 2016, Wiley-VCH Verlag GmbH.)

Co, Ni, Cu, Ru, Pd, Ag, Pt, and Au in aqueous solution by using a bidentate ligand (*tris*(triazolyl)-polyethylene glycol (*tris*-trz-PEG) (**2**)) to stabilize the TMNPs while leaving part of TMNPs' surfaces "naked" [66]. The resulting TMNPs served as efficient, stable catalysts for reduction, the Suzuki–Miyaura, transfer hydrogenation, and click reactions (Figure 17.6).

As most catalysis takes place at specific surface sites (edges and corners), the synthesis of 0D nanocatalysts with controlled crystal planes or reactive planes is of paramount importance [72–74]. For example, for monoclinic tenorite-structured CuO nanomaterials, which are active catalysts for CO oxidation, the ones with less close-packed (111) planes, which have lower free energies, higher stability, and thus less catalytic active sites than open (001) and (01$\bar{1}$) planes, are less desirable [75]. Furthermore, since the catalytic activity of nanocatalysts directly correlates to their surface area per unit mass/volume, further nanoengineering, by adding corrugation and hollow/porous structures in the 0D nanocatalysts, via various synthetic methods, is appealing [76]. In some cases, this can be done via galvanic reactions or etching [77,78]. In other cases, the so-called Kirkendall effect, the difference in diffusion rates of two metals, can be taken advantage of. For example, if Zr(IV) ions are added into a solution of ceria nanospheres, the Kirkendall process involving the diffusion of Zr(IV) into CeO_2 takes place and CeO_2–ZrO_2 ($Ce_{1-x}Zr_xO_2$) with nanocages form (Figure 17.7) [79,80]. By varying the morphology of the original nanoparticles, the reaction types and the reaction times, different hollow, shaped

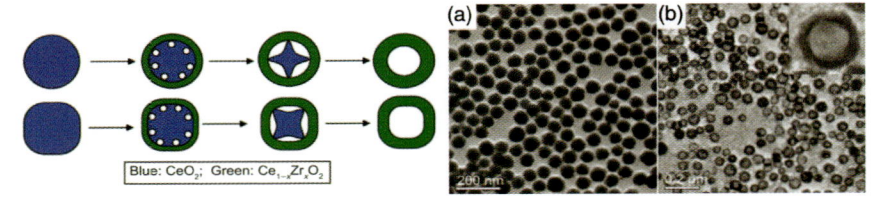

Figure 17.7 Schematic illustration of the Kirkendall process-based synthetic route leading to for $Ce_{1-x}Zr_xO_2$ nanocages from CeO_2 (a) and the TEM images of the resulting particles (b). (Adapted with permission from Ref. [79]. Copyright 2008, American Chemical Society.)

(spherical, cubic, etc.) 0D nanocatalysts can then be synthesized. Besides their higher surface areas, such nanocatalysts can display the so-called confinement effect, where a steady state of higher concentration concentration of reactants are confined in their cavities, giving better catalytic outcomes [81].

Dendrimer-encapsulated nanoparticles (DENs), which are synthesized by using the ligands present within dendrimers as surface capping agents, are notable 0D type nanocatalysts [82–85]. In their typical synthesis, metal ions are let to complex within the inner parts of the dendrimers, and then reduced with a reducing agent (Figure 17.8). This process can be repeated with the same or different type of metal

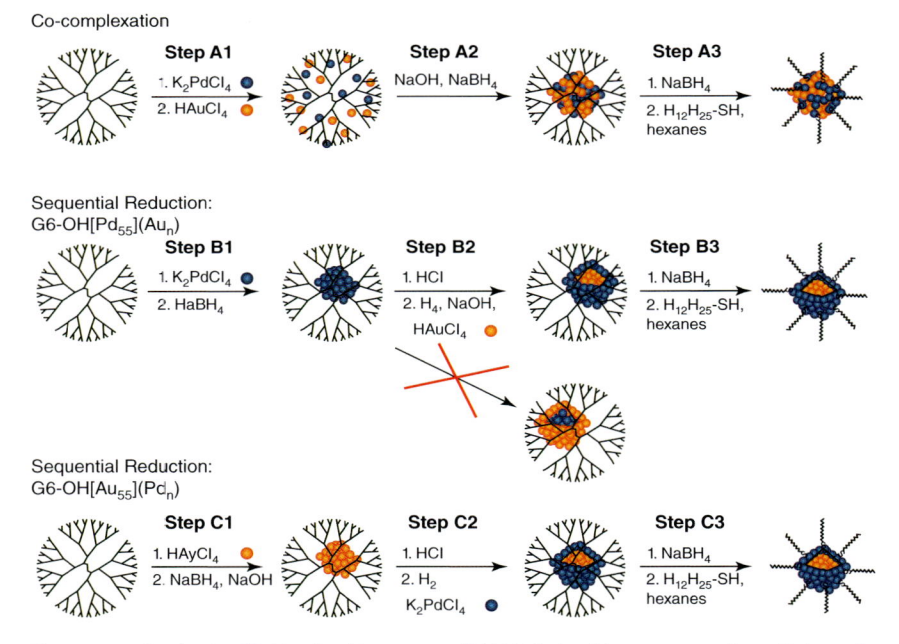

Figure 17.8 Synthesis of PdAu dendrimer-encapsulated nanoparticles (DENs) prepared via sequential reduction of the two metals using hydroxyl-terminated poly(amidoamine) (PAMAM) dendrimers as templates. (Adapted with permission from Ref. [91]. Copyright 2010, American Chemical Society.)

ions to grow monometallic, bimetallic/multimetallic alloy or core–shell DENs [86,87]. For example, catalytically active PdPt bimetallic alloy DENs were synthesized by cocomplexing Pd(II) and Pt(II) ions within hydroxyl-terminated poly (amidoamine) (PAMAM) dendrimer from the respective solutions of K_2PdCl_4 and K_2PtCl_4 [87–89]. Using a similar approach, many other catalytically active bimetallic (e.g., PdAu DENs) were also successfully synthesized [89,90]. But, interestingly, in the latter case, regardless of which metal is deposited first, the Au atoms always migrate into the core and the Pd atoms go to the surface (Figure 17.8), as evidenced by EXAFS and Au-sensitive catalytic reaction involving resazurin to resorufin transformation.

17.3.2
One-Dimensional (1D) Nanocatalysts

1D nanomaterials, such as nanowires and nanorods, are also highly attractive in catalysis especially due to their inherent anisotropies, faceted surfaces, and high surface area/volume ratios. Moreover, since their structures offer interesting charge transport, unique absorption, and other electronic properties, they sometimes have some advantages in photocatalysis and electrocatalysis [92,93]. Just like many shaped 0D nanocatalytic systems, the synthesis of nanorods and nanowires requires reaction conditions and surface passivating agents that favor the preferential and disproportionately faster growth of certain facets. For example, by using relatively higher concentrations of $AgNO_3$ and elemental S in the presence of octadecylamine (ODA), certain facets were favored to grow faster than others, resulting in 1D Ag_2S (nanowires), which can have catalytic uses [94], (Figure 17.9) [95,96].

17.3.3
Two-Dimensional (2D) Nanocatalysts

Owing to their significant proportion of exposed areas and defect sites, and thus large density of potentially active catalytic sites, two-dimensional (2D)

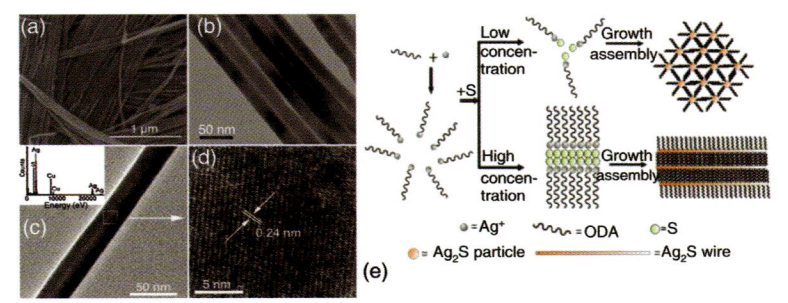

Figure 17.9 (a) SEM and (b) TEM (c) high-resolution TEM along with EDS analysis, and (d) crystal lattice images of Ag_2S nanowires. (e) Schematic illustration showing the effect of experimental parameters on the shape and size of Ag_2S crystals. Adapted with permission from Refs [95,96]. Copyright 2008, Wiley-VCH Verlag GmbH; Copyright 2009, Tsinghua University Press and Springer Berlin Heidelberg.)

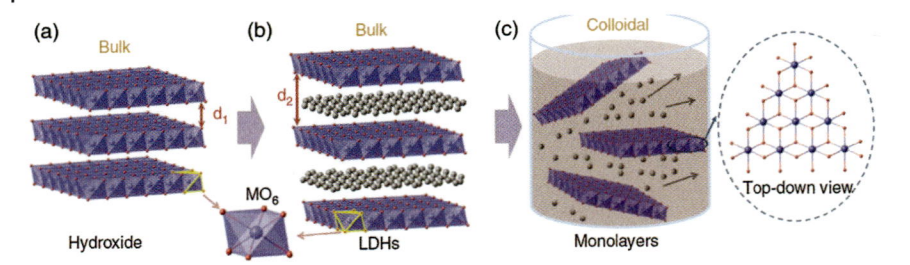

Figure 17.10 Schematic representation of the structures of layered nanomaterials synthesized for catalysis. (a) Layered hydroxides; d_1 is the inter-layer distance. (b) LDHs with inter-layer anions and water molecules; d_2 is the inter-layer distance, $d_2 > d_1$. (c) Exfoliated LDH monolayers dispersed in a colloidal solution. Each single layer is composed of edge-sharing octahedral MO_6 moieties (M denotes a metal element). Metal atoms: purple spheres; oxygen atoms: red spheres; inter-layer anions and water molecules: gray spheres. Hydrogen atoms are omitted. (Reprinted with permission from Ref. [98]. Copyright 2014, Nature Publishing Group.)

nanomaterials have also been intensely investigated for their applications in catalysis. In many 2D nanocatalysts, their thickness often matters more than their overall size in catalysis. They are often synthesized by exfoliating their corresponding bulk materials by various methods [97]. For example, exfoliated WS_2 electrocatalyzes the evolution of hydrogen from water better than bulk WS_2 [97]. In another example, single layered NiFe and NiCo double hydroxides (LDHs), which could be obtained by exfoliating their bulk counterparts via anion exchange chemistry (Figure 17.10) [98], showed higher electrocatalytic activity toward water oxidation than their respective bulk counterparts. Their higher catalytic activity was accounted by their high density of catalytic active edge sites and higher electronic conductivity.

Another good example of 2D nanocatalysts is an ultrathin Co_3O_4 that could be made via fast-heating at 320 °C for 5 min in O_2 of $Co(CO_3)_{0.5}(OH) \cdot 0.11 H_2O$. The latter could be hydrothermally synthesized from a reaction between Co(acac)$_3$ and CTAB. Compared with the bulk material, 2D Co_3O_4 electrocatalyzed CO_2 reduction to formate much more efficiently (Figure 17.11) [99]. A 1.72 nm layer Co_3O_4 nanocatalyst gave a reduction current density of 0.68 mA/cm^2 at -0.88 V versus saturated calomel electrode (SCE), which is 1.5 and 20 times more than those given by 3.51 nm thick Co_3O_4 and bulk Co_3O_4, respectively [99]. The Faradaic efficiency of the ultrathin Co_3O_4 nanocatalyst toward formate product was also higher (Figure 17.11b). When the current density was normalized with respect to electrochemical surface area (ECSA), the order of catalytic activity for the reaction was 1.72 nm $Co_3O_4 >$ 3.51 nm $Co_3O_4 \gg$ bulk Co_3O_4 (Figure 17.11d). The higher catalytic activity exhibited by the thinner Co_3O_4 materials was attributed to their higher ESCA, better 2D electronic conductivity and higher fraction of surface atoms with low coordination number and lots of dangling bonds, all of which allowed the intermediates to be stabilized during the CO_2 reduction reaction [99].

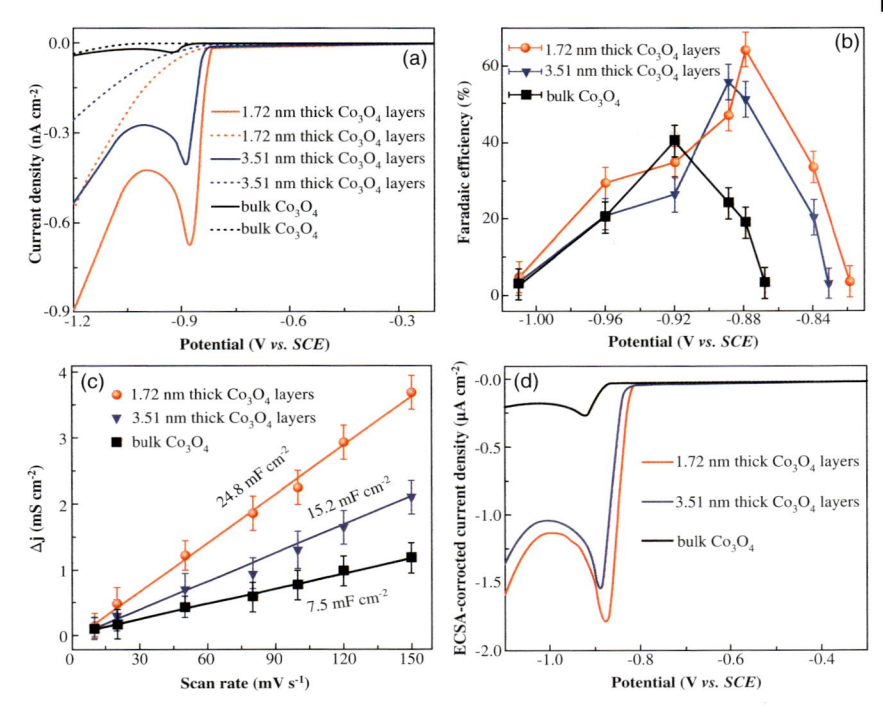

Figure 17.11 Electroreduction of CO_2 to formate catalyzed by layered Co_3O_4 nanomaterials with different thicknesses. (a) Linear sweep voltammetric curves in the CO_2 saturated (solid line) and N_2 saturated (dash line) 0.1 M $KHCO_3$ aqueous solutions. (b) Faradaic efficiency for formate at each applied potentials for 4 h. (c) Charging current density differences plotted against scan rates. (d) ECSA-corrected current densities versus applied potentials. (Reprinted with permission from Ref. [99]. Copyright 2016, Wiley-VCH Verlag GmbH.)

17.3.4
Three-Dimensional (3D), Supported, Core–Shell, and Multifunctional Nanocatalysts

Catalytic active nanoparticles are generally relatively hard to separate in stable form at the end of reactions. This is true especially when they are made with smaller sizes in order to produce in them as much catalytic activity per unit mass as possible. To overcome this conundrum, they are often supported on high surface area support materials, with a process sometimes referred to as "heterogenization of nanoparticles." The resulting 3D nanocatalytic materials possess many of the advantages of nanoparticle catalysts (e.g., high activity and selectivity) as well as many of the advantages of heterogeneous catalysts (e.g., good stability and ease of recoverability/recyclability) [100].

Generally, 3D nanocatalysts include 3D nanostructured materials that have nanosized features (e.g., possess nanopores, supported nanoparticles with different size and morphology) and catalytic properties. In 3D nanocatalytic systems, two components are required: active nanocatalytic groups and good solid-state support materials. A single nanomaterial can also offer both. The active nanocatalytic groups can be metallic, metallic alloy, inorganic, and other nanorelated species as long as the support material has some nanoscale features. The solid-support materials that can be used for making nanocatalysts range from polymers to zeolites, mesoporous metal oxides, and modified metal oxides. Among them, nanoporous silica and carbon have been widely used as support materials for making nanocatalysts because of their inexpensiveness, availability from various sources, excellent stability, and relative inertness toward various reagents [101]. They are also easy to synthesize. For instance, mesoporous silica materials are typically synthesized with surface areas as high as $1000 \, m^2/g$ or greater and tunable nanometer pores (depending on the templates used to make them) by applying a combination of the sol–gel process and supramolecular self-assembly in the presence of surfactant templates [102]. The two most common ones, MCM-41 and SBA-15, possess micron and nanoscale sizes and average pores of around 3–4 nm and 8–9 nm, respectively. Other mesoporous oxides with similar structural features can also be synthesized in the same way [103].

Using these as support materials, catalytic groups can then be immobilized or supported either by physical attachment (physisorption) or covalent attachment (chemical tethering) [104,105]. The former can be accomplished simply by letting the catalytic species present solutions or in gas phase to adsorb onto the surface hydroxyl or metaloxane groups present on the surfaces of the support materials. The second method, which is sometimes called grafting, involves chemical covalent modification of the surfaces of the support material by contacting it with various agents (e.g., organosilanes, metalalkyl complexes, halogenated silanes), which can form covalent bonding with it [106]. This process ultimately leads to covalently tethered groups, along with any ligands and catalytic active groups they may carry (amines, sulfonic acids, carboxylic acids, etc.), as shown for mesoporous silica in Figure 17.12 [107,108]. If necessary, the grafted surface groups can

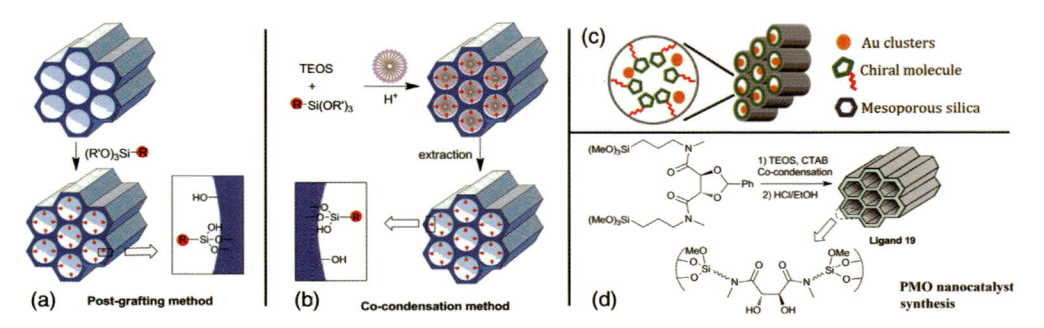

Figure 17.12 Modification of the surfaces of mesoporous silicas with various catalytic groups by (a) post-grafting method, (b) cocondensation method, (c) post-grafting/cocondensation, followed by functionalization with metal nanoclusters, and (d) assembling bridging organosilanes (synthesis of PMOs). (Reprinted with permission from Refs [105,107,110]. Copyright 2015, The Royal Society of Chemistry; Copyright 2013, American Chemical Society; Copyright 2008, American Chemical Society.)

be used as linkers for other, desired catalytic groups that may be hard to graft directly. However, this method (i.e., post-grafting) has one major drawback: it can inherently lead to a nonhomogeneous distribution of catalytic groups within the nanostructured catalyst [109].

To overcome this issue (i.e., the nonhomogeneous distribution of catalytic groups) that the post-grafting synthesis suffers from, the molecules containing the functional groups (such as (R'O)$_3$SiR) are alternatively coassembled or cocondensed with other structure formers (e.g., tetraethoxysilane, TEOS) and the amphiphilic molecules (Figure 17.12b and c) [111]. Similarly, bridging organosilanes ((R'O)$_3$Si-R-Si(OR'O)$_3$) can be assembled with surfactant templates to produce periodic mesoporous organosilicas (PMOs) containing catalytic R groups on their pore walls (Figure 17.12d) [110,112–115].

The immobilized groups (e.g., organoamine) supported in the mesoporous materials can serve as the catalytic or cocatalytic sites, or even as ligands to support other catalytic groups (e.g., Lewis acids such as Fe(III) ions) [116–119]. The supported metallic ions can also be further converted to supported metallic or multimetallic nanoparticle catalytic sites, and their type and structural features can be tailored by changing the synthetic conditions [120]. Many of the resulting materials can serve as effective, recyclable nanocatalysts for ring opening, C—C and C-heteroatom bond coupling, hydroamination, and oxidation reactions [121–123]. While many of the catalytic systems have interesting properties, they often do not form isolated species that allow precise elucidation of the structure–property of individual groups directly though. To address this issue, and enable meticulous structure–property studies, many research groups perform the grafting on thermally dehydroxylated silica and other metal oxides that possess isolated hydroxyl to produce controlled, isolated grafted catalytic moieties [124].

A different synthetic route to make 3D nanocatalysts involves the assembly of catalytic active 0D nanoparticles with metal oxide precursors in one-pot. In one of the earliest such works, Reetz and Dugal synthesized Pd nanocatalysts supported on nanoporous silica with MgO promotor by mixing premade $R_4N^+Br^-$-stabilized Pd nanoparticles with $CH_3Si(OCH_3)_3$ and $Mg(OC_2H_5)_2$ in the presence of fluoride and water, followed by washing (Figure 17.13) [125]. The resulting nanocatalysts could efficiently catalyze chemoselective olefin hydrogenation, and be recycled afterwards. Budroni and Corma reported a similar nanocatalyst

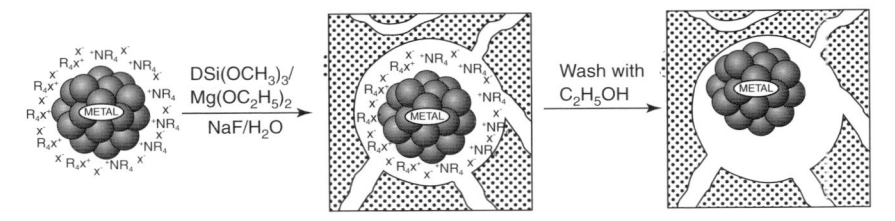

Figure 17.13 Schematic representation of the synthesis of stable metal oxide-supported Pd nanoparticle catalyst. (Reprinted with permission from Ref. [125]. Copyright 1999, Kluwer Academic Publishers.)

by mixing TEOS with 3-mercaptopropyltrimethoxysilane (MPTS)-capped Au nanoparticles synthesized in reverse micelles from the reaction between Au(III) with hydrazine in the presence of 1-dodecanethiol and MPTS [126]. The resulting silica-encapsulated Au nanocatalysts exhibited good, recyclable catalytic properties for the Suzuki coupling reaction.

Carbon nanomaterials have long been used as support materials for making various 3D nanocatalysts. In many of such catalysts, the catalytic sites are metallic, bimetallic, and metal oxide nanoparticles and the support materials are typically high surface area, nanostructured carbons (e.g., activated carbon) [127–131]. Recently, just like mesoporous silicas and organosilicas, nanoporous carbons have been a center of attention in the synthesis of 3D nanocatalysts because they have large surface area and stable structure [132–134], and also because their surfaces can easily be modified by a variety of functional groups [135]. Mesoporous carbons are generally synthesized by pyrolysis (carbonization) of different precursors (small molecules, polymers, biomass, biomolecule, etc.) using hard or soft templates [136–138]. In the first case, carbon precursors are mixed with or impregnated into porous inorganic hard templates (such as mesoporous silicas, colloidal silica, anodic alumina), sometimes along with metallic ions that help their carbonization and/or graphitization [139,140]. After carbonization of the resulting composite materials under inert conditions at high temperatures, the templates are removed by various means (e.g., NaOH solution which dissolves silica but not the carbon material) [141,142]. An example illustrating these steps is depicted in Figure 17.14 [143], in which sucrose was used as a precursor and amine-functionalized SBA-15 was used as a hard template. If metal ions are added with the precursor, the synthesis produces a high surface area mesoporous carbon with metallic nanoparticles in it [144]. Interestingly, even if they do not have metallic or metal oxide species, but are doped with heteroatoms, many of the resulting

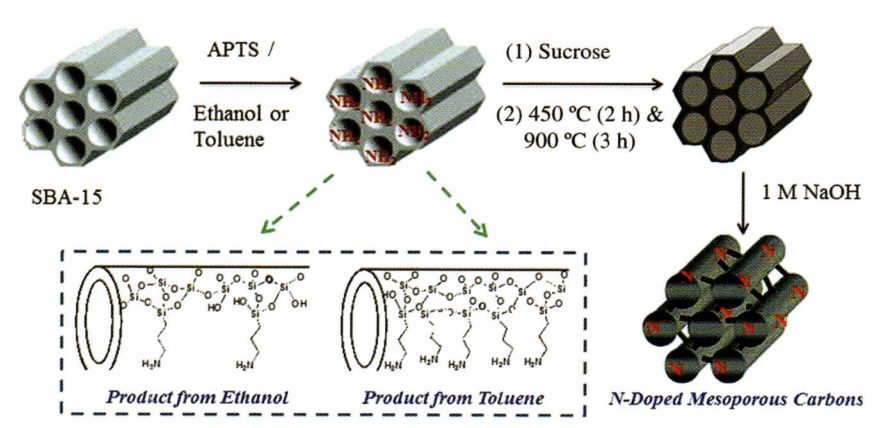

Figure 17.14 Schematic illustration of the synthesis of N-doped mesoporous carbons using sucrose as precursor and amine-grafted SBA-15 mesoporous silicas prepared by grafting in ethanol or toluene as templates. (Reprinted with permission from Ref. [143]. Copyright 2014, The Royal Society of Chemistry.)

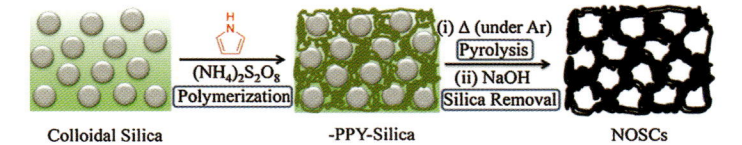

Figure 17.15 Synthetic procedures leading to N-, O-, and S-doped, polypyrrole-derived nanoporous carbons that can serve as efficient metal-free nanocatalysts. (Reprinted with permission from Ref. [149]. Copyright 2014, American Chemical Society.)

nanostructured carbons can still catalyze a number of reactions by themselves [145,146]. Due to this unprecedented catalytic property, these types of metal-free catalysts have recently become a subject of intense research, especially for reactions such as oxygen reduction reaction, hydrogen evolution reaction and so on [147,148].

In a related but different procedure, polypyrrole and persulfate were used as precursors and colloidal silica particles were used as templates to produce metal-free 3D carbon nanocatalyts (Figure 17.15) [149]. In this case, the amount and size of the template and the carbon precursors were both found to influence the structures of the resulting metal-free N-, O-, and S-doped carbon nanomaterials as well as their catalytic activities toward reactions such as the oxygen reduction and alcohol oxidation reactions.

In the case of soft-templating synthetic route to nanoporous carbons, the synthesis involves unstable soft templates such as block copolymers (e.g., polystyrene-*block*-poly(ethylene oxide) (PS-*b*-PEO)), which spontaneously form micelles (with about 20 nm in size) and which can be removed at the end of reactions by a simple

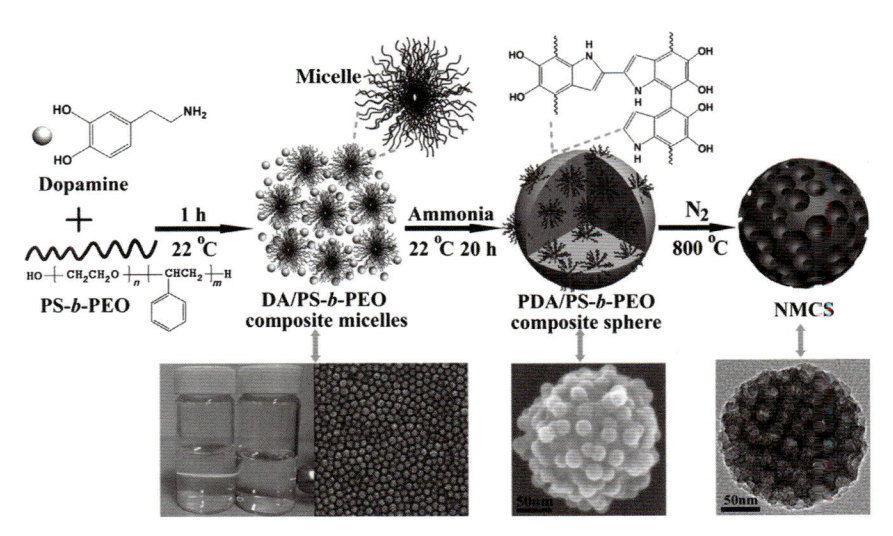

Figure 17.16 Soft-templating synthetic route for making mesoporous N-doped carbon nanospheres. (Reprinted with permission from Ref. [150]. Copyright 2015, Wiley-VCH Verlag GmbH.)

procedures (e.g., heat treatment) (Figure 17.16) [150]. When molecules such as dopamine (DA) are added into them, DA/PS-*b*-PEO micelles (with an average size of 185 nm) can form. When pyrolyzed, the composite micelles become N-doped nanoporous carbon nanoparticles with 16 nm pores due to the decomposition of the block copolymers (i.e., PS-*b*-PEO). The resulting nanoporous carbons can serve as metal-free nanocatalysts for various reactions.

Besides the above methods, incipient wetness impregnation, metal ion deposition followed by *in situ* reduction, chemical vapor deposition (CVD), atomic layer deposition (ALD), and electroless deposition methods and their combinations have been applied on different nanoporous metal oxide, carbon, zeolite, and so on to produce a variety of supported nanocatalytic systems [151–154].

17.3.5
Core–Shell Type Nanostructured Catalysts

A core–shell nanomaterial generally contains a core and a shell (multiple shells), with some nanometer dimensions. The composition of the core and shell include metals, metal oxides, polymers, dendrimers, and so on [155], and they may have different shapes (spheres, rods) [156]. Owing to their many interesting properties, core–shell nanomaterials have attracted a broad interest for catalysis applications, besides many others [157].

Core–shell nanocatalysts are usually synthesized by stepwise synthetic protocols, starting with the cores and then depositing shells around them. Alternatively, they can be synthesized from a mixture of reagents in one step by taking advantage of the faster reaction that one of them may undergo with respect to the other(s) [158–162]. Using a typical multistep synthetic route [163], thermally stable Pt/mesoporous silica (Pt@mSiO$_2$) type core–shell nanocatalysts were synthesized by coating tetradecyltrimethylammonium bromide (TTAB)-stabilized Pt nanoparticles with silica shells via a standard sol–gel process, and then removing the TTAB molecules by calcination afterwards (Figure 17.17) [164]. This results in Pt@mSiO$_2$ core–shell nanoparticles that exhibit efficient catalytic activity for ethylene hydrogenation and CO oxidation reactions. These kinds of synthetic procedures can also be applied to polymers and dendrimers, as long as compatible chemistry and interaction between the systems used as cores and shell are

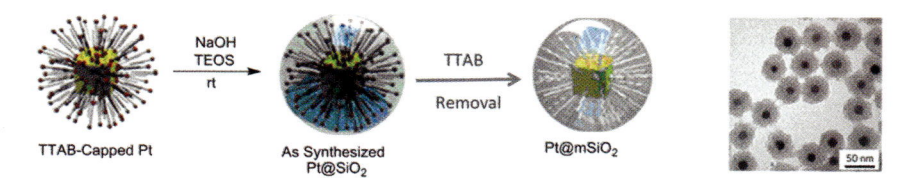

Figure 17.17 (a) Schematic representation of the synthesis of Pt@mSiO$_2$ core–shell nanocatalysts and (b) their TEM image after calcination at 350 °C. (Adapted with permission from Ref. [164]. Copyright 2009, Nature Publishing groups.)

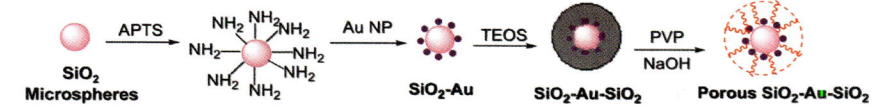

Figure 17.18 Synthesis procedure to make silica/Au nanoparticles/silica core–shell-shell nanospheres containing catalytically active Au nanoparticles. These nanomaterials were shown to serve as effective heterogeneous catalysts for a number of reactions. (Reprinted with permission from Ref. [180]. Copyright 2012, Springer Science + Business Media, LLC.)

present [89,90,165–167]. Using combinations of synthetic methods various core–shell nanoparticles containing SPR active nanoparticles have been reported by several groups, including the Calzaferri [168,169], Mulvaney [170–172], Liz-Marzan [173,174], and Kamat [89,90,175] groups for plasmonic photocatalysis and fundamental investigations in photocatalysis in nanocatalytic systems [175].

For most of the core–shell nanocatalysts discussed above, and others, the core–shell structure typically contains only one metal per particle, mostly as a core (which means a relatively low loading of metallic catalytic sites per nanoparticle). To increase the loading of metal nanoparticles, thus produce more catalytic activity per unit particle, the catalytically active nanoparticles can be placed on the outer shell of each particle, and then even coated with a nanoporous metal oxide shell, as demonstrated first independently by the Asefa group and the Yin group (Figure 17.18) [176–180]. The size of the inner silica core and the thickness of the outer shell can all be controlled by changing the concentrations of reagents and their deposition times, and the pore size of the latter can be controlled by varying the concentration of the base and the etching time used to create the pores. Importantly, the nanoporous shells around the metal nanoparticles can keep the nanoparticles not to aggregate or come off of the materials, while allowing the reactants to reach the catalytic sites [180]. Hence, the core–shell nanocatalyst exhibits more stable catalytic activity than the corresponding catalyst containing no shells around the nanoparticles.

A standard templating synthesis involving surfactants such as CTAB, Brij-30, and Pluronics as templates, instead of etching, could also be used to produce the nanoporous shells with different pores sizes around the catalytic active nanoparticles [181–184]. Other direct synthetic methods that can lead to related core–shell type nanocatalysts have also been reported by using metal-glycolates nanospheres and subjecting them to galvanic reactions, which have ultimately produced materials such as Ag/TiO$_2$ core–shell nanocatalysts possessing many Ag nanoparticles within nanoporous TiO$_2$ [185]. So, by using variants of these synthetic strategies, various core–shell nanocatalysts with different compositions and catalytic properties can be produced [186]. Additionally, multifunctional nanocatalysts can be synthesized by adding other nanomaterials, for example, magnetic active nanoparticles, into the core or shells of the nanocatalysts to easily recover them after reactions by using simple bar magnets (Figure 17.19) [187–191].

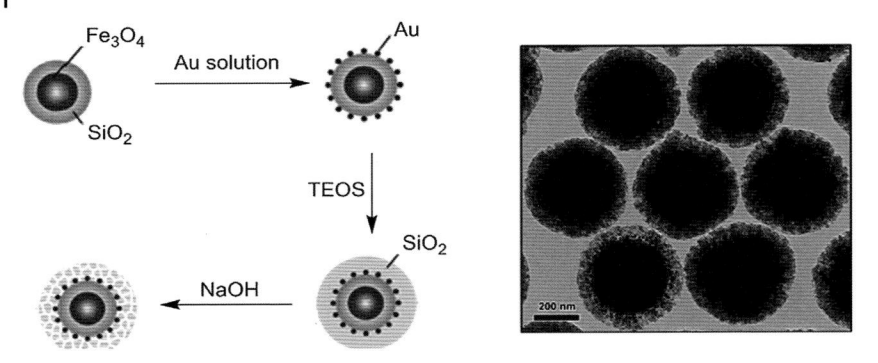

Figure 17.19 A synthetic procedure used to make of a magnetically-active $Fe_3O_4/SiO_2/Au/$ nanoporous silica core–shell-shell nanocatalyst and the TEM image of the particles. (Reprinted with permission from Ref. [192]. Copyright 2008, Wiley-VCH Verlag GmbH.)

17.3.6
Metal Organic Framework and Covalent Organic Framework Nanostructured Catalysts

Metal organic frameworks (MOFs), which have nanostructures made from the linkages of metal cluster secondary building units and organic linkers, are relatively new comers to the catalysis arena [193–198]. As they often have crystalline, microporous structures, MOFs mimic zeolites in some ways [199–201]. Similarly, covalent organic frameworks (COFs), which have periodic/ordered open covalent networks that are entirely built from organic units [202,203], are also new in catalysis [204]. In both cases, as the catalytic activities stem from the molecular species present in the frameworks, their catalytic properties resemble those of the corresponding homogenous, coordination compounds in solutions. Generally, compared with many other 3D nanocatalysts, their structural integrity and stability against various solvents, humidity, and acidic and basic solutions are relatively less though.

Nevertheless, many MOFs and COFs have recently been shown to catalyze a number of reactions. In one example, a layered MOF [Co(Hoba)$_2$.2H$_2$O] (1) (H$_2$oba = 4,4′-oxybis(benzoic acid)) is found to catalyze olefin epoxidation with 96% conversion, 96% selectivity, and good recyclable property [205]. In another example, a series of robust, porous [Ir(COD)(OMe)]$_2$-metallated bipyridyl- and phenanthryl-based MOF catalysts were synthesized [206]. The catalysts showed high catalytic activity as single-site solid catalysts for tandem hydrosilylation of aryl ketones and aldehydes, followed by dehydrogenative orthosilylation of benzylic silyl ethers as well as C—H borylation of arenes (Figure 17.20). They gave TONs up to 17 000 for C—H borylation reaction, and could be recycled over 15 times.

An example of a robust COF nanocatalyst is shown in Figure 17.21 [207]. This COF, which is prepared with high surface and chiral linkages containing water stable groups such as imine, hydrazone, triazine, phenazine, and azines on its

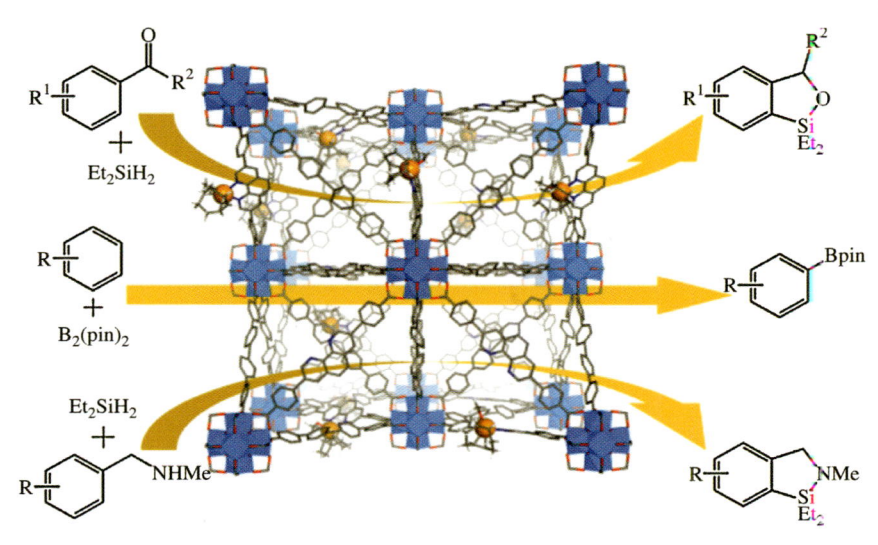

Figure 17.20 An example of a highly active, recyclable MOF catalyst for various reactions including tandem reactions. (Reproduced with permission from Ref. [206]. Copyright 2015, American Chemical Society.)

channel walls, was demonstrated to serve as a metal-free, easily recyclable efficient enantioselective catalyst for asymmetric Michael reaction in water.

17.3.7
Bifunctional- and Multifunctional Nanocatalysts

In many catalytic processes, multiple catalysts are often required before a raw material or a reactant can reach the ultimate, desired product. So, if a catalyst contains two or more than two catalytic groups, it enables such multistep, cascade catalytic reactions to take place in one-pot [208–213]. Furthermore, two or

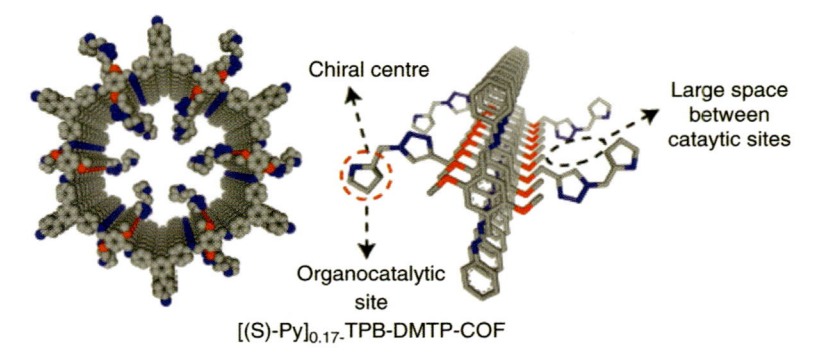

Figure 17.21 Example of a chiral COF nanocatalyst. (Reproduced with permission from Ref. [207]. Copyright 2015, Nature Publishing Group.)

more than two different types of catalytic groups (e.g., metal complexes, metal nanoclusters, organocatalysts, ligands, chiral auxialliary groups) are known to produce cooperative or synergistic catalytic effects and better catalytic outcomes in many reactions [214]. Hence, the development of such types of multifunctional nanocatalysts is appealing [215].

By supporting or immobilizing two or more precursors using the different approaches mentioned above, such types of multifunctional 3D nanocatalysts containing multiple types of catalytic groups could be synthesized [216–218]. During the synthesis of these types of nanocatalysts, it is however often necessary to avoid the potential segregation of the different phases or functional groups involved, in order to obtain highly active multifunctional nanocatalysts [219]. For example, if the rate of reduction of two metals differs substantially and if they are subject to reduction at the same time, they will likely form core–shell type nanoparticles, as opposed to an alloy type supported nanoparticles, which may be desired.

Besides conventional multifunctional catalysts, a number of bifunctional nanoelectrocatalysts have been reported, for example, for fuel cells and electrochemical water splitting [220]. Using the right bifunctional electrocatalyst, two electrochemical reactions can be performed at the same time. Zou and Asefa groups [221] have recently developed Ni_3S_2/Ni foam that can serve as a binder-free bifunctional electrocatalyst for both hydrogen evolution reaction (HER) at the anode and oxygen evolution reaction (OER) at the cathode with very good activity and stability. Similarly, Lin and coworkers reported the synthesis of hierarchically ordered, hollow nickel cobaltite ($NiCo_2O_4$) microcubes that can electrocatalyze both HER and OER [222], and Tour and coworkers reported Co phosphide/phosphate that can serve as a bifunctional nanocatalyst for both HER and OER [223].

17.4
Effect of Nanostructure (Size, Shape, Grain Boundary, Pores, etc.) on their Catalysis

17.4.1
Effect of Size

The nanoscale dimension of a nanocatalyst is unquestionably of utmost importance to nanocatalysts [224]. This can be exemplified by the fact that bulk gold is quite inert or inactive for catalysis, but nanosized gold is a very active catalyst, even at subambient temperatures [225,226]. As a specific example, Haruta and coworkers reported that metal-oxide (Al_2O_3, ZrO_2, TiO_2, CeO_2) supported Au nanocatalysts show higher catalytic activity and turn-over-frequency (TOF) for the glucose oxidation reaction as the diameter of the Au particles decreases to the range of 2–6 nm [227]. Since the surface is generally the one that affect(s) the chemisorption and the chemical reactivity of reactants over the catalytic material the most, the material shows relatively higher catalytic activity at least due to the availability of larger surface area at nanoscale size (Table 17.1) [228,229]. It could also be due to quantum size effects or inherent nanoscale-related catalytic activity.

17.4 Effect of Nanostructure (Size, Shape, Grain Boundary, Pores, etc.) on their Catalysis

Table 17.1 Number of surface atoms with respect to total number of atoms in full shell nanoclusters.

Full-shell clusters	Total number of atoms	Number of surface atoms	Surface atoms (%)
1 shell	13	12	92
2 shells	55	42	76
3 shells	147	92	63
4 shells	309	162	52
5 shells	561	252	45
6 shells	923	362	39

Decrease in surface/volume ratio, surface atom/total atom ratio, corner + edge/total atom ratio and their facets as the size or total atom containing nanoclusters increases. (Reproduced with permission from Refs [230,231]. Copyright 2013, Wiley-VCH Verlag GmbH and Copyright 2015, Elsevier B.V.)

Generally, the smaller the size of catalytically active nanomaterials (0D, 1D, etc.), the higher their catalytic activity per unit mass would be. For example, in Pd nanoparticles-catalyzed CH_4-O_2 thermocatalytic reaction (producing CO_2 and H_2O as the only detected products, except when the amount of O_2 was nearly depleted, in which a trace amount CO was formed), higher CH_4–O_2 turnover rates and lower C—H bond activation rate constants for O^*–O^* site pairs were obtained when the size of the Pd nanoparticles was decreased from 21.3 to 4.8 nm [232]. This is because the O^* atoms bind more strongly with the more unsaturated, exposed atoms present on the small nanoparticles, thus decreasing their reactivity for H-abstraction [232]. Similar size dependent effects have also been widely documented for many nanophotocatalysts and nanoelectrocatalysts. For example, for Cu nanocatalysts reported for electrochemical activation of CO_2, while Cu nanoparticles that have 1.9 and 2.3 nm size showed higher catalytic activity than a Cu foil electrode, those that have 4.8–15.1 nm size exhibited poor or only similar catalytic activity as the latter [233]. The size of nanoparticles can dictate the types of products forming from the reaction too. In the above case, Cu nanoparticles with approximately 2 nm in size and below favored the formation of H_2 and CO over hydrocarbons (CH_4, C_2H_4) due to their higher proportion of low coordinated (<8) surface atoms.

17.4.2
Effect of Shape of Nanocatalysts on Catalysis

The shape of nanocatalytic systems also plays an important role on their catalytic properties [228,234–238], because certain surface atoms at edges and corners of nanoparticles have higher surface energy, or reactivity, and thus interact better with reactants and catalyze their conversions [239]. This has actually been among the reasons why efforts to make nanocatalysts with not only controlled size but also different shapes have intensively continued [240–242]. In other words, the catalytic properties of nanocatalysts can be tailored, or improved, by controlling their shape and morphology [243,244] using shape-controlling agents, such as CTAB or polymers (e.g., poly(vinylpyrrolidone) or PVP) [245–247].

One of the earlier reports showing the effects of shaped nanoparticles on catalysis was performed by El-Sayed and coworkers [248,249]. In one of their works, they synthesized Pt nanoparticles with different shapes (tetrahedral, cubic, and spherical) and then tested their catalytic activity toward electron transfer reactions [248]. While the tetrahedral Pt nanoparticles were the most catalytically active, with a minimum activation energy of 14 kJ/mol, the cubic Pt nanoparticles were the least catalytically active, with a higher activation energy of 26.4 kJ/mol. The spherical Pt nanoparticles, on the other hand, showed catalytic activity somewhere in between those of the tetrahedral and the cubic ones. This is because the tetrahedral ones have (111) facets and a large density of surface atoms on edges and corners, and thus exhibit a higher catalytic activity. But, the cubic Pt nanoparticles possess (100) facets and a relatively smaller density of surface atoms on edges and corners. A similar trend was also observed for the same set of nanoparticles in the Suzuki coupling reaction.

In some cases, the shapes and chemical states of the nanocatalysts can evolve, and display different catalytic activity and selectivity, depending on the reaction conditions [250]. For example, in the partial and total oxidation of 2-propanol over about 0.7 nm Pt nanoparticles supported on γ-Al_2O_3, the chemical state of the Pt nanoparticle changes, and the coordination around them evolves as a function of the reaction temperature (Figure 17.22) [251,252]. Even at low

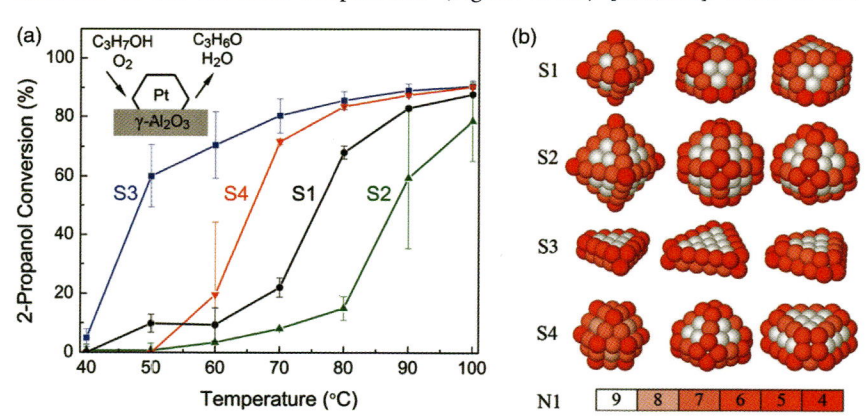

Figure 17.22 (a) Percent (%) conversion of 2-propanol during oxidation reaction catalyzed by nanocrystalline γ-Al_2O_3-supported Pt nanoparticles with different shapes but similar average sizes (<0.8 nm for S1 and ~1 nm for S2, S3, and S4). (b) Pt NPs with three different shapes; their coordination numbers are obtained from EXAFS and the colors indicate the different number of first nearest neighbors (N1) of surface atoms in each nanoparticle. (Reprinted with permission from Ref. [252]. Copyright 2010, American Chemical Society.)

temperatures (<100 °C, or within the partial oxidation regime), a significant decrease in Pt–Pt coordination number, and a concomitant increase in the Pt–O coordination number or in the amount of PtO_x, was documented for these nanoparticles. These nanoparticles favored acetone as product (>97%) during the catalytic reaction. On the other hand, at higher temperatures (>100 °C) or complete oxidation regime producing CO_2, the nanoparticles were initially in metallic state but then quickly covered with PtO_x (Figure 17.22).

17.4.3
Effect of Porosity

In case of nanoporous materials, apart from their overall particle size, their pore size as well as pore shape can play important roles in catalysis. Porosity has actually been highly relevant and critical in microporous materials such as zeolites, whose highly monodisperse micropores have been taken advantage of to do a variety of size and shape selective catalysis of different reactants, and generate specific products. This is also true for mesoporous materials that have well-controlled pore structures. As a notable example, Iwamoto et al. reported that the catalytic activity for the acetalization of cyclohexanone with methanol by mesoporous silica-based catalysts is dependent on the pore size in the materials [253]. The material with 1.9 nm pores particularly gave the best catalytic activity while those with either smaller or bigger pores showed lower activity (Figure 17.23). This difference in catalytic properties among these different material was reasoned based on the difference in mass diffusion for reactants, intermediates and products; the synergistic effects resulting from the orientation and local concentration of the reactants; and the position of the acidic silanol groups in the pores of these materials. Notably, the material with 1.9 nm pores was suggested to have these features in the most optimal manner to catalyze the reaction the best.

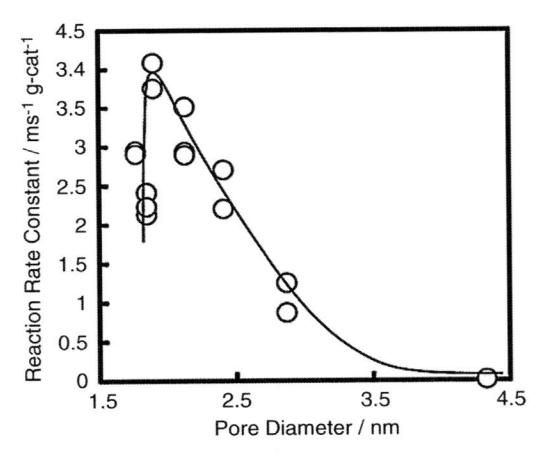

Figure 17.23 Dependence of the catalytic activity for the acetalization of cyclohexanone as a function of the pore diameter of MCM-41-based catalyst. (Reprinted with permission from Ref. [253]. Copyright 2003, American Chemical Society.)

17.5
Characterizations Methods for Nanocatalysts and Nanocatalysis

Just like many traditional catalysts and materials, nanocatalysts rely on analytical methods that allow elucidation of their structures, compositions, and changes in structures and compositions before, after, and even during catalytic reactions [254,255]. While many of the common techniques used for characterization of nanomaterials are also applicable to nanocatalyts, probing the active sites in nanocatalysts may require extra efforts or additional techniques [256]. In terms of characterization of the properties and performance of nanocatalysts, typical analytical methods used for common organic and inorganic reactions can be applied. Probing the products of nanocatalytic reactions is actually generally relatively easier to do, as the nanocatalysts are easier to separate from the reaction products, unlike common homogeneous catalysts. Similarly, as the nanomaterials can be recovered afterwards, their compositions/structures, recyclability, catalytic activity, and so on, are also easier to characterize.

The most common analytical methods used for characterization of nanostructured catalysts include those that allow: (1) structural analyses, such as gas porosimetry, X-ray diffraction (XRD), scanning electron microscopy and transmission electron microscopy and (2) compositional analysis, such as solid-state NMR, Fourier transform infrared, and X-ray photoelectron spectroscopy, and elemental analysis and thermogravimetric analysis.

However, much work still remains to elucidate the specific sites of many nanocatalysts, especially those with smaller dimensions (<1 nm). In particular, observing the effects of each species in a given nanocatalyst is a tremendous challenge as they often exist as ensembles of many species and catalytic sites, unlike in homogeneous catalysts [257,258]. The development of powerful molecular and *in situ*/operando probing techniques that allow probing the catalysts in action during catalysis has, therefore, been increasingly sought after [259].

17.6
Applications of Nanocatalysts or Nanocatalysis

Several examples of the applications or potential applications of nanocatalysts have already been discussed in preceding sections. While it is possible to add a separate section detailing the many reactions that nanocatalysts can catalyze, it would beyond the scope of this chapter. Briefly still, the applications of nanocatalysts have been shown in various common carbon–carbon reactions such as the Henry, Knoevenagel, Hiyama, Negishi, Stille, Kumada-Corriu, aldol, and various other carbon-heteroatom coupling reactions. Other reactions where nanocatalysts have been demonstrated to catalyze include skeletal isomerization, oxidation, and reduction reactions. These types of reactions are extensively used in industrial settings for the production of various synthetic materials, polymers, pharmaceuticals, and so on. Numerous nanocatalytic reactions that are highly relevant for renewable energy production and environmental remediation have also been successfully tested and being developed [260].

17.7 Future Prospects and Outlooks and Conclusions

The chapter has introduced the field of nanocatalysis, discussed the various common synthetic methods used to make different kinds of nanocatalytic materials, the unique structure and properties of nanocatalytic materials, and the effects of the structure, composition and shape of nanocatalysts on their catalytic properties. Special topics have also been dedicated to a wide array of supported and core–shell type nanocatalytic systems. Although great progress has already been made in nanocatalysis, research on the development of nanocatalysts and investigation of their properties is still intensively continuing. Some of the major research topics being pursued include the following.

Plasmonic photocatalysis and upconversion catalysis would be major areas that are still expected to bring in lots of interesting results and make great impacts [24]. In particular, nanomaterials with tunable spectral properties, improved SPR active groups and different geometries that can serve as efficient "nanoantenna" for plasmonics in a wide range of electromagnetic radiation would be of great interest. Similarly, upconversion nanocatalysts that can produce higher quantum efficiency in reactions such as C–H activation or other organic reactions would be of tremendous interest [40].

Research on nanocatalysis in any form that can contribute to the solutions of the so-called grand challenge areas such as energy, environment, and water is expected to continue. This includes the discovery of solar activated nanocatalytic processes that allow solar to chemical energy conversions, such as CO_2 reduction to synthetic fuels, water splitting to produce hydrogen, and so on.

Similarly, since petroleum or fossil fuels most likely remain major components of our energy landscape for years to come, nanocatalytic systems that allow conservation of atoms and energy, give high yield of desired products per unit time at less energy cost, allow fuel upgrading, and so on, would still be needed. In light of the recent major discoveries of shale gas/natural gas, nanocatalysis that can employ CH_4 as a feedstock and convert it to value added chemical products, pharmaceuticals, and even better upgraded hydrocarbons and liquid fuels would be a growing research area [261].

One area that has recently witnessed explosive research activities, and continues to do so, is the development metal- and noble metal-free nanocatalysts that can replace conventional noble metal-based catalytic systems for many reactions. In particular, these types of nanocatalysts that integrate photo- and electrochemical activations under full direct solar energy are of high interest [262].

The mechanisms by which many types of nanocatalysts activate reactions have not been fully understood and would still need lots of investigations. So the development of nanomaterials that can be used as stable, "ideal" model systems for such studies and that can increase our fundamental understanding of structure–property relationships of nanocatalysts can make a difference [263]. In particular, research involving new and more powerful operando/*in situ* characterization techniques, simulation, supercomputers and software, and powerful spectroscopic methods that allow spatial resolution and time resolved studies would always remain interesting research areas. Combination of theoretical and experimental studies to answer many of the remaining questions and enable precisely the correlations between structure and activity would likely to continue in the field as well.

These challenging research areas and questions yet to be addressed would certainly provide an opportunity as well as an impetus for researchers to continue to make various nanocatalysts using different methods. These may however require both the imagination of the scientific community as well as resources (funding) to the scientific community who are involved to tackle these issues.

References

1 Coperet, C., Comas-Vives, A., Conley, M.P., Estes, D.P., Fedorov, A., Mougel, V., Nagae, H., Nunez-Zarur, F., and Zhizhko, P.A. (2016) *Chem. Rev.*, **116**, 323.

2 Hansen, T.W., Delariva, A.T., Challa, S.R., and Datye, A.K. (2013) *Acc. Chem. Res.*, **46**, 1720.

3 Das, S., Goswami, A., Hesari, M., Al-Sharab, J.F., Mikmekova, E., Maran, F., and Asefa, T. (2014) *Small*, **10**, 1473.

4 Li, K., Zhao, J., Zhang, Y., Wu, P., and Zhang, Z.M. (2016) *RSC Adv.*, **6**, 11829.

5 Zhao, Z., Tian, J., Sang, Y., Cabot, A., and Liu, H. (2015) *Adv. Mater.*, **27**, 2557.

6 Xu, J., Xiao, X., Ren, F., Wu, W., Dai, Z., Cai, G., Zhang, S., Zhou, J., Mei, F., and Jiang, C. (2012) *Nanoscale Res. Lett.*, **7**, 239.

7 Tsydenova, O., Batoev, V., and Batoeva, A. (2015) *Int. J. Environ. Res. Pub. Health*, **12**, 9542.

8 Linic, S., Christopher, P., and Ingram, D.B. (2011) *Nat. Mater.*, **10**, 911.

9 Zhuang, H., Zhang, Y., Chu, Z., Long, J., An, X., Zhang, H., Lin, H., Zhang, Z., and Wang, X. (2016) *Phys. Chem. Chem. Phys.*, **18**, 9636.

10 Chen, D., Gao, Y.F., Wang, G., Zhang, H., Lu, W., and Li, J.H. (2007) *J. Phys. Chem. C*, **111**, 13163.

11 Tian, J., Zhao, Z.H., Kumar, A., Boughton, R.I., and Liu, H. (2014) *Chem. Soc. Rev.*, **43**, 6920.

12 Li, Z.T., Ren, Z.Y., Qu, Y., Du, S.C., Wu, J., Kong, L.J., Tian, G.H., Zhou, W., and Fu, H.G. (2014) *Eur. J. Inorg. Chem.*, **2014**, 2146.

13 Viswanathan, B. and Krishanmurthy, K.R. (2012) *Int. J. Photoenergy*, **2012**, 269654.

14 Fang, J., Wang, F., Qian, K., Bao, H.Z., Jiang, Z.Q., and Huang, W.X. (2008) *J. Phys. Chem. C*, **112**, 18150.

15 Mitoraj, D. and Kisch, H. (2008) *Angew. Chem., Int. Ed.*, **47**, 9975.

16 Han, B. and Hu, Y.H. (2015) *J. Phys. Chem. C*, **119**, 18927.

17 Cheng, X.W., Yu, X.J., and Xing, Z.P. (2013) *J. Phys. Chem. Solids*, **74**, 684.

18 Zhu, G., Shan, Y., Lin, T., Zhao, W., Xu, J., Tian, Z., Zhang, H., Zheng, C., and Huang, F. (2016) *Nanoscale*, **8**, 4705.

19 Yan, J., Liu, P., Ma, C., Lin, Z., and Yang, G. (2016) *Nanoscale*, **8**, 8826.

20 Rong, X.S., Qiu, F.X., Yan, J., Zhao, H., Zhu, X.L., and Yang, D.Y. (2015) *RSC Adv.*, **5**, 24944.

21 Geng, Y.L., Zhang, P., and Kuang, S.P. (2014) *RSC Adv.*, **4**, 46054.

22 Zhu, G., Lin, T., Cui, H., Zhao, W., Zhang, H., and Huang, F. (2016) *ACS Appl. Mater. Interfaces*, **8**, 122.

23 Tsukamoto, D., Shiraishi, Y., Sugano, Y., Ichikawa, S., Tanaka, S., and Hirai, T. (2012) *J. Am. Chem. Soc.*, **134**, 6309.

24 Awazu, K., Fujimaki, M., Rockstuhl, C., Tominaga, J., Murakami, H., Ohki, Y., Yoshida, N., and Watanabe, T. (2008) *J. Am. Chem. Soc.*, **130**, 1676.

25 Chen, Y.H., Bian, J.J., Qi, L.L., Liu, E.Z., and Fan, J. (2015) *J. Nanomater*, **2015**, 905259.

26 Sun, X.J., Zhang, H., Wei, J.Z., Yu, Q., Yang, P., and Zhang, F.M. (2016) *Mat. Sci. Semicon. Proc.*, **45**, 51.

27 Lopez, S.M., Hidalgo, M.C., Navio, J.A., and Colon, G. (2011) *J. Hazard. Mater.*, **185**, 1425.

28 Colon, G., Lopez, S.M., Hidalgo, M.C., and Navio, J.A. (2010) *Chem. Commun.*, **46**, 4809.

29 Kostis, I., Vasilopoulou, M., Papadimitropoulos, G., Stathopoulos, N., Savaidis, S., and Davazoglou, D. (2013) *Surf. Coat. Tech.*, **230**, 51.

30 Wang, G., Huang, B.B., Ma, X.C., Wang, Z.Y., Qin, X.Y., Zhang, X.Y., Dai, Y., and Whangbo, M.H. (2013) *Angew. Chem., Int. Ed.*, **52**, 4810.

31 Liu, Z.E., Wang, J., Li, Y., Hu, X.X., Yin, J.W., Peng, Y.Q., Li, Z.H., Li, Y.W., Li, B.M., and Yuan, Q. (2015) *ACS Appl. Mater. Inter.*, **7**, 19416.

32 Wang, W., Huang, W., Ni, Y., Lu, C., and Xu, Z. (2014) *ACS Appl. Mater. Interfaces*, **6**, 340.

33 Xu, R., Li, J., Wang, J., Wang, X.F., Liu, B., Wang, B.X., Luan, X.Y., and Zhang, X.D. (2010) *Sol. Energy Mat. Sol. C*, **94**, 1157.

34 Wang, J., Zhang, G., Zhang, Z.H., Zhang, X.D., Zhao, G., Wen, F.Y., Pan, Z.J., Li, Y., Zhang, P., and Kang, P.L. (2006) *Water Res.*, **40**, 2143.

35 Zhang, P., Wang, T., and Gong, J. (2015) *Adv. Mater.*, **27**, 5328.

36 Jiang, R., Li, B., Fang, C., and Wang, J. (2014) *Adv. Mater.*, **26**, 5274.

37 Ng, C., Cadusch, J.J., Dligatch, S., Roberts, A., Davis, T.J., Mulvaney, P., and Gomez, D.E. (2016) *ACS Nano*, **10**, 4704.

38 DuChene, J.S., Sweeny, B.C., Johnston-Peck, A.C., Su, D., Stach, E.A., and Wei, W.D. (2014) *Angew. Chem., Int. Ed. Engl.*, **53**, 7887.

39 Yu, S., Kim, Y.H., Lee, S.Y., Song, H.D., and Yi, J. (2014) *Angew. Chem., Int. Ed. Engl.*, **53**, 11203.

40 Baffou, G. and Quidant, R. (2014) *Chem. Soc. Rev.*, **43**, 3898.

41 Wang, J., Zhang, G., Zhang, Z., Zhang, X., Zhao, G., Wen, F., Pan, Z., Li, Y., Zhang, P., and Kang, P. (2006) *Water Res.*, **40**, 2143.

42 Wang, J., Wen, F.Y., Zhang, Z.H., Zhang, X.D., Pan, Z.J., Zhang, L., Wang, L., Xu, L., Kang, P.L., and Zhang, P. (2005) *J. Environ. Sci. (China)*, **17**, 727.

43 Liu, Z.E., Wang, J., Li, Y., Hu, X., Yin, J., Peng, Y., Li, Z., Li, Y., Li, B., and Yuan, Q. (2015) *ACS Appl. Mater. Interfaces*, **7**, 19416.

44 Baldauf, W., Jagers, G., Kaufmann, D., and Zimmermann, H. (1994) Proceedings of the 6th Ethylene Producers' Conference, Vol 3, 635.

45 Hou, J.T., Li, Y.Z., Mao, M.Y., Yue, L.Z., Greaves, G.N., and Zhao, X.J. (2015) *Nanoscale*, **7**, 2633.

46 Mao, M.Y., Li, Y.Z., Hou, J.T., Zeng, M., and Zhao, X.J. (2015) *Appl. Catal. B Environ.*, **174**, 496.

47 Zeng, M., Li, Y.Z., Mao, M.Y., Bai, J.L., Ren, L., and Zhao, X.J. (2015) *ACS Catal.*, **5**, 3278.

48 Jin, Q.L., Yamamoto, H., Yamamoto, K., Fujishima, M., and Tada, H. (2013) *Phys. Chem. Chem. Phys.*, **15**, 20313.

49 Li, M., Cullen, D.A., Sasaki, K., Marinkovic, N.S., More, K., and Adzic, R.R. (2013) *J. Am. Chem. Soc.*, **135**, 132.

50 Cui, Z., Chen, H., Zhao, M., and DiSalvo, F.J. (2016) *Nano Lett.*, **16**, 2560.

51 Yang, Z., Moriguchi, I., and Nakashima, N. (2015) *ACS Appl. Mater. Interfaces*, **7**, 9800.

52 Silva, R. and Asefa, T. (2012) *Adv. Mater.*, **24**, 1878.

53 Morozan, A., Jegou, P., Pinault, M., Campidelli, S., Jousselme, B., and Palacin, S. (2012) *ChemSusChem*, **5**, 647.

54 Li, J., Zhang, Y., Zhang, X., Han, J., Wang, Y., Gu, L., Zhang, Z., Wang, X., Jian, J., Xu, P., and Song, B. (2015) *ACS Appl. Mater. Interfaces*, **7**, 19626.

55 Tan, S.M. and Pumera, M. (2016) *ACS Appl. Mater. Interfaces*, **8**, 3948.

56 Pachatouridou, E., Papista, E., Delimitis, A., Vasiliades, M.A., Efstathiou, A.M., Amiridis, M.D., Alexeev, O.S., Bloom, D., Marnellos, G.E., Konsolakis, M., and

Iliopoulou, E. (2016) *Appl. Catal. B Environ.*, **187**, 259.
57 Rao, C.N.R., Vivekchand, S.R.C., Biswas, K., and Govindaraj, A. (2007) *Dalton Trans.*, 3728.
58 Makgwane, P.R. and Ray, S.S. (2014) *J. Nanosci. Nanotechnol.*, **14**, 1338.
59 Baddour, C. and Briens, C. (2005) *Int. J. Chem. React. Eng.*, **3**, R3.
60 Liu, H., Feng, Y., Chen, D., Li, C.Y., Cui, P.L., and Yang, J. (2015) *J. Mater. Chem. A*, **3**, 3182.
61 El-Toni, A.M., Habila, M.A., Labis, J.P., Alothman, Z.A., Alhoshan, M., Elzatahry, A.A., and Zhang, F. (2016) *Nanoscale*, **8**, 2510.
62 Chen, X.J., Smith, N.M., Iyer, K.S., and Raston, C.L. (2014) *Chem. Soc. Rev.*, **43**, 1387.
63 Zhang, C., Yan, Y.L., Zhao, Y.S., and Yao, J.N. (2013) *Annu. Rep. Prog. Chem. C*, **109**, 211.
64 Haruta, M., Kobayashi, T., Sano, H., and Yamada, N. (1987) *Chem. Lett.*, **16**, 405.
65 Hutchings, G.J. (1985) *J. Catal.*, **96**, 292.
66 Wang, C.L., Ciganda, R., Salmon, L., Gregurec, D., Irigoyen, J., Moya, S., Ruiz, J., and Astruc, D. (2016) *Angew. Chem., Int. Ed.*, **55**, 3091.
67 Zhu, B.W. and Lim, T.T. (2007) *Environ. Sci. Technol.*, **41**, 7523.
68 Lee, Y.C., Kim, C.W., Lee, J.Y., Shin, H.J., and Yang, J.W. (2009) *Desalin Water Treat.*, **10**, 33.
69 Pradhan, N., Pal, A., and Pal, T. (2002) *Colloid Surf. A*, **196**, 247.
70 Hudson, R., Hamasaka, G., Osako, T., Yamada, Y.M.A., Li, C.J., Uozumi, Y., and Moores, A. (2013) *Green Chem.*, **15**, 2141.
71 Mori, K., Yoshioka, N., Kondo, Y., Takeuchi, T., and Yamashita, H. (2009) *Green Chem.*, **11**, 1337.
72 Zhou, S.Q., Wen, M., Wang, N., Wu, Q.S., Wu, Q.N., and Cheng, L.Y. (2012) *J. Mater. Chem.*, **22**, 16858.
73 Wang, X.Z., Ding, L., Zhao, Z.B., Xu, W.Y., Meng, B., and Qiu, J.S. (2011) *Catal. Today*, **175**, 509.
74 Zhang, W.Y., Kong, C., and Lu, G.X. (2015) *Chem. Commun.*, **51**, 10158.
75 Zhou, K.B., Wang, R.P., Xu, B.Q., and Li, Y.D. (2006) *Nanotechnology*, **17**, 3939.

76 Vidal, C., Wang, D., Schaaf, P., Hrelescu, C., and Klar, T.A. (2015) *ACS Photonics*, **2**, 1436.
77 Macdonald, J.E., Bar Sadan, M., Houben, L., Popov, I., and Banin, U. (2010) *Nat. Mater.*, **9**, 810.
78 Yan, C.L. and Rosei, F. (2014) *New J. Chem.*, **38**, 1883.
79 Liang, X., Wang, X., Zhuang, Y., Xu, B., Kuang, S., and Li, Y. (2008) *J. Am. Chem. Soc.*, **130**, 2736.
80 Mel, A.E., Tessier, P.Y., Buffiere, M., Gautron, E., Ding, J., Du, K., Choi, C.H., Konstantinidis, S., Snyders, R., Bittencourt, C., and Molina-Luna, L. (2016) *Small*, **12**, 2885.
81 Yu, C. and He, J. (2012) *Chem. Commun. (Camb.)*, **48**, 4933.
82 Scott, R.W., Sivadinarayana, C., Wilson, O.M., Yan, Z., Goodman, D.W., and Crooks, R.M. (2005) *J. Am. Chem. Soc.*, **127**, 1380.
83 Garcia-Martinez, J.C., Lezutekong, R., and Crooks, R.M. (2005) *J. Am. Chem. Soc.*, **127**, 5097.
84 Gates, A.T., Nettleton, E.G., Myers, V.S., and Crooks, R.M. (2010) *Langmuir*, **26**, 12994.
85 Yeung, L.K., Lee, C.T. Jr., Johnston, K.P., and Crooks, R.M. (2001) *Chem. Commun. (Camb.)*, 2290.
86 Wilson, O.M., Scott, R.W., Garcia-Martinez, J.C., and Crooks, R.M. (2005) *J. Am. Chem. Soc.*, **127**, 1015.
87 Anderson, R.M., Zhang, L., Loussaert, J.A., Frenkel, A.I., Henkelman, G., and Crooks, R.M. (2013) *ACS Nano.*, **7**, 9345.
88 Weir, M.G., Knecht, M.R., Frenkel, A.I., and Crooks, R.M. (2010) *Langmuir*, **26**, 1137.
89 Scott, R.W., Wilson, O.M., Oh, S.K., Kenik, E.A., and Crooks, R.M. (2004) *J. Am. Chem. Soc.*, **126**, 15583.
90 Scott, R.W., Wilson, O.M., and Crooks, R.M. (2005) *J. Phys. Chem. B*, **109**, 692.
91 Weir, M.G., Knecht, M.R., Frenkel, A.I., and Crooks, R.M. (2010) *Langmuir*, **26**, 1137.
92 Chandrasekaran, S., Nann, T., and Voelcker, N.H. (2015) *Nano Energy*, **17**, 308.
93 Wang, Y.J., Wang, Q.S., Zhan, X.Y., Wang, F.M., Safdar, M., and He, J. (2013) *Nanoscale*, **5**, 8326.

94 Wang, J.L., Chen, K.M., Gong, M., Xu, B., and Yang, Q. (2013) *Nano Lett.*, **13**, 3996.

95 Wang, D.S., Hao, C.H., Zheng, W., Peng, Q., Wang, T.H., Liao, Z.M., Yu, D.P., and Li, Y.D. (2008) *Adv. Mater.*, **20**, 2628.

96 Wang, D.S., Xie, T., and Li, Y.D. (2009) *Nano Res.*, **2**, 30.

97 Voiry, D., Yamaguchi, H., Li, J.W., Silva, R., Alves, D.C.B., Fujita, T., Chen, M.W., Asefa, T., Shenoy, V.B., Eda, G., and Chhowalla, M. (2013) *Nat. Mater.*, **12**, 850.

98 Song, F. and Hu, X.L. (2014) *Nat. Commun.*, **5**, 4477.

99 Gao, S., Jiao, X.C., Sun, Z.T., Zhang, W.H., Sun, Y.F., Wang, C.M., Hu, Q.T., Zu, X.L., Yang, F., Yang, S.Y., Liang, L., Wu, J., and Xie, Y. (2016) *Angew. Chem., Int. Ed.*, **55**, 698.

100 Astruc, D., Lu, F., and Aranzaes, J.R. (2005) *Angew. Chem., Int. Ed.*, **44**, 7852.

101 Perego, C. and Millini, R. (2013) *Chem. Soc. Rev.*, **42**, 3956.

102 Ren, Y., Ma, Z., and Bruce, P.G. (2012) *Chem. Soc. Rev.*, **41**, 4909.

103 Gu, D. and Schuth, F. (2014) *Chem. Soc. Rev.*, **43**, 313.

104 Villa, A., Schiavoni, M., Chan-Thaw, C.E., Fulvio, P.F., Mayes, R.T., Dai, S., More, K.L., Veith, G.M., and Prati, L. (2015) *ChemSusChem*, **8**, 2520.

105 Cheng, T.Y., Zhao, Q.K., Zhang, D.C., and Liu, G.H. (2015) *Green Chem.*, **17**, 2100.

106 Kumar, P. and Guliants, V.V. (2010) *Microporous Mesoporous Mater.*, **132**, 1.

107 Gross, E., Liu, J.H., Alayoglu, S., Marcus, M.A., Fakra, S.C., Toste, F.D., and Somorjai, G.A. (2013) *J. Am. Chem. Soc.*, **135**, 3881.

108 Johnson, B.J. and Stein, A. (2001) *Inorg. Chem.*, **40**, 801.

109 Lim, M.H. and Stein, A. (1999) *Chem. Mater.*, **11**, 3285.

110 Garcia, R.A., van Grieken, R., Iglesias, J., Morales, V., and Gordillo, D. (2008) *Chem. Mater.*, **20**, 2964.

111 Han, Y.R., Park, J.W., Kim, H., Ji, H., Lim, S.H., and Jun, C.H. (2015) *Chem. Commun. (Camb.)*, **51**, 17084.

112 Mizoshita, N., Tani, T., and Inagaki, S. (2011) *Chem. Soc. Rev.*, **40**, 789.

113 Asefa, T., MacLachlan, M.J., Coombs, N., and Ozin, G.A. (1999) *Nature*, **402**, 867.

114 Melde, B.J.H., Holland, B.T., Blanford, C.F., and Stein, A. (1999) *Chem. Mater.*, **11**, 3302.

115 Inagaki, S., Guan, S., Fukushima, Y., Ohsuna, T., and Terasaki, O. (1999) *J. Am. Chem. Soc.*, **121**, 9611.

116 Mallick, S., Rana, S., and Parida, K. (2011) *Dalton Trans.*, **40**, 9169.

117 Grunberg, A., Yeping, X., Breitzke, H., and Buntkowsky, G. (2010) *Chemistry*, **16**, 6993.

118 Biradar, A.V., Patil, V.S., Chandra, P., Doke, D.S., and Asefa, T. (2015) *Chem. Commun. (Camb.)*, **51**, 8496.

119 Balakrishnan, U. and Velmathi, S. (2013) *J. Nanosci. Nanotechnol.*, **13**, 3079.

120 Fukuoka, A. and Dhepe, P.L. (2009) *Chem. Rec.*, **9**, 224.

121 Yang, S. and He, J. (2012) *Chem. Commun. (Camb.)*, **48**, 10349.

122 Xu, J., Cheng, T., Zhang, K., Wang, Z., and Liu, G. (2016) *Chem. Commun. (Camb.)*, **52**, 6005.

123 Van Der Voort, P., Vercaemst, C., Schaubroeck, D., and Verpoort, F. (2008) *Phys. Chem. Chem. Phys.*, **10**, 347.

124 Papaefthimiou, V., Dintzer, T., Lebedeva, M., Teschner, D., Havecker, M., Knop-Gericke, A., Schlogl, R., Pierron-Bohnes, V., Savinova, E., and Zafeiratos, S. (2012) *J. Phys. Chem. C*, **116**, 14342.

125 Reetz, M.T. and Dugal, M. (1999) *Catal. Lett.*, **58**, 207.

126 Budroni, G. and Corma, A. (2006) *Angew. Chem., Int. Ed. Engl.*, **45**, 3328.

127 Park, C. and Keane, M.A. (2003) *J. Colloid Interface Sci.*, **266**, 183.

128 Padhye, L., Wang, P., Karanfil, T., and Huang, C.H. (2010) *Environ. Sci. Technol.*, **44**, 4161.

129 Tseng, H.H., Wey, M.Y., Lin, C.L., and Chang, Y.C. (2002) *J. Air Waste Manag. Assoc.*, **52**, 1281.

130 Espinal, L., Suib, S.L., and Rusling, J.F. (2004) *J. Am. Chem. Soc.*, **126**, 7676.

131 Yu, F., Ji, J., Xu, Z., and Liu, H. (2005) *Ultrasonics*, **44** (Suppl 1), e389.

132 Lu, A.H., Nitz, J.J., Comotti, M., Weidenthaler, C., Schlichte, K., Lehmann, C.W., Terasaki, O., and Schuth, F. (2010) *J. Am. Chem. Soc.*, **132**, 14152.

133 Hu, G., Nitze, F., Gracia-Espino, E., Ma, J., Barzegar, H.R., Sharifi, T., Jia, X.,

Shchukarev, A., Lu, L., Ma, C., Yang, G., and Wagberg, T. (2014) *Nat. Commun.*, **5**, 5253.

134 Li, C.S., Melaet, G., Ralston, W.T., An, K., Brooks, C., Ye, Y., Liu, Y.S., Zhu, J., Guo, J., Alayoglu, S., and Somorjai, G.A. (2015) *Nat. Commun.*, **6**, 6538.

135 Stein, A., Wang, Z., and Fierke, M.A. (2008) *Adv. Mater.*, **21**, 265.

136 Nam, K., Lim, S., Kim, S.K., Peck, D., and Jung, D. (2011) *J. Nanosci. Nanotechnol.*, **11**, 5761.

137 Ma, T.Y., Liu, L., and Yuan, Z.Y. (2013) *Chem. Soc. Rev.*, **42**, 3977.

138 Ren, Z.F., Huang, Z.P., Xu, J.W., Wang, J.H., Bush, P., Siegal, M.P., and Provencio, P.N. (1998) *Science*, **282**, 1105.

139 Meng, Y., Zou, X., Huang, X., Goswami, A., Liu, Z., and Asefa, T. (2014) *Adv. Mater.*, **26**, 6510.

140 Huang, X., Zou, X., Meng, Y., Mikmekova, E., Chen, H., Voiry, D., Goswami, A., Chhowalla, M., and Asefa, T. (2015) *ACS Appl. Mater. Interfaces*, **7**, 1978.

141 Zeng, L.Z., Zhao, S.F., and Li, W.S. (2015) *Appl. Biochem. Biotechnol.*, **176**, 978.

142 Yuan, W., Feng, Y., Xie, A., Zhang, X., Huang, F., Li, S., Zhang, X., and Shen, Y. (2016) *Nanoscale*, **8**, 8704.

143 Almeida, V.C., Silva, R., Acerce, M., Pezoti, O., Cazetta, A.L., Martins, A.C., Huang, X.X., Chhowalla, M., and Asefa, T. (2014) *J. Mater. Chem. A*, **2**, 15181.

144 Marchi, M.C., Acuna, J.J., and Figueroa, C.A. (2012) *J. Nanosci. Nanotechnol.*, **12**, 6439.

145 Cheon, J.Y., Kim, T., Choi, Y., Jeong, H.Y., Kim, M.G., Sa, Y.J., Kim, J., Lee, Z., Yang, T.H., Kwon, K., Terasaki, O., Park, G.G., Adzic, R.R., and Joo, S.H. (2013) *Sci. Rep.*, **3**, 2715.

146 Li, J.C., Zhao, S.Y., Hou, P.X., Fang, R.P., Liu, C., Liang, J., Luan, J., Shan, X.Y., and Cheng, H.M. (2015) *Nanoscale*, **7**, 19201.

147 Zhang, J., Qu, L., Shi, G., Liu, J., Chen, J., and Dai, L. (2016) *Angew. Chem. Int., Ed. Engl.*, **55**, 2230.

148 Shanmugam, S. and Osaka, T. (2011) *Chem. Commun. (Camb.)*, **47**, 4463.

149 Meng, Y., Voiry, D., Goswami, A., Zou, X., Huang, X., Chhowalla, M., Liu, Z., and Asefa, T. (2014) *J. Am. Chem. Soc.*, **136**, 13554.

150 Tang, J., Liu, J., Li, C.L., Li, Y.Q., Tade, M.O., Dai, S., and Yamauchi, Y. (2015) *Angew. Chem., Int. Ed.*, **54**, 588.

151 Kim, H., Jeong, S., Kim do, H., Park, Y.K., and Jeon, J.K. (2012) *J. Nanosci. Nanotechnol.*, **12**, 6074.

152 Gong, T., Qin, L., Lu, J., and Feng, H. (2016) *Phys. Chem. Chem. Phys.*, **18**, 601.

153 Tominaka, S., Shigeto, M., Nishizeko, H., and Osaka, T. (2010) *Chem. Commun. (Camb.)*, **46**, 8989.

154 Martins, A.R., Salviano, A.B., Oliveira, A.A., Mambrini, R.V., and Moura, F.C. (2016) *Environ. Sci. Pollut. Res. Int*, doi: 10.1007/s11356-016-6692-3.

155 Carregal-Romero, S., Buurma, N.J., Perez-Juste, J., Liz-Marzan, L.M., and Herves, P. (2010) *Chem. Mater.*, **22**, 3051.

156 Mohanta, J., Satapathy, S., and Si, S. (2016) *ChemPhysChem*, **17**, 364.

157 Banin, U., Ben-Shahar, Y., and Vinokurov, K. (2014) *Chem. Mater.*, **26**, 97.

158 Gawande, M.B., Goswami, A., Felpin, F.X., Asefa, T., Huang, X., Silva, R., Zou, X., Zboril, R., and Varma, R.S. (2016) *Chem. Rev.*, **116**, 3722.

159 Nandan, B. and Horechyy, A. (2015) *ACS Appl. Mater. Interfaces*, **7**, 12539.

160 Reiss, P., Protiere, M., and Li, L. (2009) *Small*, **5**, 154.

161 El-Toni, A.M., Habila, M.A., Labis, J.P., ALOthman, Z.A., Alhoshan, M., Elzatahry, A.A., and Zhang, F. (2016) *Nanoscale*, **8**, 2510.

162 Moraes, J., Ohno, K., Maschmeyer, T., and Perrier, S. (2013) *Chem. Commun. (Camb.)*, **49**, 9077.

163 Parvulescu, V.I., Parvulescu, V., Endruschat, U., Granger, P., and Richards, R. (2007) *ChemPhysChem*, **8**, 666.

164 Joo, S.H., Park, J.Y., Tsung, C.K., Yamada, Y., Yang, P., and Somorjai, G.A. (2009) *Nat. Mater.*, **8**, 126.

165 Nemanashi, M. and Meijboom, R. (2015) *Langmuir*, **31**, 9041.

166 Bhattacharya, P., Nasybulin, E.N., Engelhard, M.H., Kovarik, L., Bowden, M.E., Li, X.H.S., Gaspar, D.J., Xu, W., and Zhang, J.G. (2014) *Adv. Funct. Mater.*, **24**, 7510.

167 Astruc, D. and Chardac, F. (2001) *Chem. Rev.*, **101**, 2991.

168 Wang, Y.G., Li, H.R., Gu, L.J., Gan, Q.Y., Li, Y.N., and Calzaferri, G. (2009) *Microporous Mesoporous Mater*, **121**, 1.
169 Ramachandra, S., Popovic, Z.D., Schuermann, K.C., Cucinotta, F., Calzaferri, G., and De Cola, L. (2011) *Small*, **7**, 1488.
170 Nann, T. and Mulvaney, P. (2004) *Angew. Chem., Int. Ed.*, **43**, 5393.
171 van Embden, J., Jasieniak, J., Gomez, D.E., Mulvaney, P., and Giersig, M. (2007) *Aust. J. Chem.*, **60**, 457.
172 Mulvaney, P., Liz-Marzan, L.M., Giersig, M., and Ung, T. (2000) *J. Mater. Chem.*, **10**, 1259.
173 Lombardi, A., Grzelczak, M.P., Crut, A., Maioli, P., Pastoriza-Santos, I., Liz-Marzan, L.M., Del Fatti, N., and Vallee, F. (2013) *ACS Nano*, **7**, 2522.
174 Kobayashi, Y., Katakami, H., Mine, E., Nagao, D., Konno, M., and Liz-Marzan, L.M. (2005) *J. Colloid Interfaces Sci.*, **283**, 392.
175 Hirakawa, T. and Kamat, P.V. (2005) *J. Am. Chem. Soc.*, **127**, 3928.
176 Zhang, Q., Zhang, T., Ge, J., and Yin, Y. (2008) *Nano Lett.*, **8**, 2867.
177 Hu, Y., Zhang, Q., Goebl, J., Zhang, T., and Yin, Y. (2010) *Phys. Chem. Chem. Phys.*, **12**, 11836.
178 Shi, Y.L. and Asefa, T. (2007) *Langmuir*, **23**, 9455.
179 Asefa, T., Duncan, C.T., and Sharma, K.K. (2009) *Analyst*, **134**, 1980.
180 Das, S. and Asefa, T. (2012) *Top. Catal.*, **55**, 587.
181 Wang, Y., Biradar, A.V., and Asefa, T. (2012) *ChemSusChem*, **5**, 132.
182 Ahmed, A., Myers, P., and Zhang, H. (2014) *Langmuir*, **30**, 12190.
183 Hu, P., Zhuang, J., Chou, L.Y., Lee, H.K., Ling, X.Y., Chuang, Y.C., and Tsung, C.K. (2014) *J. Am. Chem. Soc.*, **136**, 10561.
184 Prabhakar, N., Nareoja, T., von Haartman, E., Karaman, D.S., Jiang, H., Koho, S., Dolenko, T.A., Hanninen, P.E., Vlasov, D.I., Ralchenko, V.G., Hosomi, S., Vlasov, I.I., Sahlgren, C., and Rosenholm, J.M. (2013) *Nanoscale*, **5**, 3713.
185 Zou, X., Silva, R., Huang, X., Al-Sharab, J.F., and Asefa, T. (2013) *Chem. Commun. (Camb.)*, **49**, 382.
186 Yin, D., Zhang, L., Cao, X., Chen, Z., Tang, J., Liu, Y., Zhang, T., and Wu, M. (2016) *Dalton Trans.*, **45**, 1467.
187 Yan, J.M., Zhang, X.B., Akita, T., Haruta, M., and Xu, Q. (2010) *J. Am. Chem. Soc.*, **132**, 5326.
188 Li, X., Wang, X., Liu, D., Song, S., and Zhang, H. (2014) *Chem. Commun. (Camb.)*, **50**, 7198.
189 Shokouhimehr, M., Kim, T., Jun, S.W., Shin, K., Jang, Y., Kim, B.H., Kim, J., and Hyeon, T. (2014) *Appl. Catal. A Gen.*, **476**, 133.
190 Wang, D., Deraedt, C., Salmon, L., Labrugere, C., Etienne, L., Ruiz, J., and Astruc, D. (2015) *Chem. Eur. J.*, **21**, 6501.
191 Ge, J., Zhang, Q., Zhang, T., and Yin, Y. (2008) *Angew. Chem., Int. Ed. Engl.* **47**, 8924.
192 Ge, J.P., Zhang, Q., Zhang, T.R., and Yin, Y.D. (2008) *Angew. Chem., Int. Ed.*, **47**, 8924.
193 Gu, J.M., Kim, W.S., and Huh, S. (2011) *Dalton Trans.*, **40**, 10826.
194 Luz, I., Xamena, F.X.L.I., and Corma, A. (2010) *J. Catal.*, **276**, 134.
195 Zhang, L.J., Han, C.Y., Dang, Q.Q., Wang, Y.H., and Zhang, X.M. (2015) *RSC Adv.*, **5**, 24293.
196 Guillerm, V., Kim, D., Eubank, J.F., Luebke, R., Liu, X., Adil, K., Lah, M.S., and Eddaoudi, M. (2014) *Chem. Soc. Rev.*, **43**, 6141.
197 Czaja, A.U., Trukhan, N., and Muller, U. (2009) *Chem. Soc. Rev.*, **38**, 1284.
198 Liu, J., Chen, L., Cui, H., Zhang, J., Zhang, L., and Su, C.Y. (2014) *Chem. Soc. Rev.*, **43**, 6011.
199 Manna, K., Zhang, T., Carboni, M., Abney, C.W., and Lin, W.B. (2014) *J. Am. Chem. Soc.*, **136**, 13182.
200 Manna, K., Zhang, T., and Lin, W.B. (2014) *J. Am. Chem. Soc.*, **136**, 6566.
201 Garcia-Garcia, P., Muller, M., and Corma, A. (2014) *Chem. Sci.*, **5**, 2979.
202 Goncalves, R.S.B., de Oliveira, A.B.V., Sindra, H.C., Archanjo, B.S., Mendoza, M.E., Carneiro, L.S.A., Buarque, C.D., and Esteves, P.M. (2016) *ChemCatChem*, **8**, 743.
203 Ding, S.Y., Gao, J., Wang, Q., Zhang, Y., Song, W.G., Su, C.Y., and Wang, W. (2011) *J. Am. Chem. Soc.*, **133**, 19816.

204 Diaz, U. and Corma, A. (2016) *Coordin. Chem. Rev.*, **311**, 85.
205 Zhang, J.M., Biradar, A.V., Pramanik, S., Emge, T.J., Asefa, T., and Li, J. (2012) *Chem. Commun.*, **48**, 6541.
206 Manna, K., Zhang, T., Greene, F.X., and Lin, W.B. (2015) *J. Am. Chem. Soc.*, **137**, 2665.
207 Xu, H., Gao, J., and Jiang, D.L. (2015) *Nat. Chem.*, **7**, 905.
208 Ostrowski, K.A., Fassbach, T.A., and Vorholt, A.J. (2015) *Adv. Synth. Catal.*, **357**, 1374.
209 Chen, Y.Z., Zhou, Y.X., Wang, H.W., Lu, J.L., Uchida, T., Xu, Q., Yu, S.H., and Jiang, H.L. (2015) *ACS Catal.*, **5**, 2062.
210 Long, J.L., Yin, B.L., Li, Y.W., and Zhang, L.J. (2014) *Aiche J.*, **60**, 3565.
211 Gao, J., Zhang, X., Lu, Y., Liu, S., and Liu, J. (2015) *Chemistry*, **21**, 7403.
212 Shiju, N.R., Alberts, A.H., Khalid, S., Brown, D.R., and Rothenberg, G. (2011) *Angew. Chem., Int. Ed. Engl.*, **50**, 9615.
213 Zhang, F., Jiang, H., Wu, X., Mao, Z., and Li, H. (2015) *ACS Appl. Mater. Interfaces.*, **7**, 1669.
214 Dickschat, A.T., Behrends, F., Surmiak, S., Weiss, M., Eckert, H., and Studer, A. (2013) *Chem. Commun. (Camb.)*, **49**, 2195.
215 Motokura, K., Tada, M., and Iwasawa, Y. (2009) *Catal. Today*, **147**, 203.
216 Deng, Y.H., Cai, Y., Sun, Z.K., Liu, J., Liu, C., Wei, J., Li, W., Liu, C., Wang, Y., and Zhao, D.Y. (2010) *J. Am. Chem. Soc.*, **132**, 8466.
217 Sun, Z.B., Cui, G.J., Li, H.Z., Tian, Y.X., and Yan, S.Q. (2016) *Colloid Surf. A*, **489**, 142.
218 Tsai, C.H., Chen, H.T., Althaus, S.M., Mao, K.M., Kobayashi, T., Pruski, M., and Lin, V.S.Y. (2011) *ACS Catal.*, **1**, 729.
219 Zafeiratos, S., Piccinin, S., and Teschner, D. (2012) *Catal. Sci. Technol.*, **2**, 1787.
220 Zhang, J., Zhao, Z., Xia, Z., and Dai, L. (2015) *Nat. Nanotechnol.*, **10**, 444.
221 Feng, L.L., Yu, G., Wu, Y., Li, G.D., Li, H., Sun, Y., Asefa, T., Chen, W., and Zou, X. (2015) *J. Am. Chem. Soc.*, **137**, 14023.
222 Gao, X., Zhang, H., Li, Q., Yu, X., Hong, Z., Zhang, X., Liang, C., and Lin, Z. (2016) *Angew. Chem., Int. Ed.*, **55**, 6.
223 Yang, Y., Fei, H.L., Ruan, G.D., and Tour, J.M. (2015) *Adv. Mater.*, **27**, 3175.
224 El-Sayed, M.A. (2004) *ACC Chem. Res.*, **37**, 326.
225 Perks, B. (2010) *Chem. World UK*, **7**, 48.
226 Wang, C.L. and Astruc, D. (2014) *Chem. Soc. Rev.*, **43**, 7188.
227 Ishida, T., Kinoshita, N., Okatsu, H., Akita, T., Takei, T., and Haruta, M. (2008) *Angew. Chem., Int. Ed.*, **47**, 9265.
228 Li, Y.F. and Liu, Z.P. (2011) *J. Am. Chem. Soc.*, **133**, 15743.
229 Peter, M., Flores Camacho, J.M., Adamovski, S., Ono, L.K., Dostert, K.H., O'Brien, C.P., Roldan Cuenya, B., Schauermann, S., and Freund, H.J. (2013) *Angew. Chem., Int. Ed. Engl.*, **52**, 5175.
230 Philippot, K. and Serp, P. (2013) *Nanomaterials in Catalysis*, 1st edn (eds P. Serp and K. Philippot), Wiley-VCH GmbH & Co. KGaA, p. 1.
231 Cuenya, B.R. and Behafarid, F. (2015) *Surf. Sci. Rep.*, **70**, 135.
232 Chin, Y.H. and Iglesia, E. (2011) *J. Phys. Chem. C*, **115**, 17845.
233 Reske, R., Mistry, H., Behafarid, F., Cuenya, B.R., and Strasser, P. (2014) *J. Am. Chem. Soc.*, **136**, 6978.
234 Sanjeeva Gandhi, M. and Mok, Y.S. (2014) *Chemosphere*, **117**, 440.
235 Cheong, S., Graham, L., Brett, G.L., Henning, A.M., Watt, J., Miedziak, P.J., Song, M., Takeda, Y., Taylor, S.H., and Tilley, R.D. (2013) *ChemSusChem*, **6**, 1858.
236 Hua, Q., Cao, T., Bao, H., Jiang, Z., and Huang, W. (2013) *ChemSusChem*, **6**, 1966.
237 Tsao, Y.C., Rej, S., Chiu, C.Y., and Huang, M.H. (2014) *J. Am. Chem. Soc.*, **136**, 396.
238 da Silva, A.G., Rodrigues, T.S., Wang, J., Yamada, L.K., Alves, T.V., Ornellas, F.R., Ando, R.A., and Camargo, P.H. (2015) *Langmuir*, **31**, 10272.
239 Narayanan, R., Tabor, C., and El-Sayed, M.A. (2008) *Top. Catal.*, **48**, 60.
240 Li, Y. and Shen, W. (2014) *Chem. Soc. Rev.*, **43**, 1543.
241 Kotani, H., Hanazaki, R., Ohkubo, K., Yamada, Y., and Fukuzumi, S. (2011) *Chemistry*, **17**, 2777.
242 Fan, Z.X. and Zhang, H. (2016) *Chem. Soc. Rev.*, **45**, 63.
243 Bai, S., Li, X., Kong, Q., Long, R., Wang, C., Jiang, J., and Xiong, Y. (2015) *Adv. Mater.*, **27**, 3444.

244 Grabowska, E., Diak, M., Marchelek, M., and Zaleska, A. (2014) *Appl. Catal. B Environ.*, **156**, 213.
245 Axet, M.R., Philippot, K., Chaudret, B., Cabie, M., Giorgio, S., and Henry, C.R. (2011) *Small*, **7**, 235.
246 Shavel, A. and Liz-Marzan, L.M. (2009) *Phys. Chem. Chem. Phys.*, **11**, 3762.
247 Huczko, A. (2000) *Appl. Phys. A Mater.*, **70**, 365.
248 Narayanan, R. and El-Sayed, M.A. (2005) *J. Phys. Chem. B*, **109**, 12663.
249 Ahmadi, T.S., Wang, Z.L., Green, T.C., Henglein, A., and El-Sayed, M.A. (1996) *Science*, **272**, 1924.
250 Narayanan, R. and El-Sayed, M.A. (2004) *J. Am. Chem. Soc.*, **126**, 7194.
251 Matos, J., Ono, L.K., Behafarid, F., Croy, J.R., Mostafa, S., DeLaRiva, A.T., Datye, A.K., Frenkel, A.I., and Roldan Cuenya, B. (2012) *Phys. Chem. Chem. Phys.*, **14**, 11457.
252 Mostafa, S., Behafarid, F., Croy, J.R., Ono, L.K., Li, L., Yang, J.C., Frenkel, A.I., and Cuenya, B.R. (2010) *J. Am. Chem. Soc.*, **132**, 15714.
253 Iwamoto, M., Tanaka, Y., Sawamura, N., and Namba, S. (2003) *J. Am. Chem. Soc.*, **125**, 13032.
254 Zhang, J.P., Liao, P.Q., Zhou, H.L., Lin, R.B., and Chen, X.M. (2014) *Chem. Soc. Rev.*, **43**, 5789.
255 Somorjai, G.A. and Li, Y.M. (2010) *Top. Catal.*, **53**, 832.
256 Roduner, E. (2014) *Chem. Soc. Rev.*, **43**, 8226.
257 Yang, F., Deng, D.H., Pan, X.L., Fu, Q., and Bao, X.H. (2015) *Natl. Sci. Rev.*, **2**, 183.
258 Li, H., Li, L.E., and Li, Y.D. (2013) *Nanotechnol. Rev.*, **2**, 515.
259 Chmelik, C. and Karger, J. (2010) *Chem. Soc. Rev.*, **39**, 4864.
260 Zhang, Y.T., Arancon, R.A.D., Lam, L.Y.F., and Luque, R. (2014) *Chim. Oggi*, **32**, 36.
261 McFarland, E. (2012) *Science*, **338**, 340.
262 Zhang, X.W., Meng, F., Mao, S., Ding, Q., Shearer, M.J., Faber, M.S., Chen, J.F., Hamers, R.J., and Jin, S. (2015) *Energy Environ. Sci.*, **8**, 862.
263 Reuter, K. and Scheffler, M. (2006) *Phys. Rev. B*, **73**, 045433.

18
Heterogeneous Asymmetric Catalysis

Ágnes Mastalir and Mihály Bartók

University of Szeged, Department of Organic Chemistry, Dóm tér 8, 6720 Szeged, Hungary

18.1
Introduction

One of the most important areas in synthetic organic chemistry is the preparation of chiral organic compounds through various asymmetric syntheses. Ever since the publication of the first monograph of this field [1], progress in research has been continuously monitored by handbooks and monographs [2–8]. In the past two decades, reviews mainly focused on the application of efficient heterogeneous catalytic procedures and systematic studies of chiral reactions mediated by heterogeneous catalysts [9–18]. We have been studying the stereochemistry of heterogeneous catalytic reactions [19], including enantioselective hydrogenations and other asymmetric reactions [20–22].

The aim of this chapter was to summarize the results on asymmetric reactions induced by supported or immobilized chiral organocatalysts, collected in reviews before 2008 and also in more recent papers published between 2009 and 2015. There are several ways of organizing the results available in literature, for example, catalysts (chiral molecules, supports), or interactions between the chiral compounds and the supports (covalent, electrostatic, or adsorption interactions). We selected the reactions for the basis for classification. The most extensively studied asymmetric reactions are as follows: aldol addition, Michael reaction, Diels–Alder reaction, diethylzinc addition, epoxidation, and hydrogenation.

The most important support materials used for the fabrication of immobilized chiral organocatalysts are natural or synthetic polymers, amorphous and ordered silica, cross-linked polymers, clays, inorganic materials, nanoparticles (NPs) and metal-organic frameworks (MOFs) with various structural features. The immobilization methods include covalent bonding, ligand grafting, microencapsulation, ion exchange, and electrostatic interactions [8,17,18,23–26].

According to the database of the Web of Science, about 200 papers have been published between 2008 and 2015 on asymmetric heterogeneous catalytic

Handbook of Solid State Chemistry, First Edition. Edited by Richard Dronskowski, Shinichi Kikkawa, and Andreas Stein.
© 2017 Wiley-VCH Verlag GmbH & Co. KGaA. Published 2017 by Wiley-VCH Verlag GmbH & Co. KGaA.

reactions. One-third of these papers focused on enantioselective hydrogenations, another third discussed the reactions listed above and the remainder involved various other reactions (Mannich reaction, Henry reaction, Friedel–Crafts reaction, alkylation, carboxylation, dihydroxylation, cyclopropanation, hydroformylation, and ring opening reactions of ethers), which, despite their obvious importance, could not be included in the present chapter, as related to page limitations.

18.2
Aldol Addition

The asymmetric aldol reaction is one of the most efficient synthetic methods for obtaining optically active β-hydroxy carbonyl compounds, which may be applied in organic syntheses [27]. A large number of variations have been reviewed for this reaction and new catalysts have been developed, including organocatalysts and supported chiral catalysts. Proline has been found to be capable of catalyzing the direct asymmetric intra- and intermolecular aldol reactions with high yields and enantioselectivities [8,14,28–40].

(2S,4R)-4-Hydroxyproline immobilized onto both ends of polyethylene glycol (PEG) was used as a catalyst for the aldol reaction of acetone and aldehyde to give β-hydroxyketones (Figure 18.1) [41]. When cyclohexanecarbaldehyde was employed as a reactant, an enantiomeric excess (ee) of 98% was obtained in 81% yield. In addition, when aldimine was used instead of aldehyde, the asymmetric iminoaldol reaction took place to give β-aminoketones with high enantioselectivity [41].

Proline was also attached to cross-linked polystyrene (PS) through a benzyl thioether linkage, as shown in Figure 18.2. The PS-supported proline was used as an organocatalyst in the asymmetric aldol reaction between cyclohexanone and substituted benzaldehyde in water [42]. The reaction with *p*-cyanobenzaldehyde afforded the aldol adduct with 98% conversion and 98% ee.

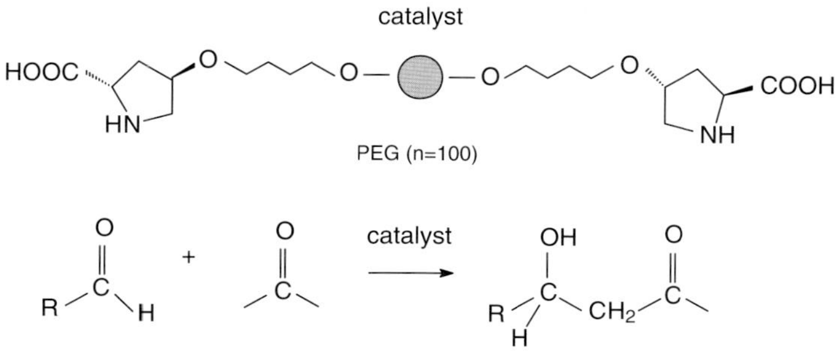

Figure 18.1 The aldol reaction of an aldehyde with acetone.

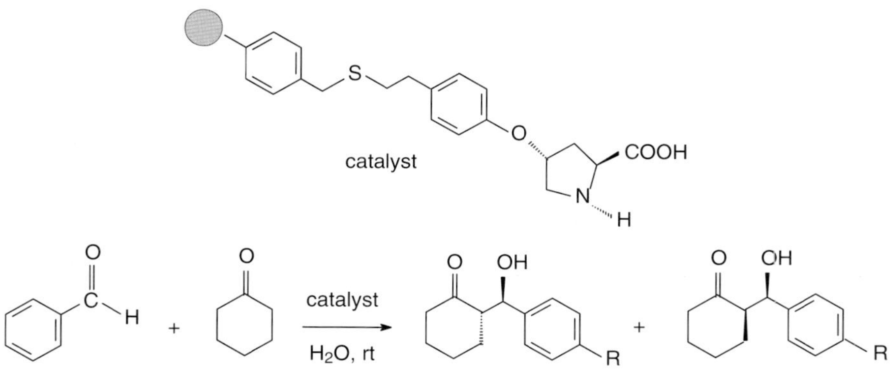

Figure 18.2 PS-supported proline as an organocatalyst in asymmetric aldol reaction.

Figure 18.3 Asymmetric direct aldol reaction of benzaldehyde and acetone.

He et al. [43] synthesized layered double hydroxide (LDH) supported proline by a calcination-reconstruction method [44] and examined its catalytic ability for the asymmetric direct aldol reaction of acetone and benzaldehyde (Figure 18.3), which was found to proceed with a yield of 90% and an ee of 94% [25,43].

Kudo and coworkers reported on the application of proline tripeptides (D-Pro–Tyr–Phe) immobilized in polyethylene glycol grafted on cross-linked polystyrene (PEG-PS) resins for aldol reactions in a triphasic solvent media (water/acetone/THF, 1/1/1) [45]. The catalyst provided quantitative yields and 71% ee and could be recycled four times without any loss of activity.

Zhang et al. applied (4S)-phenoxy-(S)-proline as an asymmetric catalyst for the aldol addition of acetone to 2-nitrobenzaldehyde [25]. By using 10 mol% of catalyst, the product was formed with 90% yield and 83% ee in 16 h in the presence of β-cyclodextrin. The existence of the hydrophobic phenol group on the catalyst allowed the formation of a host–guest complex with β-cyclodextrin. The heterogeneous character of the complex afforded catalyst separation by filtration and reuse up to four cycles with only a slight loss of activity and constant enantioselectivity [16,46]. Proline immobilized on various support materials was also found to be an effective organocatalyst in direct aldol reactions [47–49].

Recently, Luo et al. described a new strategy to immobilize pyrrolidine-type organocatalysts. The authors constructed a library by combining a PS resin bearing sulfonic acid groups with several chiral amines to achieve immobilization by acid–base interactions. The support worked as a catalyst anchor and also as a regulator for the activity and stereoselectivity of the chiral catalyst. These heterogeneous organocatalysts were successfully applied in the aldol reactions of cyclic ketones and acetone with benzaldehyde and the products were obtained with 97% and 83% ee for cyclohexanone and acetone, respectively. The catalyst retained most of its activity up to four cycles [50].

A similar effect of the support material has been reported for chiral primary amino acid-derived diamines immobilized on sulfonated solid supports through acid–base interactions. The heterogeneous catalysts were investigated in direct asymmetric aldol reactions of linear ketones and aromatic aldehydes [51].

Usually, polar solvents, such as dimethyl sulfoxide (DMSO), N,N-dimethylformamide (DMF) or water are applied for aldol reactions. The use of chloroform, in which proline is insoluble, and the immobilization of proline on silica gel resulted in decreased enantioselectivities, as compared to the homogeneous reaction performed in DMSO [14]. On the other hand, the application of ionic liquids for L-proline-catalyzed direct aldol reactions afforded yields and enantioselectivities, which were similar to those obtained in organic solvents [52–54].

The aldol reaction of benzaldehyde with acetone in 1-butyl-3-methylimidazolium hexafluorophosphate [bmim][PF$_6$] (Figure 18.3) afforded the corresponding chiral hydroxy ketone with 71% ee, which was higher than that obtained in DMSO (60%). After the reaction has been completed, the immobilized proline in the ionic liquid could be recovered and reused without any significant loss of activity or enantioselectivity [53].

The catalytic performance of L-proline in ionic liquid was improved by the addition of DMF as a cosolvent, which was attributed to the increased mass transfer in the presence of DMF. Thus, the use of only 5 mol% L-proline was sufficient to accomplish the cross-aldol reactions of aliphatic aldehydes, affording α-alkyl-β-hydroxyaldehydes with extremely high enantioselectivities (>99% ee) in moderate to high diastereoselectivities [54].

(S)-Proline has been supported on the surface of modified silica gels with a monolayer of covalently attached ionic liquid, with or without additional adsorbed ionic liquid. These materials (e.g., 4-methylpyridinium modified silica gel) were used as catalysts for the aldol reaction between acetone and various aldehydes, and provided the respective products with high yields and ee values, which were comparable to those obtained under homogeneous conditions. The supported catalysts could be recovered by simple filtration and reused up to seven times [55–57]. Chiral MOFs were also applied as catalysts for asymmetric aldol reactions [58–60].

The catalytic asymmetric aldol reaction was also investigated in continuous flow systems [51]. As related to the first organocatalytic asymmetric aldol reaction performed in a microreactor [61], Pericas and coworkers reported on the application of polystyrene-immobilized proline-based catalysts for the catalytic enantioselective aldol reaction of benzaldehydes with cyclohexanone under continuous flow conditions. The enantioselective α-aminoxylations of aldehydes were also examined [62], which afforded the facile preparation of β-aminoxy alcohols with up to 96% ee. Recently, another novel organocatalyst, 4-(1-triazolyl)proline, supported

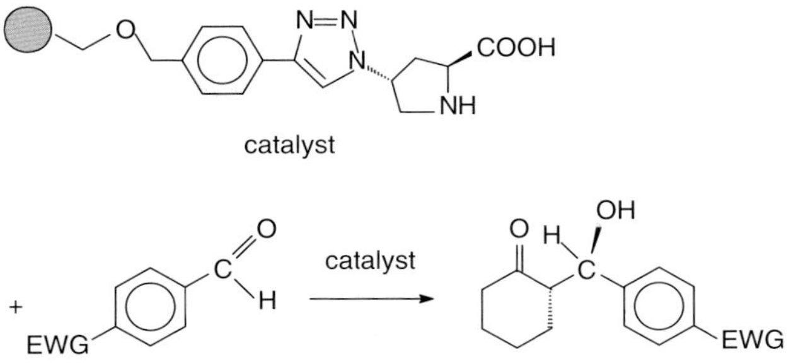

Figure 18.4 Asymmetric aldol reaction catalyzed by a heterogeneous organocatalyst.

on a divinylbenzene-polystyrene (DVB-PS) copolymer (Figure 18.4), was applied for continuous-flow enantioselective aldol reactions [63]. The high activity and stability of the supported catalyst allowed its application in packed-bed reactors for continuous flow processing. Almost enantiomerically pure (98% ee) adducts with a diastereomeric ratio of 97:3 could be produced under continuous flow conditions with a residence time of 26 min.

Asymmetric aldol reaction in a continuous flow reactor was also investigated by Fülöp and coworkers, by using a heterogeneous peptide, synthesized and immobilized by solid-phase peptide synthesis (SPPS) on a swellable polymer support in a single step [64]. After optimization of the reaction conditions, the yields and stereoselectivities obtained for β-hydroxyketone products (85% ee), were comparable with batch results (Figure 18.4). The lack of peptide cleavage from the resin resulted in no workup, no purification, and no product loss. Moreover, the residence time on the catalyst bed was as low as 6 min and, thus, suggested a very promisingly high productivity.

Massi *et al.* reported on the catalytic enantioselective aldol reaction of *p*-nitrobenzaldehyde with cyclohexanone under flow conditions, by using (2S,4R)-hydroxyproline immobilized on 3-mercaptopropyl-functionalized silica in packed-bed microreactors. The conversion was 82% and the moderate stereoselectivity could be improved by increasing the reaction temperature [65,66].

Asymmetric nitroaldol reaction was also performed under flow conditions by using a self-assembled catalyst [67]. As for the stereochemistry of the reactions, the formation of a higher ee of (R)-β-hydroxyketones was generally observed. However, reversed ee values were also obtained under certain reaction conditions [68,69].

18.3
Michael Addition

The enantioselective Michael addition is one of the most frequently studied reactions in asymmetric organocatalysis, which is suitable for the construction of optically active compounds that are useful intermediates for a number of natural and synthetic products. For this reaction, impressive levels of conversions and enantioselectivities have been achieved by using a variety of organocatalysts and Lewis acid catalysts [14,16,70–72].

Hydrogen-bonding activation, combining supramolecular recognition with chemical transformations, has been recognized as a powerful activation mode in organocatalysis [70,73,74]. Theoretical calculations revealed that the immobilization of cinchona alkaloid-derived chiral modifiers on the internal surface of mesoporous hosts can switch the activation routes. The hydrogen-bonding interaction between the substrate and the pore wall has an effect on the energy gap between R/S transition states, accounting for the dependence of enantioselectivity on the pore size experimentally observed in the heterogeneous organocatalytic Michael addition [70]. As reported by Liu and coworkers, chiral bis(cyclohexyldiamine)-based Ni(II) complexes immobilized in periodic mesoporous organosilica exhibited high catalytic activity and excellent enantioselectivity in the asymmetric Michael addition of 1,3-dicarbonyl compounds to nitroalkenes, which were comparable to those of the homogeneous catalyst [75]. Natural quinine was also successfully immobilized on mesoporous SBA-15 and the heterogeneous catalyst exhibited high catalytic activity and enantioselectivity for the asymmetric Michael addition of malononitrile to chalcones [76].

In contrast to aldol reactions, only a few Michael additions were conducted in ionic liquids, and in most cases, decreased enantioselectivities were observed in comparison with the reactions performed in organic solvents [14,77–79]. Nevertheless, imidazolium-tagged pyrrolidines showed extremely high activities in Michael additions. The reaction of cyclohexanone with β-nitrostyrene in the presence of chiral ionic liquids and trifluoroacetic acid (Figure 18.5) proceeded quantitatively with excellent diastereoselectivity (syn/anti = 99/1) and enantioselectivity (99% ee) within 8 h. The imidazolium-tagged catalyst could be recovered and reused four times, and displayed slightly decreased catalytic activities [80].

By taking advantage of the imidazolium phase transfer function, Liu and coworkers synthesized a bifunctional heterogeneous catalyst by immobilizing a chiral cinchona-based squaramide organocatalyst on imidazolium-functionalized organic–inorganic hybrid silica. For the asymmetric Michael addition of 1,3-dicarbonyl compounds to β-nitroalkenes, the catalyst exhibited excellent catalytic activity and high enantioselectivity, which was attributed to the synergistic

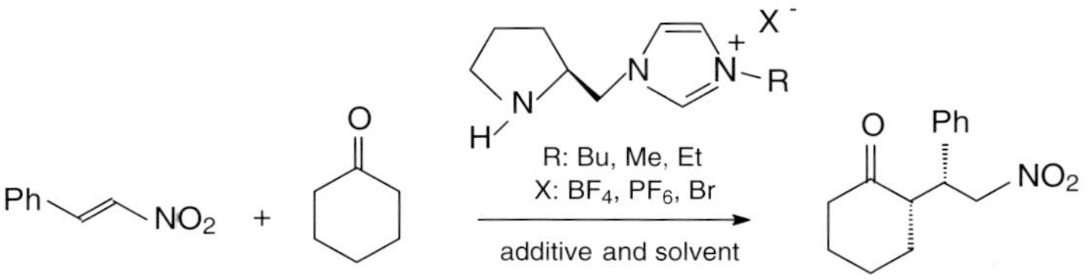

Figure 18.5 Asymmetric Michael addition of cyclohexanone to β-nitrostyrene using ionic liquid-supported chiral pyrrolidines as a catalyst.

effect of the imidazolium phase-transfer function and the confined site-isolated cinchona-based squaramide organocatalyst [81]. Likewise, a chiral squaramide supported on a polystyrene resin through a copper-catalyzed azide–alkyne cycloaddition reaction proved to be a highly active and reusable organocatalyst for the asymmetric Michael addition of 1,3-dicarbonyl compounds to β-nitrostyrenes [82].

The enantioselective Michael reaction of cyclohexanone with β-nitrostyrene (Figure 18.5) was also investigated by using an organocatalyst obtained via immobilization of a pyrrolidine derivative of proline in a polystyrene resin bearing alkyne groups. The reaction was performed in water in the presence of an additive, polyethylene glycol dimethyl ether, and quantitative conversions were obtained, with ee's up to 99%. The catalyst was recycled three times without yield and selectivity loss. When the reactions were conducted by using ketone as a solvent, the recyclability achieved 10 cycles [83,84].

Two series of polymer-bound bifunctional organocatalysts, based on chiral diamine scaffolds, were prepared by solid-phase synthesis and applied for the enantioselective Michael reaction of acetone and nitrostyrene. The best catalysts displayed high enantioselectivities, which were found to surpass those reported previously for polymer-bound catalysts [83,85,86]. The catalytic activity and the enantioselectivity were found to depend on the number of H-bond donors in the catalyst and the catalytic performance also varied with the length of the tether between the polymer and the catalytic module. The addition of an external carboxylic acid was found to increase the catalytic activity without affecting the enantioselectivity [85].

As related to their polar natures, liquid polyethylene glycols can also be used as support materials for the biphasic immobilization of organocatalysts. Xu *et al.* developed a host–guest complex of PEG with organocatalysts, which proved to be highly efficient for the asymmetric Michael addition reactions of unmodified ketones to nitroalkenes, displaying up to 97% yields and 99% enantioselectivities. The existence of a host–guest complex was confirmed by spectroscopic studies [25,87–89].

Recently, the high potential of peptide catalysis for the asymmetric Michael addition has been demonstrated by Kudo and coworkers [90,91]. The application of an aqueous solvent system was found to be essential for an effective reaction, and the hydrophobic helical part of the peptide significantly affected the reaction rate and the enantioselectivity. With the optimized peptide catalyst, enantioenriched products with all-carbon-substituted quaternary centers were obtained. This reaction provided a simple way of synthesizing sterically congested β-disubstituted γ-amino acids, for which a conventional low-molecular weight catalyst was reported to be ineffective [90].

Crystalline metal-organic frameworks (MOFs) with microporous structures [92] can also be applied for asymmetric heterogeneous catalytic reactions including Michael additions [93]. However, the utilization of chiral linkers derived from small chiral organic molecules often resulted in low enantioselectivities. Therefore, more rational strategies have been developed, such as the

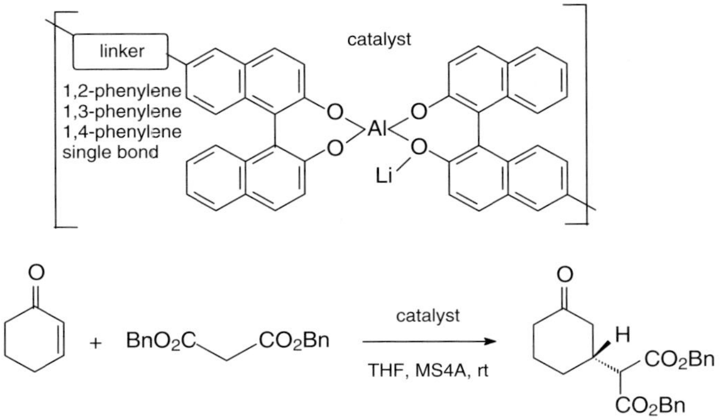

Figure 18.6 Enantioselective Michael addition catalyzed by Al-bridged hybrid polymers.

incorporation of organocatalytic units into the framework either directly or by postsynthetic modification [18]. Al–Li-bis(binaphthoxide) (ALB), a chiral heterobimetallic complex with Al and Li centers supported by two units of 1,1′-bi-2-naphtol (BINOL), was found to catalyze asymmetric reactions with high efficiency and enantioselectivity [94]. Studies indicated that a synergistic cooperation between the two types of active sites was required for the reactivity of this bifunctional catalyst. Therefore, it was essential to maintain the structural integrity of the active species during immobilization of this catalyst.

Sasai and coworkers employed bis-BINOL ligands in combination with $LiAlH_4$ and butyllithium to generate insoluble aluminum bridged homochiral polymers and examined their catalytic properties in the Michael reaction between cyclohex-2-enone and dibenzyl malonate (Figure 18.6) [95]. The Michael adduct was obtained with good yields and up to 96% ee, which was comparable to that obtained for the homogeneous ALB catalyst (97% ee). In contrast, an ALB catalyst supported on polystyrene gave the same Michael adduct in essentially racemic form.

A novel chiral MOF organocatalyst has been developed, based on a chromium terephthalate-based mesoscopic MOF (MIL-101), and the chiral primary diamine (1R,2R)-1,2-diphenylethylenediamine, by postsynthetic modification. The chiral heterogeneous catalyst was applied for the catalytic enantioselective one-step synthesis of a widely used anticoagulant, (S)-warfarin, on a gram scale, with an excellent yield (92%) and high enantioselectivity [96].

Asymmetric Michael additions have been extensively studied under homogeneous and heterogeneous conditions in batch systems. However, there are only few reports on such reactions performed under continuous flow conditions. Catalysts based on cinchona alkaloids, proline derivatives (polypeptides), and chiral Ca species obtained from $CaCl_2$ have been investigated [66,97].

Hodge *et al.* disclosed that polymer-supported cinchonidine was an efficient catalyst for the 1,4-addition reaction of a ketoester with methyl vinyl ketone

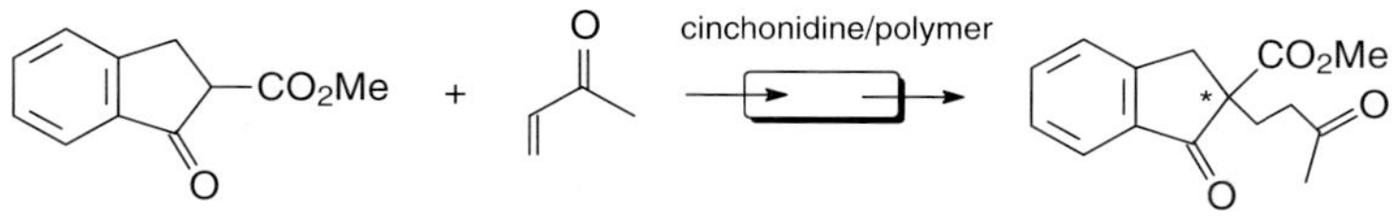

Figure 18.7 Asymmetric Michael reactions catalyzed by polymer-supported cinchonidine in a bench-top flow system.

under continuous flow conditions. Cinchonidine attached to a polystyrene resin was used as a catalyst for the continuous Michael reaction of methyl-1-oxoindan-2-carboxylate and methyl vinyl ketone (Figure 18.7), by passing solutions of the reactants through a fluid bed of polymer-supported tertiary amine at 50 °C [57,98]. The Michael product was obtained with high yield and 51% ee for polymer-supported cinchonidine and the results were comparable to those obtained by cinchonidine. The application of a fluid bed reactor made it possible to use gel-type beads in a flow system.

Fülöp and coworkers. developed another method for the first continuous-flow asymmetric organocatalytic conjugate addition of aldehydes to nitroolefins [99]. A solid-supported peptide was used as a catalyst, which was readily synthesized and immobilized in one single step. As shown in Figure 18.8, synthetically useful chiral γ-nitroaldehydes could be generated with excellent yields and stereoselectivities within short reaction times. The efficiency of this method was enhanced by the reusability of the catalyst, the ease of product separation and a facile scale-up.

The leaching of precious metals can be a serious problem in chiral metal complex-catalyzed continuous flow reactions, as it may result in the loss of expensive metals and in the contamination of products with metals. This can be circumvented by using ubiquitous and nontoxic metals such as Ca. As reported by Kobayashi *et al.*, a chiral calcium chloride complex immobilized on cross-linked polystyrene by copolymerization was an efficient catalyst for the 1,4-addition reactions of 1,3-dicarbonyl compounds with nitroalkenes, under continuous flow conditions. The same reaction was investigated by using a chiral heterogeneous Ca catalyst based on $CaCl_2$ and polymer-supported pyridinebisoxazoline, which exhibited a significant catalytic performance, maintained for 8.5 days without loss of activity [97].

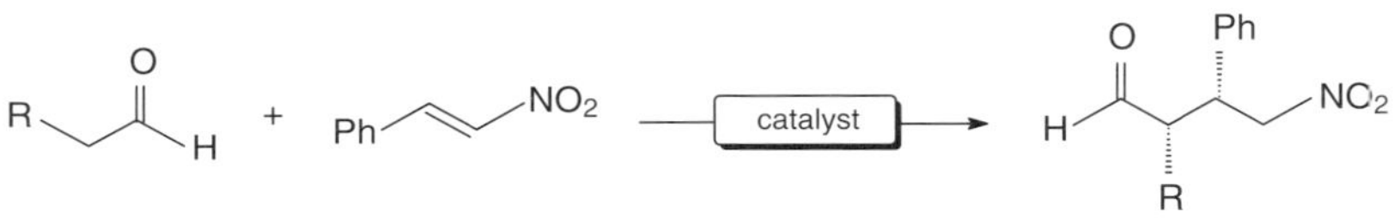

Figure 18.8 Michael reaction of aldehydes to nitroolefins by solid supported peptide catalysts.

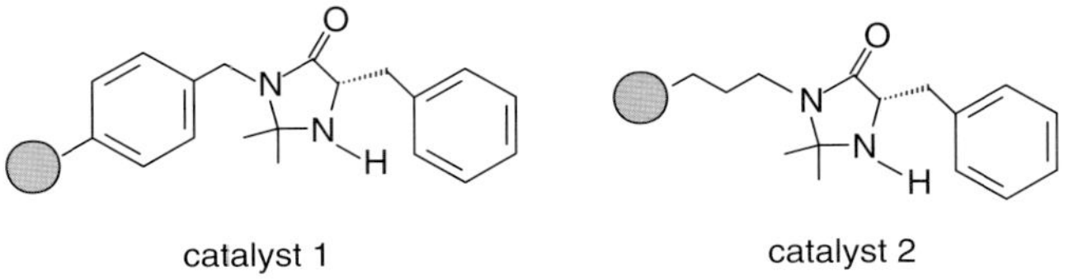

catalyst 1   catalyst 2

Figure 18.9 Supported MacMillan type catalysts.

18.4
Diels–Alder Reactions

Since their discovery, chiral imidazolidinones, developed by MacMillan and coworkers, demonstrated their enormous potential to induce the formation of new asymmetric centers in a short time. These organocatalysts can be successfully applied for several reactions, including Diels–Alder cycloadditions [16,100,101].

Pihko and coworkers reported on the immobilization of a MacMillan type catalyst on silica and JandaGel resins. The heterogeneous catalysts were prepared by solid-phase synthesis (Figure 18.9). By applying cyclopentadiene as a reactant, JandaGel-supported catalyst 1 proved to be efficient for reactions with low polarity transition states and provided good yields (60–73%) and high enantioselectivities (83–99% ee), which sometimes even surpassed those obtained for the homogeneous catalyst. Silica-supported catalyst 2 was found to be more appropriate for the Diels–Alder reactions of 1,3-cyclohexadiene or 2-methylbutadiene with acrolein and afforded high yields (79–83%) and 90% ee [102].

Metal-based asymmetric catalysts act as Lewis acids activating the dienophile in cycloaddition reactions. Various chirally modified Lewis acids have been developed for asymmetric Diels–Alder cycloadditions [103,104].

Some of these have been attached to crosslinked polymers, for example, polymeric monoliths containing (4R,5R)-2,2-dimethyl-($\alpha,\alpha,\alpha',\alpha'$)-tetraphenyl-1,3-dioxolane-4,5-dimethanol (TADDOL) subunits. The treatment of the monolith afforded Ti-TADDOLates, used for the asymmetric Diels–Alder reaction of cyclopentadiene and 3-[(E)-but-2-enoyl]oxazolidine-2-one (Figure 18.10). The major product of this reaction was the endo adduct with 43% ee. The supported Ti catalysts were reported to have excellent long-term stabilities [14,105,106]. The efficiency of a chiral MOF was also demonstrated for a hetero Diels–Alder reaction between Danishefsky's diene and benzaldehyde [107].

The same reaction was investigated in the presence of BINOL-functionalized chiral dendrimer ligands, by using an oligo(arylene) framework as a support. Hexyl groups were grafted onto the surface of the dendrimers to improve their solubility in organic solvents. Their aluminum complexes, generated *in situ* by mixing the dendrimer ligands with Me$_2$AlCl, exhibited slightly better reactivities and enantioselectivities than that of a monomer catalyst. The reactivity and enantioselectivity were found to be independent of dendrimer generation, which was related to the absence of intramolecular interactions among the catalytic centers [108].

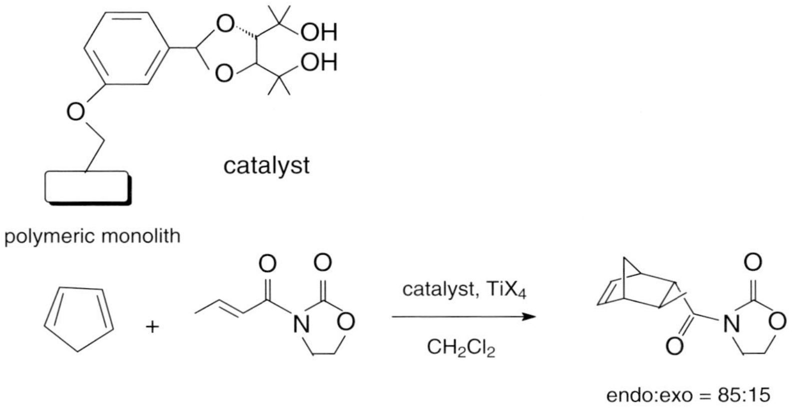

Figure 18.10 Asymmetric Diels–Alder cycloaddition.

Chiral secondary amines proved to be efficient organocatalysts in the Diels–Alder reaction of cyclopentadiene and α,β-unsaturated aldehydes [100]. A tyrosine-derived imidazolidin-4-one was immobilized on PEG to provide a soluble, polymer-supported catalyst, which was successfully employed for the cycloaddition of acrolein to 1,3-cyclohexadiene (Figure 18.11). The reaction proceeded with high endoselectivity and up to 92% ee [109].

Diels–Alder reactions performed in ionic liquids usually give high stereoselectivities at room temperatures, unlike those in conventional organic solvents, for which low temperatures are required to achieve good selectivities. The Diels–Alder reaction of cyclopentadiene and a dienophile (Figure 18.10), investigated with a Cu(II)-bisoxazoline complex as a catalyst in 1-butyl-3-butylimidazolium tetrafluoroborate ([dbim][BF$_4$]) showed higher regioselectivity and endoselectivity (endo/exo = 93/7) than those observed in dichloromethane (endo/exo = 79/21) [110].

For the same reaction, improved enantioselectivities were obtained in the presence of platinum complexes of 2,2′-bis(diphenylphosphino)-1,1′-binaphtyl (BINAP) and conformationally flexible diphosphines in ionic liquids [111].

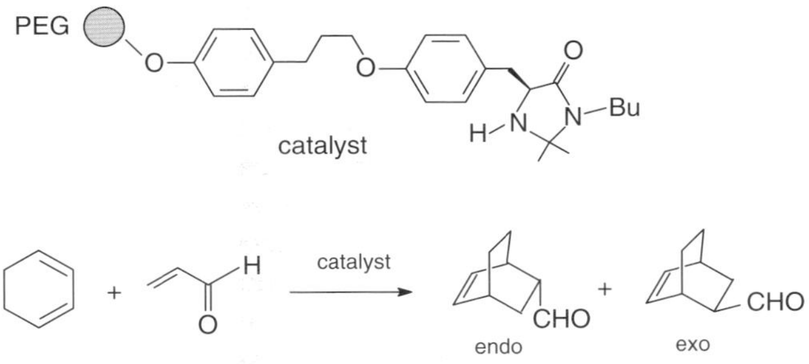

Figure 18.11 Diels–Alder cycloaddition of acrolein to 1,3-cyclohexadiene.

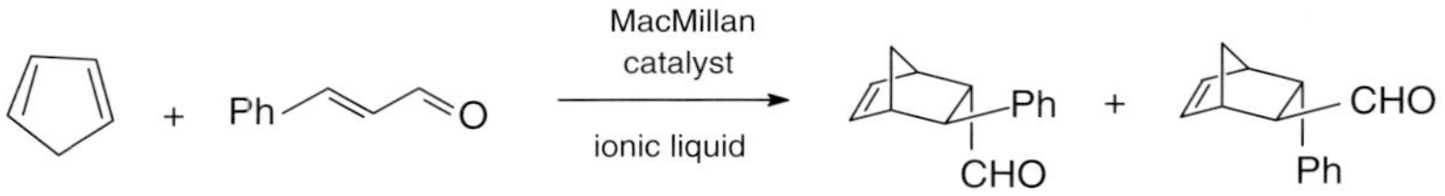

Figure 18.12 Effect of ionic liquids in the enantioselective Diels–Alder reaction.

Asymmetric Diels–Alder reactions were also investigated by using an air- and moisture-stable chiral BINOL-indium complex as a catalyst, applied in ionic liquids. The reactions of various dienes with 2-metacrolein or 2-bromoacrolein were found to proceed with good yields and excellent enantioselectivities (up to 98% ee) [112].

Recently, a noncovalently supported ionic liquid on SiO_2 bed was applied for the immobilization of MacMillan catalysts, employed for the asymmetric Diels–Alder reaction of cyclopentadiene and cinnamaldehyde (Figure 18.12) [25,113], by using a MacMillan imidazolidinone organocatalyst, immobilized in the pores of silica gel by means of the ionic liquid 1-butyl-3-methylimidazolium bis(trifluoromethylsulfonyl) imide [bmim]NTf_2. The heterogenized organocatalyst was found to provide 81% yield and 87% ee for endo- and 80% ee for exo-products [113].

Wang and coworkers reported on a novel method for the synthesis of hollow-structured phenylene-bridged periodic mesoporous organosilica (PMO) spheres with a uniform particle size of 100–200 nm, by using α-Fe_2O_3 as a hard template. The hollow structured phenylene PMO was functionalized with the MacMillan catalyst by a cocondensation process and a "click chemistry" postmodification. The catalyst was found to exhibit high catalytic activity (98% yield, 81% ee for endo and 81% ee for exo) in the Diels–Alder cycloaddition of cyclopentadiene and trans-cinnamaldehyde (Figure 18.12) and could be reused seven times without significant loss of activity. These catalysts were found to be more efficient than those prepared by a grafting method [114].

Endo/exo-selectivity in the Diels–Alder reaction strongly depends on the substrates, based on Woodward and Hoffmann's frontier molecular orbital interactions and steric interactions between dienes and dienophiles via the formation of [4 + 2] pericyclic transition states. Therefore, it is difficult to control enantioselectivity and endo/exo-selectivity at the same time, as conventional chiral catalysts are unable to distinguish between dienes. The reactions of dienes with dienophiles generally follow the endo rule, based on second-order orbital interactions. In some cases, however, the latter may be overcome by steric interactions and the predominant formation of less familiar exo adducts can be observed [115–118].

As reported by Fujita and coworkers, conformationally rigid self-assembled Pd_6 cages can be applied as a nanosized molecular flask to promote the catalytic Diels–Alder reactions of aromatic hydrocarbons with N-cyclohexylmaleimide. The induced proximity of the reaction centers of the reactants in the confined cavity reduced the entropy of the Diels–Alder reactions, and hence unreactive aromatic compounds, for example, naphthalene, anthracene, and triphenylene, underwent unusual Diels–Alder reactions in a regio- and stereoselective manner under thermal or photoirradiation conditions [119–121].

Diels–Alder reactions with anomalous endo/exo selectivities were investigated by Hatano and Ishihara [122]. Novel supramolecular catalysts were prepared *in situ* from chiral 3,3'-disubstituted binaphthols and biphenols, arylboronic acid, and $B(C_6F_5)_3$, to be employed for both anomalous endo- and exo-selective Diels–Alder reactions, including cycloadditions between cyclopentadiene and α-haloacroleins [123,124]. The tailor-made chiral supramolecular catalysts were claimed to be fine-tuned for each substrate to establish high anomalous substrate-selectivity and/or stereoselectivity [122].

18.5
Diethylzinc Addition

The enantioselective addition of organometallic reagents to aldehydes results in the formation of optically active secondary alcohols. For this reaction, a number of organometallic compounds can be used as alkyl donors, of which dialkylzincs are regarded as the most favorable alternatives. One of the most important asymmetric alkylation reactions is the enantioselective addition of diethylzinc to benzaldehyde, which is typically performed in the presence of chiral auxiliaries, for example, β-aminoalcohols (Figure 18.13) [125–127].

The immobilized derivatives of the three main types of chiral ligands (amino alcohols, BINOL, and TADDOL) have been successfully applied for the above transformation (Figure 18.14) [16,128]. TADDOL derivatives with intrinsic C2

Figure 18.13 Enantioselective addition of diethylzinc to benzaldehyde.

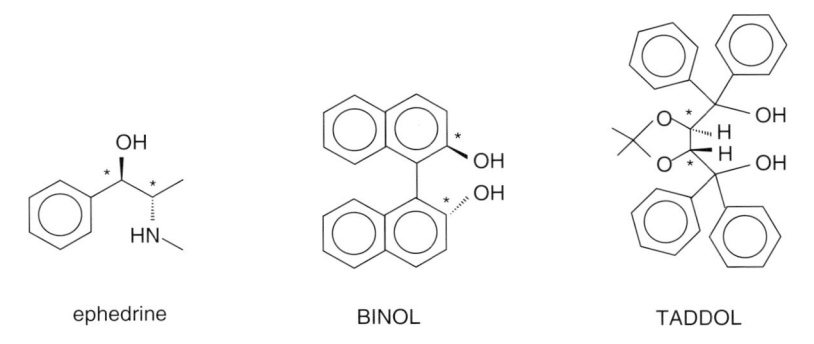

Figure 18.14 Chiral ligands for the enantioselective reaction of diethylzinc.

symmetry have been considered as particularly promising ligands in heterogeneous asymmetric syntheses [129–131]. Several polymeric ligands including chiral porous polymers (CPP) afforded high enantioselectivities and proved to be recyclable. Ephedrine was anchored to siliceous supports to achieve recyclability. The immobilization provided ligands that proved to be as enantioselective as ephedrine itself [16,128]. The structure–property–activity relationship of TADDOL-CPP-based materials was studied in the asymmetric addititon of Et_2Zn with aromatic aldehydes [8].

Immobilization of the chiral cationic surfactant (1R,2S)-(−)-N-dodecyl-N-methylephedrinium bromide in montmorillonite resulted in the formation of an organophilic clay organocomplex, which was employed as a heterogeneous catalyst for the enantioselective addition of diethylzinc to benzaldehyde. The application of the immobilized catalyst was found to result in higher conversions and lower enantioselectivities than those observed for the homogeneous reaction [132].

Soai and coworkers reported on the synthesis of chiral dendrimer amino alcohol ligands based on polyamidoamine (PAMAM), hydrocarbon and carbosilane dendritic backbones, used as catalysts for the enantioselective addition of dialkylzincs to aldehydes. The molecular structures of the dendrimer supports were found to have a significant effect on the catalytic properties [133–135].

Chiral ionic liquids derived from α-pinene have been used either as additives or as cosolvents in the copper-catalyzed enantioselective addition of diethylzinc to various enones. The addition of diethylzinc to cyclohex-2-enone by using 3 mol% $Cu(OTf)_2$ in the presence of a chiral ionic liquid afforded 76% ee, thus providing a potential application of chiral ionic liquids as asymmetric agents in symmetric reactions [14].

Gros and coworkers described the application of a pyridine-based tridentate chiral ligand in the enantioselective addition of diethylzinc to benzaldehydes supported on Merrifield resin. This heterogeneous ligand provided the respective phenylpropanols with excellent levels of enantioselectivities (up to 93% ee) and proved to be fully recyclable during five cycles [136].

Moreau and coworkers used a trialkoxylated camphorsulfonamide for synthesis of functional organosilicas as a highly ordered material using template-directed hydrolysis–polycondensation with tetraethyl orthosilicate (TEOS). This heterogeneous ligand was tested in the titanium-promoted addition of diethylzinc to benzaldehyde. The presence of uncapped free silanol groups made this catalyst less enantioselective than the unsupported analogues [137]. Highly porous cross-linked polymers [138] and 3D nanoporous silicas [139] as supports were also applied for alkylation of aldehydes.

Homochiral porous MOFs are considered as particularly attractive candidates as heterogeneous asymmetric catalysts and enantioselective adsorbents for the production of optically active organic compounds, as related to the lack of homochiral inorganic porous materials, such as zeolites. The application of MOF catalysts with high crystallinities may be advantageous for structural elucidation, and also for the "single-site" nature of the active species in the solid. The uniformly distributed active sites in an identical microenvironment may help understand the catalytic events in the solid phase on the molecular level and facilitate the structure–activity relationship study in the development of a catalyst [140,141].

Lin and coworkers investigated the synthesis and characterization of several BINOL-bearing crystalline homochiral MOF compounds, which, combined with an excess of Ti(OiPr)$_4$, exhibited network–structure-dependent catalytic activities in the enantioselective diethylzinc addition to a range of aldehydes [142–144].

By applying the catalyst [Cd$_3$(BINOL)$_3$Cl$_6$] × 4DMF × 6 MeOH × 3H$_2$O – Ti (OiPr)$_4$, the enantioselectivity was found to be excellent and comparable to that obtained by its homogeneous counterpart BINOL-Ti(OiPr)$_4$. Moreover, a remarkable substrate size dependent reactivity was observed in the reactions of the tested aldehydes [142].

By using the molecular assembly approach, Harada and Nakatsugawa immobilized a BINOLate/Ti(OiPr)$_4$ catalyst for the asymmetric addition of diethylzinc to aldehydes [145]. Treatment of the chiral tris-BINOL ligand with 3 equiv of Ti (OiPr)$_4$ in CH$_2$Cl$_2$/toluene gave an insoluble amorphous aggregate solid. It was proposed that such an aggregate might undergo partial dissociation in a reversible manner to generate active sites, thus exhibiting an activity similar to that of its parent homogeneous catalyst while maintaining a heterogeneous state. For the reaction of benzaldehyde, the catalyst recovered by filtration maintained its activity (>98% conversion in 18 h) even after six cycles without lowering of the enantioselectivity (72–75% ee). However, the potential catalyst leaching still requires further investigations [145].

The enantioselective addition of diethylzinc to benzaldehyde has also been extensively studied in continuous flow systems, by using supported catalysts [146–149]. In such systems, the reaction and the separation of the catalysts from the products can be performed simultaneously, which is highly advantageous in terms of industrial applications. Further, no mechanical stirring is required, which may result in higher overall productivity [149]. On the other hand, the mechanical degradation of the support material may result in significantly shortened lifetimes of the supported reagents.

Itsuno et al. reported on the continuous asymmetric ethylation of aldehydes by using polymeric chiral α-amino alcohols [146]. Further, a polymer-supported camphor derivative (catalyst 1, Figure 18.15) was applied by Hodge et al. [147] in

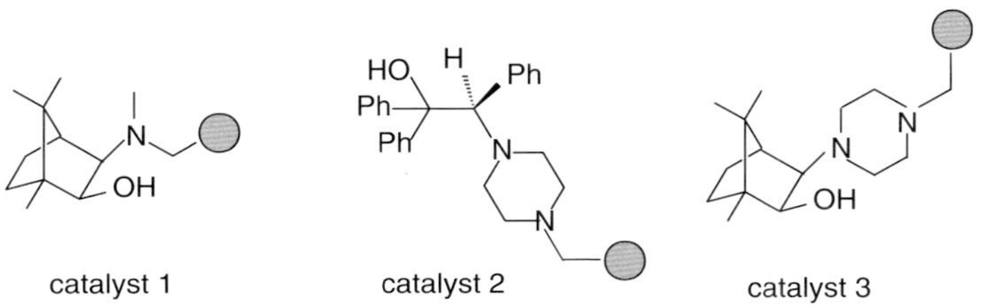

Figure 18.15 Polymer-supported chiral aminoalcohols, employed in asymmetric alkylations in the flow system.

a flow system to catalyze the reaction of benzaldehyde with diethylzinc to give 1-phenylpropanol in >95% yield and >94% ee. However, for this system, both the yield and ee were significantly deteriorated after prolonged reaction time, which was attributed to the gradual chemical degradation of the active sites.

Pericàs et al. extended the scope of dialkylzinc reagents by developing efficient heterogeneous catalysts for asymmetric alkylations performed in flow reactors. The functional polymer (catalyst 2, Figure 18.15), obtained by the reaction of (R)-2-(1-piperazinyl)-1,1,2-triphenylethanol with a Merrifield resin, has been used as a catalyst for the continuous enantioselective formation of 1-arylpropanols through a single-pass operation. The high catalytic activity of catalyst 2 afforded complete conversion of the reactants and up to 94% ee of the target products within short residence times (2.8 min). The authors also reported on the synthesis of enantiopure 1-arylpropanols by the sequential operation of the flow system [57,150]. Recently, Pericàs et al. developed a new polystyrene-supported (2S)-(−)-3-exopiperazinoisoborneol (catalyst 3, Figure 18.15), which was successfully applied as a ligand in the continuous flow asymmetric alkylation of aldehydes with diethylzinc. A broad scope of aldehyde substrates (R = aryl, alkyl, and alkenyl) could also be efficiently transformed to the corresponding alcohols with up to 99% ee [151].

18.6
Epoxidation

The catalytic asymmetric epoxidation of alkenes offers a powerful strategy for the synthesis of enantiomerically enriched epoxides. Metalated chiral Salen ligands were first introduced by Jacobsen and Katsuki as highly enantioselective catalysts for the asymmetric epoxidation of unfunctionalized alkenes. Several types of immobilized chiral complexes have been employed for this reaction, of which Mn(III)-Salen complexes (Figure 18.16) proved to be particularly efficient [3,14,152,153].

However, the large size of the complex makes its encapsulation into the cavities of porous materials rather difficult. Jacobsen-like catalysts of smaller sizes

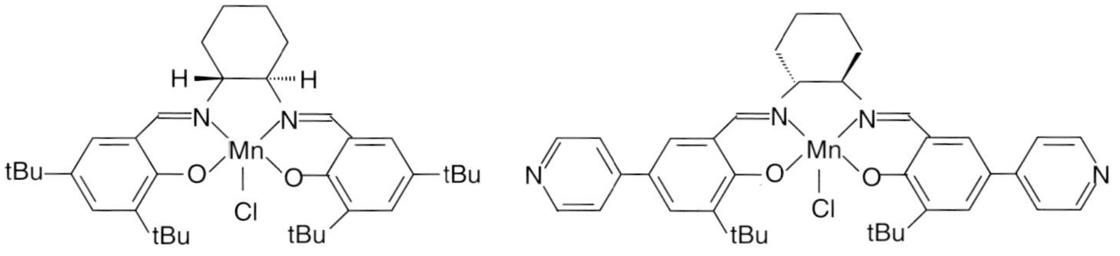

Figure 18.16 Mn(III)-Salen complexes for asymmetric epoxidations.

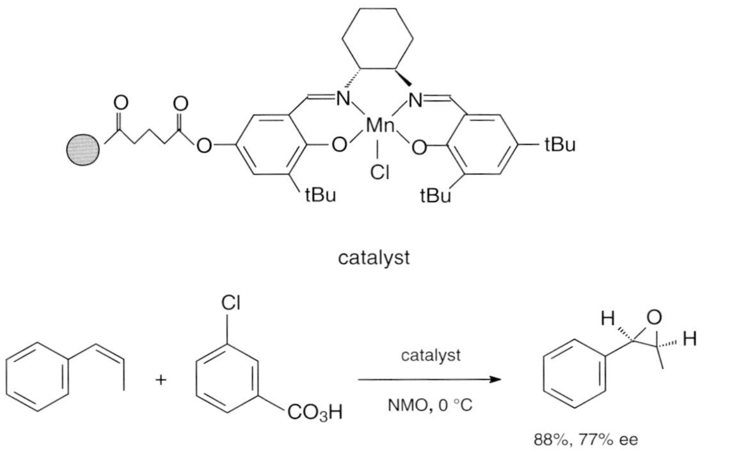

Figure 18.17 Epoxidation of cis-β-methylstyrene in the presence of a polymer Mn(III)-Salen catalyst.

can be more readily encapsulated, but tend to exhibit lower enantioselectivities [154,155]. Therefore, heterogenization of the original complex should be accomplished. Large complexes can be immobilized by bonding onto the external surface of a suitable support material.

Freire and coworkers performed the coordination of hydroxyl groups in activated carbon to the axial position of chiral Mn(III)-Salen complex and its application in asymmetric epoxidation. However, it was found that the heterogeneous catalyst lost its efficiency upon recycling [156,157]. The homogeneous Mn(III)-Salen catalyst was also immobilized in various clays. Despite revealing the same activities, the heterogeneous catalysts provided lower values of enantioselectivities [158].

Because of the large size of the free complex, Balkus et al. used the zeolite synthesis method for the encapsulation of Mn(III)-Salen complexes into the pores of MCM-22. The immobilized catalysts exhibited somewhat higher selectivities than the free complex [159]. The above procedure is regarded as an efficient method for the immobilization of large complexes [160]. MCM-41 as a support was also applied for encapsulation [161].

Polymer-supported chiral Mn(III)-Salen complexes were also employed in asymmetric epoxidation reactions (Figure 18.17). cis-β-Methylstyrene was efficiently epoxidized in the presence of a polymer catalyst. For most of the substrates investigated, the enantioselectivities were similar to those obtained for the low molecular weight catalyst [162].

Likewise, polymer-supported catalysts were applied for Sharpless asymmetric epoxidation. As a polymeric chiral ligand, methoxy PEG (MeO-PEG) was used. In the reaction performed under Sharpless epoxidation conditions (Figure 18.18), the (2S,3S)-trans product was formed with 93% ee by using a low molecular weight catalyst (Mw = 750), whereas the (2R,3R)-trans product was obtained by

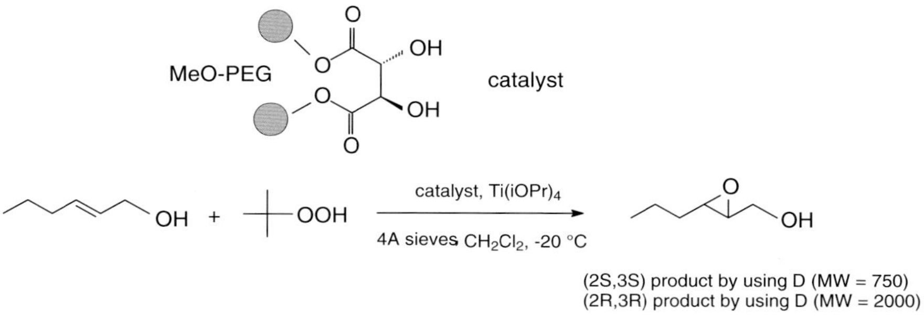

Figure 18.18 Sharpless asymmetric epoxidation.

(2S,3S) product by using D (MW = 750)
(2R,3R) product by using D (MW = 2000)

using the same catalyst of high molecular weight (Mw = 2000) [163]. Janda investigated the effect of the molecular weight of the PEG chain on the same reaction and found that the enantioselectivity could be reversed solely as a function of the molecular weight of the polymer [164].

Song and Roh reported on the application of an ionic liquid for the asymmetric epoxidation of various alkenes, investigated by using catalyst 1 (Figure 18.16) and NaOCl as a co-oxidant in a mixture of 1-butyl-3-methylimidazolium hexafluorophosphate ([bmim][PF$_6$]) and dichloromethane. For these reactions, good conversions and enantioselectivities were obtained [165]. Gaillon and Bedioui reported on the electroassisted biomimetic activation of molecular oxygen by using a chiral Mn(III)-Salen complex in [bmim][PF$_6$], where the highly reactive manganese-oxo intermediate [(Salen)Mn(V)=O]$^+$ was capable of transferring its oxygen to an alkene [166]. This method permitted to carry out electrocatalytic asymmetric epoxidations by using molecular oxygen in ionic liquids.

Liu and coworkers reported on the Mn(III)-Salen catalyzed asymmetric epoxidation of unfunctionalized alkenes in an MCM-48 supported ionic liquid phase. The Mn(III)-Salen catalysts were immobilized by mixing their acetone solutions with ionic liquid-anchored MCM-48 and the ionic liquid [bmim][PF$_6$]. For the epoxidation of α-methylstyrene and 1-phenylcyclohexene, high yields and enantioselectivities (92–99%) were obtained and the immobilized catalysts were recycled at least three times without any significant decrease in either activity or enantioselectivity [167,168].

Recently, novel types of organic polymer–inorganic hybrid material layered crystalline zinc poly(styrene-phenylvinylphosphonate)-phosphate (ZnPS–PVPA) have been applied as support materials of a chiral Mn(III)-Salen complex [169,170]. The heterogeneous chiral Mn(III)-Salen catalysts exhibited comparable or even higher enantioselectivities than that of the homogeneous catalyst for the asymmetric epoxidation of unfunctionalized olefins. The supported chiral Mn(III)-Salen catalysts were relatively stable and could be recycled nine times in the asymmetric epoxidation of α-methylstyrene. This organocatalyzed asymmetric epoxidation reaction could also be performed on a large scale, whereas the catalytic activity was maintained at the same level [169].

Asymmetric epoxidation was also studied by using other types of catalysts including amino-functionalized mesoporous silica. As reported by Park and coworkers, a chiral copper proline diamide complex has been immobilized on the surface of SBA-15. The heterogenized complexes were applied as asymmetric catalysts for the epoxidation of α,β-unsaturated carbonyl compounds by hydrogen peroxide, *tert*-butyl hydroperoxide and urea hydroperoxide as oxidants under solvent-free conditions. Enantiomeric excesses up to 84% were obtained and the catalysts could be reused five times without significant loss of activity [171].

For asymmetric epoxidation reactions, the application of MOFs as solid catalysts may also be of interest, as the pore size and functionality of the framework can be adjusted over a wide range [18,24,172].

Nguyen and coworkers reported on the immobilization of a chiral Mn(III)-Salen complex in a crystalline microporous MOF [173]. The chiral complex with 4-pyridyl groups (catalyst 2, Figure 18.16) was used as catalyst, which, upon solvothermal reaction with zinc ions and biphenyldicarboxylic acid (bpdc) in DMF, formed a crystalline coordination network. X-ray diffraction (XRD) data indicated that all Mn(III) sites were accessible to the channels. PLATON calculations showed that this network had a highly porous structure, with solvent molecules occupying the micropores and XRD measurements revealed that the evacuated sample retained its crystallinity. The catalytic activity of this framework was studied in the asymmetric epoxidation of 2,2-dimethyl-2H-cromene (Figure 18.19) and the reaction took place with an ee of 82%. The enhanced stability of the catalyst was attributed to the rigid structure of the catalytic complex and the confinement effect of the metal-organic framework, which prevented oxidation of the Salen moiety [173].

Van der Woort and coworkers studied the selectivity of the Mn(III)-Salen complex in the epoxidation of unfunctionalized olefins by using a density functional theory (DFT) methodology by considering the full Salen complex as a model and dihydronaphthalene as a substrate [174,175].

The authors presented a combined theoretical and experimental approach to elucidate the factors affecting enantioselectivity. The importance of different substituents was determined by analyzing the transition state for the oxygen transfer. Analysis of the asymmetric complex revealed an inherent tendency for a decreased selectivity due to the lack of specific bulky groups. In order to design a heterogeneous Salen catalyst with a minimal loss of selectivity, electrostatic grafting has been suggested, where the support material is charged and acts as a

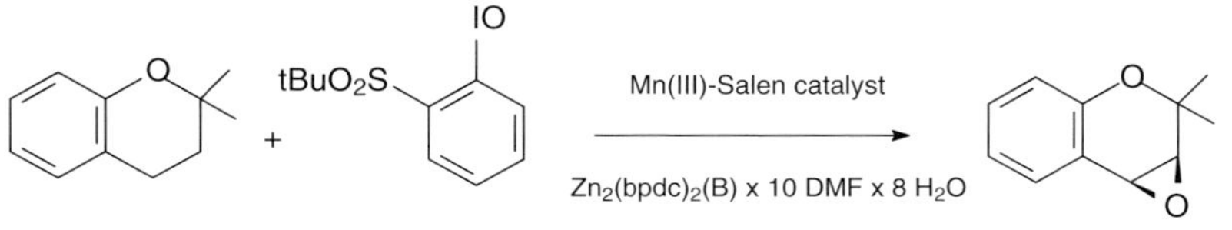

Figure 18.19 Epoxidation of 2,2-dimethyl-2H-cromene.

counter ion on the axial position of the Salen complex [176,177]. Another possibility is the encapsulation of the complex in the cages of the support material confining it by small window sizes [178].

Efficient methods were developed for enhancing the enantioselectivities in vanadium- and niobium-catalyzed asymmetric epoxidations [179–182]. This strategy, which involves pristine α-amino acid as a chelating moiety and synthetic or naturally occurring inorganic brucite- or LDH-like layers as a planar substituent, was successfully used in the epoxidation of allylic alcohols. The crucial role of the inorganic nanosheets as planar substituent in improving the enantioselectivity was revealed by relating the observed enantiomeric excess with the distribution of the catalytic centers and the accessibility of the reactants to the catalytic sites. DFT calculations indicated that the brucite layers improved the enantioselectivity by influencing the formation and stability of the catalytic transition states, both in terms of steric resistance and H-bonding interactions [179,180].

18.7
Hydrogenation

The modification of an active metal surface by an adsorbed chiral modifier proved to be very efficient in heterogeneous asymmetric catalysis. This method involves the application of an achiral solid (a metal surface), in combination with a suitable chiral organic compound (a chiral modifier) as a heterogeneous asymmetric catalyst. The metal is responsible for the catalytic activity, whereas the chiral modifier adsorbed on the surface induces stereochemical control of the reaction via interaction with the substrate. This strategy proved to be particularly efficient for the enantioselective hydrogenations of prochiral substrates. When using a chirally modified metal surface as a heterogeneous catalyst in hydrogenations, a high substrate selectivity was observed, as only a limited number of metal–modifier–substrate combinations provided enhanced enantioselectivities, suitable for synthetic applications.

The most important catalytic systems are the nickel tartrate–NaBr system for the hydrogenation of β-ketoesters, β-diketones, and methyl ketones, as reported by Izumi *et al.* [183]; the platinum–cinchona alkaloid system for the hydrogenation of α-ketoesters, α-ketoacids, and lactones, as reported by Orito *et al.* [184], (Figure 18.20); and palladium catalysts modified with cinchona alkaloids for the hydrogenation of activated alkenes. Progress on the research performed on the above catalytic systems has been reported in recent reviews [14,20,185–187]. This chapter deals solely with the Orito reaction.

The most significant model compound for investigations of Orito-reaction is ethyl pyruvate (EtPy), which may be hydrogenated to ethyl lactate (EtLt) with a high enantioselectivity (Figure 18.21).

In order to achieve high catalytic activities and enantioselectivities, the most important factors are the selection and the particle size of the metal and the

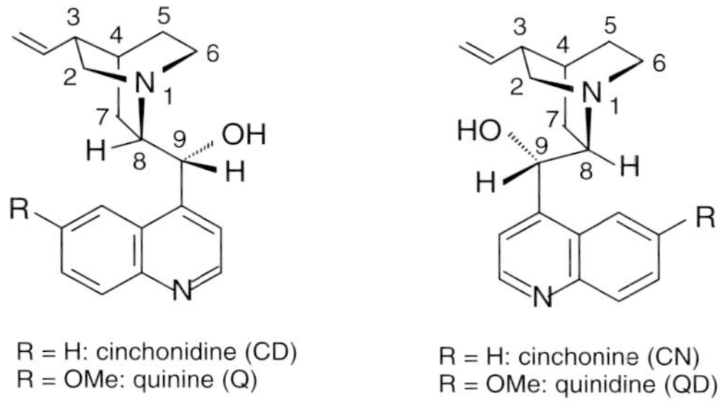

Figure 18.20 Cinchona alkaloids.

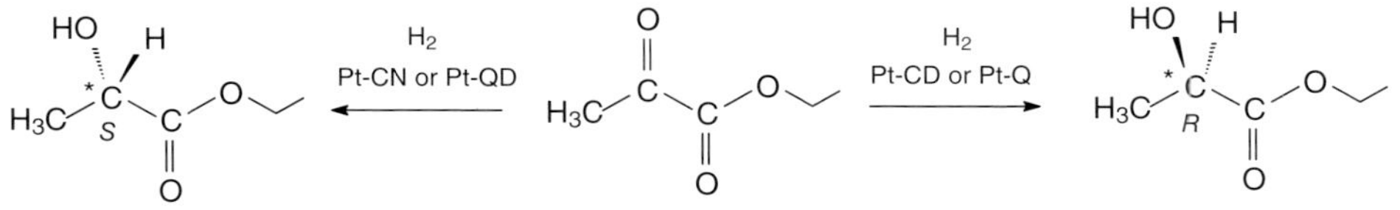

Figure 18.21 Hydrogenation of ethyl pyruvate in the presence of cinchona alkaloids.

solvents. Enantioselective hydrogenation is a structure-sensitive reaction, which implies that the simultaneous adsorption of the reactant and the modifier takes place on the metal surface [185–187].

For the hydrogenation of EtPy, the suitable Pt particle size was larger than 2 nm [188], whereas for the hydrogenation of a larger molecule, for example, 1-phenyl-1,2-propanedione, the optimum Pt particle size was 4 nm and 3.8 nm for Pt/Al$_2$O$_3$ and Pt/SiO$_2$ catalysts, respectively [189]. The presence of relatively large Pt particles ensure that the flat adsorption of CD takes place via the quinoline ring. The particle size and the morphology of the metal should be optimized in order to favor the adsorption of a modifier-reactant complex on the surface of the catalyst. For a more precise explanation of the effect of the metal particle size and morphology, the local metal environment is also to be taken into consideration [186].

The most convenient support materials, for example, Al$_2$O$_3$, SiO$_2$, and carbon, have been reported to be suitable for the hydrogenation of EtPy and have no particular effect on enantioselection [190]. However, CD adsorption experiments revealed differences between alumina and silica supports, as silica adsorbed CD more efficiently. Consequently, during the catalytic reaction, CD was adsorbed on silica and not on the metal particles, thereby decreasing enantioselectivity [186].

The beneficial effect of ultrasonic treatment on the enantioselective hydrogenation of EtPy over supported cinchona-modified Pt catalysts was reported for the E4759 catalyst and the effect of ultrasound on the surface structure of this catalyst was also investigated [191,192]. It was found that ultrasonic treatment increased the amount of chiral modifier adsorbed on the catalyst, both on the metal and on the support material. Experimental evidence indicated that ultrasound decreased the mean particle size of the metal and provided a more homogeneous size distribution. It was also found to decrease the modifier concentration in the solution and restricted the hydrogenation of the modifier. The effect of ultrasonic treatment on the catalytic systems was interpreted by the synergetic effect of a more uniform catalyst surface and an increased amount of adsorbed modifier. Accordingly, the sonochemical treatment of Pt-CD systems afforded enhanced enantioselectivities (up to 98% ee) for the hydrogenation of α-ketoesters under mild experimental conditions [20].

Li and coworkers investigated the catalytic performance of a CD-modified Pt nanocatalyst encapsulated in carbon nanotubes (CNT) for the hydrogenation of EtPy. Adsorption experiments indicated that open CNT channels exerted a unique enrichment effect on both CD and the reactant, which was attributed to the intense capillary effect of the CNT channels when their diameters were in the nanoscale range. The enrichment of both CD and EtPy in the CNT channels was the proposed reason of the high reaction rates and enhanced enantioselectivities [193]. The role of shape selective nanoparticles and metal–support interaction in the enantioselective hydrogenation of ketones on Pt was investigated by Baiker and coworkers [194,195].

The mechanism of asymmetric catalysis on chirally modified metal surfaces is complicated and has not yet been fully understood. The cinchona alkaloid modified platinum system displayed an excellent catalytic performance for the enantioselective hydrogenation of α-ketoesters and therefore has been the subject of an extensive research. It is generally accepted that the chiral modifier adsorbed on the Pt surface not only induces the enantioselectivity of the reaction, but also accelerates the reaction rate for the asymmetric hydrogenation of α-ketoesters [196]. Chiral induction was found to occur in the chemisorption layer through the formation of individual 1:1 modifier–substrate diastereomeric complexes, which may be characterized by experimental methods [185–187]. Experimental evidence for the Pt-β-CN/Al_2O_3 system, obtained by using toluene, AcOH or their mixtures as solvents, confirmed that the structure of the intermediate responsible for the enantioselection was actually modifier:substrate = 1:1 [197,198].

The application of selective *in situ* attenuated total reflection (ATR) studies for the investigation of the active species in heterogeneous asymmetric catalysis was a major step towards the understanding of chirality transfer on metal surfaces [199]. For the hydrogenation of activated ketones, the stereogenic center at the C9 atom of the cinchona modifier was found to affect both the reaction rates and the enantioselection [196,200].

The influence of the configuration at the C8 and C9 positions of cinchona alkaloids (Figure 18.20) was investigated by comparing the efficiency of CD, CN, and 9-epi-cinchonidine as chiral modifiers, in the hydrogenation of activated ketones (methyl benzoylformate, ketopantolactone, (Figure 18.22) methylglyoxal dimethylacetal, 2,2,2-trifluoroacetophenone) on Pt. The catalytic observations combined with DFT calculations on the adsorption of all three alkaloids on Pt require a revision of the widely accepted traditional concept on the role of C8

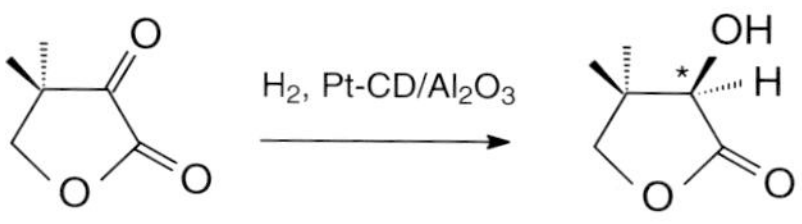

Figure 18.22 Asymmetric hydrogenation of ketopantolactone to (R)-pantolactone on cinchonidine-modified supported Pt particles.

and C9 stereogenic centers. For the hydrogenation of ketones on Pt, it was proposed that both stereogenic centers were involved in the enantioselection [200].

The role of the C9-OH group of the chiral modifier in the enantiodifferentiating surface process was examined by using *in situ* attenuated total reflectance infrared (ATR IR) spectroscopy combined with modulation excitation spectroscopy (MES) and phase-sensitive detection (PSD). Experimental evidence was found for the formation of an O—H—O bonded surface diastereomeric intermediate complex between ketopantolactone and CD under reaction conditions. The results extended the generally accepted model for enantiodifferentiation based on the formation of a H-bond between the quinuclidine N atom of the cinchona alkaloid and the oxygen of the α-carbonyl group of the ketone (N—H—O type H-bonding) and indicated that for some ketones, the C9—O···H···O=C interaction has to be taken into account for explaining the enantiodifferentiation [201].

The surface processes occurring during the catalytic enantioselective hydrogenation of activated ketones on a CD-modified Pt catalyst were monitored under real working conditions [20]. The dynamic molecular interactions between reactants, products and modifier at the catalytic solid–liquid interface were studied by using operando ATR-IR spectroscopy. Essential information was obtained on the interrelation between the catalytic performance and the molecular structure at the chiral catalytic solid–liquid interface. The spectroscopic data demonstrated the formation of a transient diastereomeric intermediate surface complex between adsorbed CD (chiral site) and the reactant ketone at the surface of the Pt catalyst. The formation of a N—H—O hydrogen bond interaction between coadsorbed CD and ketone was also confirmed, which was considered as the origin of enantioselection [202].

In another study, experimental evidence was obtained that the use of a protic solvent was not essential for the latter interaction. In an aprotic medium, such as toluene, hydrogen dissociated on Pt was involved in the N—H—O interaction between the chiral modifier, cinchonidine, and the ketone. In the absence of Pt, by using pure alumina support, no such interaction was observed, which indicated the crucial role of dissociated hydrogen in the formation of the diastereomeric transition complex [203].

The formation of bimolecular complexes on metal surfaces through interaction between a single chemisorbed chiral molecule and a single chemisorbed prochiral substrate molecule can be considered as a preorganization step toward chirality transfer. Svane *et al.* applied a combination of van der Waals corrected DFT calculations and scanning tunneling microscopy (STM) measurements to determine the structure and stability of bimolecular complexes formed between ketopantolactone and (R)-1-(1-naphthyl)ethylamine on Pt(111) [204]. The study revealed several distinct complexation geometries at the submolecular level as well as the stereodirecting forces operating in the most abundant bimolecular assemblies. The comparison of theoretical and experimental data suggested that partial hydrogenation of ketopantolactone occurred under the experimental conditions and that some of the most abundant complexes were formed by the hydroxy intermediate.

Recent results by Kaushik *et al.* provided an opportunity to use highly crystalline cellulose directly in asymmetric catalysis. The authors employed Pd patches deposited onto cellulose nanocrystals (CNCs) as catalysts for the hydrogenation of prochiral ketones in water at room temperature [205]. Their system, where CNCs acted both as a support and a chiral source, achieved an ee of 65% with 100% conversions. The 3-dimensional structure of the metal-functionalized hybrid material was characterized by cryoelectron microscopy, high-resolution transmission electron microscopy and tomography. The direct contact between the Pd patches and the support was confirmed and the presence of sub-nanometer-thick Pd patches on the surface of CNCs was found to be essential for the chiral induction mechanism. Pd patches activated the hydrogenation reactions and the CNCs provided enantioselection [205].

The database on the Orito reaction is exceedingly complex, including various examples of solvent and substituent dependent stereoinversion, different optimization conditions for different substrates and possible surface modifications through side-reactions [199]. Subtle changes of the structure of the cinchona modifier or the activated ketone may also affect the stereochemical outcome of the reaction. The complete understanding of the reaction mechanism still requires further investigations, which may promote the development of novel asymmetric reactions with high enantiomeric excesses.

18.8
Summary

Based on the current state of heterogeneous catalytic asymmetric syntheses, this chapter discloses results on chiral reactions including the aldol reaction, the Michael addition, the Diels–Alder cycloaddition, diethylzinc addition, epoxidation, and hydrogenation. The above reactions may be performed on a variety of catalysts containing chiral ligands, immobilized on supports with different structures.

The papers involved in this study discuss the syntheses and applications of catalysts more or less in detail. An unfavorable property of these catalysts, though, (except for those applied for the Orito reaction), is their low activity, as related with reaction times 2–16 h (Table 18.1). Therefore, future research should focus on the development of novel catalyst systems, with low catalyst concentrations, capable of delivering products of high optical purity within short

Table 18.1 Experimental data on heterogeneous asymmetric reactions in batch systems.

	Reaction	Reactants[a] (mmol/mmol)	Catalyst[b]	c[b] (mol%)	Time (h)	Conv (%)	Ee (%)	References
1	Aldol	p-NBA/Ac	H-Pro-Pro-Asp-NH-TG	1	2	93	80	[64,206]
2	Aldol	p-NBA/Cy	PS-supported-Pro	10	5	91	98	[63]
3	Michael	β-NS/Bu	H-D-Pro-Glu-NH-TG	3	4	100	91	[99,207]
4	Michael	β-NS/DMM	DACH-Ni-PMO	1/1.2	5	>99	94	[75]
5	Diels-Alder	CA/CPD	PMO-Mac/Fe$_2$O$_3$	0.25/1.25	12	98	81	[114]
6	Dialkylzinc	BA/ZnEt$_2$	NSAA-PS/Ti(OiPr)$_4$	0.5/0.75	16	95	92	[208]
7	Epoxidation	α-MS/MCPBA	DACH-S-Mn-MCM-48	1	2	30	>99	[209]
8	Epoxidation	α-MS/MCPBA	DACH-S-Mn-ZnPS-PVPA	1/2	5	99	>99	[170]
9	Hydrogenation	EtPy, others	Pt-CD/Al$_2$O$_3$	0.02	0.5	>90	90–98	[20,186,187]

a) Ac = acetone, BA = benzaldehyde, Bu = butanal, CA = cinnamaldehyde, CPD = cyclopentadienes, Cy = cyclohexanone, DMM = dimethylmalonate, EtPy = ethyl pyruvate, MCP BA = m-chloroperbenzoic acid, MS = methylstyrene, NS = nitrostyrene, p-NBA = p-nitrobenzaldehyde.
b) CD = chinconidine, DACH = chiral-diaminocyclohexane, Mac = MacMillan catalyst, NSAA = chiral-N-sulfonylamino alcohol, PMO = periodic mesoporous organisilica, PS = polystyrene, S = Salen, TG = tentagel, ZnPS-PVPA = zinc poly(styrene-phenylvinylphosphonato)-phosphate.

reaction times. Obviously, the experimental conditions, including the catalyst concentration, reaction temperature and additives, have a crucial effect on the reaction rate and enantioselectivity.

Despite the abundance of results reported in the literature so far, it seems that current research on heterogeneous asymmetric syntheses is mostly based on data collection. Hence, the interpretation of the relations between the structures of catalysts and their activities and enantioselectivities still requires further investigations. In order to consider industrial applications of the working catalysts, experimental data should be analyzed more in detail and further work should be accomplished, including the variation of the catalytic activity and enantioselectivity with the experimental conditions.

Acknowledgment

This work was supported by the Hungarian National Science Foundation (OTKA Grant K109278). The authors wish to thank Árpád Molnár for valuable discussions.

References

1 Morrison, J.D. (ed.) (1985) *Asymmetric Synthesis*, Academic Press, New York.
2 Noyori, R. (1994) *Asymmetric Catalysis in Organic Synthesis*, John Wiley & Sons, Inc, New York.
3 Jacobsen, E.N., Pfaltz, A., and Yamamoto, H. (eds) (1999) *Comprehensive Asymmetric Catalysis*, vols. **1–3**, Springer, Berlin.
4 Bolm, C. and Gladysz, J.A. (eds) (2003) Enantioselective catalysis, *Chem. Rev.*, **103**, 2761–3400.
5 Berkessel, A. and Gröger, H. (eds) (2005) *Asymmetric Organocatalysis*, Wiley-VCH Verlag GmbH, Weinheim.
6 Bartók, M. (2010) *Chem. Rev.*, **110**, 1663–1705.
7 List, B. and Maruoka, K. (eds) (2012) *Asymmetric Organocatalysis*, Workbench Edition, Thieme.
8 Dalko, P.I. (ed.) (2013) *Comprehensive Enantioselective Organocatalysis*, Wiley-VCH Verlag GmbH, Weinheim.
9 Jannes, G. and Dubois, V. (eds) (1995) *Chiral Reactions in Heterogeneous Catalysis*, Plenum Press, New York.
10 De Vos, D.E., Vankelecom, I.F.J., and Jacobs, P.A. (eds) (2000) *Chiral Catalyst Immobilization and Recycling*, Wiley-VCH Verlag GmbH, Weinheim.
11 Gladysz, J.A. (Guest ed.) (2002) Recoverable catalyst and reagents, *Chem. Rev.*, **102**, 3215–3837.
12 Li, C. (2004) *Catal. Rev. Sci. Eng.*, **46**, 419–492.
13 Heitbaum, M., Glorius, F., and Escher, I. (2006) *Angew. Chem., Int. Ed.*, **45**, 4732–4762.
14 Ding, K. and Uozumi, Y. (ed.) (2008) *Handbook of Asymmetric Heterogeneous Catalysis*, Wiley-VCH Verlag GmbH, Weinheim.
15 Benaglia, M. (2009) *Recoverable and Recyclable Catalysts*, John Wiley & Sons, Ltd, Chichester, UK.
16 Trindade, A.F., Gois, P.M.P., and Afonso, C.A.M. (2009) *Chem. Rev.*, **109**, 418–514.
17 Gruttadauria, M. and Giacalone, F. (eds) (2011) *Catalytic Methods in Asymmetric Synthesis: Advanced Materials, Techniques, and Applications*, John Wiley & Sons, Inc, Hoboken.

18 Yoon, M., Srirambalaji, R., and Kim, K. (2012) *Chem. Rev.*, **112**, 1196–1231.
19 Bartók, M., Czombos, J., Felföldi, K., Gera, L., Göndös, G., Molnár, Á., Notheisz, F., Pálinkó, I., Wittmann, G., and Zsigmond, Á.G. (1985) *Stereochemistry of Heterogeneous Metal Catalysis*, J. Wiley and Sons, Ltd, Chichester, pp. 1–632.
20 Bartók, M. (2006) *Curr. Org. Chem.*, **10**, 1533–1567.
21 Balázsik, K., Szőri, K., Szőllősi, G., and Bartók, M. (2011) *Chem. Commun.*, **47**, 1551–1552.
22 Török, B., Balázsik, K., Felföldi, K., and Bartók, M. (2001) *Ultrason. Sonochem.*, **8**, 191–200.
23 Liu, Y., Xuan, W.M., and Cui, Y. (2010) *Adv. Mater.*, **22**, 4112–4135.
24 Corma, A., Garcia, H., and Llabrés i Xamena, F.X. (2010) *Chem. Rev.*, **110**, 4606–4655.
25 Zhang, J., Luo, S., and Cheng, J.P. (2011) *Catal. Sci. Technol.*, **1**, 507–516.
26 Barbaro, P. and Liguori, F. (2010) *Heterogenized Homogeneous Catalysis for Fine Chemicals Production*, Springer, Dordrecht, The Netherlands.
27 Trost, B.M. (1995) *Angew. Chem., Int. Ed.*, **34**, 259–281.
28 Dalko, P.I. and Moisan, L. (2004) *Angew. Chem., Int. Ed.*, **43**, 5138–5175.
29 Wu, Y.Y., Zhang, Y.Z., Yu, M.L., Zhao, G., and Wang, S.W. (2006) *Org. Lett.*, **8**, 4417–4420.
30 Cordova, A., Zou, W., Dziedzic, P., Ismail, I., Reyes, E., and Xu, Y. (2006) *Chem. Eur. J.*, **12**, 5383–5397.
31 Mukherjee, S., Yang, J.W., Hoffmann, S., and List, B. (2007) *Chem. Rev.*, **107**, 5471–5569.
32 Zlotin, S.G., Kucherenko, A.S., and Beletskaya, I.P. (2009) *Russ. Chem. Rev.*, **78**, 737–784.
33 Geary, L.M. and Hultin, P.G. (2009) *Tetrahedron Asymmetry*, **20**, 131–173.
34 Trost, B.M. and Brindle, C.S. (2010) *Chem. Soc. Rev.*, **39**, 1600–1632.
35 Bisai, V., Bisai, A., and Singh, V.K. (2012) *Tetrahedron*, **68**, 4541–4580.
36 Mahrwald, R. (ed.) (2013) *Modern Methods in Stereoselective Aldol Reactions*, Wiley-VCH Verlag GmbH, Weinheim.
37 Mlynarski, J. and Bas, S. (2014) *Chem. Soc. Rev.*, **43**, 577–587.
38 Bartók, M. (2015) *Catal. Rev. Sci. Eng.*, **57**, 192–255.
39 List, B., Lerner, R.A., and Barbas, C.F. (2000) *J. Am. Chem. Soc.*, **122**, 2395–2396.
40 Sakthivel, K., Notz, W., Bui, T., and Barbas, C.F. (2001) *J. Am. Chem. Soc.*, **123**, 5260–5267.
41 Benaglia, M., Cinquini, M., Cozzi, F., and Puglisi, A. (2002) *Adv. Synth. Catal.*, **344**, 533–542.
42 Giacalone, F., Gruttadauria, M., Marculescu, A.M., and Noto, R. (2007) *Tetrahedron Lett.*, **48**, 255–259.
43 An, Z., Zhang, W., Shi, H., and He, J. (2006) *J. Catal.*, **241**, 319–327.
44 Nakayama, H., Wada, N., and Tsuhako, M. (2004) *Int. J. Pharm.*, **269**, 469–478.
45 Akagawa, K., Sakamoto, S., and Kudo, K. (2005) *Tetrahedron Lett.*, **46**, 8185–8187.
46 Shen, Z.X., Ma, J.M., Liu, Y.H., Jia, C.J., Li, M., and Zhang, Y.W. (2005) *Chirality*, **17**, 556–558.
47 Angeloni, M., Piermatti, O., Pizzo, F., and Vaccaro, L. (2014) *Eur. J. Org. Chem.*, 1716–1726.
48 An, Z., Guo, Y., Zhao, L.W., Li, Z., and He, J. (2014) *ACS Catal.*, **4**, 2566–2576.
49 An, Z., He, J., Dai, Y., Yu, C.G., Li, B., and He, J. (2014) *J. Catal.*, **317**, 105–113.
50 Luo, S.Z., Li, J.Y., Zhang, L., Xu, H., and Cheng, J.P. (2008) *Chem. Eur. J.*, **14**, 1273–1281.
51 Demuynck, A.L.W., Peng, L., de Clippel, F., Vanderleyden, J., Jacobs, P.A., and Sels, B.F. (2011) *Adv. Synth. Catal.*, **353**, 725–732.
52 Loh, T.P., Feng, L.C., Yang, H.Y., and Yang, J.Y. (2002) *Tetrahedron Lett.*, **43**, 8741–8743.
53 Kotrusz, P., Kmentová, I., Gotov, B., Toma, S., and Solcániova, E. (2002) *Chem. Commun.*, 2510–2511.
54 Cordova, A. (2004) *Tetrahedron Lett.*, **45**, 3949–3952.
55 Gruttadauria, M., Riela, S., Aprile, C., Lo Meo, P., D'Anna, F., and Noto, R. (2006) *Adv. Synth. Catal.*, **348**, 82–92.

56 Giacalone, F., Gruttadauria, M., Marculescu, A.M., and Noto, R. (2007) *Tetrahedron Lett.*, **48**, 255–259.

57 Zhao, D. and Ding, K. (2013) *ACS Catal.*, **3**, 928–944.

58 Dang, D.B., Wu, P.Y., He, C., Xie, Z., and Duan, C.Y. (2010) *J. Am. Chem. Soc.*, **132**, 14321–14323.

59 Liu, Y., Xi, X.B., Ye, C.C., Gong, T.F., Yang, Z.W., and Cui, Y. (2014) *Angew. Chem., Int. Ed.*, **53**, 13821–13825.

60 Bonnefoy, J., Legrand, A., Quadrelli, E.A., Canivet, J., and Farrusseng, D. (2015) *J. Am. Chem. Soc.*, **137**, 9409–9416.

61 Odedra, A. and Seeberger, P.H. (2009) *Angew. Chem., Int. Ed.*, **48**, 2699–2702.

62 Cambeiro, X.C., Martin-Rapun, R., Miranda, P.O., Sayalero, S., Alza, E., Llanes, P., and Pericàs, M.A. (2011) *Beilstein J. Org. Chem.*, **7**, 1486–1493.

63 Ayats, C., Henseler, A.H., and Pericàs, M.A. (2012) *ChemSusChem*, **5**, 320–325.

64 Ötvös, S.B., Mándity, I.M., and Fülöp, F. (2012) *J. Catal.*, **295**, 179–185.

65 Massi, A., Cavazzini, A., Zoppo, L.D., Pandoli, O., Costa, V., Pasti, L., and Giovannini, P.P. (2011) *Tetrahedron Lett.*, **52**, 619–622.

66 Tsubogo, T., Ishiwata, T., and Kobayashi, S. (2013) *Angew. Chem., Int. Ed.*, **52**, 6590–6604.

67 Hashimoto, K., Kumagai, N., and Shibasaki, M. (2014) *Org. Lett.*, **16**, 3496–3499.

68 Kandel, K., Althaus, S.M., Peeraphatdit, C., Kobayashi, T., Trewyn, B.G., Pruski, M., and Slowing, I.I. (2013) *ACS Catal.*, **3**, 265–271.

69 Szőllősi, G., Fekete, M., Gurka, A.A., and Bartók, M. (2014) *Catal. Lett.*, **144**, 478–486.

70 Zhao, L., Li, Y., Yu, P., Han, X., and He, J. (2012) *ACS Catal.*, **2**, 1118–1126.

71 Sibi, M.P. and Manyem, S. (2000) *Tetrahedron*, **56**, 8033–8061.

72 Christoffers, J. and Baro, A. (2003) *Angew. Chem., Int. Ed.*, **42**, 1688–1690.

73 Schreiner, P.R. (2003) *Chem. Soc. Rev.*, **32**, 289–296.

74 MacMillan, D.W.C. (2008) *Nature*, **455**, 304–308.

75 Jin, R., Liu, K., Xia, D., Qian, Q., Liu, G., and Li, H. (2012) *Adv. Synth. Catal.*, **354**, 3265–3274.

76 Chen, Q., Xin, C., Lou, L.L., Yu, K., Ding, F., and Liu, S. (2011) *Catal. Lett.*, **141**, 1378–1383.

77 Kotrusz, P., Toma, S., Schmalz, H.G., and Adler, A. (2004) *Eur. J. Org. Chem.*, **2004**, 1577–1583.

78 Meciarova, M., Toma, S., and Kotrusz, P. (2006) *Org. Biomol. Chem.*, **4**, 1420–1424.

79 Hagiwara, H., Okabe, T., Hoshi, T., and Suzuki, T. (2004) *J. Mol. Catal. A-Chem.*, **214**, 167–174.

80 Luo, S., Mi, X., Zhang, L., Liu, S., Xu, H., and Cheng, J.P. (2006) *Angew. Chem., Int. Ed.*, **45**, 3093–3097.

81 Xu, X., Cheng, T., Liu, X., Xu, J., Jin, R., and Liu, G. (2014) *ACS Catal.*, **4**, 2137–2142.

82 Kasaplar, P., Riente, P., Hartmann, C., and Pericas, M.A. (2012) *Adv. Synth. Catal.*, **354**, 2905–2910.

83 Alza, E., Cambeiro, X.C., Jimeno, C., and Pericas, M.A. (2007) *Org. Lett.*, **9**, 3717–3720.

84 Miao, T. and Wang, L. (2008) *Tetrahedron Lett.*, **49**, 2173–2176.

85 Tuchman-Shukron, L. and Portnoy, M. (2009) *Adv. Synth. Catal.*, **351**, 541–546.

86 McCooey, S.H. and Connon, S.J. (2007) *Org. Lett.*, **9**, 599–602.

87 Lee, J.W., Shin, J.Y., Chun, Y.S., Jang, H.B., Song, C.E., and Lee, S. (2010) *Acc. Chem. Res.*, **43**, 985–994.

88 Xu, D.Q., Luo, S.P., Wang, Y.F., Xia, A.B., Yue, H.D., Wang, L.P., and Xu, Z.Y. (2007) *Chem. Commun.*, 4393–4395.

89 Luo, S.P., Zhang, S., Wang, Y.F., Xia, A.B., Zhang, G.C., Du, X.H., and Xu, D.Q. (2010) *J. Org. Chem.*, **75**, 1888–1891.

90 Akagawa, K. and Kudo, K. (2012) *Angew. Chem., Int. Ed.*, **51**, 12786–12789.

91 Akagawa, K., Sakai, N., and Kudo, K. (2015) *Angew. Chem., Int. Ed.*, **54**, 1822–1826.

92 Rowsell, J.L.C. and Yaghi, O.M. (2004) *Microporous Mesoporous Mater.*, **73**, 3–14.

93 Wang, Z., Chen, G., and Ding, K. (2009) *Chem. Rev.*, **109**, 322–359.

94 Shibasaki, M., Sasai, H., and Arai, T. (1997) *Angew. Chem., Int. Ed.*, **36**, 1236–1256.

95 Takizawa, S., Somei, H., Jayaprakash, D., and Sasai, H. (2003) *Angew. Chem., Int. Ed.*, **42**, 5711–5714.

96 Shi, T., Guo, Z., Yu, H., Xie, J., Zhong, Y., and Zhu, W. (2013) *Adv. Synth. Catal.*, **355**, 2538–2543.

97 Tsubogo, T., Yamashita, Y., and Kobayashi, S. (2012) *Chem. Eur. J.*, **18**, 13624–13628.

98 Bonfils, F., Cazaux, I., Hodge, P., and Caze, C. (2006) *Org. Biomol. Chem.*, **4**, 493–497.

99 Ötvös, S.B., Mándity, I.M., and Fülöp, F. (2012) *ChemSusChem*, **5**, 266–269.

100 Ahrendt, K.A., Borths, C.J., and MacMillan, D.W.C. (2000) *J. Am. Chem. Soc.*, **122**, 4243–4244.

101 Dalko, P.I. and Moisan, L. (2001) *Angew. Chem., Int. Ed.*, **40**, 3726–3748.

102 Selkala, S.A., Tois, J., Pihko, P.M., and Koskinen, A.M.P. (2002) *Adv. Synth. Catal.*, **344**, 941–945.

103 Josephsohn, N.S., Snapper, M.L., and Hoveyda, A.H. (2003) *J.Am. Chem. Soc.*, **125**, 4018–4019.

104 Nakano, H., Takahashi, K., and Fujita, R. (2005) *Tetrahedron Asymm.*, **16**, 2133–2140.

105 Seebach, D., Beck, A.K., and Heckel, A. (2001) *Angew. Chem., Int. Ed.*, **40**, 92–138.

106 Altava, B., Burguete, M.I., Verdugo, E.G., Luis, S.V., and Vicent, M.J. (2006) *Green Chem.*, **8**, 717–726.

107 Jeong, K.S., Go, Y.B., Shin, S.M., Lee, S.J., Kim, J., Yaghi, O.M., and Jeong, N. (2011) *Chem. Sci.*, **2**, 877.

108 Chow, H.F. and Wang, C.W. (2002) *Helv. Chim. Acta*, **85**, 3444–3453.

109 Benaglia, M., Celentano, G., Cinquini, M., Puglisi, A., and Cozzi, F. (2002) *Adv. Synth. Catal.*, **344**, 149–152.

110 Meracz, I. and Oh, T. (2003) *Tetrahedron Lett.*, **44**, 6465–6468.

111 Doherty, S., Goodrich, P., Hardacre, C., Luo, H.K., Rooney, D.W., Seddon, K.R., and Styring, P. (2004) *Green Chem.*, **6**, 63–67.

112 Fu, F., Teo, Y.C., and Loh, T.P. (2006) *Org. Lett.*, **8**, 5999–6001.

113 Hagiwara, H., Kuroda, T., Hoshi, T., and Suzuki, T. (2010) *Adv. Synth. Catal.*, **352**, 909–916.

114 Shi, J.Y., Wang, C.A., Li, Z.J., Wang, Q., Zhang, Y., and Wang, W. (2011) *Chem. Eur.J.*, **17**, 6206–6213.

115 Hoffmann, R. and Woodward, R.B. (1965) *J. Am. Chem. Soc.*, **87**, 4388–4389.

116 Garcia, J.I., Mayoral, J.A., and Salvatella, L. (2000) *Acc. Chem. Res.*, **33**, 658–664.

117 Barba, C., Carmona, D., Garcia, J.I., Lamata, M.P., Mayoral, J.A., Salvatella, L., and Viguri, F. (2006) *J. Org. Chem.*, **71**, 9831–9840.

118 Wannere, C.S., Paul, A., Herges, R., Houk, K.N., Schaefer, H.F., and von Ragué Schleyer, P. (2007) *J. Comput. Chem.*, **28**, 344–361.

119 Yoshizawa, M., Tamura, M., and Fujita, M. (2006) *Science*, **312**, 251–254.

120 Nishioka, Y., Yamaguchi, T., Yoshizawa, M., and Fujita, M. (2007) *J. Am. Chem. Soc.*, **129**, 7000–7001.

121 Murase, T., Horiuchi, S., and Fujita, M. (2010) *J. Am. Chem. Soc.*, **132**, 2866–2867.

122 Hatano, M. and Ishihara, K. (2012) *Chem. Commun.*, **48**, 4273–4283.

123 Hayashi, Y., Rohde, J.J., and Corey, E.J. (1996) *J. Am. Chem. Soc.*, **118**, 5502–5503.

124 Corey, E.J., Shibata, T., and Lee, T.W. (2002) *J. Am. Chem. Soc.*, **124**, 3808–3809.

125 Noyori, R. and Kitamura, N. (1991) *Angew. Chem., Int. Ed. Engl.*, **30**, 49.

126 Bellocq, N., Brunel, D., Laspéras, M., and Moreau, P. (1997) *Stud. Surf. Sci. Catal.*, **108**, 485.

127 Abramson, S., Laspéras, M., and Brunel, D. (2002) *Tetrahedron Asymm.*, **13**, 357.

128 Fan, Q.H., Li, Y.M., and Chan, A.S.C. (2002) *Chem. Rev.*, **102**, 3385.

129 Seebach, D. and Beck, A.K. (2001) *Angew. Chem., Int. Ed.*, **40**, 92–138.

130 Wang, X., Zhang, J., Liu, Y., and Cui, Y. (2014) *Bull. Chem. Soc. Jpn.*, **87**, 435440.

131 An, W.K., Han, M.Y., Wang, C.A., Yu, S.M., Zhang, Y., Bai, S., and Wang, W. (2014) *Chem. Eur. J.*, **20**, 11019–11028.

132 Mastalir, Á. and Király, Z. (2008) *Catal. Commun.*, **9**, 1404–1409.

133 Sato, I., Kodaka, R., Shibata, T., Hirokawa, Y., Shirai, N., Ohtake, K., and Soai, K. (2000) *Tetrahedron Asymm.*, **11**, 2271–2275.

134 Sato, I., Hosoi, K., Kodaka, R., and Soai, K. (2002) *Eur. J. Org. Chem.*, **2002**, 3115–3118.

135 Malhotra, S.V. and Wang, Y. (2006) *Tetrahedron Asymm.*, **17**, 1032.

136 Kelsen, V., Pierrat, P., and Gros, P.C. (2007) *Tetrahedron*, **63**, 10693.

137 Gadenne, B., Hesemann, P., Polshettiwar, V., and Moreau, J.J.E. (2006) *Eur. J. Inorg. Chem.*, **2006**, 3697–3702.

138 Ma, L.Q., Wanderley, M.M., and Lin, W.B. (2011) *ACS Catal.*, **1**, 691–697.

139 Balakrishnan, U., Ananthi, N., Velmathi, S., Benzigar, M.R., Talapaneni, S.N., Aldeyab, S.S., Ariga, K., and Vinu, A. (2012) *Microporous Mesoporous Mater.*, **155**, 40–46.

140 Wang, Z., Chen, G., and Ding, K. (2009) *Chem. Rev.*, **109**, 322–359.

141 Ma, L.Q., Wu, C.D., Wanderley, M.M., and Lin, W.B. (2010) *Angew. Chem., Int. Ed.*, **49**, 8244–8248.

142 Wu, C.D., Hu, A., Zhang, L., and Lin, W. (2005) *J. Am. Chem. Soc.*, **127**, 8940–8941.

143 Wu, C.D. and Lin, W. (2007) *Angew. Chem., Int. Ed.*, **46**, 1075–1078.

144 Ngo, H.L., Hu, A., and Lin, W. (2004) *J. Mol. Catal. A Chem.*, **215**, 177–186.

145 Harada, T. and Nakatsugawa, M. (2006) *Synlett*, 321–323.

146 Itsuno, S., Sakurai, Y., Ita, K., Maruyama, T., Nakahama, S., and Frechet, J.M.J. (1990) *J. Org. Chem.*, **55**, 304–310.

147 Hodge, P., Sung, D.W.L., and Stratford, P.W. (1999) *J. Chem. Soc. Perkin Trans.*, **1**, 2335–2342.

148 Burguete, M.I., García-Verdugo, E., Vicent, M.J., Luis, S.V., Pennemann, H., von Keyserling, N.G., and Martens, J. (2002) *Org. Lett.*, **4**, 3947–3950.

149 Mak, X.Y., Laurino, P., and Seeberger, P.H. (2009) *Beilstein J. Org. Chem.*, **5**, 19. doi: 10.3762/bjoc.5.19

150 Pericas, M.A., Herrerías, C.I., and Solà, L. (2008) *Adv. Synth. Catal.*, **350**, 927–932.

151 Osorio-Planes, L., Rodríguez-Escrich, C., and Pericàs, M.A. (2012) *Org. Lett.*, **14**, 1816–1819.

152 Irie, R., Noda, K., Ito, Y., Matsumoto, N., and Katsuki, T. (1990) *Tetrahedron Lett.*, **31**, 7345–7348.

153 Larrow, J.F. and Jacobsen, E.N. (2004) *Top. Organomet. Chem.*, **6**, 123–152.

154 Corma, A., Iglesias, M., del Pino, C., and Sámchez, F. (1991) *J. Chem. Soc. Chem. Commun.*, 1252–1253.

155 Ogunwumi, S.B. and Bein, T. (1997) *J. Chem. Soc. Chem. Commun.*, 901–902.

156 Silva, A.R., Freire, C., and de Castro, B. (2004) *Carbon*, **42**, 3027–3030.

157 Silva, A.R., Budarin, V., Clark, J.H., de Castro, B., and Freire, C. (2005) *Carbon*, **43**, 2096–2105.

158 Fan, Q.H., Li, Y.M., and Chan, A.S.C. (2002) *Chem. Rev.*, **102**, 3385–3466.

159 Gbery, G., Zsigmond, Á., and Balkus, K.J. (2001) *Catal. Lett.*, **74**, 77–80.

160 De Vos, D.E., Dams, M., Sels, B.F., and Jacobs, P.A. (2002) *Chem. Rev.*, **102**, 3615–3640.

161 Hashimoto, K., Kumagai, N., and Shibasaki, M. (2015) *Chem. Eur. J.*, **21**, 4262–4266.

162 Reger, T.S. and Janda, K.D. (2000) *J. Am. Chem. Soc.*, **122**, 6929–6934.

163 Guo, H., Shi, X., Qiao, Z., Hou, S., and Wang, M. (2002) *Chem. Commun.*, 118–119.

164 Reed, N.N., Dickerson, T.J., Boldt, G.E., and Janda, K. (2005) *J. Org. Chem.*, **70**, 1728–1731.

165 Song, C.E. and Roh, E.J. (2000) *Chem. Commun.*, 837–838.

166 Gaillon, L. and Bedioui, F. (2001) *Chem. Commun.*, 1458–1459.

167 Lou, L.L., Yu, K., Ding, F., Zhou, W., Peng, X., and Liu, S. (2006) *Tetrahedron Lett.*, **47**, 6513–6516.

168 Lou, L.L., Yu, K., Ding, F., Peng, X.J., Dong, M.M., Zhang, C., and Liu, S.X. (2007) *J. Catal.*, **249**, 102–110.

169 Huang, J., Fu, X., and Miao, Q. (2011) *Appl. Catal. A Gen.*, **407**, 163–172.

170 Huang, J., Cai, J.L., Feng, H., Liu, Z.G., Fu, X.K., and Miao, Q. (2013) *Tetrahedron*, **69**, 5460–5467.

171 Prasetyanto, E.A., Khan, N.H., Seo, H.U., and Park, S.E. (2010) *Top. Catal.*, **53**, 1381–1386.

172 Ma, L., Abney, C., and Lin, W. (2009) *Chem. Soc. Rev.*, **38**, 1248–1256.

173 Cho, S.H., Ma, B.Q., Nguyen, S.B.T., Hupp, J.T., and Albrecht-Schmitt, T.E. (2006) *Chem. Commun.*, 2563–2565.

174 Bogaerts, T., Wouters, S., Van der Woort, P., and Van Speybroek, V. (2015) *J. Mol. Catal. A Chem.*, **406**, 106–113.
175 Khavrutskii, I.V., Musaev, D.G., and Morokuma, K. (2004) *Proc. Natl. Acad. Sci. USA*, **101**, 5743–5748.
176 De Decker, J., Bogaerts, T., Muylaert, I., Delahaye, S., Lynen, F., Van Speybroeck, V., Verberckmoes, A., and Van der Voort, P. (2013) *Mater. Chem. Phys.*, **141**, 967–972.
177 Maia, F., Mahata, N., Jarrais, B., Silva, A.R., Pereira, M.F.R., Freire, C., and Figueiredo, J.L. (2009) *J. Mol. Catal. A Chem.*, **305**, 135–138.
178 Bogaerts, T., Van Yperen-De Deyne, A., Liu, Y.Y., Lynen, F., Van Speybroeck, V., and Van Der Voort, P. (2013) *Chem. Commun.*, **49**, 8021–8023.
179 Zhao, L.W., Shi, H.M., Wang, J.Z., and He, J. (2012) *Chem. Eur. J.*, **18**, 9911–9918.
180 Wang, J., Zhao, L., Shi, H., and He, J. (2011) *Angew. Chem., Int. Ed.*, **123**, 9337–9342.
181 Egami, H., Oguma, T., and Katsuki, T. (2010) *J. Am. Chem. Soc.*, **132**, 5886–5895.
182 Wang, J.Z., Zhao, L.W., Shi, H.M., and He, J. (2011) *Angew. Chem., Int. Ed.*, **50**, 9171–9176.
183 Izumi, Y., Imaida, M., Fukawa, H., and Akabori, S. (1963) *Bull. Chem. Soc. Jpn.*, **36**, 155–160.
184 Orito, Y., Imai, S., and Niwa, S. (1979) *J. Chem. Soc. Jpn.*, **8**, 1118–1120.
185 Studer, M., Blaser, H.U., and Exner, C. (2003) *Adv. Synth. Catal.*, **345**, 45–65.
186 Murzin, D.Y., Maki-Arvela, P., Toukoniitty, E., and Salmi, T. (2005) *Catal. Rev.*, **47**, 175–256.
187 Mallat, T., Orglmeister, E., and Baiker, A. (2007) *Chem. Rev.*, **107**, 4863–4890.
188 Webb, G. and Wells, P.B. (1992) *Catal. Today*, **12**, 319–337.
189 Toukoniitty, E., Maki-Arvela, P., Kalantar Neyestanaki, A., Salmi, T., Sjöholm, R., Leino, R., Laine, E., Kooyman, P.J., Ollonqvist, T., and Vayrynen, J. (2001) *Appl. Catal. A Gen.*, **216**, 73–83.
190 Blaser, H.U., Jalett, H.P., Müller, M., and Studer, M. (1997) *Catal. Today*, **37**, 441–463.
191 Török, B., Felföldi, K., Szakonyi, G., Balázsik, K., and Bartók, M. (1998) *Catal. Lett.*, **52**, 81–84.
192 Török, B., Balázsik, K., Török, M., Felföldi, K., and Bartók, M. (2002) *Catal. Lett.*, **81**, 55–62.
193 Chen, Z., Guan, Z., Li, M., Yang, Q., and Li, C. (2011) *Angew. Chem., Int. Ed.*, **50**, 4913–4917.
194 Schmidt, E., Vargas, A., Mallat, T., and Baiker, A. (2009) *J. Am. Chem. Soc.*, **131**, 12358–12367.
195 Hoxha, F., Schmidt, E., Mallat, T., Schimmoeller, B., Pratsinis, S.E., and Baiker, A. (2011) *J. Catal.*, **278**, 94–101.
196 Blaser, H.U. and Studer, M. (2007) *Acc. Chem. Res.*, **40**, 1348–1356.
197 Bartók, M., Felföldi, K., Török, B., and Bartók, T. (1998) *Chem. Commun.*, 2605–2606.
198 Bartók, M., Sutyinszki, M., Felföldi, K., and Szöllősi, G. (2002) *Chem. Commun.*, 1130–1131.
199 Goubert, G. and McBreen, P.H. (2013) *ChemCatChem*, **5**, 683–685.
200 Schmidt, E., Bucher, C., Santarossa, G., Mallat, T., Gilmour, R., and Baiker, A. (2012) *J. Catal.*, **289**, 238–248.
201 Meemken, F., Maeda, N., Hungerbühler, K., and Baiker, A. (2012) *Angew. Chem., Int. Ed.*, **51**, 8212–8216.
202 Meemken, F., Hungerbühlen, K., and Baiker, A. (2014) *Angew. Chem., Int. Ed.*, **53**, 8640–8644.
203 Maeda, N., Hungerbühler, K., and Baiker, A. (2011) *J. Am. Chem. Soc.*, **133**, 19567–19569.
204 Svane, K., Dong, Y., Groves, M.N., Demers-Carpentier, V., Lemay, J.C., Ouellet, M., Hammer, B., and McBreen, P.H. (2015) *Catal. Sci. Technol.*, **5**, 743–753.
205 Kaushik, M., Basu, K., Benoit, C., Cirtiu, C.M., Vali, H., and Moores, A. (2015) *J. Am. Chem. Soc.*, **137**, 6124–6127.
206 Revell, J.D., Gantenbein, D., Krattiger, P., and Wennemers, H. (2006) *Biopolymers*, **84**, 105–113.
207 Arakawa, Y. and Wennemers, H. (2003) *ChemSusChem*, **6**, 242–245.
208 Hui, X.P., Chen, C.A., Wu, K.H., and Gau, H.M. (2007) *Chirality*, **19**, 10–15.
209 Yu, K., Lou, L.L., Lai, C., Wang, S., Ding, F., and Liu, S. (2006) *Catal. Commun.*, **7**, 1057–1060.

19
Catalysis by Metal Carbides and Nitrides

Connor Nash,[1] Matt Yung,[1] Yuan Chen,[2] Sarah Carl,[3] Levi Thompson,[3] and Josh Schaidle[1]

[1]*National Bioenergy Center, National Renewable Energy Laboratory, 15013 Denver West Parkway, Golden, CO 80401, USA*
[2]*Institute for Integrated Catalysis, Pacific Northwest National Laboratory, 902 Battelle Blvd, Richland, WA 99354, USA*
[3]*University of Michigan, Department of Chemical Engineering, Phoenix Memorial Lab, 201 Bonisteel Blvd, Ann Arbor, MI 48109, USA*

19.1
Introduction

During the last 150 years, much of the research in the field of heterogeneous catalysis has focused on transition metals and their oxides, to the extent that we are now beginning to be able to predict some of their catalytic properties [1]. Over this time period, most of the research and development efforts have been driven by "market pull," such as the industrialization of petroleum refining and fuels production in the mid-to-late 1800s (which then led to the growth of petrochemicals) and emergence of the transportation and environmental business sectors in the 1900s, in particular with regard to meeting new regulatory standards [2]. Today, catalyst development is driven by interests in biomass, sustainability, emissions control, and energy [2]. As the market pull changes, new catalysts must be developed for those specific applications. Accordingly, new catalyst formulations beyond transition metals and their oxides are being explored. Of particular interest is the development and application of early transition metal carbide and nitride (ETMCN) catalysts.

ETMCNs are materials in which carbon or nitrogen occupies interstitial sites within a parent metal lattice, and they often adopt crystal structures that differ from the parent metal. Moreover, the incorporation of carbon or nitrogen results in unique physical and chemical properties, thus driving development of these materials as catalysts for a variety of applications. During the last 40 years, drivers for investigation of these materials include:

Handbook of Solid State Chemistry, First Edition. Edited by Richard Dronskowski, Shinichi Kikkawa, and Andreas Stein.
© 2017 Wiley-VCH Verlag GmbH & Co. KGaA. Published 2017 by Wiley-VCH Verlag GmbH & Co. KGaA.

- Reports of catalytic activities similar to Pt group metals for some reactions [3–5].
- Opportunity to be a cheaper alternative to noble metals (Mo: 15 USD/kg [6] versus Pt: 35,000 USD/kg [7]).
- Ability to be synthesized with high surface areas, thus facilitating their use as support materials [5,8–11].
- Potential to tune catalytic reactivity based on selection of metal and nonmetal atoms, metal/nonmetal atom stoichiometry and crystal structure [12–15].
- Possession of multifunctional catalytic properties (i.e., metallic sites, acidic sites, and basic sites) [16–23].
- Broad applicability for a variety of reactions [4,5,23–30].
- Enhanced performance compared to existing industrial catalysts for some reactions [30].

Despite considerable advances in these materials, there appear to be few, if any, commercial applications.

As the time period from discovery to commercialization of a catalyst is reported to be on the order of 10 years [31] and the first report of metal carbide catalytic reactivity occurred in 1973 [3], it is reasonable to assume that industrial adoption of an ETMCN catalyst could have occurred by now. Obviously, this lack of commercialization could be due to inferior performance (e.g., activity, selectivity, lifetime) for the ETMCN catalysts compared to the incumbent industrial catalyst for a specific application. Beyond performance, other challenges to commercialization exist for these materials. First, these low-cost ETMCN catalysts typically come at the expense of being synthetically challenging, that is, their synthesis often requires high temperatures and reactive gases, C- and N-containing precursors must be sourced and can be in the liquid, solid, or gas form, and excess surface carbon is often encountered for carbide materials. As this chapter will not discuss methods for synthesis of ETMCN catalysts in detail, readers are directed to a number of excellent reports focused on this aspect [11,32,33].

Second, ETMCNs are pyrophoric in their native, reduced state, thus they (1) oxidize relatively easily under some reaction conditions and (2) require specialized handling. Third, the high surface areas of bulk ETMCNs are achieved by creating highly porous structures, and the mechanical stability of these porous structures has not yet been proven in industrial reactors. Thus use of these bulk catalysts may not be feasible in reactor systems that lead to significant attrition, that is, entrained flow and fluidized bed systems. To reduce attrition rates, these materials can be supported on high surface area supports such as Al_2O_3 or SiO_2; however, these supported carbide/nitride particles (nanoparticles) are even more susceptible to oxidation (especially bulk oxidation) under some reaction conditions [34].

Finally, although reported to behave similarly to noble metals, ETMCNs have complex surface structures and compositions (e.g., metal, nonmetal, or mixed termination; oxygen incorporation), which are dynamic under reaction conditions, resulting in significantly different behavior than noble metals toward certain reactants and reactions; (noble metal surfaces also vary with conditions; consider Pd and Pt under oxidizing conditions.) [35]. The complexity of the surface provides these materials their multifunctionality and diverse catalytic properties, but it also likely hinders their adoption as industrial catalysts. There is limited understanding of how the surface structure changes under reaction conditions and the identity of the active sites. In conjunction, the lifetime of these materials and the regeneration protocols are not well established.

This chapter will focus on recent investigations into the catalytic properties of ETMCN materials, including both experimental and computational efforts, and will attempt to link catalyst performance to active site identity/surface structure in order to elucidate the present level of understanding of structure–function relationships for these materials. Specifically, we will focus on research published since 2010. This chapter will first broadly describe the surface structure and potential active site identities of ETMCNs, followed by a thorough review of recent literature divided into four topical areas based on application: (1) biomass conversion, (2) syngas and CO_2 upgrading, (3) petroleum and natural gas refining, and (4) electrocatalytic energy conversion, energy storage, and chemicals production.

19.2
Surface Structure and Active Site Identity

The multifunctionality (i.e., possession of multiple types of catalytic sites, including acidic, basic, and metallic sites) of ETMCN catalysts is due in part to their complex surface structure, which is dictated by their bulk crystal structure, metal and nonmetal atom type, metal:nonmetal atomic ratio, exposed surface facet(s), and surface coverages of adsorbates/intermediates, the last two of which are controlled by reaction conditions. ETMCN materials are interstitial compounds in which carbon or nitrogen resides in interstices in a metal lattice. Based on the Hägg rule, stable structures for interstitial compounds will exist when the ratio of nonmetal to metal radii is between 0.41 and 0.59 [5,36]. Accordingly, based on the metal and nonmetal atom type, ETMCNs can adopt a variety of bulk crystal structures, as shown in Figure 19.1. Based on these bulk crystal structures, an idealized surface structure can be realized depending upon the exposed surface facet, and these surfaces can be metal terminated, nonmetal atom terminated, or mixed terminated. For example, the (001) facet of orthorhombic Mo_2C, which is one of the most thermodynamically stable surfaces [37], can be either C-terminated or Mo-terminated, as shown in Figure 19.2a and b. Thus, just knowing the dominant exposed facet does not fully describe the surface structure. Moreover, if the (011) facet of Mo_2C is exposed (this surface is also predicted to be relatively stable [38]), the surface can be both C- and Mo-terminated, as shown in Figure 19.2c. The dominant exposed facet(s) is controlled primarily by reaction conditions (i.e., temperature, pressure, and composition of the surrounding gas/liquid phase; the latter of which is quite difficult to characterize as it can change

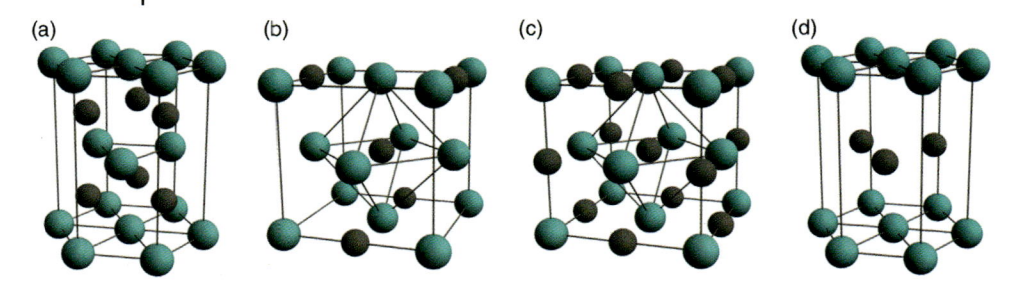

Figure 19.1 Structures for several early transition metal carbides and nitrides: (a) Mo_2C, W_2C (hcp)*, (b) MoC_{1-x}, WC_{1-x} (fcc), (c) TiC, HfC, VC, NbC (fcc), and (d) MoC, WC (hex). Teal spheres correspond to metal atoms and gray spheres correspond to nonmetal atoms. (Adapted from Ref. [39].) *There is some discrepancy in the literature about the crystal structure of this phase, whether it is hcp or orthorhombic (slightly distorted hcp). In this chapter, we refer to the orthorhombic structure [37].

during the course of a reaction) and particle size [37]. It should also be noted that the density of the nonmetal atoms on the surface is dependent not only on the exposed facet of a given crystal structure but also on the nonmetal:metal atomic ratio, as can be observed for the (001) surfaces of the fcc crystal structures shown in Figure 19.1b and c.

For a given surface termination, a specific ETMCN surface facet has a collection of potential catalytically active sites, stemming from the combination of metal and nonmetal atoms in a single structure. Often, the number of specific active sites on a given surface of an ETMCN material is much higher than a comparable surface on a transition metal. For example, the C-terminated Mo_2C (001) surface has eight distinct sites compared to only four on a Pt(111) surface, as shown in Figure 19.2a and d. The sites on the C-Mo_2C(001) surface include Mo and C atop sites, threefold hollow sites, and bridging sites; the hollow sites are distinguished based on the number of nearest neighbor C atoms (see hol2

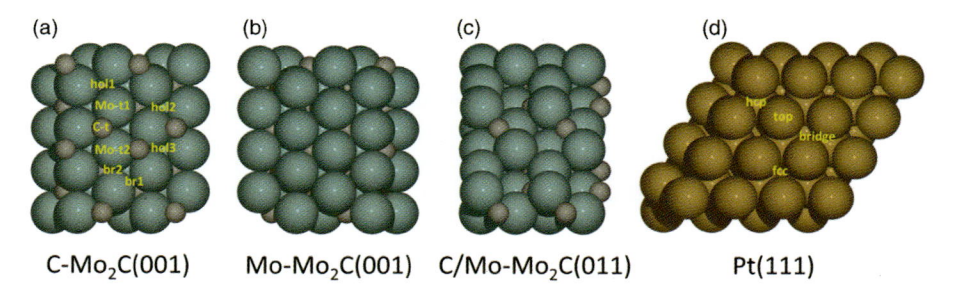

Figure 19.2 Top views of the minimum energy geometries for (a) C-Mo_2C(001), (b) Mo-Mo_2C(001), (c) C/Mo-Mo_2C(011), and (d) Pt(111). For (a)–(c), teal spheres correspond to Mo atoms and gray spheres correspond to carbon atoms. Adsorption sites on C-Mo_2C(001) and Pt(111) are labeled in yellow.

versus hol3 in Figure 19.2a) and the subsurface structure (i.e., whether a carbon or Mo atom is directly below the site). Also contributing to the number of distinct sites is the varying coordination numbers of the Mo atoms. For example, Mo atop site Mo-t1 is coordinated to one surface and two subsurface C atoms, whereas Mo atop site Mo-t2 is coordinated to two surface and one subsurface C atoms. Calculated adsorption energies for a given reactant on these various sites can vary by as much as, if not more than, 0.5 eV/molecule (about 50 kJ/mol) [11]. Furthermore, experimental and computational results show that both metal and nonmetal atoms on the surface may participate in binding reactants to the surface [20,40]. Accordingly, each of these distinct sites on the surface of an ETMCN material could play a role in the catalysis.

Thus far, this section has been focused on idealized surface models derived from the bulk ETMCN crystal structures. However, these surfaces are dynamic under reaction conditions, specifically in regards to C- and O-containing reactants. Under high temperature conditions in the presence of light hydrocarbons (e.g., methane, ethane), the surface of an ETMCN can possess a higher density of carbon atoms than anticipated from the bulk crystal structure. It has been shown that the extent of surface carburization has a strong effect on catalytic reactivity [41]. In the presence of O-containing reactants (e.g., H_2O, biomass-derived species), the ETMCN surface may become partially covered by oxygen due to the oxophilicity of these materials. It has recently been demonstrated that this surface oxygen can form hydroxyl species (i.e., M—OH) and that these hydroxyl species can act as Brønsted acid sites, facilitating hydrogenolysis, isomerization, and dehydration reactions [17,19,20,23]. Based on the extent of oxygen coverage, exposed metal sites, which are capable of activating H_2 and performing hydrogenation reactions, may act like oxygen-vacancy sites similar to reducible metal oxides [20]. While it is difficult to unambiguously determine structure–function relationships for these ETMCN materials due to their complex surface structure and dynamic nature under reaction conditions, the remainder of this chapter will define the current level of understanding of structure–function relationships for these materials based on their catalytic applications.

19.3
Catalytic Applications

19.3.1
Biomass Conversion

Biomass feedstocks have the potential to serve as a renewable source of fuels and chemicals [42]. Due to the dramatic difference in composition between biomass-derived and petroleum or natural gas-derived feedstocks, there are unique challenges associated with development of heterogeneous catalysts for conversion of biomass feedstocks [43]. For example, bio-oils produced from fast pyrolysis contain more than 400 compounds [44,45] with relatively low concentrations of sulfur and nitrogen, but have high oxygen contents, typically near 40 wt% [46,47].

These biomass derivatives cannot be "dropped-in" to existing petroleum refinery processes due to these major chemical differences, and thus, must be upgraded. It is desirable to remove oxygen via selective cleavage of C—O and/or C=O bonds such that carbon is not lost as CO or CO_2 and carbon efficiency is maximized. Often, complete deoxygenation requires the addition of hydrogen, a process known as hydrodeoxygenation (HDO). Another important reaction in biomass conversion is catalytic reforming to produce H_2 and/or CO. In contrast to HDO, C—C bond scission is required. Controlling reforming chemistry is desirable as it may occur in parallel to HDO and hydrogenation (HYD) reactions [48]. The following sections will focus on reforming and HDO reactions of biomass derivatives using ETMCN-based catalysts.

19.3.1.1 Reforming Reactions for H_2 and CO Production

Biomass-derived compounds can be used to produce hydrogen and/or syngas, both of which can be used in subsequent processes for fuel and chemical production. Recent work has focused on reforming simple model oxygenates over metal carbide catalysts. Studies of methanol interactions with Pt-modified WC polycrystalline films using density functional theory (DFT) calculations, high resolution electron energy loss spectroscopy (HREELS) and temperature-programmed desorption (TPD) of methanol in ultrahigh vacuum (UHV) conditions have shown significant selectivity shifts compared to WC; these shifts are similar to changes observed in prior work with single crystal Pt-modified C/W (111) [49,50]. Through C—O bond scission of the methoxy surface intermediate, the native WC surface favored methane production from methanol. The Pt modification eliminated C—O bond scission, hence favoring CO and complete decomposition. The clean polycrystalline Pt surface favored the same reaction pathways, but was less than half as active as Pt/WC on a per metal atom basis. DFT calculations and HREELS studies were used to support the TPD results and to conclude that Pt modification of WC resulted in synergistic effects that made Pt/WC more selective for methanol decomposition than the native Pt or WC materials. The Pt modification resulted in C—H bond scission of the methoxy intermediate to form CO as opposed to C—O scission observed over clean WC [49]. Subsequent work has shown than Ni- or Rh-modified WC catalysts have even higher activities and selectivities for H_2 and CO production from methanol compared to Pt/WC [51]. In addition, Ni-modified WC displayed similar reaction pathways to a Ni-modified Pt catalyst for the decomposition of ethanol [52], ethylene glycol [53], and glycolaldehyde [54]. It is important to note that, while UHV techniques using clean carbide surfaces provide valuable insights into reactivity and structure–function relationships, the carbide surface under UHV conditions may not be rigorously representative of the surface under reaction conditions at ambient or elevated pressure, especially in regards to oxygen coverage.

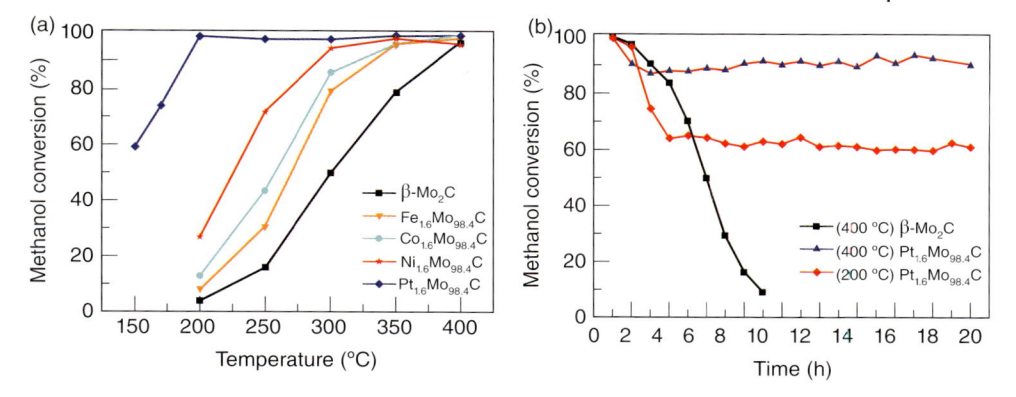

Figure 19.3 Methanol conversion in steam reforming reactions over (a) a series of metal-doped Mo$_x$C catalysts at 150–400 °C and (b) over Pt modified Mo$_x$C at 200 and 400 °C as a function of time-on-stream. (Reprinted with permission from Ref. [55]. Copyright 2014, Elsevier.)

There are several studies of carbide materials for methanol reforming to produce CO, CO$_2$, and H$_2$ [55–59]; however, recent work has focused specifically on Mo$_2$C-based materials. In a study of transition metal-doped (Pt, Fe, Co, and Ni) Mo$_2$C catalysts, Pt-Mo$_2$C (synthesized by a temperature programmed reaction resulting in a mixed α-MoC$_{1-x}$/β-Mo$_2$C phase) showed >70% selectivity to H$_2$ and the highest catalytic activity of the prepared catalysts (Figure 19.3a) [55]. The Pt-Mo$_2$C catalyst achieved 100% methanol conversion at 200 °C, and significantly increased the lifetime of the unmodified β-Mo$_2$C catalyst (Figure 19.3b). No evidence of molybdenum oxide species, which may be indicative of oxidation-induced deactivation [29], were present in post-reaction X-ray diffraction (XRD) patterns; however, the activity of the unmodified β-Mo$_2$C was recovered following regeneration in 15% CH$_4$/H$_2$ at 590 °C for 2 h.

The same authors prepared Ni- and Cu-doped Mo$_2$C (β and mixed phase, respectively) for methanol steam reforming [60,61]. In both cases, the loading of Ni or Cu had a significant effect on the catalytic performance of the catalyst, with specific formulations outperforming the bulk β-Mo$_2$C. Interestingly, X-ray photoelectron spectroscopy (XPS) of pre- and post-reaction catalysts showed low Ni loadings increased the resistance to surface oxidation, which was inferred to be the cause of deactivation [61]. Based on changes in the product distribution and activity as a function of time-on-stream, the authors hypothesized that Ni extended the time the catalyst was active for the water-gas shift reaction (WGSR; as evidenced by CO$_2$ production), thus reducing H$_2$O-induced oxidation of the surface and/or irreversible binding of CO to active sites. In the case of Cu-Mo$_2$C, the enhanced lifetime relative to β-Mo$_2$C was attributed to the presence of Cu^I species [60]. Based on XPS analysis, the Cu^I species was presumed to exist due to a charge transfer from the Mo$_2$C matrix to the Cu metal. The results from biomass-derived model compounds show that transition metal carbide surfaces can be modified with various admetals to control both activity and selectivity in catalytic reforming processes. Moreover, metal modification has been shown to improve catalyst lifetime in reforming reactions by facilitating WGSR, preventing the ETMCN surface from permanently oxidizing and blocking active sites. The role of admetals in WGSR will be discussed further in the section "Water-Gas Shift Reaction."

19.3.1.2 Hydrodeoxygenation Reactions

The removal of oxygen is one of the major challenges in converting biomass feedstocks into value-added fuels and chemicals [62]. ETMCN materials have been shown to possess bifunctional properties [16,17,19,22,63,64] that are necessary for HDO [42], favoring C—O/C=O cleavage (HDO and direct deoxygenation (DDO)) over C—C cleavage (decarbonylation and decarboxylation (DCO)) [18,53,65–69]. Biomass-derived model compounds are commonly used to gain a fundamental understanding of reaction mechanisms and allow for comparisons of catalyst performance.

Vapor Phase HDO

Vapor phase HDO of acetic acid with co-fed H_2 showed β-Mo_2C to favor acetaldehyde and ethylene at 250–400 °C [20]. Following hydrogen pretreatment, the carbide surface possessed metallic-like H-adsorption sites, (attributed to exposed Mo and C) and Brønsted acidic hydroxyl sites in a 1:8 metallic:acidic site ratio [20]. Schaidle et al. [20] proposed that C—O bond cleavage occurred over acidic or oxygen vacancy sites (similar to the mechanism proposed over MoO_3 [70]) and the deoxygenation rate was limited by the availability of adsorbed hydrogen. That is, adsorbed hydrogen functioned primarily in the removal of surface-bound oxygen (removed as H_2O) to regenerate O-vacancy sites. Similar to the acetic acid HDO results over Mo_2C, for propanol and propanal HDO, C—O/C=O bond cleavage to form propylene was favored over C—C bond scission with WC and Mo_2C catalysts [66,71,72]; however, the propanal conversion was 10 times higher over Mo_2C than WC [71]. The high selectivity (>60% to propylene from propanal) of WC and Mo_2C was attributed to strong reactant binding through the oxygen atom (similar to what was observed by Schaidle et al. [20] for acetic acid adsorption on β-Mo_2C). It was suggested that the presence of oxygen species on the WC catalyst surface provided acidic character. This oxygen-induced acidity likely contributed to the higher rate of propanol deoxygenation relative to the rate of propanal HDO, contrary to what was expected based on DFT calculations of a clean WC surface. That is, direct dehydration of propanol over acidic sites was proposed to be faster than propanal HDO (a reaction requiring H_2 incorporation).

Computational and experimental investigations of the HDO of C_3 oxygenates (1-propanol, 2-propanol, propanal, and acetone) with and without co-fed H_2 showed that the Mo_2C catalyst deactivated quickly without co-fed H_2. Ren et al. concluded that H_2 was required to remove surface oxygen (as H_2O) in order to prevent deactivation of the catalyst surface [66]. The effect of surface bound oxygen on the activity and selectivity of Mo_2C is well documented for other model

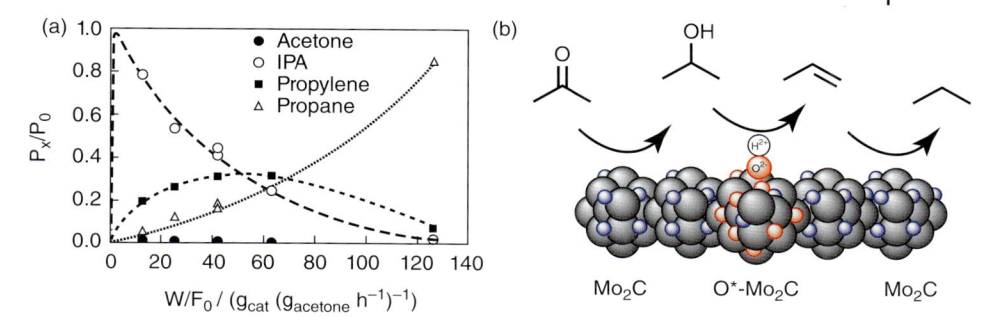

Figure 19.4 (a) Effluent partial pressures as a function of contact time during acetone HDO experiments over Mo_2C ($T = 96\,°C$). (b) Proposed acetone HDO reaction pathway, whereby rapid carbonyl HYD occurs over metallic sites. Sequential alcohol dehydration occurs over Brønsted acid sites and alkene HYD occurs over metallic sites to form the alkane. (Reprinted with permission from Ref. [19]. Copyright 2016, American Chemical Society.)

compounds [69,73], and has been shown to control the Brønsted acidity of the catalyst surface [17]. In a detailed study of acetone HDO over Mo_2C, Sullivan *et al.* reported that acetone was first hydrogenated to isopropyl alcohol over metallic-like sites and subsequently dehydrated to form propylene over Brønsted acid sites (Figure 19.4) [19]. It was shown that increased surface oxidation led to higher Brønsted acidic activity but suppressed (and eventually eliminated) the hydrogenation functionality. Similar observations have been made over tungsten carbide catalysts [74,75]. These biomass-derived model compound studies illustrate the importance of controlling surface oxidation in order to balance the availability of Brønsted acid, O-vacancy, and H_2-activation sites to achieve effective HDO.

There are limited examples of vapor phase HDO of more complex model compounds (e.g., C_{5+}, lignin-derived oxygenates) over carbide catalysts. Specifically, the HDO of furfural [67,69] and anisole [18,76] was recently studied over Mo_2C catalysts. Low temperature (150 °C), ambient pressure HDO of furfural resulted in high selectivity toward C—O/C=O bond cleavage products (50–60% selectivity to 2-methylfuran, <1% selectivity to furan) [69]. This product distribution has been attributed to a strong interaction between the carbide surface and the C=O bond of furfural [67]. It was proposed that the catalyst deactivation was due to the build-up of carbonaceous species on the catalyst surface reducing the number, but not the character, of active sites. Kinetic data suggested that two distinct sites were required for HDO of furfural, with metallic-like sites playing an important role in the observed selectivity [69]. However, the authors did not explicitly define the role of each individual site in the reaction mechanism [69].

In anisole upgrading experiments at 150–250 °C and ambient pressure, Mo_2C initially formed cyclohexane, with benzene becoming the dominant steady-state product at extended times-on-stream (>90% selectivity, Figure 19.5) [18]. The

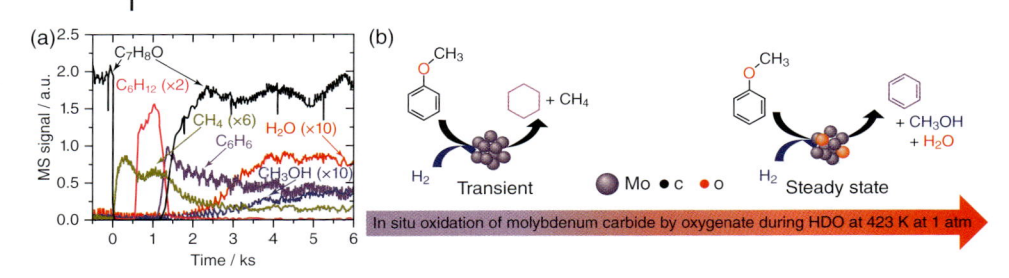

Figure 19.5 (a) Transient mass spectrometer (MS) signals of benzene (C_6H_6, m/z = 78), anisole (C_7H_8O, m/z = 108), cyclohexane (C_6H_{12}, m/z = 84), methane (CH_4, m/z = 18), methanol (CH_3OH, m/z = 31), and water (H_2O, m/z = 18) for 2.1 g of Mo_2C (BET surface area ~ 96 m^2/g, CO chemisorption ~ 199 μmol/g) tested in the vapor phase anisole HDO reaction at 150 °C under ambient pressure with reaction feed composition: anisole/Ar/H_2 (vol %) = 0.75/16.7/bal (total flow rate ~2 cm^3/s). (b) The effect of surface oxidation on the product distribution in vapor phase anisole HDO over Mo_2C. (Reprinted with permission from Ref. [18]. Copyright 2016, American Chemical Society.)

authors concluded that surface oxidation led to the suppression of the sequential HYD of benzene to form cyclohexane. Similar to the work with furfural, kinetic studies suggested that two distinct catalytic sites exist on the carbide surface and contribute to the catalysis [18,69]. In the same study, Lee *et al.* performed *in situ* CO titration to show that <20% of the sites titrated via *ex situ* methods were active for anisole HDO. Furthermore, the product selectivities, apparent activation energies and benzene synthesis turnover frequencies (TOF, based on *in situ* site titration) were similar for an oxygen-treated and freshly prepared Mo_2C, suggesting the role of oxygen was to reduce the number of active sites for vapor phase anisole HDO [18].

Upgrading of aromatic oxygenate mixtures (i.e., anisole, *m*-cresol, guaiacol, and 1,2-dimethoxybenzene) at 280 °C and ambient pressure over Mo_2C catalysts produces benzene and toluene with >90% yields [68]. These investigations demonstrate that molybdenum carbide catalysts are effective for aryl-oxygen (C—O) bond cleavage.

Liquid Phase HDO

In addition to vapor phase HDO, there have been several recent studies of liquid phase HDO using ETMCNs [34,65,77–86]. Mortensen *et al.* studied HDO of a phenol/1-octanol mixture over a Mo_2C/ZrO_2 catalyst at 280–380 °C and 100 bar H_2 [83]. Aromatic ring HYD selectivity remained low (benzene:cyclohexane = 3.3 at 340 °C), and Brønsted acid-catalyzed transalkylation selectivities to products including octylphenol and octylphenyl ether were high, particularly at lower temperatures. It is unclear, however, if the observed acid catalysis was primarily due to the acidic ZrO_2 support or to the Brønsted acidity of Mo_2C. The authors also concluded that catalyst deactivation was caused by severe oxidation of the Mo_2C. Similar work by Jongerius *et al.* showed that Mo_2C supported on carbon nanofibers (CNFs) produced transalkylation products during guaiacol HDO in a batch autoclave system at 300–375 °C and 55 bar H_2 [84]. Collectively, liquid phase HDO studies suggest that even under high hydrogen pressures, aromatic ring HYD remains difficult (likely due to thermodynamic limitations [87,88]) and Brønsted acid sites play a major role in the observed performance of metal carbide catalysts. Mo_2N materials have also been studied for liquid phase HDO of guaiacol [81,85,86]. The primary product over all of the Mo_2N formulations tested was phenol, which was presumed to be formed via demethoxylation [85]. In a related study using supported Mo_2N, higher Mo_2N dispersion was observed on supports with more oxygen surface groups and higher mesoporosity, and was correlated to increased HDO activity [82].

While the role of oxidation-induced Brønsted acidity in metal carbide and nitride catalysis is well documented, it is also clear that severe oxidation will ultimately cause catalyst deactivation. The liquid phase HDO literature shows similar trends to vapor phase HDO studies: oxidation results in less HYD activity and more acid site catalysis; however, Brønsted acidity alone is insufficient for successful HDO.

19.3.1.3 Knowledge Gaps

Despite extensive investigations, several knowledge gaps must be addressed to gain a comprehensive understanding of biomass conversion catalysis over these ETMCNs.

- The majority of recent studies present physical and chemical characterization data generated using *ex situ* methods. Work to probe *in situ/in operando* catalyst surfaces is needed.
- Detailed deactivation kinetics and mechanisms during biomass upgrading processes need to be determined. The role of oxidation is well documented; however, detailed studies probing the effect of reaction conditions on surface oxidation are limited, and the role of other deactivation mechanisms (e.g., carbon accumulation) is not well understood.
- The deoxygenation performance of ETMCNs should be evaluated with authentic biomass streams. Although model compound studies are useful in understanding fundamental reaction kinetics and mechanisms, commercial scale operation will ultimately depend on performance with whole biomass streams.
- Recent biomass conversion work has focused heavily on W, V, and Mo carbides and nitrides; however, studies with other parent-metal ETMCN (e.g., Ti, Cr, Ta, Nb), bimetallic (e.g., NiMoC), and metal-promoted (e.g., K) materials may yield useful results.

19.3.2
Syngas and CO_2 Upgrading

As mentioned previously, biomass can go through a reforming process to produce syngas, that is, a mixture of CO and H_2 with small amounts of CO_2. Early transition metal carbide materials have been demonstrated to convert syngas through Fischer–Tropsch Synthesis (FTS) for the production of hydrocarbons and oxygenates [28,89,90]. Syngas can also be generated through steam/dry reforming of alkanes and alcohols over transition metal carbide catalysts, which will be discussed in the section "Dry Reforming" of this chapter. Additionally, there has also been growing interest in using early transition metal carbides to hydrogenate CO_2 to produce syngas, hydrocarbons, and alcohols. Besides being a component in the syngas, CO_2 is also a renewable building block. The direct conversion of CO_2 to useful fuels/chemicals not only provides a sustainable route for chemical syntheses, but also helps balance greenhouse gas emissions that are linked to global climate change. For the carbide materials, the interconversion between CO_2 and CO in these hydrogenation reactions has also been frequently reported [91,92]. The following subsections will highlight the key recent findings and compare studies on CO and CO_2 hydrogenation using ETMCNs.

19.3.2.1 CO Hydrogenation

Catalytic Performance
One of the most studied metal carbide catalysts for CO hydrogenation is Mo_2C, due to its high hydrogenation activity. Table 19.1 summarizes the reaction rate and selectivity data from recent studies on bulk or metal-modified/promoted Mo_2C catalysts. In general, over bulk or supported Mo_2C, the products are predominantly hydrocarbons; the low selectivity to alcohols is partially attributed to the high operating temperatures, as alcohol production is an exothermic reaction. The selectivities shifted from hydrocarbons to alcohols when appropriate promoters, such as Rb, Ni, or K were employed. This enhancement in alcohol selectivity was suggested to be a consequence of the metal promoter inhibiting hydrocarbon formation or the formation of a alkali metal-Mo-C phase that promoted alcohol chain propagation [91]. It should be pointed out that the Mo_2C-based catalysts exhibit relatively low chain propagation probabilities ($\alpha \approx 0.3-0.4$) compared to those of more industrially-relevant materials such as Fe/Al_2O_3 (0.6–0.8) [93]. This result indicates that Mo_2C is less viable for C—C coupling than the commercial Fe-based catalysts, likely a consequence of the low site density for the Mo_2C materials as suggested by Shou *et al.* [94]. The WGSR also occurred, resulting in the formation of CO_2 with selectivities ranging from 14 to 50%. Schaidle *et al.* [15] systematically studied CO hydrogenation over a series of carbide and nitride catalysts. Arrhenius plots for the catalysts are shown in Figure 19.6. The gravimetric rates decreased in the following order: $Mo_2C \sim W_2C \sim VN \sim NbN > Mo_2N \sim W_2N \gg VC > NbC$. There also appeared to be significant activity differences between carbide and nitride materials of the same parent metal. This difference has also been observed in other types of reactions, such as acetone condensation [16] and hydrogenolysis, dehydrogenation, and isomerization [95].

Active Sites and Mechanistic Insights
Three possible mechanisms have been proposed for FTS over ETMCN materials: they are designated as the carbide, oxygenate, and CO insertion pathways [15].

Table 19.1 Reaction conditions, rates, selectivities, and chain propagation probabilities for CO hydrogenation over bulk, supported or metal promoted Mo_2C catalysts in recent studies.

Catalysts	Conditions (°C/bar/H_2 to CO ratio)	Rates		Selectivity				Chain propagation probability (α)	References
		TOF (s^{-1})	µmol/m^2/s	CH_4	C_{2+} hydrocarbons	CH_3OH	C_{2+} alcohols		
$5Mo_2C/Al_2O_3$	300/30/1:1	0.003	Not reported	31	67	1.5	0.9	0.36	[91]
$1.5Rb-5Mo_2C/Al_2O_3$	300/30/1:1	0.0007	Not reported	15	29	16	40	0.41	[91]
MoC	240/70/2:1	—	0.63	75	16	8	0.7	—	[63]
$Ni_{0.5}MoC$	240/70/2:1	—	0.24	56	12	28	4	—	[63]
$\beta-Mo_2C$	340/70/2:1	—	12.8	95	4	0.8	—		[96]
$K-\beta-Mo_2C$	340/70/2:1	—	13.6	65	14	21	—		[96]
Mo_2C	290/25/2:1	0.36	0.91	55	44	1	0.31		[15]

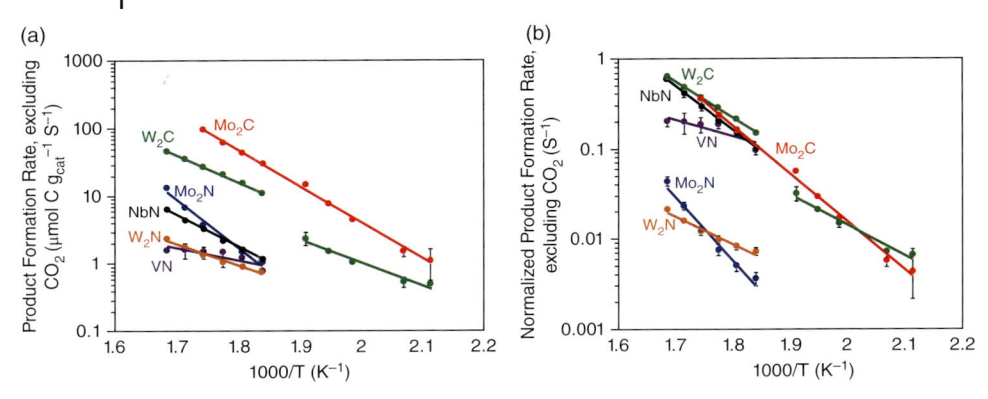

Figure 19.6 Arrhenius plots of the FTS product formation rates on a C1 basis (excluding CO_2) for the carbide and nitride catalysts on (a) a gravimetric basis and (b) normalized by the CO uptakes. (Reprinted with permission from Ref. [15]. Copyright 2015, Elsevier.)

Schaidle et al. [15] used a combination of temperature-programmed desorption (TPD) and temperature-programmed reaction (TPRxn) studies to demonstrate that at least two different types of catalytic sites were present on the Mo_2C surface. These included CO dissociation sites and molecular CO adsorption sites that facilitate C—C coupling reactions. However, the authors were unable to provide additional details about the actual identity of these sites. Interestingly, two types of active sites (Brønsted acidic and metallic (which could be both exposed C or Mo sites) [20]) were also proposed to participate in vapor phase HDO reactions over Mo_2C, as described in the section "Vapor Phase HDO." Jointly, these reports highlight not only the multifunctionality of Mo_2C, but also the need to link active site identity/structure to specific adsorbate interactions. For FTS, Schaidle et al. [15] also indicated that the oxygenate pathway (for C—C coupling) was dominant over Mo_2C, where CH_XO species were the key intermediates and alcohols were formed as the side products. It was suggested that the carbide mechanism was responsible for chain growth over Mo_2N, as illustrated in Figure 19.7. The lower CO dissociation barrier for the carbide compared to the nitride helps explain the 10 times higher TOF for Mo_2C (0.36 s^{-1}) as compared to Mo_2N (0.04 s^{-1}) for FTS at 300 °C. Interestingly, the DFT study by Medford et al. [35] also suggested that carbon–carbon coupling was kinetically facile on Mo-terminated $Mo_2C(001)$. They found that coupling involving HCO was more favorable than coupling CH_x species, which supported the hypothesis by Schaidle et al. [15] that Fischer–Tropsch type reactions occurred via the oxygenate mechanism on Mo_2C. Shou et al. [91] demonstrated that the addition of metal onto Mo_2C could introduce/modify the active sites. They found that deactivation from surface passivation was minimal for the Rb-promoted Mo_2C, suggesting that the metal-native carbide interaction controlled the reaction activity compared to the addition of oxygen via passivation.

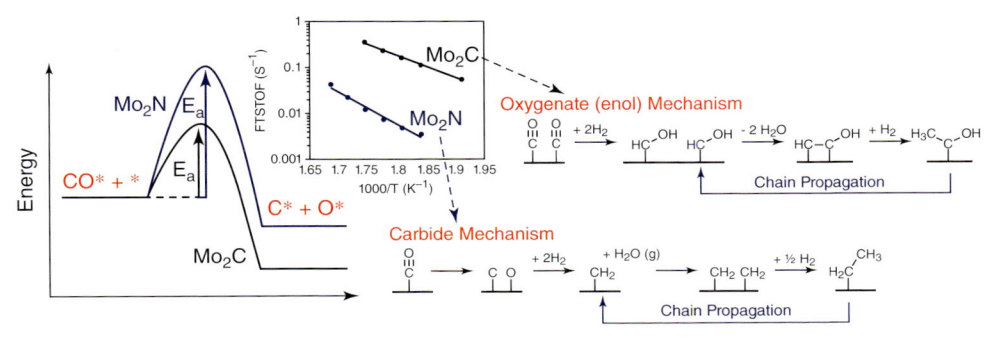

Figure 19.7 Comparison of activation energies, turnover frequencies for FTS, and reaction mechanisms on Mo_2C and Mo_2N. (Reprinted with permission from Ref. [15]. Copyright 2015, Elsevier.)

19.3.2.2 CO$_2$ Hydrogenation

Catalytic Performance
Early transition metal carbides, such as Mo_2C and TiC, are active for CO_2 conversion to fuels and chemicals. Vidal *et al.* [40] used Cu or Au promoted TiC (001) to produce CH_3OH from CO_2 hydrogenation at 250–320 °C. The Cu/TiC (001) material exhibited a CH_3OH production rate that was an order of magnitude higher than that for the pure TiC(001) material. The authors suggested the improved rate for CH_3OH synthesis was a consequence of a charge polarization on the Cu or Au, where electrons are distorted from the nuclei. This perturbation made the admetal-carbide interface effective for activating CO_2. The study by Xu *et al.* using Mo_2C-based catalysts suggested that the reaction selectivity was highly dependent on the carbide phase [14]. Their results indicated that β-MoC_y was more active than α-MoC_{1-x} and produced mainly CH_4 and CO as reaction products. On the other hand, α-MoC_{1-x} was less active but more selective for CH_3OH production. They attributed this difference in catalytic performance to the different Mo/C ratios possessed by the α-MoC_{1-x} and β-MoC_y. As α-MoC_{1-x} is a C-rich catalyst, it is proposed that the CO_2 adsorbs molecularly and cleavage of the first C—O bond only occurs through hydrogenation forming a formate (HCOO*) intermediate [14,97], which eventually decomposes into CO or goes through a subsequent hydrogenation to yield CH_3OH. In contrast, β-MoC_y is a Mo-rich catalyst, which likely forms a Mo-C bond upon the adsorption of CO_2; this bond formation can lead to the $CO_2 \rightarrow CO \rightarrow C \rightarrow CH_4$ transformation [40,97], yielding more CH_4 as the final product. This work demonstrates that the active sites of Mo carbide for hydrogenation can be tuned by modifying the Mo/C ratio. Porosoff *et al.* [92] reported that bulk Mo_2C outperformed CeO_2-supported Pt or Pd catalysts for CO_2 hydrogenation, producing primarily CO (~93% selectivity) at 300 °C and 1 atm via the reverse water–gas shift (RWGS) reaction. They also showed that the addition of Co onto Mo_2C resulted in a moderate increase in the CO selectivity, which was attributed to

Table 19.2 Reaction conditions, rates, and selectivity for CO_2 hydrogenation over bulk, supported or metal promoted Mo_2C catalysts in recent studies.

Catalyst	Conditions (°C/bar/H_2 to CO_2 ratio)	TOF (s^{-1})	Selectivity (%)					References
			CO	CH_3OH	C_2H_5OH	CH_4	C_{2+}	
Mo_2C	300/1/2:1	4.3	94	0	0	6	0	[92]
Co-Mo_2C	300/1/2:1	2.7	98	0	0	2	0	[92]
Mo_2C	200/40/3:1	0.002	5	53	16	17	9	[68]
Cu/Mo_2C	200/40/3:1	0.004	9	62	14	10	6	[68]
α-Mo_2C	200/20/5:1	0.001	52	28.0	1.0	11.0	5	[14]
β-Mo_2C	200/20/5:1	0.01	39	21	1	29	8	[14]
TiC	227/5/9:1	8.3	0	100	0	0.0	0	[40]
Cu/TiC	227/5/9:1	0.015	0	100	0	0.0	0	[40]

the ability of Co/Mo_2C to dissociate CH_4 to C and H_2, thus decreasing the selectivity to the side product CH_4. More recently, Chen et al. [68] reported the use of Cu/Mo_2C supported catalysts to produce methanol from CO_2 via a cascade hydrogenation sequence involving a formate intermediate. The CH_3OH selectivities and production rates were higher for the Cu/Mo_2C catalyst compared to Mo_2C for the hydrogenation of CO_2, formic acid, and ethyl formate (at 135 °C, 30 bar H_2). The influence of a series of admetals on the reactivity of Mo_2C for CO_2 hydrogenation was also reported [98]. The results revealed that the addition of Cu or Pd onto Mo_2C enhanced methanol formation, while the addition of Co or Fe facilitated C—C coupling to produce C_{2+} hydrocarbons. Table 19.2 summarizes the reaction performance for the aforementioned studies.

Active Sites and Mechanistic Insights
Combined experimental and theoretical studies also provide insight regarding the active sites for CO_2 hydrogenation over carbide-based catalysts. Rodriguez et al. [40] reported that the use of a TiC-based catalyst for CO_2 hydrogenation resulted in nearly 100% methanol selectivity. In contrast, other carbides, such as Mo_2C and Fe_3C, produced mainly CO and CH_4 and only a small amount of methanol (about 20%). Due to the symmetric and stable adsorption geometry of CO_2 on TiC, the adsorption energy of CO_2 on TiC(001) is relatively large. Interesting, for the calculated CO_2 adsorption mode in Figure 19.8b, the O and C (from CO_2) were bound to the Ti and C sites (on TiC surface), respectively. This bidentate binding mode is similar to that described by Schaidle et al. [20] for acetic acid HDO ("Vapor Phase" section), where O from the acetic acid was shown to only bind to surface Mo atoms. In addition, the complete dissociation of CO_2 is highly endothermic, making it difficult to cleave the C—O bond over TiC, which prevents the formation of CH_4 and CO. A small amount of metal

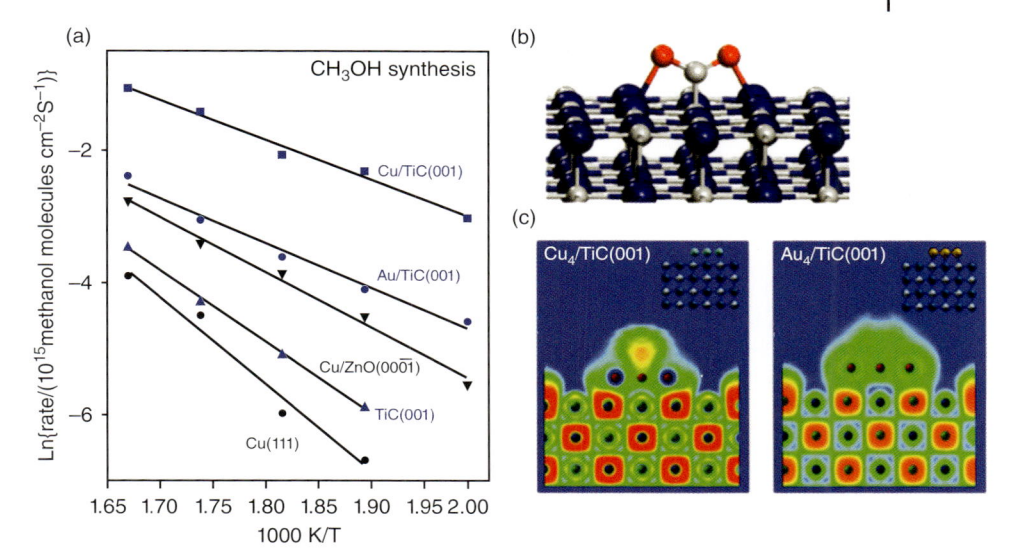

Figure 19.8 Activity measurement and surface study of TiC-based catalysts for CO_2 hydrogenation. (a) Arrhenius plot for methanol synthesis on Cu(111), a ZnO(000$\bar{1}$) surface precovered with 0.2 ML of Cu, clean TiC(001), and titanium carbide precovered with 0.1 ML of Au or Cu. (b) Calculated adsorption geometry for CO_2 on TiC(001). Color code: red (oxygen), light gray (carbon), and blue (titanium). (c) Electron-polarization plots for Cu^4 and Au^4 supported on TiC(001). The metal atoms are adsorbed on C sites of TiC(001). (Reprinted with permission from Ref. [40]. Copyright 2012, American Chemical Society.)

(Cu or Au) deposition (0.1–0.2 monolayer) on TiC(001) led to a significant enhancement in performance. For example, the amount of CH_3OH generated on Cu/TiC(001) was at least two orders of magnitude higher compared to those for TiC and Cu as shown in Figure 19.8a. The electron-polarization plots suggest that the deposited metal created significant electronic perturbations, which induced the substantial enhancement in CO_2 hydrogenation activity. These results also suggest that active sites for methanol synthesis from CO_2 hydrogenation over metal-modified carbides reside at the metal-carbide interface.

For both CO and CO_2 hydrogenation, the interconversion between CO and CO_2 is usually observed through WGSR and RWGS, which suggests that the mechanisms for CO_2 and CO hydrogenation are also interdependent. Chen et al. [98] recently investigated the reaction pathway for CO_2 hydrogenation in liquid solvents by investigating the reactivity of potential intermediates including CO and methanol. They reported that under the same reaction conditions, the use of CO as the starting material produced comparable amounts of hydrocarbons to that produced from CO_2 hydrogenation, with similar chain propagation properties. In contrast, only small amounts of CH_3OH were produced from CO hydrogenation, accounting for 3–7% of the methanol generated from CO_2 hydrogenation. This finding was consistent with the data reported by Schaidle et al. [15], where low selectivities to alcohols were observed during the FTS over

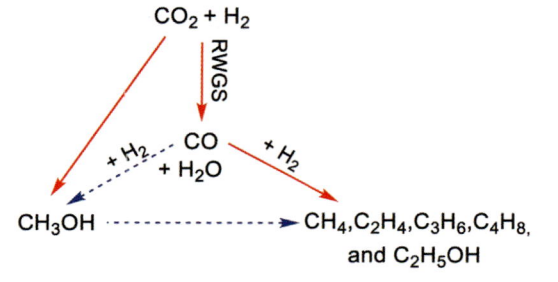

Figure 19.9 Proposed reaction pathways to produce alcohols and hydrocarbons from CO_2 and H_2. The solid arrows denote major pathways and the dashed arrows denote minor pathways. The pathways are applicable to the following experimental conditions: 200 °C, 10 bar CO_2, and 30 bar H_2 in 1,4-dioxane. (Reprinted with permission from Ref. [98]. Copyright 2016, Elsevier.)

a series of ETMCN catalysts. Accordingly, CO appeared to be a key intermediate for hydrocarbon formation during CO_2 hydrogenation, but not the primary intermediate for CH_3OH production. Surface formates and/or aldehydes are the most plausible intermediates for CH_3OH synthesis. The proposed reaction pathways for CO_2 hydrogenation over Mo_2C-based catalysts are summarized in Figure 19.9.

19.3.2.3 Knowledge Gaps

Despite recent progress using ETMCNs as catalysts for syngas and CO_2 conversion, significant unknowns remain. We suggest that future studies focus on the following key aspects/topics.

- Efforts should be made to tune CO hydrogenation product selectivities for olefins over paraffins. The bulk carbide materials typically favor the formation of paraffins, most of which are in the C_2–C_4 range due to the low chain propagation probabilities observed for ETMCN materials. Future research should focus on investigating strategies (e.g., by modifying the catalyst synthesis/pretreatment or reaction conditions) to prevent/minimize the subsequent hydrogenation of olefin intermediates to paraffins.
- Detailed mechanistic studies on ETMCN deactivation during reaction processes involving H_2, CO, CO_2, and H_2O should be performed. Although information about possible deactivation pathways exists, there have been limited systematic investigations of the long-term stability of ETMCN materials for FTS and WGSR. *In situ* spectroscopic analyses, such as FTIR and XPS, can be combined with activity measurements to understand the deactivation mechanisms.
- Identify the effect of acid/base sites on CO_2 hydrogenation rates and selectivities. Typically, basic sites are required to stabilize a CO_2 molecule, which

is weakly acidic. Future work should focus on controlling acid/base sites to elucidate this structure-function relationship.
- Investigations into the CO_2 hydrogenation performance of ETMCNs should be expanded beyond Mo_2C- and TiC-based materials.
- Perform kinetic and mechanistic investigations of alcohol and hydrocarbon formation during CO_2 hydrogenation. Future work could use FTIR spectroscopy with ^{13}C labeled CO_2 to provide useful information on reaction intermediates.

19.3.3
Petroleum and Natural Gas Refining

There are many important catalytic reactions involved in petroleum refining, which are employed to meet commercial specifications of finished products, or to improve their suitability for downstream processing. These may include hydrogenolysis to remove impurities (e.g., S, N, O), hydrogenation to add hydrogen to unsaturated products such as olefins and aromatics, and hydrocracking to break large hydrocarbons into smaller molecules. In this section we will focus on the applications of carbide and nitride catalysts towards petroleum and natural gas refining processes in the recent literature, as reported starting in 2010.

19.3.3.1 Reaction Type and Catalytic Performance

Hydrotreating and Hydrogenolysis (HDS, HDN, HDO)
The removal of heteroatoms such as sulfur, nitrogen, and oxygen is an industrially important process to upgrade petroleum-derived compounds prior to their end-use. This may be accomplished via the hydrogenolysis/hydrogenation processes of hydrodesulfurization (HDS), hydrodenitrogenation (HDN), and hydrodeoxygenation (HDO) to produce H_2S, NH_3, and H_2O, respectively. Recent studies on carbides and nitrides have focused on HDS, as it is a more significant issue than HDN and HDO for petroleum feed stocks. The majority of recent HDS studies have focused on molybdenum carbides and nitrides, though some work has also investigated multicomponent carbides/nitrides. Additionally, improving the H/C ratio by hydrogenation of aromatics and olefins is an industrially important process. *Chu et al.* studied the hydrogenation of benzene in the presence of thiophene on Ni/Mo/N catalysts at 250 °C and 3 MPa [99]. On the most active Ni/Mo/N catalyst, the presence of Ni_2Mo_3N, Mo_2C, and metallic Ni phases were observed by XRD on post-reaction catalysts. Without thiophene, benzene was hydrogenated to cyclohexane at nearly 100% selectivity. The addition of thiophene led to a decrease in activity, attributed to MoS_2, suggesting that the sulfide phase is less active for hydrogenation than other phases that were initially observed. This result suggests that at least the surface, if not the bulk, carbide/nitride structure is susceptible to sulfidation.

Villasana *et al.* studied thiophene HDS and also hydrotreated a crude oil using Ni-promoted Mo carbide and nitride catalysts supported on alumina, and found that Ni had a promoting effect for both carbide and nitride catalysts (for both hydrogenation of crude oil and sulfur removal from thiopene) [100]. Pinto-Castilla *et al.* looked at vanadium catalysts supported on carbon for HDN, HDS, and hydrogenation reactions, and observed evidence of vanadium carbide forming during synthesis [101]. These vanadium catalysts were slightly superior to a commercial NiMoS for HDS and hydrogenation, but not for HDN. Additionally, a mixture of vanadium sulfide and carbide species was observed under HDS reaction conditions. Rodriguez *et al.* studied HDS reactions on metal (Ti, V, or Mo) carbides using high-resolution photoemission spectroscopy and DFT calculations and determined that the C sites moderate the reactivity of the metal centers and provide bonding sites for adsorbates, and concluded that they should not be considered as merely spectators [102]. This finding is consistent with the adsorption geometry proposed for CO_2 on TiC (Figure 19.8b), in which the C in CO_2 binds to a surface C site [40]. Hydrogen has also been observed using IR spectroscopy to bind to surface C sites upon the adsorption of acetic acid [20]. Regarding HDS, adsorption of S on TiC(001) results in a large positive shift (1.0–1.3 eV) in the C 1s core, and correlations exist between S and thiophene adsorption energies and shifts in the carbon d-band centroid induced by metal–carbon binding in metal carbides.

Chouzier *et al.* studied HDS of thiophene on Mo_2N and Mo_2C catalysts, and their results indicate that the nitride and carbide particles themselves are not the active phase, but rather a layer of MoS_2 that is formed on the surface. The high surface areas of these particles allow for a higher surface area of MoS_2 on nitride/carbide cores, providing more activity than conventional unsupported sulfides with lower surface areas. A supported catalyst, $CoMo/Al_2O_3$, was studied for HDS experimentally and with DFT; carburization and sulfidation of Co and Mo reduced the interaction with the support and led to increased dispersion of active sites (Co, Mo) and higher HDS activity [103].

Florez *et al.* [102] suggested that the metal–carbide interface is essential for H_2 activation. They showed that the Au_4/TiC(001) operated in a synergistic way, where the active Au easily broke the H—H bond, and the TiC(001) behaved as a reservoir for dissociated H atoms to prevent H blocking of the Au sites. This spillover produced a reservoir of H on the carbide substrate that, in principle, made Au_4/TiC(001) an excellent catalyst for olefin HYD or HDS. A similar enhancement for CO_2 hydrogenation over Cu/TiC and Cu/Mo_2C was also observed due to the admetal-carbide interface, as discussed in the section "Catalytic Performance." A study by Rodriguez [104] *et al.* further examined the adsorption of Au on a TiC(001) surface. They found that the Au/TiC(001) interactions were much stronger than typical Au/oxide interactions. In addition, there was a substantial polarization of electrons around Au. This perturbation from the carbide support enhances the chemical activity of this metal [105–107] and its capability to interact with the thiophene molecule [108] and facilitate the sulfur removal process. The strong binding of metals with the native carbide surface (TiC) has also been reported on other metal/carbide systems [109]. Schweitzer et al. reported a strong metal support interaction of Pt on a native Mo_2C system, resulting in the formation of raft-like particles [23].

Under hydrotreating reaction conditions, the catalyst surfaces can be dynamic in nature and undergo phase transformations. An example system illustrating the dynamic surface has been the *in situ* carburization of HDS sulfide catalysts. Studies on MoS_2 for HDS have reported improved activity when activated by a carbon-containing sulfiding agent as opposed to H_2S. The enhanced activity is proposed to be the result of elemental carbon replacing sulfur on the edges of MoS_2. Tuxen et al. used scanning tunneling microscopy, XPS experiments, and DFT to probe the treatment of MoS_2 nanocluster HDS catalysts with organosulfur compounds used for sulfidation [110]. Their results reveal that the sulfiding agent can strongly influence the dispersion and morphology of the active sulfide phase, but the formation of a carbide phase is not favorable. Using *in situ* HRTEM to study the reactive phase of a Co/MoS_2 catalyst, Ramos et al. probed the same question as to the formation of an active carbide phase on the sulfide catalyst [111]. They reported that carbon addition to the edge sites, via addition from a C TEM grid, led to bending of MoS_2 slabs, resulting in cracks that changed morphology and increased MoS_2 exposure. This bending of MoS_2 slabs did not occur when a SiO_2 TEM grid was used to support the MoS_2. These studies illustrate the dynamic nature of the ETMCN catalysts, in that they can undergo *in situ* phase changes in the presence of C-, O-, and S-containing compounds that result in dramatic changes in catalytic activity, and the activation of these materials must be carefully controlled.

Water-Gas Shift Reaction
The water-gas shift reaction ($CO + H_2O \leftrightarrow H_2 + CO_2$) is industrially important due to its ability to generate additional H_2 during hydrocarbon steam reforming and adjust H_2/CO ratios for syntheses of alcohols and hydrocarbons. This reaction also reduces CO levels in processes where it is a contaminant (e.g., Haber–Bosch process for ammonia synthesis). The WGSR has an equilibrium constant that varies significantly over the standard temperatures of operation (with CO conversion being more highly favored at lower temperature). For this reason, the WGSR has often been conducted in two-stages, in order to take advantage of faster kinetics at higher temperatures and a favorable equilibrium for high CO conversion at lower temperatures. Recent studies of carbide catalysts for the WGSR have focused on the low temperature regime (150-250 °C) [30,112–118].

Schaidle et al. [113] reported that Mo_2C was active for the WGSR and that its activity could be improved by an order of magnitude with the addition of Pt. The Pt/Mo_2C catalyst demonstrated activity higher than a commercial $Cu/Zn/Al_2O_3$ catalyst based on TOF (active sites determined by CO uptake for carbides and N_2O uptake for the Cu-based catalyst) or normalized by surface areas. The authors also studied the stability of these carbide-based catalysts in the presence of H_2S, where both Mo_2C and Pt/Mo_2C deactivated to approximately the same activity level. Mo_2C deactivation was attributed mainly to surface sulfur adsorption to form MoS_2 as confirmed by XPS, which still maintained some activity following the treatment of H_2S. The Mo_2C active sites could be regenerated by treatment with 15% CH_4/H_2 at 590 °C. The Pt/Mo_2C catalyst contained Pt nanoparticles, which were associated with the high activity sites, and their deactivation was primarily attributed to poisoning of Pt by sulfur. This sulfur poisoning of highly active Pt sites was irreversible both when H_2S was removed from the stream and following a carburization in 15% CH_4/H_2.

The reaction mechanism and active sites for the Pt/Mo_2C WGSR catalyst have been well elucidated by Schweitzer et al. [30]. The experimental rates were most consistent with the perimeter model, indicating that active sites for the rate-determining step for WGSR were at the perimeters of the Pt particles. The proposed mechanism involves the adsorption of CO to Pt sites and then reaction at the interface between Pt and the Mo_2C, with hydrogen released during the reduction of hydroxyls on Mo_2C sites. Again, the admetal-carbide interface plays an important catalytic role, as was discussed for HDS and CO_2 hydrogenation. Furthermore, the XPS results reported by Schaidle et al. indicated that the presence of water resulted in oxidation of the Mo_2C surface, which limited the CO adsorption capacity of the Mo_2C. This oxidation, however, greatly enhanced the hydroxyl density on the catalyst surface. The deposition of Pt provided sites for CO, which led to the improved activities compared to other oxide supported Pt catalysts. Using a combination of spectroscopic analysis and theoretical calculations, they found a strong metal-carbide interaction between Pt and Mo_2C, as indicated by the X-ray absorption near edge structure (XANES) analysis showing Pt was reduced on the Mo_2C surface during the synthesis of the catalyst. A micrograph illustrating Pt supported on Mo_2C is shown in Figure 19.10. According to the extended X-ray absorption fine structure (EXAFS) spectra, these Pt particles were found to be raft-like particles. Particles with such morphology exhibited higher perimeter to volume ratios than more symmetric particles, which likely enhanced the rate-limiting step at the perimeter and contributed to the higher WGSR rate exhibited on Pt/Mo_2C. Using XANES, Schaidle et al. extended the study to a series of metals deposited onto Mo_2C catalysts and found that Cu, Pd, and Pt precursors were all reduced over the Mo_2C surface, likely accompanied by the oxidation of Mo species in Mo_2C. They also suggested that the metal-carbide interaction is strongly correlated with their standard reduction potentials [109].

Because Pt/Mo_2C had a higher WGS activity per mole of Pt than any oxide-supported Pt-catalyst, Sabnis et al. [117] sought to advance the understanding on the active sites. By preparing Pt/Mo_2C on multiwalled carbon nanotubes, they were able to perform detailed characterization and conclude that Pt–Mo alloy nanoparticles in contact with Mo_2C formed the active sites, not merely the formation of the Pt–Mo alloy. Specifically, Mo_2C activated water to form hydroxyls. These hydroxyl groups can participate in WGSR on Pt-Mo alloy nanoparticles in contact with the Mo_2C phase. The authors also studied other metals (Pt, Au, Pd, and Ni) supported on Mo_2C for their WGSR performance

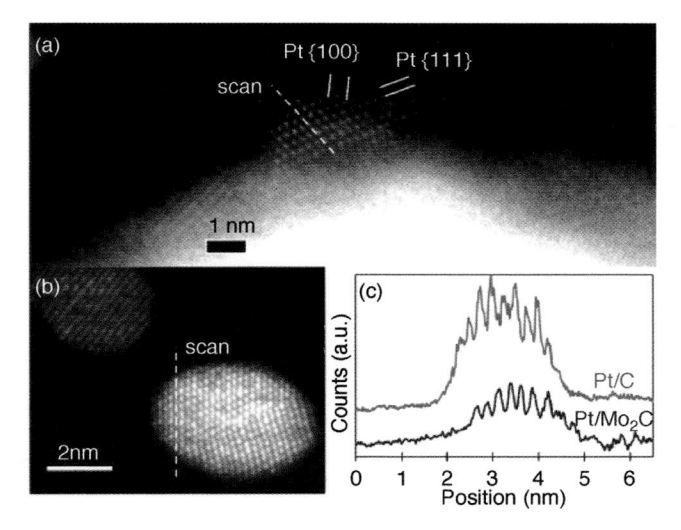

Figure 19.10 High resolution HAADF-STEM electron micrographs of (a) a Pt particle supported on Mo$_2$C and (b) a Pt particle supported on carbon. (c) Intensity line scans for the Pt particle supported on Mo$_2$C and carbon in (a) and (b), respectively. (Reprinted with permission from Ref. [30]. Copyright 2011, American Chemical Society.)

and found their activities to be 4–8 times higher than a commercial Cu/ZnO/Al$_2$O$_3$ catalyst. They concluded that the function of the metals was to enhance the surface concentration of CO, which led to enhancements in the overall rate. They determined that the kinetically relevant steps occurred either at the admetal surface or the admetal–Mo$_2$C interface. This interface is important to many of the reactions on carbide catalysts, such as WGSR, HDS, and CO$_2$ hydrogenation (as discussed previously).

Wyvratt et al. [118] studied the addition of Pt on passivated and unpassivated Mo$_2$C to understand the role of synthesis on structure, composition, and activity for WGSR on Pt/Mo$_2$C catalysts. They found that passivation of Mo$_2$C prior to Pt-addition resulted in significantly larger Pt particles (Figure 19.11), which resulted in a substantially lower WGSR activity. While the addition of Pt to unpassivated Mo$_2$C significantly improved its activity, the addition of Pt to passivated Mo$_2$C did not result in a significant difference in WGSR activity (Figure 19.12). The reason for this significant difference in activity is related to the mechanism of electrostatic adsorption on anionic Pt species to the carbide surface, followed by reduction of the Pt, as proposed by Schaidle et al. [109]. The dominant Pt species in solution are [PtCl$_6$]$^{2-}$ and [PtCl$_5$(H$_2$O)]$^-$ and result in a solution with pH ~2.2. The point of zero charge (PZC) for Mo$_2$C was 5 and the PZC for passivated Mo$_2$C was 3. This indicates a less favorable electrostatic adsorption of Pt on passivated-Mo$_2$C than unpassivated Mo$_2$C, which results in larger and/or fewer Pt particles and fewer admetal–carbide interface sites. This study further confirms that the native carbide surface is essential to achieve the

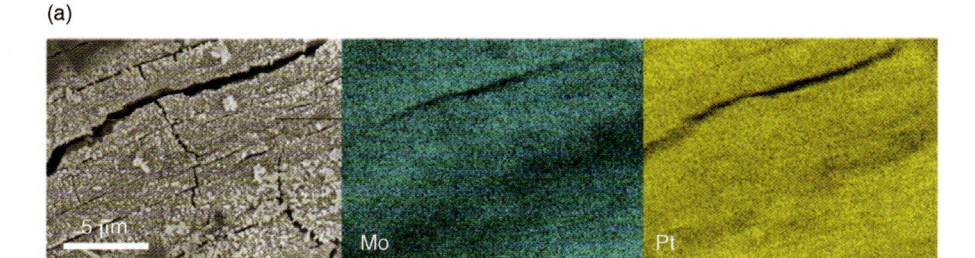

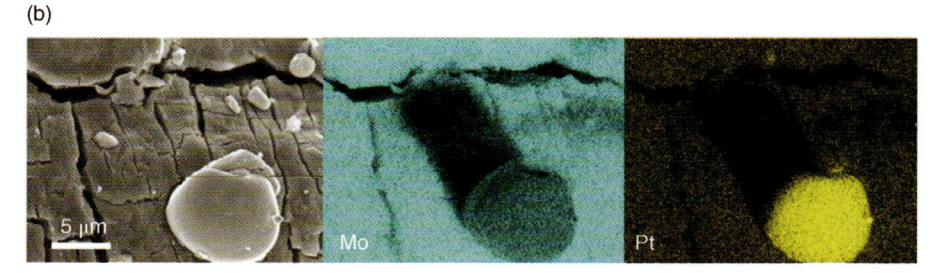

Figure 19.11 SEM micrographs and energy dispersive spectroscopy maps for Pt/Mo$_2$C catalysts prepared by Pt-addition to (a) unpassivated Mo$_2$C resulting in high dispersion and (b) passivated Mo$_2$C resulting in larger Pt particles. (Reprinted with permission from Ref. [118]. Copyright 2015, Elsevier.)

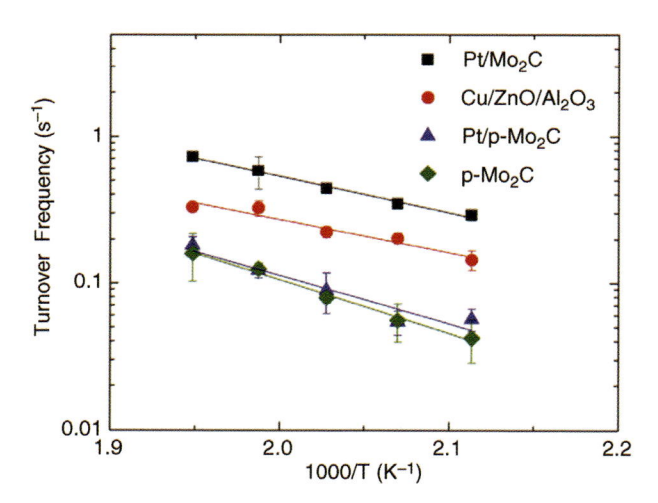

Figure 19.12 Arrhenius plots illustrating TOF (normalized by CO uptake) of Mo$_2$C-based catalysts and a commercial Cu/ZnO/Al$_2$O$_3$. (Reprinted with permission from Ref. [118]. Copyright 2015, Elsevier.)

synergistic effect between the admetal and the carbide support for WGSR and several other aforementioned reactions.

Dry Reforming
The production of hydrogen can be achieved by the reforming of hydrocarbons, particularly methane from natural gas, which can be carried out via steam or CO_2 ("dry") reforming. Dry reforming of methane ($CH_4 + CO_2 \rightarrow 2\ CO + 2\ H_2$) creates synthesis gas which can be utilized for a number of applications, including Fischer–Tropsch synthesis. In addition to generating a useful building block product (syngas), the dry reforming reaction has the benefit of utilizing CO_2 and methane, which are abundant and major contributors to greenhouse gas emissions. There have been a number of recent studies on carbide catalysts for dry reforming reactions [35,68,76,103,119–129], with most studies focusing on Ni–Mo carbides on various supports.

Zhang et al. [130] looked at the role of Mo_2C on Ni- and Mo-containing catalysts supported on La_2O_3 for dry reforming of methane (DRM) at 800 °C. They found that the presence of Mo_2C was critical in reducing the carbon deposition in the form of deactivating coke. During the DRM reaction, Ni-Mo_2C can be formed *in situ* by carburization of a freshly oxidized $NiMoO_x$ catalyst. This occurs by reduction of NiO species with methane, and the metallic Ni can then enhance the dissociation of methane and produce active carbon and hydrogen species that are capable of reducing MoO_3 to MoO_2. The replacement of oxygen atoms in MoO_2 by carbon produces β-Mo_2C. While Mo_2C was itself active for DRM, it deactivated rapidly due to oxidation by CO_2. The modification by Ni, however, resulted in stable activity by establishing a redox cycle. This scheme, proposed by Zhang et al., is shown in Figure 19.13 [131]. Similar findings were observed when Ni and Mo were added to CeZr-MgAl(O) [68] and SBA-15 [125] supports – the formation of a $NiMo_2C$ phase improved dry reforming activity and stability.

Aromatization of Hydrocarbons and Olefins
The aromatization of hydrocarbons is a promising application that allows for the upgrading of methane to produce gasoline-range molecules such as benzene, particularly via dehydroaromatization (DHA): $6CH_4 \rightarrow C_6H_6 + 9H_2$. There have recently been several studies on DHA on Mo-promoted zeolites, particularly

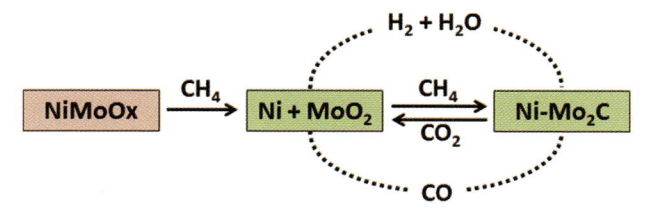

Figure 19.13 Scheme for catalytic oxidation-recarburization cycle of $NiMo_2C$ during reaction in CH_4-CO_2 feed. (Reprinted with permission from Ref. [131]. Copyright 2011, Elsevier.)

ZSM-5 [121,124,132–143], but also MCM-22 [143,144] and other unique aluminosilicates [142]. In addition to methane DHA, hexane aromatization on Mo-modified ZSM-5 has also been studied [133,141]. It has been generally established that Mo_2C is involved with methane/hydrocarbon activation. During the aromatization reactions, there is generally an induction period in which MoO_3 species are reduced by CH_4 (or other hydrocarbons) to form Mo_2C species, which is accompanied by carbonaceous deposits associated with Mo. The active Mo_2C species can then continuously activate methane to produce active intermediates, which can then undergo condensation on acid sites to produce benzene and other aromatics [122]. The size and proximity of Mo_2C sites on the catalyst with respect to Brønsted acid sites on zeolites was shown to affect the overall aromatization activity [121]. Namely, small Mo species that can be found within zeolite pores and interact with internal Brønsted acid sites are the precursors to forming the most active Mo carbide phase [121]. These studies show the importance of tuning the ratio of active carbide and acid sites for aromatization reactions, and that in addition to the ratio, the positioning of these active sites will influence the overall aromatization activity.

19.3.3.2 Knowledge Gaps

As mentioned previously, bulk and metal-modified carbide/nitride catalysts provide active sites for a series of key reactions in petroleum and natural gas refinery. However, given the complex chemistry involved in fossil fuel refining, additional research efforts are required to gain a more thorough understanding on the use of ETMCNs for these applications. The following knowledge gaps are identified to help direct future research:

- Identify the effect of admetals on the catalytic properties of ETMCN materials for hydrotreating reactions. Current literature examples only focus on several model catalytic systems, for example, Au/TiC(001) for the HDS reaction. Future research should look into other admetals, especially those that are known to have strong interactions with ETMCN materials (e.g., Pt, Cu, and Pd), and evaluate their effects on the catalytic activity and stability for HDO, HDN, and HDS reactions.
- Evaluate the catalyst activity and stability using reaction conditions similar to those used in commercial petroleum/natural gas refinery applications. These results would provide helpful insights on assessing the commercial viability of ETMCN materials.
- Perform mechanistic studies on C–H activation during methane DHA using Mo_2C-based model catalysts. Although the activity data appeared to suggest that Mo_2C provided the active sites for methane DHA, the role and identity of these sites have not been elucidated. Future work should investigate the role of Mo_2C by comparing the results for Mo-promoted zeolite with just Mo_2C. Spectroscopic analyses, such as FTIR and XPS, and computational studies should also be used for active site identification.

19.3.4
Electrocatalytic Energy Conversion, Energy Storage, and Chemicals Production

This section reviews prior investigations of ETMCNs for electrocatalytic reactions involved in energy conversion, energy storage, and chemicals production applications. The performance of bulk and supported ETMCNs, as well as ETMCN supported materials, will be discussed.

19.3.4.1 Electrocatalytic Energy Conversion
Desirable properties for an electrocatalyst include high catalytic activities and selectivities, and high electronic conductivities [145,146]. ETMCNs have been reported to have electrocatalytic activities for some reactions that are similar to those of noble metal catalysts including Pd and Pt, and electronic conductivities that approach and in some cases exceed those of metals [11,145]. Some other attractive properties of ETMCNs are their resistance to poisoning by CO, hydrocarbons, and sulfur containing compounds as well as their stabilities in acidic and alkaline electrolytes [145,147,148]. Furthermore, ETMCNs are of low cost compared to other types of catalysts [149,150].

Oxygen Reduction
The most widely used catalysts for oxygen reduction reactions (ORR) are based on Pt due to their high activities, stability at oxidative potentials, and ability to be used in acidic electrolytes [11]. EMTCNs offer similar properties to noble metals and are stable in corrosive conditions [11,145]. While these materials are active, their oxophilic nature requires modification to increase their resistance to bulk oxidation [151–153]. A variety of methods have been investigated to reduce susceptibility to bulk oxidation. Subtle changes to the bulk structure can result in increased resistance to oxidation without sacrificing electrochemical activity. Nitrogen doping has been reported to improve carbon electron donor capacities, leading to greater stabilities and higher ORR activities [154]. For example, when VC is doped with nitrogen via treatment with ammonia, the half wave potential increased from −0.12 to −0.10 V_{SCE} [155]. Nitrogen was reported to occupy interstitial sites that were vacant or previously occupied by carbon; however, the authors did not offer an explanation for the enhanced stability. Altering the morphology can also affect the activity and stability. Ko *et al.* reported that mesoporous tungsten carbide nanoplates demonstrated improved ORR activities in alkaline electrolytes compared to those of more conventional WC materials [156]. The highest activity material contained mostly WC (compared to mixtures of WC_{1-x} and WC) and a higher carbide to oxide ratio, suggesting that the active species were concentrated on the WC domains. Furthermore, the mesoporous WC had a higher number of electrons exchanged (~3.9 electrons per O_2) in O_2 saturated 0.1M NaOH than that for conventional WC catalysts (~3.25 electrons per O_2) [156]. The porous electrode structure also permits greater access of the reactants to the surface, resulting in increased activities [156].

While the structures of carbides and nitrides of the same metal can be similar, carbides typically possess higher ORR activities [157]. In addition, carbides and nitrides are more active than their parent metals. For example, TiN has twice the limiting current density of Ti, and the onset for ORR occurs at 300 mV compared to 800 mV for Ti [157]. As one might expect and as discussed in Section 19.2, facets exposed at the surface influence performance. Seifitokaldani et al. used DFT calculations to predict the ORR performance of TiN(200) and TiN (111) [158]. The TiN(200) facet was predicted to be most effective because it is active for the dissociative adsorption of oxygen and repels H_2O. Conversely, while TiN(111) is also active for the dissociative adsorption of oxygen, it has strong interactions with hydroxide, a poison that binds to the Ti sites [158]. For TiN(200), the rate determining steps for hydroxide production and water desorption were on bridge and top sites, respectively. These findings are similar to those reported earlier in the chapter regarding the strong binding of oxygen to exposed metal sites and the formation of surface hydroxyls.

ETMCNs have been supported on various types of carbons; some have been shown to enhance ORR activities due to synergistic effects between the ETMCNs and carbon [11,155,159–162]. For example, WC nanoparticles encapsulated in graphite-carbon (GC-WC) have been reported to possess enhanced ORR activities in alkaline media [162]. These materials also demonstrated enhanced stabilities. When held at 0.6 V for 10 000 s in O_2-saturated 0.1 M KOH electrolyte, the GC-WC deactivated by ~12% while the WC nanopowder and Pt/C materials experienced sharp decreases in activity by 20% and 36%, respectively [162]. Huang et al. concluded that the stability of GC-WC was due to activation of the graphite-carbon support by the WC nanoparticles and shielding of the WC from the highly oxidizing conditions by the graphite-carbon shell [162]. Nitrogen doped carbon nanotubes with encapsulated Fe_3C (Fe_3C@NCNTs) have also proved to be highly active for ORR with rates that are higher than those of the NCNTs and a commercial 20 wt% Pt/C catalyst in acidic and alkaline electrolytes (see Figure 19.14) [161]. The onset potential for ORR was 0.098 $V_{Ag/AgCl}$ in alkaline media and 0.719 $V_{Ag/AgCl}$ in acidic media, which is slightly higher than and on par with those for Pt/C [161].

The addition of precious metals (e.g., Ag [163], Au [164,165], Pd [165], and Pt [150,166–174]) and nonnoble metals (e.g., Co [154]) can improve the activities and stabilities of EMTCNs, as noted previously in this chapter. The improved performance has often been attributed to synergistic effects between the admetal and the carbide or nitride surface. The admetals can enhance the activity by limiting oxidation of the surface, while the carbides and nitrides can improve stability of the admetal in acidic and alkaline environments. For example, strong interactions between Pt and WC improved the stability of Pt compared to that for Pt/C [171–174]. DFT calculations predict nearly identical ORR activities for a catalyst with a monolayer of Pt on WC(001) surface and Pt (111) [172]. While experimental results indicated that indeed the addition of Pt improved the activity of WC (Figure 19.15a and b), even thick films did not

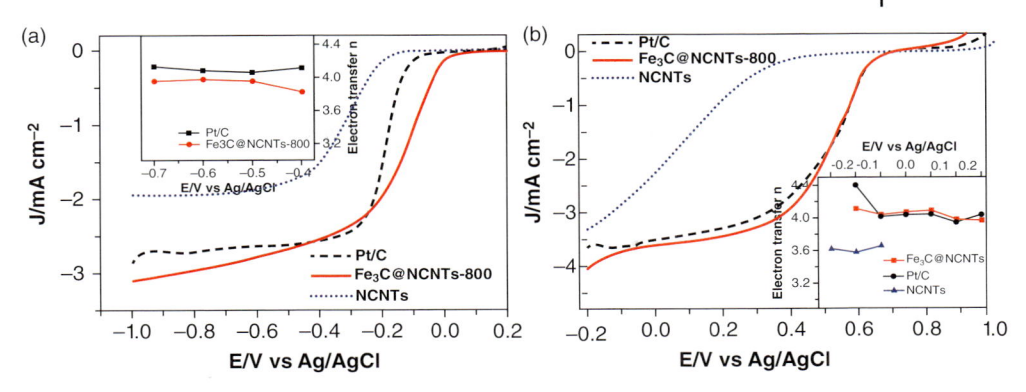

Figure 19.14 Linear sweep voltammograms and electron transfer value n for Pt/C, Fe$_3$C@NCNTs-800 in oxygen-saturated (a) 1.0 M HClO$_4$ or (b) 0.1 M KOH at 1600 rpm rotation speed and 10 mV/s scan rate. (Reprinted with permission from Ref. [161]. Copyright 2015, Elsevier.)

match the performance of bulk Pt [172]. Nevertheless, when considering the cost, catalysts consisting of Pt overlayers on WC are attractive.

Hydrogen Oxidation

Hydrogen oxidation reaction (HOR) and its reverse, the hydrogen evolution reaction (HER), are key reactions in H$_2$-O$_2$ fuel cells and H$_2$O electrolyzers [175,176]. These reactions typically occur under acidic conditions and high water contents [11]. As was the case for ORR, the structure and composition of ETMCNs affects their activities. A WC-YSZ (yttrium stabilized zirconia) composite, for example, degraded rapidly causing a drastic change to the structure

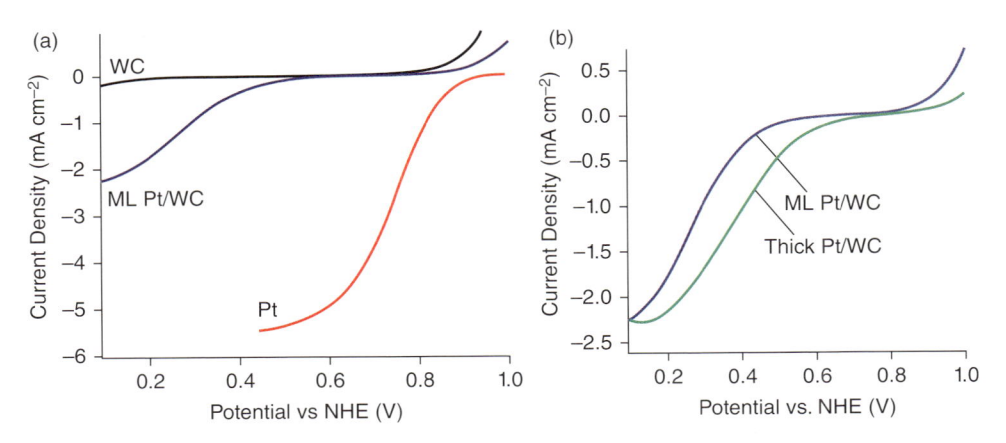

Figure 19.15 Oxygen reduction polarization curves in oxygen-saturated 0.5 M H$_2$SO$_4$ for comparison among (a) WC, ML Pt/WC, and bulk Pt, and between (b) ML Pt/WC and thick Pt/WC. (Reprinted with permission from Ref. [172]. Copyright 2012, Elsevier.)

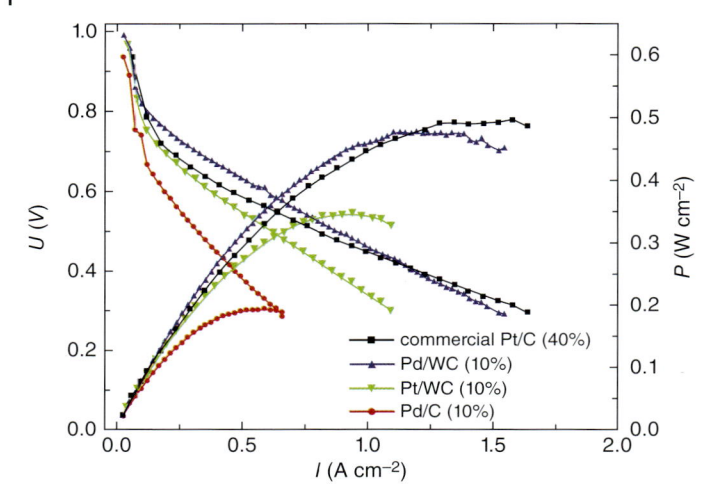

Figure 19.16 Cell voltage and power density of proton exchange membrane fuel cell with 0.2 mg/cm² of commercial Pt/C (40%), Pd/WC (10%), Pt/WC (10%), and Pd/C (10%) as anode and 0.2 mg/cm² Pt as cathode, recorded in hydrogen and oxygen at relative humidity of 100% at room temperature. (Reprinted with permission from Ref. [185]. Copyright 2013, Elsevier.)

while a WC infiltrated YSZ maintained its activity [177]. Oxidative pretreatment of multiwalled carbon nanotube (MWCNT) supports improved attachment of the tungsten carbide precursor, resulting in increased yields of WC and better overall performance for the electrocatalyst [178].

Among the carbides, WC has received the most attention [166,179–183]; however, alone it does not have sufficient HOR activity to compete with Pt [184]. Metal additives are effective in improving the activity. Tang et al. reported that WC microsphere-supported Pt electrocatalysts had higher exchange current densities than the bare microspheres [184]. Nikolic et al. reported that Pd/WC was very active for HOR and performed similar to a commercial 40% Pt/C as shown in Figure 19.16 [185]. Interestingly, the enhancement observed for the Pd/WC electrocatalyst was not obtained on a comparable Pt/WC electrocatalyst. It has been proposed that the enhancement for Pd/WC is a consequence of hydrogen spill-over on to the WC support [185]. Nevertheless, Pt on WC supported on carbon (Pt/WC/C) has demonstrated improved HOR activities compared to Pt/C [182]. The relative enhancement was a function of temperature. At 85 °C, the Pt/C and Pt/WC/C materials performed similarly; however, at 105 °C, the activity of Pt/WC/C is slightly better than that of Pt/C [182]. At either temperature with CO present, the Pt/WC/C catalyst performed better due to its tolerance of CO [182]. The CO tolerance for ETMCNs has been attributed to their activity for CO oxidation by surface hydroxyls produced via water dissociation [182]. A variety of nonnoble metals have been supported onto carbides and nitrides; however, their activities are typically low [147,186,187].

Oxidation of Organics

The electrocatalytic oxidation of liquid organics, for example in a direct alcohol fuel cell (DAFC), has garnered interest as an alternative to the oxidation of H_2 [188–190]. The best performing bulk ETMCNs possess very high surface areas and/or are promoted with metal nanoparticles. Xiao et al. developed a highly active electrocatalyst for methanol oxidation consisting of Pt supported on a porous, high surface area, molybdenum-doped titanium nitride ($Ti_{0.8}Mo_{0.2}N$) support [152]. After normalizing reaction currents to the electrochemical surface areas (ECSA), the $Pt/Ti_{0.8}Mo_{0.2}N$ catalyst was reported to have a specific activity of $1.07\,mA\,cm^{-2}$ compared to $0.35\,mA\,cm^{-2}$ and $0.6\,mA\,cm^{-2}$ for Pt/C and Pt/TiN, respectively [152]. The enhanced methanol oxidation activity was attributed to alteration of the Pt electronic structure by Mo ultimately weakening the interactions between Pt and CO. Binary carbides (combinations of two different monometallic carbides) have been reported to possess attractive electrocatalytic properties [11,153,191]. For glycerol oxidation, a tungsten-molybdenum carbide supported Pd catalyst (Pd@WC-Mo_2C) was nearly twice as active as a comparable Pd/C catalyst [153].

Supports can enhance the stabilities and activities of ETMCNs. For example, carbon supports can protect the ETMCNs from poisoning and/or oxidation [160,192–198]. Oh et al. investigated WC supported on carbon nanofibers (WC/CNFs) for alkaline direct ethanol fuel cells [199]. The WC/CNFs had superior performance compared to pure CNFs due to a significant increase in the ECSA [199]. The active ECSA for pure CNF ($73\,m^2/g$) was increased by nearly 40% on the addition of WC particles [199]. A somewhat interesting morphology has the ETMCN core encapsulated in a carbon shell. A TiN@C catalyst (TiN core and carbon shell) exhibited activity and excellent corrosion resistance [196].

Some ETMCN supports can lead to better metal dispersions than can be achieved with carbons, the most common supports used in electrocatalysts [148,153,194,200]. Xiao et al. described the performance of Pt on TiN nanotubes (Pt/TiN NT); very high Pt dispersions were achieved due to the nature of the TiN surface and the hollow, porous structure of the TiN nanotubes ($82\,m^2/g$) [194]. The onset potential for methanol oxidation for the Pt/TiN NT electrocatalyst (0.34 V) was significantly lower than that for a Pt/C electrocatalyst (0.43 V).

19.3.4.2 Energy Storage

Given their high electronic conductivities, ability to be produced in high surface area form, and stability in aqueous and nonaqueous electrolytes, ETMCNs have attracted attention for use in supercapacitors [11,145,183,201,202]. Supercapacitors are a class of electrochemical energy-storage devices that complement batteries and conventional capacitors in terms of their specific energy and specific power [203]. Particular attention has been focused on carbides and nitrides of V, Mo, and W because of their high capacitances [183].

Vanadium nitride (VN) has been the most widely studied material; there are reports of specific capacitances as high as 1340 F/g in 1.1 V aqueous KOH electrolytes [204]. While the capacitance is high, VN materials display limited rate capabilities [205]. Composites including electronically conductive materials such as VN/CNT have been reported to have better rate capabilities than pure VN. TiN generally has a higher electronic conductivity compared to VN, but its capacitance is relatively low (238 F/g in 1.0 V aqueous KOH electrolyte) [206]. Thompson and co-workers have completed systematic investigations of ETMCN electrodes in aqueous KOH and H_2SO_4 electrolytes. The highest capacitances have been reported for Mo_2N (346 F/g over 0.8 V) in H_2SO_4 electrolytes and VC, Mo_2C, and W_2N demonstrated relatively poor electrochemical properties [183]. Despite its low surface area (16 m^2/g), W_2C exhibited a high areal specific capacitance (494 $\mu F/cm^2$ over 0.7 V) in aqueous KOH electrolytes [183]. Capacitances of less than 100 F/g have been reported for NbN and WN in aqueous KOH electrolytes and for WC in aqueous H_2SO_4 electrolytes [207–209]. Most investigations of ETMCNs have utilized aqueous electrolytes; however, there have been some recent efforts to use nonaqueous electrolytes [210] that offer higher operating voltages, and consequently higher energy densities. Commercial supercapacitors employing carbon electrodes utilize acetonitrile and other nonaqueous electrolytes.

The mechanism for charge storage has been the subject of some debate, but helps inform investigations of the mechanisms for catalytic reactions. Most materials including the high surface area carbon-based materials used in commercial supercapacitors store energy via the formation of an electrical double layer at the electrode/electrolyte interface [211]. These carbon-based materials yield specific capacitances of up to 300 F/g [212]. The ETMCNs as well as metal oxides and hydroxides [213–220], and conducting polymers [221–223] can store energy via pseudocapacitive mechanisms. Pseudocapacitive storage involves fast redox reactions at or near the surface of the materials. For the ETMNCs, this mechanism involves insertion of protons (for acids) or hydroxyls (for alkaline) into interstitial sites within the lattice. For example, Thompson and coworkers determined, based on results from *in operando* X-ray absorption spectroscopy and small angle neutron scattering (SANS), that the half reaction for storage in Mo_2N is:

$$Mo_2^{\delta+}N + 2H^+ + 4e^- \leftrightarrow Mo_2^{(\delta-1)+}NH_2 \tag{19.1}$$

Note that Mo in the material is reduced on the insertion of H^+. This type of mechanistic information has facilitated the design of ETMNCs for supercapacitor and other applications.

19.3.4.3 Chemicals Production

ETMCNs have been used as thermocatalysts for a variety of industrially relevant, chemicals and fuels production reactions; however, there are also a few examples of their use as electrocatalysts for chemicals production. These examples include reduction or hydrogenation reactions and oxidation reactions.

Reduction or Hydrogenation Reactions

A reaction of particular interest is the synthesis of NH_3 [224,225]. Ammonia synthesis consumes 1–2% of the world's energy and about 3% of the natural gas produced. Horanyi and co-workers investigated tungsten carbides with HNO_3,

HNO$_2$, and NH$_2$OH as potential electrocatalysts for NH$_3$ synthesis [226]. Computational results were used to predict the stabilities, activities, and selectivities of ETMNCs for NH$_3$ synthesis [227,228]. Abghoui *et al.* reported that (100) facets of ZrN, NbN, CrN, and VN (rocksalt structures) have the highest potential as electrocatalysts because they are predicted to have stable nitrogen-vacancy sites at the surface and resist oxygen and hydrogen poisoning in aqueous electrolytes [227]. These catalysts are also predicted to suppress the competing hydrogen evolution reaction as well as require low onset potentials for NH$_3$ formation.

Nanorod WC has been reported to be active for nitromethane electroreduction [229]. With Pt foil as a counter electrode, the nitromethane was electrochemically reduced to methyl hydroxylamine at −0.89 V$_{SCE}$ with a peak current of 14.9 mA/cm^2. The electrocatalytic activity of this catalyst was slightly lower than that for the Pt foil electrode, which according to the authors was due to mass transfer effects [229].

Thompson and co-workers used solid polymer electrolyte (SPE) reactors to investigate high surface area early transition metal carbide based electrocatalysts for the electrochemical hydrogenation of polyunsaturated triglycerides [230]. Pintauro and coworkers first reported high triglyceride hydrogenation rates and better product selectivities than conventional thermocatalytic reactors when using SPE reactors with conventional electrocatalytic materials including Pt and Pd [231–233]. While ETMCNs are active, metals typically have to be added to achieve competitive rates. In some cases, there can also be synergistic effects between the ETMCN and admetal. For example, surface area normalized rates for Pd/W$_2$C electrocatalysts were much higher than those for pure Pd or W$_2$C electrocatalysts (shown in Figure 19.17) [230]. Variations in the rates with potential suggested that the rate determining step was electrochemical. In addition, selectivities for the Pd/W$_2$C catalysts were superior to those for the other catalysts. The authors suggested that the results were evidence of synergy between Pd and W$_2$C.

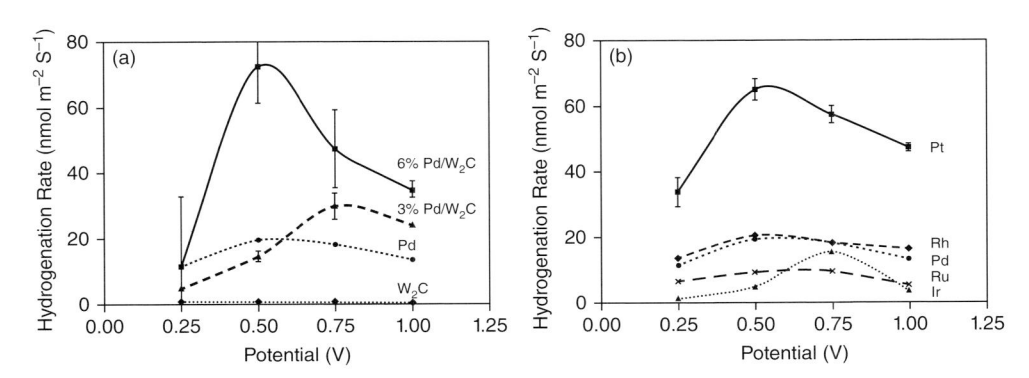

Figure 19.17 Triglyceride hydrogenation rates at various potentials for (a) W$_2$C and Pd/W$_2$C catalysts, and (b) noble metal catalysts. (Reprinted with permission from Ref. [230]. Copyright 2012, Elsevier.)

Oxidation Reactions

ETMCNs are active for the oxidation of a variety of alcohols. Li *et al.* investigated the use of Mo_2C as an anode in proton-conducting solid oxide fuel cells to dehydrogenate ethane [234]. Ethane was dehydrogenated selectively to ethylene with co-generation of electricity [234]. The ethane conversion increased from 7.7 to 42.6% when the temperature increased from 650 to 750 °C; however, the corresponding selectivity to ethylene decreased from 97.5 to 87.2%. The maximum power density also increased with temperature from 65 to 215 mW/cm^2. The power density with ethane is much lower than that with hydrogen, but there is added value with the production of ethylene [234]. Recall that carbides are active for C—H bond activation at high temperatures as demonstrated by DHA activity as discussed in the section "Aromatization of Hydrocarbons and Olefins"; the activities of these materials for electrocatalytic dehydrogenations could be due to the same types of active sites.

19.3.4.4 Knowledge Gaps

ETMCNs have been demonstrated to possess attractive electrocatalytic activities, which in some cases are similar to those for noble metals, but one of the most attractive features is their resistance to CO, hydrocarbons, and sulfur containing compounds. To achieve broader use, the activities of ETMCNs will have to be enhanced perhaps through modification of the structure (e.g., increased active surface areas or use of supports), and/or the deposition of metals onto their surfaces. Considering work cited in this section, we have identified the following knowledge gaps/opportunities with regard to the use of ETMCNs for electrocatalytic applications:

- Synthesis conditions should be optimized to preferentially expose the most active facets and increase surface area. This will not only enhance the properties of bulk ETMCNs but also benefit their use as supports.
- Refine the theoretical models of ETMCNs so that closer correspondence between experimental and computational results can be obtained. This would allow for greater confidence in the computational investigations and assist in determination of reaction mechanisms and optimal conditions.
- Expand our use of ETMCNs for electrocatalytic production of chemicals. With the prospect of low cost renewable electricity, electrocatalytic processes could be favored over thermocatalytic processes for the production of a variety of chemicals. The rich surface chemistries of ETMCNs could be exploited.

19.4
Conclusions and Focus Areas for Future Research

Due to their unique physical and chemical properties, ETMCNs are broadly applicable as catalysts for a variety of applications, ranging from biomass upgrading to petroleum refining to electrochemical energy conversion and storage. This broad applicability stems from the combination of metal and nonmetal atoms into a single structure, resulting in multiple types of sites, including metallic-like hydrogenation sites (or oxygen-vacancy type sites if the surface is oxygen-covered), acidic sites (primarily due to surface hydroxyls), and basic sites. Typically, only one of these types of sites is encountered on the surface of more common materials, for example, hydrogenation sites on noble metals (Pt, Pd), oxygen-vacancy sites on reducible metal oxides (TiO_2), acidic sites on zeolites (HZSM-5), and basic sites on metal oxides (MgO). This multifunctionality is both a benefit and a challenge of these ETMCN materials, but should be leveraged as an opportunity. As a benefit, multiple types of sites can be located near each other on a single surface, thus facilitating cascade reactions or reactions that require two separate functionalities, for example, hydrogenation and dehydration. For more common catalysts, this multifunctionality is only achieved by combining two separate catalytic materials and requires close attention to synthetic methods to produce co-located active sites. However, it is challenging to develop discrete structure-function relationships for these ETMCN materials due to the complex nature of their surfaces, and it is further complicated by their dynamic nature under reaction conditions. To date, the majority of research has focused on evaluating the catalytic performance of ETMCNs for a number of different reactions instead of on identifying and exploiting their unique properties to *design* improved catalysts.

An opportunity exists to tailor and enhance the catalytic performance of ETMCN materials through a multidisciplinary effort dedicated to the determination of structure-function relationships, similar to the types of efforts that have been ongoing within the catalysis community for acid site chemistry within zeolite pores and reactivity trends of transition metal surfaces, such that new ETMCN catalyst formulations (structure, composition) can be developed to target specific reactions. While a number of specific research opportunities were identified in Section 19.3, there are a few key future focus areas that will help researchers take advantage of this opportunity:

- Theoretical (DFT) modeling of ETMCN surfaces should be coupled with advanced catalyst characterization techniques that probe the catalyst surface and bulk structure under reaction conditions so that the modeled surfaces are representative of the working catalyst
- *In-situ/in-operando* catalyst characterization methods (e.g., site titration) need to be developed and given preference over *ex-situ* methods because the ETMCN surfaces are dynamic under reaction conditions
- Reactivity and selectivity trends for ETMCN materials should be developed, similar to those that have been developed for transition metals
- For some reactions, admetals enhance the catalytic performance of ETMCN materials; future studies should investigate this metal-ETMCN interaction and identify routes to exploit this interaction to develop improved catalysts.

Lastly, as this chapter illustrates, we have learned a considerable amount about these materials since the initial discovery of their catalytic activity over 40 years ago, but there is still much to be revealed, especially in regards to structure-function relationships. Continued research will be required to establish these relationships and assess the viability of ETMCNs for commercial catalytic applications.

References

1. Norskov, J.K., Bligaard, T., Rossmeisl, J., and Christensen, C.H. (2009) *Nat. Chem.*, **1**, 37.
2. Armor, J.N. (2011) *Catal. Today*, **163**, 3.
3. Levy, R.B. and Boudart, M. (1973) *Sci.*, **181**, 547.
4. Weigert, E.C., Stottlemyer, A.L., Zellner, M.B., and Chen, J.G. (2007) *J. Phys. Chem. C*, **111**, 14617.
5. Oyama, S.T. (1992) *Catal. Today*, **15**, 179.
6. The London Metal Exchange (2016) Molybdenum Prices, London UK. Available at http://www.lme.com/metals/minor-metals/molybdenum.
7. Matthey, J. (2014) Platinum Group Metal Price Tables, London UK. Available at http://www.platinum.matthey.com/prices/price-tables.
8. Volpe, L. and Boudart, M. (1985) *J. Solid State Chem.*, **59**, 332.
9. Volpe, L. and Boudart, M. (1985) *J. Solid State Chem.*, **59**, 348.
10. Claridge, J.B., York, A.P.E., Brungs, A.J., and Green, M.L.H. (2000) *Chem. Mater.*, **12**, 132.
11. Lausche, A.C., Schaidle, J.A., Schweitzer, N., and Thompson, L.T. (2013) *Comprehensive Inorganic Chemistry II*, 2nd edn, Elsevier, Amsterdam, p. 371.
12. Vojvodic, A., Hellman, A., Ruberto, C., and Lundqvist, B.I. (2009) *Phys. Rev. Lett.*, **103**, 146103.
13. Vojvodic, A., Ruberto, C., and Lundqvist, B.I. (2010) *J. Phys. Condens. Matter*, **22**, 375504.
14. Xu, W., Ramirez, P.J., Stacchiola, D., and Rodriguez, J.A. (2014) *Catal. Lett.*, **144**, 1418.
15. Schaidle, J.A. and Thompson, L.T. (2015) *J. Catal.*, **329**, 325.
16. Bej, S.K. and Thompson, L.T. (2004) *Appl. Catal. A*, **264**, 141.
17. Sullivan, M.M., Held, J.T., and Bhan, A. (2015) *J. Catal.*, **326**, 82.
18. Lee, W.-S., Kumar, A., Wang, Z., and Bhan, A. (2015) *ACS Catal.*, **5**, 4104.
19. Sullivan, M.M. and Bhan, A. (2016) *ACS Catal.*, **6**, 1145.
20. Schaidle, J.A., Blackburn, J., Farberow, C.A., Nash, C., Steirer, K.X., Clark, J., Robichaud, D.J., and Ruddy, D.A. (2016) *ACS Catal.*, **6**, 1181.
21. McGee, R.C.V., Bej, S.K., and Thompson, L.T. (2005) *Appl. Catal. A*, **284**, 139.
22. Bej, S.K., Bennett, C.A., and Thompson, L.T. (2003) *Appl. Catal. A*, **250**, 197.
23. Ribeiro, F.H., Dalla Betta, R.A., Boudart, M., Baumgartner, J., and Iglesia, E. (1991) *J. Catal.*, **130**, 86.
24. Aegerter, P.A., Quigley, W.W.C., Simpson, G.J., Ziegler, D.D., Logan, J.W., McCrea, K.R., Glazier, S., and Bussell, M.E. (1996) *J. Catal.*, **164**, 109.
25. Neylon, M.K., Bej, S.K., Bennett, C.A., and Thompson, L.T. (2002) *Appl. Catal. A*, **232**, 13.
26. Claridge, J.B., York, A.P.E., Brungs, A.J., Marquez-Alvarez, C., Sloan, J., Tsang, S.C., and Green, M.L.H. (1998) *J. Catal.*, **180**, 85.
27. Lee, J.S., Locatelli, S., Oyama, S.T., and Boudart, M. (1990) *J. Catal.*, **125**, 157.
28. Ranhotra, G.S., Bell, A.T., and Reimer, J.A. (1987) *J. Catal.*, **108**, 40.
29. Patt, J., Moon, D.J., Phillips, C., and Thompson, L. (2000) *Catal. Lett.*, **65**, 193.
30. Schweitzer, N.M., Schaidle, J.A., Ezekoye, O.K., Pan, X., Linic, S., and Thompson, L.T. (2011) *J. Am. Chem. Soc.*, **133**, 2378.
31. Mitchell, S., Michels, N.-L., and Perez-Ramirez, J. (2013) *Chem. Soc. Rev.*, **42**, 6094.
32. Alexander, A.-M. and Hargreaves, J.S.J. (2010) *Chem. Soc. Rev.*, **39**, 4388.
33. Leclercq, G., Kamal, M., Giraudon, J.M., Devassine, P., Feigenbaum, L., Leclercq, L., Frennet, A., Bastin, J.M., Löfberg, A., Decker, S., and Dufour, M. (1996) *J. Catal.*, **158** 142.

34 Stellwagen, D.R. and Bitter, J.H. (2015) *Green Chem.*, **17**, 582.
35 Medford, A.J., Vojvodic, A., Studt, F., Abild-Pedersen, F., and Nørskov, J.K. (2012) *J. Catal.*, **290**, 108.
36 Hägg, G. (1931) *Z. Phys. Chem. B*, **12**, 33.
37 Wang, T., Liu, X., Wang, S., Huo, C., Li, Y.-W., Wang, J., and Jiao, H. (2011) *J. Phys. Chem. C*, **115**, 22360.
38 Politi, J.R.d.S., Vines, F., Rodriguez, J.A., and Illas, F. (2013) *Phys. Chem. Chem. Phys.*, **15**, 12617.
39 King, T.E. (2007) University of Michigan.
40 Vidal, A.B., Feria, L., Evans, J., Takahashi, Y., Liu, P., Nakamura, K., Illas, F., and Rodriguez, J.A. (2012) *J. Phys. Chem. Lett.*, **3**, 2275.
41 Choi, J.-S., Bugli, G., and Djéga-Mariadassou, G. (2000) *J. Catal.*, **193**, 238.
42 Ruddy, D.A., Schaidle, J.A., Ferrell, J.R., Wang, J., Moens, L., and Hensley, J.E. (2014) *Green Chem.*, **16**, 454.
43 Shanks, B.H. (2010) *Ind. Eng. Chem. Res.*, **49**, 10212.
44 Branca, C., Giudicianni, P., and Di Blasi, C. (2003) *Ind. Eng. Chem. Res.*, **42**, 3190.
45 Milne, T., Agblevor, F., Davis, M., Deutch, S., and Johnson, D. (1997) *Developments in Thermochemical Biomass Conversion*, Springer, Netherlands, p. 409.
46 Maggi, R. and Delmon, B. (1994) *Biomass Bioenergy*, **7**, 245.
47 Bridgwater, A.V. (2012) *Environ. Prog. Sust. Energy*, **31**, 261.
48 Xiong, K., Yu, W., Vlachos, D.G., and Chen, J.G. (2015) *ChemCatChem*, **7**, 1402.
49 Stottlemyer, A.L., Liu, P., and Chen, J.G. (2010) *J. Chem. Phys.*, **133**, 104702.
50 Liu, N., Kourtakis, K., Figueroa, J.C., and Chen, J.G. (2003) *J. Catal.*, **215**, 254.
51 Kelly, T.G., Stottlemyer, A.L., Ren, H., and Chen, J.G. (2011) *J. Phys. Chem. C*, **115**, 6644.
52 Ren, H., Hansgen, D.A., Stottlemyer, A.L., Kelly, T.G., and Chen, J.G. (2011) *ACS Catal.*, **1**, 390.
53 Yu, W., Mellinger, Z.J., Barteau, M.A., and Chen, J.G. (2012) *J. Phys. Chem. C*, **116**, 5720.
54 Yu, W., Barteau, M.A., and Chen, J.G. (2011) *J. Am. Chem. Soc.*, **133**, 20528.
55 Ma, Y., Guan, G., Shi, C., Zhu, A., Hao, X., Wang, Z., Kusakabe, K., and Abudula, A. (2014) *Int. J. Hydrogen. Energy*, **39**, 258.
56 Koós, Á., Barthos, R., and Solymosi, F. (2008) *J. Phys. Chem. C*, **112**, 2607.
57 Széchenyl, A. and Solymosi, F. (2007) *J. Phys. Chem. C*, **111**, 9509.
58 Setthapun, W., Bej, S.K., and Thompson, L.T. (2008) *Top. Catal.*, **49**, 73.
59 Cao, J., Ma, Y., Guan, G., Hao, X., Ma, X., Wang, Z., Kusakabe, K., and Abudula, A. (2016) *Appl. Catal. B Environ.*, **189**, 12.
60 Ma, Y., Guan, G., Hao, X., Zuo, Z., Huang, W., Phanthong, P., Kusakabe, K., and Abudula, A. (2014) *RSC Adv.*, **4**, 44175.
61 Ma, Y., Guan, G., Phanthong, P., Hao, X., Huang, W., Tsutsumi, A., Kusakabe, K., and Abudula, A. (2014) *J. Phys. Chem. C*, **118**, 9485.
62 Wang, H., Male, J., and Wang, Y. (2013) *ACS Catal.*, **3**, 1047.
63 Zhang, W., Zhang, Y., Zhao, L., and Wei, W. (2010) *Energy Fuels*, **24**, 2052.
64 Sullivan, M.M., Chen, C.-J., and Bhan, A. (2016) *Catal. Sci. Technol.*, **6**, 602.
65 Han, J., Duan, J., Chen, P., Lou, H., Zheng, X., and Hong, H. (2011) *Green Chem.*, **13**, 2561.
66 Ren, H., Yu, W., Salciccioli, M., Chen, Y., Huang, Y., Xiong, K., Vlachos, D.G., and Chen, J.G. (2013) *ChemSusChem*, **6**, 798.
67 Xiong, K., Lee, W.S., Bhan, A., and Chen, J.G. (2014) *ChemSusChem*, **7**, 2146.
68 Chen, Y., Choi, S., and Thompson, L.T. (2015) *ACS Catal.*, **5**, 1717.
69 Lee, W.S., Wang, Z., Zheng, W., Vlachos, D.G., and Bhan, A. (2014) *Catal. Sci. Technol.*, **4**, 2340.
70 Prasomsri, T., Shetty, M., Murugappan, K., and Román-Leshkov, Y. (2014) *Energy Environ. Sci.*, **7**, 2660.
71 Ren, H., Chen, Y., Huang, Y.L., Deng, W.H., Vlachos, D.G., and Chen, J.G.G. (2014) *Green Chem.*, **16**, 761.
72 Lee, J.S., Oyama, S.T., and Boudart, M. (1987) *J. Catal.*, **106**, 125.
73 Choi, J.S., Krafft, J.M., Krzton, A., and Djéga-Mariadassou, G. (2002) *Catal. Lett.*, **81**, 175.

74 Iglesia, E., Baumgartner, J.E., Ribeiro, F.H., and Boudart, M. (1991) *J. Catal.*, **131**, 523.
75 Keller, V., Wehrer, P., Garin, F., Ducros, R., and Maire, G. (1995) *J. Catal.*, **153**, 9.
76 Lee, W.S., Wang, Z.S., Wu, R.J., and Bhan, A. (2014) *J. Catal.*, **319**, 44.
77 Lu, M., Lu, F., Zhu, J., Li, M., Zhu, J., and Shan, Y. (2015) *React. Kinet. Mech. Catal.*, **115**, 251.
78 Mai, E.F., Machado, M.A., Davies, T.E., Lopez-Sanchez, J.A., and Teixeira Da Silva, V. (2014) *Green Chem.*, **16**, 4092.
79 Qin, Y., He, L., Duan, J., Chen, P., Lou, H., Zheng, X., and Hong, H. (2014) *ChemCatChem*, **6**, 2698.
80 Monnier, J., Sulimma, H., Dalai, A., and Caravaggio, G. (2010) *Appl Catal A Gen*, **382**, 176.
81 Sepúlveda, C., Leiva, K., García, R., Radovic, L.R., Ghampson, I.T., DeSisto, W.J., Fierro, J.L.G., and Escalona, N. (2011) *Catal. Today*, **172**, 232.
82 Ghampson, I.T., Sepúlveda, C., Garcia, R., Radovic, L.R., Fierro, J.L.G., DeSisto, W.J., and Escalona, N. (2012) *Appl. Catal. A*, **439–440**, 111.
83 Mortensen, P.M., De Carvalho, H.W.P., Grunwaldt, J.D., Jensen, P.A., and Jensen, A.D. (2015) *J. Catal.*, **328**, 208.
84 Jongerius, A.L., Gosselink, R.W., Dijkstra, J., Bitter, J.H., Bruijnincx, P.C.A., and Weckhuysen, B.M. (2013) *ChemCatChem*, **5**, 2964.
85 Ghampson, I.T., Sepúlveda, C., Garcia, R., Frederick, B.G., Wheeler, M.C., Escalona, N., and Desisto, W.J. (2012) *Appl. Catal. A*, **413–414**, 78.
86 Tyrone Ghampson, I., Sepúlveda, C., Garcia, R., García Fierro, J.L., Escalona, N., and Desisto, W.J. (2012) *Appl. Catal. A*, **435–436**, 51.
87 Griffin, M.B., Baddour, F.G., Habas, S.E., Ruddy, D.A., and Schaidle, J.A. (2015) *Top. Catal.*, **59**, 124.
88 Griffin, M.B., Ferguson, G.A., Ruddy, D.A., Biddy, M.J., Beckham, G.T., and Schaidle, J.A. (2016) *ACS Catal.*, **6**, 2715.
89 Kim, H.-G., Lee, K.H., and Lee, J.S. (2000) *Res. Chem. Intermed.*, **26**, 427.
90 Kojima, I., Miyazaki, E., and Yasumori, I. (1980) *J. Chem. Soc. Chem. Commun.*, 573.
91 Shou, H., Ferrari, D., Barton, D.G., Jones, C.W., and Davis, R.J. (2012) *ACS Catal.*, **2**, 1408.
92 Porosoff, M.D., Yang, X., Boscoboinik, J.A., and Chen, J.G. (2014) *Angew. Chem.*, **126**, 6823.
93 Eliason, S.A. and Bartholomew, C.H. (1999) *Appl. Catal. A*, **186**, 229.
94 Shou, H. and Davis, R.J. (2013) *J. Catal.*, **306**, 91.
95 Neylon, M.K., Choi, S., Kwon, H., Curry, K.E., and Thompson, L.T. (1999) *Appl. Catal. A*, **183**, 253.
96 Wang, N., Fang, K., Jiang, D., Li, D., and Sun, Y. (2010) *Catal. Today*, **158**, 241.
97 Rodriguez, J.A., Evans, J., Feria, L., Vidal, A.B., Liu, P., Nakamura, K., and Illas, F. (2013) *J. Catal.*, **307**, 162.
98 Chen, Y., Choi, S., and Thompson, L.T. (2016) *J. Catal.*, **343**, 147.
99 Chu, Q., Feng, J., Li, W., and Xie, K. (2013) *Chin. J. Catal.*, **34**, 159.
100 Villasana, Y., Escalante, Y., Rodríguez Nuñez, J.E., Méndez, F.J., Ramírez, S., Luis-Luis, M.Á., Cañizales, E., Ancheyta, J., and Brito, J.L. (2014) *Catal. Today*, **220–222**, 318.
101 Pinto-Castilla, S., Marrero, S., Diaz, Y., Brito, J.L., Silva, P., and Betancourt, P. (2012) *React. Kinet. Mech. Catal.*, **107**, 321.
102 Florez, E., Gomez, T., Liu, P., Rodriguez, J.A., and Illas, F. (2010) *ChemCatChem*, **2**, 1219.
103 Ge, H., Wen, X.-D., Ramos, M.A., Chianelli, R.R., Wang, S., Wane, J., Qin, Z., Lyu, Z., and Li, X. (2014) *ACS Catal.*, **4**, 2556.
104 Rodriguez, J.A., Viñes, F., Illas, F., Liu, P., Takahashi, Y., and Nakamura, K. (2007) *J. Chem. Phys.*, **127**, 211102.
105 Rodriguez, J.A., Liu, P., Takahashi, Y., Nakamura, K., Viñes, F., and Illas, F. (2009) *J. Am. Chem. Soc.*, **131**, 8595.
106 Rodriguez, J.A., Liu, P., Viñes, F., Illas, F., Takahashi, Y., and Nakamura, K. (2008) *Angew. Chem., Int. Ed.*, **47**, 6685.
107 Ono, L.K. and Roldán-Cuenya, B. (2007) *Catal. Lett.*, **113**, 86.
108 Liu, P., Lightstone, J.M., Patterson, M.J., Rodriguez, J.A., Muckerman, J.T., and White, M.G. (2006) *J. Phys. Chem. B*, **110**, 7449.

109 Schaidle, J.A., Schweitzer, N.M., Ajenifujah, O.T., and Thompson, L.T. (2012) *J. Catal.*, **289**, 210.
110 Tuxen, A., Gobel, H., Hinnemann, B., Li, Z., Knudsen, K.G., Topsoe, H., Lauritsen, J.V., and Besenbacher, F. (2011) *J. Catal.*, **281**, 345.
111 Ramos, M., Ferrer, D., Martinez-Soto, E., Lopez-Lippmann, H., Torres, B., Berhault, G., and Chianelli, R.R. (2013) *Ultramicroscopy*, **127**, 64.
112 Nagai, M., Zahidul, A.M., Kunisaki, Y., and Aoki, Y. (2010) *Appl. Catal. A Gen.*, **383**, 58.
113 Schaidle, J.A., Lausche, A.C., and Thompson, L.T. (2010) *J. Catal.*, **272**, 235.
114 Namiki, T., Yamashita, S., Tominaga, H., and Nagai, M. (2011) *Appl. Catal. A Gen.*, **398**, 155.
115 Stottlemyer, A.L., Kelly, T.G., Meng, Q., and Chen, J.G. (2012) *Surf. Sci. Rep.*, **67**, 201.
116 Halevi, B., Lin, S., Roy, A., Zhang, H., Jeroro, E., Vohs, J., Wang, Y., Guo, H., and Datye, A.K. (2013) *J. Phys. Chem. C*, **117**, 6493.
117 Sabnis, K.D., Akatay, M.C., Cui, Y., Sollberger, F.G., Stach, E.A., Miller, J.T., Delgass, W.N., and Ribeiro, F.H. (2015) *J. Catal.*, **330**, 442.
118 Wyvratt, B.M., Gaudet, J.R., and Thompson, L.T. (2015) *J. Catal.*, **330**, 280.
119 Shi, X.-R., Wang, S.-G., and Wang, J. (2016) *J. Mol. Catal. A Chem.*, **417**, 53.
120 Myint, M., Yan, B., Wan, J., Zhao, S., and Chen, J.G. (2016) *J. Catal*, **343**, 168.
121 Song, Y., Xu, Y., Suzuki, Y., Nakagome, H., Ma, X., and Zhang, Z.-G. (2015) *J. Catal.*, **330**, 261.
122 Khavarian, M., Chai, S.-P., and Mohamed, A.R. (2014) *Chem. Eng. J.*, **257**, 200.
123 Chiang, S.-W., Chang, C.-C., Shie, J.-L., Chang, C.-Y., Ji, D.-R., and Tseng, J.-Y. (2012) *J. Taiwan Inst. Chem. Eng.*, **43**, 918.
124 Cui, Y., Xu, Y., Lu, J., Suzuki, Y., and Zhang, Z.-G. (2011) *Appl. Catal., A*, **393**, 348.
125 Huang, T., Huang, W., Huang, J., and Ji, P. (2011) *Fuel Process. Technol.*, **92**, 1868.
126 Horváth, A., Stefler, G., Geszti, O., Kienneman, A., Pietraszek, A., and Guczi, L. (2011) *Catal. Today*, **169**, 102.
127 Hirose, T., Ozawa, Y., and Nagai, M. (2011) *Chin. J. Catal.*, **32**, 771.
128 Cheng, J. and Huang, W. (2010) *Fuel Process. Technol.*, **91**, 185.
129 Arkatova, L.A. (2010) *Catal. Today*, **157**, 170.
130 Zhang, S., Shi, C., Chen, B., Zhang, Y., Zhu, Y., Qiu, J., and Au, C. (2015) *Catal. Today*, **258**, 676.
131 Zhang, A., Zhu, A., Chen, B., Zhang, S., Au, C., and Shi, C. (2011) *Catal. Commun.*, **12**, 803.
132 Velebná, K., Horňáček, M., Jorík, V., Hudec, P., Čaplovičová, M., and Čaplovič, L.u. (2015) *Microporous Mesoporous Mater.*, **212**, 146.
133 Tshabalala, T.E. and Scurrell, M.S. (2015) *Catal. Commun.*, **72**, 49.
134 Tempelman, C.H.L. and Hensen, E.J.M. (2015) *Appl. Catal. B Environ.*, **176–177**, 731.
135 Karakaya, C., Zhu, H., and Kee, R.J. (2015) *Chem. Eng. Sci.*, **123**, 474.
136 Abdelsayed, V., Shekhawat, D., and Smith, M.W. (2015) *Fuel*, **139**, 401.
137 Majhi, S. and Pant, K.K. (2014) *J. Ind. Eng. Chem.*, **20**, 2364.
138 Aoki, Y., Tominaga, H., and Nagai, M. (2013) *Catal. Today*, **215**, 169.
139 Wong, K.S., Thybaut, J.W., Tangstad, E., Stöcker, M.W., and Marin, G.B. (2012) *Microporous Mesoporous Mater.*, **164**, 302.
140 Aboul-Gheit, A.K., El-Masry, M.S., and Awadallah, A.E. (2012) *Fuel Process. Technol.*, **102**, 24.
141 Maldonado-Hódar, F.J. (2011) *Appl. Catal., A*, **408**, 156.
142 Liu, H., Wu, S., Guo, Y., Shang, F., Yu, X., Ma, Y., Xu, C., Guan, J., and Kan, Q. (2011) *Fuel*, **90**, 1515.
143 Smiešková, A., Hudec, P., Kumar, N., Salmi, T., Murzin, D.Y., and Jorík, V. (2010) *Appl. Catal. A*, **377**, 83.
144 Yin, X., Chu, N., Yang, J., Wang, J., and Li, Z. (2014) *Catal. Commun.*, **43**, 218.
145 Lavacchi, A., Miller, H., and Vizza, F. (2013) *Nanotechnology in Electrocatalysis for Energy*, Springer.

146 Jäger, R., Kasatkin, P.E., Härk, E., and Lust, E. (2013) *Electrochem. Commun.*, **35**, 97.
147 Izhar, S., Yoshida, M., and Nagai, M. (2009) *Electrochim. Acta*, **54**, 1255.
148 Hu, F.P. and Shen, P.K. (2007) *J. Power Sources*, **173**, 877.
149 Toth, L. (1971) *Mechanical Properties, in Refractory Materials*, Academic Press Inc., **7**, 141–182.
150 Yang, X.G. and Wang, C.Y. (2005) *Appl. Phys. Lett.*, **86**, 1.
151 Shen, P.K., Yin, S., Li, Z., and Chen, C. (2010) *Electrochim. Acta*, **55**, 7969.
152 Xiao, Y., Fu, Z., Zhan, G., Pan, Z., Xiao, C., Wu, S., Chen, C., Hu, G., and Wei, Z. (2015) *J. Power Sources*, **273** 33.
153 Zhang, X. and Shen, P.K. (2013) *Int. J. Hydrogen. Energy*, **38**, 2257.
154 Bukola, S., Merzougui, B., Akinpelu, A., and Zeama, M. (2016) *Electrochim. Acta*, **190**, 1113.
155 Yu, J., Gao, X., Chen, G., and Yuan, X. (2016) *Int. J. Hydrogen. Energy*, **41**, 4150.
156 Ko, A.-R., Lee, Y.-W., Moon, J.-S., Han, S.-B., Cao, G., and Park, K.-W. (2014) *Appl. Catal. A*, **477**, 102.
157 Giner, J. and Swette, L. (1966) *Nature*, **211**, 1291.
158 Seifitokaldani, A., Savadogo, O., and Perrier, M. (2014) *Electrochim. Acta*, **141**, 25.
159 Liu, Y.L., Xu, X.Y., Sun, P.C., and Chen, T.H. (2014) *Int. J. Hydrogen. Energy*, **40**, 4531.
160 Yan, Z., Zhang, M., Xie, J., and Shen, P.K. (2015) *J. Power Sources*, **295**, 156.
161 Zhong, G., Wang, H., Yu, H., and Peng, F. (2015) *J. Power Sources*, **286**, 495.
162 Huang, K., Bi, K., Xu, J.C., Liang, C., Lin, S., Wang, W.J., Yang, T.Z., Du, Y.X., Zhang, R., Yang, H.J., Fan, D.Y., Wang, Y.G., and Lei, M. (2015) *Electrochim. Acta*, **174**, 172.
163 Meng, H. and Shen, P.K. (2006) *Electrochem. Commun.*, **8**, 588.
164 Elumalai, G., Noguchi, H., Lyalin, A., Taketsugu, T., and Uosaki, K. (2016) *Electrochem. Commun.*, **66**, 53.
165 Nie, M., Shen, P.K., and Wei, Z. (2007) *J. Power Sources*, **167**, 69.
166 Wang, Y., Song, S., Maragou, V., Shen, P.K., and Tsiakaras, P. (2009) *Appl. Catal. B Environ.*, **89**, 223.
167 Nie, M., Shen, P.K., Wu, M., Wei, Z., and Meng, H. (2006) *J. Power Sources*, **162**, 173.
168 Meng, H. and Shen, P.K. (2005) *Chem. Commun.*, 4408.
169 Elbaz, L., Kreller, C.R., Henson, N.J., and Brosha, E.L. (2014) *J. Electroanal. Chem.*, **720–721**, 34.
170 Jing, S., Luo, L., Yin, S., Huang, F., Jia, Y., Wei, Y., Sun, Z., and Zhao, Y. (2014) *Appl. Catal. B Environ.*, **147**, 897.
171 Schlange, A., Dos Santos, A.R., Hasse, B., Etzold, B.J.M., Kunz, U., and Turek, T. (2012) *J. Power Sources*, **199**, 22.
172 Hsu, I.J., Kimmel, Y.C., Dai, Y., Chen, S., and Chen, J.G. (2012) *J. Power Sources*, **199**, 46.
173 Chhina, H., Campbell, S., and Kesler, O. (2007) *J. Power Sources*, **164**, 431.
174 Chhina, H., Campbell, S., and Kesler, O. (2008) *J. Power Sources*, **179**, 50.
175 Zalitis, C.M., Sharman, J., Wright, E., and Kucernak, A.R. (2015) *Electrochim. Acta*, **176**, 763.
176 Strmcnik, D., Uchimura, M., Wang, C., Subbaraman, R., Danilovic, N., Van Der Vliet, D., Paulikas, A.P., Stamenkovic, V.R., and Markovic, N.M. (2013) *Nat. Chem.*, **5**, 300.
177 Torabi, A. and Etsell, T.H. (2012) *J. Power Sources*, **212**, 47.
178 Rahsepar, M., Pakshir, M., Nikolaev, P., Piao, Y., and Kim, H. (2014) *Int. J. Hydrogen. Energy*, **39**, 15706.
179 Meyer, S., Nikiforov, A.V., Petrushina, I.M., Köhler, K., Christensen, E., Jensen, J.O., and Bjerrum, N.J. (2015) *Int. J. Hydrogen. Energy*, **40**, 2905.
180 McIntyre, D.R., Burstein, G.T., and Vossen, A. (2002) *J. Power Sources*, **107**, 67.
181 Antolini, E. and Gonzalez, E.R. (2010) *Appl. Catal. B Environ.*, **96**, 245.
182 Hassan, A., Paganin, V.A., and Ticianelli, E.A. (2015) *Appl. Catal. B Environ.*, **165**, 611.
183 Pande, P., Rasmussen, P.G., and Thompson, L.T. (2012) *J. Power Sources*, **207**, 212.

184 Tang, C., Wang, D., Wu, Z., and Duan, B. (2015) *Int. J. Hydrogen. Energy*, **40**, 3229.
185 Nikolic, V.M., Zugic, D.L., Perovic, I.M., Saponjic, A.B., Babic, B.M., Pasti, I.A., and Marceta Kaninski, M.P. (2013) *Int. J. Hydrogen. Energy*, **38**, 11340.
186 Tackett, B.M., Kimmel, Y.C., and Chen, J.G. (2016) *Int. J. Hydrogen. Energy*, **41**, 5948.
187 Xing, Z., Li, Q., Wang, D., Yang, X., and Sun, X. (2016) *Electrochim. Acta*, **191**, 841.
188 Litster, S. and McLean, G. (2004) *J. Power Sources*, **130**, 61.
189 Yu, E.H., Wang, X., Liu, X.T., and Li, L. (2012) Challenges and Perspectives of Nanocatalysts in Alcohol-Fuelled Direct Oxidation Fuel Cells, p. 227.
190 EG&G Technical Services I. (2004) *Fuel Cell*, **7**, 1.
191 Zhang, X., Tian, Z., and Shen, P.K. (2013) *Electrochem. Commun.*, **28**, 9.
192 Hu, F., Cui, G., Wei, Z., and Shen, P.K. (2008) *Electrochem. Commun.*, **10**, 1303.
193 Zhao, Z., Fang, X., Li, Y., Wang, Y., Shen, P.K., Xie, F., and Zhang, X. (2009) *Electrochem. Commun.*, **11**, 290.
194 Xiao, Y., Zhan, G., Fu, Z., Pan, Z., Xiao, C., Wu, S., Chen, C., Hu, G., and Wei, Z. (2014) *Electrochim. Acta*, **141**, 279.
195 Zheng, H., Chen, Z., Li, Y., and Ma, C.A. (2013) *Electrochim. Acta*, **108**, 486.
196 Lee, J.-M., Kim, S.-B., Lee, Y.-W., Kim, D.-Y., Han, S.-B., Roh, B., Hwang, I., and Park, K.-W. (2012) *Appl. Catal. B Environ.*, **111–112**, 200.
197 Ma, C.A., Xu, C., Shi, M., Song, G., and Lang, X. (2013) *J. Power Sources*, **242**, 273.
198 Lu, J.L., Li, Z.H., Jiang, S.P., Shen, P.K., and Li, L. (2012) *J. Power Sources*, **202**, 56.
199 Oh, Y., Kim, S.K., Peck, D.H., Jang, J.S., Kim, J., and Jung, D.H. (2014) *Int. J. Hydrogen. Energy*, **39**, 15907.
200 Dai, H., Chen, Y., Lin, Y., Xu, G., Yang, C., Tong, Y., Guo, L., and Chen, G. (2012) *Electrochim. Acta*, **85**, 644.
201 Hanumantha, P.J., Datta, M.K., Kadakia, K., Okoli, C., Patel, P., and Kumta, P.N. *Electrochim. Acta*, **207**, 37.
202 Xu, Y., Wang, J., Shen, L., Dou, H., and Zhang, X. (2015) *Electrochim. Acta*, **173**, 680.
203 Miller, J.R. and Burke, A.F. (2008) *Electrochem. Soc. Interface*, **17**, 53.
204 Choi, D., Blomgren, G.E., and Kumta, P.N. (2006) *Adv. Mater.*, **18**, 1178.
205 Zhong, Y., Xia, X., Shi, F., Zhan, J., Tu, J., and Fan, H.J. (2016) *Adv. Sci*, **3**, 1500286.
206 Choi, D. and Kumta, P.N. (2006) *J. Electrochem. Soc.*, **153**, A2298.
207 Choi, D. and Kumta, P.N. (2011) *J. Am. Ceram. Soc.*, **94**, 2371.
208 Choi, D. and Kumta, P.N. (2007) *J. Am. Ceram. Soc.*, **90**, 3113.
209 Wixom, M., Tarnowski, D., Parker, J., Lee, J., Chen, P.-L., Song, I., and Thompson, L. (1997) *MRS Proceedings*, Cambridge University Press, vol. **496**, p. 643.
210 Djire, A., Ishmwe, J.Y., Choi, S., and Thompson, L.T. (2016) *ChemElectroChem*, Under Review.
211 Conway, B. E. (1999) *Electrochemical Supercapacitors: Scientific Fundamentals and Technological Applications*, Springer Science+Business Media, pp. 1–698.
212 Simon, P. and Gogotsi, Y. (2008) *Nat. Mater.*, **7**, 845.
213 Zheng, J.P. (1999) *Electrochem. Solid State Lett.*, **2**, 359.
214 Zheng, J., Cygan, P., and Jow, T. (1995) *J. Electrochem. Soc.*, **142**, 2699.
215 Hu, C.-C., Chang, K.-H., Lin, M.-C., and Wu, Y.-T. (2006) *Nano Lett.*, **6**, 2690.
216 Hu, C.C. and Huang, Y.H. (1999) *J. Electrochem. Soc.*, **146**, 2465.
217 Hu, C.-C., Chen, W.-C., and Chang, K.-H. (2004) *J. Electrochem. Soc.*, **151**, A281.
218 Trasatti, S. and Buzzanca, G. (1971) *J. Electroanal. Chem. Interfacial Electrochem.*, **29**, A1.
219 Juodkazis, K., Juodkazytė, J., Šukienė, V., Grigucevičienė, A., and Selskis, A. (2008) *J. Solid State Electrochem.*, **12**, 1399.
220 Ma, Z., Zheng, J.P., and Fu, R. (2000) *Chem. Phys. Lett.*, **331**, 64.
221 Rudge, A., Davey, J., Raistrick, I., Gottesfeld, S., and Ferraris, J.P. (1994) *J. Power Sources*, **47**, 89.
222 Naoi, K., Suematsu, S., and Manago, A. (2000) *J. Electrochem. Soc.*, **147**, 420.
223 Mastragostino, M., Arbizzani, C., Meneghello, L., and Paraventi, R. (1996) *Adv. Mater.*, **8**, 331.

224 Liu, H. (2014) *Cuihua Xuebao/Chin. J. Catal.*, **35**, 1619.
225 Giddey, S., Badwal, S.P.S., and Kulkarni, A. (2013) *Int. J. Hydrogen. Energy*, **38**, 14576.
226 Horanyi, G. and Rizmayer, E.M. (1982) *J. Electroanal. Chem.*, **132**, 119.
227 Abghoui, Y. and Skúlasson, E. (2015) *Procedia Comput. Sci.*, **51**, 1897.
228 Howalt, J.G. and Vegge, T. (2013) *Phys. Chem. Chem. Phys.*, **15**, 20957.
229 Zheng, H., Ma, C., Wang, W., and Huang, J. (2006) *Electrochem. Commun.*, **8**, 977.
230 Lausche, A.C., Okada, K., and Thompson, L.T. (2012) *Electrochem. Commun.*, **15**, 46.
231 An, W., Hong, J.K., Pintauro, P.N., Warner, K., and Neff, W. (1999) *J. Am. Oil Chem. Soc.*, **76**, 215.
232 An, W., Hong, J.K., and Pintauro, P.N. (1998) *J. Appl. Electrochem.*, **28**, 947.
233 Pintauro, P., Gil, M.P., Warner, K., List, G., and Neff, W. (2005) *Ind. Eng. Chem. Res.*, **44**, 6188.
234 Li, J.H., Fu, X.Z., Luo, J.L., Chuang, K.T., and Sanger, A.R. (2012) *Electrochem. Commun.*, **15**, 81.

20
Combinatorial Approaches for Bulk Solid-State Synthesis of Oxides

Paul J. McGinn

University of Notre Dame, Department of Chemical and Biomolecular Engineering, 178 Fitzpatrick Hall, Notre Dame, IN 46556, USA

20.1
Introduction

Traditionally, new or improved materials have been developed in laboratories by repetitive serial synthesis and analysis of individual samples. This process is time and effort intensive because of the number of iterations required to optimize a material or process. In an attempt to accelerate and improve this process, combinatorial approaches have been developed that involve three main steps: (i) high-speed automated materials synthesis, (ii) automated serial or parallel characterization or screening, and (iii) processing and analysis of large data sets [1]. High-throughput experimentation is complemented by computational methods to aid in the identification of candidate materials systems or compositions.

The majority of reports describing high-speed automated materials synthesis involve thin film deposition. A primary reason for this is that existing laboratory thin film deposition systems were easily adapted for composition spread (gradient) library synthesis. Combinatorial thin film library synthesis is the subject of several reviews [1–3]. However, bulk (non-thin film) laboratory samples are preferred in many industries, particularly where they are dealing with higher material volumes. In addition, some bulk properties of a material can be quite different from thin film forms. For example, in films strain caused by substrate–lattice mismatch may be very important, grain size may be vastly different from the bulk, and surface and skin electrical effects can dominate. In certain fields (e.g., catalysis), powder samples are the norm; so laboratory-scale powder synthesis and testing more easily scales to manufacturing operations than do thin film samples. The focus of this chapter is on techniques (step 1 above) that have been utilized to process inorganic solid-state materials, particularly in *bulk or particulate form.*

Handbook of Solid State Chemistry, First Edition. Edited by Richard Dronskowski, Shinichi Kikkawa, and Andreas Stein.
© 2017 Wiley-VCH Verlag GmbH & Co. KGaA. Published 2017 by Wiley-VCH Verlag GmbH & Co. KGaA.

20.2
Powder Synthesis from Liquid Precursors

Combinatorial powder synthesis from liquid precursors is a commonly used route in bulk high-throughput materials studies. Liquid deposition routes are attractive to researchers because implementation is usually straightforward and relatively inexpensive. Widely available pipetting robots or low-cost consumer-level inkjet printers can be employed for library deposition. Techniques based on dispensing of liquids (e.g., pipettes, inkjets) are typically used for the synthesis of discrete composition libraries, rather than composition spread libraries that are popular in thin film processing. Metal nitrate salts are common precursors for inkjet printing of many compounds since they are highly soluble and metal oxides can be formed through pyrolysis at temperatures below 550 °C. For more complex systems, other synthesis routes including sol–gel, hydrothermal, polymer complexing reactions, combustion synthesis, and flame spray pyrolysis have all been reported, as described in Section 20.2.1.

20.2.1
Sol–Gel

Sol–gel routes are well established for synthesis of a wide variety of material. The reader is referred to several excellent handbooks that cover the technology in great detail [4–6]. The reactants in most sol–gel processes are the corresponding metal alkoxide compounds that are dissolved in an alcohol. This solution is termed the sol. When water is added, a hydrolysis reaction replaces alkoxide groups (OR) with hydroxyl groups (OH), as shown in Eq. (20.1) (R is an organic group and Me is the metal of interest). Subsequently, a condensation reaction (Eq. (20.2)) results in metal–oxygen bonding in three directions, yielding formation of a 3D network structure. The resulting suspension of the alkoxide in alcohol is termed the gel.

$$(RO)_2 - Me - OR + H(OH) = (RO)_2 - Me - OH + (OH)R \qquad (20.1)$$

$$\begin{aligned}(RO)_2 - Me - OH + (RO)_2 - Me - OR \\ = (RO)_2 - Me - O - Me - (RO)_2 + ROH\end{aligned} \qquad (20.2)$$

The characteristics and properties of a particular sol–gel inorganic network are related to a number of factors that affect the rate of hydrolysis and condensation reactions, with pH, nature and concentration of catalyst, H_2O/Me molar ratio, and temperature being the most important. By controlling these factors, it is possible to vary the structure and properties of the sol–gel-derived inorganic network over wide ranges.

The sol–gel process is relatively easy to practice, it requires simple equipment, it yields high-surface-area powder, and it is powerful in allowing precise tuning of materials compositions, product homogeneity, and is applicable to a wide variety of chemical systems [6]. In addition it is readily compatible with both pipetting and inkjet technology. As a result the sol–gel technique has been widely used for parallel synthesis of powder libraries [7–11].

20.2.2
Polymer Complex Processing

Polymer complex methods ("Pechini" processing) have also been used for combinatorial powder synthesis [12–15]. These methods are not as popular as sol–gel for combinatorial studies, but are widely used in the synthesis of complex oxides because they are simple to utilize and synthesis occurs at lower temperature and shorter reaction times than solid-state routes. The process is based on the formation of a polymer resulting from polyesterification of a chelated metallic complex, through the use of a hydroxycarboxylic acid (e.g., citric acid or maleic acid) and a polyhydroxylic alcohol (e.g., ethylene glycol). Polymer complex methods allow one to achieve intimate mixing of cations at the molecular level, and the immobilization of the metal complexes in the polymeric network reduces phase segregation, giving compositional homogeneity at the molecular scale. If appropriate for the system under investigation, this route typically uses less expensive precursors than sol–gel, but the amounts of the organic components (commonly citric acid and ethylene glycol) need to be optimized to yield high-surface-area powder.

20.2.3
Combustion Synthesis

The use of combustion synthesis for parallel powder synthesis has also been reported [15,16]. The combustion synthesis technique involves mixing oxidizing reagents (e.g., metal nitrates) and a fuel (e.g., glycine, urea). The reaction is highly exothermic, so is self-sustaining. The high temperatures produced ensure crystallization of oxides in a very short time. Because the temperature excursion is brief, agglomeration of particles is avoided, so small particle sizes result. The challenge in using this route in parallel reactions is to limit the violence of the reactions so that the fine synthesized powder does not float out of one reactor and cross-contaminate adjacent samples.

20.2.4
Hydrothermal Synthesis

20.2.4.1 Parallel Hydrothermal Synthesis
Powder synthesis by hydrothermal or solvothermal processing is well known in the inorganic synthesis community. Hence, parallel combinatorial hydrothermal techniques were adopted from the onset of combinatorial research [17–22].

A schematic of the approach is shown in Figure 20.1

They all share a common design principle of machining parallel autoclave cells in a Teflon block. Akporiaye *et al.* described an autoclave capable of carrying out

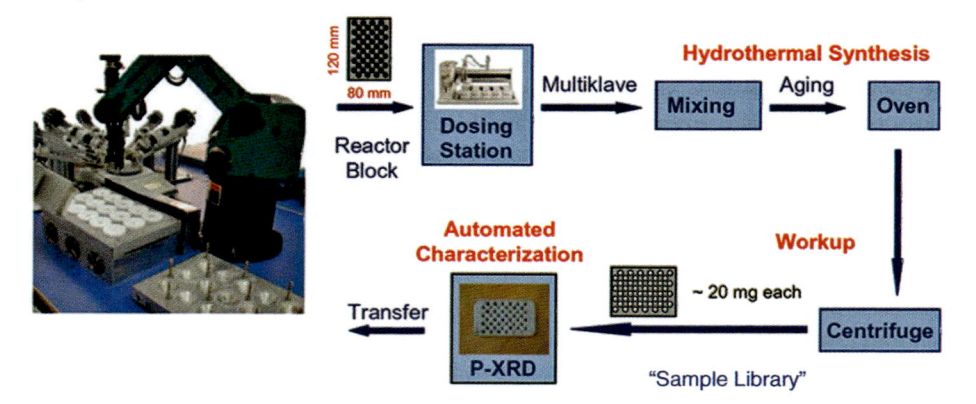

Figure 20.1 Robotic system for parallel gel preparation (*left*) [21], and typical nonrobotic HT synthesis and characterization workflow (*right*) for reactions under hydrothermal conditions comprising automated preparation of reaction mixture, synthesis, isolation, characterization, and storage of the compounds. (Reproduced with permission from Ref. [23]. Copyright 2010, Elsevier Inc.)

more than 100 crystallizations under hydrothermal conditions at temperatures up to 200 °C [24]. Their design was based on a Teflon block with 100 reaction chambers (holes) that are sealed when sandwiched between steel plates. By stacking identical blocks, the synthesis of up to 1000 compositions in one experiment is possible. The autoclave was designed to be compatible with commercial pipetting robots so that filling of the reaction wells is readily automated. After the hydrothermal synthesis, the isolation of the products is accomplished by parallel centrifugation and/or a washing. They used this for research on zeolites and then later on perovskites. The product library was characterized with a diffractometer with an area detector and XYZ stage, enabling rapid automated crystallographic characterization. Chemical analysis was performed with an automated microprobe [25]. Miniaturization was further improved by Klein *et al.* [18], with a bottom plate Si wafer being used for compatibility with their postsynthesis characterization. The reader is referred to an excellent recent review on high-throughput investigations employing solvothermal syntheses for more details [23].

20.2.4.2 Serial Hydrothermal Synthesis

Serial hydrothermal synthesis was demonstrated with a novel high-throughput continuous hydrothermal (HiTCH) flow synthesis reactor described by Weng *et al.*, which uses serial hydrothermal reactions (Figure 20.2) [26]. This is a modified continuous hydrothermal flow synthesis reactor, which allows individual samples to be made without contamination between samples.

Other designs of continuous hydrothermal flow synthesis reactors with varying levels of automation have been used to rapidly produce metal oxide nanoparticles [27,28]. The HiTCH reactor described above was subsequently

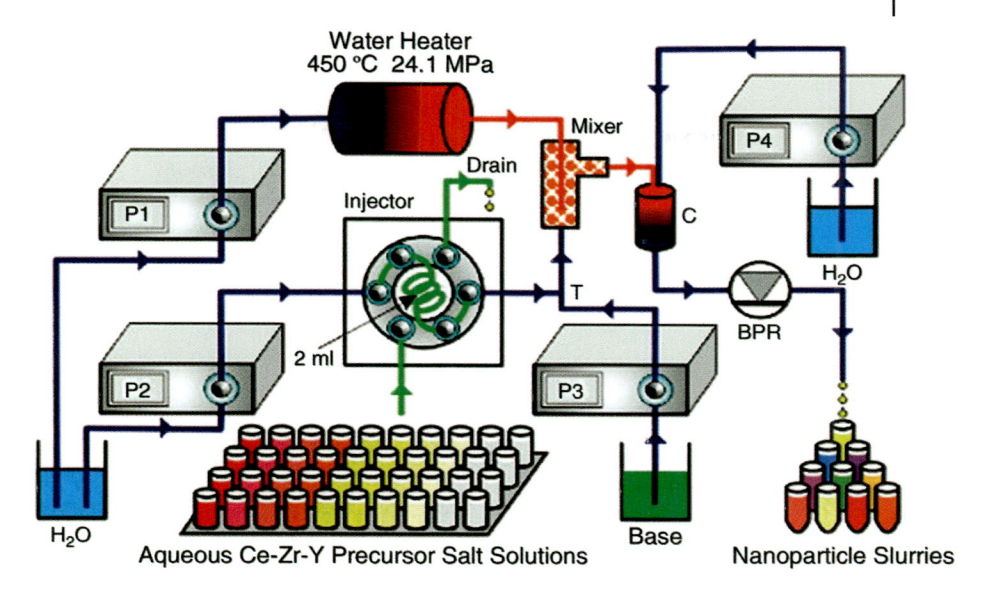

Figure 20.2 A schematic layout of the high-throughput continuous hydrothermal (HiTCH) flow synthesis system for rapid synthesis of compositionally unique ceramic samples at high pressures without contamination between consecutive samples. Each metal salt solution is filled into the injection loop, and injection of cold water flow from pump P2 carries the contents of the loop to a T, where it is mixed with a flow of KOH and H_2O_2 solution (from pump P3) to give a pH 14 mixture. The mixture then meets a flow of supercritical water (from pump P1) that is superheated in a heat exchanger to reach 450 °C at a pressure of 24.1 MPa) in the countercurrent mixer (labeled "mixer"). The reaction of the solution of metal salts with the flow of supercritical water at 450 °C and 24.1 MPa results in rapid nucleation and crystallization of nanoparticles. The nanoparticle slurries are then cooled (C) and collected at the exit of a back pressure regulator (BPR). The system is purged with H_2O for several minutes between samples. The nanoparticle slurries from HiTCH reactor are freeze-dried and heat-treated (at 1000 °C for 1 h for a $Ce_xZr_yY_zO_{2-\delta}$ library) to produce powders for further characterization. (Reproduced with permission from Ref. [26]. Copyright 2009, American Chemical Society.)

incorporated as a unit into a rapid automated materials synthesis instrument (RAMSI), which automated the handling and processing of the HiTCH output. The RAMSI instrument automates sample cleanup and enables printing of samples as dots [29,30].

20.2.5
Flame Spray Pyrolysis

A less widely practiced technique for high-throughput combinatorial powder synthesis is based on flame spray pyrolysis. Flame spray synthesis, reviewed by Teoh *et al.* [31], has been shown to be well suited for powder synthesis cases where several components are present, for example, in multimetal catalysts [32].

One advantage claimed for flame spray pyrolysis is that it can produce kinetic rather than thermodynamic phases, with such phases being difficult or impossible to produce by other methods [33]. Weidenhof *et al.* used ethanol-based precursor solutions that were atomized through a nozzle with O_2. The fine mist generated is ignited by a gas torch (typically methane–oxygen), giving combustion temperatures >1500 °C. The nanosized powders are collected downstream in an electrostatic precipitator. In their experiments they collected 30–50 g of each of the 46 samples they synthesized [33]. However, they only produced one to three different compositions per day. These quantities are likely better suited for an industrial lab where large powder batches are the norm, but the sample production rate may be too low for most university labs, where smaller batches with a rapid turnaround are preferred.

20.3
Combinatorial Inkjetting

Inkjet dispensing of liquids is easily implemented and can be readily adapted to a wide array of precursor chemistries [34]. Both consumer and technical inkjet devices have been utilized by researchers. Commercial inkjet printers designed specifically for materials research and development are now available (e.g., Fujifilm Dimatix Materials Printer DMP-2831). The application of inkjet technology for combinatorial oxide synthesis has been reviewed in detail [35–37] so will only be briefly highlighted here. Precursor inks can be made from metal salt solutions, insoluble colloidal suspensions of nanoparticles, or nanoparticle colloids produced via sol–gel synthesis. An example of a drop-on-demand inkjet printer is shown in Figure 20.3 [38].

In the case of soluble precursors, the liquid precursors (e.g., nitrates, acetates) are dispensed into well plates, onto paper or other absorbing substrates, or coated glass, silicon, or other bulk substrates or supports [39]. To make multicomponent compounds, the component inks are sequentially dispensed, with interdiffusion of the constituents readily occurring in the liquid state. After drying, a heat treatment is used to form the final compound. After heating, the synthesized sample may be in the form of particles or a film that can be controlled by the formulation of the ink, with the preference depending on the nature of the subsequent characterization or screening for which the samples are intended. Inkjetting is a relatively high-speed process, so large libraries with a thousand or more samples can be rapidly synthesized [40,41], even if multiple passes are required to deposit enough material to satisfy characterization needs.

Ultrasonic misting of larger solution volumes [42–44] has also been used. This is not capable of printing fine geometries as with inkjetting. Syringe pumps feed precursors into a stainless steel nozzle charged with a high voltage, leading to atomization. The resulting mist deposits onto a substrate heated to 400 °C. The powder deposits are subsequently heated to 700 °C to form the final compounds,

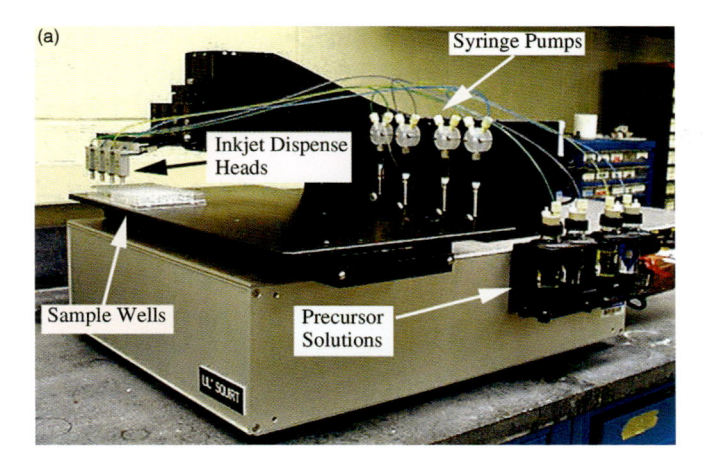

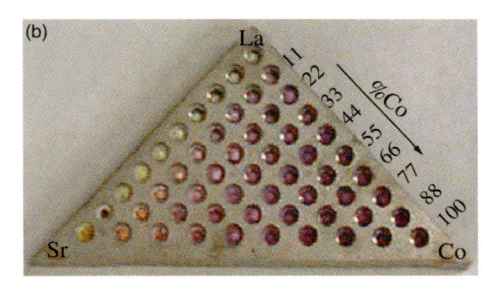

Figure 20.3 A drop-on-demand inkjet printer is shown in part (a). It combines syringe pumps with inkjet dispensers, and incorporates a computer controlled x–y–z table. In part (b), a ternary library of La-Sr-Co-O compounds is shown. The sample holder is made from a heat-resistant alloy to permit high decomposition temperatures for powder formation. (Reproduced with permission from Ref. [38]. Copyright 2003, Elsevier Science B.V.)

and then characterized by X-ray diffraction (XRD), energy-dispersive X-ray spectroscopy (EDXS), and so on.

20.4
Paste and Slurry Routes

Dispensing of powder slurries by pipetting has been used for discrete library synthesis in a number of investigations [45–53]. The process is straightforward, with the main challenge being ensuring that the components are suitably mixed and that separation does not occur [54]. One example is provided by Fujimoto *et al.*, who described pipetting combinations of metal salt solutions and nanoparticles [49,52]. In one study the starting materials were aqueous solutions of LiOH-H_2O, Cr(OCOCH$_3$)$_3$, and Ni(NO$_3$)$_2$, and slurries of Fe_2O_3 and TiO_2 nanoparticles suspended in water [49]. These starting materials were pipetted into mixing vials using a robotic pipetting system. The pipetted liquids were mixed by repetitively (50×) sucking and injecting them. These mixtures were then dispensed onto a reaction plate, where they were dried. This process was repeated to build up the volume of deposited material as desired. The reaction was then transferred to a furnace and heated for 5 h between 900 and 1100 °C to form the final compounds.

Inkjet devices have also been used to deposit nanoparticle suspensions and more viscous ceramic inks [55]. The powders for inkjetting can be prepared by a variety of approaches including microwave synthesis, sol–gel processing, hydrothermal synthesis, and high-energy ball milling (HEBM). The inks made with the powders typically consist of a nonaqueous solvent for rapid drying, a viscosity control agent (e.g., ethylene glycol), a surfactant for wetting and stabilization of the nanoparticles, and possibly an acid to control the surface charge on particles [56].

Printing of more viscous ceramic slurries is more complex because of the difficulty in achieving complete mixing. Mixing can occur on the substrate, but this is undesirable because it necessarily involves large diffusion distances during sintering. Alternatively, mixing can be accomplished in a chamber located ahead of a dispensing nozzle [57]. This requires cleaning of the chamber between compositions to avoid cross-contamination. The third approach, used in the London University Search Instrument (LUSI) robot (Figure 20.4), is to use a two-stage process where inks are dispensed into a well plate for mixing, and then the contents of the well plate are printed [48,58,59]. This approach is not as fast as the liquid inkjetting described above, but allows one to make "bulk" samples more rapidly. Their printed samples are 1 mm diameter dots, with a volume of 0.2 µl. Because this process involves printing and firing on a substrate, interactions with the substrate can be challenging [53]. Further developments of the system included the capability of robotically loading and firing the combinatorial libraries after printing.

Pullar further simplified this approach [60]. He deposits a strip that has a stepped gradient composition. This is accomplished by using two variable-rate

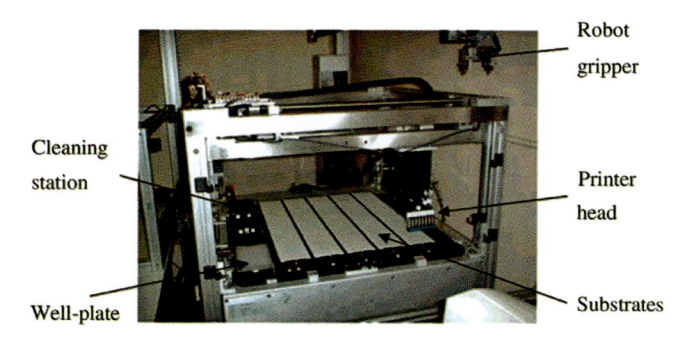

Figure 20.4 London University Search Instrument (LUSI) inkjet printer table within the robot gantry, populated with alumina substrates. The pick and place robot arm is seen in the top right. (Reproduced with permission from Ref. [48]. Copyright 2005, American Chemical Society.)

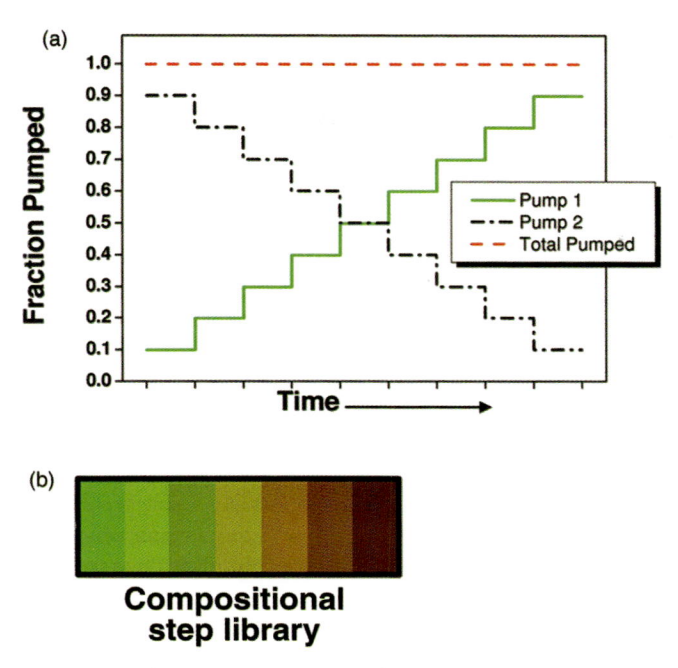

Figure 20.5 Schematic representation of (a) the variable fraction–constant total rate pumping process and (b) the step compositional gradient combinatorial libraries produced. (Reproduced with permission from Ref. [60]. Copyright 2012, American Chemical Society.)

syringe pumps that simultaneously feed two slurries into a commercial mixer tip, yielding a deposited line of mixed paste that has a varying composition along its length (Figure 20.5a). As the rate of one pump is reduced, the rate of the other is increased to keep a continuous overall feed rate. The resulting tubular deposit can be doctor-bladed perpendicular to its length, to produce a thick film (Figure 20.5b). The doctor-bladed thick film library is cut into strips parallel to the gradient, and then the five strips are simultaneously fired in a five-zone tube furnace with 25 °C increments in sintering temperature. In the demonstrated example, composite libraries with 10 mol% steps of $SrFe_{12}O_{19}$–$BaTiO_3$ were fired at 1100, 1125, 1150, 1175, and 1200 °C.

20.5
Combinatorial Synthesis from Dry Powders

20.5.1
High-Throughput Ball Milling

Combinatorial processing from dry powders is appealing because it scales directly to common industrial practice. A significant advantage compared to some liquid routes is that one can realize a high yield from inexpensive reagents, and it can be easily scaled to large quantities. However, it is more challenging to accomplish high-throughput powder processing compared with starting materials synthesis from liquids.

Ball milling is a more common materials preparation technique. This was "parallelized" by Stegk et al. [61], and by Wildcat Discovery Technologies, a US company specializing in materials discovery as part of an effort to discover hydrogen storage materials [62]. High-energy ball milling is a powerful technique popularized in the 1990s for synthesis of a wide variety of materials, including metal alloys, ceramics, and intermetallics. The appeal of HEBM is that the small particle size and intimate mixing achieved facilitates subsequent thermal processing steps such as sintering. It helps to reduce sintering temperatures and shortens annealing times needed for crystallization, phase formation, or pellet densification. Many manufacturers offer planetary mills where two or four samples can be milled simultaneously. Thus, with a line of such machines, this approach is readily scaled.

Stegk et al. developed technology for high-throughput synthesis of bulk ceramics using fine-grained ($d_{50} \sim 0.5\,\mu m$), dry powders as starting materials [61,63,64]. The advantage of such an approach is that the results are directly comparable to results from conventionally synthesized bulk samples. They use a laboratory robot fitted with a commercial solid doser. Powders are dosed into 10 mm-diameter wells machined in ultrahigh-molecular-weight polyethylene containers. Each 30 mm × 120 mm container has five wells (Figure 20.6a). A steel fixture that holds up to four containers is placed in a conventional planetary mill (Figure 20.6b). The mill can hold 2 fixtures, so up to 40 samples can be mixed in parallel. After mixing, the powders are transferred in parallel manner into silicone dies that allow pressing five samples (i.e., one container) at once (Figure 20.6c). The limit of five samples was due to the small size of the press available to them. These samples can then be sintered at elevated temperatures to densify the pellets. Because of the use of silicone dies, only relatively low compaction loads can be utilized. For denser compaction of the green pellets, an isostatic or other pressing operation is needed.

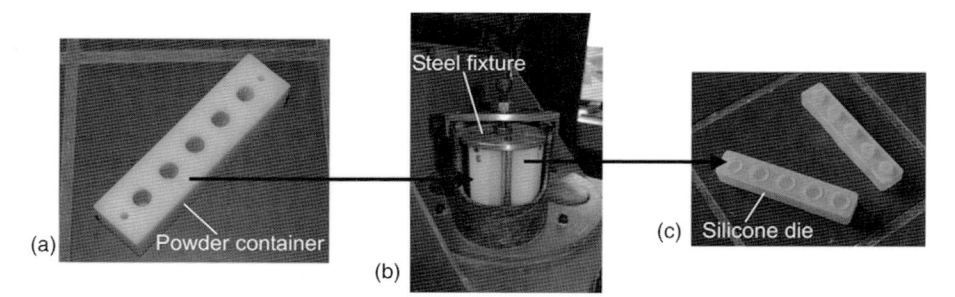

Figure 20.6 Four powder containers (a) are placed in a steel fixture (b) in a conventional planetary mill. After milling the powders are transferred to silicon dies (c) for parallel pressing. (Reproduced with permission from Ref. [61]. Copyright 2008, American Chemical Society.)

Wildcat Discovery Technologies, a US start-up, improved on this approach and developed at multiwall ball milling fixture to mill multiple samples in parallel [65]. A potential disadvantage of high-energy milling is that there can be contamination from wear of the milling jar and media. Thus, care must be taken in selecting media and in specifying milling parameters.

One complicating factor in applying HEBM for combinatorial work is that process control agents (PCAs) are often used as part of milling operations. These are additives that are used to inhibit formation of particle agglomerate and enhance grinding and particle size reduction. They are another variable that must be considered in optimizing processing, and the constancy of their effect across a compositional region must be verified. In the Wildcat work, a variety of PCAs were examined to find the best one for their systems [62]. After powder synthesis is complete, structure characterization is relatively straightforward with automated powder diffractometers. If it is desired to characterize powders in pelletized form, parallel pressing is needed to avoid a processing bottleneck. Wildcat developed technology to hot press multiple pellets simultaneously [66].

Related high-throughput powder processing technology has also been developed for *glass processing* by Inoue et al. They described a combinatorial research system for glasses based on robotic handling and weighing of powder precursors [67–69]. This setup, shown schematically in Figure 20.7, enables synthesis of up to 24 compositions of 20 g in a four-component system. The 24 samples are melted and cooled in carbon crucibles at one time. One test takes about 1 h, so in a day they can screen about 200 samples. The glass formation of the cooled

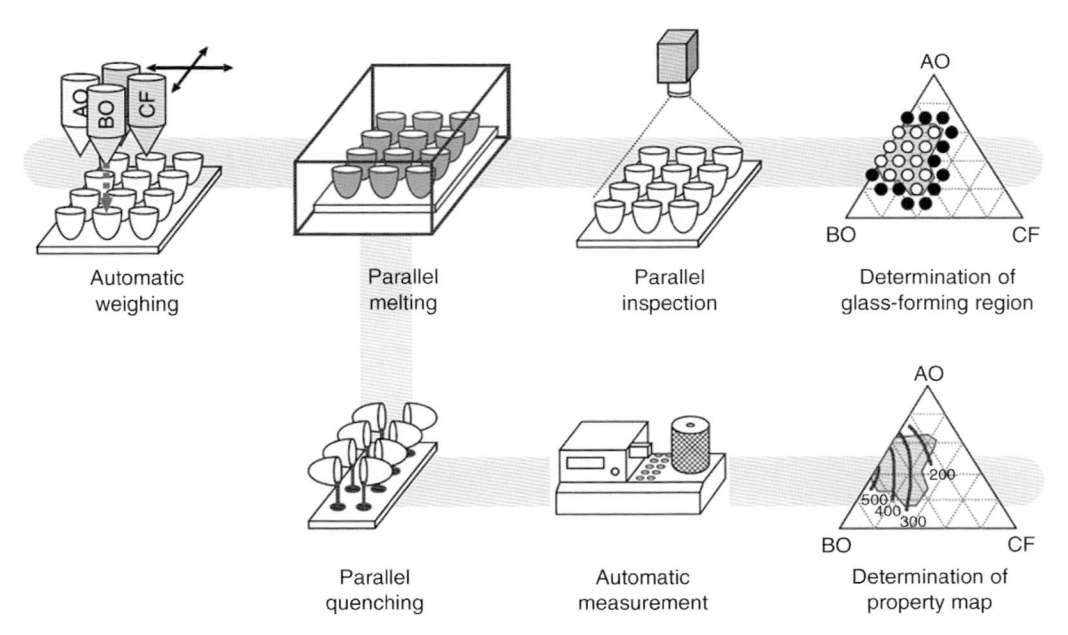

Figure 20.7 Illustration of a combinatorial glass research system. (Reproduced with permission from Ref. [70]. Copyright 2006, Taylor & Francis Group, LLC.)

melts is identified using a CCD camera system. The basic properties of the glasses such as glass transition temperature, softening temperature, crystallization temperature, and liquidus temperature are measured serially using a conventional differential thermal analysis (DTA) apparatus with an automatic sample changer. The difference between the crystallization temperature and the glass transition temperature is a simple measure of the thermal stability of the glass, with a larger difference signifying a more stable the glass. The usefulness of this was demonstrated in an exploration of borate glasses for compositions possessing high hardness and low melting temperature to serve as sealing materials in joining metals and ceramics for electronic applications. Ultimately they identified glasses in the Fe_2O_3-ZnO-B_2O_3, Co_3O_4-ZnO-B_2O_3, and REO-MnO-B_2O_3 glass systems that showed hardness values >6 GPa while having $T_g < 550\,°C$ [71].

20.5.2
Injection Molding

A quite unique approach applying injection molded powder wedges was used to generate ternary libraries of ceramic powders [72]. Following an analogous concept to that used in thin film processing, three triangular wedges produced by injection molding are stacked on top of one another with a 120° rotation to form a triangular prism with 120 mm wide sides and a 20 mm height. The stacked wedges were then heated to 100 °C to soften the binder, and the layers were then pressed into a mold with a honeycomb structure (6.4 mm cell size). This forms separated library members in the honeycomb cells. Their mold produces 171 compositions from a ternary system. After the individual cells are defined, the cell contents are mixed by transferring to wells, where they are heated high enough to liquefy the binder and then they are shaken until homogeneous. The cooled mixtures are calcined in air to remove the organics, ground in an agate mortar, and then manually pelletized prior to sintering. The powdered materials were manually pressed in a steel die and sintered. This approach was demonstrated with layers composed of Fe_2O_3, TiO_2, or Al_2O_3. The researchers used this approach to study the subsolidus phase relationships and the dielectric properties in the CuO–TiO_2–CaO ternary system [73]. The injection molding process described above is beyond the scope of typical solid-state processing laboratories

20.6
Diffusion Couples

The use of diffusion couples has a long history in metallurgy for phase diagram development. In combinatorial investigations, the use of diffusion multiples in metals and intermetallics was advanced by Zhao [74–76]. This approach relies on the formation of interdiffusion zones between multiple materials through long-term annealing, and has been used primarily for metals. In such cases,

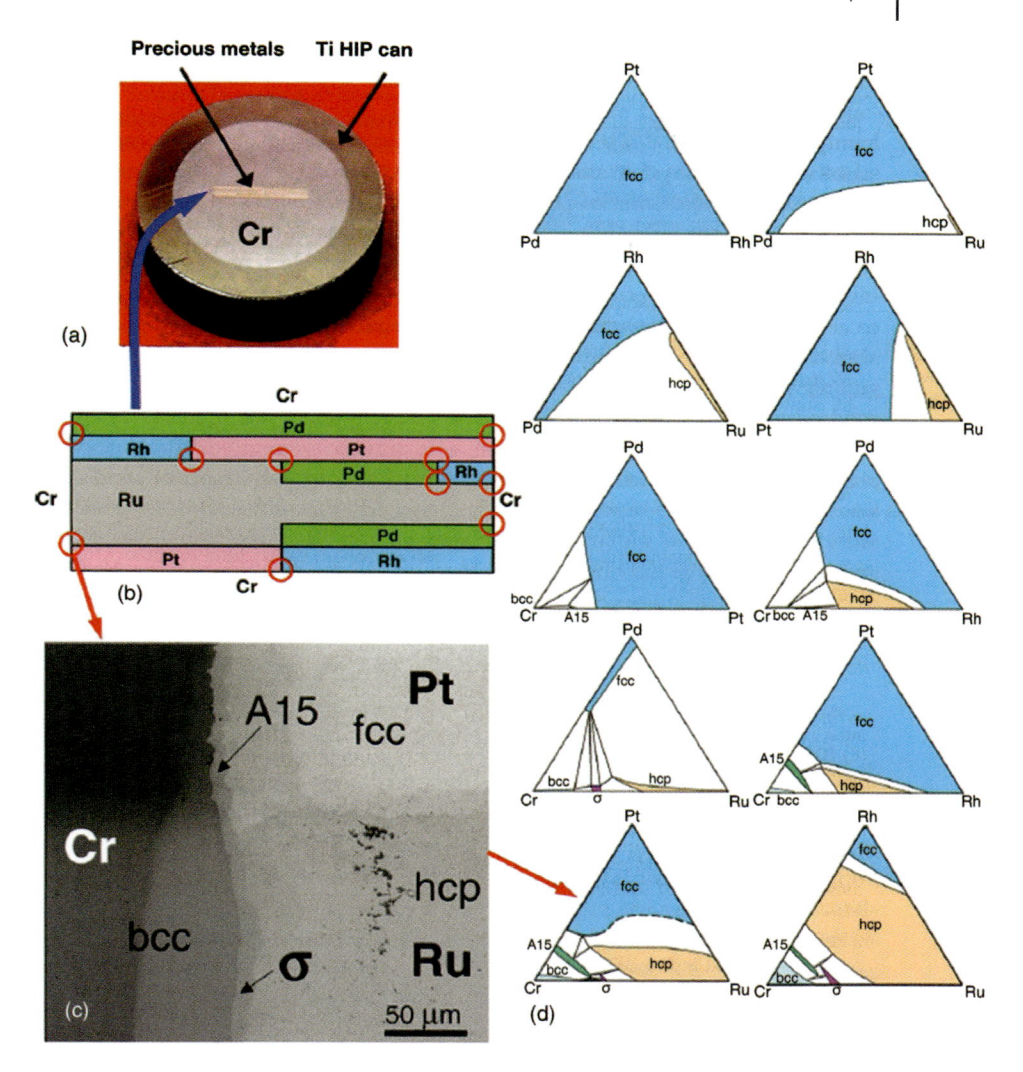

Figure 20.8 A diffusion multiple for rapid mapping of ternary phase diagrams in the Pd–Pt–Rh–Ru–Cr system. (a) An optical image of the reacted sample. (b) A schematic of the arrangement of the precious metal foils in the diffusion multiple used to create many trijunctions shown in circles. (c) A back scatter electron image of the Cr–Pt–Ru trijunction showing the formation of the A15 and σ phases due to interdiffusion of Pt, Ru, and Cr. (d) Ten ternary phase diagrams (isothermal sections at 1200 °C) obtained from this single diffusion multiple. The phase diagrams are plotted in atomic percent axes with the scales removed for simplicity. (Reproduced with permission from Ref. [36]. Copyright 2015, Schweizerische Chemische Gesellschaft (open access).)

dense metal specimens cut to size comprise the diffusion multiple (Figure 20.8). Such samples offer a very efficient way to study phase diagrams. These samples can only be studied by chemical probing techniques with high spatial resolution, such as an electron probe microanalysis (EPMA) or EDXS. Melting has to be avoided in diffusion multiples, so the highest heat treating temperature must be limited to the lowest solidus temperature of the multicomponent system, which may not be known before the investigation. In systems where the solidus is low, extended annealing times may be necessary to form intermetallic regions thick enough for a reliable composition evaluation.

This approach has also been employed in oxides, albeit in a simpler fashion. Rather than beginning with dense versions of the materials, binary diffusion couples have been fabricated by pressing two half pellets. Upon sintering, interdiffusion between the two halves yields a concentration gradient for further analysis. Jones *et al.* prepared 10 mm-diameter pellets where a thin metal separator was used during filling of the die [77]. The first half of the pellet die was filled with $Pb_{0.97}La_{0.02}(Zr_{0.6}Ti_{0.4})O_3$ and the other half was filled with $Pb_{1.01}[(Zr_{0.6}Ti_{0.4})_{0.98}Fe_{0.02}]O_3$. During sintering La and Fe ions diffused across the spatial boundary separating the two regions, giving a composition gradient that was visually distinct. The gradient was then probed by microdiffraction on a synchrotron and fluorescence spectroscopy. They were able to study the ferroelectric behavior of the graded bulk ceramic during electrical loading. They later used this same approach to study electric-field-induced phase-change behavior in $(Bi_{0.5}Na_{0.5})TiO_3$-$BaTiO_3$-$(K_{0.5}Na_{0.5})NbO_3$ [78].

20.7
Summary

High-throughput experimentation methods can offer an important advantage in terms of overall data quality since samples are processed and data are collected in an automated fashion under identical conditions. To realize this advantage one must ensure in using the techniques discussed above that only one parameter (typically composition) changes in small increments across a library.

For example, in powder processing it is important that particle size not vary along with composition. In most cases judicious choice of process conditions can help minimize such variation. However, since particle size may not be perfectly uniform across a library, this suggests that particle-based libraries are best suited for searching for relatively large changes in performance, based on some figure of merit, that exceed any effects that minor materials variations might have. With a primary characterization screen, the researcher is hoping to spot trends in a property with a parameter (e.g., composition) variation that can be further explored by more traditional routes employing real application conditions.

All of the above techniques offer the capability to accelerate the synthesis portion of a research program. As such they should be accompanied by appropriate high-throughput characterization techniques to evaluate the properties of the generated materials. Unlike the case for miniaturized libraries, for bulk materials automated serial characterization tools are often available, making unattended characterization of large numbers of samples possible. However, these tools are typically expensive as the automation option adds to the cost of sophisticated characterization instruments. Still from the viewpoint of a company, this can be attractive as it means less labor cost and 24 h operation.

Finally, these approaches are most effective if they can be coupled with good theoretical guidance and experimental design. Moving forward, combinatorial processing routes can be expected to become increasingly utilized for verifying predicted materials properties and trends rather than just for more rapid screening to accelerate materials optimization. This shift is being driven by a combination of the emergence of projects like the Materials Genome Initiative (www.mgi.gov), and the wider availability of computational data bases such as the Materials Project (materialsproject.org) and continued improvement in materials modeling and simulation. In the future one may see a transition from "automation," where experiments are preplanned and performed via robotic operations, to "autonomy" where the system makes intelligent decisions about the direction of the experiments. In that case the system will execute, evaluate, and plan new experiments in an iterative fashion with minimal human intervention.

References

1 Maier, W.F., Stowe, K., and Sieg, S. (2007) Combinatorial and high-throughput materials science. *Angew. Chem., Int. Ed.*, **46**, 6016–6067.
2 Green, M.L., Takeuchi, I., and Hattrick-Simpers, J.R. (2013) Applications of high throughput (combinatorial) methodologies to electronic, magnetic, optical, and energy-related materials. *J. Appl. Phys.*, **113**, 53.
3 McGinn, P.J. (2015) Combinatorial electrochemistry: processing and characterization for materials discovery. *Mater. Discov.*, **1**, 38–53.
4 Brinker, C.J. and Scherer, G.W. (1990) *Sol-Gel Science: The Physics and Chemistry of Sol-Gel Processing*, Academic Press, Boston.
5 Sakka, S. (2005) *Handbook of Sol-Gel Science and Technology*, Kluwer Academic Publishers.
6 Levy, D. and Zayat, M. (2015) *The Sol-Gel Handbook*, Wiley-VCH Verlag GmbH, Weinheim.
7 Holzwarth, A., Schmidt, P.W., and Maier, W.E. (1998) Detection of catalytic activity in combinatorial libraries of heterogeneous catalysts by IR thermography. *Angew. Chem., Int. Ed.*, **37**, 2644–2647.
8 Scheidtmann, J., Weiss, P.A., and Maier, W.F. (2001) Hunting for better catalysts and materials: combinatorial chemistry and high throughput technology. *Appl. Catal. A*, **222**, 79–89.
9 Saalfrank, J.W. and Maier, W.F. (2004) Doping, selection and composition spreads, a combinatorial strategy for the discovery of new mixed oxide catalysts for low-temperature CO oxidation. *C. R. Chim.*, **7**, 483–494.
10 Olong, N.E., Stowe, K., and Maier, W.F. (2007) A combinatorial approach for the discovery of low temperature soot oxidation catalysts. *Appl. Catal. B*, **74**, 19–25.
11 Reiser, M., Stowe, K., and Maier, W.F. (2012) Combinatorial development of novel Pd based mixed oxide catalysts for the CO hydrogenation to methanol. *ACS Comb. Sci.*, **14**, 378–387.

12 Reichenbach, H.M. and McGinn, P.J. (2001) Combinatorial synthesis of oxide powders. *J. Mater. Res.*, **16**, 967.

13 Vess, C.J., Gilmore, J., Kohrt, N., and McGinn, P.J. (2004) Combinatorial synthesis of oxide powders with an autopipetting system. *J. Comb. Chem.*, **6**, 86–90.

14 Henderson, S.J., Armstrong, J.A., Hector, A.L., and Weller, M.T. (2005) High-throughput methods to optically functional oxide and oxide-nitride materials. *J. Mater. Chem.*, **15**, 1528.

15 Bluthardt, C., Fink, C., Flick, K., Hagemeyer, A., Schlichter, M., and Volpe, A. (2008) Aqueous synthesis of high surface area metal oxides. *Catal. Today*, **137**, 132–143.

16 Luo, Z.L., Geng, B., Bao, J., and Gao, C. (2005) Parallel solution combustion synthesis for combinatorial materials studies. *J. Comb. Chem.*, **7**, 942–946.

17 Akporiaye, D.E., Dahl, I.M., Karlsson, A., and Wendelbo, R. (1998) Combinatorial approach to the hydrothermal synthesis of zeolites. *Angew. Chem., Int. Ed.*, **37**, 609–611.

18 Klein, J., Lehmann, C.W., Schmidt, H.W., and Maier, W.F. (1998) Combinatorial material libraries on the microgram scale with an example of hydrothermal synthesis. *Angew. Chem., Int. Ed.*, **37**, 3369–3372.

19 Choi, K., Gardner, D., Hilbrandt, N., and Bein, T. (1999) Combinatorial methods for the synthesis of aluminophosphate molecular sieves. *Angew. Chem., Int. Ed.*, **38**, 2891–2894.

20 Lai, R., Kang, B.S., and Gavalas, G.R. (2001) Parallel synthesis of ZSM-5 zeolite films from clear organic-free solutions. *Angew. Chem., Int. Ed.*, **40**, 408–411.

21 Moliner, M., Serra, J.M., Corma, A., Argente, E., Valero, S., and Botti, V. (2005) Application of artificial neural networks to high-throughput synthesis of zeolites. *Microporous Mesoporous Mater.*, **78**, 73.

22 Caremans, T.P., Kirschhock, C.E.A., Verlooy, P., Paul, J.S., Jacobs, P.A., and Martens, J.A. (2006) Prototype high-throughput system for hydrothermal synthesis and X-ray diffraction of microporous and mesoporous materials. *Microporous Mesoporous Mater.*, **90**, 62–68.

23 Stock, N. (2010) High-throughput investigations employing solvothermal syntheses. *Microporous Mesoporous Mater.*, **129**, 287–295.

24 Akporiaye, D., Dahl, I., Karlsson, A., Plassen, M., Wendelbo, R., Bem, D.S., Broach, R.W., Lewis, G.J., Miller, M., and Moscoso, J. (2001) Combinatorial chemistry: the emperor's new clothes? *Microporous Mesoporous Mater.*, **48**, 367–373.

25 Wendelbo, R., Akporiaye, D.E., Karlsson, A., Plassen, M., and Olafsen, A. (2006) Combinatorial hydrothermal synthesis and characterisation of perovskites. *J. Eur. Ceram. Soc.*, **26**, 849–859.

26 Weng, X.L., Cockcroft, J.K., Hyett, G., Vickers, M., Boldrin, P., Tang, C.C., Thompson, S.P., Parker, J.E., Knowles, J.C., Rehman, I. et al. (2009) High-throughput continuous hydrothermal synthesis of an entire nanoceramic phase diagram. *J. Comb. Chem.*, **11**, 829–834.

27 Chaudhry, A.A., Goodall, J., Vickers, M., Cockcroft, J.K., Rehman, I., Knowles, J.C., and Darr, J.A. (2008) Synthesis and characterisation of magnesium substituted calcium phosphate bioceramic nanoparticles made via continuous hydrothermal flow synthesis. *J. Mater. Chem.*, **18**, 5900–5908.

28 Marchand, P., Makwana, N.M., Tighe, C.J., Gruar, R.I., Parkin, I.P., Carmalt, C.J., and Darr, J.A. (2016) High-throughput synthesis, screening, and scale-up of optimized conducting indium tin oxides. *ACS Comb. Sci.*, **18**, 130–137.

29 Lin, T., Kellici, S., Gong, K., Thompson, K., Evans, J.R.G., Wang, X., and Darr, J.A. (2010) Rapid automated materials synthesis instrument: exploring the composition and heat-treatment of nanoprecursors toward low temperature red phosphors. *J. Comb. Chem.*, **12**, 383–392.

30 Alexander, S.J., Lin, T., Brett, D.J.L., Evans, J.R.G., Cibin, G., Dent, A., Sankar, G., and Darr, J.A. (2012) A combinatorial nanoprecursor route for direct solid state chemistry: discovery and electronic properties of new iron-doped lanthanum

nickelates up to $La_4Ni_2FeO_{10}$-delta. *Solid State Ion.*, **225**, 176–181.
31 Teoh, W.Y., Amal, R., and Madler, L. (2010) Flame spray pyrolysis: an enabling technology for nanoparticles design and fabrication. *Nanoscale*, **2**, 1324–1347.
32 Hannemann, S., Grunwaldt, J.D., Lienemann, P., Gunther, D., Krumeich, F., Pratsinis, S.E., and Baiker, A. (2007) Combination of flame synthesis and high-throughput experimentation: the preparation of alumina-supported noble metal particles and their application in the partial oxidation of methane. *Appl. Catal. A*, **316**, 226–239.
33 Weidenhof, B., Reiser, M., Stowe, K., Maier, W.F., Kim, M., Azurdia, J., Gulari, E., Seker, E., Barks, A., and Laine, R.M. (2009) High-throughput screening of nanoparticle catalysts made by flame spray pyrolysis as hydrocarbon/NO oxidation catalysts. *J. Am. Chem. Soc.*, **131**, 9207–9219.
34 Hoath, S.D. (2016) Fundamentals of inkjet printing: the science of inkjet and droplets.
35 Liu, X.N., Tarn, T.J., Huang, F.F., and Fan, J. (2015) Recent advances in inkjet printing synthesis of functional metal oxides. *Particuology*, **19**, 1–13.
36 Lesch, A., Cortes-Salazar, F., Bassetto, V.C., Amstutz, V., and Girault, H.H. (2015) Inkjet printing meets electrochemical energy conversion. *Chimia*, **69**, 284–289.
37 Zhao, J.C. (2004) Reliability of the diffusion-multiple approach for phase diagram mapping. *J. Mater. Sci.*, **39**, 3913–3925.
38 Reichenbach, H.M. and McGinn, P.J. (2003) Combinatorial solution synthesis and characterization of complex oxide catalyst powders based on the $LaMO_3$ system. *Appl. Catal. A*, **244**, 101.
39 Woodhouse, M., Herman, G.S., and Parkinson, B.A. (2005) Combinatorial approach to identification of catalysts for the photoelectrolysis of water. *Chem. Mater.*, **17**, 4318–4324.
40 Liu, X.N., Shen, Y., Yang, R.T., Zou, S.H., Ji, X.L., Shi, L., Zhang, Y.C., Liu, D.Y., Xiao, L.P., Zheng, X.M. *et al.* (2012) Inkjet printing assisted synthesis of multicomponent mesoporous metal oxides for ultrafast catalyst exploration. *Nano Lett.*, **12**, 5733–5739.
41 Haber, J.A., Cai, Y., Jung, S.H., Xiang, C.X., Mitrovic, S., Jin, J., Bell, A.T., and Gregoire, J.M. (2014) Discovering Ce-rich oxygen evolution catalysts, from high throughput screening to water electrolysis. *Energy Environ. Sci.*, **7**, 682–688.
42 Fujimoto, K., Kato, T., Ito, S., Inoue, S., and Watanabe, M. (2006) Development and application of combinatorial electrostatic atomization system "M-ist Combi": high-throughput preparation of electrode materials. *Solid State Ion.*, **177**, 2639–2642.
43 Fujimoto, K., Onoda, K., and Ito, S. (2007) Exploration of layered-type pseudo four-component Li-Ni-Co-Ti oxides. *Appl. Surf. Sci.*, **254**, 704–708.
44 Kim, K.W., Jeon, M.K., Oh, K.S., Kim, T.S., Kim, Y.S., and Woo, S.I. (2007) Combinatorial approach for ferroelectric material libraries prepared by liquid source misted chemical deposition method. *Proc. Natl. Acad. Sci. USA*, **104**, 1134–1139.
45 Yanase, I., Ohtaki, T., and Watanabe, M. (2002) Combinatorial study for ceramics powder by X-ray diffraction. *Solid State Ion.*, **154**, 419–424.
46 Yanase, I., Ohtaki, T., and Watanabe, M. (2002) Combinatorial study on nano-particle mixture prepared by robot system. *Appl. Surf. Sci.*, **189**, 292–299.
47 Yanase, I., Ohtaki, T., and Watanabe, M. (2002) Application of combinatorial process to $LiCo_1$-$XMnXO_2$ ($0 \leq X \leq 0.2$) powder synthesis. *Solid State Ion*, **151**, 189–196.
48 Wang, J. and Evans, J.R.G. (2005) London University Search Instrument: a combinatorial robot for high-throughput methods in ceramic science. *J. Comb. Chem.*, **7**, 665–672.
49 Fujimoto, K., Takada, K., Sasaki, T., and Watanabe, M. (2004) Combinatorial approach for powder preparation of pseudo-ternary system $LiO_{0.5}$-X-TiO_2 (X: $FeO_{1.5}$, $CrO_{1.5}$ and NiO). *Appl. Surf. Sci.*, **223**, 49–53.
50 Fujimoto, K. and Watanabe, M. (2005) Preparation of pseudo-ternary library by combinatorial robot system based on wet

and dry processes. *Meas. Sci. Technol.*, **16**, 41.

51 Fujimoto, K., Ito, S., Suehara, S., Inoue, S., and Watanabe, M. (2006) Effective algorithm for material exploration in ceramics with combinatorial technology. *J. Eur. Ceram. Soc.*, **26**, 731–734.

52 Suehara, S., Konishi, T., Fujimoto, K., Takeda, T., Fukuda, M., Koike, M., Inoue, S., and Watanabe, M. (2006) A combinatorial sample-preparation robot system using the volumetric-weighing method. *Appl. Surf. Sci.*, **252**, 2456–2460.

53 Zhan, Y., Chen, L.F., Yang, S.F., and Evans, J.R.G. (2007) Thick film ceramic combinatorial libraries: the substrate problem. *QSAR Comb. Sci.*, **26**, 1036–1045.

54 Wang, J. and Evans, J.R.G. (2006) Segregation in multicomponent ceramic colloids during drying of droplets. *Phys. Rev. E*, **73**, 8.

55 Evans, J.R.G., Edirisinghe, M.J., Coveney, P.V., and Eames, J. (2001) Combinatorial searches of inorganic materials using the ink jet printer: science, philosophy and technology. *J. Eur. Ceram. Soc.*, **21**, 2291–2299.

56 Teng, W.D. and Edirisinghe, M.J. (1998) Development of ceramic inks for direct continuous jet printing. *J. Am. Ceram. Soc.*, **81**, 1033–1036.

57 Mohebi, M.M. and Evans, J.R.G. (2002) A drop-on-demand ink-jet printer for combinatorial libraries and functionally graded ceramics. *J. Comb. Chem.*, **4**, 267.

58 Wang, J., Mohebi, M.M., and Evans, J.R.G. (2005) Two methods to generate multiple compositions in combinatorial ink-jet printing of ceramics. *Macromol. Rapid Commun.*, **26**, 304.

59 Chen, L., Zhang, Y., Yang, S., and Evans, J.R.G. (2007) Protocols for printing thick film ceramic libraries using the London University Search Instrument (LUSI). *Rev. Sci. Instrum.*, **78**, 6.

60 Pullar, R.C. (2012) Combinatorial bulk ceramic magnetoelectric composite libraries of strontium hexaferrite and barium titanate. *ACS Comb. Sci.*, **14**, 425–433.

61 Stegk, T.A., Janssen, R., and Schneider, G.A. (2008) High-throughput synthesis and characterization of bulk ceramics from dry powders. *J. Comb. Chem.*, **10**, 274–279.

62 Li, B., Kaye, S.S., Riley, C., Greenberg, D., Galang, D., and Bailey, M.S. (2012) Hydrogen storage materials discovery via high throughput ball milling and gas sorption. *ACS Comb. Sci.*, **14**, 352–358.

63 Stegk, T.A., Mgbemere, H., Herber, R.P., Janssen, R., and Schneider, G.A. (2009) Investigation of phase boundaries in the system $(K_xNa_{1-x})_{1-y}Li_y(Nb_{1-z}Ta_z)O_3$ using high-throughput experimentation (HTE). *J. Eur. Ceram. Soc.*, **29**, 1721–1727.

64 Mgbemere, H.E., Janssen, R., and Schneider, G.A. (2015) Investigation of the phase space in lead-free $(K_xNa_{1-x})_{1-y}Li_y(Nb_{1-z}Ta_z)O_3$ ferroelectric ceramics. *J. Adv. Ceram.*, **4**, 282–291.

65 Downs, R.C. (2010) High throughput mechanical alloying and screening. U.S. Patent 7,767,151, filed Aug. 3, 2005 and issued Aug. 3, 2010.

66 Micklash, K.J.C. II and Bailey, M.S. (2016) Hot pressing apparatus and method for same. U.S. Patent 9,314,842, filed Nov. 28, 2012 and issued Apr. 19, 2016.

67 Inoue, S., Todoroki, S., Matsumoto, T., Hondo, T., Araki, T., and Tsuchiya, T. (2002) Development of the combinatorial glass formation tester. MRS Online Proceedings. doi: https://doi.org/10.1557/PROC-700-S6.6

68 Inoue, S., Todoroki, S.I., Konishi, T., Araki, T., and Tsuchiya, T. (2004) Combinatorial glass research system. *Appl. Surf. Sci.*, **223**, 233.

69 Inoue, S., Todoroki, S., Suehara, S., Konishi, T., Chu, S.Z., Wada, K., Kikkojin, T., Isogai, M., Katsuta, Y., Sakamoto, T. et al. (2006) Advanced approaches to functional glasses. *J. Non-Cryst. Solids*, **352**, 632–645.

70 Tomoya, K., Shigeru, S., Shin-ichi, T., and Satoru, I. (2006) Combinatorial study of new glasses, in *Combinatorial and High-Throughput Discovery and Optimization of Catalysts and Materials*, CRC Press, pp. 61–83.

71 Segawa, H., Igarashi, T., Sakamoto, T., Mizuno, K., Konishi, T., and Inoue, S. (2010) Exploration of high-hardness/low-

melting borate glasses. *Int. J. Appl. Glass Sci.*, **1**, 378–387.

72 Ren, S., Kochanek, W., Bolz, H., Wittmar, M., Grobelsek, I., and Veith, M. (2008) Combinatorial preparation of solid-state materials by injection moulding. *J. Eur. Ceram. Soc.*, **28**, 3005–3010.

73 Veith, M., Ren, S.H., Wittmar, M., and Bolz, H. (2009) Giant dielectric constant response of the composites in ternary system CuO–TiO_2–CaO. *J. Solid State Chem.*, **182**, 2930–2936.

74 Zhao, J.C., Zheng, X., and Cahill, D.G. (2005) High-throughput diffusion multiples. *Mater. Today*, **8**, 28–37.

75 Zhao, J.C. (2006) Combinatorial approaches as effective tools in the study of phase diagrams and composition–structure–property relationships. *Prog. Mater. Sci.*, **51**, 557–631.

76 Zhang, Q.F. and Zhao, J.C. (2014) Impurity and interdiffusion coefficients of the Cr-X (X=Co, Fe, Mo, Nb, Ni, Pd, Pt, Ta) binary systems. *J. Alloy Compd.*, **604**, 142–150.

77 Jones, J.L., Pramanick, A., and Daniels, J.E. (2008) High-throughput evaluation of domain switching in piezoelectric ceramics and application to $PbZr_{0.6}Ti_{0.4}O_3$ doped with La and Fe. *Appl. Phys. Lett.*, **93**, 3.

78 Daniels, J.E., Jo, W., Rodel, J., Honkimaki, V., and Jones, J.L. (2010) Electric-field-induced phase-change behavior in $(Bi_{0.5}Na_{0.5})TiO_3$-$BaTiO_3$-$(K_{0.5}Na_{0.5})NbO_3$: a combinatorial investigation. *Acta Mater.*, **58**, 2103–2111.

Index

a

ABCABC sequence, 5
absolute permittivity, 25
AC. *See* activated carbon (AC)
accessible pore volume, 250
accumulation, 394
acetone
– asymmetric direct aldol reaction, 481
– condensation, 522
acetonitrile, 542
acidic H-ZSM-5 shell, 430
activated carbon (AC), 30, 263, 274
– based masks, 274
– materials, 30, 40
– production of, 31
activation energies, 525
admetals, 518
– carbide interface sites, 533
adsorbate–adsorbate interactions, 251, 252
– energy of interaction, 256
– nature, 256
adsorbate compression, 258
adsorbate critical temperature, 254
adsorbate gravimetric density, 252
adsorbate liquefaction energy, 253
adsorbate molecules, 251
adsorbate–surface interaction, 258
adsorbents, 250
– adsorbate electron cloud distance, 257
adsorption affinity (AAf), 250, 274
adsorption data
– evaluation, 252
adsorption energy, 250
adsorption enthalpies, 251
adsorption forces, 259
adsorption heats, 256, 259
adsorption isotherms, 251, 253, 255, 267
adsorption potential, 256
aerogels, 443

Ag/AgCl electrode, 77
agarose hydrogel, 137
agglomeration, 555
Ag_2S nanowires, crystal lattice images, 453
air purification
– materials, 275
– toxic chemicals adsorption, 273
Al-bridged hybrid polymers, enantioselective Michael addition catalyzed by, 486
alcohols, 528
aldol reaction, 480
– of aldehyde with acetone, 480
– of benzaldehydes, 482
– – with cyclohexanone, 482
alicyclic ketones, 347
– hydrogenation, 344
aliphatic carboxylic acids, 348
alkali earth ions, doping, 286
alkali metal ions, 290
alkaline earth metals, 262
alkane oxidation, 397, 403
– selective, 394
alkene hydrogenation, 341
alkenes/alkanes, catalytic processes, 395
alkoxide, in alcohol, 554
alkynes
– to alkenes, reaction heats of partial hydrogenation, 344
Al-Li-bis(binaphthoxide) (ALB), 486
γ-Al_2O_3, 313
Al-rich Beta zeolite, 435
alternating strips, 142
alumina, 207
aluminum
– in $LiNi_{0.8}Co_{0.15}Al_{0.05}O_2$, 13
aluminum hydrides, 304
ambient light, 123
ambient-pressure bulk modulus, 181
amine derivatives, 351

Handbook of Solid State Chemistry, First Edition. Edited by Richard Dronskowski, Shinichi Kikkawa, and Andreas Stein.
© 2017 Wiley-VCH Verlag GmbH & Co. KGaA. Published 2017 by Wiley-VCH Verlag GmbH & Co. KGaA.

amine-functionalized SBA-15, 458
amine-grafted SBA-15 mesoporous silicas, 458
amine hardener, 212
3-aminopropyl triethoxysilane (APTES), 157
– immunoassay, 158
– linker, 159
– – functionalized sensor systems, 158
– oxygen interaction-based sensor, 158
amphiphilic bidentate ligand (tris(triazolyl)-polyethylene glycol (tris-trz-PEG)
– synthesis of, 451
Anderson–Schulz–Flory (ASF), 418
anion exchange chemistry, 454
anisotropic materials, 143
anisotropic particles, magnetization of, 126
anisotropy, 257
aqueous salt EGTFT structure, schematic of, 84
area-normalized capacitance, 36
aromatic amines, 270
aromatic compounds, 348
aromatic oxygenate mixtures, 520
aromatization reactions, 535, 536
Arrhenius plots
– of product formation rates, 524
– TOF, illustrating, 534
art capacitor technology, 47
artificial DX DNA nanostructure, schematic diagram, 82
asymmetric epoxidation reactions, 497
asymmetric hybrid capacitors, 36
asymmetric Michael additions, 486
– in water, 463
atmospheric pressure, 189
atomic layer deposition (ALD), 155, 460
atomization, 558
attenuated total reflection (ATR), 500
– infrared (ATR IR), 501
$Au(CH_3)_2(acac)$, 317
Au nanoparticles, 446
Au-sensitive catalytic reaction, 453
average adsorption energy, 254
azines, 462

b

back pressure regulator (BPR), 557
ball milling, 316
band gap nanomaterials, 446
batteries, electrical energy storage
– cationic and anionic substitutions, 12
– computational high-throughput calculations, 21
– electrolyte, 3
– experimental combinatorial synthesis, 18
– high-throughput chemistry, 18
– implementation of electric vehicles, 1
– $LiCoO_2$/graphite electrodes, 2
– mechanisms during cycling, 13
– nanometric carbon coating syntheses, 10
– negative electrode materials, 6
– particle morphology/coatings, 15
– positive electrode materials, 5
– soft chemistry/discovery of, 11
– solid-state chemistry, 12
– solid-state electrolytes, 7, 16
– solid-state synthesis, 9
– state-of-the-art Li-ion batteries, 2
– structures/properties of interest, 4
battery materials, 9
BC_3 phase in the cubic diamond structure (d-BC_3), 182
benzaldehyde, 488
– asymmetric direct aldol reaction, 481
benzylic silyl ethers, dehydrogenative orthosilylation of, 462
benzyl Ti(III) grafting, for H_2 uptake
– mesoporous silica, 287
beryllium benzene tribenzoate (Be-BTB), 290
BET. *See* Brunauer–Emmett–Teller (BET)
Beta zeolite, 427
Bi-containing compounds, 105, 112
bidentate ligand (*tris*(triazolyl)-polyethylene glycol (*tris*-trz-PEG), 451
bifuncational catalysts, 426
bimetallic nanocatalysts, 360
Bi–Mo catalysts, 395
Bi–Mo–O catalyst, 397, 402, 403
1,1′-bi-2-naphtol (BINOL), 486
BINOLate/Ti(OiPr)$_4$ catalyst, for asymmetric addition of diethylzinc to aldehydes, 493
biocompatibility, 30
biodegradable, 75
bioelectronics interfaces, 75, 78
bio-fats, 430
biomass
– chemocatalytic conversion, 431
– conversions, 373, 432, 515
– – catalysis, 521
– derived compounds, 516
– feedstocks, 515
biomaterials, functional surfaces, 227
bioreceptors, 154
biosensitive semiconducting layers, 81
biosensors, using MOx EGTFTs, 82
biphenyldicarboxylic acid (bpdc) in DMF, 497
bipolar electrodes, 46
birefringent material, 143

Index | 575

2,2′-bis(diphenylphosphino)-1,1′-binaphtyl (BINAP), 489
BiSe-containing compound, 115
black dye, 66
block copolymers, 459
blood clotting, 203
boiling point (BP), 274
Boltzmann constant, 94, 237
Boltzmann transport theory, 94
bone-like crystal, 229
– growth, on polarized ceramic HAp
– – acceleration and deceleration of, 229
bone prostheses, 227
borate glasses, 564
borohydrides (tetrahydroborates), 305
boroxine linkages, 270
Bradley isotherm, 255
Bragg diffraction, 133, 135
– magnetic tuning, 133–142
Brønsted acidity, 435
– hydroxyl sites, 518
Brønsted acid sites, 515
Brunauer–Emmett–Teller (BET), 253
– isotherm equation, 253
bulk stoichiometric $TiSe_2$, 111
buoyancy effect, 251
1-butyl-3-butylimidazolium tetrafluoroborate ([dbim][BF_4]), 489
1-butyl-3-methylimidazolium hexafluorophosphate ([bmim][PF_6]), 496

c

$CaCO_3$ crystallization of, 240
calcination, 360
calcite crystals, scanning electron micrographs, 241
calcite film, X-ray diffraction pattern, 242
calcite thin films, 241
calcium apatite, 229
calcium carbonate crystallization, on polarized hydroxyapatite substrates, 237
– $CaCO_3$ crystallization, on polarized HAp substrates, 240
– heterogeneous nucleation, 237
calcium lactate, 210
capacitance, 25, 27
capacitor, 25
– electrodes, 43
carbide-derived carbons (CDC), 32
carbon aerogels, 34
carbon based materials, 263, 285
– carbon–carbon composites/hierarchical porous, 33
– functionalization and doping, 285
carbon–carbon coupling, 524
carbon-containing resins, 263
carbon dioxide capture, 260–270
– chemisorption, 261
– clathrates, 271
– oxy-fuel combustion, 261
– porous solids adsorption, 262
– post-combustion, 260
– pre-combustion, 260
carbon dioxide separation, 269
carbon dioxide uptake, 268
carbon electrodes, 34
carbonization, 458
– of precursor, 31
carbon monoliths, 34
carbon nanofiber (CNF), 520
carbon nanomaterials, 458
carbon nanotube (CNT), 27, 500
carbon nitride (CN), 180
– ligands, 293
– related compound nanoplatelets, 181
carbon pool parallel mechanism, 422, 423
carburization, 515, 530, 532
cardiovascular devices, micro/nanostructure of, 228
Carnot efficiency, 93
catalysis, 444
– characterization techniques, 370, 371
– deactivation, 331
– homogeneous and heterogeneous, 444
– particle sizes, 316
– performance, structure–property relationships, 325
– – catalyst regeneration, 331
– – catalytic activity on crystal faces, 326
– – effects of ligands, 329
– – insensitivity in catalysis by supported metal clusters and particles, 325
– – ligands and modifiers of metals, 327
– – metal nuclearity, 328
– – metal oxidation state, 330
– – roles of structural features, 326
– – sintering and poisoning, 330
catalysts characterization, supported, 313, 318
– case studies, 322
– chemisorption equilibrium measurements, 321
– industrial catalysts, 313
– transmission electron microscopy, 318
– X-ray absorption fine structure (XAFS), 319
catalysts preparation, supported, 316

– industrial catalyst preparation methods, 316
– synthesis of, 317
catalysts reactivities, supported, 322
– adsorption, 322
– oxidation, 324
– sulfidation of, 325
– supported catalysts, restructuring of, 324
catalyst support, 369
catalytic active nanoparticles, 455
catalytic asymmetric aldol reaction, 482
catalytic asymmetric epoxidation, 494
catalytic chemistry, 341, 355
catalytic hydrogenation, 339, 369
catalytic methodology, 393
catalytic oxidation activity, 402
catalytic oxidation-recarburization cycle, 535
catalytic performance, 530
cathode ray tube (CRT), 123
cation doping, 65
C—C bond formation, 424
CDC. See carbide-derived carbons (CDC)
C_{dl}, for DSP linker, 159
CD-modified Pt catalyst, 501
– nanocatalyst, 500
Ce^{4+}/Ce^{3+} redox processes, 448
cell adhesion, on biomaterials, 237
cell potential, 45
cells sense, 228
cellulose nanocrystal (CNC), 502
cell voltage, 540
ceramics
– biomaterials, 235
– from chemical solutions, 79
– material surfaces, with polarization treatment, 234
– methods, 10
– powders, ternary libraries of, 564
ceria films, 161, 164
$Ce_{1-x}Zr_xO_2$ nanocages
– schematic illustration, 452
CH_4
– adsorption forces, 282
CHA cage well, 434
charge-balancing cations, 262
charge carriers
– density, electric field control, 160
– photogeneration, schematic illustration of, 445
charge density wave (CDW), 100
charge–dipole interactions, 258, 259
charge storage, 42
– device, 25

– mechanisms, in electrochemical capacitors, 28
charge transfer reactions, 42
charge within graphite (C_{SC}), 34
chemical activation, 31
chemical agents, 365
chemical bath deposition (CBD), 67
chemical liftoff lithography (CLL), 83
chemical poisoning, 331
chemical purity, 31
chemicals production, 537, 542
chemical transformation, 340
chemical vapor deposition (CVD), 460
chemical warfare agents, 250
chemical weapons, 273
– convention, 250, 273
chemisorption process, 261
C-heteroatom bond coupling, 457
c-Hf_3N_4
– Hf—N—O phase, 190
chiral auxialliary groups, 464
chiral bis(cyclohexyldiamine)-based Ni(II) complexes, 484
chiral imidazolidinones, 488
chiral ionic liquids, 492
chiral Mn(III)-Salen catalysts, 495, 496
– in [bmim][PF_6], 496
chiral MOFs, 482
chitosan (CHIT), 156
CH_4 molecule, electron cloud polarization, 282
cholesterol oxidase (ChOx)
– on ZnO nanoflower structures, 156
CH_4 storage, 306
cigarette filters, 273
cinchona alkaloids, 499
– ethyl pyruvate, hydrogenation of, 499
cinchona-modified Pt catalysts, 499
citric acid, 555
clathrate formation, 271, 302
Clausius–Clapeyron equation, 260
climate change, 93
CMP. See conjugated microporous polymer (CMP)
CNC. See colloidal nanocluster (CNC)
β-C_3N_4 structure, 180
CNT. See carbon nanotube (CNT)
CO_2, 521
– adsorption materials
– – carbon-based materials, 262
– – isotherms, 265
– – metal–organic frameworks, 262
– – solid materials, 262
– capture materials, 275

– coordination interaction
– – with partially naked transition metal, 258
– electroreduction of, 455
– emissions, 249
– hydrogenation, 527
– – active sites for, 526
– – catalytic performance, 525
– – mechanistic insights, 526
– – reaction conditions, rates, and selectivity for, 526
CO_2 adsorption materials
– coordination polymers, 262
coarse grain simulation, 193
coatings, 15
cobalt catalysts, 358
cobalt promoters, 325
cobalt redox couple, disadvantage, 69
CO hydrogenation, 418, 522
– active sites and mechanistic insights, 522
– catalytic performance, 522
– reaction, 426
– – rates, selectivities, and chain propagation probabilities for, 523
coke formation, 331
CO ligands, 329
colloidal nanocluster (CNC), 134
combinatorial approaches, 553
combinatorial glass research system, 563
combinatorial inkjetting, 558
combinatorial libraries, 560
combinatorial powder synthesis, from liquid precursors, 554
combinatorial research, 555
combinatorial synthesis, from dry powders, 561
– high-throughput ball milling, 561
– injection molding, 564
combustion synthesis, 554
combustion synthesis technique, 555
Co/meso-H-ZSM-5, 427
composite film, 136
composite materials, 42
composites formation, 261
concrete, self-healing, 210
conducting polymer (CP), 75
conduction band minimum (CBM), 79
conductive polymers, 41
– and functionalized carbons as active materials, 41
conjugated microporous polymer (CMP), 270
contact angles, 233
conventional capacitors, 47

conventional catalysts, 369
Co_3O_4 nanomaterials, 455
CoO_2 slabs, 15
CO oxidation reaction, 330
copper catalysts, 358
copper Prussian blue analogues $Cu_3[T_i(CN)_6]_2$
– H_2 adsorption, 293
core–shell nanocatalysts, 460, 461
core–shell nanomaterial, 460
core-shell particles
– SEM images, 10
core–shell-structured bifunctional catalyst, 429
core/shell ZSM-5 catalyst, 429
cosensitization, 67
cost-effective synthesis, 12
Coulombic/spin interactions, 102
counter electrode (CE), 27, 62
covalent organic framework (COF), 270, 462
– chiral nanocatalyst, 463
covalent triazine-based frameworks (CTFs), 270
critical temperature, 252
Cr–Pt–Ru trijunction, 565
CRT. See cathode ray tube (CRT)
crystal formation
– controlling factors, 406
– scheme, 406
crystalline $H_{3-x}Cs_xPMo_{12-y}V_yO_{40}$
– methacrolein oxidation to methacrylic acid at polyoxomolybdate cluster, 400
crystalline Mo–V–O, 401
crystalline $Mo_3VO_{11.2}$ catalyst, 400, 401, 407, 410
crystalline solids, 299
crystallization temperature, 562, 564
crystallized divanadyl pyrophosphate catalyst, 398
C_{SC}. See space charge within graphite (C_{SC})
C-terminated $Mo_2C(001)$ surface, 514
Cu-Beta catalysts, 434
cubic BC_3 (c-BC_3) phase, 183
cubic boron nitride (c-BN), 175
cubic unit cell, 272
Cu nanocatalysts, 465
Cu nanoparticles, 465
$CuO–TiO_2–CaO$ ternary system, 564
Cu-TEPA-templated Cu-SSZ-13 zeolites, 434
Cu-zeolite, 435
$Cu-ZnO/Al_2O_3$ catalysts, 317
Cu-ZSM-5
– NO decomposition activity, 433
CV. See cyclic voltammetry (CV)

cyclic voltammetry (CV), 36
– measurements, 156
β-cyclodextrin, 481
cyclododecatriene, 341
cyclohexane, 519
cyclohexanecarbaldehyde, 480
cyclohexanone
– catalytic activity for acetalization, 467
– to cyclohexanone oxime, ammoximation of, 397
– derivatives, 347
– to β-nitrostyrene, asymmetric Michael addition, 484
cyclopentane, 304

d

Dalian Institute of Chemical Physics (DICP), 417
Danishefsky's diene, 488
data acquisition board (DAQ), 167
2D Co_3O_4 electrocatalyzed CO_2 reduction, 454
dead volume, 251, 252
defect-wurtzite dwur-C_2N_2(NH), 181
dehydrogenation, 522
demagnetizing tensor, 126
dendrimer-encapsulated nanoparticle (DEN), 452
density functional theory (DFT), 423, 497, 516
– calculations, 21
density of state (DOS), 94
deoxygenation, 516
diagnostic sensors, 151
– label-free biosensors, basic concept of, 152
– nanotextured semiconductor zinc oxide, as functional material, 154
– ZnO nanostructures
– – for biosensing applications, 155, 157
– – wearable biosensors, 159
dialkylzinc reagents, 494
diamond anvil cell (DAC), 183
diamond/c-BN line positions, 186
diamond-like carbon (DLC), 183
diamond-like t-BC_2 (dlt-BC_2), 182
diastereomeric transition complex, 501
dibenzothiophene, 355
1,3-dicarbonyl compounds, 1,4-addition reactions of, 487
dichloromethane, 496
dielectric capacitors, 25, 26
dielectric properties, 564
Diels-Alder reaction (DA), 214, 479, 488, 489, 490
– with anomalous endo/exo selectivities, 491

– of aromatic hydrocarbons with N-cyclohexylmaleimide, 490
– cycloadditions, 488
– – of acrolein to 1,3-cyclohexadiene, 489
– – asymmetric, 489
– – of cyclopentadiene and trans-cinnamaldehyde, 490
– of 1,3-cyclohexadiene/2-methylbutadiene, 488
– of cyclopentadiene and cinnamaldehyde, 490
– of cyclopentadiene and dienophile, 489
– cyclopentadiene and α,β-unsaturated aldehydes, 489
– endo/exo-selectivity, 490
– shape-memory-assisted self-healing, 214
diesel-range hydrocarbons, 418
diethylzinc
– addition, 479, 491
– to benzaldehyde
– – enantioselective addition of, 491
– – enantioselective addition of, 493
– chiral ligands for enantioselective reaction, 491
– to cyclohex-2-enone, 492
differential thermal analysis (DTA), 564
diffusion couples, 564
diffusion method, 241
3-dimensional materials, 399
2,2-dimethyl-2H-cromene, epoxidation, 497
dimethyl oxonium methyl ylide (DOMY), 424
dimethyl sulfoxide (DMSO), 482
dipole–dipole interactions, 125, 138, 259
direct alcohol fuel cell (DAFC), 541
discharge capacity, 3
disordered WSe_2-layered compounds, 109
dispersive forces, 267
dispersive interactions (DFT-D), 423
dispersive-type interactions, 257
displays
– emissive, 123
– – cathode ray tube (CRT), 123
– – field emission display (FED), 123
– – light-emitting diode (LED), 123
– – plasma display panel (PDP), 123
– evolution of, 123
– nonemissive, 123
– – functioning of, 124
– – liquid crystal display (LCD), 123
dithiobis(succinimidyl-propionate) (DSP), 157
divanadyl pyrophosphate, 398
– catalyst, 397

dlt-BC$_2$ phase, 182
DMTO process, 417
DNA backbone, 81
DNA detection method, 81
2D nanocatalysts, 454
1D nanomaterials, 453
DNA solution, 81
doctor-blade methods, 65
1-dodecanethiol, 450
DOE ultimate targets, 306
dopamine (DA), 460
– sensitive In-O TFTs, 83
dopants, 162
doped ceria thin films
– layout of constant voltage setup, 162
doping, 41, 101
– with transition metals, 286
double-crossover (DX), 81
double-layer capacitance, 28
drying, 317
dry processes, 316
dry reforming, 522, 535
– of methane, 535
Dubinin–Astakhov equation, 253
Dubinin–Astakhov (DA) isotherm, 254, 265
Dubinin's theory, 254
dye-sensitized nanocrystalline solar cells (DSSCs), 61
– challenges, 70
– counter electrodes, 68
– dye sensitizers, 66
– electrolytes, 68
– elements of, 62
– energy level diagram of, 62
– history, 61
– IPCE spectrum o, 64
– I–V curve of, 64
– materials and methods, for fabrication of, 64
– mesoscopic metal oxide electrodes, 64
– operational principles, 62
– parameter of, 63
– schematic energy level diagram of, 62

e
early transition metal carbide and nitride (ETMCN), 511
– catalysts, 513
– catalytic properties of, 513
– CO$_2$ hydrogenation performance of, 529
– deoxygenation performance of, 521
– industrial adoption of, 512
– surface areas of, 512

– surface facet, 514
– surface structure and active site identity, 513
3-(E)-but-2-enoyl]oxazolidine-2-one, 488
EC. See electrochemical capacitor (EC)
EDLC. See Electric Double-Layer Capacitors (EDLC)
EDX data, 10
EDX mapping, 10
effluent partial pressures, 519
EGFR antibodies, 157
elastic moduli, 176
elastic properties, 179
electrical biosensors, 152
electrical conductivity, 32
electrical double layer (EDL), 77
electrical/electrochemical signal, 152
electric double-layer capacitors (EDLC), 27
– operation, schematic illustration of, 29
electroadsorption, 45
electrocatalysis, 449, 543
– energy conversion, 537
electrocatalytic oxidation, 541
electrochemical capacitor (EC), 27, 47
electrochemical doping–dedoping, 41
electrochemical energy, 27
electrochemical impedance spectroscopy (EIS), 7, 157
electrochemical surface areas (ECSA), 541
electrocorticography (ECoG)
– fabricating, 88
electrode-electrolyte interface, 15
electrode materials, 32, 40
electrodeposition, 68
electrodes, 25, 36
electrohealing, 205
electroless deposition, 450
electrolyte-gated thin-film transistor (EGTFT), 76, 81
– MOx-based low-voltage operations, 82
– schematic cross-sectional illustration of, 77
electrolyte-gated transistor (EGT), 76, 78
– operation mechanism, 76
electrolytes, 29
– classification, 68
– decomposition, 46
– degradation, 46
– electrode interface, 21
– oxidation, 46
– reduction, 46
– selection, 44
electrolytic capacitor, 25
electromagnetism, 238

electron-based signals, 75
electron clouds, 257, 268
– polarizability, 289
electron diffraction (ED), 183
electron energy loss spectroscopy (EELS), 183
electron–hole pairs, 446
electronic conductivity, 95
electronics interfaces, 75
electronic transport
– in thermoelectric materials, 97
electron-induced surface energy modifications, 235
electron microscopy, 14, 32
electron paramagnetic resonance (EPR), 14
electron–phonon behavior, 103
electron-polarization plots, 527
electron probe microanalysis (EPMA), 190, 566
electron transfer value, 539
electrosorption, 39
electrostatic adsorption forces, 270
electrostatic force density, 239
electrostatic interactions, 267
electrowetting equation, 240
enantiodifferentiating surface process, 501
enantioselective Diels-Alder reaction
– ionic liquids, 490
energy consumption, 394
energy density, 3, 44
energy diagram, 2
energy dispersive spectroscopy maps, 534
energy-dispersive X-ray spectroscopy (EDXS), 559
energy-independent scattering, 95
energy storage, 537, 541
– devices, 25
– system, 48
engineered cementitious composite (ECC), 210
entropy change, 259
environmental gas
– sensing, 152
– sensors, 151
environmental oxygen, 161
environmental sensors
– hydrocarbon sensors, 164
– oxygen sensors, 161
enzymatic sensors, 86
epoxidation, 479

epoxy resin, 212
equations of state (EoS), 188
ethylene glycol, 516, 555, 560
ethyl lactate (EtLt), 498
ethyl pyruvate (EtPy)
– hydrogenation of, 498
exfoliated WS_2 electrocatalyzes, 454
exothermic reaction, 348
extended X-ray absorption fine structure (EXAFS), 319, 532
– characterization, 322
– data, 322
– – for $Pt/\gamma-Al_2O_3$, 322
– spectra, 330
external magnetic fields, 141, 146

f

faujasite zeolite, 324
Fe-Au-B-N alloys, 205
Fe-based zeolite catalysts, 435
FED. *See* field emission display (FED)
Fe_3O_4 microplates, 131
Fe_3O_4 nanoclusters, 127
$Fe_3O_4/SiO_2/Au$/nanoporous silica core–shell-shell nanocatalyst, 131
– synthetic procedure, 462
FER-based catalysts, 430
ferecrystals, 109
Fermi levels, 3, 12, 35, 101
ferroelectric behavior, 566
ferroelectric-like properties, 229
Fe-zeolite catalysts
– for NH_3-SCR, 435
Fe-ZSM-5 catalysts, 435
FFT analysis, frequency spectra, 170
field emission display (FED), 123
figure of merit, 85, 95
filled skutterudite structure
– polyhedral and ball-and-stick schematics of, 98
filtered cathodic arc (FCA), 190
First World War, 250
Fischer–Tropsch synthesis (FTS), 339, 417, 521, 535
– hydrocarbons, 531
– reactions, 524
flame spray pyrolysis, 554, 557, 558
flexible substrate-based biosensor, schematic representation, 160
fluid catalytic cracking(FCC)
– catalyst, 420
– yields, 421
fluorine-doped tin oxide (FTO), 62

fluorine, in HA, 236
focal adhesion kinase (FAK), 237
fossil-based energy conversion, 430
fossil fuels, 281
– combustion, 249, 261
– exploitation, 249
– refining, 536
Fourier transformations, 169
Friedel-Crafts reaction, 480
full concentration gradient (FCG) cathode material, 16
F–T *See* Fischer–Tropsch (F-T)

g

gas adsorption, 250
– adsorbate–surface interaction, 282
– measurement techniques, 250
gaseous pollutants, 249
gases/vapors
– adsorption
– – evaluation of, 250–256
– adsorption–desorption, 259
– physical adsorption, 250
– purification system, 251
gasification, 31
gasoline-range molecules, 535
gas sensing, 151
gas sensors, 151
gas storage, 252, 281
– clathrates, 273
g-$BC_{1.6}$ phase, 183
GBM. *See* graphene-based material (GBM)
gelation/flocculation, 316
germanium
– binary spinel nitrides of, 188
glass transition temperature, 564
glucose oxidation reaction, 464
3-glycidyloxypropyl)trimethoxysilane (GOPS), 86
glycolaldehyde, 516
gold microplates, 128, 130
gold plates dispersion, 129
Gouy–Chapman–Stern (GCS) model, 77
grain size, 553
– distribution, 175
graphene-based material (GBM), 32
graphene, 33
graphic coding, 138
graphite, 2
– based electrode, 35
graphite-carbon (GC-WC), 538
graphitization, 458
gravimetric capacitance, 44
gravimetric rates, 522
gravimetric storage density, 259, 267
gravimetric techniques, 250
guest–host interactions, 281
– coordination, 250
guest–host systems, 271

h

HA
– adhered osteoblasts morphology, 236
– polarized, TSDC curves, 230
HAADF-STEM images, 408
H_2, adsorption, 252, 290
– forces, 282
– heats
– – guest–host interactions, 297
– spectroscopic/structural studies, 292
Hall–Petch effect, 179
halogenated silanes, 456
hardness, defined, 177
Hashin–Shtrickman approach, 190
H_2 atmospheres, 327
H_2 clathrates, 282
HDS, reaction schemes, 356
healing, 214
health-monitoring devices, 152
heat of adsorption, 250
H_2 electron cloud
– by charge center, 296
Helmholtz layer, in EDL regions, 77
hemoglobin blocking, 274
Henry isotherm, 252
heteroatoms, 35, 399, 529
heterogeneous asymmetric catalysis, 479
– aldol addition, 480
– Diels-Alder reactions, 488
– diethylzinc addition, 491
– epoxidation, 494
– hydrogenation, 498
– Michael addition, 483
heterogeneous asymmetric reactions
– in batch systems, 503
heterogeneous catalysts, 482
heterogeneous organocatalyst
– asymmetric aldol reaction, 483
heteropolyanion, 399
heteropoly compounds, 397
heterostructures, 108
hexagonal boron nitride (h-BN), 185
hexagonal channel structure, 408
HF scavenging, 15
Hägg rule, 513
H–H bond scission, 329

high-energy ball milling (HEBM), 560
highest occupied molecular orbital (HOMO), 63
high-pressure high-temperature (HP-HT), 185
high-pressure volumetric determinations, 251
high resolution electron energy loss spectroscopy (HREELS), 516
High resolution HAADF-STEM electron micrographs, 533
high surface-indexed nanocatalysts, 370
high-throughput continuous hydrothermal (HiTCH), 556, 557
H_2 Kubas complex, 296
HKUST-1, 301
H_2 loading, 285
H_2 molecules combining coordination model
– for pore filling, 291
hole transporting p-type materials (HTM), 69
hollow glass fibers, 212
hollow nickel cobaltite ($NiCo_2O_4$) microcubes, 464
H_2O/Me molar ratio, 554
homochiral porous MOFs, 492
homogenous catalysts, 444
host-guest interactions, 216
hot electrons, 446
HP-HT-synthesized B–C–N samples
– TEM images of, 186
H-SAPO-34, 422, 426
H_2 storage, 296, 306
– adsorbent material
– – carbon-based materials, 285
– – ionic MOFs
– – H_2 adsorption through electrostatic interactions, 289
– – metal-organic frameworks (MOFs), 288
– – in MOFs, 288
– – silica/alumina/zeolites, 287
– adsorbent materials, 284
– capacity, 288, 297, 304
– carbon-based materials, 286
– current state, targets, and perspectives, 296
Human serum (HS), 157
H_2 uptake capacity, 289
$H_{4+x}PMo_{12-x}V_xO_{40}$, 399
hybrid bifunctional FT catalyst, 426
hybrid capacitors, 38, 39
hydrates, 301
hydrazone, 462
hydroamination, 457
hydrocarbon formation, 528
hydrocarbon fuel, 353

hydrocarbon oils, 233
hydrocarbon pool species, 423
hydrocarbons, 528
– aromatization, 535
– detection of, 165
– response frequencies, harmonic repetition, and frequency ambiguity, 167
– sensors, 165
hydrocracking, 356
hydrodenitrogenation (HDN), 331, 529
hydrodeoxygenation (HDO), 516, 529
hydrodeoxygenation reactions (HDO), 518
– biomass, 351
– knowledge gaps, 521
– liquid phase, 520
– vapor phase deoxygenation, of acetic acid, 518
hydrodesulfurization (HDS), 331, 529
– of petroleum crude, 353
hydrogel film, 137
hydrogenate aromatic ketones, 347
hydrogenation, 339, 344, 351
– of aldehydes, 347
– – and ketones, 344
– of alkenes, 341
– alkynes, 344
– applications, 365
– – of common nonnoble metal catalysts, 357
– of aromatic compounds, 358
– of carboxylic acids, esters, and ethers, 348
– catalysts
– – types, 357
– fatty acids, 341
– kinetics, 341
– by metals, 340
– nitroaromatics, 348
– of nitroaromatics, 351
– nitroaromaticso, 348
– nitrobenzene, 348
– oils, 341
– of unsaturated aldehydes, 348
hydrogenation reactions, 347, 357, 360, 369, 373, 542
– classification of, 340
– in petrochemical and fine chemical processes, 342
hydrogen bonding interactions, 216, 271
hydrogen clathrates, 301
hydrogen evolution reaction (HER), 464, 539
hydrogen, gas storage, 281
hydrogen hydrates, 303, 304

hydrogenolysis, 515, 522, 529
– of biomass derived feedstocks to fuels and chemicals, 354
– of C–O bond, 351
– glycerol, 352, 355
– of sorbitol and xylitol, 353
hydrogen oxidation reaction (HOR), 539
hydrogen storage, 292, 305, 306
– in clathrates, 303
– advances, 304
hydrogen sulfide, 355
hydrophobic fragments, 271
hydroprocessing, 340
hydrothermal conditions, 556
Hydrothermal growth method, 65
hydrothermal synthesis, 316, 555, 560
hydrotreating, 529
hydroxyapatite (HA), 229, 232, 240
hydroxycarboxylic acid, 555
β-hydroxyketone products
– yields and stereoselectivities, 483
hydroxylic functionalization approach, 169
5-hydroxymethylfurfural (5-HMF), 431
hypercrosslinked microporous polymers, 295
hypercrosslinked polystyrene material, 295
HY zeolite-supported $Ir(C_2H_4)_2$ complexes, 324
H-ZSM-5 shell, 421, 423, 425, 429
H-ZSM-5 zeolite, 423

i

IL. *See* ionic liquid (IL)
imidazole, 288
4,5-imidazoledicarboxylic acid (H_3ImDC), 290
imidazolium phase transfer function, 484
imidazolium-tagged pyrrolidines, 484
immobilized chiral organocatalysts, 479
immobilizing, 464
impregnation method, 359
impurities adsorption, 251
incident monochromatic photon-to-current conversion efficiency (IPCE), 63
incipient wetness impregnation (IWI), 359
inductive effect, illustration of, 13
In–Ga–Zn–O metal oxide TFT, 79
In–Ga–Zn–O surface, 81
injection molding process, 564
inkjet devices, 560
inner Helmholtz plane (IHP), 84
in situ/operando probing techniques, 468
in situ reduction, 360

instrument technologies, 370
interaction energy, 257
interaction pseudocapacitive materials, 43
intercalation pseudocapacitance, 42
interference of light, 131
– constructive, 131
– destructive, 131
Intergovernmental Panel on Climate Change (IPCC), 249
ion exchangers, 36, 267
ionic conductivities, 7
– solid electrolytes, 8
ionic liquid (IL), 31, 76
– cations and anions of, 83
– medium, 12
ionic mobility, 44
Ir atoms
– penetrating electrode, 89
– Ru bands, 14
– in SSZ-53 channels, 320
$Ir(C_2H_4)_2(acac)$, 317
iron-based particles, 444
IR spectroscopy, 372
isoelectric point (IEP), 155
isomerization, 522
isosteric heat, 260
isosteric method, 251
isotropic homogeneous solid
– elastic properties, 176
IWI. *See* incipient wetness impregnation (IWI)

j

Jacobsen-like catalysts, 494

k

α-ketoacids, 498
– asymmetric hydrogenation, 500
ketone
– α-carbonyl group of, 501
ketopantolactone, to *(R)*-pantolactone
– asymmetric hydrogenation, 501
kinetic isotope effect (KIE), 425
Kirkendall process, 451, 452
knowledge gaps, 536, 544
Kubas type complexes, 282
– of H_2 with transition metals, 290

l

label-free biosensor, schematic of, 153
lactones, 498
Langmuir constant, 252
Langmuir–Freundlich model, 254

– isotherm, 252, 256, 259
Langmuir surface areas, 295
Langmuir type monolayer, 298
laser ultrasonic (LU) technique, 183
La-Sr-Co-O compounds, ternary library, 559
layered $LiCoO_2$, 6
LCD. *See* liquid crystal display (LCD)
LED. *See* light-emitting diode (LED)
Lewis acid, 258, 431, 488
lewisite, 275
Li-air batteries, 1
LIB. *See* Li-ion batteries (LIB)
Li-Co-Mn-Ni-O pseudoquaternary system, 19
$LiCoO_2$, 2, 17
– band structure, schematic of, 2
– energy diagram of, 2
– graphite, 3, 4
Li_3PS_4/Li, interphase formed in particular cell, 18
– structure, O_3 stacking, 16
– use of, 12
– with $Li_4Ti_5O_{12}$, 7
$LiCoO_2$ cycle, 3
$LiCoO_2$-LiPON-graphite microbatteries, 20
$LiCoO_2$-LiPON interface, 20
$LiFePO_4$, 15
– with carbon coating, 11
– cycling data, 11
$LiFeSO_4F$, 11
ligands (2,6-bis(1′-methylbenzimidazolyl) pyridine), 216
$Li_{10}GeP_2S_{10}$, 8
light-emitting diode (LED), 123
light interference
– magnetic modulation of, 131
– magnetic tuning of, 129–133
light modulation, 123, 124, 128, 143
light polarization, magnetic tuning of, 142–146
light scattering, 124, 126, 133, 146
– magnetic tuning of reflection, 126–129
light transmission modulation, 128. *See also* light modulation
lignocellulose, 431
Li-ion batteries (LIB), 1, 36
Li-ion capacitors, 36
$Li[Li_{1/9}Ni_{1/3}Mn_{5/9}]O_2$, 5
Li-Mn-Co face, 19
Linear sweep voltammograms, 539
$LiNi_{0.5}Mn_{1.5}O_4$, 5
Li_3PO_4
– solid electrolytes, structures of, 8

LiPON, 17
liquid crystal display (LCD), 123, 146
– design of, 143
liquid crystals, 151
– facial magnetic control of, 145
– magnetic, 145
– molecules, 123
– – as light modulator, 123
– optical tuning of, 143
– transmittance of, 145
liquid electrolytes, 7, 69
liquid organics, 541
liquid phase hydrogenation, 340
liquidus temperature, 564
Li-rich oxide, band structure, 14
Li-S batteries, 1
Li-Ta-Ti perovskite, 8
$Li_4Ti_5O_{12}$ negative electrode, 9
London University Search Instrument (LUSI), 560
Lorenz number, 96, 97
lowest unoccupied molecular orbital (LUMO), 63
L-proline
– catalyzed direct aldol reactions, 482
– in ionic liquid, 482
lubricants, 430

m
MacMillan catalysts, 488, 490
macroporous materials, 252
Madelung potential, 79
magnetic dipole–dipole force, 134
magnetic dipole moment, 125
magnetic field
– nonhomogeneous, 126
– strength, 134
magnetic ink, 138, 139
magnetic interactions
– dipole–dipole interactions, 124
– principle of, 125–126
magnetic liquid crystal film, 144
magnetic manipulation, 146
magnetic materials
– behavior of, 124, 125
– magnetic fields, role in, 125
magnetic nanoparticle deposition, 130
magnetic shape anisotropy, 126
magnetic tuning, 138, 140
– Bragg Diffraction, 133–142
– light interference, 129–133
– light polarization, 142–146
– light scattering and reflection, 126–129

magnetization, 126
magnetophoretic force, 126
maleic acid, 555
Mars Krevelen mechanism, 395
mass diffusion, 444
mass transfer, 340
material density, 250
material science, 370
material storage capacity, 286
MAX phases, 207
MEA. *See* monoethanol amine (MEA)
Menten constant, 156
3-mercaptopropyl-functionalized silica, 483
3-mercaptopropyltrimethoxysilane (MPTS)-capped Au nanoparticles, 458
Merrifield resin, 492
meso-/macroporous zeolites
– in situ synthesis of, 419
mesoporosity, 357
mesoporous carbons, 458
– materials, 369
mesoporous silica materials, 443, 456
mesoporous solids, 252
mesoporous tungsten carbide nanoplates, 537
metal alkoxide compounds, 554
metal-based asymmetric catalysts, 488
metal catalysts, 339
– characterization techniques, 372
– for hydrogenation of carboxylic acids, esters, and ethers, 349
metal centers, 268
metal chalcogenide, 450
metal conductor, 27
metal-containing species, 317
metal dispersion, 356
metal dispersions
– by EXAFS spectroscopy, 321
metal electrodes, 34
metal-free 3D carbon nanocatalyts, 459
metal-free triphenylamine-based organic dyes, 67
metal hydroxide (M−OH)
– hydrolysis reaction, 79
metal-ligand bond, 12
metal nitrate salts, 554, 555
metal nitrides, 39, 41
metal organic framework (MOF), 258, 267, 288, 317, 479, 485
– supported iridium complexes, 328
– type solids, 267
– typical MOFs, porous framework, 266

metalorganic precursors, 360
metal oxide (MOx), 39, 40, 75, 460
– catalysis/selective oxidation, 393
– – crystalline $Mo_3VO_{11.2}$, 400
– – development of selective oxidation catalysts, 394
– – distinct complex metal oxide catalysts, types of, 395
– – divanadyl pyrophosphate, 397
– – heteropoly compounds, 399
– – industrial chemical processes, 393
– – multicomponent Bi−Mo−O, 401
– – titanosilicates, 397
– catalysts, 411
– – catalytic characteristics, 396
– – with multifunctionality, 412
– film fabrication steps
– – solution-processed, 79
– mixed oxides, 404
– nanoparticles, 556
– semiconducting materials, 78, 161
– – sol–gel approaches to fabricate high-quality, 78
– solution-dispensing robot, 19
– transistors, 78
– – bioelectronic applications, on electronic materials, 81
– – and bioelectronic interfaces, 78
– – sol–gel processed metal oxide electronic materials, 79
– promotor, 457
– supported triosmium clusters, 320
metal–oxygen bonds, 315
metal precursors, 359, 360
metal sintering, 330
metal support interaction, 356
2-methacryloyloxyethyl phosphorylcholine (MPC), 228
– polymer, 228
methane
– clathrates, 301
– gas storage, 281
– natural gas, 301
– purification, 270
– storage, 298
– – in $Cu_3(BTC)_2$, 301
methanol
– conversion, 517
– decomposition, 516
– power spectral density plot, 168
– steam reforming, 517
methanol to olefin (MTO), 417

methanol-to-propylene (MTP), 421
– demo unit of, 421
2-methoxyethanol, 425
methoxymethyl cation, 425
methoxy PEG (MeO-PEG), 495
methylene blue (MB)
– normalized concentration, 447
methylglyoxal dimethylacetal, 500
cis-β-Methylstyrene, 495
Michael additions, 483, 484
– of 1,3-dicarbonyl compounds to β-nitroalkenes, 484
– hydrogen-bonding activation, 484
Michael reaction, 486
– of cyclohexanone with β-nitrostyrene, 485
– of methyl-1-oxoindan-2-carboxylate and methyl vinyl ketone, 487
microbalance reading, 251
microcracks, 205
microelectrode arrays, 166
– fabricated, 166
– process layout, 166
microexplosions, 31
microplates
– gold, reflectance of dispersion of, 129
– reorientation of, 129
micropore filling, 283, 284
micropore volume, 253
microporosity, 288, 399
microporous carbons, 44
microporous organic polymer (MOP), 275
microspheres
– diffraction of, 140
– fabrication of, 140
microstructural properties, 179
microvascular network, test specimen, 213
microwave synthesis, 560
Mie scattering, 126
– of nanoparticles, 126
miniaturization, 556
minimum energy geometries, 514
misfit layer compound (MLC), 104
mitogen-activated protein kinase (MAPK), 237
Mitsubishi Chemicals, 400
γ-M_3N_4 changes, 188
Mn(III)-Salen complexes
– for asymmetric epoxidations, 494
– catalyzed asymmetric epoxidation, 496
Mobil Chemical Research, 421
mobile phase, 203
Mo–Bi oxide catalysts, 401, 402

modulation excitation spectroscopy (MES), 501
modulation frequency, 127
molar adsorption work, 254
molecular beam epitaxy (MBE), 163
molecular-type catalysts, 411
molecule adsorption, 259
molecule polarizability, 258
molybdenum carbide catalysts, 520
monobranched-chain unsaturated fatty acid (MoBUFA), 430
monoethanol amine (MEA), 261
monometallic nanocatalysts, 360
– synthesis of, 364
monometallic Pd nanocatalysts, synthesis of, 361, 362
mononuclear metal complexes, 329
MOP. See microporous organic polymer (MOP)
morphological observation, 235
MoS_2 clusters, 325
Mott formula, 94, 95
Mo–V–(Nb)–(Te)-oxides, 409
Mo–V–O-based catalyst system, 400
– oxide catalysts for oxidative dehydrogenation
– – of ethane to ethene, 404
– selective oxidation catalyst with high dimensionally structures, 403
$Mo_3VO_{11.2}$ catalyst
– ethane oxidative dehydrogenation, 408
Mo–V–O catalysts, 409, 413
$Mo_3VO_{11.2}$ materials, 405
– crystal structures, 405
– well-crystallized single phasic, 405
Mo_3VO_x
– crystal structures, formation mechanism of, 405
– oxidation of ethane to ethene, 407
Mo–V–Te–Nb–O catalysts, 405
$([MSe]_{1+\delta})_1(TiSe_2)_1$
– atomic plane positions, 113
– carrier densities, 113
– in-plane diffraction patterns, 114
– Seebeck coefficients and power factors, 115
multicomponent system, 566
– Bi–Mo–O catalysts, 402
– carbides/nitrides, 529
– Mo–Bi oxide catalysts, 401
multifunctional nanocatalysts, 461
multifunctional properties, complexity, 394
multiwalled carbon nanotube (MWCNT), 165, 532, 540

- functionalization, 167, 169
- material, 168
mustard agent, 274

n

nanocatalysis, 443
- catalysis, 444
- future prospects, 469
nanocatalysts, 360, 369, 444
- applications, 468
- characterizations methods for, 468
- nanoscale dimension, 464
- synthetic methods, 449
- - bifunctional- and multifunctional nanocatalysts, 463
- - core–shell type nanostructured catalysts, 460
- - metal–organic frameworks (MOF), 462
- - one-dimensional (1D), 453
- - three-dimensional (3D), 455
- - two-dimensional (2D), 453
- - zero-dimensional (0D), 450
- types of, 445
nanochain
- magnetic dipole–dipole energy of, 141
nanocrystalline composite materials (NCs), 179
nanocrystalline Hf_3N_4, 190
nanocrystalline γ-Si_3N_4, 188
nanoelectrocatalysts, 449
nanolaminate material, 104
nanomaterials, 449, 468
- synthesized for catalysis, schematic representation of, 454
nanometer-scale MOx semiconducting channels, 83
nanoparticle (NP), 156, 479
- heterogenenization of, 455
- magnetic, 127
- - light modulation, use in, 127
- photocatalyst adjacent, 447
- in solution, 450
nanophotocatalysis, 445
nanorods, 453
- arrangement of, 145
- external magnetic field, upon application of, 143
- orientation of, 146
nanoscale filler, 211
nanoscale shape-memory nanoalloys, 205
nanoscience/nanotechnology, defined, 443
nanosized carbon-coated particles, 10

nanosized powders, 558
nanostructure, effect of, 464
- porosity, 467
- shape of nanocatalysts on catalysis, 465
- size, 464
nanotechnology, 152
nanowires, 453
NaOH solution, 458
natural dyes, 67
natural gas clathrates, 302
natural gas hydrates, 301
natural gas refining, 529
natural gas storage, 302
n-butane oxidation to maleic anhydride, 397, 398
n-butane, selective oxidation of, 403
N-doped carbon nanospheres
- soft-templating synthetic route, 459
N-doped mesoporous carbons
- schematic illustration, 458
N-doped nanoporous carbon nanoparticles, 460
negative electrode materials
- properties of commercialized and potential, 7
nerve agents, 275
neutralization reactions, 274
NH_4^+ ions, 408
nickel tartrate-NaBr system, 498
NiCo double hydroxides (LDHs), 454
$NiCoMnO_4$ nanomaterial, 446
Ni–Nb–O catalyst, 404
niobium, 65
Ni-rich core, 15
Ni_3S_2/Ni foam, 464
NiTi SMA wires, 205
nitrides, 514
nitroaromatics, 340
nitrogen–containing functionalities, 269
nitrogen doping, 537
nitromethane electroreduction, 543
nitroolefins
- Michael reaction of aldehydes, 487
N,N-dimethylformamide (DMF), 482
N,N,N-trimethyl-1-adamantammonium hydroxide (TMAdaOH), 434
N-, O-, and S-doped, polypyrrole-derived nanoporous carbons
- synthetic procedures leading, 459
nonaqueous electrolytes, 542
noncatalytic reaction, 351
noncondensable adsorbates, 254
nongravimetric effects, reduction, 165

nonuniform particle materials
– use of gradient particles, 16
normal hydrogen electrode (NHE), 68
normalized activation energy
– theoretical graph of, 239
normalized electric potential
– theoretical graphs of, 239
novel structured bimetallic nanocatalysts, preparation of, 367
nucleation theory, 238

o

oceanic geological reservoirs, 249, 276
octahedral hexacyanometallate anions, 264
OLC. See onion-like carbon (OLC)
olefin hydrogenation, 325
olefins, aromatization, 535
oleic acid over ferrierite, isomerization of, 431
olivine $LiFePO_4$, 6
onion-like carbon (OLC), 32
optical modulation, 145
optical polarizations, 145
optimal adsorption heat, 259
organic compounds produces energy, oxidation of, 394
organic electrochemical transistor (OECT), 76
– based enzymatic biosensors
– – for saliva testing, 87
– based enzymatic sensor, 88
– – schematic diagrams of, 87
– for bioelectronic applications, 86
– and bioelectronic interfaces, 83
– fundamentals of, 84
– integrated with HL-1 cells
– – schematic image of, 89
organic electrolytes, 44
organic porous polymers, design, 270
organic synthesis, 355
organometallic precursors, 317
organosulfur compounds, 355, 531
Orito reaction, 498, 502
orthogonal polarizations, 143
orthorhombic $Mo_3VO_{11.2}$ catalysts
– ethane at room temperature, 410
orthorhombic Mo_3VO_x catalysts, 409
orthorhombic/trigonal $Mo_3VO_{11.2}$ catalysts
– formation of, 407
osmotic adsorption isotherm, 255
osmotic coefficient, 256
osmotic theory of adsorption, 254

osteoblasts, quantitative analysis, 235
oxidation-induced deactivation, 517
oxidation reactions, 544
oxidation–reduction mechanism, 402
oxide catalysts, 404
oxidized dye, 63
oxidizing agent, 31
oxonium ylide mechanism, 424
oxygen evolution reaction (OER), 464
oxygen plasma, 163
oxygen reduction reaction (ORR), 537
– polarization curves, 539
oxygen sensors, 161

p

packing density, 33
paraffins, 420, 528
parallel combinatorial hydrothermal techniques, 555
parallel hydrothermal synthesis, 555
parallel powder synthesis, use of combustion synthesis for, 555
parallel synthesis, 555
partial hydrogenation of alkynes to alkenes, 341
particle size, 555, 562, 563
pattern magnet, color of ink, 142
Pb/Bi-containing $([MSe]_{1+\delta})_1(TiSe_2)_1$ ferecrystals, 115
$([PbSe]_{1+\delta})_m(TiSe_2)_n$ compounds
– electronegativity and coordination environments, 112
$([PbSe]_{1+\delta})_4(TiSe_2)_4$ isomers
– structures of, 110, 111
PbTe
– density function theory (DFT), 101
– schematic g(E), 101
PCB substrate, 158
PdAu dendrimer-encapsulated nanoparticles (DEN) synthesis, 452
PDMS. See polydimethylsiloxane (PDMS)
Pd nanoparticle catalyst, schematic representation of, 457
Pd nanoparticles-catalyzed CH_4-O_2 thermocatalytic reaction, 465
PDP. See plasma display panel (PDP)
Pd–Pt–Rh–Ru–Cr system, 565
Pechini processing, 555
PEDOTPSS
– channels, 86
– chemical structure of, 85
– Pt NP- based gate electrodes, 87

pellets resembling pharmaceutical pills, 316
pentacyanonitrosylferrate anion [Fe(CN)$_5$NO]$^{2-}$, 294
pentagonal dodecahedron, 271
periodic mesoporous organosilicas (PMOs), 457, 490
– synthesis, 456
petroleum, 529
– crude, hydrodesulphurization of, 353
– refining
– – industrialization of, 511
phase-sensitive detection (PSD), 501
– analysis, 170
– for ethanol, 169
phase transformations, 15
phenazine, 462
phosphate buffered saline (PBS), 82, 157
photocatalysis, 449
photocatalytic reactions, 445
photomask, 145
photonic chains, 138, 140
photonic crystal film, 138
photonic films, light paths in, 140
photonic inks, displays, 142
photonic nanochains, 142
photonic structures
– angular-dependent, 138
– – fabrication of, 138
– nanostructures, 142
– – design of, 146
photooxidizes air pollutants, 448
photopolymerization, 137
photovoltage spectroscopy, 235
phthalocyanine dyes, 67
Pisarenko relationship, 95, 115
planetary mills, 562
plasma display panel (PDP), 123
plasmon-assisted photochemistry
– physical mechanisms, 447
platinum group catalysts, 358
– metal catalysts, applications of, 358
platinum nanoparticles (Pt NPs), 87
Pmma structures, 182
Point of zero charge (PZC), 533
polarizability, 257
polarizations, 230
– of ceramic biomaterials, 233
– hydroxyapatite, 229
– modulated patterns, 147
polarized optical microscope (POM), 143
polyacrylic acid (PAA), sodium salt, 240
– Ca^{2+} complex, morphology of, 243
polyamidoamine (PAMAM), 492

polyanionic materials, 12, 13
polycrystal, quasi-isotropic elastic moduli, 177
polydimethylsiloxane (PDMS), 134
– film, structural color
– – under nonuniform magnetic field, 136
– – under uniform magnetic field, 136
– network, 214
polyesterification, 555
poly(3,4-ethylenedioxythiophene)poly (styrenesulfonate) (PEDOTPSS), 83
polyethylene glycol (PEG), 69
– grafted on cross-linked polystyrene (PEG-PS), 481
– (2S,4R)-4-hydroxyproline, 480
poly(ethylene oxide) (PEO) polymers, 69
polyhydroxylic alcohol, 555
poly-L-lysine (PLL)-coated glass, 237
polyMB cycle, 422, 423
polymer-based film, 165
polymer-bound bifunctional organocatalysts, 485
polymer channels, 85
polymer complexing reactions, 554, 555
polymer electrode, 41
polymer film, 137
polymerizable monomers, 210
polymerization, UV-induced, 138
polymer Mn(III)-Salen catalyst
– cis-α-methylstyrene epoxidation, 495
polymers of intrinsic microporosity (PIM), 295
polymer-supported chiral aminoalcohols, 493
polymer-supported chiral Mn(III)-Salen complexes, 495
polymer-supported cinchonidine, 486
– asymmetric Michael reactions, 487
polymethylbenzenes (polyMBs)-based aromatic cycle, 422
poly(methyl methacrylate), 69
polyoxoanions, 3-dimensional arrangement of, 399
polyoxomolybdates, 405
polypropylene (PP), 210
polypyrrole, 459
polystyrene-block-poly(ethylene oxide) (PS-b-PEO), 459
poly(vinyl acetate), 211
poly(vinyl alcohol) (PVA), 210
poly(vinyl chloride), 69
poly(vinylpyrrolidone) (PVP), 466
POM. *See* polarized optical microscope (POM)
pore system, features, 252
pore volume, 48

porosity, 31, 356, 467
porous γ-Al_2O_3 (Pt/γ-Al_2O_3), 322
porous aromatic framework (PAF), 270
porous [Ir(COD)(OMe)]$_2$-metallated bipyridyl- and phenanthryl-based MOF catalysts, 462
porous organic polymer (POP), 269, 270, 274, 295
porous Prussian blue analogue, $(T_e)_3[T_i(CN)_6]_2$, 292
porous solids, functional, 269
positive electrode materials, commercialized/ potential next-generation, 5
postcombustion processes, 263
PO_4 tetrahedra, 229
powder containers, 562
powdered catalyst, 418
power density, 45, 540
power factor, 95
precipitation agent, 360
precipitation method, 359
– conventional, 359
process control agent (PCA), 563
proline, 480
propane ammoxidation, 411
– catalytic material Mo–V–(Nb)–(Te)-oxides for, 409
propanol
– dehydration of, 518
– deoxygenation, 518
– power spectral density, 168
2-propanol oxidation catalyzed
– by nanocrystalline γ-Al_2O_3-supported Pt nanoparticles, 466
propene, hydrogen abstraction, 403
propene oxidation reaction, 402
propylene, 518
protein adsorption, 228
Prussian blue, 264
– analogs, 264, 293
– – H_2 adsorption in Cu-containing, 283
– – $(T_e)_3[T_i(CN)_6]_2$ anhydrous phase, 293
pseudocapacitance, 33, 34, 35, 36, 40, 42, 43
– cathode, 37
– electrodes, 39
– material, 38
– metal oxides, 40
– storage, 542
pseudocapacitors, 36, 39
– cathode, 44
– electrodes, 36
– types of, 37
PS resin bearing sulfonic acid, 481

PS-supported proline, 480
– organocatalyst in asymmetric aldol reaction, 481
Pt(II) ions within hydroxyl-terminated poly (amidoamine) (PAMAM) dendrimer, 453
Pt@mSiO_2 core–shell nanocatalysts
– schematic representation, 460
Pt@mSiO_2 core–shell nanoparticles, 460
pulsed laser deposition (PLD), 155, 159
pure cerium oxide, flourite structure, 162
pure nanostructured TiO_2 film
– SEM images of, 447
phonon glass/electron crystal (PGEC) concept, 98
4,6-pyrimidinedicarboxylic acid, 290
pyrolysis, 554

q

quadrupole moment, 257, 282
– interactions, 258
quantum capacitance measurement setup, schematic of, 35
quantum dot (QD), 67
quantum/space charge capacitance, 34
quantum theory, 124
quinine, natural, 484

r

Raman bands, 405
Raman spectra, 183, 406
Raney nickel, 314
rapid automated materials synthesis instrument (RAMSI), 557
rare earth (RE), 420
rare earth oxide (REO), 421
rattling atom, 98
(R)-β-hydroxyketones, 483
recombination processes, 65
recyclable MOF catalyst, 463
redox-active impurities, 46
redox electrolytes, 68
redox-induced capacitance, 39
redox reactions, 36, 42
reductive alkylation, 341
– of aromatic amines, 352
– of aromatic diamines, 353
reductive amination and alkylation reactions, 351
reflection high-energy electron diffraction (RHEED), 163
reforming reactions, 516
refraction, 124

refractive indices, 143
- extraordinary (n_e), 143
- ordinary (n_o), 143
reverse water–gas shift (RWGS), 525
reversible optical modulation, 146
reversible oxygen redox, 14
RF magnetron sputtering, 155
Rh(C_2H_4)$_2$(acac), 317
Rh–O bonds, 324
rhodium catalysts, 347
rhodium particles
- on model silica spheres annealed in H_2, 319
(R)-1-(1-naphthyl)ethylamine, on Pt(111), 502
$R_4N^+Br^-$-stabilized Pd nanoparticles, 457
robotic pipetting system, 560
robotic system, 556, 560
- pipetting system, 560
rock-salt (MX)
- schematic of structures, 105
ROMP-based healing agents, 212
roughness
- macro-, micro-, and nanoscale, 228
(R)-2-(1-piperazinyl)-1,1,2-triphenylethanol
- with Merrifield resin, 494
(4R,5R)-2,2-dimethyl-($\alpha,\alpha,\alpha',\alpha'$)-tetraphenyl-1,3-dioxolane-4,5-dimethanol (TADDOL), 488
(1R,2R)-1,2-diphenylethylenediamine, 486
(1R,2S)-(−)-N-dodecyl-N-methylephedrinium bromide
- immobilization of the chiral cationic surfactant, 492
Ru(C_2H_4)$_2$(acac), 328
Ru nanoparticles, on meso-ZSM-5, 428
ruthenium-complex dyes, chemical structures of, 66
ruthenium metal, 358
Rutherford backscattering (RBS), 163

s

Salen complex, 497
SALT compounds, 99
samarium, 162
- doping concentration, adsorption regimes, 163
sample cell, 250
sample swelling, 251
SAPO-34 cages, 422
SAPO-34 zeolite, 421
sarin, 275
saturated calomel electrode (SCE), 454
SBA-15, 456
scanning electron microscopy, 30
scanning transmission electron microscopy (STEM), 318
scanning tunneling microscopy (STM), 325, 501
screen-printing method, 65
Seebeck coefficients, 94, 95, 97, 102, 103, 115
selected area electron diffraction (SAED), 186
selective catalytic reduction (SCR), 418
self-assembled monolayer (SAM), 228
self-discharge, 46
self-healing materials, 201, 202, 205, 217
- asphalt, 204, 211
- autonomic and nonautonomic, 203
- ceramics, 207, 217
- commercial applications, 217
- concrete, 209
- – schematic representation, 209
- damaged specimen, 203
- damage prevention, 201
- definitions/principles, 202
- extrinsic and intrinsic, 202
- load displacement curves, 215
- overview, 206
- polymers, 211
- potential healing mechanisms, 204
- schematic representation, 202
self-healing metallopolymers, 216
self-healing polymers, 217
- schematic representation, 212
self-healing process, 212
self-healing supramolecular polymers, 217
self-healing supramolecular rubber, 216
semiclathrates, 272
semiconductor materials, 152
semi-hydrogenation, 344
semi-hydrogenation of alkynes, 345
sensor design, 169
sensor stack, to determine oxygen concentration utilizing doped ceria, 163
separator resistance, 46
serial hydrothermal synthesis, 556
(2S)-(−)-3-exopiperazinoisoborneol, 494
shape-memory alloy, 205
Sharpless asymmetric epoxidation, 496
shell nanoclusters, surface atoms number, 465
Si/Al zeolites, 431
sigmoid type isotherms, 268
signal coupling devices, 25
silica–alumina catalyst, 418
silica/Au nanoparticles/silica core–shell-shell nanospheres
- synthesis procedure of, 461
silicon
- binary spinel nitrides of, 188
- ionic oxide semiconductors
- – schematic electronic structures, 80

single-stranded DNA, 86
single-wall carbon nanotubes (SWCNT), 170, 285
sintering, 560, 562
– temperature, 561
skutterudites, 98
skyscrapers, 201
Slack's criteria, 98
Slack's framing, 98
Slack's guidelines, 100
Slack's original concept, 99
small angle neutron scattering (SANS), 542
shape-memory alloy (SMA) wires, 205
Sm-doped ceria thin films, 164
S-Mo-S trilayer, 107
Sn-Beta zeolite, 431, 432
Snell's law, 131
– of refraction, 143
SnSe-containing compound, 113
SodZMOF, adsorption capacity, 290
softening temperature, 564
S–OH bonds, 169
solar energy conversion, physicochemical principles, 61
solar light, schematic illustration, 448
solar photovoltaic, renewable energy technologies, 61
solar spectrum, UV region, 446
sol–gel method, 9, 360, 554, 560
– for combinatorial studies, 555
– derived inorganic network, 554
– metal oxide semiconductors (MOx), 76
solid catalysts, classes of, 314
– bulk solid catalysts, 314
– supported catalysts, 314
solid electrolyte interphase (SEI), 3
solid electrolytes
– use of combinatorial sputtering, 20
solid–liquid–solid (SLS), 155
solid-phase peptide synthesis (SPPS), 483
solid polymer electrolyte (SPE), 543
solids evaluation, surface properties of, 257
solid-state heteropoly compounds, oxidizing ability, 399
solid-state materials, 94
solid-state metal, 373
solid-state semiconductor sensors, 152
solid-support materials, 456
solid surfaces, properties, 227
solution-dispensing robot, 19
solvothermal synthesis, 12

space-charge polarization, 230
space charge within graphite (C_{SC}), 34
spatial color mixing, 138
specific capacitance, 34
specific surface area (SSA), 30
spectroscopy technique, 373
speculative band structure, 96
(4S)-phenoxy-(S)-proline, 481
spin-coating method, 65
spinel $LiMn_2O_4$, 6
spinel nitrides, 189
spray pyrolysis, 68
(S)-proline, 482
sputtering, 68
SrTe nanoprecipitates, 100
SSA. See specific surface area (SSA)
stabilizers, 34
state-of-the-art Li-ion battery, 17
– radar plot of, 4
– schematic of, 2
steel cord, 210
stoichiometric aluminum oxide (Al_2O_3), 163
stoichiometric reagent, 339
storage technologies, 47
store energy, 27
strong metal–support interaction (SMSI), 327
structure-directing agents (SDA), 434
structure-insensitive reactions, 325, 326
subnanometer, 44
successive ionic layer adsorption and reaction (SILAR) method, 67
sulfidation, 529, 530
– treatments, 325
sulfide $Li_{10}GeP_2O_{12}$, 8
sulfur-containing compounds, 331
sulfur mustard, 275
Sun's radiation, 448
supercapacitors, 27, 36, 38, 44, 541
– characteristics of, 46
– construction of, 48
– devices, 48
– – configurations, 43
– electrodes, 41
– self-discharge of, 45
– store energy, 542
supercritical gases, 259
superhard materials, 175
– ball-and-stick model
– – of spinel structure, 187
– – of Th_3P_4-type structure, 189
– – of U_2S_3-type structure, 191
– B–C–N compounds, 184
– – synthesis experiments, 185

– – theoretical predictions, 184
– bulk modulus, 178
– carbon nitride (C–N) phases, 180
– – synthesis under HPHT conditions, 181
– – theoretical Predictions, 180
– c-Hf$_3$N$_4$/η-Ta$_2$N$_3$, transition metal nitrides, 189
– computational simulations, 192
– (c-Zr$_3$N$_4$) transition metal nitrides, 189
– G/B ratio, 178
– – on Poisson's ratio for homogeneous isotropic solid, 177
– hardness/strength, 176
– γ-M$_3$N$_4$, 187
– novel B–C-based compounds, 182
– role of microstructure, 179
– shear modulus, 178
– spinel nitrides γ-M$_3$N$_4$, 186
– strength, 179
– theoretical prediction of, 179
surface adsorption, 40
surface Brillouin scattering (SBS), 183
surface characterization, 357
surface free energy, 233
surface hydroxyls, 538
surface methoxy species (SMS), 424, 425
surface morphology, scanning electron micrographs, 242
surface plasmon resonance (SPR), 446
surface redox pseudocapacitance, 39
surface wettability, 233
surface wettability, 236
Suzuki coupling reaction, 466
Suzuki–Miyaura, transfer hydrogenation, 451
syngas, 521
– ETMCNs catalysts for, 528
– generation of, 522
synthesis gas, 417
synthetic carbons, 48
synthetic polymers, 31

t

TADDOL derivatives, 491
tantalum, 190
tavorite LiFePO$_4$F, 6
T$_3$[Co(CN)$_6$]$_2$
– H$_2$ adsorption isotherms, 283
β-TCP (tricalcium phosphate), 233
temperature-programmed desorption (TPD), 516, 524
– dispersions of platinum, 322
temperature-programmed reaction (TPRxn), 524

ternary B–C–N system, 184
ternary Li$_2$O-GeO$_2$-P$_2$S$_5$ system, 8
tetrabutyl ammonium fluoride, 303
tetradecyltrimethylammonium bromide (TTAB)-stabilized Pt nanoparticles, 460
tetraethoxysilane (TEOS), 457
tetraethyl orthosilicate (TEOS), 492
tetrahydrofuran (THF), 303
thermal conductivity, 95
– measurements, 116
thermal decomposition, 31
thermally stable Pt/mesoporous silica (Pt@mSiO$_2$) type core–shell nanocatalysts, 460
thermally stimulated depolarization current (TSDC), 230, 232
thermal stability, 267, 564
thermal vapor deposition, 68
thermocatalysis, 448
thermoelectric materials
– bridge to nanolaminate structures, 103
– phonon glass/electron crystal approach, 97
– kinetically trapped nanolaminates, 108
– maximizing power factor, 100
– minimizing lattice conductivity, 97
– misfit layer compounds (MLCs), 104
– [MSe]$_{1+δ}$)$_1$(TiSe$_2$)$_n$ ferecrystals, 111
– physical picture, 94
– reduce thermal conductivity, 99
– thin-film superlattice materials, 106
– transition metal dichalcogenide compounds, 104
– using 2D layers, 93
– van der Waals heterostructures, 107
thin films, 163
– fabrication of
– – lithography process for, 145
– interference, 129, 132
– patterns, 145
– with polarization patterns, 145
– resistive behavior, 164
– transistor-based biosensors (bio-TFTs), 81
thiophene molecule, 530
Ti$_2$AlC ceramic
– back-scattered scanning electron micrographs, 208
TiC-based catalysts, 527
TiC filler, 207
TiO$_2$/CeO$_2$ nanomaterial
– advantages of, 65
– material for DSSC fabrication, 65
– morphology and thickness of, 65

– nanostructures of, 65
– solar light-driven synergetic photothermocatalysis, 449
TiO_2/SiO_2 catalysts, 397
TiO_2-supported V_2O_5, 433
$TiSe_2$ layers, 111
tissue hypoxia, 274
titanosilicate catalyst
– alkene epoxidation with hydrogen peroxide at single Ti site, 397
TiX_2 intercalates, 104
Tl-doped compound, 101
– PbTe, 101
top-gate and bottom-contact (TGBC), 77
total adsorption capacity, 256
toxic gas detection, 151
transient mass spectrometer (MS) signals, 520
transition metal
– carbides, 514
– dichalcogenides, 107
transition metal nanoparticle (TMNP), 450
– catalysts, 451
transition metal nitroprussides, 264, 294
transition metals, 511
– oxides, 40
– sites, 258
transmission electron microscopy (TEM), 183, 318, 373
transparent electrodes, 146
trialkoxylated camphorsulfonamide, 492
triazine, 462
4-(1-triazolyl)proline, supported on a divinylbenzene-polystyrene (DVB-PS) copolymer, 483
2,2,2-trifluoroacetophenone, 500
triglyceride hydrogenation rates, 543
trimethyl oxonium ion (TMO), 424
troponin-T detection, 157
tungsten-molybdenum carbide, 541
turbomolecular pumps, 251
turn-over-frequency (TOF), 464

u

ultraclean fuel production, 355
ultrahard materials, 175
ultrahigh vacuum (UHV), 163, 314, 516
ultrapure gases, 251
ultrastable Y (USY), 419
– zeolite, 432
unbreakable glass, 201
unit cell size (UCS), 420, 421
UV light, 167, 216

v

vacancy solution theory, 255
valence band maximum (VBM), 79
vanadium nitride (VN), 41, 541
van der Waals gap, 113
van der Waals systems, 108
vapor phase anisole HDO, 520
vesicants, 275
Vickers hardness, 175, 188
Vickers indenter, 178
VN. See vanadium nitride (VN)
VO^{2+} cations, 407
volatile organic compound (VOC), 249, 448
volumetric equipment, 252
volumetric filling of micropores theory, 254
volumetric techniques, 250
V–P–O system, 398

w

water-gas shift reaction (WGSR), 517, 518, 531
Weibull-type distribution, 254
wettability, 233
– initial cell behaviors, 235
width–length ratio, 86
Wiedemann–Franz law, 96, 97
wurtzite crystal, 155

x

X-ray absorption fine structure (XAFS)
– spectroscopy, 323
X-ray absorption near edge spectra (XANES), 319, 532
– spectra, 324, 330
X-ray absorption spectroscopy, 15
X-ray diffraction (XRD), 181, 183, 186, 241, 468, 497, 517, 559
– yields, 19
X-ray photoelectron spectroscopy (XPS), 14, 46, 163, 314, 372, 517

y

$YbAl_3$ modeled
– density of states, 103
Young's equation, 238, 240
Young's modulus, 176, 178

z

zeolite-like hexagonal solids, $Zn_3A_2[T_i(CN)_6]_2$, 292
zeolite-like imidazolate frameworks (ZIFs), 288
zeolites, 258, 262, 318, 419, 420, 536, 556

– adsorption capacity, 262
– in biomass conversion, 430
– – role of acidity, 430
– – shape-selectivity, 432
– – stability of, 432
– catalysts, 417, 418
– – carbene mechanism, 424
– – catalyst deactivation, 425
– – core–shell-structured catalyst bifunctional F-T catalyst, 429
– – direct mechanism, 424
– – discovery of MTO, 421
– – fluid catalytic cracking (FCC), 418
– – formaldehyde/methane, 424
– – in FT synthesis, 426
– – hybrid bifunctional F-T catalyst, 426
– – hydrocarbon pool mechanism (HCP), 422
– – MFI type of, 419
– – oxonium ylide mechanism, 424
– – rare earth (RE) element, 420
– – into refinery operations, 417
– – Y skeletal diagram of, 419
– – zeolite Y, 418
– exchanged with Cu^+, 287
– HY-supported $M(CO)_2$, 323
– pores, 430
– in selective catalytic reduction (SCR), 433
– – Cu-zeolites as SCR catalyst, 433
– – Fe-zeolites as SCR catalyst, 435
– supported FT metal catalysts, 427
– ZSM-5, 419
zeolite-templated carbons (ZTC), 32
zeolite-like solids $Zn_3A_2[T_i(CN)_6]_2$
– anhydrous phase, 293
zinc oxide (ZnO), 154
– biocompatible semiconductor material, 155
– for biosensing
– – synthesis and applications of, 156
– crystal structures, 154
– electrical characteristics and surface states, 157
– nanoflowers, 156
– nanostructures
– – doping, 161
– nanostrures
– – synthesis and applications, 160
– NPs–CHIT/Pt composite, 156
– n-type semiconductor material, 159
– RF magnetron sputter deposited films, 157
– thin film, 156, 157
– – sputter, 157
zinc poly(styrene-phenylvinylphosphonate)-phosphate (ZnPS-PVPA), 496
zinc ZA-8 die-casting alloy, 205
$Zn_3A_2[T_i(CN)_6]_2$
– H_2 adsorption, 293
Zn_4Sb_3, 99
Zn–thiol interaction-based sensor, 158
$Zn_3[T_i(CN)_6]_2$
– porous framework, 292
Zr(IV) ions, 451
ZSM-5
– microspheres, 430
– skeletal diagram, 419
– zeolite, 420, 432
– – in hydrocracking, 430